PRODUCED WATER

Technological/Environmental Issues and Solutions

ENVIRONMENTAL SCIENCE RESEARCH

Series Editor:

Herbert S. Rosenkranz
Department of Environmental and Occupational Health
Graduate School of Public Health
University of Pittsburgh
130 DeSoto Street
Pittsburgh, Pennsylvania

Founding Editor:
Alexander Hollaender

Recent Volumes in this Series

Volume 39 – GENETIC TOXICOLOGY OF COMPLEX MIXTURES
Edited by Michael D. Waters, F. Bernard Daniel, Joellen Lewtas, Martha M. Moore, and Stephen Nesnow

Volume 40 – NITROARENES: Occurrence, Metabolism, and Biological Impact
Edited by Paul C. Howard, Stephen S. Hecht, and Frederick A. Beland

Volume 41 – ENVIRONMENTAL BIOTECHNOLOGY FOR WASTE TREATMENT
Edited by Gary S. Sayler, Robert Fox, and James W. Blackburn

Volume 42 – CHEMISTRY FOR THE PROTECTION OF THE ENVIRONMENT
Edited by L. Pawlowski, W. J. Lacy, and J. J. Dlugosz

Volume 43 – PRIMARY PRODUCTIVITY AND BIOGEOCHEMICAL CYCLES IN THE SEA
Edited by Paul G. Falkowski and Avril D. Woodhead

Volume 44 – SECONDARY-METABOLITE BIOSYNTHESIS AND METABOLISM
Edited by Richard J. Petroski and Susan P. McCormick

Volume 45 – GLOBAL CLIMATE CHANGE: Linking Energy, Environment, Economy, and Equity
Edited by James C. White

Volume 46 – PRODUCED WATER: Technological/Environmental Issues and Solutions
Edited by James P. Ray and F. Rainer Engelhardt

A Continuation Order Plan is available for this series. A continuation order will bring delivery of each new volume immediately upon publication. Volumes are billed only upon actual shipment. For further information please contact the publisher.

PRODUCED WATER

Technological/Environmental Issues and Solutions

Edited by

James P. Ray
Shell Oil Company
Houston, Texas

and

F. Rainer Engelhardt
Marine Spill Response Corporation
Washington, D.C.

SPRINGER SCIENCE+BUSINESS MEDIA, LLC

Library of Congress Cataloging-in-Publication Data

International Produced Water Symposium (1992 : San Diego, Calif.)
Produced water : technological/environmental issues and solutions / edited by James P. Ray and F. Rainer Engelhardt.
p. cm. -- (Environmental science research ; v.46)
"Proceedings of the 1992 International Produced Water Symposium, held February 4-7, 1992, in San Diego, California"--T.p. verso.
Includes bibliographical references and index.
ISBN 978-1-4613-6258-6
1. Oil field brines--Congresses. I. Ray, J. P. II. Engelhardt, F. R. (F. Ranier) III. Title. IV. Series.
TN871.I523 1992
622'.33819--dc20 92-44262
CIP

Proceedings of the 1992 International Produced Water Symposium, held February 4-7, 1992, in San Diego, California

ISBN 978-1-4613-6258-6 *ISBN 978-1-4615-2902-6 (eBook)*
DOI 10.1007/978-1-4615-2902-6

FOREWORD

This book represents the proceedings of the first major international meeting dedicated to discuss environmental aspects of produced water. The 1992 International Produced Water Symposium was held at the Catamaran Hotel, San Diego, California, USA, on February 4-7, 1992. The objectives of the conference were to provide a forum where scientists, regulators, industry, academia, and the enviromental community could gather to hear and discuss the latest information related to the environmental considerations of produced water discharges. It was also an objective to provide a forum for the peer review and international publication of the symposium papers so that they would have wide availability to all parties interested in produced water environmental issues.

Produced water is the largest volume waste stream from oil and gas production activities. Onshore, well over 90% is reinjected to subsurface formations. Offshore, and in the coastal zone, most produced water is discharged to the ocean. Over the past several years there has been increasing concern from regulators and the environmental community. There has been a quest for more information on the composition, treatment systems and chemicals, discharge characteristics, disposal options, and fate and effects of the produced water. As so often happens, much of this information exists in the forms of reports and internal research papers. This symposium and publication was intended to make this information available, both for open discussion at the conference, and for peer review before publication.

In the years to come as discharge options become more limited, and production operations more costly, this new information could provide a valuable technical basis upon which to make new regulatory decisions. It is hoped that the information presented will prove useful to operators in developing better and more efficient treatment technologies, and also will prove valuable to regulators in developing better, scientifically based, environmental controls.

It also should provide a clear basis upon which to plan future research to both strengthen our current knowledge base as well as fill data gaps.

James P. Ray
F. Rainer Engelhardt

ACKNOWLEDGEMENTS

We would like to thank all of the contributing authors who participated in the 1992 International Produced Water Symposium, San Diego, California, (Feb. 4-7, 1992), and their co-authors who made the papers presented in this book possible. We would also like to thank the 239 participants who contributed to the quality of the conference and these papers by their active participation in the question and answer sessions which followed each presentation.

Key to the success of the conference were Annmarie Pittman and Cathy Coyle of Courtesy Associates, Washington, D.C. They were the professionals behind the organization and execution of the conference. Mr. John Taylor and the staff of the Catamaran Hotel provided an excellent atmosphere and facilities for the conference.

The following sponsors provided funds and professional staff so that this conference could be organized:

American Petroleum Institute

Canadian Petroleum Association

E&P Forum

Environment Canada

Fisheries & Oceans Canada

Gas Research Institute

Indian & Northern Affairs Canada

Interstate Oil & Gas Compact Commission

National Energy Board of Canada

U.S. Department of Energy

U.S. DOI, Bureau of Land Management

U.S. DOI, Minerals Management Service

Representing the sponsoring organizations, and making the conference possible, were the following Steering Committee Members:

Dr. James P. Ray, (Chairman- Symposium)
Shell Oil Company
Houston, Texas USA

Dr. James Anderson
National Energy Board
Calgary, Alberta CANADA

Dr. Daniel Caudle
Conoco, Inc.
Houston, Texas USA

Dr. Sie Ling Chiang
Bureau of Land Management
Washington, D.C. USA

Ms. Susan Clarke
(Chairman, Program Committee)
Environment Canada
Hull, Quebec CANADA

Dr. Rainer Engelhardt
(Chairman, Review Committee)
Marine Spill Response Corporation
Washington, D.C. USA

Mr. James Evans
Gas Research Institute
Chicago, Illinois USA

Dr. Arthur Hartstein
U.S. Department of Energy
Washington, D.C. USA

Mr. T.A. Herenius
Nederlandse Aardolie Maatschappij
B.V.
The Hague, NETHERLANDS

Ms. Nancy Johnson
U.S. Department of Energy
Washington, D.C. USA

Mr. Rodney Paterson
Fisheries and Oceans Canada
Ottawa, Ontario CANADA

Mr. Jerry Simmons
Interstate Oil & Gas Compact Commission
Oklahoma City, Oklahoma USA

Ms. Cheryl Starke
Milpark Drilling Fluids
Houston, Texas USA

Ms. Alexis Steen
American Petroleum Institute
Washington, D.C. USA

Dr. Kenneth Turgeon
U.S. Minerals Management Service
Washington, D.C. USA

Mr. Gary Webster
(Co-Chairman Symposium)
Canadian Petroleum Association
Calgary, Alberta CANADA

Dr. Abbas Zaidi
Wastewater Technology Centre
Burlington, Ontario CANADA

In addition, Drs. R.C. Ayers, Jr., and Michael Stephenson contributed to the Review Committee. Special thanks go to our group of invited speakers that made plenary presentations to set the tone of the conference:

Dr. Alasdair McIntyre (Keynote)
University of Aberdeen
Aberdeen, Scotland

Dr. Brian Bietz
Alberta Energy Resources
Conservation Board
Calgary, Alberta CANADA

Mr. Dale Gilliam
Amoco Production Company
Houston, Texas USA

Dr. David Monk
BP Exploration, Europe
Aberdeen, Scotland

And finally, credit for the high quality of this publication go to over 150 scientists and technical specialists who served as peer reviewers and gave of their valuable time to review papers, and to Chris Noble of Noble Publishing who spent long hours preparing the final manuscripts via desktop publishing.

James P. Ray
Chairman

Gary Webster
Co-Chairman

CONTENTS

CHEMICAL CHARACTERIZATION

TOXICITY

RADIOACTIVITY

FATE AND EFFECT

METHODS DEVELOPMENT AND NEW TECHNOLOGY

A SURVEY OF PRODUCED WATER STUDIES

M. T. Stephenson

Texaco Inc.
Bellaire,Texas
USA

INTRODUCTION

Produced water is the largest volume waste stream in the exploration and production process. Over the economic life of a producing field, the volume of produced water can exceed by ten times the volume of hydrocarbon produced. During the later stages of production, it is not uncommon to find that produced water can account for as much as 98% of the extracted fluids. During 1990, Gulf of Mexico oilfield operations produced 866,514,000 barrels of water (Reilly, 1991). With volumes of this magnitude, the disposal of produced water becomes an important issue to both the operator and the environment. The optimum method for disposing of produced water is one that protects the environment while imposing the least economic burden on the operator. Often, regulatory controls limit the options available to the operator. However, the methods dictated by regulatory controls may not be necessary or appropriate to protect the environment.

COMPONENTS

In offshore petroleum production operations, produced water is typically discharged to the ocean. The discharge of large volumes of produced water into the ocean has raised fears in government agencies and non-government organizations about pollution. In an effort to allay these fears, the components of produced water and the effects of ocean discharge have been summarized from a large number of government and industry studies. These components can be grouped into the following categories:

- oil;
- heavy metals;
- radionuclides;
- treating chemicals;
- salt; and
- dissolved oxygen.

Oil

Oil is the common term applied to organic material that is dispersed or dissolved in produced water at the time of discharge. This organic material exists in two forms:

Produced Water, Edited by J.P. Ray
and F.R. Engelhart, Plenum Press, New York, 1992

- dispersed oil and;
- non-hydrocarbon organic material.

Dispersed oil consists of small discrete oil droplets suspended in the water while non-hydrocarbon organic material, or soluble oil, is present in a dissolved state. The amount of dispersed oil in a produced water stream will vary depending on the influence of several different factors. These factors include:

- the density of the oil;
- the shear history of the droplet;
- the amount of oil precipitation; and
- the interfacial tension between the water and the oil.

Oilfield produced water treating equipment in use today is designed to remove discrete droplets of oil from the water phase. These devices operate by taking advantage of the difference in density between the oil droplets and the water. If the oil density is not significantly different from the density of the water, there is little driving force to move the droplet to the surface for coalescence and subsequent recovery. As a droplet moves through chokes, valves, pumps, or other constrictions in the flow path, the droplet can be torn into smaller droplets by the pressure differential across the devices. This is especially true of low viscosity oils and condensates. Precipitation of oil from solution results in a water fraction with smaller droplets. These small droplets can be stabilized in the water by low interfacial tension between the oil and the produced water. Small droplets can also be formed by the improper use of production chemicals. Thus, the addition of excess production chemicals (such as surfactants) can further reduce the interfacial tension so that coalescence and separation of small droplets becomes extremely difficult.

The other principal parameter of concern in this separation is the size of the oil droplet. According to Stokes' Law (Sakiadis, 1984), the velocity that the oil droplet rises through the water also varies directly with the square of the droplet diameter. The size of the droplet is also a function of the viscosity of the oil and the shear history of the oil-water stream.

The Offshore Operators Committee conducted a study of 42 platforms in the Gulf of Mexico. Figure 1 depicts the results of this study, which measured the average amount of oil being discharged into the environment (Otto, 1990). These 42 platforms discharged 419 metric tons of oil into the Gulf of Mexico in 1989. These values can be extrapolated to 22,277 metric tons of oil discharged from the 2,233 platforms across the Gulf of Mexico, an average of 10.0 metric tons per platform. Compare these numbers to the 4119 metric tons of oil discharged into the North Sea from 89 oil production platforms in 1989, an average of 46.3 metric tons per platform (E&P Forum, 1990).

The amount and nature of soluble oil, or non-hydrocarbon organic materials, that may be in the produced water stream also will vary depending on several different factors. These factors include:

- type of oil;
- volume of water production;
- artificial lift technique; and
- age of production.

The only factor studied to any extent is the type of oil (paraffinic, asphaltenic, or gas condensate). The chemical characteristics of the soluble oil compounds generally place them in the following classes:

- aliphatic hydrocarbons;
- phenols;
- carboxylic acids; and
- low molecular weight aromatic compounds.

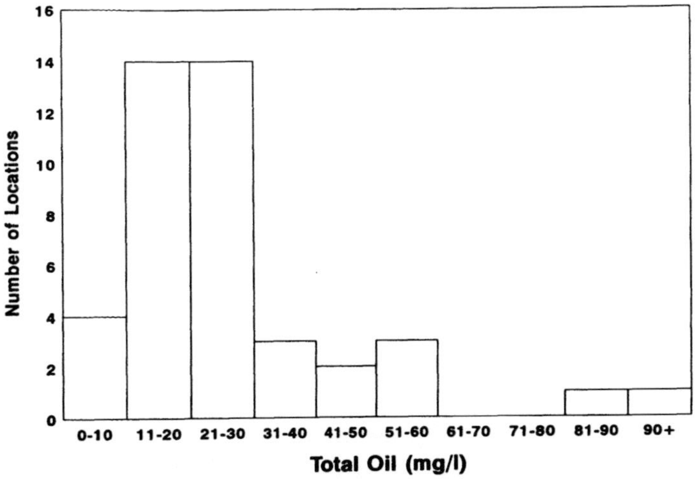

Figure 1. Oil and grease concentration: results from 42 platform study.

The quantity of non-hydrocarbon organic materials present was also determined for the same 42 platforms by the Offshore Operators Committee study (Figure 2).

Dissolved aliphatic hydrocarbons (decane through tetratriacontane) have been determined to be present in produced water from paraffinic oils in the ranges of 606 to 2,677 μg/l (OOC, 1975; Lysyj, 1981; OOC, 1982; Burns, 1983; Middleditch, 1983; Caudle, 1988; Brown, 1990). Phenols have been found in varying amounts in the water associated with all three types of oils; however, produced water from gas condensate operations tends to have higher quantities of phenols and low molecular weight aromatic compounds (Callaghan, 1990). Water produced with paraffinic oils will often have high concentrations of simple fatty acids, while water produced with asphaltenic oils can contain notable amounts of naphthenic acids.

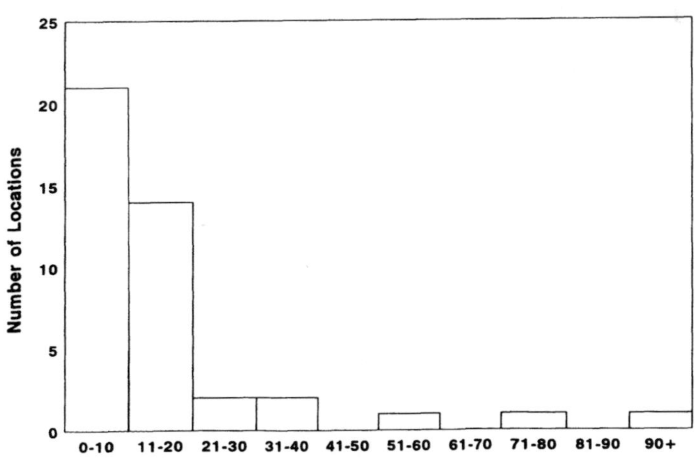

Figure 2. Magnitude of soluble oil in produced water.

Table 1. Phenols and volatile aromatic compounds in produced water (μg/l)

Gas Prod.	Phenols	Benzene	Toluene	C_2-Benz.
Average	4743	5771	5190	700
Std. Dev.	5986	4694	4850	1133
Maximum	21522	12150	19800	3700
Minimum	150	683	1010	51
Oil Prod.	**Phenols**	**Benzene**	**Toluene**	**C_2-Benz.**
Average	1049	1318	1065	221
Std. Dev.	889	1468	896	754
Maximum	3660	8722	4902	6010
Minimum	0	2	60	6

There are many studies of the fatty acid content of produced water. Early studies searching for the soluble organic constituents of produced water used current methods for the extraction of organic material from water. The acidified produced water was extracted with freon. Gas chromatography was then used to identify the types of compounds. This method was initially used because there was often a large difference in the weight of the freon extract depending on whether or not the water was acidified before analysis. Early results found that there was only a small quantity of low molecular weight fatty acid compounds present in the freon extract (Table 2).

Table 2. Organic acids found in freon extracts

Acid	mg/l	%
Butanoic (C_4)	0.4	1.9
Pentanoic (C_5)	4.2	19.1
Hexanoic (C_6)	4.6	20.8
Heptanoic (C_7)	2.7	12.4
Octanoic (C_8)	2.8	12.7
Nonanoic (C_9)	1.4	6.2

Large quantities of acetic acid (infinitely soluble in water) were found in North Sea produced water, nearly 700 mg/l (Somerville, 1987). It was soon realized that the earlier results were due to the lack of solubility of low molecular weight fatty acids in freon (Stephenson, 1990).

The content of soluble aromatic compounds in produced water varies with the type of oil. Table 3 shows the amounts of various aromatic compounds determined to be present in produced waters from the Gulf of Mexico (OOC, 1982). Larger quantities of these aromatic compounds have been reported from gas condensate platforms in the North Sea (Callaghan, 1990). This indicates that each produced water is different.

Table 3. Polynuclear aromatic hydrocarbons in produced water (μg/l)

	Naphthalenes	Other PAHs
Average	132	7
Std. Dev.	161	18
Maximum	1179	108
Minimum	0	0

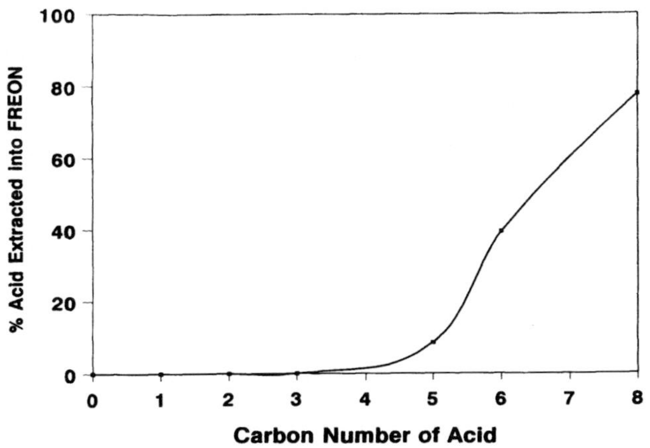

Figure 3. Solubility of low molecular weight fatty acids in freon.

Heavy Metals

There have been several studies by both industry and government to determine the quantity of various heavy metals in produced water (OOC, 1975; Jackson, 1979; Lysyj, 1981; OOC, 1982; Burns, 1983; Middleditch, 1983; Neff, 1988; Neff, et al. 1988; SAIC, 1991). The metals most commonly studied are barium, cadmium, chromium, copper, lead, mercury, nickel, silver, and zinc. Table 4 lists some of the data from these studies. This also indicates that each produced water is different.

Table 4. Heavy metals in produced water (μg/l)

Metal	Avg.	Std. Dev.	Max.	Min.	Sea Water
Cd	27	12	98	0	Trace
Cr	186	68	390	0	Trace
Cu	104	180	1455	0	45
Pb	315	670	5700	2	5
Ni	192	307	1674	0	0.5
Ag	63	17	152	12	0.3
Zn	170	253	1600	17	14

Radionuclides

Radionuclides found in oilfield production are often referred to as naturally occurring radioactive material or NORM. The source of the radioactivity in scale is from the radioactive ions, primarily radium, that co-precipitate from produced water along with other types of scale. The most common scale for this co-precipitation is barium sulfate, although radium has also been found in calcium sulfate and calcium carbonate scales.

Early studies of soluble radionuclides in produced water have been summarized (Snavely, 1989). These studies of wells in Oklahoma, the Texas panhandle, and the Gulf of Mexico coastal area show ^{226}Ra levels ranging from 0.1 to 1620 pCi/l and ^{228}Ra levels ranging from 8.3 to 1,507 pCi/l. There have been several recent studies of soluble radionuclides in produced water.

The first of these studies was initiated by the State of Louisiana Department of Environmental Quality. The second study was conducted by the Offshore Operators Committee (Stephenson, 1990). This study collected data from 42 discharge points offshore in the Gulf of Mexico and off California. The EPA also completed a detailed study of three produced water filtration systems that included the analysis of radionuclides in the produced water (SAIC, 1991). The findings of these studies are summarized in Table 5.

Table 5. Radionuclides in produced water (pCi/l)

OOC 44 Platform Study	^{226}Ra	^{228}Ra
Mean	262	277
Standard Deviation	156	146
Maximum	584	586
Minimum	4	18
Louisiana DEQ Study	^{226}Ra	^{228}Ra
Mean	68	165
Standard Deviation	144	150
Maximum	930	928
Minimum	0	0
EPA 3 Facility Study	^{226}Ra	^{228}Ra
Mean	68	29
Standard Deviation	65	19
Maximum	218	68
Minimum	4	0

Because of the lack of detailed fate and effects data of these radioactive ions, both EPA and the API have instituted additional studies to determine the environmental impacts of these ions as well as any human health effects that may be derived from those impacts.

Treating Chemicals

In addition to the chemicals that occur naturally in produced water, there may be chemicals present that were added to the production process by the oilfield operator. These chemicals are added to treat or prevent operational problems. Each individual production system is unique in its chemical requirements. Again, this indicates that each produced water is different.

Treating chemicals can generally be classified into three categories:

- production treating chemicals;
- gas processing chemicals; and
- stimulation and workover chemicals.

Production treating chemicals include:

- scale inhibitors;
- corrosion inhibitors;
- biocides;
- emulsion breakers;
- water treating chemicals;
    - — reverse emulsion breakers;
    - — coagulants; and
    - —flocculants;

- antifoams; and
- paraffin/asphaltene treating chemicals.

Gas processing chemicals are generally limited to hydrate inhibition chemicals, dehydration chemicals, and occasionally H_2S removal chemicals. Well stimulation and workover chemicals can include mineral acids, dense brines, and appropriate additives. Three reports have been compiled on the chemicals used in the Gulf of Mexico and the North Sea addressing the nature of these chemicals and the available data on aquatic toxicity (Hudgins, 1985, 1989, 1991).

ENVIRONMENTAL IMPACTS

Excessive amounts of some of the above cited substances can cause changes to an ecological system. The significance of the changes resulting from the discharge of produced water into the environment depend on the environment's capacity to absorb, use, or render harmless the constituents in the water. Some constituents of produced water are more easily dealt with by the environment than others. The ensuing discussion will focus on the offshore impacts of produced water.

Dispersed Oil

The impact of dispersed oil on a body of receiving water depends upon the location of the discharge pipe relative to the bottom and the surface as well as on the depth and energy of that body of water. If the end of the discharge pipe is close enough to the bottom for the produced water plume to strike the bottom, then there may be some local sediment contamination and an associated disturbance of the benthic community (Neff et al., 1988). The energy of the water body will strongly influence the rate of dilution of the various components of produced water into the environment. The energy of the water body involves wave motion, currents, vertical mixing, and overall rate of volume exchange of the water.

Dispersed oil droplets that do not reach the bottom may eventually rise to the surface where it spreads out and some of the volatile compounds evaporate. The more toxic compounds are among the most volatile. Thus, most of the toxic compounds are removed from the water quickly. It is interesting that researchers have concluded that:

"there is little basis for concern with hydrocarbons found in undiluted produced waters. Levels of most toxic substances in undiluted produced water are below their individual 96-hour LC_{50} concentrations. Thus, toxic levels of hydrocarbons should not be encountered in the water column beyond the immediate vicinity of the mixing zone" (Lysyj, 1981).

The primary impact of dispersed oil in discharged produced water is an increase in the biological oxygen demand (BOD), again near the mixing zone. Therefore, dispersed oil in produced water discharged from well-maintained facilities should not have a detrimental environmental impact.

Soluble Oil

Unlike dispersed oil, soluble oil is not readily removed from produced water and therefore, it is discharged to the ocean. In addition to locally increasing BOD, the components of soluble oil each have a different fate in the environment. At least one study has shown that the toxicity of soluble oil is not significant. The soluble oil fractions of two different produced waters tested for toxicity and found to have acute toxicities of 15.8% and 4.8% (Brown, 1990). One of the reported characteristics of these components is that they are easily biodegraded. Therefore, low levels of dissolved non-hydrocarbon organic materials are easily assimilated by the receiving water body.

The levels of phenolic compounds in produced water differ according to the type of production: gas or oil. As part of an industry study (Brown, 1990) of the soluble oil fraction of produced water, a literature search of the compounds found showed that phenol and alkyl phenols of the types present in produced water are readily degraded by bacterial and photo-oxidative processes in seawater and marine sediments.

The acid fraction of soluble oil has been extensively studied. The fatty acids in produced water have also been found to be naturally occurring constituents of seawater, sediment, and aquatic life (Brown, 1990). The fatty acids in produced water are produced as sodium salts of the acids (Somerville, 1987), primarily acetates and propionates (Kharaka, 1986). These salts are innocuous at the concentrations found in produced water. There is less known about the chemical character of naphthenic acids. The structure, molecular weight, solubility, and toxicity for this class of compounds is not completely known. Naphthenic acids have been shown, however, to affect marine organisms (Michel, 1986; Ponat, 1988).

The levels of polynuclear aromatic compounds soluble in produced water are extremely low (Table 3) and far below toxic levels (Neff, 1988). Volatile aromatic compounds occur at significantly higher concentrations. They escape the water column easily and exist in the water column below toxic levels (Neff, 1988). Considering these factors and the relative concentrations of the constituents of soluble oil, an adverse impact on the environment from soluble oil has not been observed.

Heavy Metals

The concentrations of the various heavy metals in produced water are often higher than those levels occurring naturally in seawater. Dilution within a mixing zone is rapid, thus diluting the concentration of this metal ion very quickly. There are no reports of elevated levels of these heavy metals in the water column around producing platforms in the Gulf of Mexico. On the other hand, higher levels of metals may exist in sediments:

"elevated levels of several metals have been reported in sediments around some producing locations. Heavy metals in these tightly adsorbed forms have much lower bioavailability to marine animals than do the metal ions in solution" (Neff, 1988).

In general, therefore, the presence of heavy metals in produced water discharged to the marine environment does not have an adverse effect.

Radionuclides

There have been several major studies of the presence of radionuclides in produced water, including a literature review (Snavely, 1989) and a screening level risk assessment (Hamilton, 1991). One of the conclusions obtained from the Offshore Operators Committee study is that the presence of radionuclides depends upon the geological formation from which the water is produced (Stephenson, 1990). Another conclusion is that the analytical method used in measurement strongly influences the accuracy and reliability of the data. Finally, a risk assessment evaluated all the available data to estimate the extent to which radium accumulates in aquatic food webs, especially the edible portions of organisms. The conclusion of the risk assessment indicated that no detectable impact on fish, mollusks, or crustaceans is likely. The number of excess human cancers predicted was comparable to that resulting from background concentrations of radium. The risks associated with the discharge of produced water offshore is extremely small.

Treating Chemicals

The effects associated with the use of treating chemicals have a great deal to do with the particular chemical and how it is used. For example, certain production treating chemicals can be used either in the continuous treatment mode or as a batch treatment. When a chemical is used as a batch treatment (including a squeeze treatment), a large amount of chemical is injected into the well. Chemicals that may be used include scale inhibitors, corrosion inhibitors, and biocides. When the treatment is completed, the well is returned to production and a large slug of fluid containing high concentrations of chemicals is immediately produced. Generally, this concentration exceeds toxic levels for that chemical. To prevent an impact on the environment, this first slug of fluid is captured separately from the basic production stream, or a continuous treatment is considered. However, in multi-well production systems, these slugs of chemical are diluted significantly by the production from the other wells.

In the continuous treatment mode, the levels of these chemicals present in the produced water stream are generally not at toxic concentrations. Other chemicals, such as emulsion breakers, reverse emulsion breakers, coagulants, flocculants, antifoams, and paraffin and asphaltene treating chemicals are not found in the final discharge stream if they are applied optimally.

Other chemicals that may have an environmental impact if they are discharged with the produced water include hydrate inhibitors (e.g., methanol), treatment acids, and dense workover brines (e.g., zinc bromide). Acids are caught separately and neutralized, toxic brines are caught and reconditioned for reuse, and hydrate inhibitors are used at minimal levels. The use of chemicals in the production process does not have adverse impacts on the environment if they are used soundly.

Salinity

The salinity of produced water can vary from very low to saturation, depending on geology and the production process. The impact of discharging fresh or brackish produced water into the ocean would be the same as rain. Observations from platforms that discharge produced water with very high salt content show that there is a lively aquatic life community present (Quigel, 1991). Dilution of a 200,000 mg/l salt water solution, such as produced water, in a 35,000 mg/l ocean occurs very quickly. Therefore, the concentration of salt in produced water discharged offshore has not been shown to cause a harmful impact on the water or aquatic life (Harper, 1986).

Temperature

The temperature of produced water varies with several geographical and geological factors. Some locations discharge water that is warm to the touch while others discharge water at 200° F. The dilution and heat transfer is so rapid that no impact is observed even at locations discharging at "high" temperatures (Quigel, 1991).

Dissolved Oxygen

Produced water does not initially contain oxygen. This means that the produced water being discharged into the environment will not contain the oxygen necessary to sustain life. Oxygen introduced into the produced water stream causes severe corrosion to the production system. This is not only costly, but may have undesirable environmental consequences if a major equipment failure occurs. Therefore, an operator will go to great lengths to prevent oxygen from getting into the produced water system. Low oxygen content in produced water has not been shown to be a problem in a high energy environment; and, in a low energy environment, it can be mitigated by allowing the discharge to occur above the surface of the receiving water. In this manner, some oxygenation takes place and dilution, though rapid, is not the sole means of contending with the lack of dissolved oxygen.

CONCLUSIONS

In the offshore environment, produced water discharged from petroleum production platforms should not have adverse effects on the receiving water or the surrounding aquatic life, especially considering that

"approximately 250,000 metric tons of petroleum hydrocarbons enter the marine environment each year from natural seepage (LLNL, 1986)."

ACKNOWLEDGEMENTS

The author wishes to acknowledge the organizations that have generated the bulk of the data: the American Petroleum Institute, the Offshore Operators Committee, the E&P Forum, and the U. S. Environmental Protection Agency. Acknowledgment is given to Texaco Inc. for allowing the publication of this paper.

REFERENCES

Brown, J. S., Neff, J. M., and Williams, J. W., 1990, The Chemical and Toxicological Characterization of Freon Extracts of Produced Water, Battelle Memorial Institute, Duxbury, Massachusetts. Report to Offshore Operators Committee.

Burns and Roe Industrial Services, 1983, Evaluation of Analytical Data Obtained from the Gulf of Mexico Sampling Program, Burns and Roe Industrial Services Corporation, Paramus, New Jersey.

Callaghan, D. and Baumgartner, W., 1990, Characterization of residual hydrocarbons in produced water discharged from gas production platforms, paper SPE 20881 presented at Europec 90, The Hague, Netherlands.

Caudle, D. D., Stephenson, M. T., 1988, The Determination of Water Soluble Organic Compounds in Produced Water, Offshore Operators Committee, New Orleans.

Hamilton, L. D., Meinhold, A. F., and Nagy, J., 1991, Produced Water Radionuclide Hazard/Risk Assessment - Phase I, Brookhaven National Laboratory, Upton, New York. Report to American Petroleum Institute.

Harper, D. E., 1986, A Review and Synthesis of Unpublished and Obscure Published Literature Concerning Produced Water Fate and Effects. Offshore Operators Committee, New Orleans, Louisiana.

Hudgins, C. M., 1985, Chemical Treatment of Produced Fluids in Offshore Oil and Gas Production Systems, Petrotech Consultants, Inc., Houston, Texas. Report to Offshore Operators Committee.

Hudgins, C. M., 1991, Chemical Usage in North Sea Oil and Gas Production and Exploration Operations, Petrotech Consultants, Inc., Houston, Texas. Report to Oljeindustriens Landsforening (OLF).

Hudgins, C. M., 1989, Chemical Treatments and Usage in Offshore Oil and Gas Production Systems, Petrotech Consultants, Inc., Houston, Texas. Report to American Petroleum Institute.

Jackson, G.F., Hume, E., Wade, M.J., and Kirsch, M., 1979, Oil Content in Produced Brine on Ten Louisiana Production Platforms, Rockwell International, Newbury Park, California. EPA Report.

Kharaka, Y.K., Law, L.M., Carothers, W.W., and Goerlitz, D.F., 1986, Role of Organic Species Dissolved in Formation Waters from Sedimentary Basins in Mineral Diagenesis, SEPM Spec. Pub., 38:111.

LLNL, 1986, (Lawrence Livermore National Laboratory, 1986, Adaptation of Marine Organisms to Chronic Hydrocarbon Exposure, Energy and Technology Review, UCRL-52000-86-6/7.

Lysyj, I., 1981, Chemical Composition of Produced Water in Selected Offshore Oil and Gas Extraction Operations, Rockwell International, Newbury Park, California. EPA Report.

Michel, W.C., Case, J.F., 1986, The effects of a water-soluble fraction on the neuroid electrical activity of the hydroid coelenterate *Tubularia crocea*, Marine Environ. Res., 19:295.

Middleditch, B. S., 1983, Ecological Effects of Produced Water Discharges from Offshore Oil and Gas Production Platforms, University of Houston, Houston, Texas. Report to American Petroleum Institute.

Neff, J. M., 1988, Bioaccumulation and Biomagnification of Chemicals from Oil Well Drilling and Production Wastes in Marine Food Webs: A Review. Battelle Ocean Sciences, Duxbury, Massachusetts. Report to American Petroleum Institute.

Neff, J. M., Sauer, T. C., and Maciolek, N., 1988, Fate and Effects of Produced Water Discharges in Near Shore Environments, Battelle Ocean Sciences, Duxbury, Massachusetts. Report to American Petroleum Institute.

Offshore Operators Committee, 1975, Environmental Aspects of Produced Waters from Oil and Gas Extraction Operations in Offshore and Coastal Waters, New Orleans, Louisiana.

Offshore Operators Committee, 1982, Determination of Priority Pollutants in Produced Water By Isotope Dilution GC-MS and Standard Addition AA, Final Report, Radian Corporation, Austin, Texas.

Oil Industry Exploration and Production Forum, 1990, North Sea Work Group Report: Inventory of North Sea Activity 1989, London.

Otto, G. H., 1990, Oil and Grease Discharge Characteristics of Methods 413.1 and 503E (42 Platform Study), University of Houston, Houston, Texas. Report to Offshore Operators Committee.

Ponat, A., 1988, Effects of water soluble crude oil fractions on cirral beat frequency in *Balanus balanoides,* Bull. Environ. Contam. Toxicol., 41:759.

Quigel, J.C., 1991, Sea-life around offshore Installations and their Role as Artificial Reefs, Southeast Asia Regional Safety and Environmental Workshop, Jakarta, Indonesia.

Reilly, W.K., O'Farrell, T., and Rubin, M.R., 1991, Development Document for 1991 Proposed Effluent Limitation Guidelines and New Source Performance Standards for the Offshore Subcategory of the Oil and Gas Extraction Point Source Category, U.S. Environmental Protection Agency, Washington, D.C.

Sakiadis, B.C., 1984, Fluid and Particle Mechanics, in: Perry's Chemical Engineers' Handbook, R.H. Perry and D.W. Green, ed., McGraw-Hill, New York.

Science Applications International Corporation, 1991, Produced Water Pollutant Variability Factors and Filtration Efficacy Assessments from the Three Facility Oil and Gas Study, EPA Contract No. 68-C0-0035.

Snavely, E. S., 1989, Radionuclides in Produced Water, A Literature Review. Report to American Petroleum Institute.

Somerville, H. J., *et al.*, 1987, Environmental Effect of Produced Water from North Sea Oil Operations, Marine Pollution Bulletin, 18:549.

Stephenson, M., and Bartels, C., 1990, Challenges in Treating Produced Water, paper presented at the AIChE National Meeting, Orlando, Florida, March 18-22.

Stephenson, M. T. and Supernaw, I. R., 1990, Offshore Operators Committee 44 Platform Study Radionuclide Analysis Results, Offshore Operators Committee, New Orleans, Louisiana.

THE COMPOSITION OF PRODUCED WATER FROM SHELL OPERATED OIL AND GAS PRODUCTION IN THE NORTH SEA

R.P.W.M. Jacobs [1], R.O.H. Grant[1], J. Kwant[2], J.M., Marquenie[3], and E. Mentzer[2]

[1] Shell Internationale Petroleum Maatschappij B.V.
The Hague, NETHERLANDS

[2] Shell U.K. Exploration and Production
Aberdeen, Scotland
UNITED KINGDOM

[3] Nederlandse Aardolie Maatschappij B.V.
Assen, NETHERLANDS

ABSTRACT

Shell U.K. Exploration and Production (Shell Expro) and the Nederlandse Aardolie Maatschappij B.V. (NAM) are both companies active in the exploration and production of oil and gas in the North Sea. They belong to the Royal Dutch Shell Group of Companies and are Shell operated.

As part of programs to implement a policy of progressive reduction in emissions, effluents, and discharges of waste, known to have a negative impact on the environment, both Companies have made a start in 1989 by determining the quantities and qualities of effluents from a total of 15 gas and 12 oil platforms. An overview of the results of this "1989 sampling and analysis programme" of produced water discharged into the North Sea is presented in this paper. The data include both organic and inorganic compounds, as well as some ecotoxicological characteristics of the effluents.

INTRODUCTION

Shell U.K. Exploration and Production (Shell Expro) and the Nederlandse Aardolie Maatschappij B.V. (NAM) are oil and gas exploration and production companies active in the North Sea. Both belong to the Royal Dutch Shell Group of Companies and are Shell operated. Half of the shares of both Companies is owned by Esso. The NAM is active in the Netherlands and is the main gas producer on the Dutch continental shelf. Shell Expro produces both oil and gas in the U.K. sector of the North Sea from oil fields in the central and northern North Sea and gas fields in the southern part.

Society's concerns and expectations regarding the environment have increased tremendously over the last ten years. In parallel, industry's attitude toward environmental protection has changed markedly. Shell companies always have valued their reputation, which has been

Produced Water, Edited by J.P. Ray
and F.R. Engelhart, Plenum Press, New York, 1992

founded on the quality of products and services. During the eighties they began to recognize the "environmental challenge" as one of the most important determinants of long-term business success. Meeting this challenge demanded the highest management standards in all operations and required a good understanding of environmental vulnerabilities and the consequences of industrial activities.

As part of policy developments to meet the "environmental challenge", Shell companies started to pursue in their operations progressive reductions of all emissions, effluents and discharges of waste materials known to have a negative impact on the environment, with the long-term goal to eliminate these, if possible. In implementing this, the first step was to gain insight into the extent of the problem by making an inventory of the quantities and qualities of waste streams. NAM and Shell Expro started an inventory of their offshore discharges of produced water in 1989, when there was also widespread public interest in the North Sea. At this time there were international political discussions to prepare for the third Ministerial Conference on the North Sea (in 1990), as well as to develop national policies to protect the quality of the Sea (Peet, 1987; Salomons et al., 1988; Newman and Agg, 1988; and National Environmental Policy Plan - The Netherlands, 1989).

This paper presents an overview of the results of sampling and analysis programs carried out at all platforms discharging produced water into the North Sea. The campaigns included NAM's effluents from eight gas/condensate platforms and Shell Expro's effluents from seven gas and 12 oil platforms. In addition to the legally required monitoring of oil-in-water content, occasionally analyses of other constituents were carried out. Although the sampling and analysis campaigns of both Companies showed large overall similarities, there were many differences in details. For reasons of length, this paper is restricted to an overview of representative data. Hence, no details are presented or discussed here on sampling, analytical methods, or difficulties associated with sampling, shipping and storage, test and analysis methodology, etc.

PRODUCTION AND EFFLUENT STREAMS

Crude oil, with associated gas, is produced from the Expro fields in the East Shetland Basin, i.e., Brent, Dunlin, Eider, Tern and Cormorant, and in the central North Sea area, i.e., Fulmar, Auk and Kittiwake. Gas is produced both by Expro and NAM from many smaller fields in the southern North Sea. Some of these gas fields also produce associated condensate.

In 1989 Expro produced offshore 27.6 x 10^6 m^3 of crude oil (or 75,600 m^3 per day), 9,628 x 10^6 m^3 gas and 2 x 10^6 tons natural gas liquids. NAM's offshore production was 9,080 x 10^6 m^3 gas and 113 x 10^6 m^3 condensate (see Table 1). The NAM figures do not include the gas/condensate produced at the offshore platform in the "onshore" Ameland concession.

Produced water discharged in 1989 by both Companies amounted to 23.4 x 10^6 m^3 for the oil platforms and 224 x 10^3 m^3 for the gas platforms, which was on average 64.7 x 10^3 m^3 per day (see Table 1). This includes the produced water from the NAM AWG (Ameland) platform.

Produced water consists of formation water, i.e., the water naturally present in the reservoir, and/or, in the case of gas production, condensed water. In addition, the effluent stream from oil production can also contain: (a) seawater that has been injected to maintain reservoir pressure and that has broken through to production wells and (b) occasionally some smaller oily streams like displacement water from oil storage facilities, process and drainage water.

The properties of produced water vary from field to field. In general, water from oil production is saturated with dissolved gases including CO_2, and possibly H_2S, as well as hydrocarbons. These gases come out of solution as process pressures are reduced. In addition to gases, the produced water contains significant amounts of both dissolved salts and organic chemical constituents, which were present in the original formation water or were introduced via in injection water and production systems as production chemicals.

Table 1. The 1989 oil, gas and produced water figures (as daily averages) for the North Sea and the Shell Expro and NAM operations

1989 Production figures	Crude oil (10^3 m^3/day)	Gas (10^6 m^3/day)	Water (10^3 m^3/day)
United Kingdom[a]	284.9	122.4	291.7
Expro contribution	75.6	24.4	64.1
	26.5%	20.0%	22.0%
Netherlands[a,b]	7.2	51.8	24.9
NAM contribution	0.0	24.7	0.6
	0.0%	47.1%	2.4%
Total North Sea	534.8	261.7	355.9
Expro/NAM contribution	14.1%	18.8%	18.2%

[a] Taylor et. al., 1991
[b] Verslag....,1990

Table 2. Produced water characteristics which demonstrate clearly the differences between distinguished oil and gas fields operated by Shell Expro and NAM

Produced water parameters	pH	Chloride (g/l)	Temperature (°C)
Oil Fields			
Brent	6 -7.7	12.4- 14.8	3-80*
Other northern	6 -7.7	14.7- 16.9	
Central North Sea	6 -7.7	81.0-100	
Gas fields			
UK sector	3.5-5.5	0.1-277	10-50
Dutch sector	3.8-5.5	0.1-189	13-45
North Sea seawater[1]	8.1	18.8- 19.5	3-17

*Note:
Temperature ranges of effluent water: concrete storage: 3-17°C
— steel platforms northern North Sea: 50-70°C
— steel, central NS: 70-80°C
Otto et al., 1990

Formation waters are in principle salt solutions, of which the cationic composition generally resembles seawater. However, they usually have higher salt concentrations and are more acidic. The extent to which this also applies to produced waters is determined by the relative contribution of formation water to the actual effluent discharged into the sea. The data in Table 2 on pH and chloride content of the effluents and of seawater illustrate that the effluents from the oil fields in the East Shetland Basin mainly consisted of "injected water". For the oil fields in the Central North Sea the situation was somewhat different: either no injection (e.g. at Auk) or no/hardly any breakthrough of injected water, so that the major part of produced water was still formation water. This appeared to be of greater consequence for the chloride content of the effluents than the pH. The chloride content was up to 5 to 6 times that of seawater (see Table 2).

Water injection is not applied in gas fields, therefore, the produced waters are primarily a mixture of formation water and condensed water. The effluents, however, could also contain drainage water. The chloride content of the discharges varied greatly: from almost fresh - the condensed - water, to salty formation water with a chloride content of about 14 times that of seawater. These produced waters also appeared to be more acidic than those discharged from the oil platforms.

ORGANIC COMPOUNDS

Under the Paris Convention special requirements exist regarding the quality of platform discharges. These refer not only to the oil content of the streams, i.e., a "target standard" of 40 mg/l dispersed oil, but also to the frequency of sampling and analysis. These requirements have been translated in different ways into the national legislations and are still under revision. Under the Prevention of Oil Pollution Act 1971 (POPA) in the U.K., oil production facilities have to monitor dispersed oil levels two times a day during discharge periods. Dissolved hydrocarbons, polar compounds and natural gas liquids discharged from gas platforms are not currently covered by POPA. According to the Dutch Mining Regulations platform effluents were regularly (16 times a month) sampled and analyzed for oil content. Dissolved hydrocarbons and polar compounds were not determined with this analytical method.

Data from sampling and analyses carried out in 1989 in accordance with the above-mentioned regulatory requirements are summarized in Table 3. From these figures it is clear that there is a structural and fundamental difference between produced water from oil and gas platforms. The Paris Convention "target standard" was easily met at oil producing platforms, whereas with the water treatment equipment installed on gas/condensate platforms the standard appeared an unapproachable target. However, due to the relatively small volumes of water discharged from gas/condensate platforms the contribution to the total oil input from this source is small.

Table 3. Dispersed oil in produced waters discharged from the Shell Expro and NAM platforms. In addition to concentrations, the calculated loadings for 1989 are given

Dispersed oil in produced water	Conc. range (mg/l)	Average (mg/l)	Total load (tonnes/year)
Oil fields			
Brent + Cormorant	2 - 10	3.9	40.80
Other northern	15 - 40	24.8	269.97
Central North Sea	20 - 40	24.1	46.00
Gas fields			
UK sector	10 - 8000	1862	28.40
Dutch sector	41 - 2617	382	79.69
Total			464.86

The total of 465 tonnes of (dispersed) oil in Shell Expro and NAM effluents was about 1.6% of the total quantity (PARCOM reportable) discharged by the offshore industry into the whole North Sea. The two Companies contributed 12% of the total quantity of oil in platform effluents on the North Sea. The total volume of water was, however, 18% of the total amount discharged by the offshore industry (see Table 1), so that the two Companies can be appraised as having above average performance among the offshore North Sea operators in this respect.

If dissolved hydrocarbons are taken into account then the total oil discharged from the oil platforms increased from 356.8 tonnes to 500.1 tonnes in 1989. About half of the dissolved hydrocarbons were monocyclic aromatics and their derivatives. For the gas platforms the situation was somewhat different: the quantity of dissolved hydrocarbons was about the same as that of the dispersed hydrocarbons. Benzene formed 50-88% of these dissolved hydrocarbons and reached concentrations up to 480 mg/l.

Organic components other than oil hydrocarbons included the polar petrogenic compounds, i.e., phenols and carboxylic acids. During analyses they are normally removed by liquid chromatography (e.g., Florisil) before IR determination of total hydrocarbons. They are, however, not removed by conventional water treatment equipment on the platforms. Some production chemicals also belong to this group. Only the phenols and organic acids are reported here; they were analyzed on an irregular basis. Based on the results, average concentrations in effluents and the total discharge over 1989 have been estimated. These results are presented in Table 4.

Acetic acid formed the greater part of the total soluble organic acids discharged from Shell Expro and NAM platforms, i.e., from 70 up to 94%.

The total amount of organics of petrogenic origin in the effluents discharged in 1989 by the two Companies consisted of 465 tonnes dispersed oil (see Table 3), 251 tonnes dissolved hydrocarbons and 3909 tonnes other organics (see Table 4): in total 4625 tonnes. The relative contribution of polar compounds was 84%.

Table 4. Average concentrations of some organics other than hydrocarbons in produced waters discharged by Shell Expro and NAM in 1989 from their 12 oil and 15 gas platforms

Some organics other than hydrocarbons in produced water	Organic acids (av. mg/l)	Phenols (av. mg/l)	Total load (tonnes/year)
Oil fields			
Brent	185.5	1.6	1982.6
Other northern	158.4	1.4	1743.2
Central North Sea	64.0	5.0	132.0
Gas fields			
UK sector	171.7	0.6	2.6
Dutch sector	227.3	6.4	48.8
Total			3909.2

INORGANIC COMPOUNDS

The produced waters contained a wide range of both cations and anions. The anionic composition was not analyzed on a regular basis; more data were available on the cationic composition. We, therefore, concentrate here on the metals and metalloids. An overview of the summarized data is presented in Table 5. The data are concentrations averaged for a number of effluents in a specific area. This appeared the only way to get an overall picture of the enormous amount of data collected. A disadvantage is, however, that differences between effluents in a specific area or concentrations below the detection limit, are no longer directly visible.

All effluents showed similar patterns of concentration for the different metals. The measured alkaline (Na, K, and Li), earth (only B) and alkaline-earth (Ba, Ca, Mg and Sr) metals showed the highest concentrations in all cases. Sodium appeared the dominant cation, but calcium, magnesium and potassium were also present in high concentrations. The concentration of calcium was in most instances higher than of magnesium, i.e., reversed compared to seawater. Although not apparent from the averaged data in Table 5, some effluents from oil platforms showed a ratio $\leq$ 1, indicating a significant breakthrough of injected seawater. All effluents also contained appreciable quantities of boron, strontium and barium. However, particularly in effluents from oil platforms with a significant "water cut," the barium content was much lower than in effluents with less or no breakthrough of seawater. This was probably a consequence of precipitation of barium sulphate scale following contact with seawater McCourt and Peers, 1988).

Although the pattern of concentrations of metals in effluents from oil platforms closely resembled the composition of (far offshore) North Sea seawater (compare data in Tables 5 and 6), the actual concentrations were a factor of up to about 1000 higher than the levels of injected seawater. This suggests that in case of water injection, a certain leaching of metals from reservoir sediments occurred. For effluents composed of only formation water, significantly higher concentrations were found.

Table 5. Average concentrations of metals in produced waters discharged in 1989 by Shell Expro and NAM into the North Sea

Element	Average concentration (mg/l)				
	Oil fields			Gas fields	
	Brent	Other Northern	Central North Sea	UK Sector	Dutch Sector
Al	0.66	0.83	2.33	4.48	nd
As	nd*	nd	nd	nd	0.005
B	20.89	26.49	81.55	1.70	82.61
Ba	1.27	1.92	3.44	0.66	6.60
Be	nd	nd	nd	nd	0.61
Ca	1167	467.9	6010	5560	7240
Cd	<0.10	0.01	<0.10	0.03	0.19
Co	<0.3	<0.3	0.4	<0.3	<0.1
Cr	0.10	0.10	0.10	0.26	0.07
Cu	<0.06	0.6	0.8	0.6	0.01
Fe	1.81	3.78	5.93	66.09	nd
Hg	<0.003	<0.003	<0.003	<0.001	0.004
K	375.4	393.8	1048.2	199.6	266.1
Li	1.44	4.21	17.15	52.38	nd
Mg	770.8	711.7	816.2	445.7	994.4
Mn	0.27	0.65	1.72	1.18	11.33
Na	8611	1060	42909	8185	35484
Ni	0.10	0.23	0.90	1.22	0.08
Pb	<0.20	<0.05	<0.05	0.38	0.84
Sb	nd	nd	nd	nd	0.26
Se	nd	nd	nd	nd	0.002
Sn	nd	nd	nd	nd	0.07
Sr	13.98	29.75	393.0	65.56	154.27
V	<0.10	<0.10	0.10	0.07	<0.1
Zn	0.49	0.84	4.59	0.20	68.96

Note:
nd = not determined

The cationic composition of produced waters from gas/condensate platforms followed a pattern generally comparable to that described above for oil platforms (see Table 5). However, here the earlier demonstrated difference between almost fresh - condensed - water and very salty formation water is confirmed: very low quantities of all metals in condensed water to indeed very high concentrations of alkaline, earth and alkaline-earth metals, but also some heavy metals, in formation water. For the NAM effluents a relationship between the depletion of the gas reservoirs and concentrations of inorganic compounds could be demonstrated (Jacobs and Marquenie, 1991). This means that the longer gas reservoirs are in production, the greater the relative contribution of formation water in the effluent streams and the higher the concentrations of inorganics therein. The data for the Expro effluents from gas platforms confirm this phenomenon. These relationships were most clear for the alkaline and earth-alkaline metals. For the other metals and the radionuclides this was not so marked.

Total radio-activity was measured (as radionuclide activity) in effluents from NAM platforms. Radioactivity was detected in six of the eight effluents with a total activity of water soluble radionuclides between 0 and 450 Bq/l. In Expro effluents only uranium concentrations were determined, which were below the detection limit.

Table 6. Background concentrations of metals in North Sea and acute toxicity levels of some metals

Element	Background level North Sea water (ranges) (mg/l)[1]	Acute toxicity LC/EC$_{50}$ values (mg/l)		
		Algae	Crustaceans	Fish
Al	0.16-1.9			
As$^+$	0.003-0.024	0.58	24.7	15
Ba	0.05			
Cd$^+$	0.00001-0.00005		0.015-11.6	0.8-50
Co	0.0001			
Cr$^+$	0.2-1.0	1-5	2-98	13.4-91
Cu$^+$	0.0001-0.001	0.05-0.5	0.5-100	0.03-0.5
Fe	0.002-0.02			
Hg$^+$	0.000001-0.0001	<0.005 - 0.4	0.004- 0.4	3-4
Li	0.1			
Mg	1.272			
Mn	0.001-0.01			
Ni$^+$	0.0001-0.0005	2	0.15-47	8-350
Pb$^+$	0.00001-0.00006		0.7- 3	
Se	0.004			
Sn	0.003			
Sr	13			
V	0.0003			
Zn$^+$	0.00055-0.0043	0.1-0.25	0.5- 4	60-70

[1]Beukema et al., 1986; Ad Hoc Working Group..., 1991; Weast and Astle, 1989
[2]Kersten et al., 1980, Metals mentioned in the "black and grey lists" of Paris Convention

From the results it appeared that most effluents contained inorganic compounds in concentrations higher than in seawater. All metals mentioned as "black and grey-listed" compounds of the Paris Convention, are present in the effluents. According to the Convention the discharge of these compounds should be prevented or controlled because of their health and/or environmental hazard. The compounds are indicated in Table 6. The black-listed compounds are mercury and cadmium. The rest, including radioactive compounds, are grey-listed. A further discussion on the actual environmental hazard posed by the discharges is presented in the following section.

TOXICITY - DISCUSSION WITH REGARD TO COMPOSITION

In order to get a better insight into potential effects on the environment following discharge, both Shell Expro and NAM had already commenced several studies prior to 1989. These studies included laboratory toxicity testing, biodegradation experiments, collection of data on bioaccumulation and determination of the chemical oxygen demand (COD). In addition to the bioaccumulation potential, field experiments were carried out by Expro by means of the "mussel watch". An overview of the acute toxicity data for NAM and Expro effluents is presented in Table 7.

Many aspects related to the assessment of potential environmental effects caused by the discharges have been presented and discussed earlier (Jacobs and Marquenie, 1991; Girling, 1987; I. Somerville et. al.,1987 Girling and Streatfield, 1988; Jacobs, 1987). The following is worth noting. Although rather different in composition, the discharges tested (i.e., from the Brent and Dunlin oil platforms and all NAM effluents) demonstrated a rather similar level of toxicity: the median toxicity of the effluents from gas/condensate platforms being about 10 times lower (i.e., more toxic) than that of effluents from oil platforms. Taking into account the levels of acute toxicity for some heavy metals (see Table 6) and the actual concentration of these metals in the effluents (see Table 5), the toxicity can hardly be ascribed to these compounds. These were apparently not present in an ionic form. This means that, although some metals are on the Paris Convention "black and grey lists", they do not pose a significant acute hazard. Jacobs and Marquenie (1991) concluded that the toxicity of the NAM gas/condensate platforms could be ascribed to a mixture of several "key pollutants", of which the dissolved hydrocarbons were of prime ecotoxicological importance.

Table 7. Results of acute toxicity tests carried out with produced waters from Shell Expro and NAM platforms

Toxicity of produced waters Species Tested	Test Conditions	Effect conc. (ml/l)
Produced waters from Expro oil fields		
Artemia salina (brine shrimp)	24h LC_{50}	160-180
Crassostrea gigas (oyster embryo)	48h LC_{50}	50
Crangon (brown shrimp)	24h LC_{50}	20
Calanus finmarchicus (a calanoid)	24h LC_{50}	100
Salmo gairdneri (rainbow trout)	24h LC_{50}	100
Campanularia flexuosa (a hydroid)	10d growth	50
Acartia tonsa (a calanoid copepod)	24h LC_{50}	50
	96h LC_{50}	20
	20d reproduction	3
Photobacterium phosphoreum ("Microtox")	15 min	50-60
Produced waters from NAM gas fields[2]		
Phaeodactylum tricornutum (an alga)	96h EC_{50}	0.9 - 36
	96h EC_{50}	3 - 60
	96h NOEC	0.3 - 18
Chaetogammarus marinus (an amphipod)	96h LC_{50}	1.8 - 32
	96h LC_{50}	5 - 71
	96h NOEC	1 - 18
Poecilia reticulata (guppy)	96h LC_{50}	7.5 - 423
	96h NOEC	5.6 - 320

EC = effect concentration; LC = lethal concentration; NOEC = No-Observed-Effect-Concentration
[1]Girling, 1987; Somerville et al., 1987; Girling et al., 1988
[2]Jacobs, 1987; Jacobs and Marguenie, 1991

The ecotoxicity of the effluents from the gas platforms is mainly accounted for by the presence of high concentrations of monocyclic aromatics, although it is possible that in a few effluents, production chemicals also contribute to the toxicity. Among these, the only product with a considerable toxicity was a corrosion inhibitor. The effluents from the oil platforms contained significantly less monocyclic aromatics. This can explain the difference in toxicity between the oil and gas discharges. However, based on the measured concentrations of these aromatics, a greater difference than the observed factor 10 would be expected. Possibly the contribution of other production chemicals, such as corrosion inhibitors, oxygen scavengers and water conditioners, was relatively more important in determining the toxicity of the effluents. It is worth noting that acute toxicity data from literature for produced water from oil platforms all indicate levels which are comparable to the data presented here (Jacobs and Marguenie, 1991).

REFERENCES

Ad Hoc Group of ICES Marine Chemistry Working Group, 1991, "A review of measurements of trace metals in coastal and shelf sea water samples collected by ICES and JMP laboratories during 1985-1987", Draft ICES Coop. Res. Report.

Beukema, A.A., G.P. Hekstra and Venema, C., 1986, "The Netherlands' environmental policy for the North Sea and Wadden Sea," Env. Mon. Ass., 7:117-155.

Girling, A.E., 1987, An assessment of the environmental hazard associated withthe discharge of production water from a North Sea oil platform based on laboratory bioassays with a calanoid copepod - *Acartia tonsa* (Dana), in: "Fate and Effects of Oil in Marine Ecosystems," J. Kuiper and W.J. van den Brink (eds), Martinus Nijhoff, Dordrecht, pp. 215-216.

Girling, A.E., and Streatfield, C.M., An assessment of the acute and chronic toxicity of production water from a North Sea oil platform based upon laboratory bioassays with a calanoid copepod - *Acartia tonsa* (Dana), 1988, in: "Environmental Protection of the North Sea," P.J. Newman and A.R. Agg (eds), Heinemann Professional Publishing, Oxford, pp. 789-791.

Jacobs, R.P.W.M., "Environmental impact assessment of effluentsdischarged from gas/condensate platforms," 1987, NOGEPA Report no. 6, The Hague, The Netherlands.

Jacobs, R.P.W.M., and Marquenie, J.M., 1991, Produced water discharges from gas/condensate platforms: environmental considerations, in: "Proceedings of the First International Conference on Health, Safety & Environment in Oil and Gas Exploration and Production," 11-14 November 1991, The Hague, The Netherlands. Society of Petroleum Engineers (SPE) Inc. Volume 2, pp. 89-96.

M. Kersten, M. Dicke, M. Kriews, K. Naumann, D. Schmidt, M. Schulz, M.Schwikowski and Steiger, N., (1988), Distribution and fate of heavy metals in the North Sea, in: "Pollution of the North Sea - an Assessment," W. Salomons, B.L. Bayne, E.K. Duursma and U. Förstner (eds), Springer-Verlag, Berlin, pp. 300-347.

McCourt, C.B., and Peers, D.M., 1988, "Environmental Monitoring of Trace Elements in Water Discharges from Oil Production Platforms," Petroanalysis 1987, Developments in Analytical Chemistry in the Petroleum Industry, John Wiley & Sons.

National Environmental Policy Plan - The Netherlands, 1989, SDU Uitgeverij, The Hague.

Newman, P.J., and A.R. Agg, 1988, (eds). "Environmental Protection of the NorthSea," Heinemann Professional Publishing, Oxford.

Otto, L., J.T.F. Zimmerman, G.K. Furnes, M. Mork, R. Saetre and Becker, G., 1990, "Review of the physical oceanography of the North Sea," Neth. J. Sea Res., 26 (2-4): 161-238.

Peet, G., 1987, "The Status of the North Sea Environment: Reasons for Concern," Proceedings of the 2nd North Sea Seminar, 1-3 October 1986, Rotterdam, The Netherlands. Werkgroep Noordzee, Amsterdam.

Salomons,W., B.L. Bayne, E.K. Duursma and Förstner, U., 1988, (eds). "Pollution of the North Sea: an Assessment," Springer-Verlag, Berlin.

Somerville, H.J., D. Bennett, J.N. Davenport, M.S. Holt, A. Lynes, A. Mahieu, B. McCourt, J.G. Parker, R.R. Stephenson, R.J. Watkinson and Wilkinson, T.G., 1987, Environmental effect of produced water from North Sea oil operations, Mar. Poll. Bull., 18:549-558.

Taylor, J., R.E. Larson and Scherer, B., 1991, Treatment of offshore producedwater - an effective membrane process, in: "The 1st International Environment Northern Seas Conference and Exhibition," 26-30 August 1991, Stavanger, Norway, Vol. 7 pp. 127-142.

Verslag Inspecteur Generaal der Mijnen over 1989, 1990, Ministerie vanEconomische Zaken, Staatstoezicht op de Mijnen, The Hague, The Netherlands.

Weast, R.C., and Astle, M.J. (eds), 1989, "CRC Handbook of Chemistry and Physics," 69th Edition, CRC Press, Boca Raton, Florida (1988-1989).

REVIEW OF THE COST-EFFECTIVENESS OF EPA'S OFFSHORE OIL AND GAS EFFLUENT GUIDELINES

J.A. Veil

Argonne National Laboratory
Washington, DC
USA

INTRODUCTION

The United States Environmental Protection Agency (EPA) published proposed Effluent Limitation Guidelines and New Source Performance Standards for the Offshore Subcategory of the Oil and Gas Extraction Point Source Category on March 13, 1991 (56 FR 10664). These guidelines include limitations for Best Practicable Technology (BPT), Best Conventional Technology (BCT), Best Available Technology (BAT), and New Source Performance Standards (NSPS) for produced water, drilling fluids, drill cuttings, well treatment fluids, deck drainage, produced sand, and sanitary and domestic waste.

As a support document to these regulations, EPA published a Cost-Effectiveness (CE) Analysis (EPA, 1991). This report calculated the CE of several different BAT and NSPS treatment options for produced water and drilling fluids and drill cuttings. Separate calculations were made for restricted versus unrestricted drilling rights and for three different oil costs per barrel. Each treatment option has a different cost and will result in removal of different levels of pollutants. The report calculates the annualized cost for each option and the pound equivalents (PE) removed and expresses the CE as $/PE. The CE value for the selected treatment option can be compared to the CE statistics from other industries for which EPA has done the analysis. While not officially required for decision-making, as long as the CE falls within some reasonable range, the selected treatment option is considered acceptable.

This paper examines the premises and assumptions used in the CE analysis and recalculates the CE statistic using alternate assumptions.

EPA'S CE METHODOLOGY

Ehrensberger and Rico (1988) describes and provides guidance on how to use EPA's CE methodology. CE is defined as the incremental annualized cost of a pollution control option in an industry or industry subcategory per incremental pound equivalent of pollutant removed annually by that control option. When calculating annual pollutant removal, the relative strengths of the different pollutants must be considered. A pound of chromium will be far more harmful to the environment than a pound of iron. One way of scaling different pollutants is to assign a weighting factor to each pollutant based on its toxicity. EPA's methodology calculates weighting factors using EPA's ambient water quality criteria for chronic aquatic life protection (fresh or salt water) and for human health protection through fish consumption. For carcinogenic

Produced Water, Edited by J.P. Ray
and F.R. Engelhart, Plenum Press, New York, 1992

pollutants, the risk factor is taken as 10^{-5}. All pollutants are normalized to a standard of 5.6 ppb, the aquatic life protection criterion for copper at the time the methodology was developed. Equation 1 shows the formula used.

$$\text{weighting factor} = 5.6/AL + 5.6/HH, \text{ where} \quad (1)$$

AL = aquatic life protection criterion and
HH = human health protection criterion.

Two examples are shown below.

Nickel 5.6/100 + 5.6/8.3 = 0.73
Mercury 5.6/.025 + 5.6/.146 = 262.4.

A more toxic pollutant will have lower criteria resulting in a higher weighting factor.

The next step in the procedure is multiplication of the actual pounds removed by the weighting factors to get PE removed. Since CE is calculated on an incremental basis, the PE must be calculated separately for each treatment option. The incremental PE is the PE for the option under consideration minus that for the next less stringent option. Costs must also be calculated on an incremental basis. The incremental CE is then calculated by dividing the incremental annualized cost by the incremental PE. Tables 4-1 through 4-14 in EPA (1991) contain the results of EPA's calculations for the different options. In the CE analysis, all costs are scaled to 1981 dollars to allow comparison to the statistics calculated for other industries.

The list of pollutants used in the CE analysis is not limited to those pollutants regulated by the effluent guidelines. The list will also contain selected other toxics or non-conventional non-toxics which will be removed by a treatment option. For example, for drilling fluids and drill cuttings, only mercury and cadmium are limited in the effluent guidelines, but many other metals are also present and will be removed by the treatment options.

EPA'S CONCLUSIONS

In the March 13, 1991 proposed effluent guidelines, EPA selected treatment options for each waste stream. For drilling fluids and drill cuttings, wells from 0 - 4 miles from shore may not discharge at all, and wells greater than 4 miles from shore have limits for mercury and cadmium of 1 mg/kg. For produced water, wells from 0 - 4 miles from shore must meet very stringent limits for oil and grease which EPA believes will require membrane filtration, and wells greater than 4 miles from shore must meet less stringent standards which do not require any additional treatment.

EPA performed the CE analysis for each waste stream and found that BAT and NSPS for drilling fluids and drill cuttings would cost $22/PE, BAT for produced water would cost $60/PE, and NSPS for produced water would cost $63/PE. CE statistics for twenty other industries range from less than one dollar to $404/PE (EPA, 1991). Therefore EPA was satisfied that the treatment options selected for drilling fluids and drill cuttings and produced water were not unreasonable from a CE perspective.

DISCUSSION OF EPA'S ASSUMPTIONS

There are several assumptions used in the CE methodology which warrant discussion. Some of these are applicable to all industries and others are only relevant to the offshore oil and gas industry. The first category of assumptions represents shortcomings in EPA's CE methodology, but since they have been consistently applied to all other industries for which a CE analysis has been done, they are not the subject of this paper and will be discussed only briefly.

GENERAL SHORTCOMINGS IN CE METHODOLOGY

This category includes: the approach of using weighting factors to account for different environmental impacts of various pollutants; the need to sum the effect of aquatic life protection criteria and human health protection criteria; the use of an incremental approach rather than a cumulative approach; and the failure to consider the form of metal used in the calculations.

Using Weighting Factors

It is desirable to equitably account for the wide range of toxicities found among chemicals, but the methodology used to weight different pollutants must be reasonable or the resulting weighted pollutant loads will not be meaningful. The approach used in EPA's CE methodology is plausible and is based on a set of generally consistently derived criteria.

One weakness is the absence of a complete set of EPA criteria. A complete set of criteria for a pollutant includes freshwater and marine acute criteria, freshwater and marine chronic criteria, and human health criteria from fish consumption and from fish consumption and drinking water intake. While EPA has developed criteria for over 100 pollutants, many of those criteria sets are incomplete, especially for the chronic or human health criteria which are used in the CE methodology. Ehrensberger and Rico (1988) advises that the person conducting the CE analysis contact EPA's Criteria and Standards Division for some less rigorously derived "criterion" or to otherwise estimate the criterion value. No specific guidance is given on how the person should estimate the criterion, lending some doubt to the reliability of the subsequent weighting factor.

Another weakness of using EPA criteria is that those criteria are technical recommendations, not adopted standards. As such, they have not faced full regulatory scrutiny. Following the Water Quality Act of 1987, as states were required to adopt water quality standards for many toxic standards, some of the EPA criteria were seriously questioned by both states and the regulated community. One important example of this is the human health criterion for arsenic. Murphy (1990) provides substantial evidence that the arsenic criterion is based on invalid conclusions and is overly stringent. This is particularly noteworthy since the actual criterion value for arsenic is one of the lowest (0.175 ppb) and it results in a relatively high weighting factor.

EPA (1991a) recommends updating human health criteria by using the latest risk assessment information from the Integrated Risk Information System (IRIS). EPA (1991) has followed this guidance and has updated many of the human health criteria for pollutants used in the CE analysis for the offshore oil and gas industry. While using up-to-date information is desirable, this practice creates two problems. The data is not fully peer reviewed and subjected to EPA's formal criteria development procedures, leaving the numbers open to criticism. By using changed criteria, newer CE analyses can no longer be equitably compared to earlier CE analyses. Table 1 shows both the published and recalculated human health criteria for all pollutants considered in the CE analysis for produced water in the offshore industry.

Summing Criteria

Ehrensberger and Rico (1988) specifies that the weighting factor should be calculated separately for both the aquatic life and human health criteria and then should be summed (see equation 1). This practice serves to make the weighting factor larger, especially if the two criteria are approximately equal. It makes more sense to use whichever of the two criteria is most restrictive and base the weighting factor on only that single number.

Incremental Approach

Ehrensberger and Rico (1988) requires that CE should be calculated on an incremental basis. When several treatment options are being considered, costs and PE removed are calculated for each option. $/PE are then incrementally calculated. This approach is dependent on the number of options under consideration. While for most industries only a few treatment options are considered, the more options that are considered, the less meaningful the incremental approach becomes. As an extreme example, consider 100 treatment options each one marginally more restrictive than the previous one. Imagine that cadmium and mercury limits for drilling fluids and drill cuttings could be established anywhere from 0.5 mg/kg to 10.5 mg/kg at increments of 0.1 mg/kg. This creates 100 possible regulatory options. The incremental $/PE are going to be very small and not very useful. A more logical approach would be to calculate the cumulative $/PE for each option. This approach would provide obvious breakpoints in CE, and allow selection of the most appropriate treatment option.

Form of Metals

As the emphasis on toxics has increased, greater attention has been focused on analytical methods for metals. Most metals can be analyzed in several ways using different sample preparation techniques. The most common metal forms in order of increasing portions of the total metal present are dissolved, acid soluble, total recoverable, and total. Most of EPA's aquatic life criteria are based on and expressed as the acid soluble form as this more closely approximates the way metals are encountered by biota in the environment. On the other hand, permit limits and probably the pounds of pollutants removed by the various treatment options considered in a CE analysis are measured as total or total recoverable metals.

Table 1. Criteria used for toxic weighting factors - Produced Water

Pollutants	Salt/Fresh	Chronic Criteria (ppb)	Published Human Health Criteria (ppb)	Updated but Not Published Human Health Criteria (ppb)
benzene	S	700 a	400	710
2-4-dimethyl phenol	F	- b	- b	-
ethylbenzene	F	- b	3,280	29000
naphthalene	F	620 a	- b	-
toluene	S	5,000 a	424,000	300,000
bis(2-EH) phthalate	S	360 a	50,000 a	5.9
phenol	F	2,560 a	3,500 a	4,600,000
m-xylene	S	- b	- b	-
p-cresol	F	- b	- b	-
n-alkanes	S	- b	- b	-
2-butanone	S	- b	- b	-
steranes	F	- b	- b	-
triterpanes	F	- b	- b	-
radium	S	- b	- b	-
aluminum	F	87	- b	-
arsenic	S	36	.175	1.4
boron	F	- b	- b	-
barium	F	- b	1,000 a	-
copper	S	2.9	- b	1,300
iron	F	1,000	300 a	-
manganese	F	- b	- b	-
titanium	F	- b	- b	-
zinc	S	86	- b	-

a - published criterion available and shown here, but EPA used another value instead
b - no published criterion available

There is very little information available about the ratio between acid soluble and total metals. Battelle Ocean Systems has conducted a sampling program in New York Harbor and the preliminary results suggest that there is very little difference between acid soluble and total recoverable metals. Brumfield (1988) studied treated municipal wastewater and sludge from eight cities in Indiana and found very little difference between acid soluble and total metals in the treated wastewater, but in the sludge analysis, the acid soluble test results ranged from 9 - 90% of the total metal results. Another source of data comes from the Hart-Miller Island Dredged Material Disposal Facility in Chesapeake Bay near Baltimore Harbor (Stone, 1991). A summary of eight samples found that acid soluble metals values are from 45 - 59% of total metals values.

This limited data suggests that when metals are associated with elevated solids levels the acid soluble measurement declines noticeably compared to the total metal measurement. The ratio will vary depending on the suspended solids and water chemistry of the receiving water, but it is clear that the biologically available pounds of metals discharged are less than the total pounds of metals. Thus a more environmentally accurate $/PE for acid soluble metals will be higher than the regular $/PE for total metals.

SHORTCOMINGS IN EPA'S CE METHODOLOGY SPECIFIC TO THE OFFSHORE OIL AND GAS INDUSTRY

The offshore oil and gas industry is unique among major industries in that all regulated operations are located away from land and are in a marine environment. No other major industrial category has all its members discharging to salt water or facing the space and logistical problems inherent to the offshore industry. These factors need to be considered in developing not only the CE analysis, but in selecting a treatment option as well. There are three shortcomings in EPA's CE analysis which are specific to the offshore oil and gas industry. These are the availability and stringency of marine chronic criteria compared to fresh water chronic criteria, failure to consider the vast dilution found at an offshore oil or gas platform, and the use of human health criteria for determining weighting factors when the discharges are short-lived. The first two issues will be discussed briefly and a detailed recalculation of weighting factors is provided for the third issue.

Marine vs. Fresh Water Criteria

EPA has developed 67 fresh water chronic criteria, but only 42 marine chronic criteria (EPA, 1986). For many of the pollutants chosen for the CE analysis, no marine chronic criterion exists, and some alternative, less structured "criterion" must be chosen to develop weighting factors.

The actual values for marine criteria are smaller than the fresh water criteria for the same pollutants in some cases. This is not necessarily due to an inherently higher toxicity of the pollutants in marine waters, but rather to a feature of the criterion derivation procedure. Calculation of chronic criteria involves statistical analysis of chronic toxicity data which is sensitive to the number of species tested (EPA, 1985). When fewer species are tested, the uncertainty that the data represents a wide range of species remains high and the criterion is increased accordingly. Historically, there have been far more aquatic toxicity tests conducted on fresh water species than on marine species. By virtue of the smaller number of available marine test results to be used in the calculations, a marine criterion will be lower than the corresponding fresh water criterion even if the most sensitive species in both sets of data show exactly the same toxicity results.

The impact on the offshore oil and gas industry is that weighting factors can be higher than they would be for an industry discharging into fresh water, due to a lack of data. It is not feasible to quantify the effect that this has on the CE statistics.

Dilution in the Offshore Environment

Normally a CE analysis does not consider dilution of the discharged pollutants, but in the offshore environment the dilution is extremely large and warrants evaluation. Discharging 10 pounds of copper into a small stream or river may be important to the local ecosystem, but discharging the same amount of copper at a well several miles offshore has virtually no impact on the local ecosystem. Dilution can be calculated by equation 2.

dilution	=	(V x D x W)/Qe, where:	(2)
V	=	ambient flow velocity	
D	=	water depth	
W	=	width of zone of dilution	
Qe	=	effluent flow	

Ambient flow velocity is variable with weather conditions and depth. Ecomar (1980) reported offshore currents near a Gulf of Mexico drilling platform from 0 - 30 cm/sec, and above the bottom, the lowest velocity was 3 cm/sec. O'Reilly et al (1989) found surface currents near a California offshore platform from 2.5 - 7.2 cm/sec and bottom currents of 1.5 - 4.1 cm/sec. A conservative velocity of 1.5 cm/sec is assumed here.

Water depth is also variable; the proposed effluent guidelines define shallow water as 20 m or less. EPA (1991b) reports that for the Gulf of Mexico only 10 - 37% of all new wells will be drilled in shallow water. Nevertheless, a depth of 20 m is used for the dilution calculation. Other wells would have even greater dilution. There is no particular convention for selecting a zone of dilution width. Two widths, 200 m and 1000 m, were used in the dilution calculation. They represent affected radii of 100 m and 500 m respectively.

The average daily effluent flow for drilling fluids and drill cuttings in the Gulf of Mexico is 163 bbl/day (EPA, 1991b). For produced water, the average daily discharge is less than 1800 bbl/day, but some large facilities may discharge as much as 157,000 bbl/day (Walk, Haydel & Associates, 1984).

These assumptions result in dilutions of 200,000:1 for a dilution zone width of 200 m and 1,000,000:1 for 1000 m width for drilling fluids and drill cuttings. For produced water, the dilutions are 18,126:1 and 90,629:1 respectively for the average discharge volume, and 208:1 and 1038:1 respectively for the maximum daily discharge.

Even the smallest of these dilution ratios is significant and the larger dilution ratios are so high that they should be considered in assessing the true environmental impact of the pounds or PE discharged. When diluted to the extent predicted by these calculations, many of the pollutants will be nondetectable and very few if any will be biologically harmful.

Applicability of Human Health Criteria for Calculation of Weighting Factors

Ehrensberger and Rico (1988) requires that the weighting factor be based on both chronic aquatic life protection and human health criteria. Clearly using the chronic criterion is appropriate for protecting the ecosystem, but when the assumptions used to derive a human health criterion are considered, its use is questionable for the offshore oil and gas industry. A human health criterion is based on exposure to a given quantity of the pollutant for a 70 year period (EPA, 1991a). For many land-based dischargers, their discharge will continue in the same location for many years, and the 70 year assumption is reasonable. In the offshore oil and gas industry, the discharges are shorter in duration, especially discharges of drilling fluids and drill cuttings.

EPA (1991b) reports that the average total volume of drilling fluids and drill cuttings from a Gulf of Mexico well is 8,379 bbl, and that the average daily discharge is 163 bbl. This yields an average period of discharge of 51.5 days. The duration of produced water discharge depends on the productivity of the well. At the start of oil or gas production, the proportion of produced water is small, but as the reservoir becomes depleted, the proportion of produced water increases until at some point the amount of oil and gas produced becomes uneconomical and the well is closed. A typical duration of produced water discharge is about 15 years per field.

The duration of both waste streams is significantly less than 70 years and therefore the weighting factor should not be calculated using the human health criterion. The weighting factors used in EPA (1991) have been recalculated using only the chronic aquatic life protection criteria whenever possible. Marine chronic criteria were used if available. If not, fresh water chronic criteria were used. If no chronic criterion was available, the value used by EPA was also used here.

Table 2 shows the EPA and revised toxic weighting factors for BAT for produced water. Table 2 also provides revised PE loads calculated using the revised weighting factors. Similar calculations were made for drilling fluids and drill cuttings (NSPS - restricted, and NSPS - unrestricted) and for produced water (NSPS - restricted, and NSPS - unrestricted). Due to space limitations, these are not reproduced in this paper.

Table 2. Toxic Weighting Factors And Pound Equivalents - Produced Water - BAT

Pollutants	Pollutants Removed (lb/yr)	EPA Toxic Weighting Factor	Revised Toxic Weighting Factor	EPA Equivalents Removed (PE/yr)	Revised Equivalents Removed (PE/yr)
benzene	58471	.036	.008	2,105	468
2-4-dimethyl phenol	458	.0056	.006	3	3
ethylbenzene	16148	.262	.001	4,231	16
naphthalene	4411	1.47	.009	6,484	40
toluene	49387	.0018	.001	89	49
bis(2-EH) phthalate	3226	.097	.016	313	52
phenol	30446	.0193	.002	588	61
m-xylene	4888	.160	.160	782	782
p-cresol	11640	.007	.007	81	81
n-alkanes	52482	.074	.074	3,884	3,884
2-butanone	53357	.0001	.0001	5	5
steranes	2012	.074	.074	149	149
triterpanes	2429	.074	.074	180	180
radium	.0135	1060000	1060000[a]	14,310	14,310[b]
aluminum	1648	.037	.064	61	105
arsenic	4172	32.16	.155	134,172	647
boron	4374	.177	.177	774	774
barium	51305	.002	.002	103	103
copper	1514	1.93	1.93	2,922	2,922
iron	42726	.0056	.0056	239	239
manganese	269	.056	.0144	15	4
titanium	114	.029	.029	3	3
zinc	78589	.065	.065	5,108	5,108
Total				176,601	29,985[c]

[a] 122,000 - recalculated toxic weighting factor for radium
[b] 1,647 - reflects recalculated toxic weighting factor for radium
[c] 17,322 - reflects recalculated toxic weighting factor for radium

Table 3 provides a comparison of EPA PE loads, revised PE loads, EPA $/PE, and revised $/PE for each option. The revised PE loads were in all cases much less than EPA's PE loads and the revised $/PE were much greater than EPA's $/PE.

It is interesting to note that the pollutant most affected by the recalculation procedure is arsenic. In the EPA PE loads, arsenic makes up from 10 to 76% of the total load, but in the revised loads, arsenic is never more than 2.1% of the total load. This reduction is attributable to the extremely low human health criterion for arsenic. As discussed earlier, Murphy (1990) points out the shortcomings of the arsenic human health criterion. If this suspect criterion is not included in the analysis, the weighting factor becomes much more moderate.

While the need to use criteria values derived from outside sources and not true criteria is understandable when no criteria are available, EPA did not use published criteria to calculate the weighting factors for a number of pollutants, even when they were available. The cases for which EPA used an amended criterion when a published criterion was available, and when no published criterion was available are indicated on Table 1. Instead, EPA relied on some outside source of information which was not documented in EPA (1991) or elsewhere in the official record. The resulting numbers were listed in several tables found in the official record (EPA, 1990) without any accompanying report or explanation of how they were derived. Most likely, the amended numbers were based on more current information which was not extensive enough to allow formal criterion development and may have come from the AQUIRE data base or from some other source. Often the data was in the form of an LC_{50}, an acute toxicity statistic, and not a chronic toxicity result.

By using the amended data for some pollutants, EPA has deviated from the specified procedures in Ehrensberger and Rico (1988). Since the amended values are almost all more stringent than the official criteria, the resulting weighting factors become higher. The use of numbers other than true criteria creates some uncertainty, especially when the pollutant in question represents a substantial portion of the total load. In the case of produced water two pollutants make up 60% of the revised total PE load — radium (47%) and n-alkane (13%). But criteria are not available for either of these so EPA relied on amended data. The radium data is solely human health data, and no information is given on the derivation of the n-alkane data. When important components of the PE load are based on either inappropriate or unexplained data, the total PE load becomes questionable.

EPA's calculation of toxic weighting factors for ^{226}Ra and ^{228}Ra was not found in the public record, but rather was an unpublished memorandum (EPA, 1990). In this document, EPA used safe drinking water concentrations of 2.1 pCi/l for ^{226}Ra and 2.8 pCi/l for ^{228}Ra. Since then, EPA has proposed National Primary Drinking Water Regulations for radionuclides which establish drinking water standards for ^{226}Ra and ^{228}Ra of 20 pCi/l. This corresponds to risk levels of 10^{-4} and 8×10^{-5} respectively. Although the CE methodology calls for a risk level of 10^{-5}, if EPA is willing to use a less stringent risk level in its national drinking water regulations, that same level should be acceptable here. The revised toxic weighting factor for radium (average

Table 3. Comparison of PE loads and $/PE

	EPA PE Load (PE/year)	Revised PE Load (PE/year)	EPA $/PE (1981 $)	Revised $/PE (1981 $)
Drilling Fluids and Cuttings				
NSPS-Unrestricted	491,192	329,818	$28	$42
NSPS-Restricted	442,493	330,227	$22	$29
Produced Water				
BAT	176,601	29,985 [17322]	$60	$355 [$615]
NSPS-Unrestricted	350,240	59,400 [34355]	$40	$237 [$410]
NSPS-Restricted	140,065	23,776 [13739]	$63	$372 [$644]

[ ] - reflects recalculated toxic weighting factor for radium

of 226 and 228) is 1.22×10^{-5}. Since radium accounted for 47% of the revised total load for produced water, an 8.7 fold reduction in its toxic weighting factor will have a marked effect. The revised PE loads decrease by an additional 42% when the radium toxic weighting factor is recalculated, as shown in Table 3.

The earlier CE analyses dealt primarily with the priority pollutants and mainly the heavy metals. But more recently, the list of pollutants considered has been expanded to include other non-priority pollutants and nonconventional non-toxics (Ehrensberger and Rico, 1988). Many of the pollutants considered in the CE analysis for the offshore oil and gas industry fall in these latter categories. This represents one more way in which the total PE load for the offshore industry is not completely comparable to some of the other industries for which the CE analysis has been conducted. Since more PE are contributed by these additional pollutants, the \$/PE will be proportionately lower than for some of the earlier analyses.

Although offshore oil and gas industry discharges do not continue for 70 years, there is some concern that the discharged pollutants may enter the food chain and be subject to bioaccumulation, the uptake and retention of a chemical in animal tissues from external sources. This could lead to human health effects through consumption of fish and shellfish exposed to offshore industry discharges. Neff (1988) reviewed the available literature concerning the potential for bioaccumulation of toxic constituents of drilling and production wastes in estuarine and marine food chains and reported that although produced water contains various alkanes and aromatic hydrocarbons which are rapidly bioaccumulated by marine organisms, these chemicals are not persistent in animal tissue. Half times for release of the accumulated hydrocarbons ranges from less than one day to ten days. Biomagnification, a type of bioaccumulation in which the concentration of a pollutant increases at each higher trophic level in the food chain, does not occur because most marine animals have only a limited ability to accumulate toxics from their food and all species are able to rapidly clear their tissues of hydrocarbons.

Neff also found that metals in drilling fluid discharges are primarily in an insoluble form. Even though they may be bioaccumulated to a limited extent, since they are in the insoluble form, they have low toxicity to marine organisms. Biomagnification of the metals does not occur. Since biomagnification does not occur for either the hydrocarbons or metals, there is little likelihood that fish and shellfish exposed to offshore industry discharges and later consumed by humans will cause any health effects.

The revised numbers presented in this paper represent an attempt to amend the CE procedure to be more representative of the specific environmental setting experienced by the offshore oil and gas industry. Even though the revised calculations must, by necessity, rely on some human health numbers or on some of EPA's amended numbers in the absence of chronic aquatic life protection numbers, the revised \$/PE still increases dramatically, especially for produced water.

The produced water revised value for BAT is \$355 which is higher than the \$/PE for all but one of the other industries which have been tested. A comparison of \$/PE for the different industries is given in Table 4. The Electronics I industry has a \$/PE of \$404, but that industry is substantially smaller than the offshore oil and gas industry. A total of only 9000 PE per year were discharged prior to BAT being implemented compared to 6.9 million PE per year for produced water. The economic impacts are disproportionate for the two industries. No other industry has a BAT \$/PE which is even close to these two. The recalculated \$/PE due to the revised radium toxic weighting factor becomes \$614 which is clearly higher than anything else considered.

For NSPS, the comparison is even more striking. The revised \$/PE for both the restricted and unrestricted options, \$372 and \$237 respectively, are higher than those for any other industry. This becomes even more distinctive when the \$/PE are recalculated using alternate toxic weighting factors for radium. The \$/PE are \$644 and \$409 respectively.

CONCLUSIONS

The CE methodology assigns weights to the different pollutants which will be removed by the treatment options under consideration. The procedure for calculating weighting factors has some overall limitations, but some of them are common to all of the CE analyses done thus far and are therefore consistent limitations. There are other features of the CE methodology which are specifically inappropriate to the offshore oil and gas industry. The easiest to quantify is the weighting factor calculation procedure. Considering that discharges from offshore platforms are relatively short-lived, the use of human health criteria, which are based on a 70 year exposure, to calculate the weighting factors is misleading and

Table 4. Comparison of Cost-Effectiveness Among Industries (based on EPA, 1991)

Industry	$/PE (BAT)	$/PE (NSPS)
Aluminum Forming	121	190
Battery Manufacturing	2	47
Can Making	10	—
Coal Mining	BAT=BPT	—
Coil Coating	49	13
Copper Forming	27	132
Electronics I	404	—
Electronics II		183
Foundries	84	—
Inorganic Chemicals I	<1	—
Inorganic Chemicals II	6	—
Iron and Steel	2	—
Leather Tanning	BAT=BPT	—
Metal Finishing	12	<1
Nonferrous Metals Forming	69	—
Nonferrous Metals Manufacturing I	4	—
Nonferrous Metals Manufacturing II	6	9
Organic Chems. Plastics & Synthetics	5	—
Pesticides	15	—
Petroleum Refining	BAT=BPT	—
Pharmaceuticals	1	—
Plastics Molding and Forming	BAT=BPT	—
Porcelain Enameling	6	38
Pulp and Paper	18	—
Textile Mills	BAT=BPT	—
Offshore Oil and Gas		—
produced water - EPA	60	40 (unrestricted) 63 (restricted)
produced water - revised	355	237 (unrestricted) 372 (restricted)
produced water - (reflects revised toxic weighting factor for radium)	615	410 (unrestricted) 644 (restricted)
drilling fluids - EPA		28 (unrestricted) 21 (restricted)
drilling fluids - revised		42 (unrestricted) 29 (restricted)

When this incongruity is coupled with the vast dilution present in the offshore environment, the costs for treatment of wastes which have limited environmental impact outside of a small zone of dilution and which will impact only a geographically minuscule percentage of the offshore ecosystem represents an unreasonable burden on the industry.

ACKNOWLEDGEMENTS

This work was supported by the U. S. Department of Energy, Office of Environmental Analysis, under contract W-31-109-Eng-38.

REFERENCES

Brumfield, L., 1988, Comparing total metals to acid soluble metals, Presented at the August 15, 1988 Indiana Water Pollution Control Association Annual Conference.

Ecomar, 1980, "Maximum Mud Discharge Study", for the Offshore Operators' Committee under the direction of Exxon Production Research Co.

Ehrensberger, K. and R. Rico, 1988, "Cost-Effectiveness Analysis for Effluent Guidelines," U.S. Environmental Protection Agency.

EPA, 1985, "Guidelines for Deriving Numerical National Water Quality Criteria for the Protection of Aquatic Organisms and Their Uses", NTIS-PB85-227049.

EPA, 1986, "Water quality criteria summary", update 1.0, U. S. Environmental Protection Agency.

EPA, 1990, unpublished tables of toxic weights dated June 28, 1990 and July 17, 1990, U. S. Environmental Protection Agency.

EPA, 1991, "Cost-Effectiveness Analysis of Proposed Effluent Limitation Guidelines and Standards for the Offshore Oil and Gas Industry", EPA 440/2-91-002, prepared by Eastern Research Group, Inc.

EPA, 1991a, "Technical Support Document for Water Quality-based Toxics Control", EPA/505/2-90-001, U. S. Environmental Protection Agency.

EPA, 1991b, "Proposed Development Document for Effluent Limitation Guidelines and Standards for the Offshore Subcategory of the Oil and Gas Extraction Point Source Category", EPA 440/1-91/055.

Murphy, D., 1990, Criteria Issue: the Necessity of a Fish Consumption Criterion for Arsenic, (unpublished report), Maryland Department of the Environment.

Neff, J., 1988, "Bioaccumulation and Biomagnification of Chemicals from Oil Well Drilling and Production Wastes in Marine Food Webs: A Review", prepared by Battelle Ocean Sciences for American Petroleum Institute.

O'Reilly, J., Sauer, T., Ayers, R., Jr., Brandsma, M., and Meek, R., 1989, Field verification of the OOC Mud Discharge Model, in: "Drilling Wastes," Elsevier Applied Science, London and New York, pp. 647 - 665.

Stone, E., 1991, unpublished monitoring data for Hart-Miller Island Dredged Disposal Facility, Maryland Department of the Environment.

Walk, Haydel & Associates, 1984, "Potential Impact of Proposed EPA BAT/NSPS Standards for Produced Water Discharges from Offshore Oil and Gas Extraction Industry", report to Offshore Operators Committee, New Orleans, LA.

PROCEDURE FOR DEVELOPMENT OF CONTINGENCY PLANS TO MITIGATE PRODUCED WATER RELEASES ON BLM LANDS

J.L. Rogers[1], R.T. Hicks[1], B. Shaw[2], and J. Jensen[2]

[1] H+GCL
Albuquerque, New Mexico
USA

[2] Amoco Production Company
San Juan Operations Center
Farmington, New Mexico
USA

INTRODUCTION

The development of a produced water spill contingency plan to assess and remediate impacted soils and vegetation is an important aspect of the Comprehensive Environmental Analysis and Spill Contingency Plan now required for oil and gas development conducted on Federal lands administered by the Bureau of Land Management (BLM) in the San Juan Basin of northwestern New Mexico and southwestern Colorado (Figure 1). Amoco Production Company (Amoco) is developing several well fields in the Fruitland Formation coal seam play in the north central portion of the San Juan Basin. These wells produce large volumes of water that contain elevated levels of total dissolved solids, sodium and chloride (Ayers et al., 1988). If released, the chloride and sodium content of this produced water can create elevated soluble salt, exchangeable sodium and pH levels in soils. After such a release, soils may not be able to sustain plant life until further action is taken to reduce the saline and sodic conditions of the soils.

In 1990, Amoco produced an estimated 500,000 bbl of coal seam water compared to a total basin production of approximately 3,500,000 bbl. Disposal wells will be completed at central locations to handle the increased volumes of water discharged from the Fruitland Formation gas wells. Produced water will be separated from the gas at the well head and transported via pipeline or truck to the disposal wells. As a part of the permitting procedure, Amoco and other operators in the basin were required to prepare spill contingency plans to address spill assessment and remediation.

The goal of the BLM in requiring the development of area-specific spill contingency plans is to facilitate proper management of public lands by examining potential threats to sensitive environmental receptors (soils, floral and faunal populations including threatened and endangered species, surface water, ground water, visual impacts, and cultural resources). To obtain this goal, specific regulatory concerns must be clearly defined at the outset of the study and addressed throughout plan development. The most expedient method commonly employed by H+GCL in achieving this and similar goals for clients in the minerals extraction industry is

Produced Water, Edited by J.P. Ray
and F.R. Engelhart, Plenum Press, New York, 1992

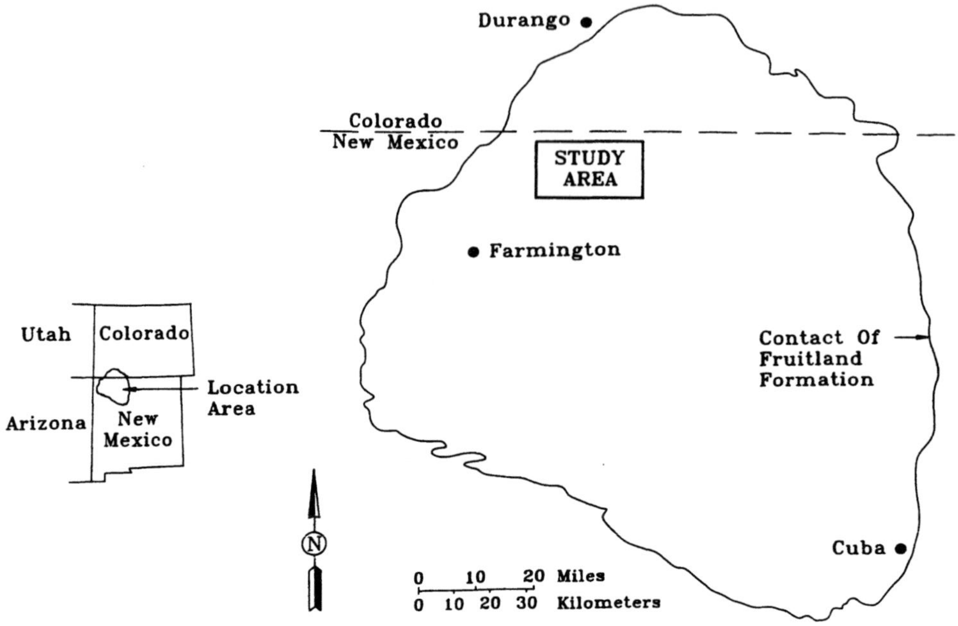

Figure 1. Study area location map in San Juan Basin of northwestern New Mexico and southwestern Colorado after Ayers et al., 1988.

through open dialogue with regulatory agencies throughout all phases of plan development specifically in defining the scope of the spill contingency plan and the objectives of remedial activities for a specific location. This proactive approach in scope development reduces additional time and costs encountered in preparing multiple drafts and acquiring additional data late in the project schedule.

This manuscript addresses the concerns expressed by the BLM for impacts of produced water releases to soils and vegetation from the recently intensive development of natural gas from coalbeds in the Fruitland Formation within the San Juan Basin. The following presents an overview of the regulatory environment associated with produced water spills on BLM land and a case history of spill contingency plan development.

REGULATORY OVERVIEW

The management of produced water developed on Federal lands is regulated under several environmental acts, regulations, and notices on specific topics. An exhaustive survey of produced water regulations is beyond the scope of this text. The major broad-scope legislation leading to the requirement of spill contingency plans on BLM lands is the National Environmental Policy Act (NEPA; 42 U.S.C. Sections 4321-4347). The BLM has interpreted NEPA to include an environmental analysis or assessment that examines "... the short term, long term, and cumulative effects ...." (Fellows, 1988) of produced water management operations in the basin. Specifically, the nature of potential impacts to soils and vegetation and the design of appropriate mitigation scenarios (Fellows, 1989) were expressed as concerns to be addressed in this study.

The categorization of produced water spills into "Major Undesirable Events" (greater than 100 barrels [bbl]) and "Other-Than-Major Undesirable Events" (10 to 100 bbl) was promulgated in Notice to Lessees and Operators of Onshore Federal and Indian Oil and Gas Leases 3A (NTL-3A, 1979). This Notice provides the action level of 10 bbl for cumulative produced water spills at a single site. With this notice as a regulatory basis, assessments of spill sites are required by the BLM when a cumulative spill of 10 barrels (bbl) or more occurs at a site. Finally, in a letter to operators in the basin, the BLM requested that cleanup scenarios for water spill impacts be addressed. Requiring the development of spill contingency plans assures regulators that operators are able to handle any release in the most effective manner to minimize or eliminate impacts to the environment through assessment and remediation of impacted sites.

WASTE STREAM CHARACTERIZATION

The initial step in preparing a spill contingency plan is to determine the characteristics of the waste stream, in this case, produced water from coalbeds in the Fruitland Formation. This information will provide insights into potential impacts to soils and vegetation and determine the deleterious characteristics requiring consideration during spill assessment and remediation.

The Fruitland Formation in the study area produces water that is predominantly brackish to saline and of the $NaHCO_3$ type (Figure 2 - Ayers et al., 1988). The only brines identified to date were produced from coalbeds to the south of the study area (Ayers et al., 1988). Water analyses from wells located in the study area (Florance C and Schwerdtferger #1) are presented in Table 1. Standard cation and anion, heavy metal, hydrocarbon, and radiogenic analyses were performed on the produced water samples to develop a thorough profile of the average chemical composition of the produced water. These data indicate that produced water contains high values of total dissolved solids (TDS; 16,300-19,600 mg/l), bicarbonate (14,700-17,000 mg/l), chloride (905-1,500 mg/l), and sodium (5,370-6,532 mg/l). Without treatment this produced water is not suitable for domestic, livestock, or irrigation use.

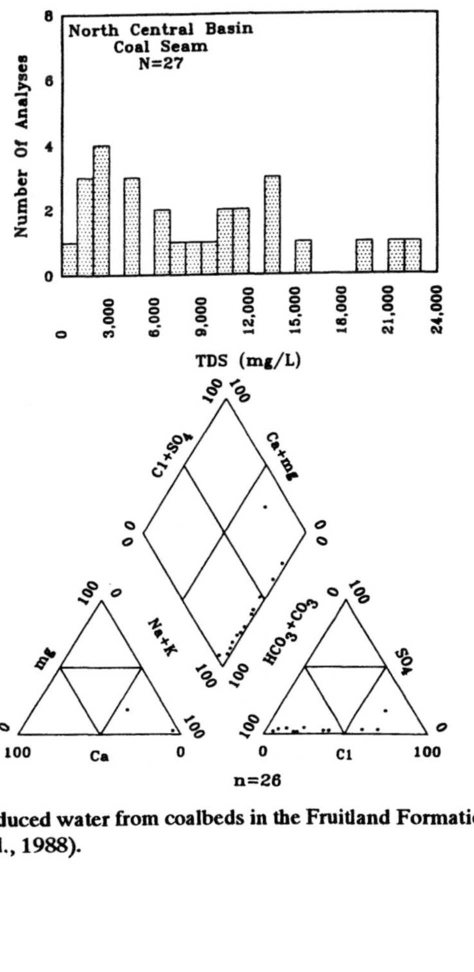

Figure 2. Chemistry of produced water from coalbeds in the Fruitland Formation north-central San Juan Basin (Ayers et al., 1988).

ENVIRONMENTAL SETTING

Soils

The Amoco well fields are located in an area containing twelve soil associations and a total of seventeen unique soil types (Keetch et al., 1980). The soils underlying these fields are predominantly thin to thick loams, sandy loams, and silty and sandy clay loams (Keetch et al., 1980). Representative samples of each soil type were collected and analyzed to determine baseline soil properties. In the event of a produced water spill, this data will be compared to soil analyses from spill impacted areas to assess the magnitude and extent of soil impacts.

Composite soil samples from a vertical profile of each soil type were collected in undisturbed locations from a depth of approximately three feet, or to bedrock or a caliche horizon if encountered at depths of less than three feet. The soil association, texture, vertical interval, bottom hole lithology, surrounding vegetation, and total depth were recorded for each sample. A total of seven parameters were selected for soil characterization analyses (Table 2). In general, the data presented in Table 2 indicate that the thin clay loams derived from shales tend to have higher levels of sodium and chloride and lower levels of calcium and magnesium than the thin sandy loams derived from sandstones. The natural pH, electrical conductivity (EC), sodium adsorption ratio (SAR), exchangeable sodium (ES), and exchangeable sodium percentage (ESP) consequently tend to be higher for the thin clay loams compared to the thin sandy loams. The thicker loams, clay loams and sandy loams tend to have ionic concentrations, pH, conductance, and sodium exchange and adsorption ratios that are intermediate to the thin sandy loams and thin clay loams. These intermediate chemical properties reflect the mixed sources of these soil textures. The cation exchange capacity (CEC) tends to be greater in clay loams compared to sandy loams because clays contain more cation exchange sites than sands. Exchange sites are also prevalent in organic matter as well as clay minerals (Richards, 1954).

Salt-impacted soils are characterized and classified on the basis of chemical properties. Saline soils are defined as having an electrical conductivity (EC) equal to or greater than 4.0 mmhos/cm (richards, 1954). This conductivity value is the level of soluble salt that adversely affects the growth of a wide range of common agricultural plants (Richards, 1954). Sodic soils contain excess sodium (relative to other cations) on exchangeable surfaces in the soil. Sodium causes clay to deflocculate lowering the permeability of soil to air and water (Richards, 1954). Sodic soils are defined as having an exchangeable sodium percentage (ESP) greater than 15 (Richards, 1954). Saline-sodic soils are soils which have both saline and sodic conditions present.

Flora

The study area is located in an upland area in the desert southwest characterized by pinon pine and juniper forests with shrubs (sagebrush, greasewood, etc.) and mixed grasslands (Elmore, 1976). *Gilia formosa* is the only threatened or endangered species known to occur within the study area, however, it was not noted to be growing at any of the locations where soil samples were collected. The spill contingency plan was designed to reestablish vegetative cover using native species. Certain types of plants, termed halophytes, are more resistant to the deleterious effects of saline and sodic soils. Local plant populations throughout the study area were evaluated to determine if native salt tolerant species existed in the area and if they were appropriate for revegetating spill sites following soil amendment applications. Alkali sacaton, Western wheatgrass, Fourwing saltbush, and Indian ricegrass are naturally occurring salt tolerant species in the study area (Aldon, 1978; Keetch, 1980; Fuller et al., 1982). These native salt tolerant grasses will be used to re-establish vegetative cover in areas impacted by produced water spills.

Table 1. Fruitland coal produced water chemistry characterizations from two wells within study area. Results reported in mg/l except where noted

Parameters	Florance C	Schwerdtferger #1
Aluminum - Al	<1	<1
Arsenic - As	0.011	0.003
Barium - Ba	57.9	61.9
Cadmium - Cd	<0.011	<0.005
Calcium - Ca	56.1	29.6
Chromium - Cr	<0.01	<0.01
Cobalt - Co	<0.02	<0.02
Copper - Cu	<0.25	<0.25
Iron - Fe	3.7	4.2
Lead - Pb	0.07	0.27
Magnesium - Mg	33.9	18.4
Mercury - Hg	<0.001	<0.001
Nickel - Ni	<0.05	<0.03
Potassium - K	37.8	48.4
Selenium - Se	0.001	<0.001
Sodium - Na	6532.0	5370.0
Vanadium - V	<0.01	<0.1
Zinc - Zn	0.16	0.39
Bicarbonate as $CaCO_3$	17000.0	14700.0
Carbonate as $CaCO_3$	0.0	0.0
Boron - B	2.0	1.6
Chloride - Cl	1500.0	905.0
Fluoride	0.78	0.94
Nitrate/Nitrite - N	<0.04	0.14
Oil and Grease	5.0	70.0
Phenols	0.028	0.027
Sulfate	<20	<20
Benzene	0.072	<0.01
Toluene	0.152	0.03
Ethyl Benzene	0.074	<0.01
Total Xylene	0.017	<0.01
Alkalinity as $CaCO_3$	17000.0	14700.0
Conductivity (µmhos/cm)	7.49	8.07
pH (units)	19600	16300
Total Dissolved Solids	4.5 ± 0.8	4.3 ± 1.0
Radium-226 (pci/l)	2.1 ± 0.8	8.6 æ 2.2
Radium-228 (pci/l)	1.2 ± 0.7	70.7 ± 1.1
Strontium-90 (pci/l)		

IMPACTS OF PRODUCED WATER SPILLS

Excess soluble salts increase the osmotic pressure of root systems. Plants consequently dehydrate and die because the uptake of water is significantly diminished. Additionally, sodium and chloride are directly toxic to vegetation at certain concentrations. Sodium also causes clay to deflocculate and lowers the permeability of soil to air and water, hinders the emergence of seedlings, and inhibits root penetration. Excess exchangeable sodium is harmful to vegetation because it hydrolizes easily elevating soil pH which can lower the availability of nutrients such as iron, manganese, calcium, and magnesium.

Soil samples were collected from three spill sites located in the north-central San Juan Basin outside of the study area, to evaluate impacts to soils similar to those in the study area (H+GCL, 1991b). Analytical results of background and impacted soil chemistry are presented in Table 3. Approximately 80 to 100 bbl of produced water were released onto a thick clay loam. Soil samples were collected one day after the spill. Analytical results (Table 3) indicate that the impacted soils were saline-sodic (EC of 8.0 mmhos/cm, ESP of 11.4 meq/100g). The soil pH was elevated by one unit over background conditions.

Table 2. Natural soil chemistry from eight soils within the study area (H+GCL, 1991a & b)

Soil Texture	pH	EC[1] (mmhos/cm)	SAR[2]	CEC[3] (meq/100g)	ESP[4] (%)	Ca+Mg+Na (mg/l)	Cl (mg/l)
Sand	8.71	0.214	0.90	3.37	3.37	31.35	3.0
Loam Sand	8.76	0.108	0.11	7.07	7.07	20.2	1.7
Sandy Loam	7.94	0.161	0.20	11.3	11.3	33.8	6.3
Fine Sandy Loam	8.54	0.135	0.15	15.6	15.6	31.6	3.8
Clay Loam	8.10	0.336	5.21	49.4	49.4	83.6	7.8
Gravelly Loam	8.04	0.189	0.27	17.2	17.2	37.35	1.1
Loam	8.09	0.141	0.11	16.1	16.1	24.7	1.1
Silty Clay Loam	8.92	0.261	1.67	21.1	12.1	56.1	2.4

[1]EC - Electrical conductivity,
[2]SAR - Sodium adsorption ratio,
[3]CEC - Cation exchange capacity,
[4]ESP - Exchangeable sodium percentage to or greater than 4.0 mmhos/cm (Richards, 1954).

Table 3. Produced water impacts to three soil types from locations outside of study area within the San Juan Basin

Soil Texture	Spill VoL (bbl)	pH	EC[1] (mmhos /cm)	SAR[2]	CEC[3] (meq/100g)	ESP[4] %	Ca+Mg +Na (mg/l)	Cl mg/l
Thick clay loam	90							
Background Spill		8.20	0.182	0.33	16.0	0.40	35.22	5.00
Impacted		9.20	8.0	60.5	25.4	44.80	2797.2	87065.
Thin sandy loam	75							
Background Spill		7.72	1.15	24.8	17.7	1.35	368.39	68.00
Impacted		10.20	7.01	66.2	9.19	108.8	2168.3	900.0
Thin clay loam	20							
BackgroundSpill		8.50	0.216	1.78	32.1	1.10	48.86	6.0
Impacted		9.50	0.512	17.1	20.7	24.5	136.09	16.0

[1]EC - Electrical conductivity,
[2]SAR - Sodium adsorption ratio,
[3]CEC - Cation exchange capacity,
[4]ESP - Exchangeable sodium percentage

A produced water spill of approximately 75 bbl impacted thin sandy loams in an arroyo. Soil samples collected approximately one month after the spill were saline-sodic (EC of 7.0 mmhos/cm, ESP of 108.8 meq/100g) and soil pH was elevated 2.5 units above background.

Vegetation was dead and the soil was hard and crusted. Another release (approximately 20 bbl) impacted thin clay loams confined in an arroyo. Soil samples were collected approximately 8 months after the spill. Analyses indicated that sodic conditions persisted (24.5 percent ESP). However, soils were not saline at the time of sample collection (EC 0.514 mmhos/cm). The soil pH was elevated one unit higher than background. Vegetation at the spill site was dead at the time this sample was taken, but soils were not hard or crusted.

Soils with high clay contents have lower permeabilities causing the spilled salt water to accumulate in the soil and create more significant impacts. These soils also have higher natural concentrations of sodium and chloride, suggesting that a relatively small spill volume may elevate sodic levels higher than that tolerated by native vegetation. Sandy soils allow the salt water to disperse away from the spill and are, therefore, more resistent to damage. In general, steep terrain and arroyos are more susceptible due to the thin soils found in these areas which concentrate saline and sodic conditions in a thin horizon. The thick soils which develop in areas of level topography are more resistant to salt water impacts because of vertical dispersal of the spilled water through the soil profile.

SOIL RECLAMATION

Reclamation of soils impacted by releases of produced water from the Fruitland Coal requires the removal of soluble salt and exchangeable sodium to the extent necessary to return soil to a productive state. Soluble salts are rather easily removed from saline soil if the salts can be leached with water to a depth beyond the rooting zone (Auchmoody and Walters, 1988; Waisel, 1972). Saline soil can heal within a few years without mitigative treatment if abundant precipitation flushes the soil (Auchmoody and Walters, 1988). However, in arid regions, such as the study area with an annual precipitation ranging from 4.85 to 21.02 inches (U.S. Weather Service records for Navajo Dam), require the active removal of salt. Reclamation tends to proceed more rapidly if the water is impounded over the spill site rather than sprinkled (Prichard, et al., 1985). In general, a minimum of one inch of water must be applied for each inch of soil depth to be reclaimed. Successful leaching also requires adequate drainage to promote the rapid downward flow of water and the addition of a soluble calcium amendment.

Exchangeable sodium is displaced from sodic soil by adding an amendment that supplies soluble calcium which replaces excess sodium in exchangeable sites within the soil (Mustafa et al., 1985). Free sodium then becomes a part of the soil solution and can be removed by leaching. Gypsum or limestone are among the most commonly used amendments. Other amendments and their properties are listed in Table 4. Gypsum is generally preferred to limestone because of its greater solubility providing more free calcium for sodium exchange. The amendment used must be fine grained and mixed thoroughly with the soil to increase the contact of calcium with exchange surfaces in the soil. Sufficient amendment must be added to react with other soluble compounds in the soil that will compete with the sodium for the calcium. Leaching of saline-sodic soils without the addition of calcium amendments may deflocculate clay and reduce permeability to the point where further leaching is impossible. Amendment application should therefore proceed quickly prior to and during seasons of maximum local precipitation. Leaching in saline-sodic soils forms a desalinized zone that is underlain by a zone of salt accumulation (Mustafa et al., 1985). The highest salt concentrations coincide with the wetting front. Leaching is more efficient with large and less frequent applications of water (Mustafa et al., 1985).

The addition of organic matter with the amendments tends to counteract the unfavorable effects of exchangeable sodium (Richards, 1954). Organic matter such as native grass mulch or aged manure also prevents deterioration of the physical condition of the soil by interacting with inorganic cation exchange material, by serving as energy material for microorganisms that promote the stable aggregation of soil particles, and by decreasing the bulk density of the soil (Richards, 1954). Native grass mulch is preferred to aged manure since it reduces the possibility of establishing stands of non-native grasses.

Table 4. Amendments commonly applied to reduce saline and/or sodic soil conditions (after Richards, 1954 and Burley, 1988)

Type	Composition
Soluble Calcium Salts	Gypsum
	Calcium Chloride
Acids or Acid Formers	Sulfur
	Sulfuric Acid
	Iron Sulfate
	Aluminum Sulfate
Calcium Salts of Low Solubility	Ground Limestone
Miscellaneous products such as ADP-7 or SC-1 sodic convertor (Petro Green, Inc.) And Spersal (Helena Chemical Co.)	Commercially available properties
	Inexpensive, easy to apply, acts slowly
	Expensive, easy to apply, acts rapidly
	Inexpensive, lengthy application process, acts rapidly once
	Costly difficult and hazardous to apply, acts rapidly, may cause excess acidity
	Expensive, easy to apply, acts rapidly, may cause excess acidity
	Expensive, easy to apply, acts rapidly, may cause excess acidity
	Inexpensive, easy to apply, acts slowly, ineffective in soils with pH > 7.5
	Moderately expensive, apply according to manufacture's recommendations

Soil reclamation tends to proceed as a logarithmic function of time (Sommerfeldt and Chang, 1987). Improvement of the EC and ESP may occur within one month of initial reclamation efforts, but lowering these parameters to levels necessary to return soil to a productive state may require a year or longer. The rate of reclamation will depend on the amount of water leached into the soil, subsequent rainfall, the soil texture, the depth of the soil profile, and the interaction occurring on the exchange complex (Sommerfeldt and Chang, 1987). In general, fine textured soils are more adversely affected at a given ESP than coarse textured soils (Richards, 1954). Clay tends to leach slowly (because of lower permeability) and resists the exchange of sodium. Sands and loamy sands improve leaching rates and exchange sodium more readily. The rate of leaching and the exchange of sodium also increase as the soil profile deepens and erratic or shallow profiles impede leaching and sodium exchange (Sommerfeldt and Chang, 1987).

SPILL CONTINGENCY PLAN

The following is a summary of the spill contingency plan prepared during this study. The flow chart in Figure 3 presents all steps and decision points in the execution of the assessment and remedial operations.

Remedial efforts should be implemented if cumulative spills at a single location of 10 bbl or more of produced water create saline (EC than 4.0 mmhos/cm), sodic (ESP greater than 15), or both saline and sodic soil conditions (H+GCL, 1990 a & b). These factors should be determined by assessments (sampling and analysis) of impacted soils immediately following a cumulative spill above the threshold value of 10 bbl.

If saline, sodic or saline-sodic soil conditions are not present, further action is not required. However, additional soil sampling will be necessary following another spill of any volume at that site. A system of record keeping should be implemented to document locations, volumes, and dates of produced water spills and analytical results. Soil removal and in situ soil treatment are the two options suitable to remediate spill sites. Excavation will require disposal of the salt impacted soil, replacement with suitable fill material, compaction or grading of the fill, and revegetation of the reclaimed site. An excavation program would likely be effective only if it were implemented manually to remediate a very small volume of soil. Active, in situ remedial measures are recommended to mitigate impacts from relatively large spills.

In situ remediation should be conducted promptly after determining the presence of saline or sodic soil conditions to minimize deterioration of the soil structure . The recommended amendments are fine grained gypsum and native grass mulch or aged manure (GCL, 1990). The amendments should be worked into the soil. Leaching with water should follow immediately. The amount of water applied should equal the thickness of the soil profile present at the spill site up to a maximum amount of 60 inches. The water should be applied in as few separate stages as practical and impounded on the spill site is practical.

Seeding with salt tolerant vegetation (in this instance Alkali sacaton, Western wheatgrass, Fourwing saltbush, and Indian ricegrass) should be performed during the period of maximum local rainfall (GCL, 1990).

Post-amendment soil samples should be collected 6 months and one year after the amendments are applied to monitor the success of remedial measures. The parameters to be analyzed include pH, EC, SAR, CEC, ES, ESP, Ca, Mg, NA, and Cl. Verification samples should be taken at one foot intervals (0-1, 1-2, and 2-3 feet) to allow vertical profiling of soil conditions. Soil reclamation can be considered complete when soils are no longer saline (EC less than 4.0 mmhos/cm) or sodic (ESP less than 15.0). Should the second (12 months after amendment application) post-amendment sample fail any of these criteria, an additional stage of in situ remediation and verification sampling is recommended. The success of reseeding should be evaluated during the verification sampling events and on an annual basis until vegetative cover is restored.

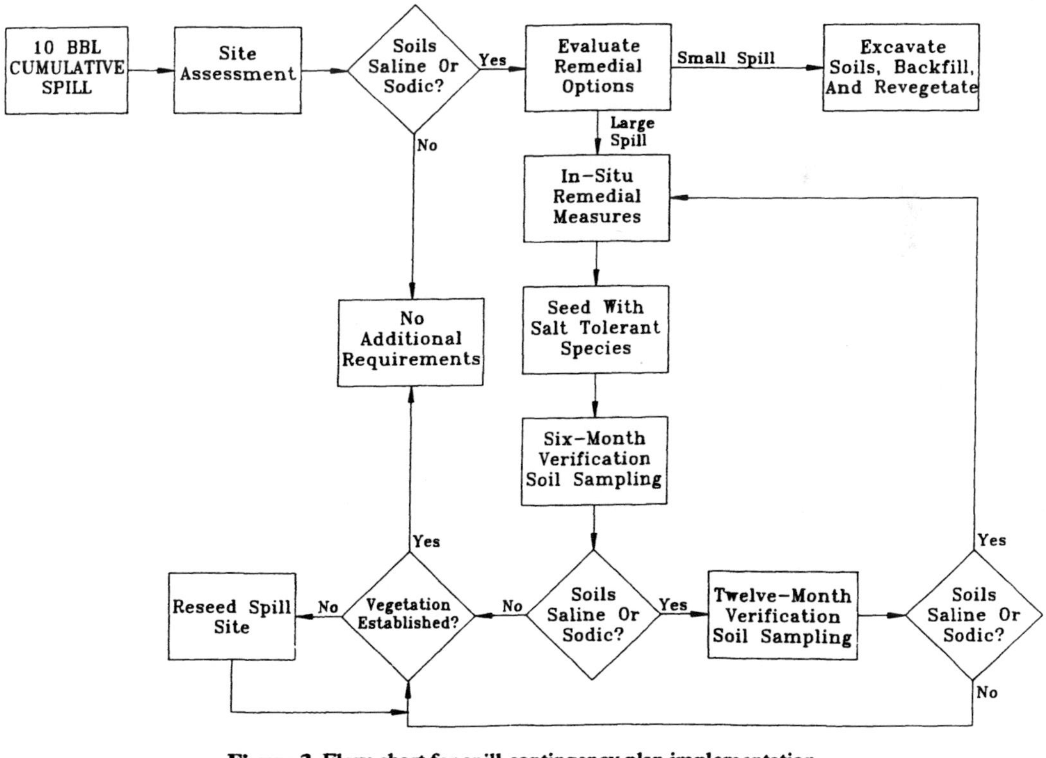

Figure 3. Flow chart for spill contingency plan implementation.

ACKNOWLEDGEMENTS

The authors would like to acknowledge Amoco Production Company and H+GCL for allowing presentation of the information contained in this manuscript. Word processing assistance from Mrs. Eileen Armagost is greatly appreciated. Drafting services were provided by Mr. Joe Nee.

REFERENCES

Aldon, E.F., 1978, Reclamation of coal-mined land in the southwest, J. Soil & Water Conservation 33:80.

Auchmoody, L.R., Walters, R.S., 1988, Revegetation of a brine killed forest site, Soil Science Society of America Journal 52:277.

Ayers, W.B., Jr., Kaiser, W.R., Swartz, T.E., Zellers, S.D. and Scanlon, A.H., 1988, "Geologic Evaluation of Critical Production Parameters for Coalbed Methane Resources, Part I, San Juan Basin," Gas Research Institute, Chicago, Il.

Burley, M.J., 1988, Reclamation of brine spill sites, Spill Technology Newsletter 13:3

Elmore, F.H., 1976, Shrubs and Trees of the Southwest Uplands, Southwest Parks and Monuments Association, Tucson, AZ.

Fellows, R., 1988, Correspondence to Public Land Users and other Interested Parties, from Mr. R. Fellows, Area Manager, Bureau of Land Management, Farmington, NM.

Fellows, R., 1989, Correspondence to Amoco Production Company, from Mr. R. Fellows, Area Manager, Bureau of Land Management, Farmington, NM.

Fuller, R.D., Nelson, E.D.P., Richardson, C.J., 1982, Reclamation of red mud (bauxite residues) using alkaline tolerant grasses with organic amendments, J. Env. Quality 11:533.

H+GCL, 1991a, Contingency Plan Fruitland Coal Wells, prepared for Amoco Production Company, H+GCL Englewood, CO.

H+GCL, 1991b, Blackwood & Nichols Contingency Plan: Produced Water Spills from Fruitland Coal Gas Wells, prepared for Blackwood & Nichols Company, Limited, H+GCL, Englewood, CO.

Keetch, C.W., 1980, "Soil Survey of San Juan County, New Mexico, Eastern Part," USDA, Soil Conservation Service, Washington, DC.

Mustafa, M.A., De Jong, R., Hayhoe, H.N., Topp, G.C., 1985, Effects of intermittent ponding on salt redistribution through a saline-sodic clay soil, Canadian Agricultural Engineering 27:105.

Prichard, T.L., Hoffman, G.J., Oster, J.D., 1985, Reclamation of saline organic soil, Irrigation Science 6:211.

Richards, L.A., (ed.), 1954, "Diagnosis and Improvement of Saline and Alkali Soils," U.S. Salinity Laboratory Staff, U.S.D.A. Handbook 60, Washington, D.C.

Sommerfeldt, T.G. and Chang, C., 1987, Desalination of an irrigated, mole drained saline loam soil, Canadian J. Soil Science 67:263.

Waisel, Y., 1972, "Biology of Halophytes," Academic Press, New York, USA.

DEVELOPMENT OF A PRODUCT SCREENING PROTOCOL FOR EVALUATING ENVIRONMENTAL EFFECTS OF OILFIELD CHEMICALS

J. D. Berg, S. Bakke, E. A. Vik and B. S. Nesgård

Aquateam - Norwegian Water Technology Centre A/S
Oslo, Norway

INTRODUCTION

North Sea operators are now faced with more stringent forthcoming regulations concerning usage and discharges of chemicals. The Paris Commission (PARCOM) Working Group on Oil Pollution (GOP) is trying to harmonize procedures for approval, evaluation, and testing of offshore chemicals and drilling muds. PARCOM has proposed that a series of tests covering toxicity testing of a marine algae and a marine copepod, biodegradation in seawater, and bioaccumulation are performed for all compounds in chemical products used in offshore production. The development of these regulations depends upon a more thorough understanding of the fate of chemical products during processing and discharge than we presently have. This affects economic and environmental aspects of chemical selection, process technology selection and daily offshore operation. The PARCOM proposed procedures, which are based on existing or proposed OECD or ISO standards, do not take into consideration that the chemicals used offshore are often slightly water soluble and complex mixtures of ill-defined components, and that modification in procedure or special sample preparation is needed. Therefore Conoco Norway (CNI), in 1989, initiated the development of a products screening protocol for evaluation of environmental effects of oil field chemicals discharged to the marine environment. The protocol includes test procedures for determination of the oil/water partitioning of chemicals, the biodegradation and bioaccumulation potential of the chemicals, and to assess and classify complex chemical products according to environmental risk. In addition to the marine environment, the oil field chemicals can also reach the air, due to volatile fractions in the compounds and they can go to marine sediments. The test protocols developed should also take this information into account. In a complete test protocol, acute and chronic toxicity also need to be addressed. However, existing published biodegradation and toxicity test procedures for surfactants and non-water soluble chemical products are not suitable, since homogeneous samples are very difficult to prepare, making reproducibility a significant problem.

Development of a standardized toxicity test system for oil field chemicals has been initiated by PARCOM, which performed a round-robin test on marine toxicity at two trophic levels: algae and copepods. The Norwegian Oil Industry Association (OLF) decided to adapt existing techniques to offshore production chemicals for the evaluation of acute and chronic marine toxicity, and to evaluate screening protocols for marine toxicity of oil field chemicals compared

Produced Water, Edited by J.P. Ray
and F.R. Engelhart, Plenum Press, New York, 1992

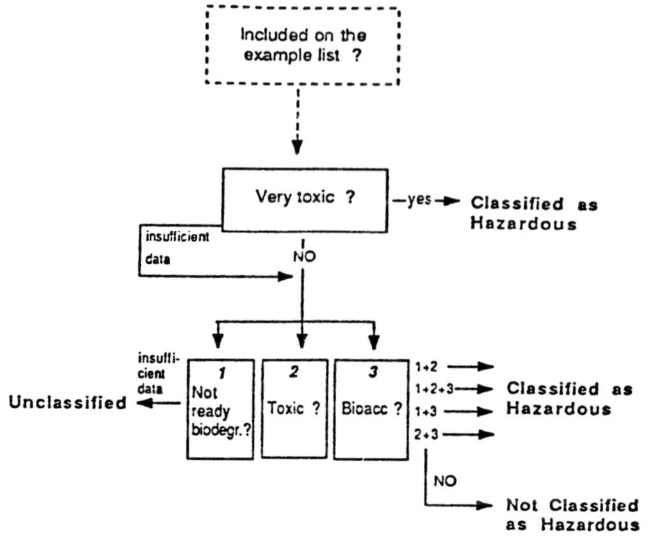

Threshold values:

	Very toxic	Toxic
Oral LD_{50}	≤ 25 mg/kg	≤ 200 mg/kg
Inhal LC_{50}	≤ 0,5 g/m^3	—
Fish LC_{50}	≤ 1,0 mg/l	≤ 100 mg/l
Daphnia EC_{50}	≤ 1,0 mg/l	≤ 100 mg/l
Algae EC_{50}	≤ 1,0 mg/l	≤ 100 mg/l

Bioaccumulating
BCF ≥ 100 or K_{ow} ≥ 1000
Not ready biodegradable
As defined in OECD Guidelines or BOD_5/COD < 0,5

Figure 1. Classification system for chemicals recommended by a Nordic Group of experts (Gjøs et al., 1989; Lundgren, 1989).

to PARCOM "authorization" testing. The test candidates included three screening tests (*Artemia salina*, Microtox, and mitochondria) and two PARCOM species (*Skeletonema costatum* and *Acartia tonsa*).

Five products which exhibited a representative and wide range of characteristics, have been selected as model compounds for evaluation of the suitability of available test protocols: a corrosion inhibitor, a biocide, a scale inhibitor, an emulsion breaker, and a flocculating agent. This paper summarizes experimental work which is presented in the context of the approach to screen the environmental effects of products. Only the results for two products of greatly different character are described herein to illustrate the screening concept. These are the corrosion inhibitor, a surfactant, oil soluble compound, which is dispersible in water and typically used in oil field operations, and the scale inhibitor which is a water soluble compound. The composition of the corrosion inhibitor tested is: imidazoline derivates, aromatic solvent, isobutanol, and etherdiamine. The composition of the scale inhibitor tested is: aminomethylene phosphonic acid and ethylene glycol.

THEORY, DEFINITIONS AND TEST METHODOLOGY

Classification of Chemicals

The North Sea countries with major offshore activities are the United Kingdom, Denmark, the Netherlands and Norway. Each of the countries have recently introduced different systems for notification and approval of oil field chemicals (production and drilling chemicals) use and discharges. Different tests are required, and GOP is trying to harmonize this. A worldwide classification system for evaluation of toxicity of chemicals discharged to the marine environment was developed by GESAMP (Joint Group of Experts on the Scientific Aspects of Marine Pollution) in 1989. This system is used to evaluate the hazards of chemicals transported on ships. A Nordic joint group of experts have agreed upon a general classification system for chemicals. This system is presented in Figure 1 to illustrate how such a classification scheme functions.

A similar classification system for chemicals, indication of danger and choice of risk phrases has been proposed by EEC (1990). All international experts agree upon the need for evaluating long-term effects of chemicals by bioaccumulation potential and biodegradability, as well as the acute toxicity. The threshold levels, test organisms and the acceptable test methodology are more difficult to agree upon. Establishment of test methodology also needs to consider oil soluble and surface active chemicals. No test method will, however, be accurate enough to determine the impact by chemicals to the environment. Simplified procedures will therefore have to be established.

Risk Assessment

For internal quality control purposes and environmental risk assessment, the Potential Environmental Concentration (PEC) should be taken into account. PEC is the highest likely concentration in the environment. By using a dilution factor for the chemicals being discharged in a mixing zone close to the discharge point, the PEC can be estimated as:

PEC = Dilution factor x use conc. x chemical fraction discharged (mg/l).

The respective regulators have to agree upon a reasonable dilution factor, considering the actual discharge surroundings. The chemical fraction discharged in the aqueous phase has to be determined. If we are working with an organic chemical, the PEC needs to be compared with the Minimum Adverse Effect Concentration (MAEC) or No Observed Concentration (NOEC) (mg/l), which is the lowest concentration at which harmful effects are likely to be observed.

$$\frac{\text{MAEC}}{\text{PEC}} = \text{the safety ratio for the chemical}$$

According to de Henau et al., (1986), if this ratio is 100, chronic toxic effects are only rarely observed. The MAEC corresponds to the lowest acute toxicity value which is reported for the chemical using for example the concentration at which 50 percent lethality (LC_{50}) or effect (EC_{50}) occurs. NOEC, on the other hand, is determined in prolonged chronic effect laboratory tests. Most toxicity data reported are, however, based on chemical dosages used in the sample preparation for the toxicity test and not on the actual chemical exposure in the toxicity test.

The purpose of an experimental program must be to enable the PEC and MAEC to be determined with sufficient accuracy to demonstrate that a reliable margin of safety exists. Once this is done, further testing is unnecessary.

Obviously, where the margin of safety is large, the estimates of PEC and MAEC need only be approximate, and a relatively simple program should include establishing test results showing that the environmental hazard is small. Where the margin of safety appears to be small, on the other hand, a more detailed program of work will usually be necessary to enable the estimates of PEC and MAEC to be refined, and the environmental hazard to be quantified with greater confidence. This system is used by many industries (e.g., the detergent industry, de Henau, 1986), and is based on many investigations. It is proposed that a similar system be used for oil field chemicals.

Oil/Water Distribution

Several researchers have proposed different extraction methods for the preparation and characterization of the aqueous phase of oil soluble products or sparingly soluble organic substances in offshore chemicals, and no standardized sample preparation procedure (Bennett et al., 1990) exists for biodegradation tests and toxicity tests. Aquateam is proposing the following sample preparation procedure for determination of the oil/water distribution factor for production chemicals used offshore.

- A mixture of 50:50 (400 ml standard oil (Exxol D 100 S) to 400 ml natural seawater) and the chemical to be tested is gently mixed in a teflon separatory funnel for 24 hours.
- After 24 hours of mixing, equilibrium between the different phases is presumably attained, and after one hour the oil and seawater fraction is separated by removing the first 50 ml of seawater and then collecting the next 250 ml of the seawater for further testing and analyses.

Natural seawater has been selected as the aqueous phase since it has similar composition as the North Sea brine. The chemical dosage used has been 10 times the normal dosage used in the North Sea. In a production well, the production water volume increases over the life time of the well, so an average oil/seawater ratio of 50:50 has been selected as an unbiased value. The mixing time, mixing speed and separation time have also been standardized with this "offshore simulation". The distribution of the chemical between the oil and aqueous phases (produced water) can then be determined.

The oil/water distribution factor (OWDF) has been defined as follows:

$$OWDF = \frac{[Chem]_o}{[Chem]_w}$$

where $[Chem]_o$ and $[Chem]_w$ means the concentration of the chemical in the oil- and seawater phases, respectively. The oil/water distribution of an unknown organic chemical can be determined using a surrogate parameter such as total organic carbon (TOC), UV or GC fingerprints.

The partitioning of the chemical between oil and water is the most critical factor in evaluating the environmental risk of chemicals used in the oil production process offshore. Both for the determination of:

1) potential environmental concentration (PEC);
2) accurate mass balance studies and evaluation of discharge limits of oil and chemicals with produced water;
3) calculation of biodegradability;
4) calculation of the potential for toxic concentrations to occur; and
5) for evaluation of the overall environmental effect.
6) Moreover, the chemical concentration in the water phase has to be known to compare the LC_{50} or the EC_{50} with the potential environmental concentration, and thus evaluate the environmental safety factor.

Bioaccumulation

For determination of the bioaccumulation potential of the chemicals, the octanol/water partition coefficient (P_{ow}) can be used for oil field chemicals. Three methods for determination of P_{ow} can be used: 1) Shake Flask Test (OECD Test Guideline 107, 1981), 2) Thin Layer Chromatography (TLC), and 3) High Performance Liquid Chromatography (HPLC). For oil field chemicals, the HPLC method has been chosen (OECD Guideline 117, 1989) in this study due to its simplicity, accuracy and reproducibility (ECETOC, 1983). The partition coefficient is deduced by linear relationship from the capacity factor, k, (derived from the retention times of the compounds) in the HPLC column. The partition coefficients of unknowns can be predicted using a correlation plot of log k and log P_{ow}. A standard curve is prepared using selected compounds covering several compound classes as typically used in oil field chemicals and proposed in the OECD's analytical procedure. All compounds are dissolved in a mobile phase consisting of methanol and water.

Ionized compounds will elute faster than the deadtime compounds (at pH = 7), and no bioaccumulation potential will be observed under normal environmental conditions. The deadtime compound is an unretained organic compound which is used to determine the deadtime, t_0. Thiourea has been selected based on OECD's proposal (OECD, 1989).

Sparingly or non-water soluble and surface active compounds, i.e., the corrosion inhibitor, will often have log P_{ow} values 6, which makes this method unsuitable. These compounds are not likely to dissolve in the water phase, but it is necessary to experimentally verify whether the low water soluble fractions with high log P_{ow} values enter the water phase. In EEC's environmental research program, recent scientific achievements on ecotoxicology have been reported by Barth (1991). The following correlation between partition coefficient (log P_{ow}), and bioaccumulation potential or bioconcentration factors in algae (BCF_A) and fish (BCF_F) have been found for chlorinated benzenes:

$$\log BCF_A = 0.928 \log P_{ow} - 0.894 \text{ (99\% significance)}$$

$$\log BCF_F = 0.820 \log P_{ow} - 0.682 \text{ (99\% significance)}$$

Very few similar data exist for typical oil field chemicals.

Biodegradation

The biodegradation procedures standardized today are not suited for most oil field chemicals, but they are required by the authorities for the biodegradation testing. However, the Modified AFNOR Shake Flask Test (OECD test 301 A, 1981) and the Closed Bottle Test (OECD 301 D, 1981) can be further modified to suit the testing of typical oil field chemicals discharged to the marine environment. For the Shake Flask Test, chemicals giving more then 70% dissolved organic carbon (DOC) reduction within 28 days, can, according to local authorities and agreed upon through PARCOM, be regarded as readily or inherently biodegradable. For the Closed Bottle Test, chemicals giving a measured biological oxygen demand (BOD) which is greater than 60% of the theoretical oxygen demand (ThOD) or chemical oxygen demand (COD) within 28 days, the same conclusion can be drawn. Ready biodegradability is determined using non-acclimated cultures, while inherent biodegradability is determined using acclimated cultures.

The formula for calculating percent (%) biodegradation in the two OECD tests are presented below:

Closed Bottle Test

$$\% \text{ biodegradation} = \frac{[BOD_0] - [BOD_{28}]}{\text{COD (or ThOD)}} \times 100$$

Shake Flask Test

$$\% \text{ biodegradation} = \frac{[DOC_0] - [DOC]_{28}}{DOC} \times 100$$

where 0 and 28 present the test day.

To reduce the labor intensity and increase the accuracy of the existing methodology, the Closed Bottle and Shake Flasks Tests can be simplified using a respirometric system with continuous and simultaneous monitoring of CO_2 and O_2. Then, in keeping with OECD Test Guidelines, either TOC or COD samples would be taken initially and at 28 days. The procedure is anticipated to have the following advantages over Closed Bottle or AFNOR Tests:

- better reproducibility;
- simpler set-up with minimal flask (reactor) requirements;
- minimal TOC/COD testing;
- continuous O_2/CO_2 (i.e., "BOD" readings);
- ability to observe continuously for abnormalities in the biodegradation process (inhibition, lag/adaption, toxicity, multiphasic degradation of individual components).

Since it seems that the Netherlands now requires the Closed Bottle Test with COD measurements, the respirometric procedure would be a simple means of conducting a Closed Bottle Test, regardless of whether COD or TOC (DOC) measurements are envisioned. The development of the respirometer procedure as an improved hybrid of the Closed Bottle and AFNOR Shake Flask Tests is recommended.

Toxicity

According to PARCOM's guidelines for harmonization procedures of approval, evaluation and testing of offshore chemicals (and drilling muds) results from acute toxicity tests on marine organisms have to be presented for at least one test species from the following trophic levels:

- algae
- herbivore
- fish

PARCOM's ringtest in 1992 and follow-up workshop in January 1992 recommended the following test species for chemical approval:

- algae: *Skeletonema costatum*
- crustacean: *Acartia tonsa*, or *Mysidopsis bahia*, or *Tisbe batagliai*.

Two of PARCOM's recommended species, *Skeletonema costatum* and *Acartia tonsa*, were compared with simpler screening toxicity tests for internal quality control purposes. These were:

Screening tests: *Artemia salina*, Microtox, mitochondria

The principles of these five tests are summarized in Table 1.

Table 1. Principles of toxicity tests

Test species	Principle
Screening	
Artemia salina	The lethality of the fresh larvae is tested
Microtox	Inhibition of light produced by the luminescent bacteria is measured. EC_{50} (effective concentration) is the sample concentration resulting in 50% light inhibition (Microbics 1990).
Mitochondria	RET (reverse electron transfer). EC-values are measured by inhibition of enzymatic process converting $NAD^+ \rightarrow NADH$ (Blondin et al. 1987; 1989).
PARCOM	
Skeltonema costatum	Growth inhibition (decrease in cell-density per unit time) is measured. EC_{50} is the test concentration of test substance which result in 50% reduction in growth rate (ISO 1991; PARCOM 1992).
Acartia tonsa	Lethality is tested

RESULTS AND DISCUSSION

Risk Assessment. In risk assessment of production chemicals used offshore, the important factor is determination of the potential exposure of the chemical to the marine organisms. By assuming a 1:100 dilution of the chemical being discharged, the PEC can be calculated based on results from the oil/water distribution, using the following equation:

$$PEC = 1/100 \times \text{Use conc.} \times 1/(1+OWDF)\ (mg/l)$$

where OWDF = Oil/Water Distribution Factor. The MAEC has been estimated from the lowest toxicity results obtained in the laboratory, taking into account the fraction of the chemical being discharged. The PEC-values, the margin of safety (MAEC/PEC) and proposed test program for the corrosion and scale inhibitors are presented in Table 2, using dosages of 15 and 20 ppm, respectively. These results indicate that chronic toxicity testing should be required for the corrosion inhibitor, but not for the scale inhibitor.

Table 2. Estimated PEC and margin of safety-values for the corrosion and scale inhibitors

Chemical	PEC (mg/l)	Margin of Safety = MAEC/PEC	Test program indicated
Corrosion inhibitor	0.5	20	Chronic toxicity of chemicals to aquatic organisms at different trophic levels should be determined
Scale inhibitor	1.0	250	No further aquatic toxicity testing is required

OWDF. Preliminary test results from the oil/water distribution factor for the corrosion and scale inhibitors according to the OWDF-procedure described, are presented in Table 3. The results in Table 3 indicate that a larger fraction of the corrosion inhibitor enters the water phase at lower dosages. The effect seen at the higher dosages can either be due to the fact that more of the chemical enters the oil phase, that more of the chemical used is dissolved in the interface, or that larger fractions of volatile compounds are lost. (The corrosion inhibitor used consists of approximately 40% volatile compounds). Normal test concentration in our experiments is, however, 150 ppm. The corrosion inhibitor is a surfactant which often shows this behavior at dosages exceeding the optimum corrosion inhibitor dosage. This can be explained by increased droplet coalescence (Anderson et al., 1974). As the droplet coalescence increases, the rate at which chemical returns to the surface increases, resulting in lower chemical concentration in the aqueous phase. The small surfactant droplets in the water phase also presents problems regarding reproducing the results when the chemical samples are to be tested with respect to biodegradation and toxicity. We are doing further experiments to confirm the results. Mass balance studies have, however, indicated similar results in the OWDF standard experiments. According to Hunton (1991), the same trend regarding results from similar mass balance experiments have been seen.

Bioaccumulation. HPLC chromatograms and log P_{ow}-values of the corrosion inhibitor are shown in Figure 2. Since the corrosion inhibitor consists of compounds, the HPLC specter shows several peaks, each having an individual log P_{ow}-value (see Figure2). The whole log P_{ow}-range observed should be reported, i.e. log P_{ow} = 1.5-4.8. For the scale inhibitor, only tow peaks are observed. The first peak has a retention time before the deadtime compound, indicating a log P_{ow} , while peak number 2 has almost the same retention time, indicating a log $P_{ow} \approx 0$. The bioaccumulation potential for the scale inhibitor is therefore reported as: log $P_{ow} = 0$.

Aquateam has evaluated various standards for calibration of the HPLC system, and has selected the following compound classes which are representative for the wide range of offshore chemicals tested: aniline (amines), nitrobenzene (nitro compounds), toluene (aromates), benzophenone (ketones), phenylbenzoate (esters), and diphenylether (ethers). The standards are selected from OECD's list of standards (OECD, 1989), and cover a log P_{ow}-range of 0.9-4.2.

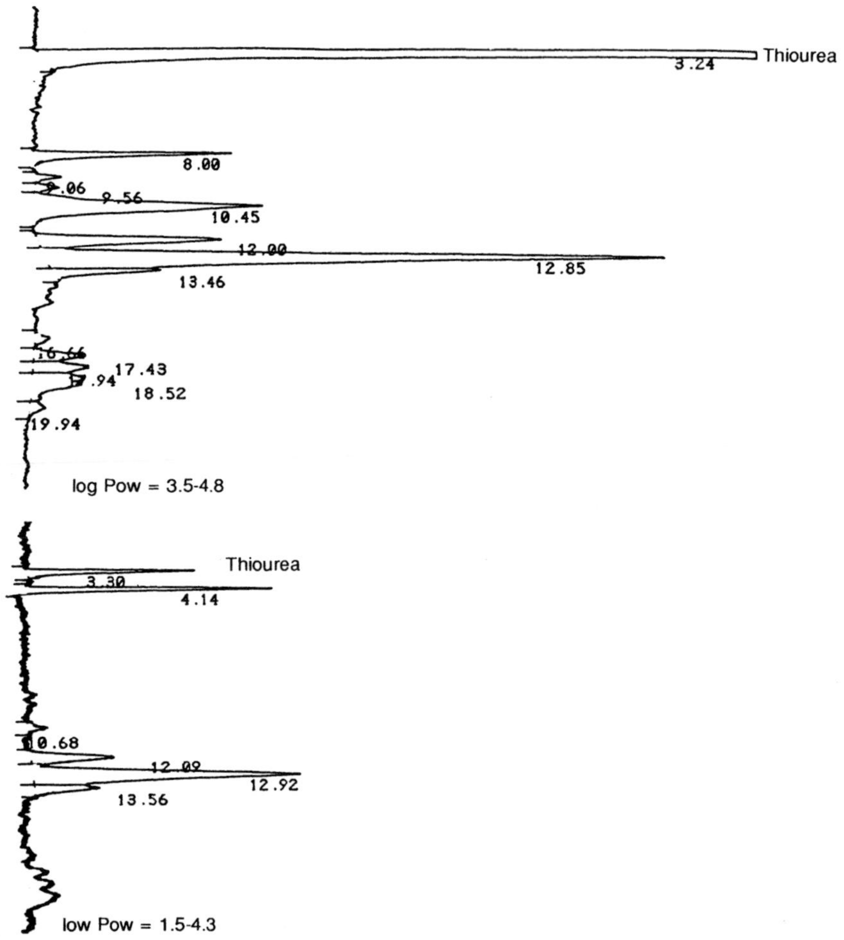

Figure 2. **HPLC chromatogram of the corrosion inhibitor using two detectors. 1) UV and 2) RI (Refractive Index). An unretained organic compound (Thiourea) has been used to determine the deadtime, t_0, of the system.**

The calibration curve for the capacity factor of the HPLC system and log P_{ow} shows a correlation coefficient of 0.9927. According to the Nordic classification system (see Figure 1), a chemical bioaccumulates if log P_{ow} 3.

The authorities in various countries do not agree on which value to use, i.e., the Dutch authorities use log $P_{ow} = 4$ as regulating value. Based on the Nordic system, and even the Dutch proposed regulatory values, the corrosion inhibitor is considered likely to bioaccumulate, while the scale inhibitor will not.

Table 3. TOC, percent chemical concentration in the water phase and calculated OWDF-values for the corrosion and scale inhibitors

Chemical	Dosage (ppm)	TOC (ppm)	% Chemical in the water phase	OWDF
Corrosion inhibitor	150	90 ± 15	33 ± 6	2-3
	500	130 ± 7	15 ± 2	5-7
	750	172	13	7
	2000	367	11	8
Scale inhibitor	200	11	50	1

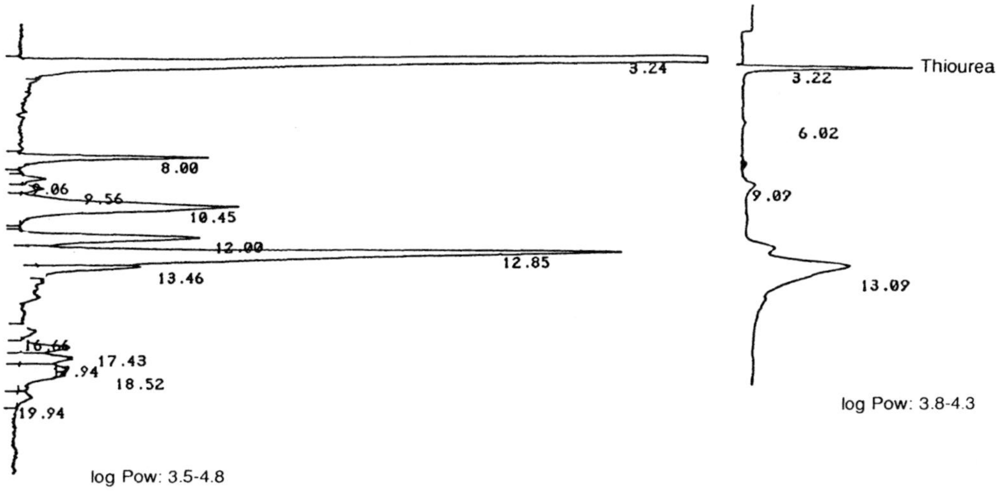

Figure 3. HPLC chromatogram of the corrosion inhibitor (UV-detector) as 1) pure chemical, and 2) seawater fraction. Thiourea is the deadtime compound.

The oil field chemicals, such as the corrosion inhibitor, are to be used in an oil/seawater mixture offshore. The regular concentration range varies between 15 and 50 ppm. Based on the results seen from the OWDF-tests, approximately 30% of the chemical enters the water phase. In preliminary tests, the log P_{ow} of the fraction of the chemical present in the water phase has been determined (at a 2000 ppm dosage), indicating a log P_{ow} range of 3.8-4.3, indicating that the fraction of the chemical with the lowest and highest log P_{ow} disappears during the mixing of oil and seawater (Figure 3). When analyzing seawater fractions of lower dosages (150-200 ppm), we are not able to detect the chemicals. The log P_{ow} can, however, be determined if a concentrating step is introduced.

Biodegradation. The chemicals have been tested using two different fractions of the chemical; 1) water soluble fraction (WSF) (after mixing with seawater), and 2) water extractable fraction (after mixing with oil and seawater). Three different cultures have been used: 1) acclimated, 2) non-acclimated, and 3) seawater with only natural bacteria present.

Cultures (1) and (2) were derived from a mixture of garden soil, a shallow marine sediment, and activated sludge. Culture (1) was further exposed to the test chemical at low doses over a period of 20 days in increasing concentrations at five day intervals.

Two contract laboratories, in addition to Aquateam's laboratory, have been used to perform the two OECD procedures. Very briefly, results from these laboratories using the two standard procedures, ranged from 15 to 99%, and 17 to 79% biodegradability between laboratories, for the corrosion and scale inhibitors, respectively, tested on the water extractable fractions (Table 4). Biodegradation test results (Shake Flask Test, Lab. 1) for the two chemicals are presented in Figure 4 to illustrate the rate and extent of biodegradation for this test. The test results show that the corrosion inhibitor is readily biodegradable according to the definition of the OECD guidelines, while the scale inhibitor is not.

However, vastly different results can be obtained if different procedures are followed, making it impossible to classify the products. The variable results presented in Table 4 are due to shortcomings in the procedures, including inappropriate use of COD determinations (in seawater samples) and ThOD calculations, non-standard acclimation procedures, and inherent problems with reproducibility when grab-sampling the water phase which includes small droplets of oil and sparingly water-soluble compounds. Tentatively, the AFNOR Shake Flask Test is preferred over the Closed Bottle Test, due to the inherent difficulties with measuring the

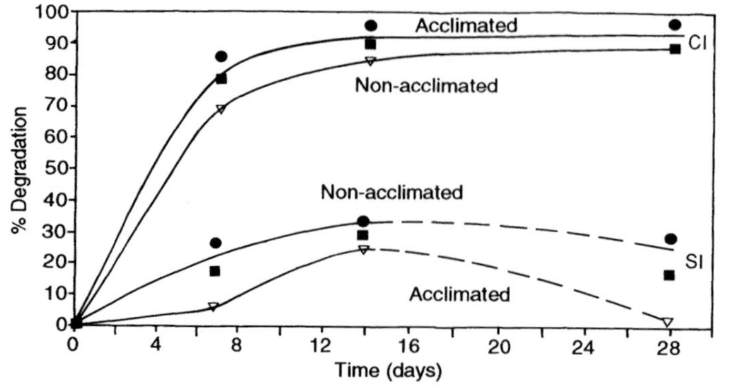

Figure 4. **Biodegradation test results from Lab 1 on the corrosion inhibitor (CI) and scale inhibitor (SI) (Shake Flask Test) using acclimated and non-acclimated cultures.**

COD of complex mixtures in seawater. The TOC (DOC) measurements required for the AFNOR-test are more reliable.

Based on preliminary results, the WSF of the chemicals using seawater with only natural bacteria present has been selected for further standardization. It is hoped that problems with reproducing the same seawater extractable fractions of the chemicals with oil droplets present in the extraction can be avoided. Further work is under way with the Shake Flask Test and a respirometric test method to develop a reliable test system.

Internal quality control has shown that the Shake Flask Test gives greater precision, and that the respirometric procedure can improve upon this, as well as broaden the capabilities of the test.

Regarding a sensitivity analysis of the percent biodegradation, neither the model compound selection chosen for the "ThOD" calculation, nor the exact composition of the product significantly influences the final value. However, the OWDF can have significant effects depending upon the assumptions made on the distribution of the chemical. For example, assuming 100% dispersion of the corrosion inhibitor in seawater will yield only 20% biodegradability. Whereas, assuming 50% dispersion in the water phase, will yield almost 70% degradation and the product can be classified as readily biodegradable.

Table 4. Percent (%) biodegradation of the corrosion inhibitor and scale inhibitor. Tests performed at three different laboratories.

Product	% Biodegradation								
	Shake Flask Test					Closed Bottle Test			
	No culture	Acclimated		Non-acclimated		Acclimated		Non-acclimated	
	Lab 3	Lab 1	Lab 3	Lab 1	Lab 3	Lab 2[1]	Lab 1	Lab 2[1]	Lab 1
Corrosion inhibitor	82	95	75	90	79	22-26	99	15-18	90
Scale inhibitor	11	31	24	25	25	73-79	17	17-26	25

[1]The results presented are based on COD and ThOD values.

Toxicity. Toxicity results (EC_{50}, LC_{50}) from preliminary range finding tests of the WSF of the corrosion and scale inhibitors are presented in Table 5. The results in Table 5 indicate that the corrosion inhibitor can be considered as acutely toxic according to the Nordic Classification model in Figure 1 for all test species, except *Acartia* which was considered to be very toxic. The scale inhibitor is not toxic according to the preliminary test results using *Artemia*, Microtox and *Acartia tonsa* tests, while toxic properties are observed performing the *Skeletonema* test, which proves to be more sensitive to this chemical.

Table 5. Toxicity results (EC_{50}, LC_{50}) from range finding tests of the WSF of the corrosion and scale inhibitors

Chemical	Toxicity results (ppm)				
	Artemia LC_{50}	Microtox EC_{50}	Mitochondria EC_{50}	*Skeletonema* EC_{50}	*Acartia* LC_{50}
Corrosion inhibitor	6	5	>33	1	0.3
Scale inhibitor	>10,000	>10,000	>10,000	15	6,000

NA = Not Available

CONCLUSIONS

In a product screening protocol for risk assessment and evaluation of environmental effects of oil field chemicals, a number of important factors need to be determined. Based on the experiences obtained through a two-year test program for these types of chemicals, the following conclusions can be drawn:

Margin of safety

- To determine the extent of a test protocol, the margin of safety should be determined.
- The definition is: MAEC/PEC = Minimum Adverse Effect Concentration/Potential Environmental Concentration

Oil/water partitioning

- Determination of the oil water distribution factor is a critical parameter for evaluation of potential environmental effect of oil field chemicals, and is defined as the ratio of the chemical concentration in the oil and seawater phases.
- The procedure needs to be standardized and verified for chemicals in all application areas offshore.
- The procedure should be developed to an international standard.
- Biodegradation
- Modified OECD's Shake Flask Test is recommended.
- The biodegradation procedure for calculation of percent biodegradation needs to be evaluated.
- The sample preparation procedure needs to be standardized, especially with respect to non-water soluble, surface active and volatile compounds.
- The test set-up can be simplified, automatized and made more reliable, especially for non-water soluble, surface active and volatile compounds, using respirometric methods.
- The respirometric methods need to be developed to a European standard.

Bioaccumulation

- OECD's HPLC procedure is recommended.
- Use of appropriate standard chemicals, verification by several types of detectors and various eluent concentrations and flow rates through the column should be performed

in order to avoid that some of the components in the chemical is adsorbed on the walls etc.

- Log P_{ow} values should be presented as a range covering relevant compounds present.
- A concentration procedure for analyzing of water fractions needs to be developed.
- The relationship between P_{ow} and bioconcentration factors for typical oil field chemicals should be developed.

Toxicity

- The toxicity test procedure should be designed depending on the margin of safety.

MAEC/PEC > 100:	No further aquatic toxicity testing required.
100 <MAEC/PEC> 1:	Chronic toxicity of chemical to aquatic organisms should be determined.
10 <MAEC/PEC> 1:	Further acute toxicity testing required. Consideration of most sensitive species may guide testing towards three trophic levels.
MAEC/PEC < 1:	This indicates an acute aquatic toxicity hazard. No further aquatic toxicity testing therefore required unless further work results in PEC being reversed downwards.

- There is a need for simplifying the toxicity testing scheme for oil field chemicals and agree upon one species at each trophic level as a standard when 10 MAEC/PEC 1, and in addition there is a need to agree upon one standard screening test which can be used.
- The sample preparation procedure for surfactants and low, non-water soluble and volatile compounds needs to be standardized.

Risk assessment

- The authorities have to agree upon test methodology, test organisms and threshold values for classification of chemicals in order to establish a completely test protocol, which also concerns non-water soluble, surface active and volatile components. This includes determination of Minimum Adverse Effect Concentration (MAEC) or No Observed Effect Concentration (NOEC).

ACKNOWLEDGEMENTS

The financial and professional contribution from Conoco Norway (CNI) and the Norwegian Oil Industry Association (OLF) is gratefully appreciated.

REFERENCES

Anderson, J.W., Neff, J.M., Cox, B.A., Tatem, E.H. and Hightower, G.M., 1974, Characteristics of dispersion and water soluble extracts of crude oils and their toxicity to estuarine crustaceans and fish, Marine Biology, 27:75-88.

Barth, H., 1991, Recent scientific achievements on ecotoxicology in the European community environmental program, PACON Workshop Ecotoxicology.

Bennett, D., Girling, A.E. and Bounds, A., 1990, Ecotoxicology of oil products: Preparation and characterization of aqueous test media, Chemosphere, 21(4-5):659-669.

Blondin, G.A., et al., 1989, An *in vitro* submitochondrial bioassay for predicting acute toxicity in fish, *in*: "Aquatic Toxicology and Environmental Fate", G.W. Suter and M.A. Lewis, eds., ASTM, Philadelphia, 551-563.

Blondin, G.A., et al., 1987, Mammalian mitochondria as *in vitro* monitors of water quality, Bull. Environ. Contam. Toxicol., 38:467-474.

de Henau, H., Lee, C.M. and Gilberg, P.A., 1986, Environmental safety of detergents, the AIS approach to assessment of the environmental safety of detergents, Tenside Detergents, 23:5, 267-271.

ECETOC, 1983, Assessment of reverse phase chromatographic methods for determining partition coefficients, Chemosphere, 12(11):1459-1475.

EEC, 1990, Classification and labelling of dangerous substances and preparations: criteria for the choice of phrases indicating special risks (R-phrases) and safety advice (S-phrases), Directive 67/548/EEC.

Gjøs, N., Møller, M., Hægh, G.S. and Kolset, K., 1989, Existing chemicals - systematic data collection and handling for priority setting, Nordic Council of Ministers, NORD 1989:87.

Hunton, A., 1991, Personal Communications.

IMO/FAO/UNESCO/WMO/WHO/IAEA/UN/UNEP Joint Group of Experts on the Scientific Aspects of Marine Pollution (GESAMP), 1989, The evaluation of the hazards of harmful substances carried by ships: revision of GESAMP reports and studies No. 17, Reports and Studies No. 35, GESAMP International Maritius Organization.

ISO, 1990a, Water quality determination of acute lethal toxicity to marine copepods (copepoda, crustacea), UK's proposal ISO/TC 147/SC 3/WG 2.

ISO, 1990b, Water quality determination of acute lethal toxicity to marine copepods (copepoda, crustacea), UK's proposal ISO/TC 147/SC 3/WG 2.

ISO, 1991, Water quality - marine algal growth inhibition test with *Skeletonema costatum* and *Phaeodactylum tricornutum*. ISO/TC 147/SC 5/WG 5/N 120.

Lundgren, A., 1989, Comparison of different models for environmental hazard classification of chemicals (report from the Nordic Expert Group), Kemi Report No. 9/89.

OECD, 1989, Partition coefficients (n-octanol/water). High performance liquid chromatography (HPLC) method, OECD Guideline 117 for testing of chemicals.

OECD, 1981, Guidelines for testing chemicals, EEC 1979 Directive 79-731.

PARCOM, 1990, Acute toxicity to the marine copepod *Acartia tonsa*, Paris Commission, Ring Test Protocol, WQI.

PARCOM, 1992, Intercalibration and comparison of tests and test laboratories for evaluation and approval of offshore chemicals and drilling muds, WQI, May.

MODELING OFFSHORE DISCHARGES OF PRODUCED WATER

M. G. Brandsma[1], J. P. Smith[2], J. E. O'Reilly[2], R. C. Ayers, Jr.[2], and A.L. Holmquist[3]

[1] Brandsma Engineering
Durango, Colorado
USA

[2] Exxon Production Research Company
Houston, Texas
USA

[3] Chevron Oil Field Research Company
La Habra, California
USA

INTRODUCTION

Industry and government have used the Offshore Operators Committee (OOC) mud and cuttings dispersion model (Brandsma and Sauer, 1983a) for years to assess the effects of drilling effluent discharges. Now, a revised version of the model simulates the dispersion of produced water discharges as well. The same fundamental predictive methods are used in either case. The revised model predicts the fate of soluble components, particulate solids and oil droplets or gas bubbles in the produced water stream. The model also predicts the decline of the plume temperature and salinity to ambient levels.

This paper describes how the model works, the modifications to adapt it for produced water discharges, and special features that make it widely applicable. We also compare predictions of the model with laboratory and field measurements.

DESCRIPTION OF OOC MODEL

Overview

The OOC model predicts the short term fate of drilling and production discharges in the marine environment. Model results, combined with biological effects data, may be used to assess the environmental effects of these discharges. The model simulates constant rate discharges of produced water, drilling mud or cuttings. The calculations are divided into three phases: initial dynamics, passive dispersion, and a transition that converts the continuous mass fluxes of the initial dynamics to independent clouds. These clouds are free to disperse according to local ambient conditions. The initial dynamics calculation is further divided into a jet/plume calculation for the initial descent and a collapse calculation. Although this paper describes the simulation of plumes that are heavier than their ambient surroundings, the model can also simulate light, ascending plumes.

The model uses two coordinate systems. For simplicity, the initial dynamics are treated in a local coordinate system with its origin at the water surface above the discharge point. Bathymetry and ambient conditions are described in reference to a global coordinate system covering the entire simulation region. The global system is also used for the dispersion phase calculation. Both coordinate systems have the +Y axis pointing downward, the +Z axis pointing east and the +X axis pointing south.

Initial Descent

Much of the total dilution of a discharge occurs during its initial descent because of entrainment of ambient fluid into the jet/plume. The densimetric Froude number, $F = U_0 / \sqrt{g'D}$, determines the classification of the initial descent as jet-like (F > 2) or plume-like (F < 2) (R. C. Y. Koh, personal communication, 1985) and selects the entrainment equation used in the calculations. The densimetric Froude number is the ratio of momentum flux to buoyancy flux (U_0 is the exit velocity, g' is the apparent gravitational acceleration $g(\rho - \rho_a)/\rho_o$, is the plume density, ρ_a is the ambient density, ρ_o is a reference density, and D is the discharge pipe diameter). Most offshore oilfield discharges are plume-like (i.e., buoyancy is more important than momentum).

The water body receiving a discharge is characterized by ambient density that is a function of depth, $\rho_a(y)$, and by an ambient current field that is a function of position in the water body and time, $\vec{U}_a(x,y,z,t)$. A top hat profile characterizes the distribution of properties across a plume; i.e., properties are uniform across the plume and have ambient values outside it. The flow quantities at any cross-section are the plume density, ρ, the plume centerline velocity vector, $\vec{U}$, the plume radius, r, the cross-sectional area of the plume, $A = \pi r^2$, and the total plume volume flux, $Q = A|\vec{U}|$. Besides fluid, the plume may contain up to 12 particulate classes, each of which is characterized by density, ρ_i, and its volume concentration, C_i and fall velocity, v_i, where i is the particle class. The density of the discharged fluid is ρ_f and its volume concentration is C_f. Equations for the conservation of mass, momentum, buoyancy (density defect times volume flux) and constituent flux describe the initial descent of the plume. The derivatives of these quantities with respect to the arc length, s, along the plume centerline are:

$$\frac{d}{ds}(\rho Q) = E\rho_a \sum_i G_i \rho_i - G_f \rho_f \tag{1}$$

$$\frac{d}{ds}(\rho Q\vec{U}) = Ag(\rho - \rho_a)\hat{j} + E\rho_a\vec{U}_a - \sum_i G_i\rho_i\vec{U} - G_f \rho_f\vec{U} + \vec{F}_D \tag{2}$$

$$\frac{d}{ds}(Q(\rho_o - \rho)) = E(\rho_o - \rho) - \sum_i G_i(\rho_o - \rho_i) - G_f(\rho_o - \rho_f) \tag{3}$$

$$\frac{d}{ds}(QC_i) = -G_i \tag{4}$$

In the above equations, E is the rate of ambient fluid entrainment, G_i is the rate at which constituent i leaves the plume, G_f is the rate at which the discharged fluid constituent leaves the plume, $\hat{j}$ is the vertical unit vector. $\vec{F}_D$ is the drag force vector accounting for the ambient current's action bending the jet toward the direction of the current flow. Its magnitude is $C_D \rho_a r (|U_a| \sin\gamma)^2$, where C_D is a drag coefficient and γ is the angle between the ambient current and the local plume trajectory.

For produced water discharges, temperature and salinity defects are also conserved:

$$\frac{d}{ds}(Q(T_d - T)) = E(T_d - T_a) \tag{5}$$

$$\frac{d}{ds}(Q(S_d - S)) = E(S_d - S_a) \tag{6}$$

where T_d and S_d are the temperature and salinity of the discharged fluid. Equation 5 assumes the discharged and ambient fluids have the same heat capacity.

The ambient fluid entrainment rate is:

$$E = 2\pi r\,(\alpha\,|\,\vec{U} - \vec{U}_a \cos\gamma\,| + \alpha_2\,|\,\vec{U}_a\,|\cos\theta\sin\theta) \tag{7}$$

The first term is momentum jet entrainment, with the entrainment coefficient α depending on the densimetric Froude number. The second term represents two-dimensional buoyant thermal entrainment with its magnitude depending on the angle between the vertical axis and the local plume velocity vector, θ, and a fixed entrainment coefficient, α_2.

Particulate constituents leave the descending plume at the greater of two rates; separation by turbulence near the mouth of the pipe (equation 8) or by gravitational settling where particles settle faster than the vertical speed of the plume (equation 9). Separation of constituents within a few diameters of the pipe mouth, which has been observed in field experiments (Ayers, et al., 1980), cannot be explained by the gravitational settling described by Koh and Chang (1973). We postulated the removal of particulates and fluid from the emerging plume by ambient turbulence in the receiving water. Using dimensional analysis, we derived the following equation that has worked well in practice:

$$G_i = 2C_i\,|\,\vec{U} - \vec{U}_a|\;rK \tag{8}$$

where K is a coefficient. Separation of discharged fluid, G_f, is handled the same except that it is proportional to the volume concentration of the fluid portion of the discharge, C_f. The settling rate is given by an equation similar to Koh and Chang's:

$$G_i = \frac{1}{2}P\,|\,v_i|\,C_i\,(1 - \beta_i)e_s \tag{9}$$

where P is the plume circumference, e_s is an enhanced settling factor and the settling coefficient depends on the vertical component, v, of the plume velocity vector and on the constituent settling speed:

$$\beta_i = \begin{cases} 1 & \textit{while} \left|\frac{v}{v_i}\right| \geq 1 \\ 0 & \textit{once} \left|\frac{v}{v_i}\right| < 1 \end{cases} \tag{10}$$

The enhanced settling factor is

$$e_s = max(1.0, min(0.013C^{4/3}, e_{\max})) \tag{11}$$

and is based on laboratory work by Bowers and Goldenblatt (1978), which showed that solids settling speeds in a concentrated suspension were much greater than characteristic settling velocities for individual particles. The coefficient e_{max}, which sets an upper limit on the enhancement, is 10. C is the local concentration of suspended solids in mg/l.

The derivatives dx/ds, dy/ds, and dz/ds, which are equivalent to the direction cosines of the plume centerline velocity vector, define the plume trajectory. The trajectory equations plus the others given above are integrated using an adaptive step size Runge-Kutta algorithm, given initial conditions for U, r, ρ, C_i and C_f.

The initial descent (or ascent) ends when the plume hits the sea bed (or the sea surface) or when the plume turns horizontal after passing through its depth of neutral buoyancy.

Dynamic Collapse

After the initial descent, the plume has a mean velocity vector roughly the same as that of the ambient fluid. A cross-section of the plume will behave like a two-dimensional thermal. If the plume hits the sea bed or the sea surface, the material remaining in it will collapse into a thin layer because of gross density differences. If the plume reaches its level of neutral buoyancy, it collapses due to the difference in density gradients between the inside and the outside of the plume, as described below. The plume cross-section in the collapse phase is elliptical within the water column, or semi-elliptical on the sea bed or surface, and is characterized by semi-major and semi-minor axes b and a, respectively.

The material coming to rest at neutral buoyancy is well mixed because of passing through the turbulence of the initial descent, so the density gradient within the plume is small. Neutral buoyancy implies that the density of the center of the plume is the same as the density of the ambient fluid at that depth. Because the ambient fluid must be stably stratified (density increasing with depth) to arrest the vertical motion of the plume, the material within the lower half of the plume is lighter than the surrounding ambient fluid so it moves upward. Similarly the fluid in the upper half of the plume is heavier than the surrounding ambient fluid, so it moves downward. Since volume is conserved, the plume must spread laterally to accommodate these motions. As the vertical dimension of the plume decreases, its density gradient increases. The collapse continues until the difference in density gradients disappears.

Consider a plume collapsing within the water column. The density gradient inside the plume is assumed to be less than that in the ambient fluid by a factor of $\lambda\, a_o/a$, where λ is a coefficient and a_o is the radius attained in the initial descent. The driving and resisting forces act on the centroid of a quarter section (quadrant) of the elliptical plume cross-section. Integrating the density gradient difference gives the magnitude of the collapse driving force:

$$F_c = \frac{g}{6}\frac{d\rho_a}{dy}\left(1 - \frac{\lambda a_o}{a}\right)a^3 \tag{12}$$

Tests have shown that $\lambda = 0.57$ produces collapse dynamics consistent with experimental observations in the field (Merritt, 1974) and laboratory (Albright, et al., 1986).

The forces resisting the collapse (per unit length of plume) are form drag, D_c, and skin friction, S_c, respectively:

$$D_c = \frac{1}{2}C_{drag}\,\rho_a\left|\frac{db}{dt}\right|\frac{db}{dt} \tag{13}$$

$$S_c = \frac{1}{2}C_{fric}\,\frac{b}{a}\,\frac{db}{dt} \tag{14}$$

where C_{drag} and C_{fric} are drag coefficients. These forces all act at the centroid of the quadrant and the resultant force is equated with the inertial force of the quadrant to form the equation for dynamic collapse:

$$I = F_c - D_c - S_c \tag{15}$$

The horizontal inertial force of the quadrant is the time rate of change in the product of its mass per unit length and the transverse horizontal velocity of its centroid (which is related to the tip velocity *db/dt* by geometry):

$$I = \frac{d}{dt}\left(\frac{ab}{3}\rho\frac{db}{dt}\right) \tag{16}$$

Each new tip velocity is obtained by finding the value of I using the old value of the tip velocity, and then solving the integral of equation (15) for the new value of *db/dt*.

As in the initial descent, mass, momentum, buoyancy, constituent fluxes, temperature and salinity must be conserved. The governing equations are the same except that they are written for an elliptical cross-section so that the cross-sectional area is $A=\pi ab$. The terms E, S_i, and $\vec{F}_D$ in equations 1-6 take different forms in the collapse phase. The entrainment rate is now the sum of terms related to convection and to spreading of the plume:

$$E = \frac{1}{2}P\left(\alpha_3 f_\alpha\left|\vec{U}-\vec{U}_a\right| + \alpha_4\frac{db}{dt}\frac{a}{b}\right) \tag{17}$$

α_3 and α_4 are entrainment coefficients, and f_α is a modification factor to attenuate convective entrainment. Particle settling is the same as in the initial descent. The drag force acts to maintain the plume direction downcurrent:

$$F_D = 2C_w a\,\rho_a \sin\phi\left|\vec{U}-\vec{U}_a\right|(\vec{U}-\vec{U}_a) \tag{18}$$

where C_w is a drag coefficient for a wedge and ϕ is the angle in the horizontal plane between the plume velocity vector and the ambient velocity vector.

Collapse on the Sea Bed or Sea Surface

If the plume hits the sea bed or sea surface, the equations described so far can still be used to describe the process if we add additional terms. There is an additional collapse driving force that arises from gross density differences. For a plume collapsing on the sea bed, its form is:

$$F_b = \frac{1}{2}(\rho - \rho_a)ga^2 \tag{19}$$

For a buoyant plume collapsing at the sea surface, the sign of the density difference is reversed to maintain a positive F_b. If a plume collapses on the sea bed, additional terms for friction at the bed enter the momentum equation and the collapse dynamics equation. No such term is needed for collapse at the surface.

Passive Dispersion

After completing the convective and dynamic collapse phases, currents and turbulence in the receiving water body disperse the discharged fluid and particulates. The passive dispersion part of the model processes each particulate class and the fluid separately. At the end of a simulation, results from each individual class are combined to form cumulative results.

The LaGrangian advection-dispersion scheme used in the OOC model represents a dispersing field of discharged materials as a set of independent clouds. These clouds are created from the history of the initial descent and dynamic collapse by a process described in the next section. Each individual cloud has a Gaussian distribution in three dimensions. The vertical extent is usually much smaller than the horizontal so that each cloud might be visualized as looking something like our galaxy. Local ambient currents transport the clouds. The change of position of each cloud in a time step is

$$\Delta x = u(x,y,z,t)\,\Delta t\,; \qquad \Delta z = w(x,y,z,t)\,\Delta t \tag{20}$$

where $u(x,y,z,t)$ and $w(x,y,z,t)$ are the x and z (horizontal) components of the ambient current, and Δt is the time increment.

The building block of the method is the concentration distribution of a single Gaussian cloud of material (Slade, 1968, for example). Given an ensemble of clouds containing solids or oil droplets, the contribution of the i-th cloud to the concentration at a point (x,y,z) is:

$$C(x,y,z)_i = (2\pi)^{-\frac{3}{2}} \frac{m_i}{\sigma_{xi}\sigma_{yi}\sigma_{zi}} \exp\left[-\frac{1}{2}\left(\frac{(x-x_i)^2}{\sigma_{xi}^2} + \frac{(y-y_i)^2}{\sigma_{yi}^2} + \frac{(z-z_i)^2}{\sigma_{zi}^2}\right)\right] \tag{21}$$

where the i-th cloud has mass m_i, coordinates x_i, y_i, z_i and variances σ_{xi}, σ_{yi}, σ_{zi}. By the principle of superposition, the total concentration at (x, y, z) is the sum of contributions from individual clouds, $C(x, y, z) = \sum_i C(x, y, z)_i$.

Turbulent eddies acting horizontally and vertically disperse each cloud. The horizontal growth of a cloud obeys the four-thirds power law of dispersion (Fischer, et al., 1979). The horizontal dispersion coefficient is $K_{x,z} = A_L L^{4/3}$, where A_L is a dissipation parameter, and L is the characteristic length of the dispersing cloud, equivalent to four times the horizontal cloud variances, σ_{xi} and σ_{zi}. The power law assumes that turbulent eddies of all sizes are available to disperse a cloud. Using the power law, Koh (1971) derived an equation for the growth of the cloud in a time interval Δt:

$$\sigma = \sigma_o \left[1 + 4^{4/3}\frac{2}{3}\frac{A_L\,\Delta t}{\sigma_o^{2/3}}\right]^{3/2} \tag{22}$$

This is the cloud growth equation used in the OOC model. Yudelson (1967) reports especially good agreement between the four-thirds power law and data observed in over 40 independent studies when L is in the range 9-9000 m. The range of values of the dispersion parameter, A_L, bracketing the observed results is approximately 0.0002-0.0009 $m^{2/3}/s$.

Vertical cloud growth relies on the Fickian expression characterizing the vertical size of a cloud, $\sigma_y = \sqrt{2K_y t}$ where K_y is the vertical dispersion coefficient and t is the time since the cloud was released from the dynamic plume. Taking differentials provides an expression for the change in vertical cloud size during a time increment:

$$\Delta\sigma_y = \frac{K_y}{\sigma_y}\Delta t \tag{23}$$

The OOC model estimates vertical dispersion coefficient profiles based on sea state and the ambient density profile. Near the water surface, the coefficient depends on wave conditions; at deeper depths, the coefficient depends on the density gradient of the water column (Koh and Fan, 1970).

Besides being advected by the ambient current and growing because of turbulence, each cloud moves vertically through the water column according to its settling velocity. Solid particles settle toward the sea bed. Oil droplets and gas bubbles, having a negative settling velocity, move toward the sea surface. When clouds reach the sea surface or sea bed, special treatments are needed.

The model treats the sea surface as an absorbing boundary for oil droplets and gas bubbles rising through the water column and as a reflecting boundary for settling solid particles. The sea bed is an absorbing boundary for settling solids and a reflecting boundary for oil droplets and gas bubbles. Both horizontal boundaries are assumed to reflect the discharged fluid and any tracer chemical contained in it. Reflecting boundaries are simulated by adding a virtual (imaginary) source located symmetrically across the boundary from the real source (e.g, a cloud). The sum of contributions from the real and virtual sources accurately model a reflecting boundary (Slade, 1968).

When the centroid of a Gaussian cloud is near the sea bed, that portion of the distribution lying below the sea bed is considered to have settled, whereas the portion above the sea bed is still associated with the cloud. The model maintains the distinction by keeping, for each cloud, its total mass and its suspended fraction. The mass is constant for each individual cloud. The suspended fraction is a function of the cloud's vertical variance, σ_{yi}, and the position of its centroid with respect to the seabed. This arrangement allows the model to properly accumulate particles on the seabed while preserving appropriate concentration distributions in the water column.

Gaussian clouds settle a discrete distance during each time step of the dispersive phase. The vertical distances from the centroid to the seabed before and after the settling step define two virtual horizontal planes through the cloud. If either of these planes are within 3 σ_{yi} of the cloud centroid we can calculate the amount of settling as the portion of the cloud mass lying between the planes. The planes are a distance $\Delta y = y_2 - y_1$ apart, where y_1 and y_2 are the distances from the cloud centroid to the sea bed at the beginning and end of a time step, measured positive when the centroid is above the sea bed and negative otherwise. Taking the triple integral of equation (21) over the vertical range Δy and infinite horizontal ranges gives the mass between planes:

$$T = \int_{y1}^{y2} \int_{x=-\infty}^{x=+\infty} \int_{z=-\infty}^{z=+\infty} C\ (x,y,z)_i\, dz dx dy \tag{24}$$

The integrals with infinite limits in the horizontal directions evaluate to 1, leaving the expression:

$$T = erf\left(\frac{y_2 - y_i}{\sigma_{yi}}\right) - erf\left(\frac{y_1 - y_i}{\sigma_{yi}}\right) \tag{25}$$

A grid of square cells on the sea bed collects the settled material. The amount in each cell is the integral of the horizontal cloud distribution over the limits of the cell. For the *i*-th cloud and a cell having boundary coordinates , x_1, z_1, x_2, z_2 the amount accumulated in the cell is:

$$\delta = T\left[erf\left(\frac{x_2 - x_i}{\sigma_{xi}}\right) - erf\left(\frac{x_1 - x_i}{\sigma_{xi}}\right)\right]\left[erf\left(\frac{z_2 - z_i}{\sigma_{zi}}\right) - erf\left(\frac{z_1 - z_i}{\sigma_{zi}}\right)\right] \tag{26}$$

A cloud with all its material deposited ceases to exist. Deposited particles of each solid class remain in place on the seabed; the model does not include resuspension effects. The distribution of the cumulative totals of particle accumulations on the sea bed is one product of the OOC model. When oil droplets are deposited at the sea surface, a cumulative total is calculated, but no horizontal distribution is formed and no submodel for surface oil transport is invoked.

Transition from Initial Dynamics to Passive Dispersion

The initial dynamics part of the model calculates constituent fluxes in a continuous plume. The passive dispersion phase models the fate of an ensemble of Gaussian clouds. This section describes the transition between the two.

The initial dynamics calculations are saved at intervals, forming a history of the dynamic plume. Each interval is a potential source of clouds for passive dispersion. The flux change of each constituent as it passes through the interval determines the number and mass content of clouds created from that interval. Separation or settling of the constituent from the dynamic plume cause the flux changes. Non-settling constituents pass out the end of the dynamic plume. Visualize the dynamic plume as a leaky pipe composed of connected intervals, fixed in space, with a leak in each connection. The inflow to each interval and its leakage rate determines the flow passed on to the next interval. Each leak in the pipe is a source of clouds to be passively dispersed. Clouds from any one interval always have the same initial position in space, but different creation times. The size of the clouds is initially based on the size of the dynamic plume at the point of cloud creation. Later, an optional wake effect process may increase the size and randomly alter the position of individual clouds.

Model Organization

The model is organized in five parts: input, dynamic plume, transition from dynamic plume to passive dispersion, passive dispersion, and output. The input processor interprets an input data file and sets up the conditions of the simulation. The input file is a series of independent sections separated by blank lines. There are seven required sections defining information needed for every simulation. Optional data sections can be used to control options and override default settings in the model.

The second and fourth parts perform the simulation using the equations presented above. The model traverses the time span of the simulation twice, once in performing the dynamic plume calculations, and once in performing the passive dispersion calculations.

MODIFICATIONS FOR PRODUCED WATER DISCHARGES

The most important change made to the OOC model released in 1983 (Brandsma, et al., 1980; Brandsma and Sauer, 1983a; Brandsma and Sauer, 1983b; Brandsma and Sauer, 1983c; O'Reilly, et al., 1989) was the addition of conservation equations for temperature and salinity defects (equations 5 and 6) to the dynamic part of the model. Temperature and salinity defects are not explicitly calculated in the dispersive part of the model; these can be derived from tracer concentrations, which the model provides.

The temperatures and salinities of some produced water discharges lie well outside the range of validity of the usual density equations for sea water, the first of which was formulated by Knudsen (U. S. Naval Hydrographic Office, 1952). Muellenhoff et al. (1985) include a routine for calculating σ_t (= 1000(ρ−1)) based on the Knudsen equations. The sodium chloride brine density table contained in the *International Critical Tables, Volume III* (Washburn, 1928) extends to 100° C temperature and 260 ppt salinity. We have combined an implementation of the Knudsen equations with a quadratic surface fit to the brine density table (using the RSREG procedure of the Statistical Analysis System from SAS, Inc.) to obtain an extended range brine/seawater density function. The RMS difference between the surface fit and the table is 0.00069 g/cm^3 (J.P. Smith, unpublished memorandum to R.C. Ayers, 1990).

A produced water plume may be lighter than the ambient sea water and will rise to the surface and stay there for some time. While the plume is on the surface, we must allow for cooling of the plume by the air above it, as well as cooling by entrainment of ambient sea water. The conservation equations for temperature defect and buoyancy (density defect) given previously are revised to account for external heat losses. The term $-Q\ (dT/ds)$ is added to the right side of equation 5, and the term $-Q\ (d\rho/ds)$ is added to the right side of equation 3. These terms are transformed for use in the model:

$$-Q\,\frac{dT}{ds} = -\frac{dT}{dt}\frac{dt}{ds}\,Q = -\frac{dT}{dt}\frac{Q}{|U|} \tag{27}$$

$$-Q\frac{d\rho}{ds} = -\frac{d\rho}{dt}\frac{dt}{ds}Q = -\frac{d\rho}{dT}\frac{dT}{dt}\frac{Q}{|U|} \tag{28}$$

A simplified expression for the heat loss rate is used:

$$\frac{dT}{dt} = \frac{hA_p(T-T_{air})}{mC_p} \tag{29}$$

where h is a surface heat transfer coefficient that varies as a function of wind speed, A_p is the area of the plume element in contact with the surface, m is the mass of the plume element and C_p is the specific heat of water. The derivative of plume density with respect to temperature, $d\rho/dT$, is found by differentiating the polynomial approximations for sea water density in the Knudsen equations.

SPECIAL FEATURES OF OOC MODEL

The OOC model incorporates several features that make it suitable for simulating marine discharges over a wide variety of conditions. A notable feature of the model is its flexibility. It will accept most discharge, bathymetric and ambient conditions. In shallow water cases, the dynamic phase calculations are reliable because the calculation adapts to local conditions at each integration step. There are additional program modules to handle the collapse phase and transition to passive dispersion when the initial plume hits these boundaries at near normal angles in zero or near zero currents. The model includes an independent program to translate current velocity fields produced by a public domain finite element hydrodynamics program to a form readable by the OOC model.

The model accepts discharges with low densimetric Froude numbers (pure plumes). The EPA plume models described by Muellenhoff et al. (1985) fail at Froude numbers less than 2.44.

The model uses the LaGrangian approach for dispersion calculations, i.e., it tracks individual clouds. This has three advantages. Numerical dispersion is eliminated. Computer memory is used more efficiently—all available memory holds data for dispersing material and none is used to denote the absence of material. Finally, output products can be obtained in whatever form and spacing desired—the output is not dependent on the geometry of a numerical grid (except for the distribution of solid particles on the sea bed).

The effects of the wake of a drilling or production platform are accounted for by introducing quasi-random fluctuations of cloud positions and sizes downcurrent of the platform, based on studies of wake intensity by Lin and Pao (1979). The strength of these fluctuations depends on ambient conditions and on the underwater configuration of the platform. The model implementation is a crude approximation of a complex phenomenon.

COMPARISON WITH LABORATORY EXPERIMENTS

Heated Water Plume into Two Layer Flow

Pierre and Viollet (1977) conducted a series of experiments on heated water plumes injected into a water flume. One of their experiments injected 50° C water upward into a flume with a layer of cool water (10° C) overlain with a layer of warm (30° C) water. The cool and warm layers were recirculated with the aid of splitter plates at the flume entrance and exit. The flow velocity was uniform except for a region of lower velocity at the interface. The discharge occurred from a vertically upward directed pipe with its end approximately 1.6 pipe diameters (D) above the flume bed, such that the densimetric Froude number, F, was 0.4. The exit velocity ratio (K = pipe exit velocity/ambient current) was 1.0. Because of the low Froude number, the initial trajectory of the discharged plume bent over about 50° at the mouth of the pipe. Figure

1(a) shows the concentration contours calculated by Pierre and Viollet from their temperature measurements and the ambient temperature profile they used. Figure 1(b) shows the concentration contours calculated by the OOC model. The coordinates are normalized by the pipe diameter, *D*. Overall, the comparison is pretty good—the model reproduces the measured behavior of the plume. The comparison for the 10% contour is very good. The lower concentration contours for 1% and 5% levels are located differently close to the discharge and come together 10 diameters from the discharge. The main flaw in the comparison is the intrusion of the upper 1% and 5% contours calculated by the model into the upper layer. This occurs because the model uses Gaussian similarity profiles to calculate water column concentrations caused by the dynamic plume (which are distinct from concentrations caused by clouds). The application of a Gaussian similarity profile to a plume calculated using a top-hat profile is acceptable because the profile shape makes little difference in plume behavior (Davidson, 1986). Because of the sharp temperature gradient, a Gaussian similarity profile does not work well in this region. In spite of this deficiency, the model clearly works well enough to give useful predictions.

Heated Water Plume into Constant Density Flow

In another series of experiments, Pierre and Viollet injected 50° C water into 10° C water in a flume. The ambient density was uniform. We selected an experiment for which the densimetric Froude number was 0.8 and the exit velocity ratio was 2.0. The ambient velocity profile in the experiment varied with distance above the flume bed according to:

$$u_a(y') = u_a(H)\left(\frac{y'}{H}\right)^{0.25} \tag{30}$$

where y' is the distance above the flume bed, H is the height of the discharge port above the flume bed, and $u_a(H)$ is the ambient velocity at the discharge port (this is the value used in calculating the exit velocity ratio). Figure 2(a) shows concentration contours calculated by Pierre

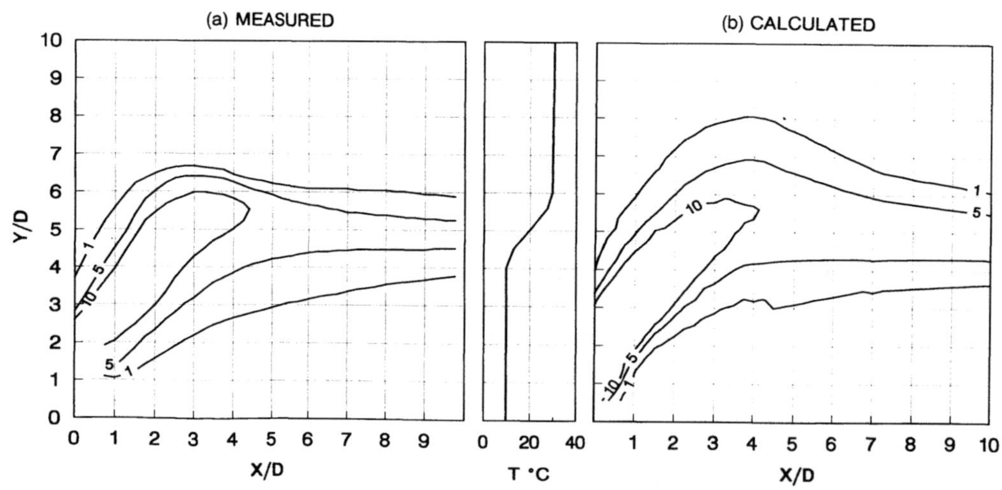

Figure 1. (a)Plume concentration contours as percent of initial concentration calculated from temperature measurements in two layer flow by Pierre and Viollet (1977). F = 0.4, K = 1. The thermocline (middle frame) shows the boundary between layers. (b). The concentration contours calculated by the OOC model.

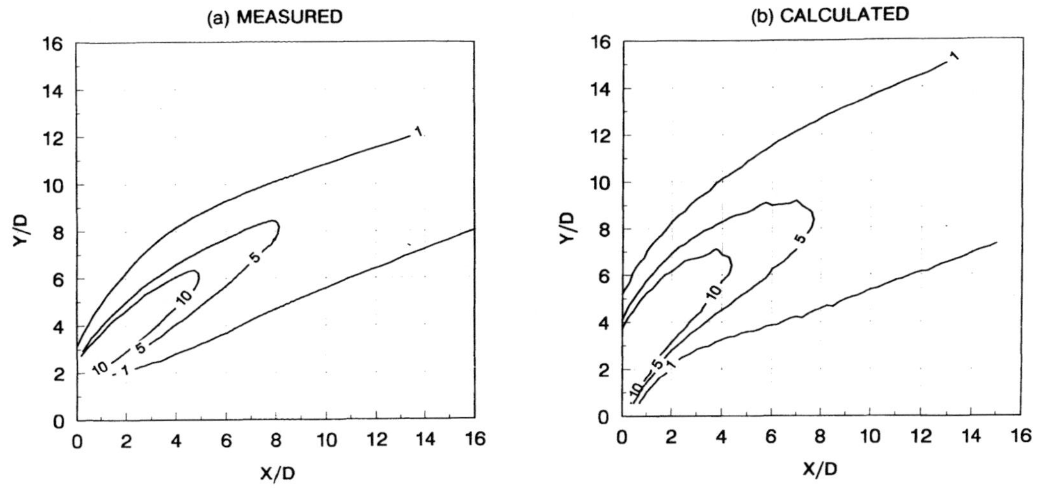

Figure 2. (a) Plume concentrations calculated from temperature measurements in constant density flow by Pierre and Viollet (1977). $F = 0.8$, $K = 2$. (b) The concentration contours calculated by the OOC model.

and Viollet from their temperature measurements. Figure 2(b) shows the contours calculated by the OOC model. The contours beneath the plume centerline compare quite well. The distances at which concentrations have declined to 10 and 5 percent of the initial value also compare well. The deformation of the laboratory plume by the ambient current causes the measured upper 1 and 5 percent contours to occur at shallower depths than calculated by the model using a Gaussian similarity profile. The contours calculated by the model are too widely distributed near the discharge pipe compared to observations. This is because the top-hat plume distribution used in the dynamic phase is converted to a Gaussian distribution when the model calculates concentrations in the water column. The constant ($\sqrt{2}$) relating the Gaussian variance to the half-width of the top-hat causes parts of the Gaussian distribution to lie outside the top-hat. The difference is pronounced near the pipe and decreases with

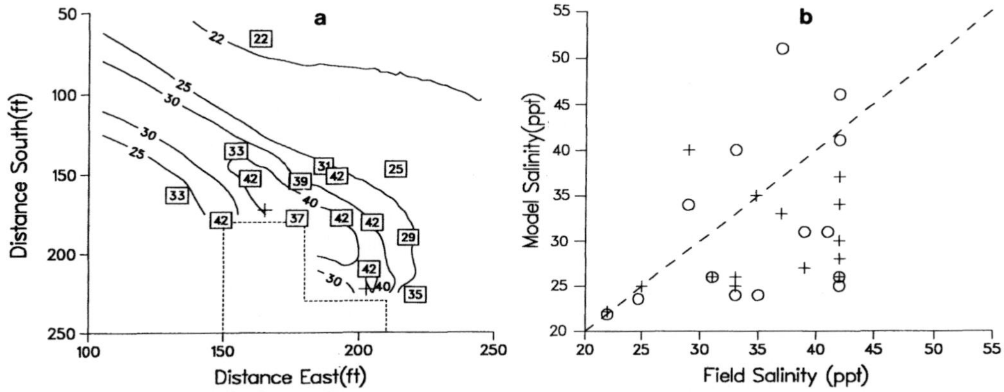

Figure 3. Near-bottom salinities (ppt) predicted by the OOC model (contour lines) compared with measured near-bottom salinities (in boxes) for a produced water discharge from a grounded barge (dotted line). Plus signs (+) mark discharge pipe locations.

distance from it. In practical use, concentrations near the pipe are of little interest compared to concentrations at regulatory boundaries.

COMPARISON WITH FIELD EXPERIMENT

OOC model predictions were compared with field observations of near-bottom water salinity around a produced water discharge from a facility on an L-shaped barge grounded in 8 ft of water. Produced water (115 ppt salinity, 35.6° C temperature) was discharged from two points on the barge at a total rate of 32,000 bbl/day. The average background salinity and temperature of the receiving water 110 ft from the discharge were 21.5 ppt and 28° C, respectively. A few current measurements made near the barge showed that the currents were mostly northwesterly with speeds ranging from 0.17-0.4 ft/s. Since the current measurements were inadequate to characterize the complex current field near the barge, input current velocity data for the OOC model were calculated with a hydrodynamic model (Froehlich, 1989) using upstream boundary conditions of 0.3 ft/s northwesterly current and 8 ft depth. The plume from each discharge pipe was modeled separately and the salinity contributions were added. This approach treats the plumes as propagating independently, which is not an unreasonable assumption since the initial dynamics phases of both discharges end within 10 ft of the discharge pipe. The field data show that the plume is transported to the northwest and is diluted to near-background salinity levels within 110 ft north of the barge. Model predictions of near-bottom salinity accurately predict these features of the field data (Figure 3). Model results tend to underestimate the breadth of the salinity plume in the area northeast of the barge.

CONCLUSION

The OOC model predicts the short term fate of drilling and production discharges. The model has been validated for drilling discharges (Brandsma and Sauer, 1983c; O'Reilly, et al., 1989). The revised model matched the limited experimental data presented here. A produced water discharge field study is planned to validate the model further.

ACKNOWLEDGEMENTS

The Offshore Operators Committee and Exxon Production Research Company supported the work reported here.

REFERENCES

Albright, S., Davis, L.R. and Herron, J.R., 1986, An experimental investigation into the dynamic collapse phase of drilling fluid discharges into a flowing stratified environment, *in:* "Proceedings of the 1986 ASME Winter Annual Meetings," ASME Pub. No. 86-WA/HT-33.

Ayers, R.C., Jr., Meek, R.P., Sauer, T.C., and Stuebner, D.O., 1980, An environmental study to assess the effect of drilling fluids on water quality parameters during high rate, high volume discharges to the ocean, *in:* "Proceedings, Symposium/Research on Environmental Fate and Effects of Drilling Fluids and Cuttings, Jan. 21-24, 1980, Lake Buena Vista, Florida".

Brandsma, M.G., Davis, L.R., Ayers, R.C., Jr., and Sauer, T.C., Jr., 1980, A computer model to predict the short-term fate of drilling discharges in the marine environment, *in:* "Proceedings, Symposium/Research on Environmental Fate and Effects of Drilling Fluids and Cuttings," Jan. 21-24, 1980, Lake Buena Vista, Florida".

Brandsma, M.G. and Sauer, T.C., Jr., 1983a, "Mud Discharge Model: Report and User's Guide," available from Exxon Production Research Company, P.O. Box 2189, Houston, TX 77252-2189.

Brandsma, M.G. and Sauer, T.C., Jr., 1983b, The OOC model: prediction of short-term fate of drilling mud in the ocean, part 1: model description, *in:* "Proceedings of Minerals Management Service Workshop on Evaluation of Effluent Dispersion and Fate Models for OCS Platforms," Santa Barbara, California, February 7-10, 1983, MBC Applied Environmental Sciences.

Brandsma, M.G. and Sauer, T.C., Jr., 1983c, The OOC model: prediction of short-term fate of drilling mud in the ocean, part 2: model results, *in:* "Proceedings of Minerals Management Service Workshop on

Evaluation of Effluent Dispersion and Fate Models for OCS Platforms," Santa Barbara, California, February 7-10, 1983, MBC Applied Environmental Sciences.
Bowers, G.W. and Goldenblatt, M.K., 1978, "Calibration of a Predictive Model for Instantaneously Discharged Dredged Material," U.S. EPA report EPA-6003-78-089. Washington, D.C.
Davidson, G.A., 1986, Gaussian versus top-hat profile assumptions in integral plume models, Atmospheric Environment, 20(3)471.
Fischer, H.B, List, E.J., Koh, R.C.Y., Imberger, J., and Brooks, N.H., 1979, "Mixing in Inland and Coastal Waters," Academic Press, New York.
Froehlich, D.C., 1989, "Finite Element Surface-Water Modeling System: Two-Dimensional Flow in a Horizontal Plane - Users Manual," U.S. Department of Transportation Publication No. FHWA-RD-88-177.
Koh, R.C.Y., 1971, Ocean sludge disposal by barges, Water Resources Research, 7(6):647.
Koh, R.C.Y. and Chang, Y.C., 1973, "Mathematical Model for Barged Ocean Disposal of Wastes," U.S. EPA report EPA-6602-73-029, Washington, D.C.
Koh, R.C.Y. and Fan, L.N., 1970, "Mathematical Models for the Prediction of Temperature Distributions Resulting from the Discharge of Heated Water into Large Bodies of Water," U.S. EPA report 16130 DWO10/70 in the Water Pollution Control Research Series, Washington, D.C.
Lin, J.T., and Pao, Y.H., 1979, Wakes in stratified fluids, Ann. Review of Fluid Mechanics, 11:317.
Merritt, G.E., 1974, Wake growth and collapse in stratified flow, American Inst. Aeronautics and Astronautics J., 12(7):940.
Muellenhoff, W.P., Soldate, A.M., Baumgartner, D.J., Schuldt, M.D., Davis, L.R., and Frick, W.E., 1985, "Initial Mixing Characteristics of Municipal Ocean Discharges," Report EPA/600/3-85/073a, U.S. Environmental Protection Agency, Newport, Oregon.
O'Reilly, J.E., Sauer, T.C., Jr., Ayers, R.C., Jr., Brandsma, M.G., and Meek, R., 1989, Field verification of the OOC mud discharge model, *in:* "Drilling Wastes,", F.R. Engelhardt, J.P. Ray, and A.H. Gillam, Eds., Elsevier Applied Science, London.
Pierre, M. and Viollet, L., 1977, "A Study of Jets in Cross-currents and Stratified media," Electricite de France, Chaton, France, translated from French in Argonne National Laboratory Report LIB-TR-41.
Slade, D.H., 1968, "Meteorology and Atomic Energy," U.S Atomic Energy Commission, Washington, D.C.
U.S. Naval Hydrographics Office, 1952, "Tables for Seawater Density," H.O. Pub. 615, Washington, D.C.
Washburn, E.W., Ed., 1928, "International Critical Tables of Numerical Data, Physics, Chemistry and Technology, Volume III," McGraw-Hill, New York.
Yudelson, J.M., 1967, "A Survey of Ocean Diffusion Studies and Data," W.M. Keck Laboratory of Hydraulics and Water Resources, California Inst. of Technology, Tech. Memo 67-2.

AN ANALYSIS OF MANAGEMENT STRATEGIES FOR PRODUCED WATERS FROM NATURAL GAS PRODUCTION

D. V. Nakles,[1] I. Ortiz,[2] and J. R. Frank[3]

[1] Remediation Technologies, Inc.
Pittsburgh, Pennsylvania
USA

[2] United Exploration & Development Company
Pittsburgh, Pensylvania
USA

[3] Gas Research Institute
Chicago, Illinois
USA

INTRODUCTION

The Gas Research Institute (GRI) has embarked on a multi-year program to conduct technology assessments of environmental control strategies in the natural gas production industry. This program builds on previous work which has been in progress over the past couple of years at GRI which conducted environmental impact assessments of this industry. The purpose of this paper is to present the initial approach which has been taken to analyze management strategies for produced waters from the natural gas industry.

BACKGROUND

GRI has assembled a multidisciplinary team to examine the regulatory, institutional, economic, and technical factors involved in determining appropriate options for waste minimization and treatment in the natural gas industry. The team was directed to identify and implement research and development opportunities that will effectively decrease environmental treatment (and therefore gas production) costs in the natural gas production industry. This team initially identified several specific environmental issues which warranted consideration for further study: (1) produced water management, (2) management of naturally occurring radioactive materials (NORM), (3) management of associated wastes, (4) management of hydrostatic test waters, (5) management of air toxics, (6) remediation of contaminated groundwaters, and (7) remediation of contaminated soils/impoundments. From this list, the team identified produced waters as the initial environmental issue upon which to focus. This high priority assignment was based upon several factors, the most important of which were the large volumes of the water which are produced and the potential impact of the costs of treatment on the profitability of the natural gas

Produced Water, Edited by J.P. Ray
and F.R. Engelhart, Plenum Press, New York, 1992

industry. The current treatment issue associated with produced water is the management of total dissolved solids and inorganic salts; however, potential future issues include the removal of trace organic constituents and the management of NORMs. These treatment requirements have the potential to represent a substantial economic burden to the natural gas industry.

The objectives of the produced water technology assessment program are:

- To provide gas producers and developers with treatment cost information to identify cost-effective, feasible environmental control strategies;
- To quantify treatment costs associated with achieving compliance with current and future environmental regulations; and
- To identify the research and development that is required to significantly reduce the costs of treatment.

To accomplish these objectives, several key questions must be answered. Some of the more important of these questions are: How is the produced water currently managed and what does it cost?; Can the current treatment costs be reduced through changes in the current management practices?; and Are there new technology applications which can provide more cost-effective treatment than the current practices? If these questions can be answered, the gas industry will have the baseline information that is required to make sound business decisions that will protect the environment while maintaining a long-term economically viable gas production operation. The industry will also have the capability to examine the cost impact of current as well as future environmental regulations on the industry. The results of this cost analysis will provide a technical basis upon which the industry can rely to respond to existing and proposed regulations and to target future research opportunities.

TECHNICAL APPROACH

The produced water study was initiated by focusing on the coalbed methane industry. This initial focus was driven by the unique characteristics of the produced water generation profiles of this industry. During coalbed methane production, peak water volumes occur during initial operation after which they decrease over time. This production profile requires an initial capital investment for produced water treatment which tends to discourage the entry into the industry by new investors and developers. As such, the produced water management costs have contributed to the depressed development of the coalbed methane industry in spite of the fact that it represents a future gas reserve of large potential to the industry. It is anticipated that this technical assessment will result in a technical approach that can be applied to produced water management in other portions of the natural gas industry as well as to the management of other environmental issues.

The collection of information on produced water management in the coalbed methane industry was initiated with region- and site-specific visits to the Black Warrior Basin in Alabama and the San Juan Basin in southern Colorado and northern New Mexico. The visits were conducted by a multidisciplinary team of experts with experience in petroleum engineering geology, environmental engineering, process engineering, and microbiology. The intent of these visits was to gather information from two of the primary coalbed methane regions on the current and future treatment strategies for produced waters and the cost of treatment. These data were then analyzed and consolidated to identify differences as well as commonalities in the treatment methods, treatment costs, and regulatory requirements that exist from region to region. On the basis of this analysis, the feasibility of developing "models" to describe and analyze produced water management within and among gas production regions as well as on a well-specific basis within a gas production field is being evaluated. In addition, targeted research and development projects are being identified and, in some cases, have already been initiated. A summary of the results-to-date from this effort and the plans for its completion are provided in the remainder of this paper.

STUDY RESULTS

The technology assessment of produced waters was initiated in April of 1991. As such, the study results are not complete; however, a partial reporting of the results are provided in this section. More specifically, this section presents results from the initial site visits, the analysis of the data which were collected during the site visits, and the preliminary research effort that was begun to investigate deep well injection options in the Black Warrior Basin. Efforts to develop produced water management "models" based upon the current findings have only just begun and are only conceptually discussed in this paper.

Summary of Site Visits

The GRI project team conducted site visits to two of the more prominent coalbed methane production areas: Black Warrior Basin in Alabama and the San Juan Basin in Colorado and New Mexico. The Black Warrior Basin visit was conducted May 1 through May 3, 1991, and included four gas production and operating companies, the Alabama Geologic Survey, and the Coalbed Methane Association. The San Juan Basin visit was conducted August 19 through August 23, 1991, and included eight gas production and operating companies, the Southern Ute Indian Tribe, three state regulatory agencies (one in Colorado and two in New Mexico), and the U.S. EPA Region VIII.

Review of Produced Water Management Strategies

The purpose of the site visits was to gain a first-hand understanding of the technical, regulatory, and economic issues associated with produced water management. As such, as much information as possible was obtained regarding the current produced water management practices, the regulatory driving forces which governed those practices, the treatment costs, and the industry and agency investigations into new and innovative treatment options.

Summary of treatment options. Table 1 provides a summary of the produced water management options that are practiced in either the Black Warrior or San Juan Basins, or both. The "X's" indicate that the option has been practiced in the region; "•'s" indicate the option is not practiced and does not represent a technically viable option for the future; and "?'s" indicate that the option may be technically possible but has not been practiced in the region as of this time. A review of the information in Table 1 reveals that only three management options are practiced in both regions. These are deep well injection, offsite commercial disposal, and re-use as frac water. Of these, offsite commercial disposal is routinely utilized in both regions whereas deep well injection is more common in the San Juan Basin. Reuse as a frac water, while utilized in both regions, is not a very common practice.

Produced water management practices that are common only to the Black Warrior Basin are land application and surface discharge although the use of wetlands and mine co-disposal, whose commercial application is somewhat questionable, have only been considered for use in this region. Alternatively, surface evaporation is the only option that is used exclusively in the San Juan Basin. This option has not been seriously considered for application in the Black Warrior Basin due primarily to climate conditions. Discharge to surface water bodies has been investigated in the San Jaun Basin and in December, 1991, EPA issued a surface discharge permit for the Colorado area. However, the regional application of this management option is complicated by issues associated with endangered species and the assignment of water rights. The discharge of the proposed water to the local/regional municipal or public water supply and/or waste treatment plants is an option which may be technically viable in both regions but wh ich has not be practiced earlier.

It is evident from this brief overview that deep well injection, treatment followed by surface discharge, offsite commercial disposal, land application, and surface evaporation define the universe of produced water management options that are currently in use in these two coalbed methane regions.

Overview of Critical Regulatory and Technical Factors. There are a variety of technical and regulatory factors which drive the selection of the produced water treatment options in any given region or area. Those factors which are most important in the Black Warrior and San Juan Basins and which have influenced the selection and costs of produced water treatment are discussed below for the primary produced water treatment options.

Table 1. Summary of produced water management options

Management Option	Gas Production Basin	
	Black Warrior	San Juan
Deep Well Injection	X	X
Surface Discharge	X	?
Surface Evaporation	•	X
Land Application	X	•
Wetlands Application	?	•
Municipal Treatment	?	?
Mine Co-Disposal	?	•
Re-Use As Frac Water	X	X
Offsite Commercial Development	X	X

Key: (X) Current Practices; (•) Not Practiced; (?) Possible option for Future Use

Deepwell Injection. In general, deep well injection is the preferred means of disposal for coalbed methane produced waters. Its use in the Black Warrior Basin is governed by the Alabama Oil and Gas Board who permits them as Class II Wells. While there are several requirements outlined in the underground injection control regulations, no specific constituent limitations exist for injection of produced water. Rather, the Oil and Gas Board (in accordance with the national Safe Drinking Water Act [SDWA]) restricts the injection of produced water into those subsurface regions where the total dissolved solids exceed 10,000 ppm (unless it can be shown that the disposal zone is not a drinking water source) and limits the maximum injection pressure based on a site-by-site maximum which is established to prevent fracturing of the injection zone. These constraints have the effect of increasing the cost of deep well injection since the water quality limit forces injection into the deeper subsurface zones and the pressure constraints can limit the useful lifetime of the well. These limitations have resulted in only sporadic success in locating subsurface geologic formations that can accept the required quantities of produced water. As such, its use by the gas industry in the Black Warrior Basin has been limited.

Similar problems face the gas producers in the San Juan Basin although for similar and different reasons. In this region, the depth of injection is driven by the requirements of the Southern Ute Indian Tribe in Colorado and the New Mexico Energy and Minerals Department (NMEMD) and also SDWA. Both of these groups favor injection into deeper geologic horizons, although no uniform water quality limitations have been specified. The Southern Utes prohibit shallow injection to protect the gas production horizons that are nearer the surface whereas the NMEMD wishes to ensure isolation of the injection zones from the drinking water sources. An additional consideration in the west is that the deep well injection of produced water is perceived to represent a loss of a potential water resource that can never be recovered. However, in spite of all of these limitations and constraints, the use of deep well injection is used widely in the San Juan Basin. The primary reason for this is the relatively high rate of success which has been achieved in the drilling of deep wells in the region. In addition, there are larger volumes of gas and produced water generated from coalbeds of the San Juan Basin, as compared to the Black Warrior Basin, which also improve the economics of a deep well disposal system.

It should be noted that operational as well as environmental limitations often require the pretreatment of produced waters prior to deep well injection. Typical pretreatment processes include gravity settling and filtration for the removal of suspended solids. However, more stringent requirements may dictate additional treatment for the removal of trace organics or dissolved solids. The nature and extent of the pretreatment that is required could have a significant impact on the future use of this produced water management option.

Treatment Followed By Surface Discharge

- Black Warrior Basin. Due to the limitations associated with deep well injection, treatment followed by surface discharge is used extensively in the Black Warrior Basin. Surface discharges from the coalbed methane industry are permitted as part of the mining industry in the State of Alabama; no surface discharge is permitted from oil production facilities. The surface discharges are permitted based upon a two-tier system which addresses both mass loadings of chloride to the surface water bodies (Tier I) as well as the ultimate chloride concentrations of these receiving water bodies (Tier II). Current regulatory trends are to totally eliminate the mass loading permits in lieu of the concentration-based standards; to reduce the acceptable chloride concentrations; and to introduce toxicity testing to the discharge criteria. All of these trends will most likely result in the increase in the cost of produced water management in the region and slow the development of this domestic energy resource.
- San Juan Basin. The surface discharge option has not been implemented in the San Juan Basin because of the extensive use of deep well injection. However, applications for general discharge permits have been filed in the Colorado portion of the basin with total dissolved solids (TDS) concentrations of 500 to 1,000 ppm as the primary discharge criteria. The general permit for the Colorado area was issued in December, 1991. There has been a great deal of controversy surrounding this permit with regards to the presence of endangered species in the lower San Juan River and the assignment of water rights. In addition, New Mexico has recently published water quality standards which may introduce additional discharge requirements beyond TDS and other inorganic salts.

  The primary treatment options for the produced waters prior to surface discharge depend heavily upon the discharge requirements that are imposed on the treated effluent. The current methods that are practiced in the Black Warrior Basin are simple aeration followed by coagulation, flocculation, and settling. Additives are often added to enhance coagulation and flocculation and to adjust the pH to near neutral. This treatment scheme is capable of achieving the current limitations of total dissolved solids, iron, and pH. However, should treatment for chlorides be forced by the upcoming regulatory trends, it is likely that more extensive treatment will be required and may include the use of technologies such as reverse osmosis, forced evaporation, electrodialysis, or ion exchange. Several gas producers have considered these options as part of their planning for bringing new wells online. In fact, some laboratory and pilot-scale projects have been implemented to generate performance and economic data for the various treatment options with a particular emphasis on reverse osmosis. Similarly, preliminary investigations in the San Juan Basin have emphasized reverse osmosis treatment to achieve the lower limit of total dissolved solids which have been proposed, i.e., 500 ppm. However, many of the produced waters in this region would require little or no treatment if the upper limit of 1,000 ppm were incorporated into the discharge limitations.

Land Application/Surface Evaporation. The application of the produced waters to the land is practiced in the Black Warrior Basin. This practice is currently limited to produced waters which have a TDS concentration of less than 2,000 ppm. However, it is likely that this treatment option will be banned in Alabama since the state recognizes the groundwaters as a receiving water of the state. As such, the land application of the waters would be governed by surface discharge permit requirements. This would introduce additional treatment requirements which will most likely eliminate this management option from further consideration.

Surface evaporation in the San Juan Basin could be considered the analog of land application in the Black Warrior Basin. The regional climate makes this a viable option with current practices using either lined or unlined surface pits. However, while this is a technically viable option, there have been regulatory and community actions which may increase the costs of this treatment option or totally eliminate it from further use. First, the Southern Ute Indian Tribe and the Bureau of Land Management have both decided to no longer permit surface evaporation due to a combination of negative visual impacts and concerns regarding the use of unlined pits and the subsequent subsurface transport of contaminants. As such, consideration has been given to specifying RCRA design requirements for all surface evaporation facilities. Secondly, there are concerns regarding the release of toxic organics into the atmosphere during the use of this "open" evaporation process. The answer to these evolving questions may require the use of closed, forced evaporation systems. These systems would collect the water that is evaporated and treat the non-condense offgas using activated carbon or other gas treatment technologies. However, forced evaporation has not been adequately tested to date and it is likely that the cost associated with such treatment may be prohibitive.

Publicly Owned Treatment Works. The discharge of produced waters into the local publicly owned treatment works (POTWs) represents a potentially cost effective treatment option which has not been thoroughly explored at this time. The use of POTWs for the management of industrial wastewaters and groundwaters is not a new concept although it often involves some pretreatment of the waters prior to their discharge to the local sewer. The primary factors which must be addressed are the fate of the contaminants in the POTW treatment plant and their potential to interfere with the overall plant performance. From a practical point of view, another issue is the proximity of the sewers to the point of produced water generation. GRI has investigated this treatment option as part of other research programs (i.e., the treatment of groundwaters from manufactured gas plant sites) and found the results to confirm the general technical viability of this management strategy. GRI believes that the same research methodology which was developed in these programs can be applied to produced waters.

Co-Disposal In Mines. The use of abandoned mines for the co-disposal of produced waters has been considered in the Black Warrior Basin. The abandoned mines provide a reservoir into which the produced waters can be discharged. These reservoirs typically contain waters from surface runoff as well as underground seepage. In addition, some mining companies have obtained permits to discharge their waters from other mining operations into these abandoned facilities. The obvious concern associated with this management technique is the ultimate fate of the inorganic and organic contaminants in the surrounding soil/water environment. However, the initial arguments supporting this disposal option are based upon its current use by the coal mining industry and its similarity, in concept, to deep well injection.

Wetlands Applications and General Use of Vegetation. There has been a great deal of discussion regarding the potential use of natural or engineered wetlands for the management of produced waters. The concept is based upon the premise that the wetland vegetation has the ability to uptake inorganic salts and to produce a water that contains less total dissolved solids for direct discharge into the surface waters. This results in the accumulation of a biomass that contains a high concentration of salts. The biomass would then require some form of management. Furthermore, if the process takes place in a natural wetlands, there are two concerns: (1) that the wetlands itself, i.e., soil and shallow groundwater, will become contaminated with the salts and pose a potentially larger environmental problem and (2) that the discharge is a regulated "surface discharge". To avoid these issues, the concept of using engineered wetlands has also been proposed where the wetlands is contained in a closed system that will not permit the uncontrolled escape of the contaminants into the environment. There is evidence of the acceptability of this option in the fact that the U.S. EPA has agreed to examine the use of artificial wetlands for the control of acid mine drainage and a similar plan has been approved to pilot-test this concept on produced waters in the Black Warrior Basin.

Another related concept which is under investigation is the use of plants to treat produced waters to yield a low-volume, concentrated brine which can then be treated using conventional means. In effect, this concept is the opposite of the wetlands approach in that the plants enhance the removal of water from the produced waters through evapotranspiration. This evaporative process can substantially reduce the volume of the water that requires treatment. Preliminary engineering calculations indicate that several species of vegetation have the potential to serve in this capacity and that bench-scale testing of the concept is warranted.

Offsite Commercial Disposal. Offsite disposal facilities exist in both the Black Warrior and San Juan Basins. The exact treatment that is provided varies but generally includes some combination of deep well injection, surface treatment and discharge, and surface evaporation. These facilities have evolved since they are able to spread the risk of development over a broad base of clients and, hence, are able to provide the service at reasonable costs. However, some key issues associated with offsite management of the waters are the cost of transportation, the risk associated with the movement of the waters from the site to the disposal facility, and the environmentally sound maintenance of the commercial disposal facility.

Treatment Economics. The economic data which are available for the produced water treatment options are typically single points which have been generated for a specific application. These data have been generated by a number of sources although the primary published data have been provided by GRI in two reports or publications that address the Black Warrior and San Juan Basins (Lee-Ryan, et. al., 1991; Decker, 1991). Figure 1 provides a brief overview of the economic data from these two reports. In this figure, the data from both studies are plotted although no attempt has been made to normalize the information on a common basis. Never-

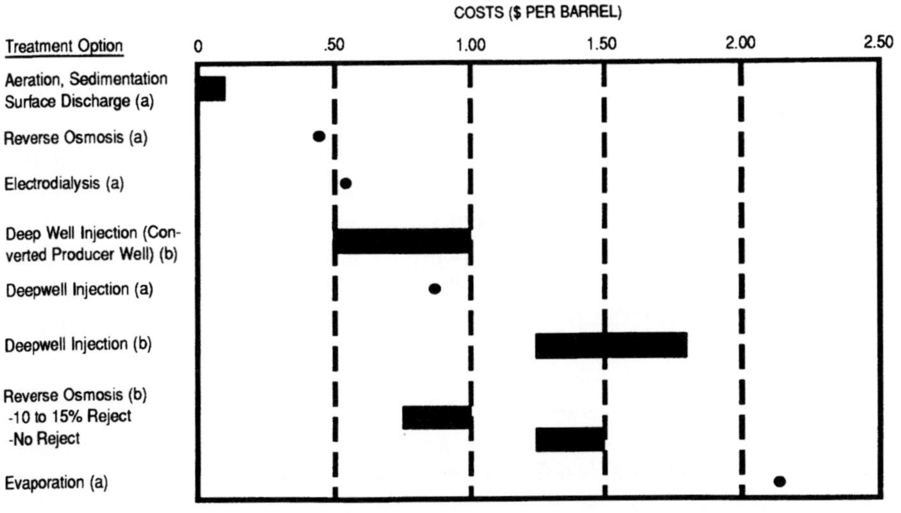

Figure 1. Preliminary costs for produced water management (a) Black Warrior Basin, (b) San Juan Basin.

theless, a brief review of this information does provide some insight on the relative cost of produced water treatment.

The range of costs presented in Figure 1 covers nearly two orders-of-magnitude, from \$0.03/bbl to \$2.15/bbl. These extremes were derived for treatment applications in the Black Warrior Basin and represent surface discharge preceded by the most basic treatment (i.e., aeration followed by sedimentation) and evaporation, respectively. The cost range for deep well injection and treatment by reverse osmosis prior to surface discharge are very similar and fall between these extremes with the former ranging from \$0.50/bbl to \$1.75/bbl and the latter from \$0.50 to \$1.60/bbl. The range in the deep well injection costs may reflect the regional differences (i.e., \$0.81/bbl in the Black Warrior Basin versus \$1.25 to \$1.75/bbl in the San Juan Basin) as well as technical differences associated with converting producer wells versus well development from scratch (\$0.50/bbl to \$1.00/bbl versus \$1.25 to \$1.75/bbl, respectively). The variation in the cost of reverse osmosis appears to reflect the percentage of reject that is produced, with the cost increasing at the percent reject decreases. There are insufficient data to make any observations regarding potential regional differences in the application of this treatment technology. Lastly, the cost of electrodialysis is at the lower end of the cost range for deep well injection and reverse osmosis. However, only one data point was available for this treatment option.

While the available data do provide some sense of the cost of produced water treatment, they only represent single points and there is no ability to determine the impact of any changes in water quality or extent of treatment on these cost estimates. Consequently, GRI anticipates moving beyond these current estimates to develop a more integrated and comprehensive cost analysis of produced water management in the natural gas production industry.

Detailed Analysis of Deep Well Injection in the Black Warrior Basin

The extensive use of deep well injection for produced water management prompted GRI to initiate a more in-depth review of its application and limitations. This review, which focused on the Black Warrior Basin, was undertaken in order to develop a data base and understanding of the cost, engineering and geological factors which govern underground disposal and to characterize potential geologic formations as candidate disposal zones. These data will then be used to compare the cost of this disposal technology to other produced water management options.

Technical Approach. The study of deep well injection in the Black Warrior Basin was completed by gathering existing data, interviewing operators, and assessing the information available from public and operator interviews. More specifically, information was gathered from the GRI coalbed methane library (University of Alabama), the State Oil and Gas Survey and local operators. This information included disposal site locations, descriptions of geologic formations, well and boring logs, and estimated disposal costs.

Summary of Results

Injection well status. A total of 12 injection wells have been constructed in the Black Warrior Basin. Two of the wells are Class I subsurface disposal wells and ten of the wells are Class II wells designated for salt water disposal.

- UIC-Class I wells. Five Class I subsurface disposal wells have been drilled and completed in Alabama. Two of the wells are in the Black Warrior Basin and are owned by the Reichhold Chemicals, Inc., Tuscaloosa County and U.S. Steel Corp., Jefferson County.

  The U.S. Steel Corporation disposal well was completed and tested by March, 1970. The well was drilled to a total depth of 6,060 feet and was completed from open hole between 4,415 and 6,060 feet. The primary injection zone, 4,440 to 4,688 feet, (Silurian-Red Mountain) is a fine grained calcareous sandstone with fracture porosity and permeability. The zone of secondary importance, 5,620 to 5,770 feet, (Knox) is a dolomite whose porosity and permeability are provided by fractures and/or cavernous vugs. The upper zone has the capacity to accept the greater volume rate at lower pressure than the lower zone, and the well is so constructed that the lower zone would begin to take fluid if the upper zone failed to function properly. The well has a tested injection rate of 300 gpm or over 10,000 barrels per day at a surface pressure of 600 psi. Pickle-liquor waste which contains hydrochloric acid or sulfuric acid was targeted for disposal in this well.

The Reichhold Chemicals, Inc., waste disposal well was completed and tested by August 1970. The well was drilled to a total depth of 8,098 feet and was completed from open hole between 5,902 and 8,098 feet. Permeability exists in two main horizons in the well. The upper zone is from 6,620 to 6,670 feet (Ordovician-undifferentiated) and the lower zone is from 7,036 to 7,230 feet. Porosity and permeability exist in the upper zone by virtue of fractures and in the lower zone by virtue of fractures and cavernous vugs. The well has a tested injection rate of 600 gpm or 20,000 barrels per day at a surface pressure of 700 psi. Chemical processing wastes containing sodium sulfite and sulfate, phenol, caustic and traces of organic compounds were targeted for disposal in this well.

After 22 years, neither of these Class I wells have been used to any extent as of the writing of this paper.

UIC-Class II wells. A list of the salt water disposal (SWD) wells in the Black Warrior Basin is provided in Table 2. A total of ten Class II injection wells have been drilled. Two wells encountered drilling problems and were plugged shortly after drilling operations commenced. Only seven have been approved and only five are currently active: Cain, USX, Brown, Cork, and Dorroh.

The results of a comparison of the relative performance of four of the operating SWD wells is shown in Figure 2. This figure presents the daily volume of injected water divided by its average injection pressure plotted against cumulative injected water volumes. Ideally, the best wells will inject large volumes with very little pressure (i.e., demonstrate a high ratio of injection volume to injection pressure). The best performance wells are the Cork, Cain, and USX wells, all of which are completed in the Pottsville formation and which demonstrate a volume to pressure ratio of between 1 and 10. The Brown well, which was completed in the Knox formation, initially performed well, although it suffered a rapid decrease in performance above a cumulative injection of 500,000 barrels of produced water where the volume to pressure ratio rapidly dropped from 1 to less than 0.1.

The costs of the SWD wells range from $400,000 to over $3,000,000. One reason for these high drilling costs is that not only does the operator have to find a suitable zone that has the permeability and porosity for injection; but the operator must also demonstrate to the Alabama State Oil and Gas Board (ASOGB) that the waters within the zone have greater than 10,000 ppm TDS. This is a costly operation since it requires the sampling of water from the zones of interest during drilling operations or the setting of casing followed by the perforation and testing of each zone.

Based upon an examination of groundwaters in Alabama, it would appear that TDS concentrations of greater than 10,000 ppm TDS exist anywhere from 500 to 2,500 feet beneath the surface. However, the water quality data are quite sparse and no additional data are currently being generated. A more intense mapping of this information is critical since a recent SWD well, located in Bibb Co., was tested having waters with less than 10,000 ppm TDS (total dissolved solids) at depths near 8,600 feet. As a result, the operator is currently back-flowing the well in an attempt to recover drilling fluid which was lost in the well during drilling operations and has petitioned the Oil and Gas Board for an aquifer exception.

There is some question as to whether produced water in the Black Warrior Basin that contains less than 10,000 mg/l TDS is recent or ancient fresh water. Age dating the waters could indicate that the potential injection formations are not hydrologically connected to fresh water sources and permit their acceptance as an injection zone for produced waters. GRI is investigating techniques to age date subsurface waters for this purpose.

Table 2. Summary of saltwater disposal wells: Black Warrior Basin

Well Name	Formation Name	PERF's (ft.)	Total Depth (ft.)	BWPD[a]	Injection Pressure (PSIG)	Comments
Cain	Pottsville	4590-4720	4,600	4,000	600	Approved 10/02/89
USX	Pottsville	4177-4412	4,300	2,221	1,830	Approved 6/01/90
Brown	Knox	9836-10924	10,000	3,351	995	Approved 12/07/89
G.S.P.	Knox	6200-8540	10,000	__[b]	—	TDS <10,000 ppm @ 8600'
Smith	Carter	3760-3886	N/A	13	1,700	Approved 04/06/90-S.I. tite
Self	Carter	2408-2416	N/A	3,600	1,050	To be plugged
Cork	Pottsville	4240-4944	4,300	9,324	787	Approved 08/24/90
Little Bear	Pottsville	5142-5224	5,400	2,160	350	Approved 10/16/90
MIW-224A	Knox	@8860	9,000	10,800	1,750	Tested only - <10,000 ppm TDS @4,400'
Dorroh	Devonian	7254-7271	7,250	2,436	1,500	Approved 11/30/90

Notes: [a]Barrels of Water Per Day; [b] No Data

Geologic Evaluation. The geologic assessment portion of the study has focused on the deeper horizons in the coalbed methane producing fields of the Black Warrior Basin and the adjacent surrounding region, including the Appalachian structural front. The recent drilling of several wells for underground disposal in these fields has greatly increased the available control for horizons below the producible coal seams. These candidate zones for injection of produced waters should exhibit porosity, matrix or fracture enhanced permeability, and continuity of reservoir. Incorporating these recent well data into the existing geologic framework will better characterize the target formations, thereby leading to a more accurate assessment of underground injection as a disposal option for produced waters in this region.

The initial focus of effort was to provide an overview of the zones that are currently being used for injection in the Black Warrior Basin. These zones range from Cambro-Ordovician dolomite found at depths greater than 10,000 feet at basin center, to shallow Mississippian sandstones found at less than 3,000 feet just north of the main coalbed producing areas.

- Knox Dolomite. Permeability was encountered in three out of six disposal wells drilled to the Knox formation. One well had an injection potential of 3,000 BWPD. Depth to the Knox ranges from 8,000 feet in the north to over 12,000 feet in the southern portion of the basin. In contrast, a 3,000 feet thick Knox section can be found at or near surface just outside the basin in the overthrust belt.
  The overthrust belt represents a region where operators can drill Knox wells at shallower depths, thereby lowering drilling costs. However, fresh water has been encountered in the Knox in the overthrust. A current area of investigation is to date this fresh water to determine whether it is indigenous to the Knox or a result of groundwater recharge along the major

thrust faulting. If it is ancient, trapped water, then the overthrust belt has considerable injection potential in the Knox.

In addition, the mapping of the dolomite and limestone facies using well logs and sample descriptions of well cuttings is underway to determine the geographic distribution of the depositional environments in the Knox. This effort, combined with the use of seismic to identify local structures, will permit the optimal selection of drilling sites in this formation.

- Paleozoic Carbonates. Injection rates vary from 2,000 to 10,000 BWPD from three wells completed in various zones in the undifferentiated Ordovician, Silurian and Devonian sections lying above the Knox formation . Fracture porosity in conjunction with the occurrence of sandy carbonate zones appears to play a significant rolė in the injection potential of these sections.
- Mississippian Sands. The Mississippian sands were deposited in narrow shoreline and shallow marine environments in the shallower part of the Black Warrior Basin primarily to the north of the coalbed methane producing areas. Except for two northern SWD tests drilled to the Carter sand at depths less than 3,000 feet, the disposal potential for Mississippian sands remains generally untested. The sands are often thinly laminated and locally developed. One sandstone, the Hartzelle, appears to trend through the northern edge of the coalbed methane producing area. Use of optimal well logging suites, including micrologs and enhanced sampling, can improve recognition of these thinly-laminated, high porosity zones which have good injection potential.
- Pennsylvanian Pottsville Sands. The disposal wells with the best injection and success rates to date have been completed in the Pottsville sands directly underlying the productive coal at average depths of 4500 feet. One well has an injection rate of 9,000 BWPD, and another is currently being tested. The depositional environment consists of upward shoaling, prograding fluvial-deltaic cycles. Well control is greater for the lower Pottsville sands than for deeper horizons, making structural and depositional factors easier to assess. In addition, a disposal test could be drilled at incremental cost

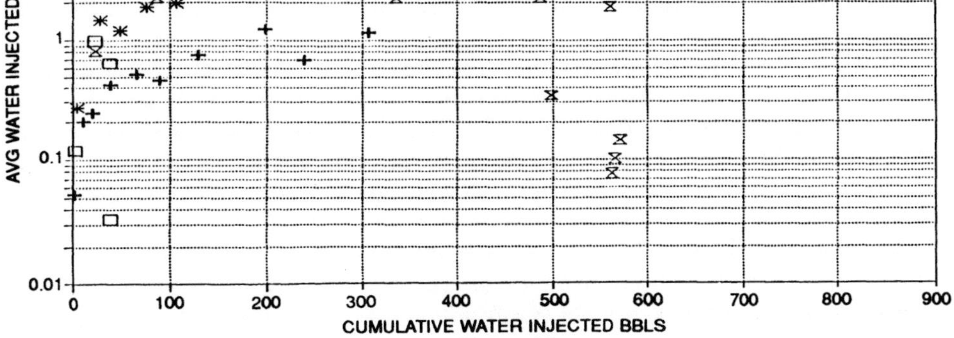

Figure 2. SWD well performance in the Black Warrior Basin, Alabama

and plugged back to a productive coal seam if porosity is not encountered or if injection rates prove insufficient. Economically, this horizon may prove optimal for underground injection in the deeper parts of the basin.

In summary, the Knox formation offers the greatest potential for deep well injection in the eastern portion of the basin where major structural features occur. Mississippian sands are present at the northern edge of the coalbed region while the undifferentiated Paleozoics and Pottsville have intermittent potential throughout the area. Pottsville sands may be the best underground option in the deeper, southern portion of the basin. The probability for drilling an adequate disposal well is currently around fifty percent. However, recently drilled injection wells have greatly increased the geological data base in the Black Warrior Basin. These data should increase the probability of drilling a successful disposal well in the future and will enable operators to target areas where underground injection may be the most cost effective option.

With regards to deep well injection, the following are potential areas for further study which were identified as a result of the current GRI investigations: (1) the problems that the operators are experiencing in running their step rate tests should be investigated further; (2) the age of the produced waters with less than 10,000 TDS at depths greater than 2500 feet should be examined, perhaps by radioactive dating of the water; (3) reservoir limits tests should be conducted on several of the wells to ascertain reservoir boundaries; and (4) integration of newer disposal wells into the regional geologic data base should occur to better define areas with optimal reservoir characteristics.

PLANS FOR ANALYSIS OF PROJECT MANAGEMENT OPTIONS

GRI intends to build upon its current work to provide the natural gas production industry with not only the proper information to evaluate produced water management options but also with a management information system, or model, to permit effective use of the information. A description of the plans in this area are provided in the remainder of this section.

Overview of Produced Water Management Information System

Figure 3 provides a conceptual overview of a simple decision tree which has been proposed to serve as the initial framework for the analysis of produced water management options and their associated costs. This framework identifies the following technical elements for investigation: (1) produced water volume and quality; (2) onsite versus offsite management; (3) transportation options; and (4) produced water treatment options.

Since it is the ultimate intent of the information system to focus on the economic aspects of the produced water management, it will be necessary to develop quantitative relationships within each element that will relate the critical process variables to the cost of produced water management.

In addition to the elements that are shown in Figure 3, it is recognized that within each region there will be specific "treatment endpoints" as well as other regulatory compliance requirements which must be met. As such, GRI anticipates the need for a continual regulatory review to occur as the information framework is developed. Another observation is that the element of transportation is really a subset of a broader technical element that involves the overall spatial description of the coalbed methane resource basins including the subsurface geology, the surface terrain, the location of surface water bodies, the location of disposal facilities, and the location of indian reservations to name a few. This element has characteristics similar to a regional geographic information system (GIS) which may be incorporated into the effort if needed.

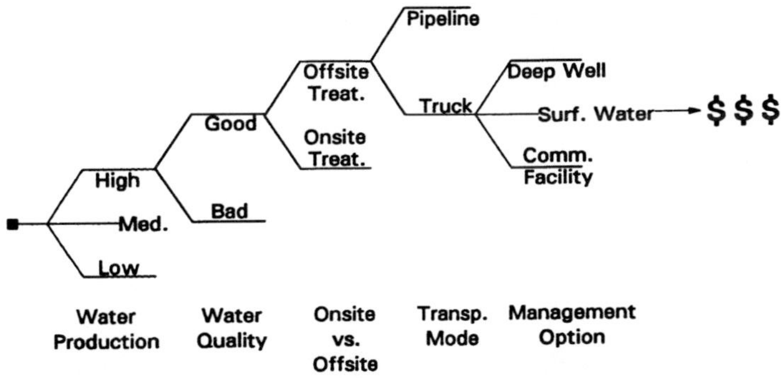

Figure 3. Example of model framework for analysis of produced water management options.

Technical Approach for Development of Management Information System

It is recognized that the gas producers have a great deal of the information that is needed to provide an initial assessment of the cost of produced water management. However, the economics associated with the various treatment options have not been well documented and probably reflect a great deal of uncertainty. As such, GRI has elected to focus on the development of this element by developing cost models for each of the primary produced water treatment options. These cost models will relate the treatment cost to the primary process variables to permit the estimation of the treatment costs under a variety of conditions. It is believed that the data that are required to address the other technical elements in Figure 3 can be initially obtained from the gas producers to permit a complete cost analysis of the produced water management options. If necessary, at a later date, a more in-depth review of each of the other technical elements will be conducted.

Definition of produced water quality data. GRI is utilizing its existing produced water quality data base to develop "representative" characteristics upon which to base the development of the process cost models. A water quality profile will be developed for each of the basins to provide a basis for comparing various treatment options from a regional perspective. The impacts of varying the water quality within a region will be considered at a later date after the general approach for the development of the cost models has been accepted.

Selection of baseline treatment options. A schematic of the management options which are available to most producers in either the San Juan or Black Warrior Basins is shown in Figure 4. This figure shows eight disposal options which can be used, each of which may require some form of treatment of either the produced waters or the frac fluids prior to their use. The type and extent of treatment that is required will depend upon the quality of the untreated water and the discharge requirements of each disposal option. Since the current discharge requirements are heavily oriented towards the removal of suspended solids and soluble inorganics, the candidate technologies include coagulation, clarification, sedimentation, and flotation for the former and electrodialysis, ion exchange, distillation, and reverse osmosis for the latter. Should the removal of trace organics be required as a result of toxicity requirements in upcoming permits, technologies such as carbon adsorption, biological treatment, and chemical oxidation may also be required.

Given the above considerations, the baseline treatment options which will be examined initially involve treatment followed by surface discharge and deep well injection as shown in Figure 5. Two levels of treatment will be examined prior to surface discharge including a minimum treatment option that includes pH adjustment, aeration, and polymer addition and

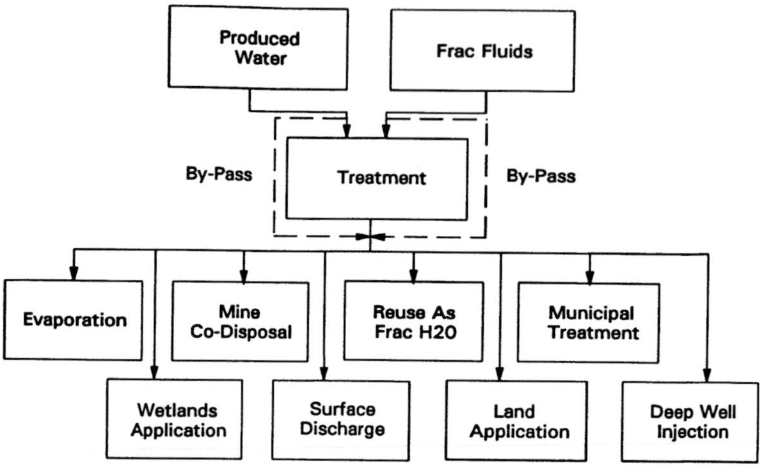

Figure 4. Schematic of produced water management options for coalbed methane facilities.

settling and a maximum treatment option that includes pretreatment (if necessary) followed by reverse osmosis. For the latter option, the concentrated reject from the reverse osmosis is disposed of using deep well injection. Prior to deep well injection, one level of treatment will be evaluated: gravity settling followed by filtration. However, as environmental and operational limitations become more stringent, other pretreatment options will be defined and evaluated.

Cost sensitivity analysis. Engineering cost estimates will be developed for the baseline treatment options for the produced water quality that is representative of the Black Warrior Basin. The cost estimates will be developed for a range of flowrates that are representative of

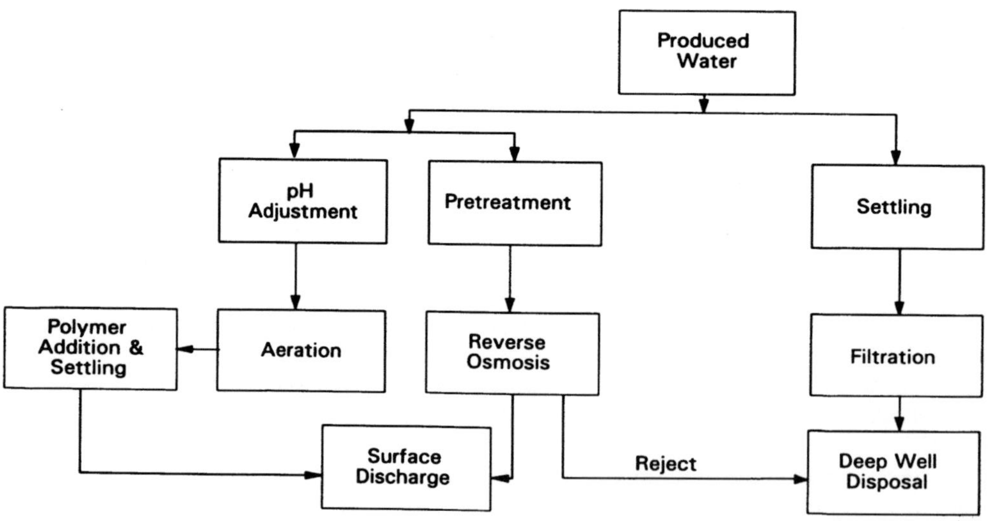

Figure 5. Baseline treatment options for produced water management.

the coalbed methane operations to determine the sensitivity of the treatment costs to the volumetric flow rate. An identical analysis will be performed based upon produced water quality from the San Juan Basin for a single flow rate. This will permit the analysis of the effect of regional differences in produced water quality on treatment costs. Lastly, cost estimates will be generated for a number of different variations in the produced water quality within a given region to investigate the impacts of selected water quality parameters on the produced water treatment costs.

A draft spreadsheet model for deep well injection has already been developed and validated using information from operators in the San Juan and Black Warrior Basins. This model estimates the disposal cost per barrel for salt water disposal wells based upon wellhead injection pressures (calculated), water disposal rates, and total injected water volume. The model predicts the effects of fracture length and threshold on the injection pressures, the estimated costs for the life of the well, and the number of injection wells that are required. The model generates a cost graph based on well conditions and fracture lengths.

SUMMARY

In summary, GRI has initiated a project to develop the necessary data base and management information system, or model, to assist the gas producers in the identification and selection of the optimal produced water management strategies. The initial efforts of this project are focusing on the better definition of the deep well disposal option and the development of cost models for the primary surface water treatment options. Using this information, gas producers can assess the costs of the integrated treatment strategies that are most commonly considered by the natural gas production industry. As needed, other technical and cost information will be gathered in other areas to support the use of these process models in examining the tradeoffs associated with the overall economics of produced water management. GRI believes that the methodology that is developed and implemented for produced waters should be applicable to the analysis of other environmental issues within the gas industry.

REFERENCES

Lee-Ryan, P. B., J. P. Fillo, J. T. Tallon, and Evans, J. M., 1991, "Evaluation of Management Options for Coalbed Methane Produced Water," Proceedings of the 1991 Coalbed Methane Symposium, The University Alabama/Tuscaloosa, Tuscaloosa, AL, May 13-16, 1991.

Decker, D., 1991, "Estimate of Water Production from Northern San Juan Basin", Natural Gas Supply Project Advisor Group Meeting - Coalbed Methane Project Area, (Internal Document - Unpublished), Birmingham, AL, Nov. 13-15, 1991.

A STATISTICAL MODEL TO PREDICT TOXICITY OF SALINE PRODUCED WATERS TO FRESHWATER ORGANISMS

D.D. Gulley[1], D.R. Mount[2], J.R. Hockett[2], and H.L. Bergman[1]

[1] University of Wyoming
Laramie, Wyoming
USA

[2] ENSR Consulting and Engineering
Fort Collins, Colorado
USA

INTRODUCTION

Changes in United States Environmental Protection Agency (USEPA) policy during the 1980's led to the increased use of whole effluent toxicity testing, or "biomonitoring," for the monitoring and regulation of effluent discharges to surface waters (Mount and Norberg, 1984; Norberg and Mount, 1985; USEPA, 1985; USEPA, 1987; Federal Register, 1989; USEPA, 1989; USEPA, 1991). More recently, many states and USEPA regions have indicated that surface discharges of co-produced waters from conventional, coalbed, or other types of natural gas facilities are subject to biomonitoring requirements. Produced water generally has high salinity relative to most fresh waters. We define salinity as the concentration of major inorganic ions (e.g., sodium, chloride, potassium). While not considered as traditional toxicants, these inorganic ions can be toxic to freshwater organisms (O'Neil et al., 1992), particularly the organisms commonly used for biomonitoring. As such, understanding the relationship between major inorganic ions and toxicity is important to properly managing the disposal of production waters.

Although elevated salinity is common to many effluents as well as produced waters, relatively little is known about the toxicity of individual ionic components of salinity. Birge et al. (1985) conducted a series of bioassays aimed at determining safe environmental concentrations of chloride in eastern rivers. While this approach had some site-specific utility it was not useful for universal application. Saline discharges (as well as receiving waters) are quite variable in ionic composition (Drever, 1982; Meyer et al., 1985; Peavy, 1977) and cannot always be accurately assessed by single-ion criteria. Furthermore, comparisons of instream toxicity data with laboratory toxicity data have suggested that laboratory toxicity tests are useful predictors of instream effects (Mount et al., 1992; O'Neil et al., 1992). What would be most useful is a model that predicts the toxicity of the complex ion mix present in produced waters.

In summer 1990, we began development of a Salinity/Toxicity Relationship (STR) model that could predict the toxicity of a given produced water to commonly used biomonitoring test organisms based on its composition of major ions. In addition to directly predicting toxic effects of the major ions present, the STR's potentially greater contribution is in determining whether a given produced water contains toxicity greater than that attributable to total dissolved solids

Produced Water, Edited by J.P. Ray
and F.R. Engelhart, Plenum Press, New York, 1992

alone. Toxicity greater than that predicted would indicate toxicity due to other constituents of the produced water. This information is invaluable to the design of toxicity characterization studies, the identification of suitable wastewater treatments, and the appropriate management of produced waters.

METHODS

Laboratory Studies

Development of the STR model focused on seven inorganic ions: bicarbonate (HCO_3^-), calcium (Ca^{++}), chloride (Cl^-), magnesium (Mg^{++}), potassium (K^+), sodium (Na^+), and sulfate ($SO_4^=$). The solutions tested were further restricted to a pH range between 6 and 9, as waters outside this range are generally prohibited from surface discharge. Although multiple carbonate species theoretically exist at these neutral pH values, bicarbonate was the only carbonate species assessed because it is the predominate species in this pH range.

Initially, only single pairs of cations and anions were tested. Test solutions were created by mixing inorganic salts in moderately hard reconstituted water (hardness = 80 to 100 mg/l as $CaCO_3$; USEPA, 1989) at concentrations ranging from 0 to 10,000 mg/l (cation + anion). These single salts included: NaCl, Na_2SO_4, $NaHCO_3$, KCl, K_2SO_4, $KHCO_3$, $MgCl_2$, $MgSO_4$, $CaCl_2$, and $CaSO_4$. Later testing included solutions with multiple cations and/or anions; these were created by combining appropriate volumes of the original 10,000 mg/l stock solutions. All salts were obtained from Sigma Chemical Company (St. Louis, Missouri) and were of reagent grade or better. Salts used included all possible cation/anion pairs except for those combinations that do not exist as salts for thermodynamic reasons (e.g., sodium + carbonate). In order to achieve some ion combinations, it was necessary to acidify stock solutions with sulfuric or hydrochloric acids (also reagent grade or better). The salts that required acidification were $CaCO_3$ and $MgCO_3$; each was acidified with both HCl and H_2SO_4. Some solutions showed visible precipitates, particularly combinations of calcium or magnesium with sulfate. These solutions were filtered using a 1 µm pore-diameter filter prior to testing. Aliquots of these and all other stock solutions were taken for chemical analyses. In cases where measured ion concentrations differed from nominal values by more than 10%, measured concentrations were used instead of nominal values in the data analysis.

All toxicity testing was conducted at ENSR's Environmental Toxicology Laboratory in Fort Collins, Colorado. The primary test organisms were *Ceriodaphnia (Ceriodaphnia dubia)*, taken from in-house cultures and less than 24 hours old at test initiation. Other test organisms included *Daphnia (Daphnia magna)* (less than 24 hours old) and one- to three-day-old fathead minnows (*Pimephales promelas*). Toxicity test methods were patterned after USEPA (1985; 1988) guidance for effluent toxicity tests. Test chambers were 30 ml plastic beakers containing 10 ml of test solution. Five test organisms were tested per chamber. For the *Ceriodaphnia* and fathead minnow tests, the temperature was maintained at 25 °C; the *Daphnia* tests were performed at 20 °C. The photoperiod was maintained at 16 hours light/ 8 hours dark. During initial testing, observations of mortality were made after 2, 4, 6, 8, 12, 24, and 48 hours of exposure; later testing involved observations at 24 and 48 hours only. Dilution water for all exposures was moderately hard reconstituted water (USEPA, 1989). Except where technical considerations dictated otherwise, a 0.5x dilution factor (e.g., 100%, 50%, 25%, 12.5%, and 6.25%) was used to prepare the salt solutions. In addition to the toxicity tests described above, sodium chloride reference toxicant tests were conducted monthly on organisms from the in-house cultures, and concurrently with each group of STR tests.

At the beginning of the project, feeding of test organisms was identified as a procedural variable to be addressed. Standard acute testing guidelines generally do not include feeding; nonetheless, withholding food during a 48 hour test can induce considerable stress on the organism. To evaluate the effect of feeding, initial experiments were conducted using two replicates, one with food added and one without. Feeding did increase survival slightly, but the effect of feeding on survival, when compared to salinity, was very small, explaining less than one percent of the total variance. However, subsequent tests included feeding to insure that no starvation-related stress was inadvertently introduced. *Ceriodaphnia* and *Daphnia* were fed the YCT food mixture as described in the USEPA chronic test manual (USEPA, 1989). Fathead minnows were fed freshly hatched brine shrimp.

Data Analysis

All toxicity data were entered into a computer database using Paradox Version 3.01a software (Borland International, 1988). This database contains all information pertinent to the toxicity testing. Over 135,000 values were entered into the database to record information pertaining to 12,500 observations on *Ceriodaphnia, Daphnia* and fathead minnows. This large quantity of data cannot be presented here, but the raw data can be found in Mount and Gulley (1992).

A statistical model relating survival to ion concentration was developed for *Ceriodaphnia* using logistic regression. Additional models are being developed for *Daphnia* and fathead minnows. The data were analyzed using the stepwise logistic regression program (program LR) in BMDP statistical software (BMDP, 1985). At each step in the logistic regression the selection of independent variables included in the model was based on the maximum likelihood ratio. All analyses were performed on the University of Wyoming's Vax mainframe computers.

During model development several data transformations (e.g., log) and independent variable interactions (e.g., anion*anion interactions) were considered. Each potential model was evaluated using the following criteria: 1) each independent variable in the model must significantly improve the fit of the model to the data; 2) the model should maximize R^2 (maximize the amount of variance in the data that is explained by the model) and minimize the number of independent variables; 3) the model should provide realistic predictions even when extrapolating outside the limits of the data used to generate the model. Thus, our goal was to create the simplest model that explained the most variance in the survival data and that made realistic predictions.

Early in the modeling effort we observed that equal concentrations of anions or cations affected *Ceriodaphnia* survival differently depending on the number of cations present in the test solution. For example, a solution of NaCl and $CaCl_2$ was less toxic than a solution of NaCl even though chloride concentrations were equal in both solutions. To account for this phenomenon, we added a new variable to the modeling effort, the number of cations in the test solution

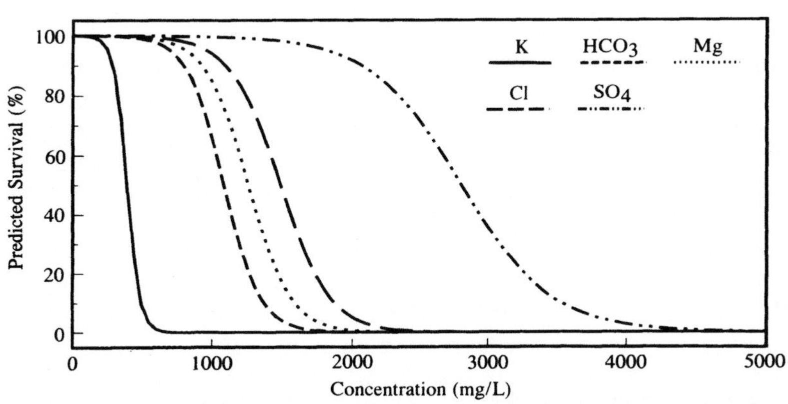

Figure 1. The predicted survival response of *Ceriodaphnia* to increasing concentrations of potassium, bicarbonate, magnesium, chloride, and sulfate. Response curves were plotted with NumCat set to one.

(NumCat). This value was calculated for all test solutions. A cation was counted only if its concentration was greater than 100 mg/L and its molar concentration was at least ten percent of the total molar concentration of cations. The addition of NumCat, NumCat*cation and NumCat*anion interactions significantly improved the *Ceriodaphnia* model.

RESULTS

We developed an STR model for *Ceriodaphnia* that predicts 48 hour survival based on ion concentrations. The best model is as follows:

Ceriodaphnia Model

Logit Survival =	-0.0299 *	K^+
	-0.00668 *	Mg^{++}
	-0.00813 *	Cl^-
	-0.00439 *	SO_4^-
	-0.00775 *	HCO_3^-
	-0.446 *	NumCat
	+0.00870 *	NumCat * K^+
	+0.00248 *	NumCat * Cl^-
	+0.00140 *	NumCat * SO_4^-
	+8.83	

NumCat may not exceed two.
Logit survival can be transformed to percent survival by the following formula:
percent survival = $[1/(1+ \exp^{-\text{logit survival}})]*100$

The variable NumCat may not exceed two in the *Ceriodaphnia* model. This limit was set as values higher than two will extrapolate outside of the range of data used to generate the model. Calculating survival with NumCat exceeding the set limit may produce survival predictions that are unrealistically high. Models for *Daphnia* and fathead minnows are still in development and their equations are not presented.

The *Ceriodaphnia* model has an R^2 value of 0.84; that is, 84 percent of the variance in survival is explained by the independent variables (ion concentrations) in the model. Thus, the *Ceriodaphnia* model adequately predicts survival based on ion concentration. Preliminary models for *Daphnia* and fathead minnows show similar R^2 values.

Figure 1 shows the *Ceriodaphnia* survival response for each ion as predicted by the model equation above. This figure has been plotted with NumCat set equal to one (only one major cation present in solution). With the exception of sodium and calcium, increased cation or anion concentrations decreased the survival of *Ceriodaphnia*. Survival responses of *Daphnia* and fathead minnows were similar to those of *Ceriodaphnia*. Potassium was obviously the most toxic ion followed by bicarbonate, magnesium, and chloride. The least toxic ion was sulfate. Sodium and calcium concentrations did not contribute directly to the regression, although they do influence the NumCat variable. Preliminary models indicate that for all ions, except magnesium, *Ceriodaphnia* was the most sensitive species followed by *Daphnia* and then fathead minnows.

Anion*anion, cation*cation, and cation*anion interactions, when entered into the *Ceriodaphnia* model, complicated the model and explained little or no variance in the survival data. Thus, these interaction terms were not included in the final model. Since the number of cations was an important variable, the number of anions was also entered into the model. However, the number of anions did not significantly improve the fit of the model to the data and was excluded from the final model.

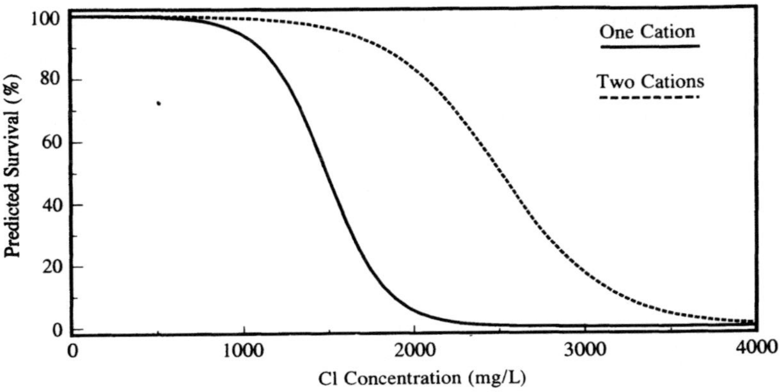

Figure 2. **The predicted survival response of *Ceriodaphnia* to increasing chloride concentrations at two different values for NumCat.**

In the *Ceriodaphnia* model, the toxicity of K^+, Cl^-, and SO_4- was moderated by the number of cations present in solution. The influence of the number of cations on *Ceriodaphnia* was especially apparent if the dominant anions were chloride or sulfate. For example, the predicted toxicity of chloride to *Ceriodaphnia* decreased as the number of cations is increased from one to two (Figure 2).

DISCUSSION

Salinity appears to be a major source of toxicity in produced waters. Collins (1975) described the ionic content of over 3,000 water samples collected from oil fields throughout the United States; the minimum ion concentrations reported were all sufficient to cause acute toxicity to *Ceriodaphnia, Daphnia,* and fathead minnows. Ion concentrations in produced water reported from typical coalbed gas operations also can exceed toxic concentrations for some species (Fillo et al., 1987; O'Neil et al., 1989). Clemens and Jones (1954) reported that produced waters from oil wells in Oklahoma were extremely toxic to freshwater organisms. Produced water releases from oil field development in Kentucky were toxic to aquatic organisms in the Green River, KY (Charles, 1964). Thus, nearly all oil, coal, and gas operations in the United States must consider the potential toxicity of produced waters managed by surface discharge.

Although high salinity can be toxic to freshwater organisms, it has some characteristics that set it apart from many traditional toxicants. For example, the toxic effect of salinity can be removed effectively by dilution alone, without fear of bioaccumulation or biomagnification which may be of concern for many toxicants. Particularly in arid regions, naturally occurring surface waters may already contain concentrations of major ions near or in excess of toxicity thresholds for sensitive species. Moreover, major ions generally cannot be treated (i.e., destroyed) by traditional wastewater treatment technologies as can many organic compounds. Technologies like reverse-osmosis and evaporation can reduce the volume of waste that must be managed, but will not increase the residual waste's acceptability for surface discharge. Thus, from both environmental protection and produced water management viewpoints, it is clearly desirable to differentiate between toxicity caused by salinity and that caused by other co-occurring contaminants. The STR approach, which predicts toxicity of ions to test organisms, could prove very useful for the management of produced waters.

A major utility of the salinity toxicity models is to help distinguish between toxicity due to salinity and toxicity due to other contaminants. In the following example we demonstrate how the

STR might be applied to toxicity data. A produced water sample was collected from an oil production operation in Wyoming. This sample was split into two sub-samples; one sub-sample was not altered, while the second was aerated vigorously to remove volatile components (e.g., sulfide). Both sub-samples were analyzed for major ion and sulfide concentrations; in addition, acute fathead minnow toxicity tests were performed on each sub-sample. The preliminary STR for fathead minnows predicted that there should be little or no mortality due to salinity. However, the observed mortality in the unaltered sample was high at the 50 and 100% concentrations (Figure 3A). Based on the higher survival predicted by the STR, one can infer that a toxicant, in addition to salinity, was present in the un-aerated sample. When the aerated sub-sample was tested, mortality values matched or were less than those predicted by the model (Figure 3B). This indicates that the additional toxicity was removed by aeration. Sulfide analyses confirmed that the unaltered sample contained toxic concentrations of sulfide, while those in the aerated sub-sample were much lower. Thus, the fathead minnow STR was a powerful investigatory tool in showing the presence of an additional toxicant in the un-aerated sample.

When tested with data from the field, the STR model for *Ceriodaphnia* performed very well. For example, the STR was used to predict toxicity for a series of water samples collected from a Wyoming watershed receiving produced water inputs (Boelter et al., 1992). When STR predictions were compared to actual results of acute *Ceriodaphnia* toxicity tests, predicted survival was within 15% of the observed survival for 28 of the 30 samples, explaining 94.6% of the overall variance (Mount et al., 1992). The STR model has also been applied to several sets of data from produced

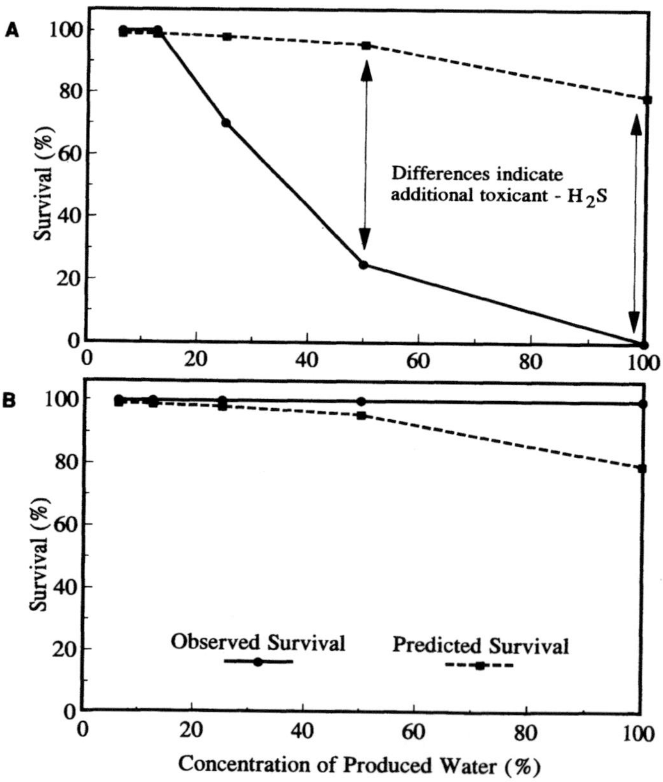

Figure 3. Predicted versus observed survival for a produced water from Wyoming. A) This sample was unaltered and the H_2S concentration was high. B) This sample, from the same produced water, was vigorously aerated removing most of the H_2S and thus the toxicity.

waters in Wyoming and Alabama and has been effective in predicting the observed survival of the test animals. Given that the prediction of field data is notoriously difficult using models based on laboratory data, the *Ceriodaphnia* model does an excellent job of predicting responses to water samples from the field. Additional field and laboratory validation of the *Ceriodaphnia, Daphnia,* and fathead minnow models are in progress.

In conclusion, modeling the survival responses of freshwater organisms to salinity, using laboratory generated data, appears to be a valid method to predict the toxicity of major ions. Models that were developed fit very well to the laboratory data and appear to make reasonable survival predictions for some produced waters. Initial models indicate that for all ions, except magnesium, *Ceriodaphnia* was the most sensitive species followed by *Daphnia* and then fathead minnows. The relative toxicity of each ion in the *Ceriodaphnia* STR was as follows:

$$SO_4^{=} << Cl^- < Mg^{++} < HCO_3^- \; K^+$$

ACKNOWLEDGEMENTS

This research was funded by the Gas Research Institute (GRI) through Contract Nos. 5084-253-1117 and 5091-253-2160 with ENSR Consulting and Engineering. We would like to thank John Fillo (ENSR) and Jim Evans (GRI) for their contributions to the study design. The following University of Wyoming personnel provided help and advice: Aïda Farag, Ann Boelter, and Joe Fernandez. Sheryl Hill (UW) collected several field samples used to validate the model while collecting samples for a produced water project funded by the Wyoming Water Research Center State Grants Program. Ms. Hill also provided technical assistance in performing some of the field validation bioassays. Tyler Garrison and Karen Barten of ENSR performed many of the laboratory toxicity tests used to develop the STR.

REFERENCES

Birge, W.J., Black, J.A., Westerman, A.G., Short, T.M., Taylor, F.B., Bruser, B.R., and Wallingford, E.D., 1985, Recommendations on numerical values for regulating iron and chloride concentrations for the purpose of protecting warmwater species of aquatic life in the commonwealth of Kentucky. Report submitted to the Kentucky Natural Resources and Environmental Protection Cabinet, Frankfort, KY. Memorandum of Agreement No. 5429.

Boelter, A.M., Lamming, F.N., Farag, A.M., and Bergman, H.L., 1992, Environmental effects of saline oil-field discharges on surface waters. Environ. Toxicol. Chem. 11:1187-1195

Borland International, 1988, Paradox Relational Database: Version 3.01a. Scotts Valley, CA.

Charles, J.R., 1964, Effects of oil field brines. "Proceedings of the 18th annual conference, Southeastern Association of Game and Fish Commissioners." Oct 18-21, 1964. Clearwater, FL.

Clemens, H.P., and Jones, W.H., 1954, Toxicity of brine water from oil wells. Amer. Fish. Soc. 84:97-109.

Collins, A.G., 1975, "Geochemistry of Oil Field Waters." Elsevier Scientific. New York, NY. 496p.

Dixon, W.J., 1985, "BMDP Statistical Software." University of California Press. Berkeley, CA. 734p.

Drever, J.I., 1982, "The Geochemistry of Natural Waters." Prentice-Hall, Englewood Cliffs, NJ. 388 p.

Federal Register, 1989, USEPA: National pollutant discharge elimination system; Surface water toxics control program. USEPA, Volume 54, No. 105, Friday June 2, 1989.

Fillo, J.P., Hamaker, T.A., Koraido, S.M., Stefanacci, R.G., Tallon, J.T., and Wesolowski, G., 1987, Production waters associated with the production, processing, transmission, and storage of natural gas: a literature survey. Gas Research Institute: Chicago, IL. GRI-87/0126.

Meyer, J.S., Sanchez, D.A., Brookman, J.A., McWhorter, D.B., and Bergman, H.L., 1985, Chemistry and aquatic toxicity of raw oil shale leachates from Piceance Basin, Colorado. Environ. Toxicol. Chem. 4:559-572.

Mount, D.R., and Norberg, T.J., 1984, A seven-day life-cycle cladoceran toxicity test. Environ. Toxicol. Chem. 3:425-434.

Mount, D.R., and Gulley, D.D., 1992, Development of a salinity/toxicity relationship to predict acute toxicity of saline waters to freshwater organisms. Gas Research Institute: Chicago, IL. (In press).

Mount, D.R., Drottar, K.R., Gulley,D.D., Fillo, J.P., and O'Neil, P.E., 1992, Use of laboratory toxicity data for evaluating the environmental acceptability of produced water discharge to surface waters. Proceedings of the 1992 International Produced Water Symposium. Plenum Publishing Corporation: New York, New York.

Norberg, T.J., and Mount, D.I., 1985, A new fathead minnow (*Pimephales promelas*) sub-chronic toxicity test. Environ. Toxicol. Chem. 4:711-718.

O'Neil, P.E., Harris, S.C., Mettee, M.F., Drottar, K.R., Mount, D.R., and Fillo, J.P., 1989, Biomonitoring of a produced water discharge from Cedar Cove degasification field, Alabama. Gas research institute: Chicago, IL. GRI-89/0073.

O'Neil, P.E., Harris, S.C., Mettee, M.F., Isaacson, H.R., and Evans, J.M., 1992, Biological fate and effect of coalbed methane produced waters discharged into streams of the Warrior Basin, Alabama. Proceedings of the 1992 International Produced Water Symposium. Plenum Publishing Corporation: New York, New York.

Peavy, H.S., 1977, Nonpoint source pollution: Problem definition. American Society of Civil Engineers National Environmental Engineering Conference, Nashville, TN, 13-15 July 1977.

United States Environmental Protection Agency (USEPA), 1985, "Methods for Measuring the Acute Toxicity of Effluents to Freshwater and Marine Organisms." EPA/600/4-85/013. U.S. Environmental Protection Agency, Washington D.C.

United States Environmental Protection Agency (USEPA), 1987, "Biomonitoring to Achieve Control of Toxic Effluents." EPA/600/8-86/002. U.S. Environmental Protection Agency, Washington D.C.

United States Environmental Protection Agency (USEPA), 1988, "Methods for Acute Toxicity Identification Evaluations: phase I Toxicity Characterization Procedures." EPA/600/3-88/034. U.S. Environmental Protection Agency, Duluth, Minnesota.

United States Environmental Protection Agency (USEPA), 1989, "Short-term Methods for Estimating the Chronic Toxicity of Effluents and Receiving Waters to Freshwater Organisms." Second edition. EPA/600/4-89/001A. U.S. Environmental Protection Agency, Washington D.C.

United States Environmental Protection Agency (USEPA), 1991, "Technical Support Document for Water Quality-based Toxics Control." Second edition. EPA/505/2-90/001 U.S. Environmental Protection Agency, Washington D.C.

A COMPREHENSIVE DETERMINATION OF PRODUCED WATER COMPOSITION

P.J.C. Tibbetts[1], I.T. Buchanan[2], L.J. Gawel[3] and R. Large[1]

[1] M-Scan Limited
Ascot, Berkshire
ENGLAND

[2] Conoco (UK) Limited
Aberdeen
SCOTLAND

[3] Conoco Inc.
Ponca City, Oklahoma
USA

INTRODUCTION

Concomitant with the production of oil from North Sea oilfields is the recovery of water associated with the oil in the reservoir. Produced oil and water are usually separated on the production platform before the oil is pumped ashore via a pipeline. The separated water is discharged overboard into the sea. In 1990 approximately 159 x 10^6 tons of produced water were discharged in the UK sector of the North Sea (HMSO, 1991). The current study is concerned with the detailed characterization of produced waters on the Murchison (block 211/19) and Hutton TLP (block 211/28) platforms in the Northern North Sea, both of which are operated by Conoco (UK) Limited. The proportion of water to oil production during the lifetime of the reservoir is not constant. In the early stages of production from the reservoir, the water cut is low (zero to a few percent) with a constant increase until production is no longer economically feasible (80% or so). During the lifetime of a reservoir water-injection (water flooding) is usually practiced during part of the production with "breakthrough" water contributing to the produced water. Both Murchison and Hutton have been under water flood for several years; Murchison has been under pressure maintenance (water flood) since 1981 and Hutton since 1985. Currently Murchison produces 180,000 barrels per day (bpd) and Hutton 130,000 bpd of water. Produced waters discharged from these platforms comprise a variable mixture of formation water, injection water and deck drainage, which may include lubricating oil, grease and diesel from leaks and spillages.

A thorough understanding of the produced water discharge stream is necessary to predict its potential impact on the marine environment. Its composition and physical properties are fundamental to its environmental impact. The treated water discharged overboard is a highly complex mixture of inorganic and organic components in various physical states (true solution, suspension, emulsion, particulate and particle-adsorbed). In addition, the composition would be

Produced Water, Edited by J.P. Ray
and F.R. Engelhart, Plenum Press, New York, 1992

expected to vary with time, partly as a result of variability in the performances of oil/water separators (Neff et al., 1989).

Routine produced water analyses carried out offshore in the North Sea are based on Freon extraction followed by the determination of "total oil" by IR. This is because legislation on the composition of discharged production water offshore has, in general, been limited to "total oil" concentration (currently 40 mg/l) and has been concerned mainly with monitoring the efficiency of the oil/water separators. With the increase in environmental concern in recent years, efforts have been made to establish the identity and concentrations of more polar, water-soluble compounds (Somerville et al., 1987), some classes of which may be more toxic than the discharged oil and which are unlikely to be removed from the produced water using the current technology available offshore. These more polar compounds include treating chemicals (process chemicals) which are required for efficient operation of the platform (Middleditch, 1984 and Neff, 1987). The treating chemicals routinely used include corrosion inhibitors, scale inhibitors, emulsion breakers, biocides, H_2S scavengers, water treating chemicals, and detergents (deck wash). The amounts and type of chemical additives vary with the individual needs of each platform. The chemical treatments are specifically selected to treat a particular problem in a given time frame. For several reasons the chemicals in question are, in general, extremely difficult to analyze. Firstly, each product is a complex mixture of crude grade chemicals with a variety of physical and chemical properties. Secondly, many of the chemicals are of high molecular weight making then intractable to most conventional analytical techniques. Thirdly, many of the chemicals are surface active and, as such, are difficult to extract from water to enable the analysis to be performed.

The first phase of the current project was to conduct a review of the literature concerning the analysis of produced waters and, on the basis of these past findings, to recommend the parameters most important for a full characterization of produced waters, and suitable analytical techniques. A summary of this review, showing the range of reported values for each chosen parameter, both for the North Sea and for the whole world, is shown in Table 2. The second phase of the project involved the collection of water samples in triplicate from four sampling points each on Murchison and Hutton TLP on three separate occasions (October 1990, January 1991 and March 1991). Similarly, 'pristine' seawater samples which served as field blanks for background concentration levels were collected from the stand-by vessel on each occasion at a position one mile up current from the Murchison platform. The present paper summarizes the analytical results obtained on the final discharge waters with occasional reference to other samples from 'upstream' sampling stations. Analytical parameters have been divided into five tasks: Task 1, Physical parameters; Task 2, Petroleum Hydrocarbons; Task 3, Minerals, Salts and Metals; Task 4, Non-hydrocarbon Organics and Task 5, Process Chemicals.

EXPERIMENTAL

A summary of the analytical methods is given in Table 1. The majority of the analyses used throughout the study have been carried out according to standard methods published by ASTM, API, CONCAWE, USEPA or one of the 'Blue Book' methods by the Standing Committee of Analysts (SCA) in a series of publications entitled "Methods for the Examination of Waters and Associated Materials" (MEWAM). Most of these standard techniques are straightforward with well documented data on repeatability and reproducibility. Repeatability in the present study appeared good for most techniques, particularly as sampling variability was included in the three replicate analyses. Hence repeatability appeared better in the final discharge waters than in the more inhomogeneous samples from the first stage separators and the slops tanks.

Variations were made to certain of the standard analytical techniques. These are summarized below, together with those techniques that have not been standardized by one of the organizations listed above.

Table 1. A Summary of the Analytical Procedures used in the Present Study

Parameter	Method	Description
Task 1		
Density	API RP45	Pyknometer method
Surface Tension		Du Nouy Tensiometer
TOC	SCA/MEWAM	Combustion-Infra-red
COD	SCA/MEWAM	Dichromate value
TSS	ASTM:D1888	Millipore filtration/gravimetry
pH		Electrode
Temperature		Thermometer
Task 2		
Total Oil	Paris Commission	IR
Volatiles	EPA 624	Purge and Trap/Thermal Desorption/GLC/MS
Base/neutrals	EPA 625	GLC/MS
Bases	M-Scan	GLC/MS (see text)
Task 3		
Chloride Sulphate	ASTM D 4327	Ion Chromatography
Bicarbonate	ASTM D 3875	Titration with acid
Ammoniacal Nitrogen	ASTM D 1426 A	Direct Nesslerization (with distillation)
Sulphide	ASTM D 4658	Ion selective electrode
Total Iron	APHA	Hach method (colorimetric)
Metals	ASTM D 4190	ICP
	ASTM D 3919	AA
Task 4		
Total Polars	M-Scan	Florisil/IR (see text)
Higher Acids/Phenols	EPA 625	Silylation GLC/MS
Volatile Fatty Acids	SCA/MEWAM	Direct GLC/FID of water
Task 5		
Process Chemicals	M-Scan	Sublation

Total Suspended Solids (TSS)

Due to low back pressure it was possible to obtain in-line samples at only one station. Samples were therefore collected in bottles and filtered within two hours of collection. If samples are taken ashore prior to filtration and determination of TSS, the results can be affected by precipitation of iron salts which would increase the measured TSS value. For comparison purposes, therefore, parallel samples were taken ashore for this determination. An experiment was also carried out to see if the addition of EDTA (30 mg/l), to complex with the iron, would prevent precipitation of iron salts, thus making analysis onshore feasible.

Purgeable Volatiles

The purge and trap procedure was performed offshore on the production platform as soon as the sample had reached ambient temperature. This overcame any problem of loss of volatiles during transport and storage of the samples. Each sample (5 ml) was spiked with two internal standards (d_{10}-xylene and bromofluorobenzene; 1g) prior to purging the sample onto Tenax TA. Analysis was performed using a Chrompack Thermal Desorption Cold Trap (TCT) injector by heating the Tenax to 200° C for 10 min with the cold trap held at -130°C. The cold trap was subsequently flash heated to 200° C for injection of the cryotrapped sample for gas-liquid chromatography/mass spectrometry (GLC/MS). GLC/MS was carried out on a VG Analytical ZAB-HF mass spectrometer with a Hewlett-Packard 5790 capillary gas chromatograph. The column was a 25 m x 0.32 mm i.d. fused silica capillary column with a 1m SE-54 stationary phase. The temperature program was 3 min at 30° C then to 280° C at 10° C/min. The mass spectrometer was scanned over the mass range 500-25 amu once a second.

Base/Neutrals

These were determined according to USEPA method 625 except that the following internal standards were added to each sample (1 liter) prior to extraction:

heptamethylnonane	40g
chlorooctadecane	40g
squalene	40g
d_8-naphthalene	10g
d_{10}-anthracene	10g
d_{10}-pyrene	10g
d_5-benzoic acid	10g
d_5-aniline	10g

GLC/MS was performed on a VG Trio-1 mass spectrometer with a Hewlett Packard 5890 gas chromatograph. Cold on-column injection was used to inject an aliquot (1l) of the sample onto a 25 m x 0.32 mm i.d. fused silica capillary column with a 0.25 m Rtx-5 (Restek Corporation) stationary phase. The temperature program was 1 min at 35° C then to 325° C at 10° C/min. The mass spectrometer was scanned over the range 25-500 amu twice a second.

Bases

Base fractions were obtained from the base/neutral fractions after they had been analyzed by GLC/MS. The base neutral fractions were diluted with dichloromethane (200 ml) and extracted with dilute HCl (pH 1-2; 2 x 100 ml). The acid extracts were combined and backwashed with dichloromethane (100 ml). The pH of the aqueous fraction was altered to 12 by the addition of potassium hydroxide (60% in water). The base fraction was obtained by extraction of the aqueous phase with dichloromethane (2 x 100 ml). The extract volume was reduced to 100 µl prior to analysis by GLC/MS. GLC/MS conditions were the same as for the base/neutrals. Quantification was by reference to the internal standard d_5-aniline.

Total Polar Compounds

Having determined 'total oil' concentration by IR the Freon 113 extract was reduced in volume to about 1 ml and added to the top of a 2 cm diameter column of Florisil (9 g). The column was eluted with Freon 113 (50 ml) and made up to 100 ml prior to further analysis by IR. The difference between the oil concentrations calculated before and after Florisil clean-up was taken as the 'total polar' concentration.

Acid Fraction

The acid fraction was examined according to USEPA 625. Prior to GLC/MS analysis the fraction was silylated using MSTFA. GLC/MS conditions were as for the base/neutrals except that the temperature program was 35° C for 30 seconds then to 60° C at 50° C/min then to 325° C at 10° C/min. Quantification was by direct reference to the internal standard d_5-benzoic acid.

Metals

Most of the metals were determined by inductively coupled plasma (ICP) emission spectroscopy (ASTM D4190). However, cadmium, lead, zinc, arsenic, antimony and silver were analyzed by atomic absorption (AA; ASTM D3919). The addition of a slotted tube atom trap was used to reduce the limit of detection for cadmium, lead, zinc and arsenic.

Samples were prepared by taking an aliquot (100 ml) from the center of the sampling bottle (avoiding oil) and adding conc. HNO_3 (5 ml). This was brought slowly to the boil and the volume allowed to reduce to 50 ml. Further conc. HNO_3 (5 ml) was added and the solution refluxed until it became clear. After cooling the solution was diluted to 100 ml in a clean volumetric flask ready for analysis.

Process Chemicals

Fast atom bombardment mass spectrometry (FABMS) was carried out on a VG Analytical ZAB-HF mass spectrometer using an M-Scan FAB gun. Xenon gas was used as the primary ionizing beam with a beam energy of 10 keV. Positive ion spectra were obtained over the range 1500-50 amu and negative ion over the range 1000-100 at a scan speed of 5 seconds/decade. Data were recorded and stored on an on-line VG Analytical 11-250J data system.

Mass spectra were obtained by placing the samples onto the stainless steel target at the end of the FAB probe. Spectra were obtained with the target coated with glycerol. Five µl of the water samples were used for analysis, but only 2 µl was used for analysis of acid, base and sublation samples.

To perform sublation, each water sample (250 ml) was shaken with ethyl acetate (100 ml) prior to transferring the water and ethyl acetate into a sintered gas washer bottle for sublation. The sample was then purged with helium at 90 ml/min for 1 h, after which time the gas flow was stopped and the ethyl acetate transferred to a clean round bottomed flask. The extract was concentrated by evaporation of the ethyl acetate to a volume of *ca.* 500 µl prior to analysis by FABMS. Quantification was achieved by comparison of peak heights with that of the internal standard tetrapentyl ammonium bromide (TPAB) which was added to the sample at a concentration of 10 mg/l prior to sublation.

RESULTS AND DISCUSSION

The summarized results from Tasks 1-4 are compared with literature values in Table 2. An additional column for the Hutton discharge waters has been included, which excludes the results obtained from the Test Vortoils discharge water in the second sampling excursion. When these samples were taken in the second excursion the well being processed through the Test Vortoils had a very high water cut (96%) and high seawater breakthrough (90%). These results are not, therefore, typical in terms of routine operations.

Physical Parameters (Task 1)

All of the discharge waters have a density similar to that of sea water. However, these determinations were carried out at 20° C, while the discharge water temperature ranged from 50° C to 94° C. This would result in a lower density, possibly causing the discharge waters to remain at the sea surface.

Surface tension may affect dispersion/partition of the discharge waters in the receiving environment and in particular particle adsorption/desorption. The range of surface tension values of the Murchison waters (50-58 dyne/cm) is lower than that of the Hutton waters (59-73 dyne/cm). Jackson et al. (1981) has suggested that there is a negative correlation between surface tension and oil content. On this basis the Hutton waters should contain lower oil concentrations than Murchison, which does not appear to be the case (below). It is, however, likely that other components (e.g. emulsion breakers) will also have a significant effect on the surface tension.

Both total organic carbon (TOC) and chemical oxygen demand (COD) show more variation in the Hutton samples than the Murchison samples. This is largely as a result of larger values at the Test Vortoils in excursions 1 and 3 and lower values in excursion 2. Greater variation would be expected at this station as the efficiency of this separator was being tested with water from a wide variety of wells. The levels of TOC and COD are significantly higher than can be accounted for by the oil concentrations (below), but the ranges are similar

Table 2. Summary of the quoted literature values of the parameters studied in the present program and comparative results for seawater and Murchison and Hutton discharge waters determined in the present study

Task	Parameter	Quoted Literature Values			Discharge Waters		
		World	North Sea	Sea Water	Murchison	Hutton	Hutton[1]
1	Density(kg/m^3)	1014-1140	1014-1085	nm	1020-1021	1019-1025	1019-1022
	Surface Tension (dynes/cm)	43-78	None quoted	nm	50-58	59-73	59-73
	TOC (mg/l)	0-1500	100-1000		142-335	15-522	139-522
	COD (mg/l)	1220	None quoted	-392*	441-869	127-2070	420-2070
	TSS (mg/l)	1.2-1000	None quoted	2.4-2.8	15-85	3-29	3-29
	pH	4.3-10	6.7-7.3	7.6-8.3	7.1-8.1	6.9-8.3	6.9-8.3
	Temperature (°C)	None quoted	None quoted	nm	50-94	66-81	66-81
2	Total Oil (IR; mg/l)	2 - 565	2 - 64		7 - 75	9 - 220	9 - 220
	Volatiles (BTX; mg/l)	0.39 - 35	4.5 - 35	nm	2.2 - 43	1.3-92.6	1.3-92.6
	Base/neutrals (mg/l) (Total non-volatile oil and grease by GLC/MS)	<140	<140	<0.1	3.2-22	4-96	7.1-96
	Bases (μg/l)	275	Detected	nm	24-4360	62-7530	69-7530
3	Chloride (mg/l)	80-200,000	12,400-81,000	20020-20820	16900-18690	16310-21035	16310-19040
	Bicarbonate (mg/l)	77-3990	420-1430	134-155	500-715	260-995	565-995
	Sulphate (mg/l)	<2-1650	18-1650	2810-2960	1295-1800	705-2895	705-1920
	Ammoniacal Nitrogen (mg/l)	10-300	10-250	nd	0-17	0-8.9	0-8.9
	Sulphide	10	Not quoted	nd	0-0.2	0-1.2	0-1.2
	Metals	Table 3	Table 3	Table 3	Table 3	Table 3	Table 3
4	Total Polars (mg/l)	9.7 - 600	25-520	0-9.6*	0-21	1-36	1-36
	Higher Acids (mg/l)	<1-63	0.1-53	1.2-11.9	3.7-9.5	1.7-9.3	3.1-9.3
	Phenols (mg/l)	0.009-23	2-23	nd	2.2-6.8	1.4-6.2	3.6-6.2
	VFA's (mg/l)	2-4900	200-930	nd	178-237	20-626	238-626

nd = not detected
nm = not measured

* Anomalous results, possibly due to contamination of sampling apparatus.
1. Excluding anomalous results at Station 9 in second excursion.

Table 3. Summary of literature values of cations studied in the present program and comparative results for seawater and Murchison and Hutton Discharge waters determined in the present study (mg/l)

	Quoted Literature Values			Discharge Waters		
Metal	World	North Sea	Sea Water	Murchison	Hutton	Hutton[c]
Calcium	13-25800	151-5700	393-427	482-560	372-491	372-491
Sodium	132-97000	5000-43600	10700-11500	10100-11000	10200-11100	10200-11100
Potassium	24-4300	160-744	472-564	330-409	212-492	212-415
Magnesium	8-6000	25-791	1180-1322	440-668	182-1100	182-811
Iron	<0.1-100	1-33	<0.02-0.5	4.6-6.4	4.5-10	4.5-10
Aluminum	310-410	None quoted	0.04-0.18	0.05-0.28	0.09-0.18	0.09-0.18
Boron	5-95	15-95	3.4-4.9	22.6-33.1	6.63-49.2	18.1-49.2
Barium	1.3-650	1.3-218	0.05-0.08	1.3-5.4	0.2-6.04	0.59-6.04
Cadmium[a]	<0.005-0.2	None quoted	<0.005-0.1	<0.005-0.01	<0.005-0.02	<0.005-0.02
Cobalt	None quoted	None quoted	<0.02	<0.02	<0.02-0.03	<0.02-0.03
Chromium	0.02-1.1	None quoted	<0.02	<0.02	<0.02	<0.02
Copper	<0.002-1.5	0.01-1.0	0.01-0.03	0.01-0.04	0.01-0.14	0.01-0.02
Lithium	3-50	None quoted	0.23-0.59	5.04-12.4	3.08-7.53	2.0-7.53
Manganese	<0.004-175	None quoted	<0.002-0.01	0.25-0.43	0.11-0.47	0.11-0.41
Nickel	None quoted	None quoted	<0.02-0.01	<0.02-0.04	<0.02-0.03	<0.02-0.03
Lead[a]	0.002-8.8	3.3-8.8	<0.03	<0.03	<0.03-0.03	<0.03
Silicon[b]	None quoted	None quoted	0.23-0.42	14.5-25.6	13-22.8	13-22.8
Strontium	0.02-1000	15-550	5.8-7.6	23.1-37.2	9.4-39.3	16.8-39.3
Titanium	<0.01-0.7	None quoted	<0.01-0.01	<0.01-0.01	<0.01-0.01	<0.01-0.01
Vanadium	None quoted	None quoted	<0.02	<0.02	<0.02-0.02	<0.02-0.02
Yttrium	None quoted	None quoted	<0.002	<0.002	<0.002-0.01	<0.002-0.01
Zinc[a]	0.01-35	0.01-35	<0.005	<0.005-0.01	<0.005	<0.005
Arsenic[a]	<0.005-0.3	None quoted	<0.01	<0.01	<0.01	<0.01
Tin	None quoted	None quoted	<0.01	<0.01	<0.01	<0.01
Antimony[a]	None quoted	None quoted	<0.01	<0.01	<0.01	<0.01
Mercury	<0.001-0.002	None quoted	<0.05	<0.05	<0.05	<0.05
Silver[a,b]	<0.001-0.15	None quoted	<0.01	<0.01	<0.01	<0.01
Phosphorus	None quoted	0.01-111	0.7-2.7	3.2-6.5	1.3-17.1	3.5-17.1
Beryllium	<0.001-0.004	None quoted	<0.002	<0.002	<0.002	<0.002
Molybdenum	None quoted	None quoted	<0.05-0.07	<0.05-0.14	<0.05-0.11	<0.05-0.11
Thorium	None quoted	None quoted	<0.1	<0.1-0.13	<0.1-0.24	<0.1-0.24

nd = not detected
a = Analyzed by Atomic Absorption (excursions 2 and 3)
b = Values should be regarded as a minimum due to poor solubilities.
discharge waters (mg/l) but were slightly higher on Hutton (mg/l).
c = Second Hutton Test Vortoils

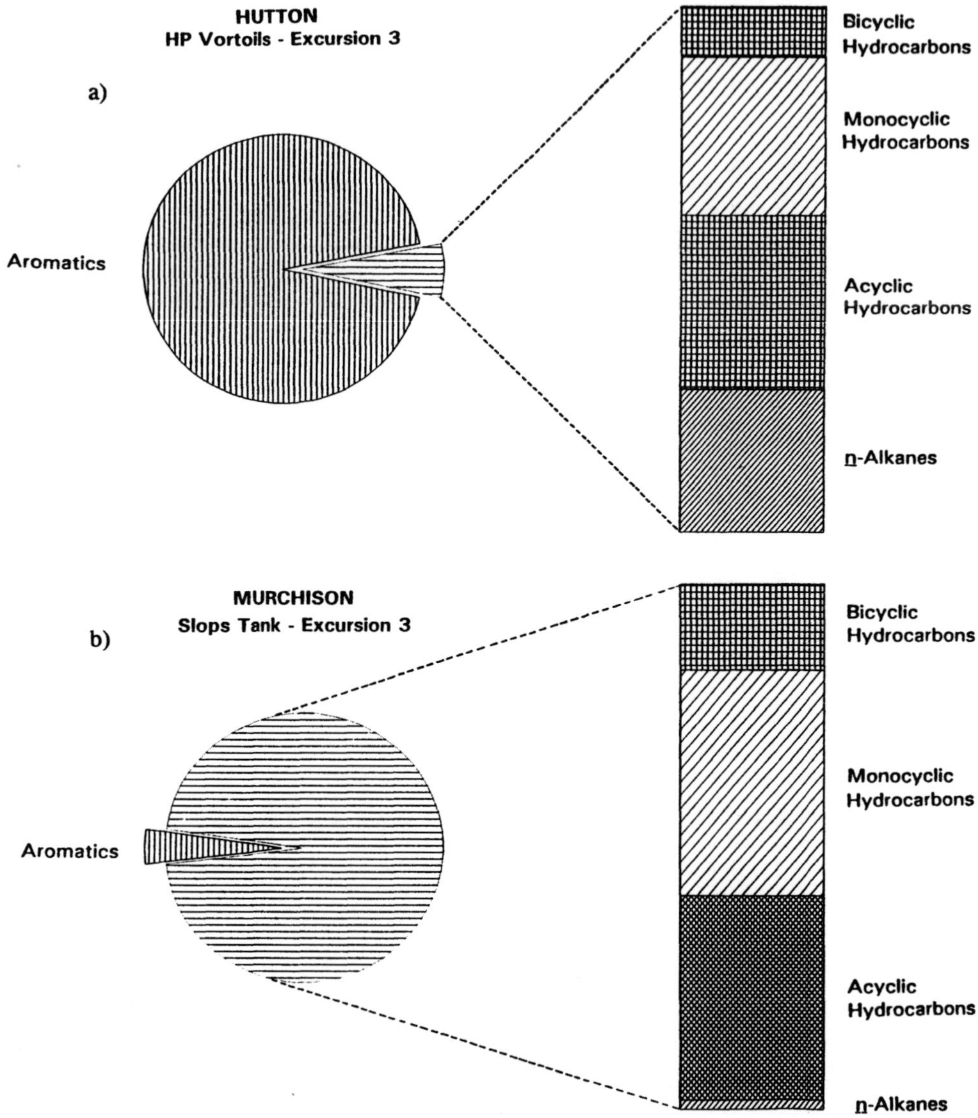

Figure 1. Pie charts showing the relative proportions of aromatic and aliphatic hydrocarbons in: a) Hutton HP Vortoils discharge water, b) Murchison Slops Tank water.

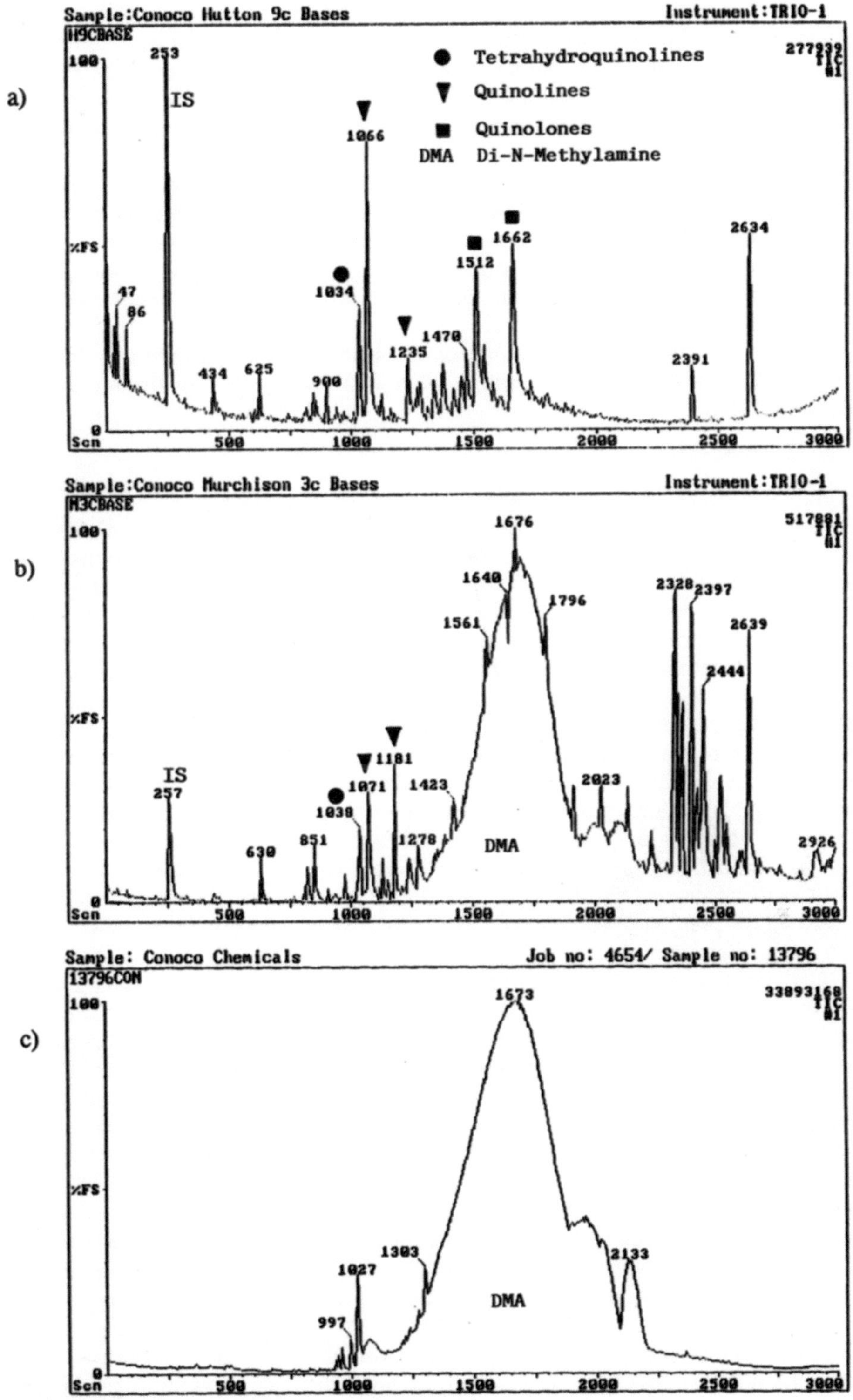

Figure 2. TIC traces of the base fractions of: a) Hutton Test Vortoil discharge waters, b) Murchison Vortoil discharge waters, c) Corrosion inhibitor.

to those reported elsewhere (Table 2). The TOC and COD concentrations appear to parallel those of the volatile fatty acids (VFAs) (below), suggesting the dominance of these components in the waters. It is worth noting that there appeared to be little decrease in the TOC and COD values through the separation facilities on both platforms from the first stage separator to the final discharge waters. This would be expected as the water soluble VFAs are unlikely to be removed by the oil/water separators.

Total suspended solids (TSS) were generally low on both platforms. Samples stored under air and analyzed onshore contained enhanced levels of TSS probably due to the precipitation of iron salts. Stabilization of the samples with EDTA reduces the TSS values determined onshore, but, in some cases, appears to reduce them below those determined in-line offshore. This suggests that the EDTA may be causing particulate matter originally present in the produced water to become dissolved.

pH values are similar on both Murchison and Hutton and are generally slightly lower than that of seawater measured in the current study and also quoted in the literature (7.5-8.4; Middleditch, 1984).

Organic Compounds (Tasks 2 and 4)

Total oil concentrations determined by IR are similar to those published for other oilfields (worldwide; Table 2). Although the range extends to higher values than recorded elsewhere for North Sea fields, this is again due to the samples from Hutton Test Vortoils, which gave particularly high values in the third excursion. The IR technique is an indirect method for the determination of oil concentrations based primarily on the IR absorbance of solvent extractable, aliphatic hydrocarbons. GLC/MS analysis of the volatile (purgeables) and base/neutral fractions, however, shows that the dominant components present are the volatile aromatic hydrocarbons, particularly benzene, toluene, ethylbenzene and xylenes (BTX). These are the more water soluble hydrocarbon components of the oil (Boylan and Tripp, 1971) and are therefore less likely to be removed from the water by physical oil/water separation. The dominance of the aromatic hydrocarbons in the discharge waters is emphasized by the pie chart in Figure 1a. For comparative purposes a corresponding chart is shown (Figure 1b) for a slops tank sample, in which there is a predominance of oil in separated or droplet form. Some of the BTX concentrations are higher than any others reported in the literature. This may be the result of the oil water separator performance at this location, and possibly due to performing the purge and trap procedure on the platform, which would minimize loss of the volatile aromatic hydrocarbons during transport and storage.

Little work has been carried out on the analysis of organic bases in produced waters. Many process chemicals, however, contain compounds that would be expected to be observed in a base fraction. The dominance of the neutral hydrocarbon components in a base/neutral fraction precludes the detection of low concentrations of bases. The present study included the separation of a base fraction after GLC/MS analysis of the base/neutrals. A range of basic compounds including anilines/pyridines, quinolines, tetrahydroquinolines, indoles/octahydroacridines, quinolines, and benzimidazoles has been detected in the present study. The discharge waters on Murchison and Hutton exhibited different patterns of basic peaks. Typical GLC/MS total ion current (TIC) traces are shown in Figure 2. The dominant components in the Hutton discharge waters are quinolinones (up to 1.4 mg/l) and quinolines (up to 1.4 mg/l) with lower levels (up to 0.5 mg/l) of anilines/ pyridines, tetrahydroquinolines and indoles/octahydroacridines. The trace of the Murchison discharge water (Figure 2b) is dominated by a broad hump, which exhibits a mass spectrum similar to an N,N-dimethylamine (DMA). From the peak shape this is thought to be a thermal degradation product, possibly from a quaternary amine salt in a process chemical. The concentrations of this component in the Murchison discharge waters ranged from 0.3 to 6.6 mg/l. Base fractions isolated from two of the process chemicals (a biocide and a corrosion inhibitor) under study showed similar GLC/MS TIC traces (Figure 2c). This suggests that the examination of a base fraction may help in the detection and quantification of process chemicals in discharge waters. The concentrations of bases appeared to drop in the second excursion and the DMA was not detected in the base fractions. These samples, together with the third excursion samples, were stabilized with acid to minimize biodegradation of the samples during transport and storage. FABMS of sublated extracts of unstabilized waters (see below), however, indicated that the quaternary salts were still present. This suggests that the acid used for stabilization may have caused a degradation of the bases. The TIC traces of the base fractions in the third excursion were also weak and DMA was again absent. d_5-Aniline was used as an internal standard for the base fraction. However, due to the extensive sample preparation necessary to produce a base fraction, it was found that there was highly variable recovery of this volatile internal standard, resulting in a wide range of calculated concentrations of the basic compounds. It is therefore recommended that a less volatile, deuterated internal standard (e.g. d_7-quinoline) is used for future work. More work is needed to validate the use of a base fraction to detect process chemicals.

Total polars were estimated by IR using the difference in the value of total oil before and after removal of 'polars' by a Florisil column. The values within each excursion were surprisingly consistent considering the inaccuracies inherent in the method. The values are generally lower than those quoted in the literature for other North Sea oilfields.

The concentrations of higher acids/phenols were consistent at all stations throughout the program. The exception was at the Hutton Test Vortoils in the second excursion, when the lower values are thought to relate to the presence in the produced water of sea water, possibly from the breakthrough of injection water. The majority of the extractable acidic compounds were accounted for by phenols with values up to 6.8 mg/l on Murchison and 6.2 mg/l on Hutton. The ranges are within those quoted for other North Sea fields. Benzoic acids ranged from 0.4 to 1.2 mg/l and acyclic aliphatic carboxylic acids ranged from 0.6 to 1.6 mg/l. Both of these ranges are within those quoted in the literature for other North Sea fields.

Volatile fatty acid (VFA) concentrations range up to 237 mg/l on Murchison and 626 mg/l on Hutton. The dominant component in every case is acetic acid, which comprises 60% or more of the VFAs. Apart from the Hutton Test Vortoils samples in the second excursion, which was affected by breakthrough of injection water, the VFAs consistently account for the majority of the organic material present in the produced water samples. An example of this is displayed in Figure 3. Although the concentrations of VFAs appear high, the values lie within the range quoted in the literature (Table 2).

Relative proportions of volatile fatty acids and other organic contaminants in Hutton Test Vortoils discharge water.

Inorganic Parameters (Task 3)

Chloride concentrations (Table 2) in the discharge waters were all relatively consistent and are slightly lower than sea water. Bicarbonate concentrations, however, are about four times greater than seawater although they lie within the range quoted in the literature. Sulphate concentrations were generally about half that of sea water. The Hutton Test Vortoil discharge samples in the second excursion, however, exhibited chloride, bicarbonate and sulphate concentrations more similar to that of seawater, which supports the suggestion that these samples contain high levels of seawater as a result of breakthrough of injection water.

Ammoniacal nitrogen was detected only in the second excursion and then only at low levels (mg/l) compared with other North Sea fields. Sulphide concentrations were low in all Murchison

The range of concentrations of metals and metalloids in the discharge waters and seawater samples along with those quoted in the literature are given in Table 3. Most parameters appeared consistent with time with the exception of Hutton Test Vortoils in the second excursion. A separate column is therefore included in Table 3 excluding these second excursion results. Most values lie within the ranges quoted in the literature. Phosphorus was present in relatively high concentrations although similar levels have been reported in other North Sea fields (McCourt and Peers, 1988). These levels have been attributed to the presence of phosphonate scale

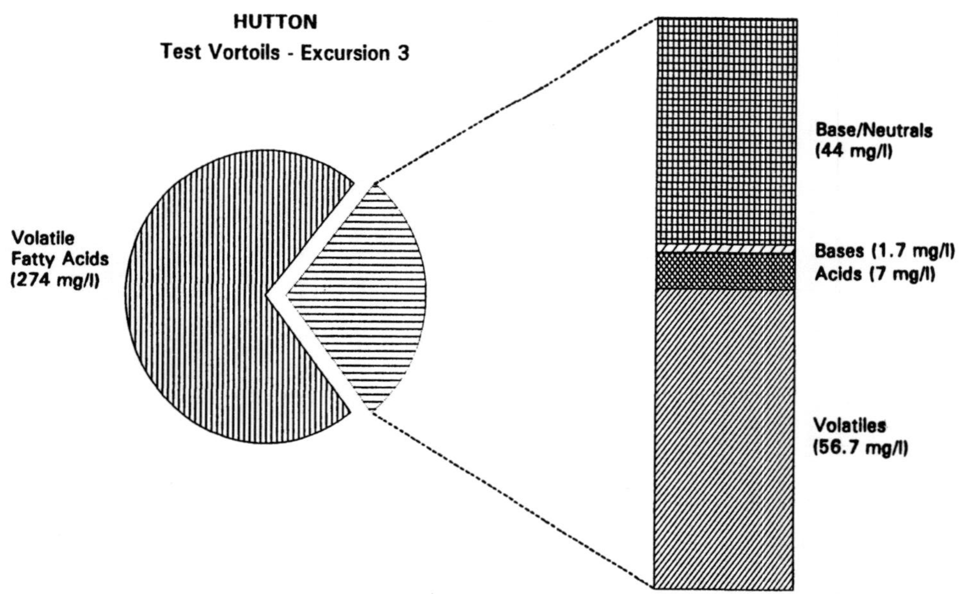

Figure 3. Relative proportions of volaitle fatty acids and other organic contaminants in Hutton Test Vortoils discharge water.

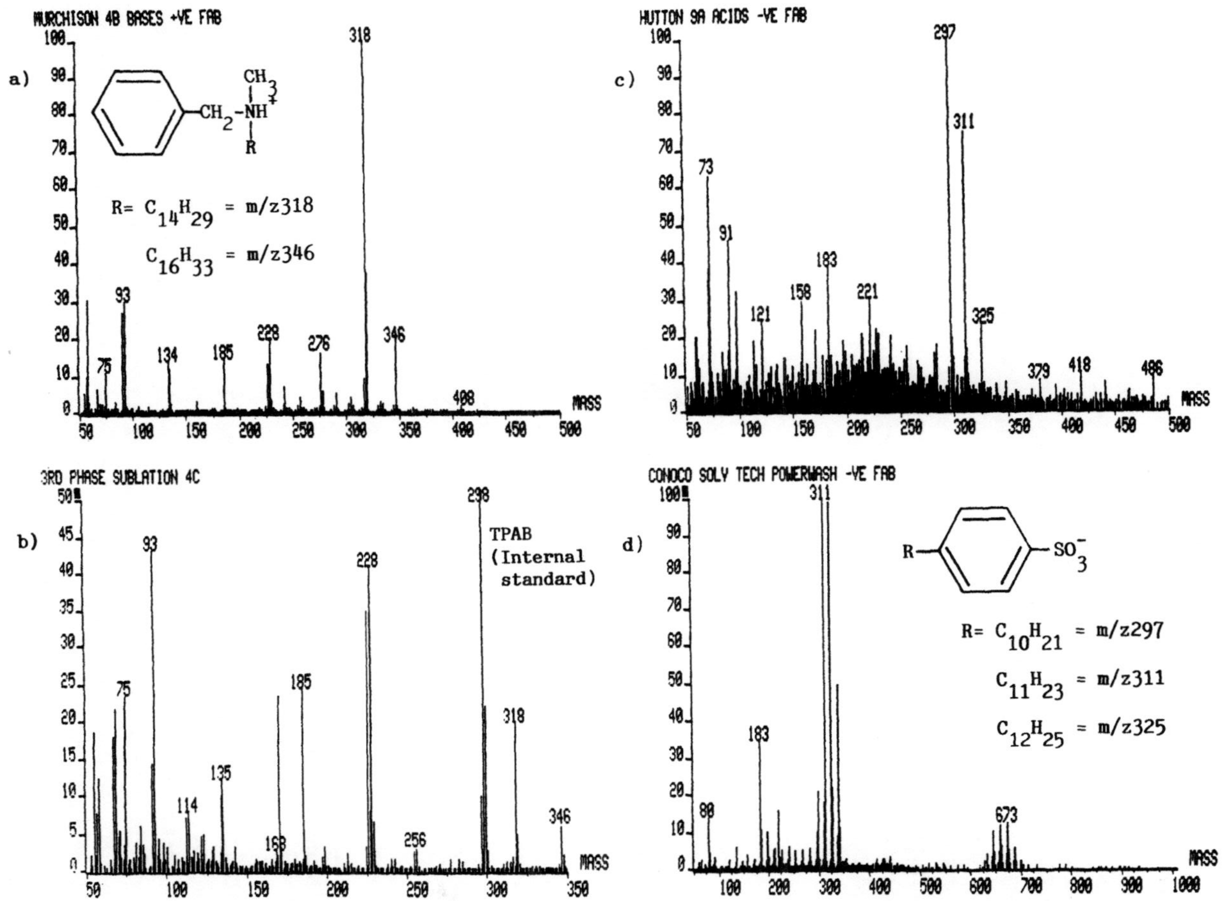

Figure 4. Fast atom bombardment mass spectra (FABMS), showing quaternary amines in: a) Murchison Wemco discharge water base fraction, and b) sublation extract (positive ion); and alkylbenzene sulphonates in, c) Hutton Test Vortoils discharge water acid fraction (negative ion) and, d) Aqueous Degreaser 2 (negative ion).

inhibitors. Such process chemicals were in use on both the Murchison and Hutton platforms during the present study.

Process Chemicals

Task 5 of the current study involved the analysis of eight process chemicals in use on the platforms to determine their composition, as far as possible, and thereafter the examination of produced water samples for the presence of process chemical components. The selection of process chemicals comprised two demulsifiers, two scale inhibitors, a corrosion inhibitor, a biocide and two aqueous degreasers. Each was examined directly by GLC/MS and by fast atom bombardment mass spectrometry (FABMS; both positive and negative ion). In addition, a base fraction was isolated from the process chemicals (excluding the aqueous degreasers) for examination by GLC/MS. A summary of the results obtained for the chemicals is given in Table 4. It can be seen that FABMS was able to identify specific compounds in the formulations of seven out of the eight process chemicals. The eighth chemical (Scale Inhibitor 2) exhibited FAB mass spectra containing apparently characteristic ions, the identity of which remain unknown. These ions could, in principle, still be used as 'tracers' for the formulation. More information was gained from FABMS analysis of the chemicals than from GLC/MS, which detected peaks in only four out of the eight process chemicals. One of these (Demulsifier 1) comprised mainly aromatic hydrocarbons that would be difficult to differentiate from those present in the produced water from crude oil.

Table 4. A Summary of the Components Detected in the Process Chemicals

	FABMS	GLC/MS of TOE[a]	GLC/MS of Bases
Demulsifier 1	Cyclic propoxylates	Heavy Aromatic Solvent	Aniline
Demulsifier 2	Amine oligomers	No Peaks +2 unidentified.	Decahydrophenazine
Scale Inhibitor 1	Diethylenetriamine pentamethylene phosphonic acid	No Peaks	No Peaks
Scale Inhibitor 2	Unknown at m/z 289	No Peaks	No Peaks
Corrosion Inhibitor	Quaternary amines	N,N-Dimethylamine	N,N-Dimethylamine
Biocide	Quaternary amines	N,N-Dimethylamine	N,N-Dimethylamine
Aqueous Degreaser 1	Diethanolamide ethoxylates C_{10} to C_{18} aliphatic acids ethoxylates	Cyclic ethoxylates	nm[b]
Aqueous Degreaser 2	Nonylphenolethoxylates alkyl benzene sulphonates	No peaks	nm

[a]TOE - Total Organic Extract
[b]nm not measured

It has been shown (Large et al., 1988) that it is possible to detect selected process chemicals directly in produced water by FABMS. To determine the detection limits for this type of analysis each process chemical was spiked into saline water at concentrations from 1000 mg/l down to 1 mg/l. It was found that four of the process chemicals could not be detected even at 1000 mg/l (both demulsifiers and both scale inhibitors). It is estimated that the aqueous degreasers could be detected down to *ca.* 20 mg/l and the corrosion inhibitor and biocide down to 5 mg/l. These detection limits are considered too high for practical purposes. It therefore appears necessary to extract and/or concentrate the waters prior to FABMS analysis. FABMS was therefore performed on the acid and base extracts of the production waters. These analyses confirmed the GLC/MS evidence (above) for the presence of quaternary amines from the corrosion inhibitor and/or the biocide (both of which gave identical FABMS and GLC/MS results, Table 4) in the base fractions of the Murchison discharge waters (Figure 4a). Negative ion FABMS of the acid fractions suggested the presence of alkylbenzene sulphonates in the Hutton discharge waters

(Figure 4c); these are also present in the formulation of Aqueous Degreaser 2 (Figure 4d). These analyses were, however, only qualitative, partly because it is difficult to extract surface active compounds quantitatively from water.

Sublation, a method designed specifically to extract surface active compounds from water, was reported by Wickbold (1966). This has more recently been developed and combined with FABMS for the detection of surfactants in water (Levsen et al., 1984; Righton and Watts, 1986). In the present study the sublation method has been investigated for the analysis of produced waters. Initially, saline water spiked with Demulsifier 1, Scale Inhibitors 1 and 2 and the Biocide was sublated and the concentrated extract analyzed by FABMS. The demulsifier and the scale inhibitors remained undetected, but the technique resulted in a 500-fold reduction in the detection limit of the biocide to a more useful 10 g/l, even in the presence of 20 mg/l of crude oil.

This procedure was then applied to the produced waters using tetrapentyl ammonium bromide (TPAB) as an internal standard for quantitative purposes. The quaternary ammonium compounds from the corrosion inhibitor and/or the biocide were detected at concentrations up to 2 mg/l in the Murchison discharge waters (Figure 4b) in both the second and third excursion samples, which had not been stabilized by acid. Negative ion FABMS of the sublation extract detected alkyl benzene sulphonates possibly from Aqueous Degreaser 2 at estimated concentrations of up to 3 mg/l.

Although sublation has resulted in the reduction of the detection limits of four of the current chemicals to useful levels, four of the process chemicals remain undetected at concentrations of 10 mg/l. Future investigations should include the study of alternative methods of extraction and detection to improve the detection limits of these chemicals in produced waters. It appears unlikely that it will be possible to develop a comprehensive method that will allow the quantification of all process chemicals in a single analysis.

ACKNOWLEDGEMENTS

The authors would like to thank Conoco (UK) Limited for funding the project and for permission to publish the data. In addition, they are grateful to the partners in Hutton TLP and Murchison for the use of these facilities. They would like to thank Simon Hird, Karl Ropkins and David Norris of M-Scan Limited for their conscientious endeavors over many months of intense analytical work. They would also like to thank Inchcape Inspection and Testing Services (UK) Limited (Caleb Brett), Aberdeen for the determination of many of the physical and inorganic parameters and, in particular, their representatives Graham Rettie and James Gray for help in field sampling and Scott Ramage for useful discussions. Metal/metalloid analyses were carried out by Dr. Nick Walsh of Royal Holloway and Bedford New College, Egham; COD and TOC analyses were carried out by the Water Research Center, Medmenham and surface tension by SGS Yarsley, Redhill, to all of whom we are grateful. Finally, the authors would like to thank all Conoco personnel, both onshore and offshore, who contributed to the project. In particular, thanks goes to Dan Caudle of Conoco Inc., Houston for guidance and many useful discussions throughout the study.

REFERENCES

Boylan, D.B. and Tripp, B.W., 1971. Determination of hydrocarbons in seawater extracts of crude oil and crude oil fractions, Nature, 230:44-47.

HMSO, 1991, "Development of the Oil and Gas Resources of the United Kingdom," HMSO, ISBN O 11 413705 6.

Jackson, G.F., Hume, E., Wade, M.J. and Kirsch, M., 1981, "Oil Content in Produced Brine on Ten Louisiana Production Platforms," Report to the U.S. Environmental Protection Agency, Municipal Environmental Research Laboratory, Cincinnati, OH. EPA-600/2-81-209.

Large, R., Tibbetts, P.J.C. and Holland, A.J., 1988, FABMS Analysis of Surfactants and Polar Petroleum Compounds, *in*: "Petroanalysis '87: Developments in Analytical Chemistry in the Petroleum Industry," G.B.Crump,ed., John Wiley and Sons, Chichester.

Levsen, K., Schneider, E., Rollgen, F.W., Dähling, P., Boerboom, A.J.H., Kistemaker, P.G. and McLuckey, S.A., 1984, Mass Spectrometric Identification of Surfactants, *in*; "Proceedings of the 3rd European Symposium on Analysis of Organic Micropollutants in Water," G. Angelotti and A. Bjorsoth, eds., Reidel.

McCourt, C.B. and Peers D.M. 1988, Environmental Monitoring of Trace Elements in Water Discharges from Oil Production Platforms, *in*: "Petroanalysis '87: Developments in Analytical Chemistry in the Petroleum Industry", G.B. Crump, ed., John Wiley and Sons, Chichester.

Middleditch, B.S., 1984, "Ecological Effects of Produced Water Discharges from Offshore Oil and Gas Production Platforms," Final Report on API Project No. 248. American Petroleum Institute, Washington, D.C.

Neff, J.M., 1987, Biological Effects of Drilling Fluids, Drill Cuttings and Produced Waters, *in*: "Long Term Environmental Effects of Offshore Oil and Gas Development," D.F. Boesch and N.N. Rabalais, eds., Elsevier Applied Science Publishers Ltd., London.

Neff, J.M., Sauer, T.C. and Maciolek, N. 1989, "Fate and Effects of Produced Water Discharges in Nearshore Marine Waters," American Petroleum Inst. Publ. No. 4472, American Petroleum Institute, Washington, D.C.

Righton, M.J.G. and Watts, C.D., 1986, "Identification of Surfactants in Water Samples using Sublation Extraction and Fast Atom Bombardment Mass Spectrometry," Water Research Center, U.K. Publication ER 1194-M.

Somerville, H.J., Bennett, D., Davenport, J.N., Holt, M.S., Lynes, A., Mahieu, A., McCourt, B., Parker, J.G., Stephenson, R.R., Watkinson, R.J. and Wilkinson, T.G., 1987, Environmental effect of produced water from North Sea oil operations, Marine Pollution Bulletin, 18:549.

Wickbold, R., 1966, Zur analyze nichtionogener ten side in wasser und abwasser, Vom Wasser, 33:229.

CHEMICAL AND TOXICOLOGICAL CHARACTERIZATION OF PRODUCED WATER FREON EXTRACTS

J.S. Brown, T.C. Sauer, Jr., M.J. Wade, and J.M. Neff

Arthur D. Little, Inc
Cambridge, Massachusetts
USA

ABSTRACT

Chemical characterizations and toxicity determinations were performed on produced water Freon extracts from three Gulf of Mexico offshore oil production platforms. The purpose of the study was to: 1) chemically characterize produced water Freon extracts prepared according to EPA Method 413.1, and additionally characterize Freon extracts after a standard silica gel cleanup (Standard Method 5520 F), and 2) determine the toxicity to *Mysidopsis bahia* of produced water and produced water Freon extracts. A comparison of oil and grease residue weights determined by both standard methods revealed that 79 to 98% of the material in the Freon extracts was polar (i.e., removed by silica gel treatment). The chemical composition of the polar organic fraction was determined by gas chromatography/mass spectrometry (GC/MS) and was found to vary from one produced water sample to another, but all samples contained a complex mixture of straight chained, branched, unsaturated and cyclic fatty acids as well as smaller amounts of aromatic acids and phenols. Although the whole produced waters were acutely toxic to mysids, the base-soluble polar material isolated from Freon extracts was not acutely toxic to mysids (LC_{50} 23%). The toxicity results indicate that the polar acid organic compounds in produced water do not contribute significantly to its toxicity.

INTRODUCTION

As mandated by Section 402 of the Clean Water Act, the U.S. Environmental Protection Agency (EPA) regulates the discharge of produced water from offshore platforms through National Pollution Discharge Elimination System (NPDES) permits. One provision of the permits limits the highest concentration of hydrocarbons, as determined by the NPDES oil and grease method (EPA, 1979), that can be discharged into the ocean with produced water. Based on data collected from a limited number of produced water discharges, EPA established a monthly mean upper limit of 48 mg/l and a daily mean upper limit of 72 mg/l oil and grease in produced water destined for ocean disposal. EPA (1985) has proposed a new source performance standard of 59 mg/l (no single sample to exceed) oil and grease for produced water based on what industry can achieve using best practical technological methods (i.e., the proposed performance standard is technology-based rather than toxicity-based).

Determination of total recoverable oil and grease is relatively simple by EPA Method 413.1. It involves extracting acidified water samples with Fluorocarbon-113 (1,1,2-trichloro-1,2,2-

trifluoroethane) or Freon equivalent, and weighing the dried extract residue to calculate total oil and grease. Because produced water is acidified before extraction, polar nonhydrocarbon compounds may be extracted with petroleum hydrocarbons. Some production platforms have been unable to comply with the best practicable technology oil and grease limits for their produced water discharges even though the best known technology is being used to separate oil from produced water.

Analysis of Freon extracts for organic compounds has revealed that the reason for noncompliance of some discharges is the high concentrations of nonhydrocarbon compounds, such as fatty acids and other polar compounds, extracted with Freon. In the present context, hydrocarbons are defined as organic compounds composed of carbon and hydrogen. Nonhydrocarbons are defined as organic compounds that contain, in addition to carbon and hydrogen, other elements such as oxygen, nitrogen, and sulfur covalently bonded to the molecule. Examples of nonhydrocarbons commonly found in petroleum are phenols, fatty acids, aldehydes, and alcohols. These compounds usually are more polar and more water-soluble than hydrocarbons of similar molecular weight. Brown et al., (1988) have demonstrated that as much as 80 percent of the oil and grease residue weight determined gravimetrically by the NPDES-approved method (EPA Method 413.1) can be attributed to water-soluble, polar components in the Freon extract.

Because the treatment technology for oil and grease control in produced water acts only on petroleum hydrocarbons, the oil industry has investigated established methods for separating nonhydrocarbons from hydrocarbon extracts to obtain a more accurate estimate of the concentration of petroleum hydrocarbons in produced water. One such method is Standard Method 5520°F with gravimetric finish (APHA-AWWA-WPCF, 1989), which involves treating the Freon extract with silica gel prior to drying the extract and weighing the treated residue. This method effectively removes the major water-soluble polar compounds from the Freon extract (Jackson et al., 1981).

The study included the following two tasks: 1) characterize the major organic compounds in the Freon extracts, and perform a mass balance for these constituents in produced water extracts before and after treatment with silica gel, and 2) determine the acute toxicity to mysids *Mysidopsis bahia* of whole produced water samples and base-soluble (polar hydrocarbon) fractions of Freon extracts of produced water.

The objectives of the chemical characterization task were to:

- Identify the major chemical components of the Freon extract of three produced water samples originating from three offshore oil production platforms;
- Produce a mass balance of the major identified chemicals in Freon extracts of produced water;
- Determine if the silica gel treatment used to remove polar organics in Freon extracts of produced water also removes aromatic hydrocarbons from the extracts; and
- Describe an analytical method for identifying and quantifying the major components in the Freon extracts of produced water.

Characterizing the major chemical components of Freon extracts of produced water involved separation of the major components by acid/base-partitioning schemes and silica gel chromatography, and identifying and quantifying the major components by gas chromatography/mass spectrometry (GC/MS). The compound classes targeted for identification were n-alkanes, polynuclear aromatic hydrocarbons (PAHs), and water-soluble polar compounds (fatty acids, aromatic acids, phenols, alcohols, and other polar compounds). The carbon, hydrogen, nitrogen, and sulfur (CHNS) composition of the Freon extracts of produced water was also investigated. A mass balance was determined gravimetrically and instrumentally throughout the analytical procedures to measure the efficiency of the separation and analytical techniques, and to allow a comparison between silica gel-treated and -untreated extracts.

The objectives of the aquatic toxicology task were to:

- Determine the acute toxicity to mysids of the polar organic compounds in the Freon extract of produced water; and

- Compare the toxicity of the Freon extract to the acute toxicity of the whole produced water. This comparison permits an assessment of the extent to which the polar organic compounds coextracted with petroleum hydrocarbons by Freon contribute to the toxicity of the whole produced water.

The 96-hour acute toxicity test with the mysid *Mysidopsis bahia* was used to evaluate the toxicity of the produced water and polar fractions of produced water Freon extracts. The mysid test has been used routinely to evaluate the toxicity of effluents such as those from municipal treatment plants, refineries and chemical manufacturing plants to marine animals. The test is not routinely used to monitor the toxicity of produced water, such methods are still under development.

MATERIALS AND METHODS

Produced water samples were obtained from three Gulf of Mexico offshore production platforms, and designated as Platforms 1, 2, 3. Twenty liters of acidified (HCl, < pH 1) and 20 l of unacidified produced water, were obtained from each platform for chemical characterization and toxicity testing. The samples were stored refrigerated (4°C) until analyzed.

All samples were homogenized by gentle stirring prior to subsampling for analysis. One 3 l subsample was withdrawn from the Platform 1 and 2 unacidified composite samples and designated as "whole" produced water for toxicity testing. One 6-l subsample was withdrawn from each acidified composite sample (Platforms 1, 2, and 3) for subsequent extraction and chemical analysis as part of the chemical characterization task. One 3 l subsample was withdrawn from the Platform 1 and 2 acidified samples for subsequent extraction and use in the toxicity tests. The Platform 3 sample was deemed unsuitable for toxicity testing due to excessive storage time.

Determination of the Chemical Characteristics of Produced Water Extracts

The analytical methods described herein were designed to meet the specific objectives of the chemical characterization task. An overview of the analytical scheme for chemical characterization of produced water is presented in Figures 1 and 2. EPA Method 413.1 was followed for determination of total residue weight [residue weight (1) in Figure 1]. Standard Method 5520 F (previously identified as Standard Method 503 E) was followed for determination of silica gel-treated residue weight [residue weight (8) in Figure 2]. In addition, residues were reweighed by electrobalance to increase accuracy of low-weight determinations.

Extraction of Produced Water. The pH of the three 6 l acidified produced water subsamples was measured to verify that it was ≤ 1.0. Any samples with a pH greater than 1.0 were adjusted to pH 1.0 with 6 N HCl. The three samples were spiked with 100 μg of fatty acid and phenol internal standards (d_{23}-dodecanoic acid and d_8-p-cresol). In addition, one spiked blank and one procedural blank were prepared by acidifying 6 l of distilled water to pH 1.0 with 6 N HCl. Both the blank and the spiked blank were spiked with 100 μg of phenol and fatty acid internal standards. In addition 100 μg of the fatty acid mixture spike solution and 100 μg of the phenol mixture spike solution were added to the spiked blank. The produced water samples, procedural blank, and spiked blank were extracted three times with 200 ml of Freon (Figure 1). The Freon extracts from each sample were combined and filtered through sodium sulfate (Na_2SO_4) and a glass fiber filter. The filter and Na_2SO_4 were then rinsed with 50 ml of Freon, and the rinse was combined with the Freon filtrate.

The combined Freon extract from each sample was concentrated using a Kuderna-Danish concentrator (K-d) to 6 ml, in a 80° C water bath. Each concentrated extract was then split into six equal fractions. One aliquot from each sample was archived, and one aliquot was saved for silica gel treatment. The remaining four aliquots from each sample were transferred to preweighed vials and concentrated to dryness under a stream of clean N_2 gas in a 80° C water bath. The vials containing the dried extracts were placed in a desiccator for 30 minutes, and then reweighed to determine the residue weight. The dried Freon extracts were redissolved in 1 ml

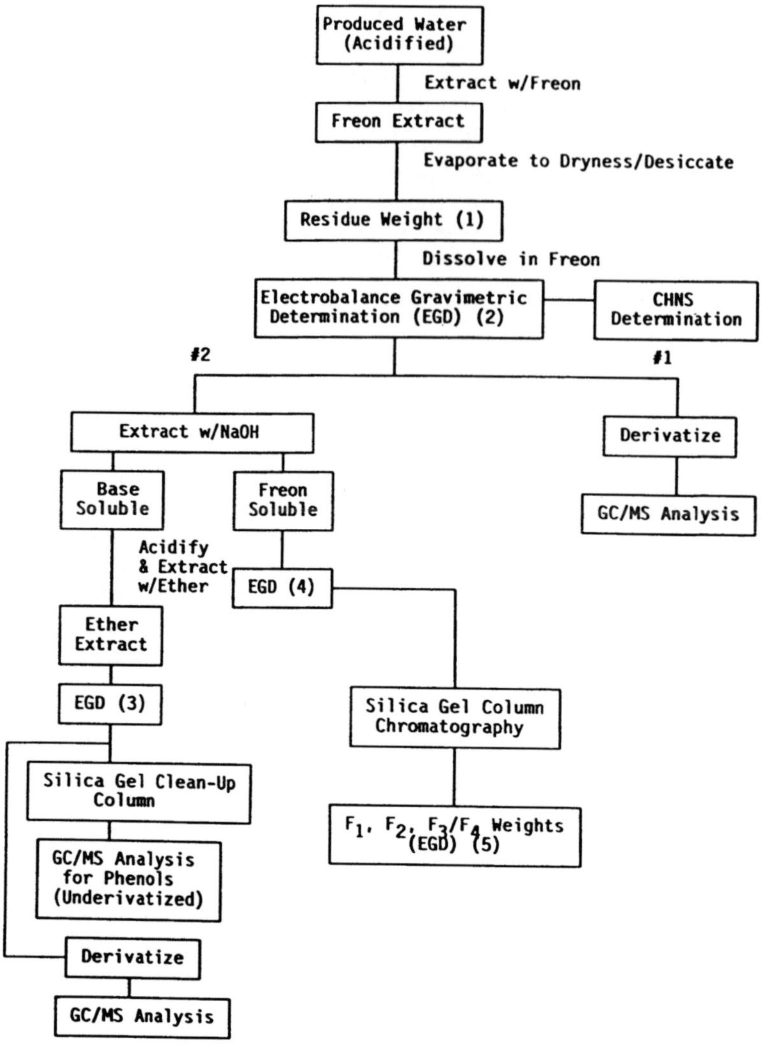

Figure 1. Produced water sample processing scheme showing individual analysis steps. Numbers in parentheses refer to the extract weights summarized in Tables 1 and 2.

of Freon and two aliquots of each extract were weighed on a Cahn electrobalance (electrobalance gravimetric determination in Figure 1) to calculate the aliquot weight.

CHNS Analysis. One Freon extract split from each sample was analyzed for carbon, hydrogen, nitrogen, and sulfur (CHNS). The sulfur concentration was determined with a Leco sulfur analyzer, which combusts the sulfur-containing organic compounds to form SO_2 which is then measured with an infrared (IR) detector. The carbon and hydrogen concentrations were determined with a Leco CHN analyzer which combusts the organic matter, forming CO_2 and H_2O, which are then measured with an IR detector. The nitrogen concentration was determined using a Perkin Elmer CHN analyzer, which combusts the organic matter, producing nitrogen oxides. The nitrogen oxides were reduced with copper, and measured with a thermal conductivity detector (TCD).

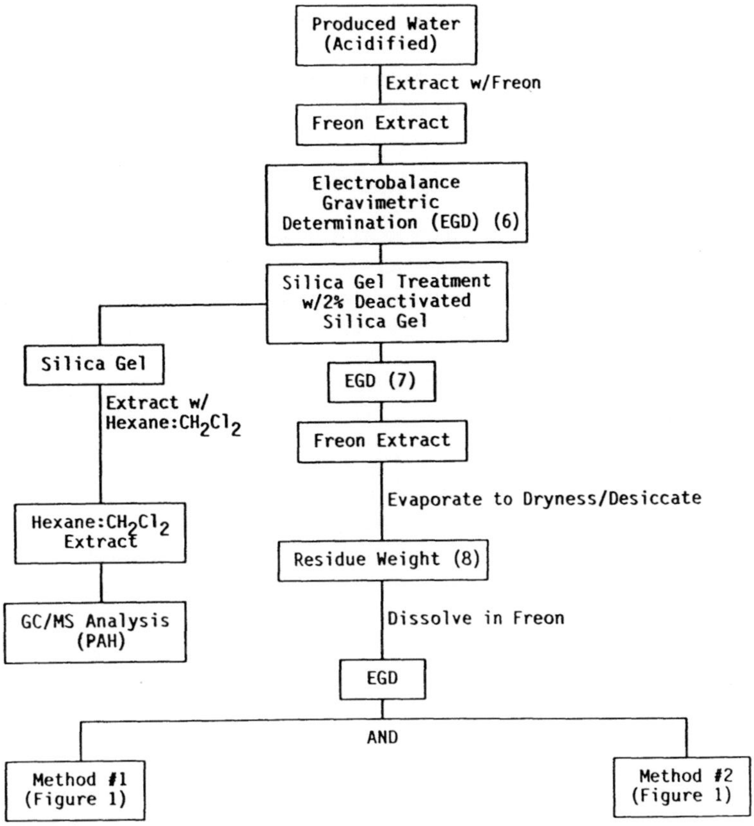

Figure 2. Produced water sample processing scheme with silica gel treatment. Numbers in parentheses refer to the extract weights summarized in Tables 1 and 2.

Method #1: Extraction/Analysis of Methyl Esters

One split from the acid extract of each produced water sample and the procedural blank were methylated to form acid methyl esters and analyzed directly by GC/MS to screen for the major fatty acids and other polar organic compounds (Figure 1).

Methylation of Extracts. The designated extract was transferred to a 25 ml vial, concentrated to 200 μl under a stream of N_2 gas, and the recovery standard d_{27}-tetra-decanoic acid was added. Two ml of BF_3/CH_3OH solution (14% w/v) was added to each sample, and the resulting solution was refluxed in a closed vial for 4 min and then cooled to room temperature. Then, 10 ml of petroleum ether and 5 ml of distilled water were added to each vial, and the solutions were shaken for 5 min. The layers were allowed to separate, and the ether layer (top layer) was decanted with a pipette and dried over 5 g of Na_2SO_4. The ether extracts were concentrated under N_2 gas to a volume suitable for injection into the gas chromatograph and were analyzed by GC/MS for fatty acid methyl esters and other major designated polar compounds.

Analysis of Methylated Fatty Acids by GC/MS. The methylated ether extracts were analyzed by full scan GC/MS using a 30 m x 0.25-mm DB-5 (J&W Scientific) fused silica capillary column. The GC oven was temperature-programmed from 40° to 290° C at 4° C/min with a 1 min initial hold and a 15 min final hold. A fatty acid standard, methylated in the same manner as the samples, was analyzed before and after the samples. An attempt was made to identify the major components (the 30 largest peaks in each sample chromatogram) in the

Method #1 sample extracts by making comparisons to National Bureau of Standards (NBS) library spectra and spectra generated by analysis of fatty acid standards. The Method #1 GC/MS analyses also provided a qualitative comparison to the subsequent analyses of the base-extractable Freon splits (Method #2) to determine if the major components were recovered throughout the sample processing scheme.

Method #2: Extraction/Analysis of Freon- and Base-Soluble Compounds

The remaining two aliquots of the Freon extract from each acidified produced water sample were combined and, with the procedural blank and the spiked blank, were brought up to a volume of 200 ml with Freon and extracted three times with 50 ml of 1.0 N sodium hydroxide (NaOH) (Figure 1). The NaOH extracts from each sample were combined in a clean separatory funnel and acidified to pH $\leq$ 1 with 6 N HCl. The acidified aqueous phase was then extracted three times with 30 ml of ethyl ether. The ether extracts were combined, dried over Na_2SO_4, and concentrated to 4 ml by K-d in a 60° C water bath. The ether extracts were then split into two equal fractions.

Characterization of the Polar Acid Fraction (Base-Soluble) Extracts by GC/MS. Analysis of fatty acids and polar compounds in extracts prepared according to Method #2 involved the identification of the major fatty acid, aromatic acid, and polar components (the 30 largest analyte peaks in each sample chromatogram). The characterization and identification of the fatty acids and other major polar compounds in the Method #2 extracts was based on computer comparisons with the NBS library spectra and spectra generated from the analysis of fatty acid standards. However, best-estimated manual identifications of groups of compounds based on key diagnostic spectral ions (e.g., 74, 87, and M-31 atomic mass units for fatty acids) were necessary to identify many analytes. The total resolved and unresolved compound concentrations for the polar fractions and the recovery of the fatty acid analytes in the spiked blank were also determined using the recovery standard d_{27}-tetradecanoic acid.

One ether extract fraction from each sample was concentrated to 200 μl under a stream of N_2 gas, and 15 μg of d_{27}-tetradecanoic acid recovery standard was added. The concentrated ether splits were then methylated according to the procedure described earlier, and analyzed by GC/MS for fatty acids and polar compounds by the same GC/MS instrumental techniques described for Method #1. Initial GC/MS analyses of the methylated polar fractions provided evidence of coelution from the GC column of many of the compounds in the extracts. Consequently, efforts to identify the individual analytes based on comparisons of mass spectra were severely hampered due to mass spectral interferences from the coeluting peaks. In an effort to reduce the coelution of analytes and obtain analyte spectra of sufficiently high purity to enable comparison to the NBS library spectra, the methylated polar fractions were reanalyzed using a 60-m x 0.25-mm DB-5 column. The GC oven was temperature-programmed from 40° to 290° C at 3° C/min, with a 1 min initial hold and a 45 min final hold. The injection port was maintained at 300° C. The modified GC/MS method enhanced the separation of coeluting analytes and enabled more accurate mass spectral interpretations and identifications.

Analysis of Phenols. The remaining ether extract splits were weighed on an electrobalance and processed through a phenol cleanup column in an attempt to remove interferences from co-extracted polar components (Figure 1). The phenol extracts were charged to a 5% deactivated silica gel/Na_2SO_4 (10 g/2 g) chromatography column (1 cm i.d.). The phenolic analytes were eluted with 22 ml of 9:1 methylene chloride:ethyl ether. After the column procedure, 15 μg of phenol recovery standard (2-isopropyl phenol) was added to each extract. The extracts were then analyzed by GC/MS for phenol and alkyl phenols on a 30 m x 0.25 mm DB-5 chromatography column. The GC oven was programmed from 40°C to 180° C at 6° C/min and from 180° C to 290° C at 10° C/min, with a 1 min initial hold and a 15 min final hold. The concentrations of the phenol and alkyl phenols were calculated by the internal standard method using the recovery standard 2-isopropyl phenol. In addition, the percent recoveries of the internal standard

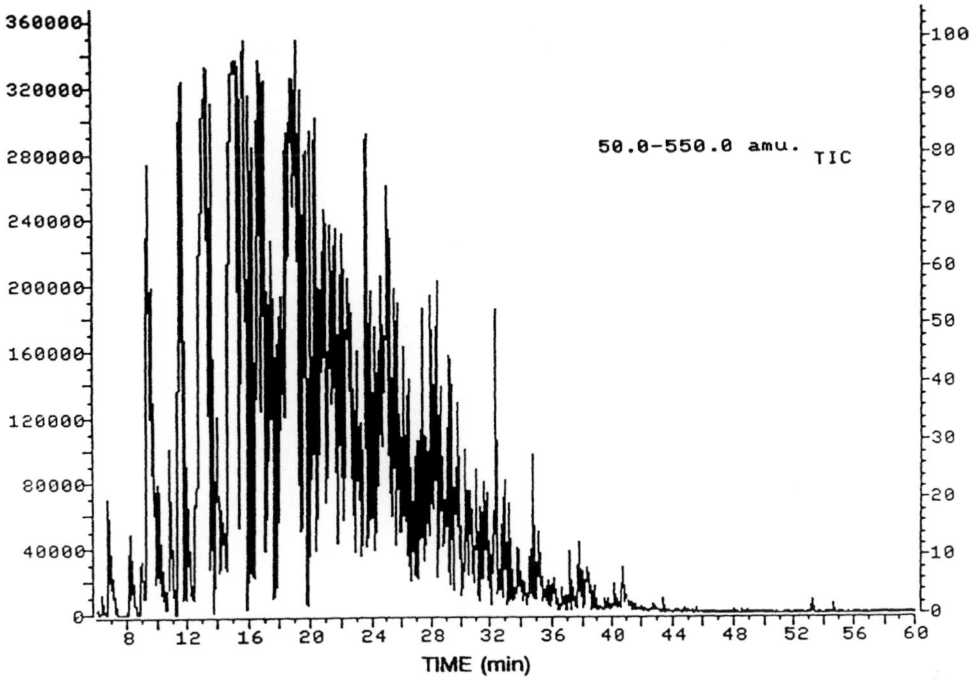

Figure 3. Total ion chromatogram (TIC) of the methylated Freon extract of produced water from Platform 3 (Method #1 analysis).

d_8-p-cresol and the analytes in the spiked blank were calculated. The presence of any phenols in the procedural blank was also determined.

Fractionation of Freon-Soluble Compounds. The base-extracted Freon extracts were concentrated to 5 ml by K-d in an 80° C water bath. The extracts were transferred to hexane under a stream of N_2 gas, and aliquots were weighed on an electrobalance to determine the extract weights, and were fractionated by silica gel column chromatography (Figure 1).

The hexane extracts were charged to a 100% activated silica gel/5% deactivated alumina (11 g/1 g) chromatography column (1 cm i.d.). The saturated hydrocarbons were eluted with 18 ml of hexane (F_1); the aromatic hydrocarbons (F_2) with 21 ml of hexane:methylene chloride (1:1); and the polar compounds (F_3) with 25 ml of methanol. An additional polar fraction (F_4) was eluted after the F_3 fraction with 30 ml of a 1 percent solution of 6N HCl in methanol (v/v). The fractions were then concentrated under a stream of N_2 gas, and aliquots were weighed on an electrobalance to determine the fraction weights.

Treatment of Extracts with Silica Gel. The Freon extract split from each sample, the procedural blank, and the spiked blank designated for silica gel treatment were weighed on an electrobalance and then diluted to 50 ml with Freon (Figure 2). To each Freon extract, 3 g of 2% deactivated silica gel was added and the resulting mixture was agitated on a shaker table for 5 min. The silica gel was allowed to settle, and the Freon was decanted into a preweighed vial. The silica gel was then rinsed with 5 ml of Freon, and the Freon was decanted and combined with the treated Freon extract.

The silica gel used to treat the Freon extracts was then extracted three times with 10 ml of hexane:methylene chloride (50:50). The hexane:methylene chloride extracts from each sample were decanted, combined, and concentrated to 250 μl under a stream of N_2 gas. Each extract

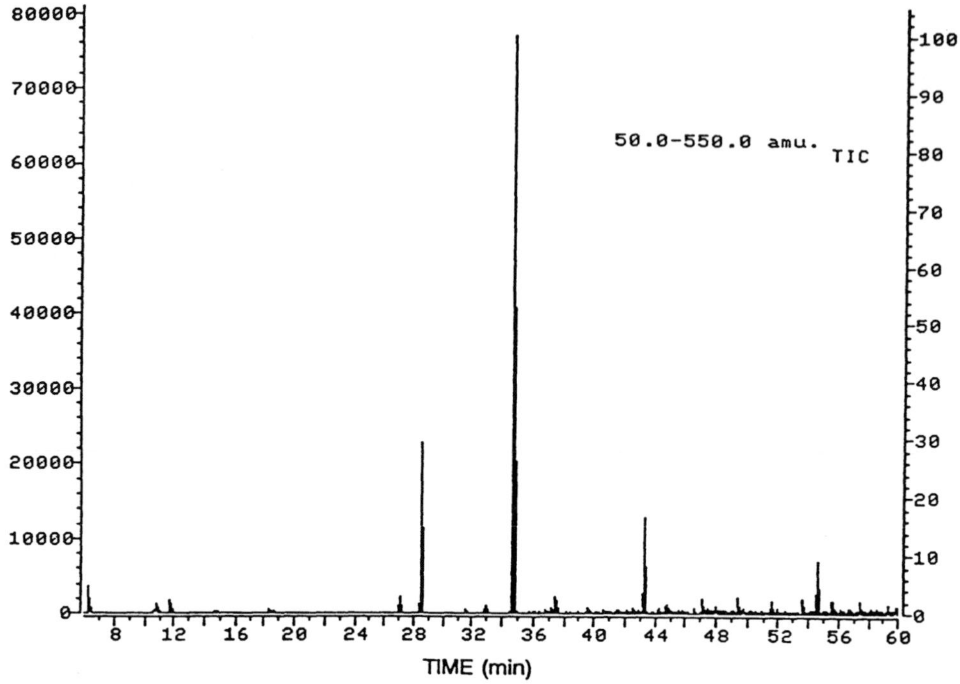

Figure 4. Total ion chromatogram (TIC) of the silica-gel treated Freon extract of produced water from Platform 3 (Method #1 analysis).

was then spiked with 10 µg of d_{12}-chrysene internal standard and analyzed for PAH by GC/MS on a 30 m x 0.25 mm DB-5 capillary column. The concentration of PAH analytes in the samples and procedural blank were calculated versus d_{12}-chrysene.

Aliquots of the silica gel-treated Freon extracts of each sample, the procedural blank, and the spiked blank were weighed on an electrobalance to calculate the extract weight, and were then concentrated to dryness under a stream of N_2 gas in a 80° C water bath. The dry extracts were placed in a desiccator for 30 min and then reweighed to determine the residue weight. The dry extracts were then redissolved in 2 ml of Freon and split into two equal fractions. One fraction was processed by Method #1 and the other fraction was processed by Method #2 (Figure 1).

TOXICITY TESTS

The acidified 3 l produced water subsamples from Platforms 1 and 2 designated for the toxicity tests, and one laboratory blank consisting of 3 l of distilled water acidified to pH $\leq$ 1 with 6 N HCl, were extracted three times with 100 ml of Freon. The three Freon extracts of each sample were combined and filtered through a glass fiber filter and Na_2SO_4. The filter and Na_2SO_4 were rinsed with 20 ml Freon and the rinses were combined with the Freon extract. The Freon extracts were then concentrated to dryness using K-d techniques and N_2 gas, in an 80° C water bath. The dry extracts were placed in a desiccator for 30 min, weighed, and then redissolved in 300 ml of Freon. The Freon was extracted three times with 50 ml of 1 N Na_2SO_4. The NaOH extracts from each sample were combined and designated as produced water base extracts for the subsequent toxicity tests.

Toxicity tests were performed on whole (unacidified) produced water samples and base extract samples from Platforms 1 and 2, and a base extract procedural blank. A reference toxicant test with sodium dodecyl sulfate (SDS) was performed concurrently with the toxicity tests for the produced water samples.

Platform 1, Unacidified Produced Water. The original salinity of this sample was 59‰ and the pH was 7.43. A 500 ml aliquot of the original 3 l sample was removed and the pH was adjusted to 8.00 with NaOH. The aliquot was then diluted 1:1 with deionized (Milli-Q) water and natural seawater to obtain a final salinity of 30‰ and a concentration of 50 percent produced water. This solution was diluted with seawater to obtain the following dilution series for testing: 50%, 25%, 12.5%, 6.2%, and 3.1% whole produced water.

Platform 2, Unacidified Produced Water. The original salinity of this sample was 126‰ and the pH was 6.32. This sample had a rust-orange color. A 250 ml aliquot of the original 3 l sample was removed and the pH was adjusted to 7.99 with NaOH. The salinity was adjusted to 31‰ and the final sample concentration was adjusted to 23% produced water by dilution with deionized water. This solution was diluted with seawater, to obtain the following dilution series for testing: 23%, 11.5%, 5.8%, 2.9%, 1.4% whole produced water.

Platform 1, Acidified Produced Water Base Extract. A 150 ml volume of base extract was diluted to a total volume of 3 l with seawater. The salinity of this sample was 29.5‰ and the pH was 10.04. A 500 ml aliquot was removed and the pH was adjusted to 8.05 with HCl. This aliquot was diluted 1:1 with seawater to obtain a final concentration equivalent to 50% produced water. This solution was diluted with seawater to obtain the following dilution series for testing: 50%, 25%, 12.5%, 6.2%, and 3.1% produced water expanded extract.

Platform 2, Acidified Produced Water Base Extract. A 150 ml volume of base extract was diluted to a total volume of 3 l with seawater. The salinity of this sample was 30‰ and the pH was 9.93. pH was adjusted to 8.06 with HCl and the sample was diluted with seawater to obtain a final salinity of 31‰ and a concentration equivalent to 23 percent produced water. This solution was diluted with seawater to obtain the following dilution series for testing: 23%, 11.5%, 5.8%, 2.9%, and 1.4% produced water expanded extract.

Procedural Blank. A 150 ml volume procedural blank was diluted to a total volume of 3 l with seawater and prepared for testing in the same manner as the acidified produced water base extracts. The salinity of this sample was 29.5‰ and the pH was 10.14. pH was adjusted to 8.08 with HCl. The solution was diluted with seawater to obtain the following dilution series for testing: 50%, 25%, 12.5%, 6.2%, and 3.1% expanded extract.

Toxicity Test Procedures. Mysids, *Mysidopsis bahia,* used for this study were obtained from cultures maintained at the toxicology facility. Mysids were held in glass aquaria at the test temperature (22° C ± 2°) and salinity (30 ± 3 ‰) prior to the tests. They were fed freshly hatched (< 24 h) brine shrimp, *Artemia salina,* nauplii at least once daily.

Test Conditions and Observations. The test conditions and procedures for the 96 hour acute mysid tests with produced water, produced water extracts, the blank, and the reference toxicant were those of Peltier and Weber (1985) and did not incorporate daily renewals. Observations of mysid survival were performed at 0, 24, 48, 72, and 96 h. The water quality parameters of temperature, salinity, dissolved oxygen, and pH were measured in the controls, low, mid, and high test solutions throughout the performance of the tests.

Toxicity data were tabulated and the median lethal concentrations (LC_{50}s) for different exposure times were calculated by using the trimmed Spearman-Karber method with Abbott's correction applied for control mortality when appropriate. Significant differences in toxicity of the samples was determined by examining the 95% confidence limits for overlap.

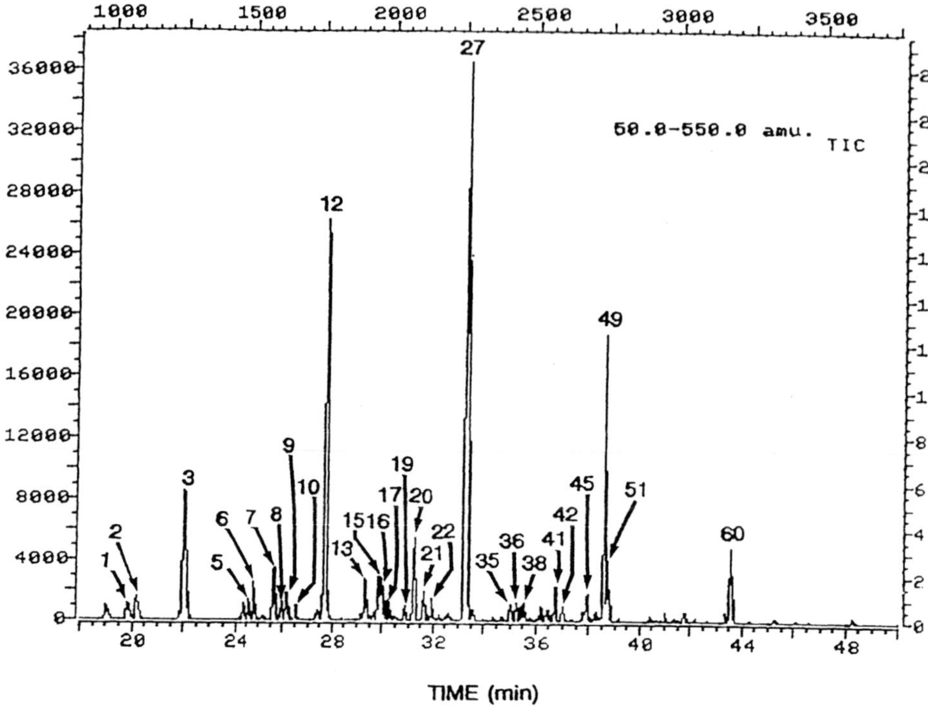

Figure 5. **Total ion chromatogram (TIC) of the methylated base-soluble fraction of the Freon extract of produced water from Platform 1 (Method #2 analysis). Peak numbers correspond to those in Table 3.**

RESULTS

CHNS

The percent composition of carbon, hydrogen, nitrogen, and sulfur in the organic matter in the produced water Freon extracts was quite consistent among the three samples. Carbon (64.4 to 73.9%) was the most abundant element, followed by hydrogen (8.33 to 10.0%). Nitrogen and sulfur combined accounted for less than 2 percent of the total organic matter in the extracts. Total recovery of the extract weights as CHNS ranged from 73.7 to 84.8%. Much of the remainder of the weight of the organic matter probably was oxygen.

Mass Balance

The weights of the silica gel-treated extracts ranged from 2 to 21% of the weights of the untreated Freon extracts (Table 1). These results indicate that a majority of the organic compounds in the Freon extracts was removed by the silica gel treatment. The residue weights of the silica gel-treated extracts, determined by the EPA gravimetric method, differed considerably from the weights determined by electrobalance. However, the electrobalance procedure is considered to be the more accurate measure, especially at the low residue weights encountered. The mass balances for the sum of the Freon-soluble fraction weights ($\Sigma F_1, F_2, F_3, F_4$) were 77, 82, and 92% for samples from Platforms 1, 2, and 3, respectively. This pattern could be related to the fact (discussed below) that the extract from Platform 1 was enriched in lower molecular weight polar compounds relative to the Platform 3 extract. The extract from Platform 2 was intermediate to Platforms 1 and 3 in polar compound molecular weight distribution.

The oil and grease weights (calculated as residue weight by the EPA gravimetric method) and the extract weights of untreated Freon extracts were similar for samples from Platforms 2 and 3 (Table 2). The oil and grease weight for the sample from Platform 1 was considerably higher than the corresponding extract weight; however, the oil and grease method appears to be subject to high variability at low weights, as indicated by a value of 18.9 for the procedural blank. The mass balances for the sum of the Freon-soluble fraction weights ranged from 33 to 61%. The F_1 and F_2 fractions accounted for more than 60% of the sum of the fraction weights in all samples. The sum of the base-soluble and Freon-soluble extract weights represented 78, 72, and 84% of the extract weights for extracts from Platforms 1, 2, and 3, respectively.

Method #1: Fatty Acid Methyl Esters

A complex mixture of fatty acids accounted for a significant portion of the total organic matter in the Freon extracts of the produced water samples. This is exemplified in Figure 3, which is a total ion chromatogram (TIC) of the methylated produced water extract from Platform 3. The TIC contains a large number of analyte peaks, as well as a large unresolved complex mixture (UCM). Extracted ion profiles (EIPs) of ion 74 (the major diagnostic ion for fatty acid methyl esters) from the chromatograms of all the Method #1 Freon extracts revealed that the major peaks identified in the TICs corresponded to fatty acid peaks in the EIPs.

Table 1. Mass balances of silica gel-treated Freon extracts of the produced water samples and the procedural blank. Numbers in parentheses refer to steps in the extraction/analysis schemes in Figures 1 and 2

	Platform			
Procedural Parameter	**1**	**2**	**3**	**Blank**
Extract Weight[a] (mg/l)				
Pre Si Gel Treatment (6)	7.28	54.7	80.2	<0.01
Post Si Gel Treatment (7)	0.26	11.4	1.44	<0.01
% Remaining after Si Gel Treatment	3.6	20.8	1.8	
Residue Weight[b] (mg/l)				
Post Si Gel Treatment (8)	3.9	<0.1	<0.1	<0.1
Extract Weight[a] (mg/l)				
Base-Soluble (3)	<0.01	<0.01	<0.01	<0.01
Freon-Soluble (4)	0.24	8.56	<0.01	<0.01
Fraction Weights[a] (mg/l) (5)				
F1	0.08	5.10	0.72	<0.01
F2	0.04	2.94	0.32	<0.01
F3	<0.01	1.04	<0.01	0.72
F4	0.08	0.32	0.28	<0.01
Total	0.20	9.4	1.32	0.72
% Si Gel-Treated Extract Recovered[c]	77%	82%	92%	—

[a] Calculated from the weight of an aliquot on an electrobalance (EGD in Figures 1 and 2).
[b] Calculated as residue weight by EPA gravimetric method.
[c] Total non-base soluble material in silica-gel-treated extract recovered after column chromatography. [Total Fraction Weight (5)/Extract Weight, Post Si Gel Treatment (7)].

A comparison of the Freon extract TICs and the base-soluble extract TICs reveals nearly identical chromatogram traces for each respective sample, indicating that the majority of polar analytes in the Freon extracts of produced water were recovered through the base-soluble sample processing scheme. The TIC of the silica gel-treated, methylated Freon extract from Platform

3 (Figure 4) was essentially devoid of peaks when compared to the untreated Freon extract (Figure 3), demonstrating that the polar analytes identified in the Freon extract are effectively removed by the silica gel treatment. This pattern of polar analyte removal by silica gel treatment was also observed in samples from Platforms 1 and 2.

Method #2: Polar Acid (Base-Soluble) Fraction

Organic Acids. The major polar analytes, other than phenols, in the polar, acid (base-soluble) fraction of the untreated Freon extracts of produced water samples from Platforms 1, 2, and 3, as determined by Method #2, are summarized in Table 3. Compound identifications for the silica gel-treated polar fractions are not provided because these samples were generally devoid of peaks, with the exception of the recovery standard and some analytes identified as contaminants from the methylation procedure. Positive identifications are provided for analytes for which NBS or fatty acid library spectra-match probabilities were greater than 80%. In a few cases, positive identifications are indicated for compounds that had spectra-match probabilities of less than 80%, but were confirmed by manual interpretations of spectra. Tentative identifications are listed for analytes with spectra-match probabilities ranging from 30 to 80% that were considered to be positive matches based on manual interpretations of spectra. Analytes for which positive or tentative identifications could not be provided were grouped into one of five classes of compounds: fatty acids, unsaturated fatty acids, cyclic fatty acids, aromatic acids, and unknowns. These identifications were based on manual reconstruction and interpretation of the analyte spectra by using key diagnostic ions: e.g., ions 74 and 57 for fatty acids, ion 74 and molecular ions for unsaturated fatty acids, ion 87 for cyclic fatty acids, and predominant molecular ions for aromatic acids.

Table 2. Mass balances of untreated Freon extracts of the produced water samples and the procedural blank. Numbers in parentheses refer to steps in the extraction/analysis schemes in Figures 1 and 2

	Platform			
Parameter	1	2	3	Procedural Blank
Oil and Grease[a] (mg/l) (1)	20.3	65.6	108.0	18.9
Extract Weight[b] (mg/l) (2)	5.48	54.2	83.0	<0.01
Extract Weight[a] (mg/l)				
Base-Soluble (3)	2.85	19.1	65.8	<0.01
Freon-Soluble (4)	1.40	20.1	3.63	<0.01
Fraction Weights[a] (mg/l) (5)				
F1	0.18	3.78	1.02	<0.01
F2	0.11	5.38	0.48	<0.01
F3	0.09	1.06	0.18	0.03
F4	0.08	2.03	<0.01	0.12
Total	0.46	12.3	1.68	0.15
% Untreated Extract Recovered[c]	33%	61%	46%	—
% Hydrocarbons[d]	5%	17%	2%	—

[a] Calculated as residue weight by EPA gravimetric method.
[b] Calculated from the weight of an aliquot on an electrobalance (EGD in Figures 1 and 2).
[c] Total non-base soluble material in untreated extract recovered after column chromatography [Total Fraction Weight (5)/Extract Weight, Freon-Soluble (4)].
[d] F1 + F2 (5)/Untreated Extract Weight (2).

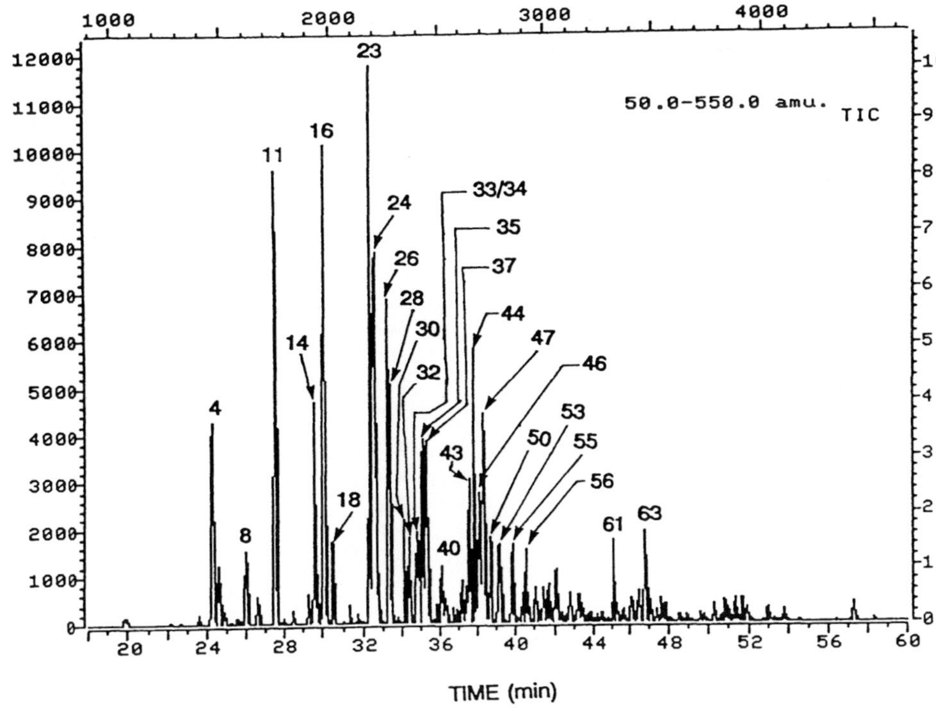

Figure 6. Total ion chromatogram (TIC) of the methylated base-soluble fraction of the Freon extract of produced water from Platform 2 (Method #2 analysis). Peak numbers correspond to those in Table 3.

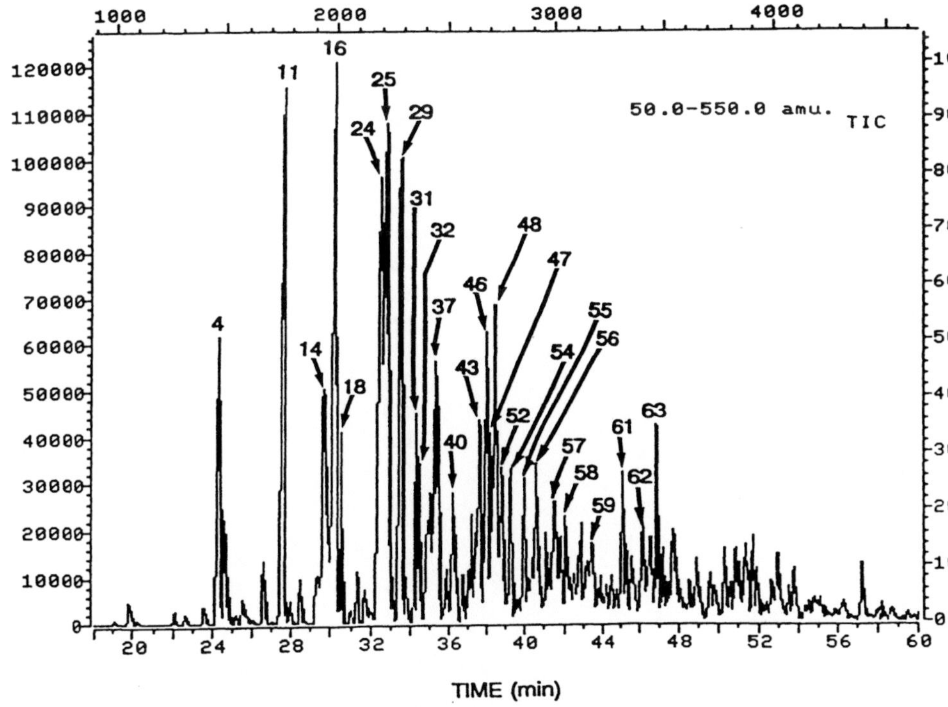

Figure 7. Total ion chromatogram (TIC) of the methylated base-soluble fraction of the Freon extract of produced water from Platform 3 (Method #2 analysis). Peak numbers correspond to those in Table 3.

Table 3. Results of mass spectra identifications of the major analytes in the polar fraction of the produced water samples

Peak No.	Identification[a]	RRI[b]	Platform 1	Platform 2	Platform 3
1	Fatty acid (FA)	561	X		
2	Pentanoic acid 4-methyl	567	X		
3	Hexanoic acid	600	X		
4	Cyclic fatty acid (CFA)	640		X	X
5	FA	645	X		
6	FA	649	X		
7	FA	664	X		
8	Aromatic acid	671	X	X	
9	FA	673	X		
10	CFA	680	X		
11	Cyclopentcarbonic acid 3-methyl	697		X	X
12	Heptanoic acid	701	X		
13	FA	730	X		
14	(6-Heptenoic acid)	735		X	X
15	FA	740	X		
16	Cyclohexanecarboxylic acid	743	X	X	X
17	FA	747	X		
18	CFA	752		X	X
19	FA	758	X		
20	FA	764	X		
21	FA	771	X		
22	(benzoic acid)	777	X		
23	FA	783		X	
24	Unsaturated fatty acid (UFA)	788		X	X
25	FA	793			X
26	(Cyclohexanecarboxylic acid 4-methyl)	800		X	
27	Octanoic acid	802	X		
28	CFA	803		X	
29	(Cyclohexanecarboxylic acid 4-methyl)	806			X
30	FA	817		X	
31	(7-Octenoic acid)	820			X
32	(6-Heptenoic acid)	823		X	X
33	UFA	829		X	
34	UFA	832		X	
35	FA	836	X	X	
36	(Hexanoic acid 2,4-dimethyl)	840	X		
37	(3-Nonanoic acid)	840		X	X
38	Unknown	843	X		
39	FA	845			
40	Unknown	856		X	X
41	FA	867	X		
42	FA	873	X		
43	(4-Octenoic acid)	884		X	X
44	UFA	888		X	
45	Benzoic acid 4-methyl	891	X		
46	FA	893		X	X
47	FA	897		X	X
48	(8-Nonanoic acid)	901			X
49	Nonanoic acid	904	X		

Table 3. Results of mass spectra identifications of the major analytes in the polar fraction of the produced water samples (Continued)

Peak			Platform		
No.	Identification[a]	RRI[b]	1	2	3
50	FA	904		X	
51	FA	907	X		
52	FA	907			X
53	UFA	913		X	
54	CFA	917			X
55	UFA	928		X	X
56	(2,8-Nonadienoic acid)	941		X	X
57	Aromatic acid	960			X
58	FA	972			X
59	Benzoic acid 3,5-dimethyl	998			X
60	Decanoic acid	1000	X		
61	CFA	1033		X	X
62	(Benzenepropanoic acid,-dimethyl)	1062			X
63	Aromatic acid	1069		X	X

[a] Analyte identifications in parentheses represent tentative identifications. Identifications without parentheses represent positive identifications. FA=fatty acid; CFA=cyclic fatty acid; UFA=unsaturated fatty acid.
[b] RRI=Relative retention index based on the analyte retention time relative to the retention time of straight chain fatty acids (C_5 fatty acid=RRI 500; C_{10} fatty acid=RRI 1000).

Figures 5, 6, and 7 show TICs of the methylated polar fractions of Freon extracts of produced water from Platforms 1, 2, and 3, respectively. The peak numbers in the figures correspond to the peak numbers listed in Table 3, allowing a qualitative assessment of the distribution and abundance of the identified peaks.

The composition of the organic acid assemblage in each of the Freon extracts was quite different for each of the three produced water samples. The extract from Platform 1 contained mostly lower molecular weight acids, from about methylpentanoic acid to benzoic acid. The extract from Platform 2 was enriched in slightly higher molecular weight acids from about heptenoic acid to nonadienoic acid. The extract from Platform 3 contained many acids between octenoic acid and dimethylbenzenepropanoic acid.

The most abundant analytes, as indicated by relative peak height, were different in each produced water sample. The most abundant polar organic compound in produced water from Platform 1 was octanoic acid. Heptanoic acid and nonanoic acid were also relatively abundant in this sample. The most abundant analyte in produced water from Platform 2 was an unidentified fatty acid (peak 23). The most abundant analyte in produced water from Platform 3 was an unidentified unsaturated fatty acid (peak 24). Cyclohexanecarboxylic acid, and 3-methyl-cyclopentanecarbonic acid were abundant in produced water from Platforms 2 and 3. 4-Methylcyclohexanecarboxylic acid was also abundant in produced water from Platform 3. Of the 63 peaks identified as major analytes in the Freon extracts of the three produced water samples, only one analyte (cyclohexanecarboxylic acid) was present in all three samples.

The concentrations, determined by GC/MS, of the total resolved and unresolved components in the untreated and silica-gel-treated Freon extracts (corresponding to residues obtained by EPA Method 413.1 and Standard Method 5520 F, respectively), and polar fractions of the respective Freon extracts, are presented in Table 4. The concentrations of total resolved and unresolved components in the untreated Freon extracts ranged from 10.8 to 39.7 mg/l. Concentrations of the polar acid fractions of untreated extracts were between 9.67 and 40.8 mg/l. The total resolved and unresolved concentrations of components of the total and polar acid fractions of silica-gel-treated Freon extracts were considerably lower, ranging from 0.18 to 5.91 mg/l.

Concentrations of total resolved and unresolved components in the procedural blanks were also low, with a maximum of 0.36 mg/l.

Phenols. Phenolic compounds were detected at trace levels in all untreated Freon extracts of produced water samples. Untreated Freon extracts from Platform 1 contained the highest concentration of individual and total phenols. The most abundant phenol in this sample was a C2-phenol at 23.2 g/l. Because the extraction method used in this study is not a recommended method for extraction of phenols from produced waters, levels of phenols found in the Freon extracts may not be representative of actual levels of phenolic compounds in produced waters. This is further supported by the low recoveries of phenols in the Freon extract of the spiked blank.

No phenolic compounds, with the exception of phenol, were detected in any of the silica gel-treated extracts of produced water. However, phenol was detected in the silica gel-treated procedural blank at a concentration greater than in any of the corresponding samples, indicating that the phenol values (in silica gel-treated samples) can be attributed to laboratory contamination.

Table 4. Concentrations (mg/l), as determined by GC/MS, of the total resolved and unresolved components in the total Freon extracts and polar fractions of untreated and silica gel-treated Freon extracts of produced water

	Platform			
Sample Type	1	2	3	Procedural Blank
Untreated Freon Extract				
Total Extract[a]	10.8	32.8	39.7	0.35
Polar Fraction[b]	9.67	22.6	40.8	0.11
Silica Gel-Treated Freon Extract				
Total Extract[c]	0.28	5.91	0.18	0.05
Polar Fraction[d]	0.63	1.08	0.66	0.36

[a] Represents total Freon extract as determined by Method #413.1 (Residue (1) in Figure 1)
[b] Represents polar acid (base-soluble) fraction of EPA Method 413.1 residue (Residue (3) in Figure 1)
[c] Represents residue as determined by Standard Method 5520F (Residue (8) in Figure 2)
[d] Represents polar acid (base-soluble) fraction of Standard Method 5520F residue (Residue (3) in Figure 1)

PAH In Extracts of Silica Gel. Polynuclear aromatic hydrocarbons (PAH) were found at trace levels in the silica gel used to treat Freon extracts of produced water. The extract of the silica gel used to treat the extract of produced water from Platform 2 had the highest concentrations of individual and total PAH. This silica gel sample contained 0.19 mg/l total resolved PAH. The PAH abundance distribution in the above sample was characteristic of a crude oil. The most abundant PAH were the alkyl naphthalenes, fluorenes, phenanthrenes, and dibenzothiophenes. The extracts of silica gel used to treat produced water samples from Platforms 1 and 3 contained 0.03 and 0.01 mg/l total PAH, respectively. These extracts also were dominated by alkyl PAH, which is characteristic of crude oil. Although PAH were removed by the silica gel, the amount removed is below the detection limit (1 mg/l) of the gravimetric method used to calculate oil and grease.

A retrospective review of the GC/MS data revealed that the three sample extracts also contained saturated hydrocarbons at concentrations below 100 ng/sample (i.e., below the gravimetric method detection limit). However, the procedural blank and the blank spike also contained a similar pattern of saturated hydrocarbons, indicating that the hydrocarbon signal most likely resulted from contamination during processing or analysis. There was no further investigation of saturated hydrocarbons in silica gel extracts.

Toxicity Tests

Water-quality parameters of temperature, salinity, dissolved oxygen concentration, and pH were all within acceptable limits during all toxicity tests. Survival of mysids in the seawater control solutions for each of the tests was ≥ 90%. The sensitivity of the mysids to the reference toxicant was similar to that reported elsewhere, indicating that the mysids were healthy at the initiation of the bioassay. Therefore, all tests were considered to be acceptable for evaluation. The median lethal concentrations at 96 hours (96-h LC_{50}) for produced water samples from Platforms 1 and 2 were 15.8 percent and 4.8 percent produced water, respectively (Table 5). The base extracts of Freon extracts of the produced water samples were not toxic to mysids at the highest concentrations evaluated, indicating that the toxicity of produced water does not reside in the polar organic compounds in the produced water. The procedural blank also was nontoxic.

Table 5. Acute toxicity of produced water and base extracts of Freon extracts of produced water to Mysids, *Mysidopsis bahia*. Results are given as percent of the original produced water sample

Sample Description	LC50	95% Confidence Limits
Platform 1 Unacidified	15.8%	11.5 - 21.6%
Platform 1 Freon Base Extract	>50%	None
Platform 2	4.8%	3.9 - 5.8%
Platform 2 Freon Base Extract	>23%	None
Procedural Blank	>50%	None
Reference Toxicant[a]	8.2 mg/l	6.9 - 9.7 mg/l

[a]Sodium dodecyl sulfate SDS

DISCUSSION

Freon extracts of produced water contain a complex mixture of petroleum hydrocarbons and nonhydrocarbon polar organic chemicals. Based on the amount of material removed by the Standard Method 5520 F silica gel treatment, 79 to 98% of the material in Freon extracts analyzed was polar (Table 3). The polar fraction includes a mixture of straight-chain, branched, unsaturated, and cyclic fatty acids, as well as smaller amounts of aromatic acids and phenols. The composition of the polar organic fraction varies from one produced water sample to another. The most abundant fatty acids in the three produced water samples analyzed in this investigation are octanoic acid, heptanoic acid, nonanoic acid, cyclohexanecarboxylic acid, 3-methylcyclopentanecarbonic acid, and unidentified saturated and unsaturated fatty acids. Because of the natural occurrence of low molecular weight fatty acids and phenols in the marine environment (Smith et al., 1983; Johns et al., 1977; Degens et al., 1964), their ease of degradation, particularly by marine bacteria (Stieb and Schink, 1985; Hill and Robinson, 1975), and their low to moderate toxicity (Cairns and Scheier, 1962; Buikema et al., 1979), the presence of these classes of compounds at low concentrations in produced water poses a minimal hazard to the receiving waters for produced water discharges.

The dominant elements in the organic matter extracted with Freon are carbon and hydrogen. Nitrogen and sulfur account for less than 2% of the organic matter in the extracts. Oxygen is probably another major constituent of the organic fraction of the Freon extracts. The amounts of individual compounds and classes of compounds in Freon extracts may not reflect actual levels in produced water, i.e., the low recovery of the phenols documented in this study indicates that the analytical procedures used are not adequate for quantitative analysis of phenols in produced water. The low recovery can be attributed to two factors: 1) the use of Freon as a solvent is inappropriate for extracting phenols from a water matrix, and 2) any phenols extracted by the Freon are subsequently lost by volatilization during the evaporation of the extracts.

Treatment of the Freon extract of produced water with silica gel effectively removes the polar compounds, and does not remove a significant amount of petroleum hydrocarbons (aromatic hydrocarbons were found to be removed at levels below the EPA Method 413.1 or Standard Method 5520 F gravimetric detection limit).

The whole produced water samples are acutely toxic to mysids. The two produced water samples evaluated have significantly different toxicities (96-h LC_{50} 4.8 to 15.8%). The more toxic of the two samples (that from Platform 2) has the higher oil and grease concentration, both before and after silica gel treatment.

Base-soluble material from produced water Freon extracts, representing the polar acid organic fraction, are not acutely toxic to mysids at any concentration examined (96h $LC_{50} > 23\%$). The results from the limited number of tests performed indicate that polar acid organic compounds in produced water do not contribute significantly to its acute toxicity. It appears that the toxicity of produced water is due largely to the petroleum hydrocarbons. This is consistent with our current knowledge about the components in crude oil that are responsible for its acute toxicity to marine organisms (Neff and Anderson, 1981).

The standard EPA gravimetric method for determining total oil and grease (EPA Method 413.1) is inappropriate for use in the regulation of petroleum hydrocarbons in produced water. Because it measures other classes of compounds in addition to petroleum hydrocarbons, it overestimates the concentration of petroleum hydrocarbons in the produced water. For example, in this study only 2 to 17% of the oil and grease in untreated Freon extracts (from EPA Method 413.1) was determined (by liquid chromatography followed by gravimetric analysis) to be hydrocarbon material (Table 2). The non-hydrocarbon constituents of produced water coextracted in the EPA Method 413.1 oil and grease analysis are fatty acids, phenols, and related compounds. They do not contribute significantly to the acute toxicity of the produced water. Therefore, a method that employs a silica gel cleanup, such as Standard Method 5520 F, should be used for measuring petroleum hydrocarbons in produced water.

Standard Method 5520 F and EPA Method 418.1, established methods for total recoverable petroleum hydrocarbons, are similar to EPA Method 413.1, currently being required for analysis of total oil and grease in produced water. They differ from the latter in that the Freon extract is cleaned up with silica gel. Total petroleum hydrocarbons in the final extract are quantified by gravimetric or infrared analysis, respectively. The methods were designed specifically for the analysis of total petroleum hydrocarbons in water (including seawater) samples. Given the use of appropriate standards (for infrared analysis), the methods produce accurate and precise results. Potential minor losses of PAH from the extract during silica gel cleanup are not a problem because, for gravimetric analysis, the amount lost is less than the method detection limit, and for infrared analysis, the wavelength used for quantification measures only aliphatic hydrocarbons. Either method would be more suitable than EPA Method 413.1 for analysis of total petroleum hydrocarbons in produced water.

ACKNOWLEDGMENTS

We wish to thank the Offshore Operators Committee, Environmental Subcommittee which funded this work as part of a produced water characterization study. We would also like to thank Edith Hutchinson who performed the chemical fractionations, and John Williams who conducted the toxicity tests.

REFERENCES

Brown, J.S., Wade, M.J., and Sauer, T.C., 1988, "Produced Water Chemical Characterization Study. Report to the Offshore Operators Committee, Environmental Subcommittee," Houston, TX. 15 pp.

Buikema, A.L., Niederlehner, B.R., and J. Cairns, Jr., 1981 The effects of a simulated refinery effluent and its components on the estuarine crustacean, *Mysidopsis bahia*. Arch. Environ. Contam. Toxicol. 10: 231-240.

Cairns, J., Jr., and Scheier, A. 1962, The effects of temperature and water hardness upon the toxicity of naphthenic acids to the common bluegill sunfish, *Lepomis macrochirus*, Raf., and the pond snail, *Physa heterostropha* Say. Novae Naturae. Acad. Nat. Sci. Philadelphia. No. 353: 1-12.

Degens, E.T., Reuter, J.H., and Shaw, K.N.F., 1964, Biochemical compounds in offshore California sediment and seawater. Geochim. Cosmochim. Acta 28: 45-66.

EPA (U.S. Environmental Protection Agency), 1979, Methods for cHemical Analysis of Water and Wastes. Environmental Monitoring and Support Laboratory, Environmental Research Center, Cincinnati, OH 45268, pp. 413.2-1 through 413.2-3.

EPA (U.S. Environmental Protection Agency), 1985, "Development Document for Effluent Limitations Guidelines and Standards for the Offshore Segment of the Oil and Gas Extraction Point Source Category." Industrial Technology Division, EPA, Washington, D.C. EPA 440/1-85/056. 408 pp.

Hill, G.Q., and Robinson, C.W., 1975. Substrate inhibition kinetics: phenol degradation by *Pseudomonas putida*. Biotechnol. Bioeng. 17: 1599-1615.

Jackson, G.F., Hume, E., Wade, M.J., and Kirsch, M. 1981, "Oil Content in Produced Brine on Ten Louisiana Production Platforms." Report to the U.S. Environmental Protection Agency, Municipal Environmental Research Laboratory, Cincinnati, OH. EPA-600/2-81-209.

Johns, R.B., Perry, G.J., and Jackson, K.S., 1977, Contribution of bacterial lipids to recent marine sediments. Estuar. Cstl. Mar. Sci. 5: 521529.

Neff, J.M., Sauer, T.C., and Maciolek, N., 1989, "Fate and Effects of Produced Water Discharges in Nearshore Marine Waters". Final Report submitted to the American Petroleum Institute, Washington, D.C. Battelle Ocean Sciences, Duxbury MA. 300 pp.

Peltier, W., and Weber, C.I., 1985, "Methods for Measuring the Acute Toxicity of Effluents to Freshwater and Marine Organisms". U.S. Environmental Protection Agency, Office of Research and Development, Cincinnati, OH. EPA 600/4-85/013. (NTIS No. PB 85-205383). 216 pp.

Smith, D.J., Eglinton, G., and Morris, R.J., 1983, The lipid chemistry of an interfacial sediment from the Peru continental shelf: Fatty acids, alcohols, aliphatic ketones and hydrocarbons. Geochim. Cosmochim. Acta 47: 2225-2232.

Stieb, M., and Schink, B., 1985, Anaerobic oxidation of fatty acids by *Clostridium bryantii* sp. nov., a sporeforming, obligately syntrophic bacterium. Arch. Microbiol. 140: 387-390.

SPECIFIC TOXIC ORGANICS IN PRODUCED WATERS FROM IN-SITU HEAVY OIL RECOVERY OPERATIONS IN WESTERN CANADA

J. Carey[1], A. Zaidi[2], and J. Ribo[3]

[1] Environment Canada
National Water Research Institute
Burlington, Ontario
Canada

[2] Wastewater Technology Centre
Burlington, Ontario
Canada

[3] Farma-lepori s.a.
Tarragona
Spain

INTRODUCTION

The work described here follows a series of four preliminary studies on produced water characterization. The first study (Zenon Environmental Inc., 1983) involved the analysis of two samples of produced water from two operating sites in Alberta. Samples from both sites were extracted with dichloromethane, then analyzed and quantified using gas chromatography/mass spectrophotometry (GC/MS). Using a sample from the first site, the recoveries of two adsorbents and an ion exchange resin were also compared. The second study (Rodgers, 1987) followed up on this work and involved the development of a rapid, general characterization scheme based on adsorbent resins. The organic compounds were adsorbed on the resin and the resulting fractions analyzed using a Total Organic Carbon (TOC) Analyzer, the Microtox toxicity assay, Gel Permeation Chromatography (GPC) and High Pressure Liquid Chromatography (HPLC). Following the first two preliminary studies, the methods published for specific organic analysis were critically reviewed (Ribo, 1987a) and a characterization scheme based on solvent extraction was recommended (Ribo, 1987b).

During this study, the characterization scheme was first experimentally validated on one sample of produced water from an in-situ heavy oil recovery site in Western Canada. Subsequently, the same scheme was used to analyze six more samples from four different in-situ heavy oil recovery sites in Western Canada. Two of the samples were extracted in the field within hours of being taken to determine whether the organic content of the sample changed during storage. The other samples were stored at 4°C for a maximum of 10 days before analysis.

The characterization scheme involved filtration of the produced water sample followed by sequential extraction using dichloromethane (DCM) and diethyl ether (EE) under acidic ($1 < pH < 2$) and basic ($pH > 11$) conditions. This was followed by adsorption and recovery of the

Produced Water, Edited by J.P. Ray
and F.R. Engelhart, Plenum Press, New York, 1992

residual dissolved organics using a macroreticular resin. Aliquots of the aqueous phase from various extraction steps were prepared and analyzed for TOC and Microtox toxicity. Aliquots of the various fractions of organic phases were evaporated to obtain masses of the total organics in each fraction. Also, these fractions were analyzed using a TOC analyzer, GC/FID, and GC/MS to obtain information on specific organics present in each fraction.

Detailed results from the validation of the characterization scheme and the analysis of six additional samples were presented in two reports (Ribo, 1987c,d). This paper presents a summary of the procedures and results for the study.

EXPERIMENTAL

Extraction

The extraction procedure used in this study is shown schematically in Figure 1. A one litre sample of produced water was passed through a glass fibre filter (GFC - 1.2 µm) to separate the solids from the filtrate. An aliquot of the aqueous filtrate was submitted for TOC analysis.

The suspended solids (SS) collected on the glass fibre filter were washed sequentially with 200 ml portions of dichloromethane (DCM) and diethyl ether (EE). The two washings were kept separate and the solvent was eliminated from them by evaporation under vacuum in a rotary evaporator. This procedure yielded three fractions: (i) DCM soluble suspended solids (SS/DCM), (ii) ethyl ether soluble suspended solids (SS/EE), and (iii) insoluble suspended solids (SS/Insoluble).

The aqueous filtrate was acidified (< pH 2) by addition of 10 ml of concentrated hydrochloric acid (HCl). At this point a precipitate formed which was removed using a GFC filter. The precipitate was washed using DCM and EE to prepare three fractions (Acid Precip/DCM; Acid Precip/EE; Acid Precip/Insoluble) as described earlier. The precipitate free and acidified aqueous filtrate was then extracted with two 50 ml portions of DCM and five 200 ml portions of EE. These DCM and EE extracts were added to the Acid Precip/DCM and Acid Precip/EE fraction respectively to produce acid plus neutral DCM and EE extractables (A+N/DCM and A+N/EE). An aliquot of the aqueous phase after DCM and EE extraction was subjected to vacuum evaporation to remove any residual solvent. After evaporation, the volume of the aliquot was adjusted to its initial volume by adding sufficient HPLC grade water. This solvent free aliquot was then used for TOC and toxicity tests as described elsewhere.

The acidified aqueous filtrate after DCM and EE extraction was treated with solid NaOH (6.6 g) to obtain a pH greater than 11. At this stage a precipitate formed which was removed on a GFC filter and washed with DCM and EE as described earlier for other solids. The washing of the base precipitate yielded three fractions, (i) Base Precip/DCM, (ii) Base Precip/EE and (iii) Base Precip/Insoluble. After the removal of basic precipitates, the filtrate was extracted sequentially with two 200 mL portions of DCM and EE. The solvent was removed from an aliquot of the extracted aqueous phase by vacuum evaporation to prepare a sample for TOC and Microtox tests.

The basic aqueous phase after DCM and EE extraction was then passed through a column (20 cm x 20 cm i.d.) packed with a macroreticular resin (XAD-4 by Rhom & Haas). The column was then eluted with 400 ml of methanol (10 ml/min) to recover the adsorbed compounds. The aqueous effluent from the XAD-4 column was acidified (pH <2), and passed through the column again. The adsorbed organic compounds were eluted with 400 ml of methanol. As a result, the XAD/Basic and XAD/Acid fractions were obtained. Finally, a 100 ml sub-sample of the acidic aqueous effluent from the XAD column was neutralized and prepared for TOC and Microtox analysis following the solvent removal procedure described earlier.

The mass of organics recovered in various fractions of DCM and EE were obtained by evaporating the solvents under vacuum (rotavap).

Another component of the characterization scheme involved the fractionation of the A+N fraction to separate the acidic from the neutral components and to derivatize the phenolics and carboxylic acids, making them more amenable to GC analysis. An aliquot of the A+N fraction

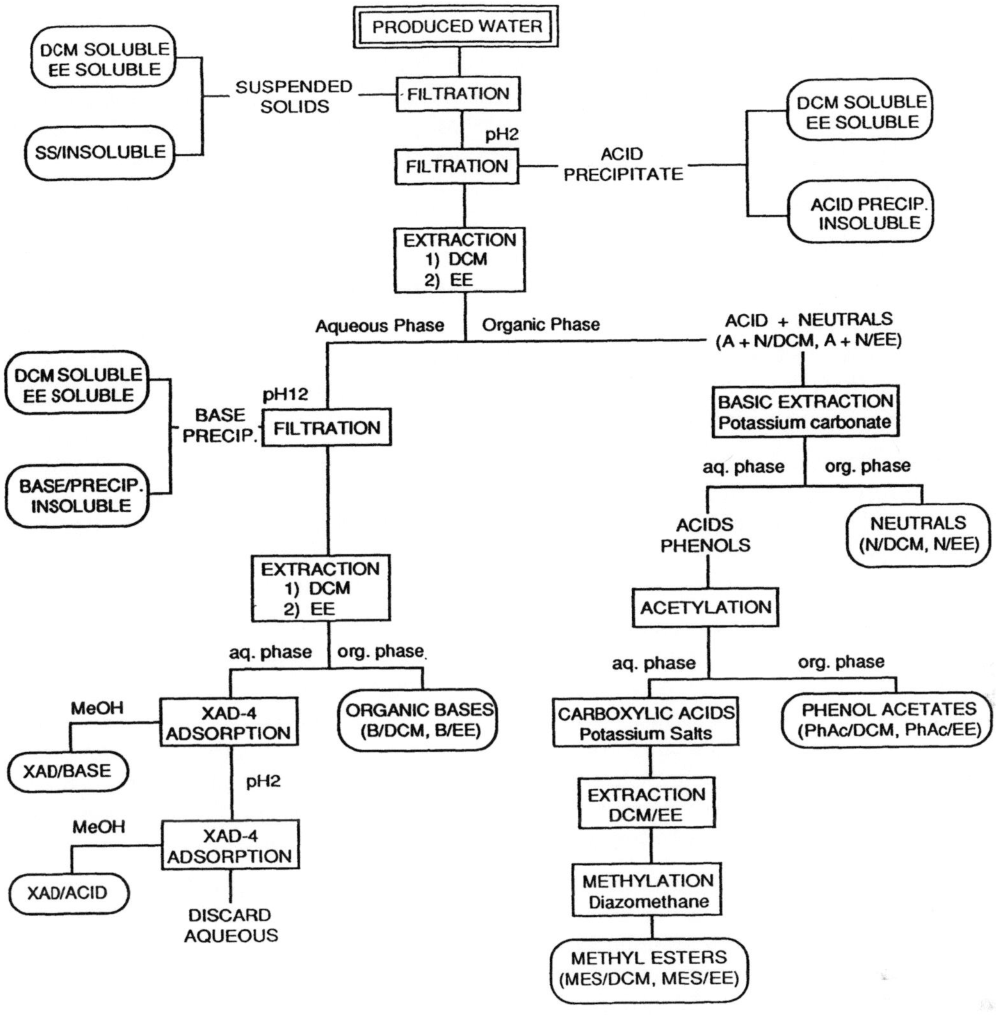

Figure 1. Produced water characterization scheme.

was redissolved in 150 ml of EE and the acidic component extracted with a 0.1 M solution of potassium carbonate (2 x 50 ml each). The organic phase containing the neutral compounds (N/DCM and N/EE), was evaporated to dryness and weighed. The aqueous phase from this step contained the carboxylic acid and phenolic compounds. To separate them, 150 ml of EE was added to this fraction which was then treated with an excess of acetic anhydride. The solution was shaken for two hours at room temperature, the organic phase was then separated and the solvent evaporated to yield acetate derivatives (PhAc/DCM and PhAc/EE) of the phenolics present in each fraction. The remaining aqueous phase was then acidified (pH <2)) with concentrated HCl and the carboxylic acids extracted with both DCM and EE until the organic phase was colourless. Both extracts were concentrated down by vacuum evaporation to 4 ml volumes and treated with an excess of freshly prepared diazomethane in EE solution to form the methyl esters of the carboxylic acids. The reaction mixtures were allowed to react overnight. The solvent was removed by evaporation under a gentle stream of dry nitrogen, yielding the fractions containing methyl esters of the carboxylic acids (MES/DCM and MES/EE).

Quality Assurance

For the first sample, procedural blanks of all fractions were obtained by repeating the extraction procedure using one litre of organic-free water as the sample. Due to time constraints, no new blanks or spiked samples were analyzed during the work on the additional six samples. Checks of solvent prints were performed for all fractions using organic-free water.

Analysis of the Extracts

The following analyses and measurements were performed on each fraction as a part of the characterization scheme: mass, TOC content, Microtox toxicity and specific organic compounds.

Mass Balance. The amount of material in each fraction was determined gravimetrically after the fraction had been extracted, water removed by filtration through anhydrous sodium sulphate and the solvent removed by vacuum distillation followed by evaporation to dryness using a gentle nitrogen stream.

Total Organic Carbon Content. TOC content was measured using a Beckman Total Carbon Infrared Analyzer. Prior to TOC analysis, the extracts were reconstituted with organic free water and acidified with concentrated HCl to eliminate inorganic carbonates. The TOC of the filtered produced water before and after the various extraction procedures, as well as after the XAD treatment, was also measured after removing the solvent as described earlier.

Microtox Toxicity Measurements. Aliquots of the different fractions were dissolved in a 10% methanol/distilled water solution at neutral pH and their toxicity measured in the Beckman Microtox Toxicity Analyzer, which uses luminescent bacteria (*Photobacterium phosphorum*) as the test organism. The toxicity of the filtered produced water before and after the extraction procedure, as well as after the XAD columns, was also measured after adjusting the pH to 7 and removing the solvent through vacuum evaporation. Results are reported as the 15 minute EC_{50} values which represent the percentage of the original sample which causes a 50% decrease in light emitted by the bacteria after 15 minutes exposure.

GC/MS and GC/FID Determination of Individual Compounds. The organic extracts were first analyzed by GC using flame ionization (GC/FID) to confirm their suitability for GC/MS analysis. Specific compounds were tentatively identified by GC/MS, although no confirmations were carried out. Individual compounds were identified on the basis of their mass spectra and the assignment of each spectrum by computerized peak matching. Known concentrations of the organic extracts were analyzed by GC/FID. The FID chromatogram and the MS total ion chromatogram were then compared, the peaks were matched according to their retention time and, using an average response calculation from the FID chromatogram, quantitative values were assigned to the identified peaks. Instrumental information and the conditions used in the GC/MS and GC/FID procedure are presented in Table 1.

RESULTS AND DISCUSSION

The results from the initial evaluation of the analytical scheme using one produced water sample will be presented in detail. The database of information subsequently generated on the six additional samples from four enhanced oil recovery sites will also be presented and any notable or anomalous results will be briefly discussed.

Mass and TOC Balance

After filtration of a 1 l sample of the produced water, 190 mg of suspended solids was collected on the filter. Of this material, 47.4% was soluble in DCM, 10.5% was soluble in EE and 42.1% remained insoluble after being rinsed with both DCM and EE.

The extraction of the acid filtrate yielded 37 mg/l of DCM extractables and 114 mg/l of EE extractables. Extraction of the base filtrate yielded only 5 mg/l of residue; 1 mg/l DCM extractables and 4 mg/l EE extractables. After adsorption on XAD-4 resin and elution from the column with methanol, a total of 33 mg/l of organics was recovered from the acid filtrate, and 6 mg/l was recovered from the basic filtrate. Most of the soluble organics (77.5%) were recovered in the A+N/DCM and A+N/EE fractions with a higher proportion extracted by EE (58.5%) than by DCM (19.0%).

Table 1. Instrumental conditions for GC analysis

Equipment and conditions used for test sample and samples 1(a) 2(a) 3 and 4		
GC / FID - Varian Gas Chromatograph Model 3700		
Column:	SE 54 30 m x 0.25 mm i.d.	
Carrier Gas:	Hydrogen	Flow Rate: 2 ml / min
Injection:	Temp:	240° C hold 5 min
	Mode:	Split 10 / 1
	Volume:	1 μl
Oven Temp:	Initial:	70° C
	Program Rate:	4° C / min
	Final:	275° C hold 15 min
Detector:	270° C	
GC / MS - Carlo Erba Fractovap Series 4160 Gas Chromatograph		
Column:	DB5 wide bore 30 m x 0.3 mm i.d.	
Carrier Gas:	Helium	Flow Rate: 15 ml / min
Injection:	Temp:	240°C
	Volume:	1 μl
Oven Temp:	Initial:	70° C hold 2 min
	Program Rate:	5° C / min
	Final:	270° C hold 10 min
Mass Spectrometry:	Ribermag R 10-10 C Quadrupole MS	
Interface:	Direct Coupling	
Mode:	Electron Impact	
Electron Energy:	70 eV	
Scan Range:	50 - 500 u.a.m.	
Equipment and Conditions Used for Samples 1(b) and 2(b)		
GC / FID - Hewlett Packard Model HP5890		
Column:	DB5 Wide Bore 30 m x 0.3 mm i.d.	
Carrier Gas:	Helium	
Oven Temp:	Initial:	70° C
	Program Rate:	5° / min
	Final:	270° C
GC/MS - Finnigan Model 4000 GC/MS		
Column:	DB1	60 m x 0.32 mm i.d.
Carrier Gas:	Helium	
Oven Temp:	Initial:	70° C
	Program Rate:	5° C min
	Final:	270° C

In summary, a total of 195 mg/l of organics was extracted from the filtered produced water sample using DCM/EE/XAD. The DCM/EE soluble fraction of the suspended solids amounted to 110 mg/l giving a total of 305 mg/l of organics removed from the test sample. A total of 395 mg/l of insoluble solids were measured in the original filtrate residue and in the acid and base precipitates from the sample filtrate. The largest portion of the insoluble solids was associated with the basic fraction (300 mg/l), with only 15 mg/l of insoluble acid precipitate. The filtrate residue contained 80 mg/l of insoluble solids.

TOC was measured on the filtered sample of produced water and on the effluent after the solvent extraction and XAD adsorption of the organic components. The initial TOC concentration was 170 mg/l and after the solvent and XAD extractions, 75 mg/l of organic carbon remained in the water. This indicated a removal of approximately 56% of the organic carbon using the described procedure.

For each of the fractions obtained, the TOC content was also measured, as shown in Table 2. The total TOC measured in the DCM and EE extracts using this approach was approximately 71 mg/l. This compared favourably with 73 mg/l extracted and analyzed by TOC analysis on the aqueous phase. However, the TOC values for the XAD step did not compare. By difference, it was determined that only part of the organics adsorbed on the XAD resin were eluted, with some 9% of the original TOC irreversibly adsorbed. The percentage of organics extracted from the aqueous matrix by the solvents and the XAD resin and recovered for analysis was therefore only 46% (measured as TOC of the original sample).

Table 2. TOC content of fractions from test sample

Fraction/Extract	TOC (mg/l)	% of recovered TOC	% of original TOC
A + N/DCM	19	24.4	11.2
A + N/EE	50	64.1	29.4
Total A + N	69	88.5	40.6
B/DCM	<1	1.3	0.6
B/EE	<1	1.3	0.6
Total Base	<2	2.6	1.2
XAD/BASE	2	2.6	1.2
XAD/ACID	5	6.4	2.9
Total XAD	7	9.0	4.1
TOC Recovered (approx)	78	100	45.9
Retained ON XAD	15		8.8

The results from the separation of the A+N fraction into its neutral, phenolic and carboxylic acid components are presented in Table 3. The methyl ester fraction contributed substantially to the A+N fraction and represented 29% of the DCM fraction and 40% of the EE fraction. This would indicate a high concentration of compounds with acidic functional groups in the sample.

Table 3. A + N fractionation results from test sample

Fraction/Extract	mg/l	% of Corresponding A+N Fraction
A + N / DCM	37	100
A + N / EE	114	100
N/DCM	7	20
N/EE	13	11
PhAc/DCM	4	11
PhAc/EE	5	4
MES/DCM	11	29
MES/EE	46	40

The mass balance data subsequently generated for the six additional produced water samples are presented in Tables 4 to 6. The level of total suspended solids present in the samples varied from 92 to 926 mg/l, as shown in Table 4. The sample from site 3 and the fresh sample

(b) from site 2 had only minor concentrations (<40 mg/l) of soluble suspended solids in the raw filtered sample, while the site four sample contained a large amount of bitumen-like material (718.3 mg/l). In all cases, most of the soluble material was extracted with DCM and only a small amount was subsequently extracted with EE. The samples that were extracted in the field (1(b) and 2(b)) showed lower levels of DCM soluble suspended solids as compared to the corresponding samples that were allowed to age before filtration and extraction.

Table 4. S uspended solid fraction of produced water samples (mg/l)

Fraction/Extract	Sample					
	1(a)	1(b)	2(a)	2(b)	3	4
SS/DCM	209.1	78.4	87.5	25.5	35.5	711.9
SS/EE	2.8	3.6	2.0	5.8	3.8	6.7
SS/Insoluble	33.0	83.8	122.6	60.8	208.0	208.1
Total Suspended Solids	244.9	165.8	212.1	92.1	247.3	926.7

Table 5 describes the soluble organics obtained from the six produced water samples. As with the test sample, most of the organics extracted from the filtered samples were recovered in the A+N fraction (85.8 to 99.4%). DCM extracted the largest proportion of the A+N fraction in the samples 1(a), 1(b), and 4, while EE extracted more than DCM from the other samples.

Table 5. Soluble organics recovered from produced water samples (mg/l)

Fraction/Extract	Sample					
	1(a)	1(b)	2(a)	2(b)	3	4
Acid Precip/DCM	0.2	2.3	2.4	1.6	0.8	1.4
Acid Precip/EE	2.6	7.7	9.8	4.9	1.3	2.2
Acid Precip/Insoluble	30.5	121.8	16.8	32.9	25.4	175.7
Total DCM/EE Soluble	2.8	10.0	12.2	6.5	2.1	3.6
Base Precip/DCM	<0.1	<0.1	<0.5	0.4	<0.1	0.6
Base Precip/EE	<0.1	2.6	1.3	15.9	<0.1	0.7
Base Precip/Insoluble	975.0	1106.0	712.2	877.2	1917.0	5808.7
Total DCM/EE Soluble	<0.2	2.7	1.8	16.3	<0.2	1.3
A + N/DCM	42.0	48.1	55.4	43.7	21.1	20.2
A + N/EE	24.1	34.2	94.1	47.5	28.6	5.1
Total A + N	66.3	82.3	149.5	91.2	49.7	25.3
B/DCM	0.8	1.1	0.7	1.8	2.8	2.0
B/EE	0.7	2.0	0.2	0.6	1.0	2.2
Total Bases	1.5	3.1	0.9	2.4	3.9	4.2
XAD/BASE	2.3		26.0		1.2	3.8
XAD/ACID	5.6		11.8		3.8	4.0
Total XAD	7.9		37.8		5.0	7.8
Total	78.7	98.1	202.2	116.4	60.9	42.2

Table 6 presents a summary of organics and insoluble solids in the six samples analyzed. The total organics recovered ranged from 100.2 to 766.4 mg/l. Lower values were measured with the fresh samples (b) than with the corresponding aged samples (a). The total insoluble solids measured in the six samples ranged from 851.6 to over 6,000 mg/l. In this case, no trend was observed between the fresh and aged samples obtained from sites 1 and 2.

Table 6. Organics and insoluble solids in the produced water samples (mg/l)

Fraction/Extract	Sample					
	1(a)	1(b)	2(a)	2(b)	3	4
Extracted Organics	78.7	98.1	202.2	116.4	60.9	42.2
DCM/EE soluble solids	211.9	82.0	89.5	31.3	39.3	718.1
Total Organics Removed	290.6	180.1	291.7	147.7	100.2	760.4
Insoluble suspended solids	33.0	83.8	122.6	60.8	208.0	208.1
Insoluble acid precip	30.5	121.8	16.8	32.9	25.4	175.7
Insoluble basic precip	975.0	1106.0	712.2	877.2	1917.0	5808.7
Total Insoluble Solids	1038.5	1311.6	851.6	970.9	2150.4	6192.5

TOC was measured on the six samples after filtration, after extraction with the two solvents and after XAD adsorption. TOC was also measured on each fraction obtained from the various samples. The results of these analyses are presented in Table 7. The TOC concentration in the raw filtered samples ranged from 87 mg/l for site 3 to 181 mg/l for site 1(b). Because of the irreversibly adsorption of some organics on the XAD resin, observed with the test sample, a TOC balance was carried out on the solvent extract fractions only. TOC values before and after extraction indicate that between 14 and 66% of the original TOC was recovered by solvent extraction. However, these values do not agree with the total TOC values measured in the DCM and EE extracts. The two sets of numbers differ by as much as 350% (for the site four sample). Part of the discrepancy may be due to the loss of organics from the post-extraction aqueous samples as they underwent vacuum distillation to remove the solvents present in the water. The samples with high post-extraction TOC values probably had a solvent residue in the analyzed solution.

Results of the fractionation of the A+N phases from the six samples into their three major components are presented in Table 8. The acetylatable compounds and the methyl esters are seen to contribute significantly to the total A+N fraction. Thus a large fraction of the dissolved organics in these produced water samples is made up of compounds with acidic functional groups such as phenol and carboxylic acids.

Microtox Toxicity

The Microtox test has proven to be a useful bioassay to determine the toxicity, not only of the whole sample, but of each of the fractional parts. In a recent review (Munkittrick et al., 1991) of Microtox and three acute lethality bioassays (namely, rainbow trout, fathead minnow, and daphnid) for determining toxicity of industrial effluents, Microtox was found to be more sensitive for pure individual organics, although somewhat less sensitive to most inorganics, and was recommended as the best available technique for rapid assessment of toxicity in environmental samples.

The 15 min EC_{50} value is defined as the effective concentration of the sample that causes a 50% reduction in the light emitted by the test organism after 15 minutes of exposure. High toxicity would therefore be indicated by low EC_{50} values. Table 9 shows the EC_{50} values for the filtered, extracted and post-XAD test sample in terms of the percentage of the original sample. In addition, the toxicity of the various extracted fractions was determined in terms of mg/l after reconstitution with water/methanol, as described in the experimental part. The toxicity of the final effluent after extraction (80%) was comparable to the final effluent from the blank run (70%). Thus, the proposed procedure effectively removed the components toxic to photobacterium in the filtered sample of produced water. Useful information may be obtained from the measurement of the individual extract's toxicity. The fractions containing acidic compounds (A+N fractions) showed high toxicity with 15 min EC_{50} values less than 10 mg/l. This would indicate a high concentration of toxic phenolic derivatives in the sample. In all cases, the fractions containing DCM extractable components were found to be more toxic than the corresponding EE fractions. It is likely that the DCM extractable compounds would be less polar, more lipophilic and perhaps more available to the test organisms.

The toxicity test results are presented in the form of rank values (total weight of the fractions divided by the EC_{50} in mg/l) in Table 10. This method allows the toxicity of each fraction to be compared on a concentration adjusted basis. Otherwise, a highly toxic fraction representing only a small portion of the original sample could be mistaken as the source of the sample's toxicity. Thus, the larger the ranking the greater the proportion of the sample's toxicity that the fraction contributes. It is important to note that Microtox test results represent the combined toxicity of all components (organics and inorganics) present in a sample, and do not consider synergistic and antagonistic effects of various components. For a first approximation, it may be assumed that the most toxic compounds in the complex mixture would make the highest contribution to the overall toxicity assuming the sample contained equal concentrations of the various components.

Table 11 presents the EC_{50} data for the six produced water samples analyzed. Five were quite toxic, exhibiting toxicity at dilutions as low as 3.4%. The site 3 sample was the exception

Table 7. TOC balance data for produced water samples (mg/l TOC).

Fraction/Extract	Sample					
	1(a)	1(b)	2(a)	2(b)	3	4
Filtered Sample	110.0	181.0	122.0	145.0	87.0	105.0
After Extraction	95.0	62.0	83.0	96.0	67.0	50.0
After XAD Column	71.0	—	30.0	—	142.2	13.0
TOC Removed by Extraction % of Original TOC	13.6%	65.7%	32.0%	33.8%	23.0%	52.4%
A + N/DCM	17.9	23.1	28.3	21.1	10.4	8.3
A + N/EE	14.3	23.6	48.0	36.3	10.0	2.7
Total A + N	32.2	46.7	76.3	57.4	20.4	11.0
B/DCM	0.34	0.50	0.27	0.60	2.2	1.2
B/EE	0.10	0.35	0.07	0.15	0.23	0.43
Total Bases	0.44	0.85	0.34	0.75	2.43	1.63
XAD/Base	0.60	—	—	—	0.38	0.93
XAD/Acid	3.80	—	4.50	—	1.90	2.0
Total XAD	4.40	—	4.50	—	2.28	3.0
N/DCM	5.1	11.1	7.9	17.2	2.5	2.0
N/EE	1.3	3.1	5.1	3.6	0.87	0.57
Total Neutrals	6.4	14.2	13.0	20.8	3.37	2.57
PhAc/DCM	—	5.8	4.7	3.2	2.0	1.5
PhAc/EE	—	6.6	16.2	10.3	3.8	1.1
Total PhAc	—	12.4	20.9	13.5	5.8	2.6
MES/DCM	—	17.1	3.5	15.9	3.8	—
MES/EE	—	15.1	7.9	14.4	5.6	—
Total MES	—	32.2	11.4	30.3	9.4	—
TOC Removed by Extraction						
Sum of Fractions (mg/l)	32.6	47.6	76.6	58.2	22.8	12.6
% of Original TOC	29.6%	26.3%	62.8%	40.1%	26.2%	12.0%

Table 8. A + N fractionation results for produced water samples (values in mg/l)

Fraction/Extract	Sample					
	1(a)	1(b)	2(a)	2(b)	3	4
A+N/DCM	42.0	48.1	55.4	43.7	21.1	20.2
A+N/EE	24.1	34.2	94.1	47.5	28.6	5.1
Total A+N	66.3	82.3	149.5	91.2	49.7	25.3
N/DCM	7.5	16.3	16.1	19.6	4.4	7.1
N/EE	2.5	10.2	17.3	5.8	3.1	5.5
Total Neutrals	10.0	26.5	33.4	25.4	7.5	12.6
PhAc/DCM	14.2	8.6	20.5	6.1	3.4	5.0
PhAc/EE	3.3	9.7	47.7	8.2	6.8	7.5
Total PhAc	17.5	18.3	68.2	14.3	10.2	12.5
MES/DCM	10.6	18.5	7.5	20.2	12.5	5.3
MES/EE	13.1	15.0	21.9	23.2	18.7	6.9
Total MES	23.7	33.5	29.4	43.4	31.2	12.2

Table 9. Toxicity of test sample based on the microtox test

Sample	15 min EC_{50} (%)
Filtered Sample	5
After Extraction	60
After XAD Column	80
Blank	70
Fractionation (mg/l)	
A + N/DCM	1.6
A + N/EE	6.1
N/DCM	4.1
N/EE	9.5
PhAc/DCM	3.2
PhAc/EE	37.7
MES/DCM	2.5
MES/EE	20.1
B/DCM	13.0
B/EE	28.5
XAD/Base	13.5
XAD/Acid	16.5

Table 10. Ranking of test sample extracts according to toxicity

Fraction	Rank Value	Class
A + N/DCM	23.1	2
A + N/EE	18.7	2
XAD/Acid	2.0	1
N/DCM	1.8	1
N/EE	1.3	1
XAD/Base	0.4	0
B/DCM	0.1	0
B/EE	0.1	0

with an EC_{50} of 52%. After extraction, there was a large drop in toxicity, with two of the six samples (2(a) and 2(b))showing no measurable toxicity. The normalized toxicity ranking for these samples is also presented in Table 11. As can be readily seen, the A+N fractions contained the vast majority of the toxicity with rank values ranging up to 79.1. The Base and XAD fractions were relatively non-toxic (rank values < 0.51). The neutral fractions appeared to contain a significant amount of the toxicity of the A+N fraction.

Table 11. Microtox toxicity results for produced water samples

Fraction/Extract	Sample					
	1(a)	1(b)	2(a)	2(b)	3	4
EC_{50} (%)						
Blank	—	—	—	—	—	—
Filtered Sample	7.4	4.1	6.7	3.4	52.0	10.4
After Extraction	65.0	21.1	—	—	55.0	38.2
After XAD Column	16.0	—	23.0	—	46.0	50.8
Ranking						
A + N/DCM	7.0	28.6	79.1	40.5	9.6	0.74
A + N/EE	2.4	2.2	67.2	7.2	2.6	0.28
B/DCM	0.10	0.37	0.08	0.19	0.38	0.06
B/EE	0.02	0.25	0.04	0.02	0.14	—
XAD/Base	0.10	—	—	—	0.08	0.05
XAD/Acid	0.51	—	2.5	—	0.38	0.27
N/DCM	2.3	10.6	14.0	8.5	2.7	2.5
N/EE	0.85	1.8	28.8	2.2	0.28	—

GC / MS Analysis

Preliminary GC/FID analysis of all the extracts from the test sample showed good peak separation indicating the suitability of the samples for GC/MS analysis without further cleanup operations. A list of the components identified and their abundance in each fraction of the test sample is presented in Table 12. The major components identified were phenols, alkyl phenol derivatives, carboxylic acids and esters of aromatic and aliphatic acids. Calculations regarding the amount of organic material detected and separated by GC/FID analysis and the percentage of compounds identified in the MS analysis are given in Table 13. About one third of the extracted material was fractionated and detected by GC and 53% of the chromatographable compounds were tentatively identified. In relation to the total organics extracted, then, only 17% have been identified as specific compounds in this sample of produced water.

Table 12. Compounds identified in the test sample (% fraction)

Neutral Compounds			
DCM		ETHER	
Methyl Cyclopentenone	0.5	Phenol	8.7
Dimethyl Cyclopentenone	1.3	Methyl-Phenol (2)	3.4
Trimethyl Cyclopentenone	0.6	C_8H_{18} Olefine	1.0
Methyl-Phenols (2)	16.1	$C_{10}H_{22}$ Olefine (2)	1.6
Dimethyl-Phenols	3.9	C-14 Acid Methyl Ester	0.4
Ethyl-Phenol (2)	16.4	C-16 Acid Methyl Ester	0.1
C-3 Alkyl Phenols	4.0	C-18 Acid Methyl Ester	0.1

Table 12. Compounds identified in the test sample (% fraction) (Continued)

Ethyl-Benzaldehyde	3.9	Dimethyl Esters Of Dioic Acids	
Butoxy-Ethoxy-Ethanol	11.5	C-5 (2)	1.3
4-Ethyl-Benzyl-Alcohol	3.1	C-8	0.7
Methyl-Alkyl-Phenol	0.7	C-9	0.4
Dihydro-inden-ol	11.2	C-10 (2)	1.1
Carboxylic Acids (as Methyl Esters)			
DCM		ETHER	
Benzoic	1.1	Methyl Benzene Dioic (3)	2.4
Methyl-Benzoic (3)	5.7	3-Hydroxy-Benzoic	12.9
Phenyl-Propanoic	1.5	Me-Hydroxy Benzoic (2)	5.1
Dimethyl Benzoic (2)	1.8	Acetoxy-Hydroxy-Benzoic	10.7
P-Methoxy Benzoic	0.7	Di-Hydroxy Benzoic	1.4
Naphtoic	2.6		
Octane Dioic	1.3		
Nonane Dioic	1.0		
Decane Dioic	2.3		
Undecane Dioic	2.2		
Dodecane Dioic	1.7		
Hexadecanoic	1.5		
Octadecanoic	0.5		
Phenolic Compounds (as Acetates)			
DCM		ETHER	
Phenol	6.0	Dihydroxybenzenes	18.0
Methyl Phenols (3)	6.2		
P-ethyl Phenol	1.3		
Methyl Heptanoate	0.1		
Bases			
DCM			
Butoxy-Ethoxy Ethanol	0.02		

Table 13. Estimated balance of organic compounds detected and identified in each extract of test sample

	Estimated % Detected In GC Analysis	% IDentified from GC Etected	From Total Extracted
A + N/DCM	98	—	—
B/DCM	17	0	0
N/DCM	82	89	73
PhAc/DCM	27	52	14
MES/DCM	6	36	24
A + N/EE	74	—	—
B/EE	14	0.13	0.02
N/EE	74	25	19
PhAc/EE	18	98	18
MES/EE	53	61	32
Total DCM	38	59	22
Total EE	31	51	16
Total	32	53	17

Tables 14 to 17 detail the results of the GC/MS analysis of the six produced water samples. Very few compounds were identified in the basic fractions (Table 14), which represented a small proportion of each sample's mass. All the samples had a very high concentration of phenolic derivatives in the neutral fractions (Table 15), most of them in the DCM extracts. The remainder of the identified organics were a mixture of neutral aromatics, long chain carboxylic acid esters and, in the case of the two field samples, long chain alkanes. The results in Table 16 show the amount of phenol derivatives. For sites 3 and 4, very little was quantified. The aged (a) sample results from sites one and two did not correlate with their corresponding fresh (b) samples. The aged samples contained various phenols, while the phenolic fractions of the fresh samples also contained various long chain carboxylic acids. The phenolic content of the fresh samples seems to have remained in the neutral fraction as shown in Table 15. Table 17 shows a number of similarities between the carboxylic acid fractions. This fraction gave the most predictable and consistent data of all of the fractions. Most noteworthy are the two benzoic acid derivatives (methoxy and hydroxy benzoic acids) which were quantified in all six samples and represented major concentrations in five of these. Most samples had a mixture of aromatic and aliphatic carboxylic acids. As shown in the mass balance discussion, a substantial proportion of the acids were quantified in the ether fraction.

Table 14. Organic bases in produced water samples (values in mg/l).

	Sample					
	1(a)	1(b)	2(a)	2(b)	3	4
DCM EXTRACTED						
Dimethyl Indole	0.04	—	—	—	—	—
Dichloro Diisopropyl Cyclopropane	—	—	—	—	—	0.71
Toluene	—	—	—	—	—	0.1
Cyclohexylamine	—	—	—	—	—	0.034
C-2 Alkyl-Piperidinone	—	—	—	—	—	0.32
C-12 Alkane	—	—	—	—	—	0.015
Total DCM Extracted	0.04	—	—	—	0.71	0.47
Ether EXTRACTED						
Dichloro Diisopropyl Cyclopropane	—	—	—	—	0.29	—
Methyl-Cyclopentanol	—	—	—	—	—	0.73
Toluene	—	—	—	—	—	0.69
C-2 Alkylpiperidinone	—	—	—	—	—	0.15
Cyclohexylacetamide	—	—	—	—	—	0.46
C-12 Alkane	—	—	—	—	—	0.034
Total EE Extracted	—	—	—	—	0.29	2.06
TOTAL DCM + EE EXTRACTED	0.04	—	—	—	1.00	2.53

Table 15. Neutral compounds in produced water samples (values in mg/l)

	Sample					
	1(a)	1(b)	2(a)	2(b)	3	4
DCM EXTRACTED						
Methyl Phenol (2)	0.55	1.37	1.69	2.74	—	0.17
Dimethyl Phenol	1.45	0.96	—	0.55	0.71	—
C-3 Alkyl Phenol	0.25	—	0.34	—	—	0.40
Ethyl Benzaldehyde	0.42	—	—	—	—	—
Methyl Allyl Phenol	0.62	—	—	—	—	—

Table 15. Neutral compounds in produced water samples (values in mg/l) (Continued)

Benzene Methanol	0.05	—	—	—	—	—
Methyl Methoxy Benzene	0.39	—	0.49	—	—	—
Benzene Dicarboxaldehyde	0.83	—	—	—	—	—
Methyl Benzofuranone	0.14	—	—	—	—	—
C-5 Alkyl Benzene	—	—	—	—	0.90	—
C-2 Alkyl Benzene Methanol	—	—	—	—	0.28	—
Ethoxy Methyl Propane	—	—	—	—	0.13	—
N-Methyl, N Benzyl Benzylamine	—	—	—	—	0.26	—
Dimethyl Pyrrol	—	—	0.98	—	—	—
C_8H_{14} Unconfirmed Olefin	—	—	1.45	—	—	—
C-2 Alkyl Phenol (2)	—	—	3.37	—	—	—
2, 3 Dihydro-1H-Idenol (2)	—	—	1.60	—	—	—
Phenol	—	2.41	—	1.41	—	0.10
Ethylphenol (2)	—	1.67	—	2.59	—	—
Trimethyl Phenol (2)	—	0.48	—	—	—	—
Methyl Ethyl Phenol	—	0.26	—	0.56	—	—
Propyl Phenol	—	0.32	—	0.57	—	—
Trimethyl Cyclopentenone	—	3.59	—	0.12	—	—
Ethyl Methoxy Benzene	—	0.25	—	0.08	—	—
4 Octyne	—	0.30	—	—	—	—
Trimethyl Cyclohexadienedione	—	0.17	—	—	—	—
C-16 Acid Methyl Ester	—	0.06	—	0.26	—	—
C-16 Acid Ethyl Ester	—	0.11	—	0.10	—	—
C-18 Acid Methyl Ester	—	0.051	—	0.36	—	—
C-18 Acid Ethyl Ester	—	0.06	—	—	—	—
C-21 Alkane	—	0.11	—	—	—	—
C-23 Alkane	—	0.015	—	0.075	—	—
C-24 Alkane	—	0.17	—	0.18	—	—
C-25 Alkane	—	0.15	—	0.27	—	—
C-26 Alkane	—	0.15	—	0.31	—	—
C-27 Alkane	—	0.12	—	0.37	—	—
Hexyl-Heptyl Ether	—	0.11	—	—	—	—
3-H Dihydro Indenol	—	—	—	0.25	—	—
Dimethyl Naphthalene (3)	—	—	—	0.33	—	—
C-18 Carboxylic Acid	—	—	—	0.12	—	—
C-18 Alkane	—	—	—	0.12	—	—
C-28 Alkane	—	—	—	0.13	—	—
C-29 Alkane	—	—	—	0.36	—	—
C-30 Alkane	—	—	—	0.51	—	—
Methyl Cyclopentanol	—	—	—	—	—	0.078
Toluene	—	—	—	—	—	0.49
C-10 Olefine (3)	—	—	—	—	—	0.037
C-9 Diene (2)	—	—	—	—	—	0.037
Dimethyl Decahydro Naphthalene	—	—	—	—	—	0.056
Methyl Octahydro Naphthalene	—	—	—	—	—	0.045
C-14 Diene	—	—	—	—	—	0.015
C-12 Olefine	—	—	—	—	—	0.026
Dimethyl Ethylidenecyclopentane	—	—	—	—	—	0.034
Total DCM Extracted	5.12	12.96	9.92	12.78	2.28	1.49
ETHER EXTRACTED						
3,4 Dihydrobenzopyranone	0.36	—	—	—	—	—
Pyridinyl Hexanone	0.08	—	—	—	—	—

Table 15. Neutral compounds in produced water samples (values in mg/l) (Continued)

Hydroxy Benzoic Acid Ethyl Ether	—	—	—	—	0.38	0.15
N-Methyl N-Benzylbenzylamine	—	—	—	—	—	0.37
Phenol	—	0.48	2.35	—	—	—
Dihydroxy Benzene	—	—	1.47	—	—	—
Methyl Phenol	—	0.18	—	—	—	—
Di-Ethoxy Ethane	—	0.15	—	—	—	—
Pentanedioic Acid Diethyl Ester	—	0.026	—	0.056	—	—
C-12 Alkane	—	0.15	—	—	—	—
Ethyl Hydroxy Benz. Acetate	—	0.32	—	—	—	—
C-10 Dioic Acid Diethyl Ester	—	0.18	—	—	—	—
C-16 Acid Ethyl Ester	—	0.17	—	0.066	—	—
C-8 Acid Ethyl Ester	—	0.043	—	—	—	—
Oxo-hexanoic Acid Ethyl Ester	—	—	—	0.037	—	—
Ethyl (OH Me Benz) Acetate	—	—	—	0.33	—	—
Decanedioic Acid Ethyl Ester	—	—	—	0.29	—	—
Cyclohexyl Phenol	—	—	—	0.056	—	—
C-16 Carboxylic Acid	—	—	—	0.22	—	—
C-23 Alkane	—	—	—	0.056	—	—
C-24 Alkane	—	—	—	0.16	—	—
C-25 Alkane	—	—	—	0.35	—	—
C-26 Alkane	—	—	—	0.35	—	—
C-27 Alkane	—	—	—	0.32	—	—
C-28 Alkane	—	—	—	0.29	—	—
C-29 Alkane	—	—	—	0.23	—	—
C-30 Alkane	—	—	—	0.23	—	—
Hexanedioic Acid Di Me Ester	—	—	—	—	—	0.052
Total EE Extracted	0.44	1.70	3.82	3.42	0.52	0.52
TOTAL DCM + EE EXTRACTED	5.56	14.66	13.74	16.20	2.80	2.01

Table 16. Phenol derivatives in produced water samples (values mg/l)

	Sample					
	1(a)	1(b)	2(a)	2(b)	3	4
DCM EXTRACTED						
Phenol	0.48	—	—	0.37	0.5	—
Methyl Phenol (2)	1.00	—	—	—	—	—
Ethyl Phenol	0.45	—	—	—	—	—
Hydroxy Benzyl Alcohol	0.21	—	—	—	—	—
Dihydroxy Benzene	—	—	7.28	—	—	—
Methoxy Phenol	—	—	1.07	—	—	—
C-17 Acid Methyl Ester	—	0.38	—	—	—	—
C-16 Carboxylic Acid	—	0.18	—	0.075	—	—
C-8 Carboxylic Acid	—	—	—	0.12	—	—
C-9 Carboxylic Acid	—	—	—	0.13	—	—
C-12 Carboxylic Acid	—	—	—	0.12	—	—
3-H-Dihydro Indazolone	—	—	—	0.84	—	—
Dihydroxy Biphenyl	—	—	—	0.22	—	—
C-18 Carboxylic Acid	—	—	—	0.047	—	—
Toluene	—	—	—	—	—	0.28
Total DCM Extracted	2.14	0.56	8.35	1.17	0.5	0.28

Table 16. Phenol derivatives in produced water samples (values mg/l) (Continued)

ETHER EXTRACTED						
Benzene Diol	1.48	4.35	—	6.41	—	—
Trimethyl Benzaldehyde	0.21	—	—	—	—	—
Phenol	—	—	0.19	0.094	—	—
Dihydroxy Benzene	—	—	39.24	—	—	—
Methoxy Phenol	—	—	1.9	—	—	—
Ethane Diol	—	0.38	—	—	—	—
C-14 Carboxylic Acid	—	0.15	—	—	—	—
C-16 Carboxylic Acid	—	0.84	—	0.037	—	—
C-18 Carboxylic Acid	—	0.85	—	—	—	—
Cyclohexane	—	—	—	—	—	5.97
Toluene	—	—	—	—	—	0.56
Total EE Extracted	1.69	5.81	41.33	6.54	—	6.53
TOTAL DCM + EE EXTRACTED	3.83	6.37	49.68	7.71	0.5	6.81

Table 17. Carboxylic acids in produced water samples (mg/l).

	Sample					
	1(a)	1(b)	2(a)	2(b)	3	4
DCM EXTRACTED						
Methyl Benzoic Acid	0.05	3.37	—	0.24	—	—
Dimethyl Benzoic Acid (2)	0.15	—	—	0.33	—	—
Trimethyl Benzoic Acid	0.14	—	—	—	—	—
Naphthalene Carboxylic Acid	0.39	0.96	—	—	—	—
C-13 Carboxylic Acid	0.23	—	—	—	—	—
C-18 Carboxylic Acid	0.10	—	—	—	—	—
Meo-dihydro-Benzofuranoic Acid	0.65	—	—	—	—	—
Ethoxy-Butanoic Acid	—	—	—	—	0.35	—
C-8 Dicarboxylic Acid	—	—	0.17	—	—	—
C-11 Unsaturated Acid (2)	—	—	1.02	—	—	—
Methyl Benzene Dioic Acid	—	0.82	—	—	—	—
Methoxy Benzoic Acid	—	0.56	—	—	—	—
Hexanoic Acid	—	—	—	0.22	—	—
C-16 Carboxylic Acid	—	—	—	0.8	—	0.35
C-18 Carboxylic Acid	—	—	—	—	—	0.2
Total DCM Extracted	1.71	5.71	1.19	1.59	0.35	0.55
ETHER EXTRACTED						
Methoxy Benzoic Acid	2.58	2.6	2.86	1.51	1.31	0.06
Hydroxy Benzoic Acid	0.23	1.89	4.4	2.07	1.88	0.019
Methoxy Benzene Acetic Acid (2)	1.56	—	—	—	—	—
Benzene Dicarboxylic Acid (2)	0.52	0.92	—	0.48	4.16	1.63
Methoxy Benzene Propanoic Acid	0.36	—	—	—	—	—
C-9 Dicarboxylic Acid	0.27	—	—	—	—	0.15
Meo, OH, Trimethyl Benzoic Acid	0.11	—	—	—	—	—
Hexane Dioic Acid	—	—	—	0.42	1.12	—
C-7 Unsaturated Acid	—	—	0.47	—	—	—
C-5 Dioic Acid	—	1.43	—	—	—	—
Acetoxy Benzoic Acid	—	2.15	—	—	—	0.3
Methyl Benzene Dicarboxylic Acid	—	1.87	—	—	—	—
Di-Methoxy Methyl-Benzoic Acid	—	0.26	—	—	—	—

Table 17. Carboxylic acids in produced water samples (mg/l) (Continued)

C-16 Carboxylic Acid	—	0.26	—	1.05	—	0.13
Methyl-Benzene Sulfonic Acid	—	—	—	1.32	—	—
Hydroxy-Methyl-Benzoic Acid	—	—	—	0.48	—	—
Benzoic Acid	—	—	—	1.03	—	—
Methyl-Benzene Dioic (3)	—	—	—	1.12	—	0.26
C-14 Carboxylic Acid	—	—	—	0.094	—	—
Benzene Tricarboxylic Acid	—	—	—	0.56	—	0.47
C-10 Carboxylic Acid	—	—	—	—	—	0.078
Acetoxybenzene Acetic Acid	—	—	—	—	—	0.41
C-18 Carboxylic Acid	—	—	—	—	—	0.13
Total EE Extracted	5.63	5.81	7.73	10.13	8.47	3.64
TOTAL ACIDS (AS ME ESTERS)	7.34	11.52	8.92	11.72	8.82	4.19

CONCLUSIONS

The proposed characterization scheme was experimentally validated on one sample and successfully applied for identifying the major toxic organic components of six produced water samples from four in-situ heavy oil recovery operations in Western Canada.

In the initial validation of the characterization scheme using one produced water sample, approximately 46% of the organic matter (measured as TOC) was extracted from the aqueous sample and available for analysis. Most of this material was associated with the DCM fractions. Approximately 32% of this extract was amenable to GC separation and subsequent identification by MS. The major components identified were phenols, alkyl phenols, carboxylic acids and esters of aromatic and aliphatic acids. Toxicity measurements using the Microtox test suggest that the extraction procedure removed essentially all the Microtox toxicity of the sample. The acid fraction of the sample was found to contain the most toxic components, specifically phenolic derivatives. Fractions containing DCM extractables were more toxic than the corresponding EE fractions perhaps because of the higher lipophilicity of the DCM extracts.

In subsequent test work on six additional produced water samples, the characterization scheme was used to assemble a database of information on organics in produced waters from in-situ heavy oil recovery operations in Western Canada. The nature and amount of organic material in each sample varied widely. Based on Microtox testing, five of the six samples were found to be quite toxic, with 15 min EC_{50} values of 3.4 to 10.4 %, and the toxicity was strongly associated with organics extracted by DCM under acidic conditions. GC/MS analysis of the six produced water samples gave similar results as with the test sample. However, significant concentrations of long chain alkanes were found in the two samples that had been extracted in the field. It appears that the storage of samples caused degradation of alkanes since field extracted and laboratory extracted samples were similar with respect to most other organic compounds identified.

ACKNOWLEDGEMENTS

The authors wish to acknowledge the extensive help that Karen Simms provided in the preparation and editing of this paper. Also, the technical review and comments provided by Mr. P. Fowlie on the first draft of this paper are gratefully acknowledged. This work was partly supported by the Federal Panel on Energy R & D (PERD).

REFERENCES

Munkittrick, K.R., Power, E.A., and Sergy, G.A., 1991, The relative sensitivity of Microtox,daphnid, rainbow trout, and fathead minnow acute lethality tests, Environmental Toxicology and Water Quality, Vol.6, 35-62.

Ribo, J.M., 1987a, "Literature Review and Evaluation of Methodologies for the Characterization of Aqueous Effluents from Oil Recovery Processes," Unpublished Report No. WTC-PC-2-1987, Wastewater Technology Centre, Burlington, Ontario.

Ribo, J.M., 1987b, "Evaluation and Development of Methodologies and Protocols for the Analysis of Aqueous Effluents from In-Situ Oil Recovery Operations: Experimental Program," Unpublished Report No. WTC-PC-3-1987, Wastewater Technology Centre, Burlington, Ontario.

Ribo, J.M., 1987c, "Characterization of Produced Water from In-Situ Oil Recovery Operations." Unpublished Report No. WTC-PC-4-1987, Wastewater Technology Centre, Burlington, Ontario.

Ribo, J.M., 1987d, "Characterization of Six EOR Produced Water Samples from In-Situ Oil Recovery Operations." Unpublished Report No. WTC-PC-5-1987, Wastewater Technology Centre, Burlington, Ontario.

Rodgers, D.R., 1987, "Characterization of the Organic Content of EOR Produced Water." Unpublished Report No. WTC-PC-1-1987, Wastewater Technology Centre, Burlington, Ontario.

Zenon Environmental Inc., 1983, "Analysis of Organics in Wastewater." Report prepared for Environment Canada, Environmental Protection Service, Ottawa, Ontario, under contract #OSE83-00259, March.

SOURCES, CHARACTERISTICS, AND MANAGEMENT OF PRODUCED WATERS FROM NATURAL GAS PRODUCTION AND STORAGE OPERATIONS

J.P. Fillo,[1] S.M. Koraido,[1] and J.M. Evans[2]

[1] ENSR Consulting and Engineering
Pittsburgh, Pennsylvania
USA

[2] Gas Research Institute
Chicago, Illinois
USA

INTRODUCTION

Produced water is the single largest volume source of waste produced by the oil and gas industry. These waters are generated as a result of crude oil and natural gas production, including both onshore and offshore sources. Unconventional sources of natural gas, such as coalbed methane also generate water upon production of the gas. Produced waters are also generated along with the withdrawal of natural gas from underground storage reservoirs.

Produced waters generally are regulated and managed in accordance with the specific disposal method used. For example, deep-well disposal of produced waters is regulated under the Safe Drinking Water Act (SDWA), via the Underground Injection Control (UIC) program, as Class II injection operations. Waters discharged to surface water require a National Pollutant Discharge Elimination System (NPDES) permit under the Clean Water Act (CWA), to the extent this disposal method is allowed. Other management methods, including livestock watering and road/landspreading, are utilized, and are typically state regulated and approved on a case-by-case basis.

There has been a significant increase in the quantity and quality of publicly available data on produced waters since conduct of the U.S. Environmental Protection Agency's (EPA) Congressionally mandated study of exploration and production (E&P) wastes under the Resource Conservation and Recovery Act (RCRA). Prior to this point in time, data on the sources, characteristics, and management of produced waters were sparse (Fillo, et al., 1987). Since the EPA study, several major technical and economic studies have been conducted and research into alternate management practices has increased.

The purpose of this paper is to review results of major studies conducted by the Gas Research Institute (GRI) and others that address the sources, characteristics, and management of produced waters from conventional on-shore natural gas operations. Results from GRI's produced water characterization studies (Wesolowski et al., 1987, 1988) are reviewed and compared with those from the EPA study (US EPA, 1987), and studies conducted by the American Petroleum Institute (ERT, 1987; Gruy, 1991). The major features of each of these studies are summarized in Table 1. Various aspects of the studies are discussed in the paper, as

Produced Water, Edited by J.P. Ray
and F.R. Engelhart, Plenum Press, New York, 1992

Table 1. Major features of U.S. EPA, API, and GRI produced water studies

Study	Industry Survey/ Volume Estimates	Produced Water Characterization	Environmental mpacts	Economic Impacts
U.S. EPA E&P Waste[1]		X	X	X
API Production Waste[2]	X	X	X	X
API - Gruy RCRA Reauthorization Economics[3]	X			X
GRI - Gas Production/ Storage[4]	X	X		

[1]Source: U.S. EPA (1987)
[2]Source: ERT (1987)
[3]Source: Gruy (1991)
[4]Source: Wesolowski, et al. (1987, 1988)

appropriate. These studies and others form the initial basis for more comprehensive GRI investigations of produced waters, which are currently in progress.

SOURCES AND VOLUMES OF PRODUCED WATERS

Produced waters are coproduced during the extraction of natural gas and crude oil from onshore and offshore production operations. The withdrawal of natural gas from underground storage reservoirs also generates produced waters. Over 75 percent of the storage fields in the U.S. are former non-associated gas producing fields (AGA, 1983). Former producing oil and gas fields and water reservoirs comprise most of the remainder. Salt domes, condensate fields, and oil fields are also used to store natural gas.

Produced waters are recovered from several locations. They are typically recovered at or near the production wellhead, but often times can be recovered via pipeline drips and/or separators located at gathering stations, processing/conditioning plants, and underground storage facilities. Figure 1 illustrates the major sources of produced waters.

Volumes of water generated from individual extraction or withdrawal wells vary widely. However, estimates of produced water generation have been developed via the previously cited American Petroleum Institute (API), EPA, and GRI studies. Nationwide estimates of total produced water generation, based on wellhead and/or state-specific data compiled in each of these studies, are presented in Figure 2.

The API conducted a survey of a cross-section of the E&P industry in 1986 in order to: (1) develop independent estimates of waste volumes and sources, and (2) analyze current waste characteristics and management practices, and associated costs. The waste volumes and facility populations in the survey were representative of 1985 E&P activity and included 16 onshore production sites, which were mostly oil producing. Four sites in the study were non-associated gas producing sites. The API survey was accepted and utilized by the EPA as a part of their June 30, 1988 regulatory determination for exploration and production waste.

More recently, Gruy Engineering conducted an economic impact study for the petroleum extraction industry for the API. Volume data for this study were compiled by Petroleum Information Corporation (PIC) for 25 of the 32 producing states. Data for the other seven states (i.e., Missouri, New York, Ohio, Pennsylvania, Tennessee, Virginia, West Virginia) were compiled from various sources including state regulatory agency reports, the 1986 API Production Waste Survey, and other sources. Water production from oil wells was reported by PIC in 15 states, as well as water production from gas wells in 13 of these states. It should be noted that these states account for 95 percent of the total U.S. oil and 92 percent of the total U.S. gas

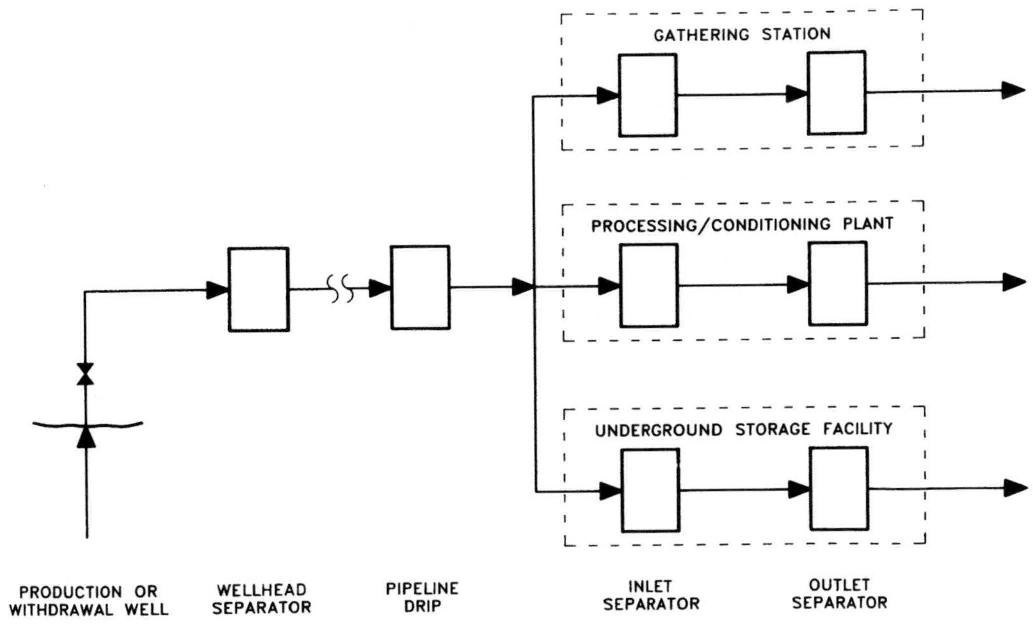

Figure 1. Major sources of produced waters.

production. Water/oil and water/gas ratios for states where water production was not reported were estimated from data for neighboring states in the API survey.

Volume estimates for gas production and underground storage operations were based on the average produced water to gas ratios for 16 production sites and 7 underground storage facilities, respectively, in the GRI study. The average ratios were extrapolated using total U.S. production and withdrawal rates for 1985 (AGA, 1986, 1987), respectively.

Estimates of produced water generation vary significantly between studies, but are of a similar order of magnitude. Based on the results of the API studies, between 14 and 21 billion barrels per year of produced water are generated from onshore oil and gas production. The difference between the two API studies can be attributed mostly to the differences in oil production for the two study years. Gas production appears to generate less produced water than oil production. Based on 1985 statistics, produced water from gas operations (GRI study) accounts for less than 10 percent of the total generated from oil and gas production (API/EPA studies). Based on the Gruy study, less than 2% of the volume of produced waters is generated from gas production operations. Gas storage operations generate approximately 51 million barrels per year, which is between 3 and 25 percent of the volume of produced waters generated from gas production operations.

PRODUCED WATER CHARACTERISTICS

Produced waters can be characterized generally as saline water. They contain other inorganic and organic constituents, largely dependent upon the nature of the producing/storage formation from which they are withdrawn and the operations associated with a given facility. The composition of produced water between different sources and formations can vary by orders of magnitude on a constituent-specific basis. However, produced water composition is qualitatively similar from oil production, gas production, and gas storage operations.

Comprehensive characterization data were generated for produced waters as a result of the exploration and production waste studies conducted by EPA (1987), API (ERT, 1987), and GRI (Wesolowski, et al., 1987, 1988). Produced water from primarily crude oil production (i.e., 4 of 16

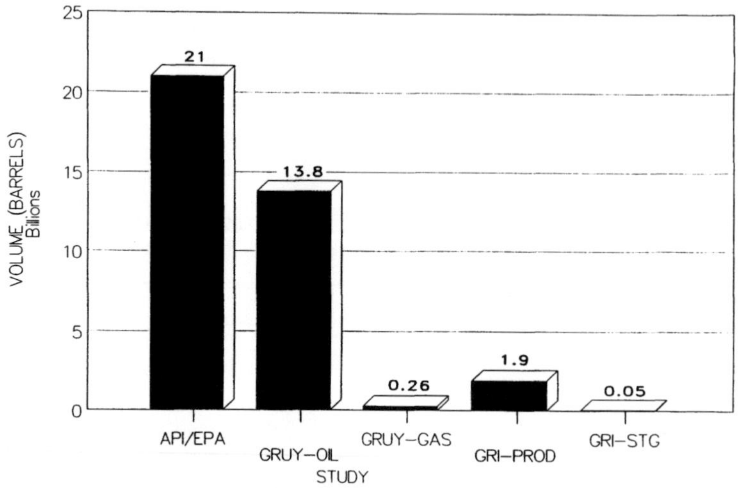

Figure 2. Produced water volume estimates reported in API's, EPA's and GRI's studies.

total sites were non-associated natural gas fields) were characterized in both the API and EPA studies. Produced water from non-associated natural gas production and gas storage reservoirs were characterized in the GRI studies. Although, a small population of samples was collected relative to the number of producing oil and gas fields and underground storage fields, samples were carefully selected to represent a wide cross-section of geographic locations and/or geologic conditions.

The slate of analytical parameters included conventional water quality parameters, minor and trace metals, volatile and semivolatile organic compounds, and RCRA criteria. The studies utilized EPA-approved protocols to ensure that quality data were developed and the results of the studies were comparable. In addition, the GRI study performed organics analyses using both conventional (i.e., EPA Methods 624 and 625) and isotope dilution (i.e., EPA Methods 1624 and 1625) gas chromatography/mass spectroscopy (GC/MS) methods. In this manner, GRI's results for all samples could be related to results from EPA's study (i.e., using isotope-dilution GC/MS methods) and results for gas production samples could be related to results from API's study (i.e., using conventional GC/MS methods). The analytical aspects of the GRI study are discussed in Shepherd et al., (1992).

Salt Constituents

The predominant constituents in produced water were dissolved salts (see Figure 3). Total dissolved solids concentrations ranged from less than 100 ppm to over 300,000 ppm. While the large majority of these salts consists of sodium chloride, the ionic composition of these waters varied considerably. Major cations included sodium, calcium, magnesium, and potassium listed in gneerally decreasing concentrations. Anions, other than chloride, included sulfate and bicarbonates.

Minor and Trace Metals

Dissolved minor and trace metals were present at substantially lower concentrations than the salt constituents (see Figure 4). Barium, boron, iron, manganese, potassium, strontium, arsenic, silicon, zinc, and lithium were detected most frequently. The maximum values for barium and strontium in the GRI production samples represent single samples taken from production fields with naturally high levels of each element.

Cadmium, cobalt, copper, lead, mercury, and nickel were detected least frequently. Priority pollutant metals were generally detected less frequently and at lower levels (i.e., generally less than ppm quantities) in storage samples.

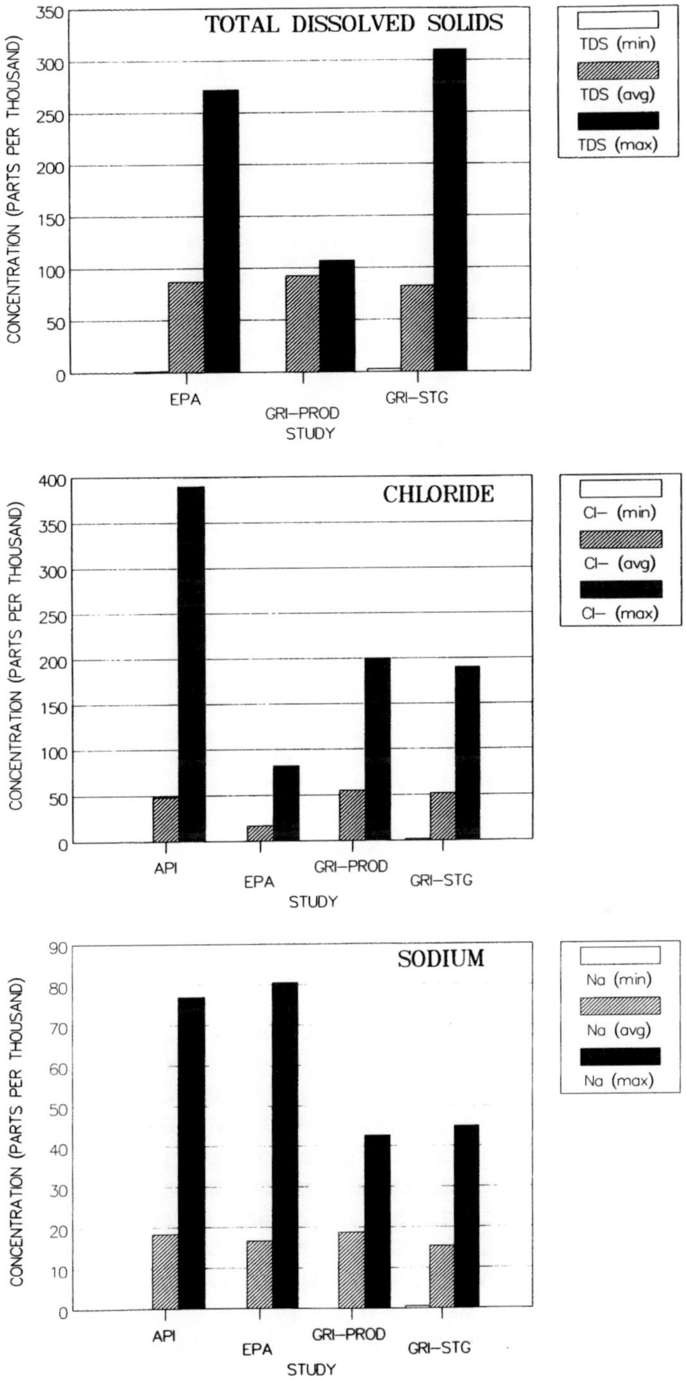

Figure 3. **Comparison of total dissolved solids, chloride, and sodium concentrations in produced waters from API's, EPA's, and GRI's gas production and gas storage studies.**

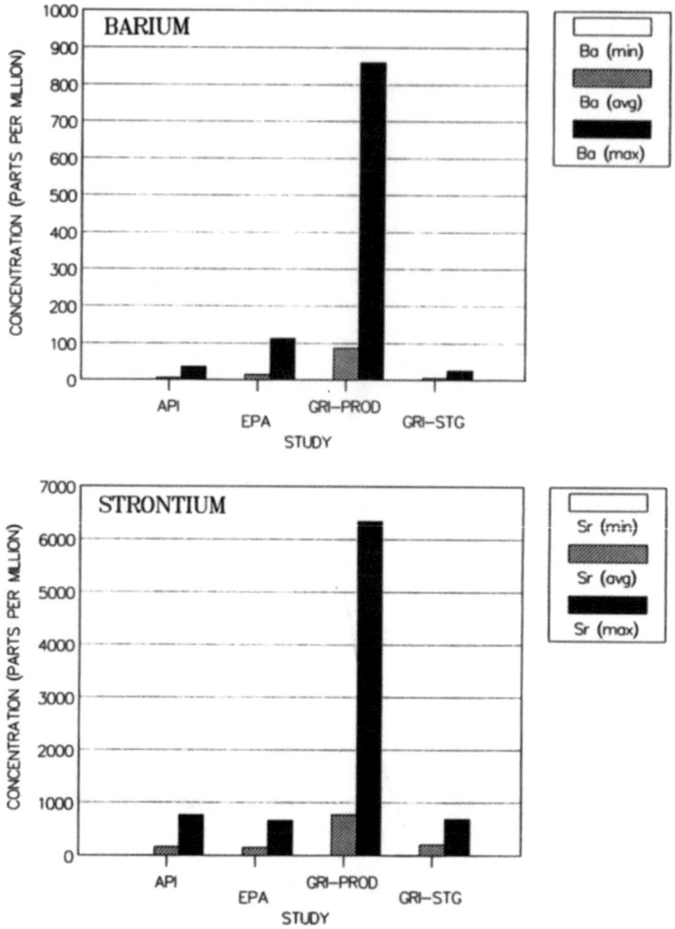

Figure 4. Comparison of barium and strontium concentrations in produced water from API's, EPA's, and GRI's gas production and gas storage studies.

Volatile Organic Compounds

The most prevalent volatile organics detected in production and storage samples were benzene, toluene, and ethylbenzene (see Figure 5). Their presence is a direct result of the nature of hydrocarbons present in crude oil and natural gas liquids that can accompany natural gas in producing formations; they were detected in roughly 75 to 80% of the production samples and all of the storage samples. Produced waters from gas production operations generally contained higher levels of benzene, toluene, and ethylbenzene than produced waters from either oil production or gas storage. This is not surprising because several of the gas production sites produced natural gas liquids along with the gas. Other volatile organics, particularly solvents (i.e., 2-butanone, chlorinated solvents), were detected infrequently in the produced water samples. The presence of certain solvents in the produced water (e.g., acetone, methylene chloride) could be related to contamination from the analytical laboratory.

Semivolatile Organic Compounds

The most prevalent semivolatile organics detected in the produced water samples were C_{10} to C_{30} straight-chain alkanes (see Figure 6). Levels of semivolatile organics generally decreased as compound molecular weight increased, and the studies that included data from oil production sites generally had higher mean, median, and maximum values. Storage sites had appreciably lower levels than either oil or gas production sites. Other semivolatile organics that were detected most frequently included phenols, carboxylic acids, and naphthalene. High molecular weight PAHs were detected infrequently and at levels lower than naphthalene.

PRODUCED WATER MANAGEMENT

There are several alternatives for effective management of produced waters. These are illustrated in Figure 7, and are summarized as follows:

- deepwell disposal,
- surface discharge, and
- recycle/reuse.

The specific methods used are a function of the geographic/geologic location, the specific regulations that govern management, and the cost of produced water disposal.

Deepwell Disposal

Based on the results of the aforementioned E&P waste studies, deepwell injection of produced waters is the most frequently practiced management alternative. This is the case because most of the gas production in the U.S. occurs in areas where underground injection is geologically feasible and cost effective.

An injection well operation can be captive to a producing site, or produced waters can be collected and hauled to a central disposal facility. For example, produced water was disposed at off-site injection wells for 17 of the 24 sites sampled in the GRI study. Four other sites utilized on-site injection wells for produced water disposal.

Surface Discharge

There are several options for effecting surface discharge. In certain locations, it is possible to treat produced waters and dispose them to surface water. This typically requires an NPDES permit for the affected facility or water can be hauled to a central facility for treatment and ultimate disposal.

Produced water can be discharged to the ocean, to inland surface waters, or to a municipal treatment facility, in accordance with applicable federal and/or state regulations. Effluent limitations guidelines for offshore oil and gas development have been reproposed (Federal Register, 1991) and produced water discharges to the outer continental shelf are regulated (Federal Register, 1986). Discharge of produced waters to the ocean or tidal wetlands is becoming increasingly more restrictive, as the State of Louisiana recently issued a ruling that mandates the deepwell injection of produced waters currently being discharged to coastal wetlands (Environmental Reporter, 1991). Discharge of produced waters to surface waters is often considered on a case-by-case basis, but can be strictly prohibited unless treatment for removal of salt constituents is conducted (e.g., New York, West Virginia, Virginia, New Mexico). One storage site, which was a depleted oil/gas producing field in the GRI study, discharged produced water to a Publicly Owned Treatment Works (POTW).

Land and/or road application can be allowed in certain states if composition criteria are met for the produced water. Examples of such states include Wyoming, Pennsylvania, and Virginia, and this typically is allowed only for low-TDS waters or for waters within a given range of TDS values (e.g., between 5,000 and 50,000 ppm in Wyoming). One site from the GRI study utilized road application for the disposal of produced water. In most cases, land application is considered on a case-by-case basis and must be conducted in accordance with a specific state's water quality regulations. Protection of ground water typically controls the extent to which land spreading can be practiced.

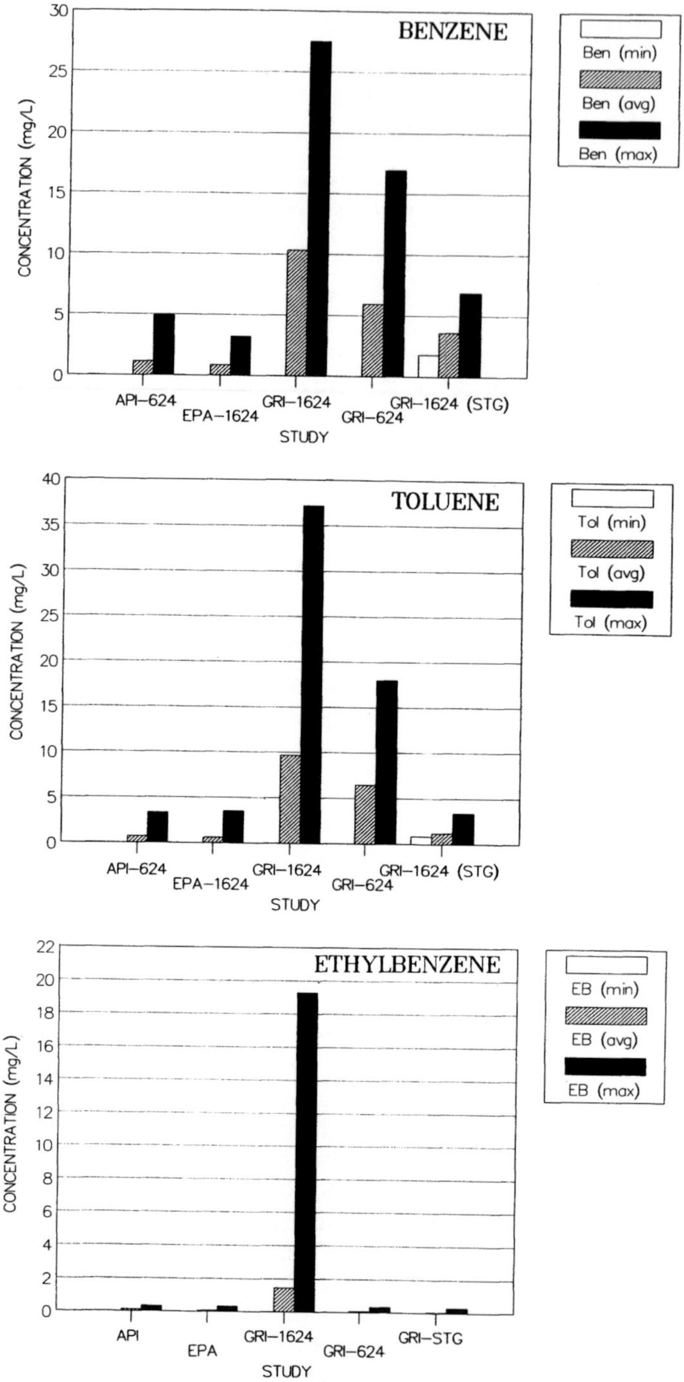

Figure 5. **Comparision of benzene, toluene, and ethylbenzene concentrations in produced waters from API's, EPA's, and GRI's gas production (EPA Method 1624 and 624) and gas storage studies.**

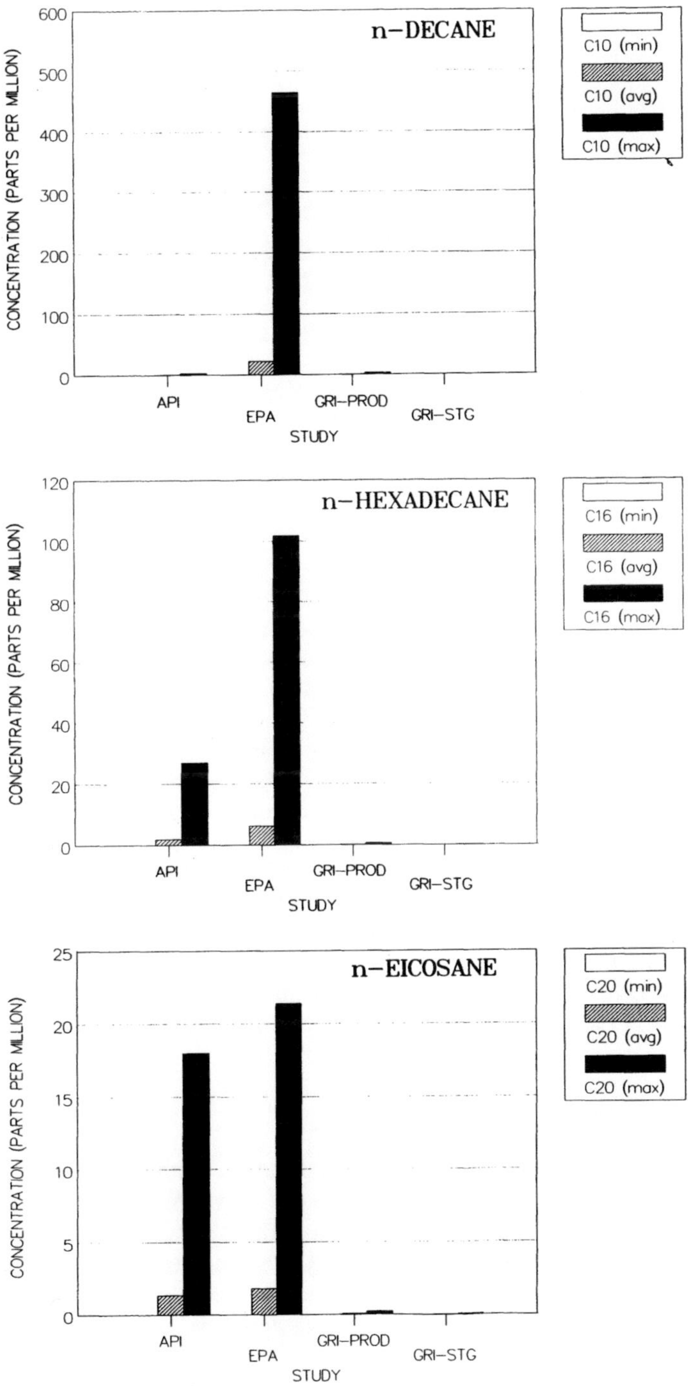

Figure 6. Comparison of n-decane, n-hexadecane, and n-eicosane concentrations in produced water from API's, EPA's, and GRI's gas production and gas storage studies.

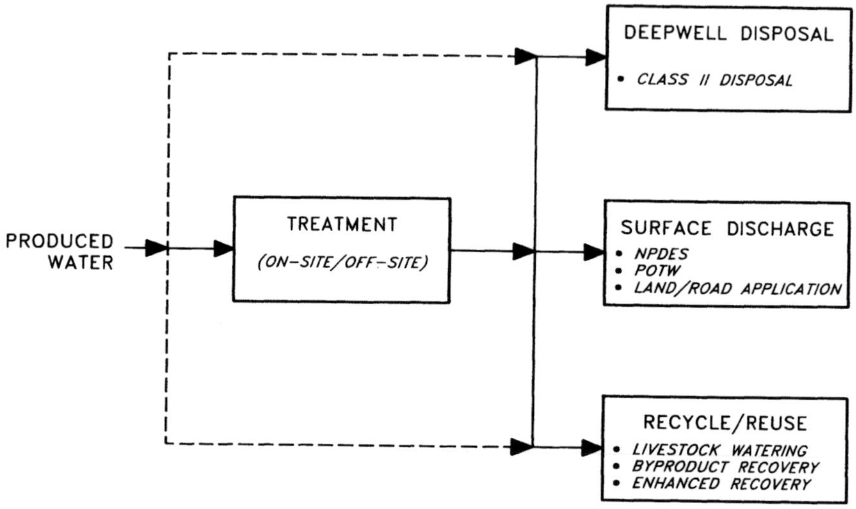

Figure 7. Schematic of major produced water management alternatives.

Recycle/Reuse

There are several options to effect produced water recycle/reuse. One alternative that is practiced in oil production operations is the use of produced water for enhanced recovery. Produced water removed from a producing formation is recovered, treated, and reinjected into the producing formation to extend the producing life of a field. This method of recycle/reuse is not feasible for gas production operations.

In a manner similar to land application, certain states allow the use of produced water for livestock watering. Specific composition criteria must be achieved to enable use of this management alternative, which is practiced predominantly in the western U.S.

Treatment of produced water can be conducted to recover salt byproducts and generate an essentially salt-free water stream. This concept has been demonstrated and uses chemical treatment combined with evaporative crystallization (Hart, 1986, 1990). Chemical treatment prior to crystallization provides for the removal of oil, ferric chloride, barium sulfate, and other heavy metals. The remaining aqueous solution is processed via evaporative crystallization to produce sodium chloride crystals and a salable calcium chloride solution. This concept potentially could be applied to produce a low-TDS water for surface discharge or possibly for process consumption.

FUTURE DIRECTIONS

Produced waters are generated in appreciable volumes from the extraction of oil and gas and the withdrawal of gas from underground storage reservoirs. While the majority of this water has been managed through use of deepwell injection, other methods can and have been used to effect proper management. Because the regulations governing management of produced waters from both conventional and unconventional sources are becoming increasingly stringent, it is essential to accurately understand the sources, characteristics, and management of produced water.

The studies discussed in this paper are being augmented to develop a more thorough understanding of produced water and feasible alternatives for its management. A more in-depth evaluation of produced water sources, characteristics, and management is in progress. The objective is to supplement existing data with more recent data, including produced waters from coalbed methane production and natural gas industry operating statistics, to develop more complete regional and national profiles for produced waters. In addition, GRI research focused to produced water treatment and disposal is presented in another paper in this volume (Nakles et al., 1992).

REFERENCES

AGA, 1983, "Survey of Underground Gas Storage Facilities in the United States and Canada." Operating Section Report, American Gas Association, Statistics Task Group, Arlington, VA.

AGA, 1986, "Gas Facts 1985, A Statistical Record of the Gas Utility Industry, American Gas Association," Department of Statistics, Arlington, VA.

AGA, 1987, "Gas Facts 1986, A Statistical Record of the Gas Utility Industry, American Gas Association," Department of Statistics, Arlington, VA.

ERT, 1987, "Oil and Gas Industry Exploration and Production Wastes." Prepared for the American Petroleum Institute. July.

Environmental Reporter, 1991, Volume 21, 1989, Bureau of National Affairs, Inc., Washington, D.C. March 8

Federal Register, 1986, Volume 51, 24897, July 9.

Federal Register, 1991, Volume 56, 10664, March 13.

Fillo, J.P., Hamaker, T.A., Koraido, S.M., Stefanacci, R.G., Tallon, J.T., and Wesolowski, D., 1987. "Production Waters Associated with the Production, Processing, Transmission, and Storage of Natural Gas" A Literature Survey: GRI Report No. GRI-87/0126. July.

Gruy Engineering Corporation, 1991, "Estimates of RCRA Reauthorization Economic Impacts on the Petroleum Extraction Industry." Prepared for the American Petroleum Institute. July 20.

Hart, L., 1986, "Crystallization of Sodium Chloride from Gas Well Brine." Presented at the Forty-sixth Annual Appalachian Gas Measurement Shoot Course, Coraopolis, PA. August 12-15.

Hart, L., 1990, Hart Chemical Company. Personal Communication with M. Conedero, ENSR Consulting and Engineering, Pittsburgh, PA. August 28.

Nakles, D., Ortiz, I., and Frank, J., 1992, An analysis of management strategies for produced waters from natural gas production. in: "Produced Water: Technological/Environmental Issues and Solutions," J.P. Ray and F.R. Engelhardt, eds., Plenum Publishing Corp., New York, New York, USA.

Shepherd, M.C., Shore, F.L., Mertens, S.M., and Gibson, J.S., 1992, Characterization of Produced Waters from Natural Gas Production and Storage Operations. in: "Produced Water: Technological/Environmental Issues and Solutions," J.P. Ray and F.R. Engelhardt, eds., Plenum Publishing Corp., New York, New York, USA.

USEPA, 1987, "Report to Congress-Management of Wastes from the Exploration, Development, and Production of Crude Oil, Natural Gas, and Geothermal Energy." NTIS Report EPA-530-SW-88-003. December.

Wesolowski, D., Broughton, A., Hansotte, C.A., Koraido, S.M., and Fillo, J.P., 1987, "Characterization of Produced Waters from Natural Gas Production Operations." GRI Report No. GRI-87/0335.1. December.

Wesolowski, D., Broughton, D., Hansotte, C.A., Koraido, S.M., and Fillo, J.P., 1988, "Characterization of Produced Waters from Natural Gas Storage Reservoir Operations." GRI Report No. GRI-88/0080.1. June

CHARACTERIZATION OF PRODUCED WATERS FROM NATURAL GAS PRODUCTION AND STORAGE OPERATIONS: REGULATORY ANALYSIS OF A COMPLEX MATRIX

M.C. Shepherd, F.L. Shore, S.K. Mertens, and J.S. Gibson

Radian Corporation
Austin, Texas
USA

INTRODUCTION

In 1986, there was a general lack of comprehensive data on the chemical characteristics of produced waters. Specifically, waters from non-associated gas production and underground storage operations were not well characterized. The Gas Research Institute (GRI) commissioned a sampling program to obtain and analyze produced water from gas production and storage operations not associated with oil production (Keith et al., 1987, 1988; Wesolowski et al., 1987, 1988) which includes the following major technical aspects:

- A total of 24 produced water samples, 17 taken from natural gas production and 7 from underground storage operations;
- Geographical distribution of the samples based on statistics for natural gas production, number of wells, and storage field output; and,
- Analytes and analytical methodology were selected to parallel those available on similar exempt production wastes.

The objective of this study was to characterize these samples with respect to existing regulations and RCRA hazardous waste characteristics. Further, that characterization was to be done using existing standard regulatory analyses. This paper focuses on the analytical methodologies as they were applied to this matrix and the analytical challenges presented by this matrix.

Experimental and Field Approach

The sampling program was designed to include samples from geographically diverse sections of the gas industry, based on industry statistics, and to include samples from both natural gas production and storage operations. The sampling effort was carried out by ENSR Consulting and Engineering. Further discussion of this aspect will be presented in a related paper (Koraido et al., 1992). A total of 24 samples were taken from either the well head separator or prior to underground injection or from storage operations.

Sampling procedures and practices were used to comply with those required by the various analytical protocols employed. An independent site audit of sampling, handling and quality control was conducted by Mason (1987). The parameters determined for each sample and the

Produced Water, Edited by J.P. Ray
and F.R. Engelhart, Plenum Press, New York, 1992

protocols used are listed in Table 1. Analytical methods employed were taken from either EPA SW 846 (EPA 530); Methods for Analysis of Municipal and Industrial Wastewater (40 CFR 136; EPA 600/4-71-020) the EPA Contract Laboratory Program (CLP); (USEPA Contract Laboratory Program); ASTM and Materials, (D56-70); or Standard Methods (1985).

Samples were analyzed to determine if they exhibited any of the characteristics of corrosivity, ignitability and reactivity as defined by RCRA. Both pH and corrosivity using the NACE standard were measured. Ignitability was measured using ASTM Method D56-70 with a selected subset reanalyzed using EPA Method 1010. Reactivity in this study refers to the ability of a sample to release cyanide or sulfide gases, vapors or fumes. The guidelines for EP Toxicity extraction were employed as at the time of this study the Toxicity Characteristic Leaching Procedure (TCLP) had not yet been promulgated. Since none of the samples contained total suspended solids at concentrations greater than 0.5% by weight, the EP leaching procedure was not required but analyses were performed on whole samples.

Samples were analyzed for trace metals using standard EPA procedures and also by sequential inductively coupled plasma emission spectrometry (ICPES) for 69 elements (Superscan). The 69 elements detected by this technique are listed in Table 2. Superscan, now a part of EPA Method 1620, is a semi-quantitative approach simultaneously analyzing these elements by rapidly scanning the emission lines for each element to determine their presence or absence. Positive results were confirmed using ICPES. The Superscan analysis and subsequent ICPES confirmation were performed by Centec Analytical Services, Inc. Due to the nature of this technique, there are some limitations. Instrument detection limits for most analytes are higher than in standard quantitative ICPES analysis and resulting method detection limits may be above regulatory limits. Additionally, spectral interferences lead to a tendency for false positives for some elements. Conventional parameters were determined following the standard EPA procedures listed in Table 1.

Organic parameters were determined using EPA methods 1624 /1625 Revision B and EPA Methods 624/625. The analytes included in each of these methods are listed in Tables 3 and 4. Each of these methods are gas chromatography/mass spectrometry methods of analysis. Methods 1624 and 1625 are isotope dilution techniques for determining volatile and semi-volatile organics respectively. Isotope dilution is a more elegant technique whereby stable isotopically labeled analogues of the compounds being determined are added to the sample prior to analysis. The compounds of interest are then quantitated with respect to their labeled analogues thereby theoretically correcting for matrix effects. A subset of samples from production sites were also analyzed by Methods 624 and 625 specifically to compare performance of these more standard methods with the isotope dilution techniques and to establish a linkage to studies sponsored by API (ERT, 1987). Furthermore, methods 1624/1625 (as well as Superscan) were selected so as to parallel those of the EPA Exploration and Production waste study (USEPA Report to Congress, 1987) performed on similar samples.

General Sample Handling and Description

All samples were collected between February 17 and March 30, 1987 by ENSR Consulting and Engineering. All sample containers were Series 300 obtained from I-Chem Research, Inc. which are cleaned to EPA protocols and QC analyzed. Samples were preserved in the field according to the requirements of the individual analytical methods. All field samples, with the exception of two samples for BOD, were analyzed within the recommended method hold times.

Gross sample composition varied widely between samples of various locations and sources but all samples were highly saline. Several samples had slight oil layers which prevented homogenous sample aliquotting for sample analysis. Most of the analytical difficulties encountered were due to the concentration of salts and organic compounds in these samples.

Table 1. Analytical methods

Parameter	Preparation Technique	Preparation Method	Analyses Technique	Analyses Method
Volatile Organics	N/A		GC/MS	EPA 1624 Rev B
Volatile Organics	N/A		GC/MS	EPA 624 & CLP
Semivolatile Organics	Liquid Extraction	EPA 1625 Rev B	GC/MS	EPA 1625 Rev B
Semivolatile Organics	Liquid Extraction	EPA 625 & CLP	GC/MS	EPA 625 & CLP
Metals:				
Aluminum	Acid Digestion	SW-846:6010	ICPES	EPA 200.7
Antimony	Acid Digestion	SW-846:3020	GFAAS	EPA 204.2
Arsenic	Acid Digestion	SW-846:3010	GFAAS	EPA 206.2
Barium	Acid Digestion	SW-846:6010	ICPES	EPA 200.7
Beryllium	Acid Digestion	SW-846:6010	ICPES	EPA 200.7
Boron	Acid Digestion	SW-846:6010	ICPES	EPA 200.7
Cadmium	Acid Digestion	SW-846:6010	ICPES	EPA 200.7
Calcium	Acid Digestion	SW-846:6010	ICPES	EPA 200.7
Chromium	Acid Digestion	SW-846:6010	ICPES	EPA 200.7
Cobalt	Acid Digestion	SW-846:6010	ICPES	EPA 200.7
Copper	Acid Digestion	SW-846:6010	ICPES	EPA 200.7
Mercury	Acid Digestion	SW-846:7470	Cold Vapor AA	EPA 245.1
Iron	Acid Digestion	SW-846:6010	ICPES	EPA 200.7
Lead	Acid Digestion	SW-846:6010	ICPES	EPA 200.7
Magnesium	Acid Digestion	SW-846:6010	ICPES	EPA 200.7
Manganese	Acid Digestion	SW-846:6010	ICPES	EPA 200.7
Molybdenum	Acid Digestion	SW-846:6010	ICPES	EPA 200.7
Nickel	Acid Digestion	SW-846:6010	ICPES	EPA 200.7
Selenium	Acid Digestion	SW-846:3020	GFAAS	EPA 270.2
Silver	Acid Digestion	SW-846:3020	GFAAS	EPA 272.7
Sodium	Acid Digestion	SW-846:6010	ICPES	EPA 200.7
Tin	Acid Digestion	SW-846:6010	ICPES	EPA 200.7
Titanium	Acid Digestion	SW-846:6010	ICPES	EPA 200.7
Thallium	Acid Digestion	SW-846:6010	ICPES	EPA 200.7
Vanadium	Acid Digestion	SW-846:6010	ICPES	EPA 200.7
Yttrium	Acid Digestion	SW-846:6010	ICPES	EPA 200.7
Zinc	Acid Digestion	SW-846:6010	ICPES	EPA 200.7
69 Elements	Acid Digestion	SW-846:3020	ICP Superscan	ICP Superscan
Conventional:				
BOD-5	N/A	N/A	Dissolve O_2	SM 507
COD	N/A	N/A	Redox Titration	EPA (1984) 410.3
TSS	N/A	N/A	Gravimetric	EPA 160.2
Purgeable Organic Carbon	N/A	N/A	UV Oxidation/IR	EPA 415.1
TOC	N/A	N/A	UV Oxidation/IR	EPA 415.1
Oil and Grease	Liquid Extraction	EPA 413.1	Gravimetric	EPA 413.1
Cyanide	N/A	N/A	Colorimetric	EPA 335.3
Sulfide	N/A	N/A	I_2 Titration	EPA 376.1

Table 1. Analytical methods (continued)

	Preparation		Analysis	
Parameter	Technique	Method	Technique	Method
Ammonia as N	N/A	N/A	Colorimetric	EPA 350.1
Nitrate/Nitrite as N	N/A	N/A	Hydrazine Reduction/Colorimetric	EPA 353.17
Conductivity	N/A	N/A	Conductimetric	EPA 120.1
Floride	Bellack Distillation	EPA 340.1	SIE	
TDS	N/A	N/A	Gravimetric	EPA 160.1
RCRA:				
Corrosivity	N/A	N/A	Coupon	SW-846:1110 (NACE)
Ignitibility	N/A	N/A	Tag closed tester Pensky-Martens	ASTM D56-70 SW846:1010
Reactivity	Toxic gas generation	USEPA	Cyanide/Sulfide	SW845:9010/376.1
EP Toxicity	N/A	N/A	See Metals Analyses	See Metals Analyses

N/A = Not applicable.

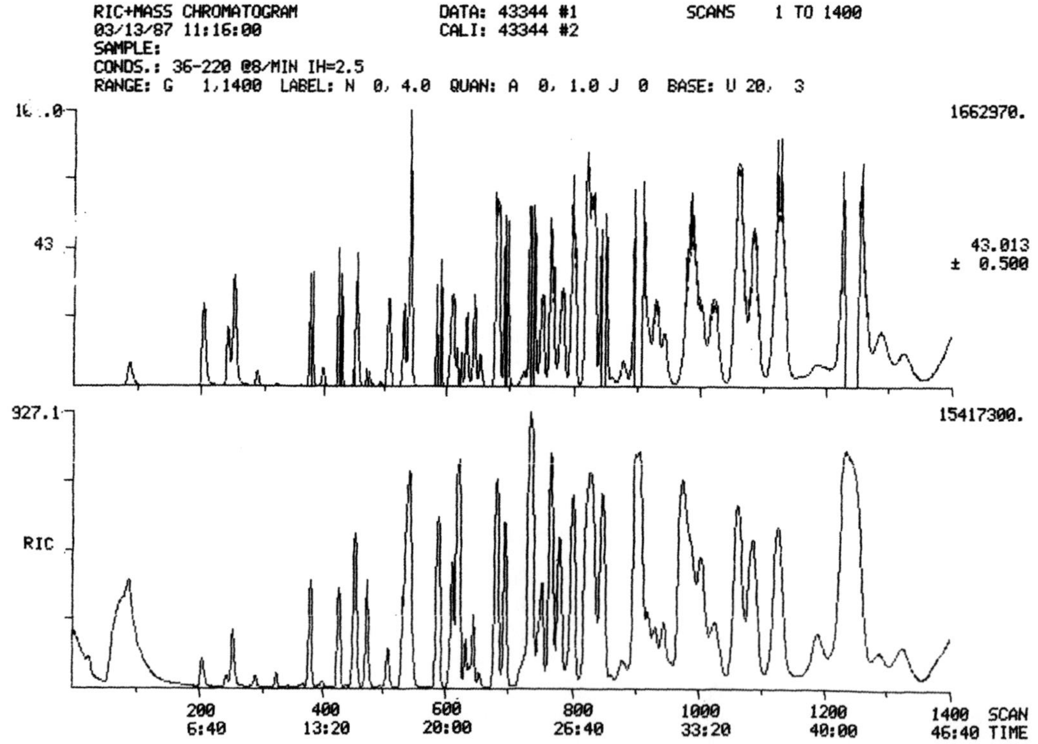

Figure 1. Reconstructed ion chromatograms of volatiles fraction of a produced water sample. The top trace shows ion fragments attributable to hydrocarbons. The bottom trace shows the total ion chromatogram.

Table 2. Metals analyzed by Superscan

Aluminum	Iodine	Samarium
Antimony	Iridium	Scandium
Arsenic	Iron	Selenium
Barium	Lanthanum	Silicon
Beryllium	Lead	Silver
Bismuth	Lithium	Sodium
Boron	Lutetium	Strontium
Cadmium	Magnesium	Sulfur
Calcium	Manganese	Tantalum
Cerium	Mercury	Tellurium
Chromium	Molybdenum	Terbium
Cobalt	Neodymium	Thallium
Copper	Nickel	Thorium
Dysprosium	Niobium	Thulium
Erbium	Osmium	Tin
Europium	Palladium	Titanium
Gadolinium	Phosphorus	Tungsten
Gallium	Platinum	Uranium
Germanium	Potassium	Vanadium
Gold	Praseodymium	Ytterbium
Hafnium	Rhenium	Yttrium
Holmium	Rhodium	Zinc
Indium	Ruthenium	Zirconium

RESULTS AND DISCUSSION

RCRA Criteria

None of the samples were found to be reactive or corrosive by RCRA standards and no particular problems were encountered in performing these tests. Ignitability was initially determined using ASTM (American Society for Testing and Materials) Method D56-70 instead of EPA Method 1010. The EPA method was not used because it requires a Pensky-Martens closed-cup flash point tester, which was unavailable during the period scheduled for sample analysis. These two methods are comparable for liquid flash point determination below 200° F. Selected samples were retested using the Pensky-Martens apparatus approximately 14 weeks after sampling. The results agreed within 20% of the original tests and did not alter the ultimate conclusions of the original testing.

Metals Analysis

The major difficulty with this set of analyses was due to the high inorganic salt content which caused spectral interference and sample crystallization during digestion. Sample dilution for the most part alleviated this problem, however some crystallization occurred even with higher dilutions. Sodium contamination in blanks due to carryover was unavoidable with samples having sodium concentrations in excess of 25,000 ppm.

The higher viscosity of these samples caused interferences with the sample nebulization and transport process in the ICPES analyses resulting in more reruns and increased instrument downtime. Salt buildup on the nebulizer tip also caused more instrument drift. Matrix modifiers were added to digestates to alleviate chloride interferences during the graphite furnace analyses.

As mentioned, the Superscan is prone to false positives due to spectral interferences. Increased iodine levels in some of these samples also caused low level interference with some of the other elements determined. Lower instrument threshold limits were raised to correct for this interference. Several elements that were detected in the Superscan analysis were not quantifiable by subsequent standard ICPES analysis in any of the samples tested. These include: cerium, neodymium, osmium, tungsten, europium, gallium, gold, hafnium, indium, palladium, praseodymium, ruthenium, samarium, tantalum, thallium and uranium. These elements were either false positives or in the samples near the threshold limit of detection. Elements that were confirmed in these samples included those on the priority pollutant and trace metal analysis performed by Radian, as well as iodine, lithium, potassium, silicon, strontium, sulfur and zirconium, which were confirmed by Centec.

Table 3. Base/neutral extractable compounds

Acenaphthene	2-Chloronaphthalene
Acenaphthylene	4-Chlorophenyl phenyl ether
Anthracene	Chrysene
Benzidine	p-Cymene
Benzo(a)anthracene	Dibenzo(a,h)anthracene
Benzo(b)fluoranthene	Dibenzofuran
Benzo(k)fluoranthene	Dibenzothiophene
Benzo(a)pyrene	Di-n-butyl phthalate
Benzo(g,h,i)perylene	1,2-Dichlorobenzene
Biphenyl	1,3-Dichlorobenzene
bis(2-Chloroethyl)ether	1,4-Dichlorobenzene
bis(2-Chloroethoxy)methane	3,3-Dichlorobenzidine
bis(2-Ethylhexyl)phthalate	Diethyl phthalate
4-Bromophenyl phenyl ether	2,4-Dimethylphenol
Butyl benzyl phthalate	Dimethyl phthalate
n-C_{10}	2,4-Dinitrotoluene
n-C_{12}	2,6-Dinitrotoluene
n-C_{14}	Di-n-octyl phthalate
n-C_{16}	Diphenylamine
n-C_{18}	Diphenyl ether
n-C_{20}	1,2-Diphenylhydrazine
n-C_{22}	Fluoranthene
n-C_{24}	Fluorene
n-C_{26}	Hexachlorobenzene
n-C_{28}	Hexachlorobutadiene
n-C_{30}	Hexachloroethane
Carbazole	Hexachlorocyclopentadiene
Isophorone	1,2,3-Trichlorobenzene
Naphthalene	1,2,4-Trichlorobenzene
B-Naphthylamine	4-Chloro-3-methylphenol
Nitrobenzene	2-Chlorophenol
N-Nitrosodimethylamine	2,4-Dichlorophenol
N-Nitrosodi-n-propylamine	2,4-Dinitrophenol
N-Nitrosodiphenylamine	2-Methyl-4,6-dinitrophenol
Phenanthrene	2-Nitrophenol
Phenol	4-Nitrophenol
d-Picoline	Pentachlorophenol
Pyrene	2,3,6-Trichlorophenol
Styrene	2,4,5-Trichlorophenol
α-Terpineol	2,4,6-Trichlorophenol

Table 4. Volatile organic compounds

Acetone	trans-1,2-Dichloroethene
Acrolein	1,2-Dichloropropane
Acrylonitrile	cis-1,3-Dichloropropene
Benzene	trans-1,3-Dichloropropene
Bromodichloromethane	Diethyl ether
Bromoform	p-Dioxane
Bromomethane	Ethyl benzene
Carbon tetrachloride	Methylene chloride
Chlorobenzene	Methyl ethyl ketone
Chloroethane	1,1,2,2-Tetrachloroethane
2-Chloroethylvinyl ether	Tetrachloroethene
Chloroform	Toluene
Chloromethane	1,1,1-Trichloroethane
Dibromochloromethane	1,1,2-Trichloroethane
1,1-Dichloroethane	Trichloroethene
1,2-Dichloroethane	Vinyl chloride
1,1-Dichloroethene	

Conventional Parameters

A great deal of variability was seen among the conventional parameters for these samples. This was particularly so for the parameters affected by organic compounds; biological oxygen demand (BOD), chemical oxygen demand (COD), oil and grease, total organic carbon (TOC) and purgeable organic carbon (POC). Aside from this variability, no particular problems were

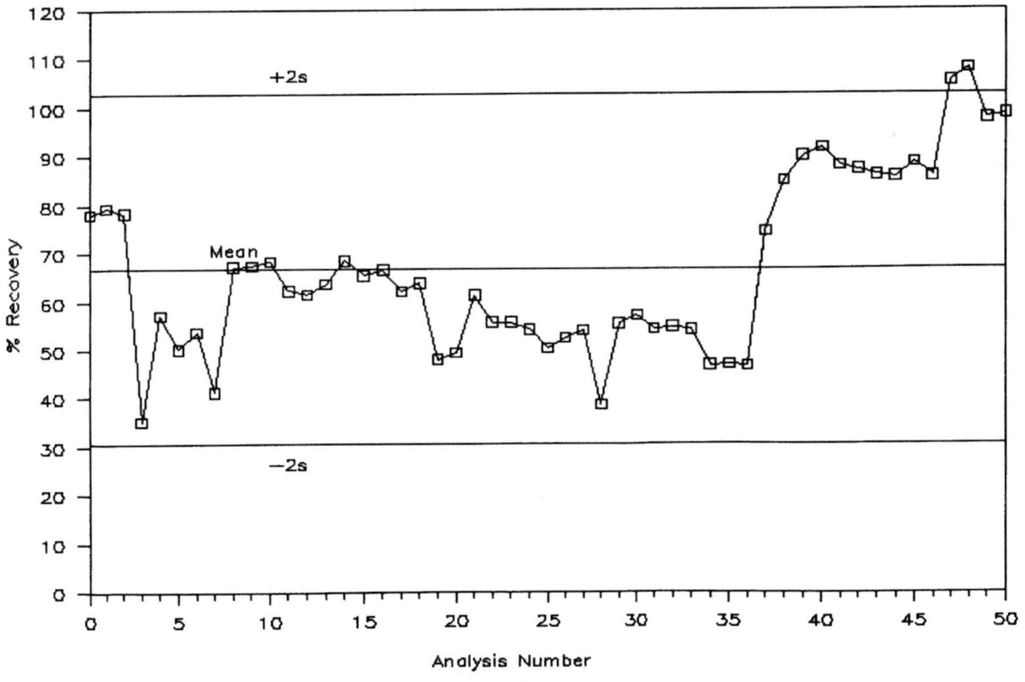

Figure 2. X-chart of naphthalene-d_8 recovery.

encountered in the analysis of BOD, COD, TOC. No problems were noted in the analysis of solids, cyanide or conductivity.

It was difficult to obtain the pH of these samples. Often it took from 3 to 5 minutes to obtain a stable reading. This is common with concentrated saline samples and probably has to do with cation interference at the electrode membrane.

Oil and grease contamination was present in field/trip blanks and duplication was outside of control limits. Method blanks for these analyses were clean. Even though the commercially available sample containers were precleaned, they may not be suitable for this analysis without additional cleaning.

Chloride analysis presented problems only in that the extremely high, varying concentrations required multiple dilutions. Initial fluoride analyses showed low spike recoveries in most cases. It was postulated that this was due to the presence of interferences, most likely organic compounds in the matrix. Consequently, samples were reprepared by Bellack distillation to remove the interfering substances before proceeding with reanalysis by specific ion electrode. Upon reanalysis, all matrix spikes were within acceptable limits.

The method employed for nitrate/nitrite determination, EPA 353.1, is a colorimetric method done by coupling reduced nitrate and nitrite to yield an azo dye and measuring color intensity at a specified wavelength. Any other agents that absorb at the specified wavelength will interfere with the analysis. Several of the samples had a background interference which resulted in a higher detection limit for these samples.

Several problems were encountered with the sulfide method. The method involves treating the sample with zinc acetate to produce zinc sulfide, if sulfide is present, followed by the addition of an excess of iodine to the zinc sulfide. Under acidic conditions, the iodine oxidizes the sulfide to sulfur. Excess iodine is then titrated with a sodium thiosulfate solution. The zinc sulfide is collected by filtering the sample. Salt from these samples also collected on the filter paper and when redissolved, clouded the solution and made the endpoint for the titration very difficult to see. An additional filtration step was incorporated between the iodine addition and the titration to separate the excess iodine from the salt. Other reducing agents that react with iodine will interfere with this method. In some cases this occurred and valid results for these samples could not be obtained.

Organics

All of the samples were analyzed using EPA Methods 1624 and 1625. A subset of the 24 samples were also analyzed using EPA Methods 624 and 625. Methods 624 and 625 are the usual required protocols for determining organic priority pollutants in water and wastewater matrices. The more elegant isotope dilution methods (1624 and 1625) are designed to improve the performance of the analytical methods on difficult matrices. The improved performance is derived from the ability of isotope dilution to automatically correct for matrix effects manifested in variable purging or extraction efficiencies or instrument variability.

In this study, high levels (relative to labeled compound spikes) of hydrocarbons were detected. Figure 1 illustrates a trace of m/z 43 over an entire chromatogram. Full scan spectral identification indicates that these peaks are primarily due to hydrocarbons. The presence of these compounds in these samples required dilutions as much as 250:1 to keep the instrument response within the calibration range. In some samples, up to six hours was required to bake off all of the compounds from the system before another sample could be analyzed. Both specific and non-specific interferences were noted in 1624 and 1625, often making quantification of target analytes or recovery of the isotopically labeled compounds difficult.

A comparison of the results from the isotope dilution methods versus the internal standard methods was performed for a set of nine samples. In general, there is good qualitative agreement between the two methods (isotope dilution and internal standard) with respect to detected compounds. Higher detection limits were necessary in the 1600 methods. This was due in part to the minimum detection limits required by these methods and the constraints on dilution procedures described in the methods. Some of the discrepancy however was due to the nature of the sampling protocol and sample matrix. Hydrocarbon levels in some samples analyzed without dilution required extensive system cleanup. Because of this and the lengthier QA/QC requirements for the 1600 methods, some predilutions based on representative samples within batches were made to meet the holding time constraints and sampling schedule. Since the 600 series tests were performed on a much smaller subset of samples, some evaluation of samples and multiple dilutions at lower levels were possible. Method 1624 tended to yield results slightly higher than Method 624 in this set of samples. Higher values may be a consequence of the isotope dilution technique which employs a stable labeled compound as a quantitation standard. For the purposes of comparison though, it must be noted that there is not a complete replication of target analytes between Methods 1624/1625 and Methods 624/625. This is particularly true in the case of 1625 vs. 625. In fact, most of the analytes found in the samples on this program were normal hydrocarbons, compounds which are not on the list of Method 625 analytes. Selected results of these comparisons are given in Tables 5 and 6. Only those compounds that were detected in at least one of the nine samples are reported. The agreement is generally good (within factors of 2). Summaries of reproducability comparisons are given in Tables 7 and 8. The tabulated RPDs represent the relative percent difference (for one sample) between the results for the same analyte as determined by the two methods, i.e. the difference between the result for benzene as determined by Method 624 versus the result for benzene as determined by Method 1624. While the target compound quantitation is comparable between the two methods, Methods 1624/1625 also require quantitation of labeled compound recoveries. The requirement to calculate recovery of the over 100 labeled compounds adds rigor to the 1600 methods not present in the 600 methods. A plot of the recovery for naphthalene-d_8 over the course of this study is shown in Figure 2. Labeled compound recoveries can be taken as a monitor of method performance, thus it can be seen that generally where data was obtainable, the method performed acceptably (for this analyte) for these samples.

In addition to interferences from hydrocarbons and oxygenates, some of the analyses suffered from sample non-homogeneity. That is, some samples were multiphasic. Floating organic layers made taking a representative aliquot for analysis extremely difficult. In those cases where an entire sample was taken for analysis, the relatively high levels of hydrocarbons resulted in elevated detection limits.

Isotope dilution analysis requires a minimum purchase of necessary isotopically labeled materials (approximately $3000US) and the set of isotopically labeled standards available exhibits some stability problems with a few compounds. Acrolein-d4 decomposes within days, while methyl ethyl ketone-d3 was found to be present as the d8-substituted compound. The presence of the d8-substituted methyl ethyl ketone is perhaps a result of deuterium exchange with the deuterated methanol solvent.

Table 5. Comparison of Method 1624 vs. 624 (μg/l)

	Sample #1		Sample #2	
Compound	1624	624	1624	624
Acetone	660	270	<3,400	<750
Methylethyl ketone	360	110	<3,400	<2,500
Benzene	<10	<8.8	10,300	8,000
Toluene	<10	<12	13,300	18,000
Ethylbenzene	<10	<14	<700	<720

Table 6. Comparison of Method 1625 vs. 625 (μg/l)

	Sample #1		Sample #2	
Compound	1625	625	1625	625
Di-n-butylphthalate	15	5.3	<10	<25
Phenanthrene	33	29	<20	<54
Acenaphthene	<10	2.3	<10	<19
Bis(2-ethylhexyl) phthalate	<10	9.4	<10	6
Phenol	<10	<1.6	442	110
Naphthalene	<10	<1.7	68	87
Pentachlorophenol	<20	<3.8	<20	<36
2,4-Dimethylphenol	<10	<2.8	305	290
Fluorene	<10	<2.0	<10	<19

Table 7. Semivolatiles analysis comparison Method 625/1625

	# Observations			
Analyte	>5 X DL	<5 X DL	Average RPD%	RPD Range
di-n-butyl-phthalate		1	65	—
Phenanthrene	3		44	11 - 61
Phenol		3	87	72 - 115
Naphthalene		2	104	22 - 185
Naphthalene	2		24	18 - 29
2,4-Dimethylphenol		1	89	—
Fluorene		1	89	—

Table 8. Volatiles analysis comparison Method 624/1624

	# Observations			
	>5 X DL	<5 X DL	Average RPD%	RPD % Range
Acetone	1		59	—
MEK		1	70	—
Benzene	5		37	2.8 - 89
Benzene		1	24	—
Toluene	4		45	9.4 - 92
Toluene		2	60	21 - 99

CONCLUSIONS

Most of the analytical methodologies employed in this study were adequate to characterize these produced waters. However, several of the methods with some of the samples suffered from interferences directly attributable to the sample matrix. In many cases complications due to the nature of the samples were overcome by use of techniques such as matrix modification. However, the constraints of EPA standard methods in many cases necessitated unacceptably high detection limits. High saline content of some samples made qualitative and quantitative determinations for several of the metals analyses extremely difficult. Other more specialized approaches could have been employed in several of the methods but this would have exceeded the bounds of the standard, routine methods and the scope of this program. Furthermore, attempts to analyze samples of this type without consideration of the kinds of matrix effects reported in this paper could yield unacceptable data. Furthermore, use of the isotope dilution techniques for the organics analysis gave generally comparable results to those generated by the more routine analytical methods.

REFERENCES

40 CFR, 136, Appendix A. Title 40 Code of Federal Regulations.

American Society for Testing and Materials Method D56-70, vols 05.01 and 06.03.

EPA 530/SW 846 "Test Methods for Evaluating Solid Waste; Physical Chemical Methods", 3rd Edition.

EPA 600/4-79-020 "Methods for Chemical Analysis of Water and Wastes".

ERT, Oil and Gas Industry Exploration and Production Wastes. Prepared for API, July 1987.

Keith, L.H., Mertens, S.K., Shore, F.L., Shepherd, M.C., Schrynemeeckers, P.J., and Gibson, J.S., 1987, Characterization of Produced Waters from Natural Gas Production Operations, Volume IIE, GRI Report 87.0335, December 1987.

Keith, L.H., Mertens, S.K., Shore, F.L., Shepherd, M.C., Schrynemeeckers, P.J.,and Gibson, J.S., 1988, Characterization of Produced Waters from Underground Natural Gas Storage Reservoir Operations, Volume IID, GRI Report 88.0080.2, June 1988.

Koraido, S., Fillo, J., Evans, J., Sources, Characterizations and Management of Produced Waters from Natural Gas Industry Operations, in: "Produced Water: Technological/Environmental Issues and Solutions," J.P. Ray and F.R. Engelhardt, eds., Plenum Publishing Corp., New York, New York, USA.

Mason, B., 1987 Audit Report for Field Audit - Grant's Pass Oregon, April 23.

Standard Methods for the Examination of Water and Wastewater, 1985, 16th ed.

USEPA Contract Laboratory Program, Statement of Work for Organics Analysis, Multimedia, Multiconcentration, 10/86.

USEPA Report to Congress, Management of Development Wastes from the Exploration and Production of Crude Oil, Natural Gas, and Geothermal Energy. EPA 530/SW 88003, December, 1987.

Wesolowski, D. Broughton, A., Hansotte, C. A., Koraido, S.M. and Fillo, J.P. , 1987, Characterization of produced Waters from Underground Natural Gas Production Operations, Volume 1, GRI Report 87.0335.1, December 1987.

Wesolowski, D., Broughton, A., Hansotte, C.A., Koraido, S.M. and Fillo, J.P., 1988, Characterization of Produced Waters from Underground Natural Gas Storage Reservoir Operations, Volume 1, GRI Report 88.0080.1, June 1, 1988.

USE OF LABORATORY TOXICITY DATA FOR EVALUATING THE ENVIRONMENTAL ACCEPTABILITY OF PRODUCED WATER DISCHARGE TO SURFACE WATERS

D.R. Mount[1], K.R. Drottar[1], D.D. Gulley[2], J.P. Fillo[3], and P.E. O'Neil[4]

[1] ENSR Consulting and Engineering
Fort Collins, Colorado
USA

[2] University of Wyoming
Laramie, Wyoming
USA

[3] ENSR Consulting and Engineering
Pittsburgh, Pennsylvania
USA

[4] Geological Survey of Alabama
Tuscaloosa, Alabama
USA

INTRODUCTION

Produced water from fossil fuel production operations are managed using a variety of disposal technologies. Even in freshwater situations, discharge to surface waters can be an attractive management option, particularly for produced waters with comparatively low (e.g., <5,000 mg/l) total dissolved solids concentrations (TDS). Not only is surface discharge an inexpensive means of disposal (assuming extensive treatment is not necessary), but it also provides water that may have other beneficial uses, especially in arid regions. However, in order to comply with Clean Water Act mandates, surface discharges must not adversely affect designated uses of the receiving waters. This generally includes assuring protection of aquatic organisms.

In most areas of the United States, effluent toxicity or "biomonitoring" tests are used as a complement to chemical-specific measures for quantifying the potential ecological effects of produced water discharges and/or determining their acceptability for discharge. USEPA (Federal Register 49:9016-9019) has placed heavy emphasis on biomonitoring tests as part of the water quality-based approach to toxics control being implemented through the National Pollutant Discharge Elimination System (NPDES). In defense of this approach, USEPA completed a series of studies attempting to link results of laboratory toxicity tests with instream effects (USEPA, 1984; 1985a, b; 1986a, c, d, e).

Produced Water, Edited by J.P. Ray
and F.R. Engelhart, Plenum Press, New York, 1992

To better understand the potential ecological effects of produced water discharges to surface waters, the Gas Research Institute (GRI) funded a series of studies examining the responses of aquatic organisms to produced water exposure. Cooperating investigators included ENSR Consulting and Engineering, the Geological Survey of Alabama (GSA), and the University of Wyoming (UW). The purpose of this paper is to summarize important findings of these studies as they apply to the surface discharge of produced water.

As mentioned above, the ability of laboratory toxicity tests to predict actual instream biological effects is a key premise underlying the application of biomonitoring requirements to produced water discharges. While USEPA has conducted studies to verify the connection, none of these studies involved produced water. Furthermore, these field studies focused on existing conditions in the receiving waters; they did not involve deliberate manipulation of the instream conditions that could then be used to establish concentration/response relationships for the instream community. The first of the studies described in this paper involved comparison of laboratory toxicity data with the responses of the native benthic community for a produced water discharged to a surface water. As is described below, the results indicated that laboratory toxicity tests could be used to predict instream effects. In addition, it seemed that a large part of the produced water toxicity was due to TDS, or more specifically, chloride.

Having demonstrated the predictive ability of laboratory toxicity tests, additional studies were conducted to evaluate potential sources of aquatic toxicity within produced waters as a group. Initial experimentation with inorganic salts quickly showed that the toxicity of produced water would likely vary depending on the concentrations of different major ions in the water. This research was then expanded and ultimately led to the development of a Salinity/Toxicity Relationship (STR) that could predict aquatic toxicity based on the concentrations of major ions (Gulley et al., 1992). In the final section of this paper, we present the results of an initial application of the STR to surface waters receiving produced water discharges.

RELATIONSHIP BETWEEN LABORATORY TOXICITY TESTS AND INSTREAM EFFECTS OF EXPOSURE TO PRODUCED WATER

From 1986 to 1988, a series of laboratory and field biomonitoring studies were conducted on produced water from a coalbed methane operation in the Cedar Cove Degasification Field in Alabama. The primary goal of these studies was to determine the environmental acceptability of discharging coalbed methane produced water to surface waters of the region. In addition, the study provided an excellent opportunity to compare the results of laboratory and field methods for making this determination. Within the Cedar Cove field, the specific study site was Pond #13, which received produced water from several wells. Chemically, Pond #13 water was dominated by sodium and chloride (typically 1,000 to 2,000 mg/l chloride), although it also contained a variety of other constituents. Extensive chemical characterizations of this water were conducted and reported by O'Neil et al. (1989). After retention in the pond, the produced water was released to Little Hurricane Creek (LHC), a small warm-water stream (1987 average flow = 0.394 m^3/s).

Laboratory Toxicity Tests

Laboratory toxicity studies on the Pond #13 produced water were conducted in three locations: (1) a mobile bioassay trailer located adjacent to Pond #13; (2) laboratory space at the GSA in Tuscaloosa; and (3) at ENSR's Environmental Toxicology Laboratory in Fort Collins, Colorado. The majority of these tests were acute and short-term chronic tests using water fleas, *Ceriodaphnia dubia*, and fathead minnows, *Pimephales promelas*. Procedures used for these tests followed the current USEPA guidance for estimating the acute and chronic toxicity of effluents (USEPA, 1985c, d). In addition, flow-through acute exposures (USEPA, 1985c) were conducted using two organisms native to LHC, rough shiners (*Notropis baileyi*) and mayflies

(*Stenacron interpunctatum*). Finally, a limited number of toxicity characterization tests (USEPA, 1988a) were conducted in an attempt to evaluate the cause(s) of acute toxicity in the produced water.

Instream Studies

Concurrent with the laboratory toxicity studies, GSA staff conducted a series of instream studies examining the effect of produced water discharge on benthic organisms native to LHC. To conduct these experiments, a portion of the LHC channel was isolated by constructing an instream experimental mesocosm. Concentrations of produced water were maintained in the mesocosm by controlled mixing of stream water and produced water. A continuous specific conductance recorder monitored the concentration of produced water in the mesocosm.

Benthic invertebrate community structure was evaluated using a modified Hester-Dendy artificial substrate sampler (Hester and Dendy, 1962; O'Neil et al., 1989). Prior to each experiment, two sampling arrays were each loaded with 48 Hester-Dendy samplers and placed in LHC upstream from the produced water discharge. After a period of colonization, one array was moved into the mesocosm and the other remained in a control section of LHC away from produced water influence. Replicate samplers were collected at pre-determined intervals from each array for a period of 20 days. Benthic organisms present on the samplers were removed, preserved, identified, and enumerated. A total of seven instream trials were conducted, each at a different produced water concentration within the mesocosm. For additional experimental details regarding both laboratory and field studies, see O'Neil et al. (1989).

Data Interpretation

Acute toxicity data were used to calculate LC_{50} values and 95 percent confidence intervals (95% CI) by either non-linear interpolation/binomial (USEPA, 1985c) or Spearman-Karber (Hamilton et al., 1977) methods; the LC_{50} is the estimated concentration of a material that will cause 50 percent mortality of test organisms within a specified time period. Chronic toxicity data are expressed in terms of the IC_{25} concentration (1988b). The IC_{25} is a regression-based value representing the sample concentration estimated to cause a 25 percent reduction in performance (e.g., young production) relative to control organisms. Recent analyses by USEPA (1989) have found the IC_{25} concentration to approximate the No Observed Effect Concentration (NOEC) for effluent toxicity tests.

Results

Acute toxicity testing showed *Ceriodaphnia* to be the most sensitive species to produced water, with 48-hour LC_{50} values of 46 percent (95% CI = 40 to 54 percent), 64 percent (57 to 72 percent), and 71 percent (50 to 100 percent) produced water (LC_{50} values of 1,063, 1,377, and 990 mg/l as chloride, respectively). Both fathead minnows and rough shiners were not acutely sensitive to the produced water (96-hour LC_{50} > 100 percent produced water containing 1655 to 2160 mg/l chloride). The acute test with mayflies was unsuccessful because of emergence occurring during the test; nonetheless, the mayflies did not appear to be more acutely sensitive than were *Ceriodaphnia*.

Although the concentrations of chloride were thought to be sufficient to account for the observed acute toxicity to *Ceriodaphnia*, several acute toxicity characterization tests (USEPA, 1988a) were conducted to help verify this suspicion. In these tests, acute toxicity was not reduced by aeration, indicating that volatile materials (e.g., sulfide) were not a primary toxicant. Likewise, filtration at ambient pH or at pH 11 did not reduce toxicity either; filtration at high pH removes toxicity caused by several cationic metals. Acute tests conducted using sodium chloride gave a LC_{50} values comparable to the chloride concentration in produced water at its LC_{50} concentration. Based on these data, it was concluded that chloride was the primary cause of acute toxicity to *Ceriodaphnia*.

Ceriodaphnia were also the most sensitive species to the chronic toxic effects of exposure to the Pond #13 produced water. Initial tests conducted in April 1986 and May 1987 showed IC_{25} values of 8 and 15 percent produced water, respectively (177 and 255 mg/l as chloride; Figure 1). In contrast, all subsequent tests showed much less chronic toxicity, with IC_{25} values of 26 to 56 percent produced water (544 to 676 mg/l as chloride; Figure 1). When tested in laboratory water, the toxicity of sodium chloride was very similar (IC_{25} = 652 mg/l chloride); when sodium chloride was tested in LHC water, a somewhat lower IC_{25} value of 394 mg/l chloride was observed (Figure 1). It was hypothesized that the relatively low pH of LHC water at the time (pH 5.5 to 6.1) could have contributed to this lower IC_{25} value, although performance of control organisms in LHC water met standard acceptance criteria (USEPA, 1989). Regardless, it is clear that chloride concentration alone can account for the chronic toxicity observed in tests conducted from June 1987 forward. The lower IC_{25} values observed in April 1986 and May 1987 (indicating greater toxicity) do not appear consistent with chronic toxicity due to chloride alone. This difference was taken to indicate the presence of one or more toxicants in addition to chloride, but the disappearance of the additional toxicity in later testing did not allow us to characterize it further.

Instream studies were conducted from June 1987 through October 1988. Although extensive benthic community data were collected (see O'Neil et al., 1989), we have summarized all seven trials based on the existence of a statistically significant difference in the number of taxa, density, or diversity of benthic organisms inhabiting control and exposed substrates after 20 days of exposure (Table 1).

A clear threshold for effect was demonstrated by these exposures, with no significant effects on native invertebrates at chloride concentrations of 519 mg/l and below, and consistent effects at concentrations of 615 mg/l chloride and above. These values agree very nicely with the concurrent laboratory studies (June 1987 to October 1988), which showed IC_{25} values of 544

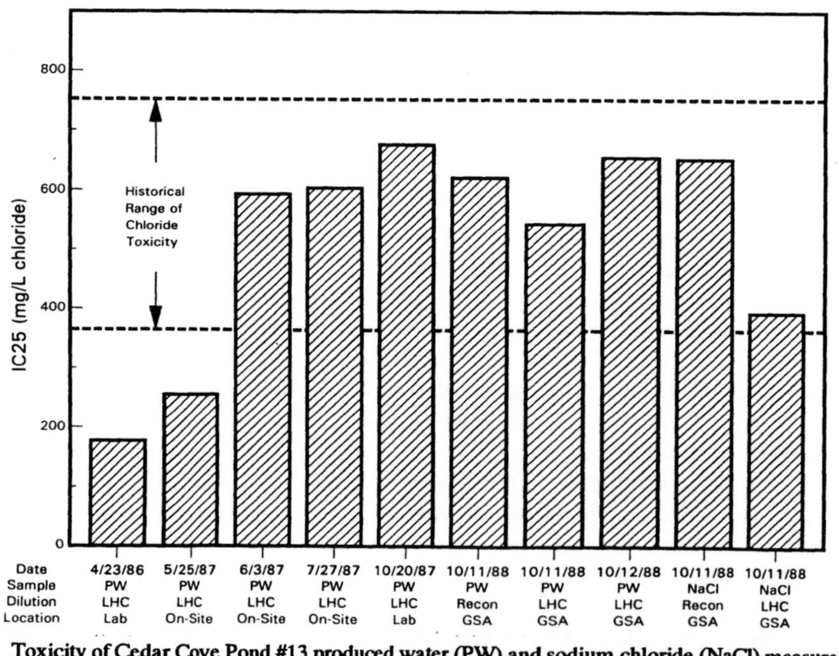

Figure 1. Toxicity of Cedar Cove Pond #13 produced water (PW) and sodium chloride (NaCl) measured during 1986 to 1988. Studies were conducted using either Little Hurricane Creek water (LHC) or moderately hard reconstituted water (Recon; USEPA, 1989) and were conducted either in a mobile laboratory at Pond #13 (On-site), at the Geological Survey of Alabama (GSA), or at ENSR's laboratory in Fort Collins, Colorado (Lab). Results are expressed as the IC_{25} value normalized to chloride concentration. Historical range is from ENSR's in-house sodium chloride data (± 2 S.D.; n = 12).

to 676 mg/l chloride for the most sensitive of the laboratory species, *Ceriodaphnia* (note that the two tests with lower IC_{25} values pre-dated the instream studies). Modeling of instream results indicated that a 25 percent reduction in number of invertebrate taxa would occur at approximately 600 mg/l chloride (O'Neil et al., 1992b). These findings confirmed a close relationship between standard biomonitoring tests and effects on instream communities.

As an epilogue to the Cedar Cove study, the GSA has conducted long-term monitoring of the fish and benthic communities in LHC during controlled discharge of produced water at concentrations below the approximately 600 mg/l chloride associated with adverse effects in previous laboratory and instream experiments (O'Neil et al., 1991; 1992a). Results of these studies have indicated no effect of these discharges on benthic invertebrate community structure, demonstrating that produced water can be discharged to a surface water without adverse effect if its potential toxicity is properly assessed and the discharge is managed and monitored accordingly (O'Neil et al., 1992b).

Table 1. Results of instream exposures of benthic macroinvertebrates to seven concentrations of produced water, normalized to chloride concentration (O'Neil et al., 1989). Significant effect was defined as a statistically significant ($p < 0.05$) reduction in density, diversity, or number of taxa

Chloride Concentration[a](mg/l)	Significant Effect?
268	No
490	No
519	No
615	Yes
641	Yes
883	Yes
944	Yes

[a]Conditions during each trial are expressed as the mean chloride concentration during the 20-day exposure.

EXTRAPOLATION OF RESULTS

While both laboratory and instream studies conducted at Cedar Cove implicated chloride as the primary cause of toxicity, there was also indication that other toxicants may have been present during the first two chronic toxicity tests. Produced waters can contain a variety of trace constituents (either natural or introduced), and these materials have the potential to influence produced water toxicity. However, virtually all produced water contains elevated concentrations of chloride or other major ions, regardless of trace constituents. Given the demonstrated importance of these major ions in determining toxicity of the Cedar Cove produced water, evaluation of toxicity of produced waters as a group must incorporate a basic understanding of toxicity caused by major ions.

Initial experimentation showed that the relationship between total dissolved solids and toxicity was not simple. Results of acute *Ceriodaphnia* toxicity tests (USEPA, 1988a) conducted on a variety of common salts dissolved in moderately hard reconstituted water (USEPA, 1989) clearly shows that the toxicity of these salts varies widely (Table 2). For example, LC_{50} values for chloride were very similar when tested as sodium chloride and calcium chloride (1,294 and 1,147 mg/l as chloride, respectively), but LC_{50} values for potassium and magnesium chlorides were markedly lower (261 and 717 mg/l as chloride, respectively). Likewise, the toxicity of sulfate salts varied, with LC_{50} values from 397 to 2,503 mg/l as sulfate (Table 2). We believe that these differences are not the result of variations in the toxicity of chloride or sulfate, but rather toxicity caused by another ion (e.g., potassium). Regardless, the overriding concept is that the specific ionic composition of the solution must be known before its toxicity can be predicted; simple measures such as TDS, salinity, or chloride concentration are not robust predictors of toxicity over a range of ionic compositions. Moreover, evaluating these simple salt

Table 2. Acute toxicity of several salts to *Ceriodaphnia dubia*

	Ceriodaphnia 48-h LC_{50} (mg/l)			
Salt	As Salt[a]	95% C.I.[b]	As Cation[c]	As Anion[d]
KCl	548	464 - 648	287	261
K_2SO_4	720	610 - 850	323	397
$KHCO_3$	501	419 - 599	196	305
NaCl	2,132	1,831 - 2,482	838	1,294
Na_2SO_4	3,701	2,683 - 4,960	1,198	2,503
$NaHCO_3$	998	857 - 1,162	273	725
$CaCl_2$	1,794	1,326 - 2,500	647	1,147
$CaSO_4$	> 2,020[e]	NA[f]	> 595[e]	> 1,425[e]
$MgCl_2$	962	688 - 1,305	245	717
$MgSO_4$	1,766	1,246 - 2,425	356	1,410

[a]Expressed in terms of the combined salt.
[b]95% confidence interval calculated by the binomial method (USEPA 1985c) for Na_2SO_4, $CaCl_2$, $MgCl_2$, and $MgSO_4$; all others (except $CaSO_4$) calculated using Spearman-Karber method (Hamilton et al., 1977).
[c]Expressed in terms of the cation concentration.
[d]Expressed in terms of the anion concentration.
[e]No acute toxicity in a saturated solution.
[f]Not applicable.

solutions does not consider the potential interactions among ions when several are present at elevated concentrations, as occurs in many produced waters.

PREDICTION OF MAJOR ION TOXICITY USING A SALINITY/TOXICITY RELATIONSHIP

While toxicity data for simple salts show that predicting toxicity of saline mixtures is not always straightforward, the ability to rapidly assess the potential toxicity of such mixtures is still clearly desirable. Not only is chemical analysis for major ions less expensive than toxicity testing, but chemical composition of future discharges might be inferred from geochemical information even before actual samples of a produced water are available. To provide a means to predict major ion toxicity for both simple and complex ion mixtures, GRI funded additional research to develop a Salinity/Toxicity Relationship (STR). In this research, multi-variate logistic regression was used to quantify the empirical relationship between major ion composition and toxicity for a wide variety of ion mixtures. Ions considered in the STR are sodium, calcium, magnesium, potassium, chloride, sulfate, and bicarbonate. At its current stage of development, predictions of acute toxicity can be made for three freshwater species, *Ceriodaphnia, Daphnia magna*, and fathead minnows. Development and details of the STR are described elsewhere in this symposium by Gulley et al. (1992).

Although the STR was only recently developed, its initial applications to actual produced waters have been very promising. As an example, the STR was applied to chemistry and toxicity data extracted from the research of Boelter et al. (1992). This study was conducted on Salt Creek and the Powder River in Wyoming, an area receiving a number of produced water discharges. Ambient water samples were collected from the watershed on four different dates; these samples were analyzed for major ion concentrations and tested for acute toxicity to *Ceriodaphnia*. A range of both toxicity and chemical composition was observed (Table 3). Some of the waters were dominated by sulfate, while others showed chloride and bicarbonate as the predominate anions; survival of *Ceriodaphnia* ranged from 0 to 100 percent.

Though the STR was not available at the time the original research was conducted, the data provide an excellent verification exercise. Major ion concentrations measured in these samples

Table 3. Chemical and toxicity data from Salt Creek, Wyoming (Boelter et al, 1992), along with STR (Gulley et al., 1992) predictions of *Ceriodaphnia* survival

			Concentration (mg/l)								*Ceriodaphnia* 48-h Survival (%)		
Date	Site	Percent Sample	Sodium	Calcium	Magnesium	Potassium	Chloride	Sulfate	Bicarbonate	Total Dissolved Solids	STR Predicted	Observed	Error
May 1988	USC	100	845	219	109	9	8	809	323	2,322	98	100	2
	LIN	100	1,255	55	52	18	1,042	383	712	3,517	1	0	-1
	LIN	50	628	28	26	9	521	192	356	1,760	86	100	14
	LIN	10	126	6	5	2	104	38	71	352	100	90	-10
	CSC	100	1,410	47	62	17	1,066	466	878	3,946	0	0	0
	HBR	100	268	71	32	4	171	152	268	966	100	100	0
	IRY	100	239	70	31	4	139	149	242	874	100	100	0
	CPR	100	73	70	24	2	8	85	157	419	100	100	0
Aug 1988	USC	100	895	235	132	11	8	914	307	2,502	99	90	-9
	LIN	100	1,420	24	24	14	1,123	204	927	3,736	0	0	0
	LIN	50	710	12	12	7	562	102	464	1,869	76	70	-6
	LIN	25	355	6	6	4	281	51	232	935	99	90	-9
	CSC	100	1,430	20	20	15	1,238	215	859	3,797	0	0	0
	CSC	50	715	10	10	8	619	108	430	1,900	75	90	15
	HBR	100	1,290	24	24	14	1,118	249	831	3,550	0	0	0
	HBR	50	645	12	12	7	559	125	416	1,776	81	80	-1
	IRY	100	1,205	37	37	15	1,006	322	599	3,221	4	0	-4
	IRY	50	603	19	19	8	503	161	300	1,613	92	100	8
	CPR	100	254	208	208	8	82	419	211	1,390	100	100	0
Oct 1988	USC	100	805	229	112	9	8	804	343	2,310	99	100	1
	LIN	100	1,390	30	30	16	1,294	242	934	3,936	0	0	0
	LIN	50	695	15	15	8	647	121	467	1,968	64	50	-14
	LIN	25	348	8	8	4	324	61	234	987	99	100	1
	CSC	100	1,395	25	33	16	1,299	260	876	3,904	0	0	0
	CSC	50	698	13	17	8	650	130	438	1,954	68	80	12
	HBR	100	845	51	42	9	777	282	604	2,610	14	50	36
	HBR	50	423	26	21	5	389	141	302	1,307	96	90	-6
	IRY	100	685	74	53	11	552	273	414	2,062	67	90	23
	CPR	100	200	161	64	5	144	320	305	1,199	100	90	-10
Sept 1989	LIN	100	1,015	21	21	39	1,301	543	899	3,839	0	0	0
Correlation with Survival[a] (r^2)			0.80	0.12	0.04	0.82	0.89	0.00	0.89	0.80	0.95	NA	NA
		Slope	(-)	+	+	(-)	(-)	(-)	(-)	(-)	+	NA	NA

[a]Coefficients of determination are from linear regressions for predicted survival and total dissolved solids; logistic regression was used for the remaining chemical parameters.

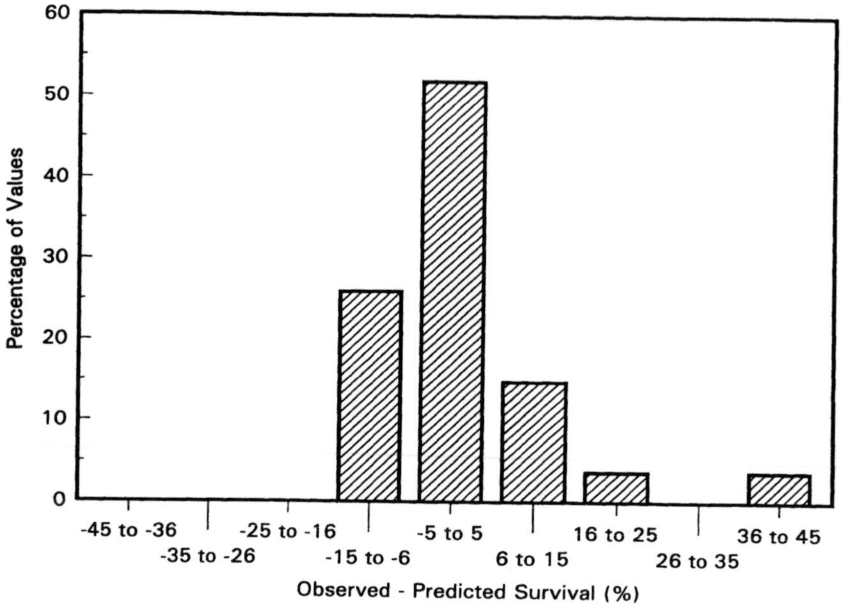

Figure 2. Frequency distribution showing differences (in percent) between the observed *Ceriodaphnia* survival (Boelter et al., 1992) and survival predicted by the STR (Gulley et al., 1992; Table 3)

were entered into the STR equation, which then predicted *Ceriodaphnia* survival for each of the samples (Table 3). Agreement between predicted and observed *Ceriodaphnia* survival was remarkable (r^2 = 0.946); more than half of the predicted values were within 5 percent of the actual observed survival and more than 90 percent of the predictions were within 15 percent (Figure 2). The correlation of observed survival with survival predicted by the STR was higher than that for any chemical variable (Table 3).

More importantly, this example application demonstrated the ability of the STR to account for the toxic interactions among ions in a complex mixture. Despite relatively high sulfate concentrations (up to 914 mg/l), there was no relationship between sulfate concentration and *Ceriodaphnia* survival (r^2 = 0.00). In contrast, both chloride and bicarbonate concentrations show clear relationships with survival (Figures 3 and 4). From these field data, one can also estimate LC_{50} values of about 710 mg/l chloride and about 515 mg/l bicarbonate (see Figures 3 and 4). Although the graphical relationships are strong, these LC_{50} values from the field data are much lower than the values derived from experiments with sodium salts (1310 mg/l chloride and 824 mg/l bicarbonate; Table 2), even though sodium was the dominant cation in the field samples. Toxicity observed in the field samples with elevated concentrations of both chloride and bicarbonate was apparently due to an additive toxicity between the two, rather than the toxicity of either one individually. By modeling all major ions simultaneously, the STR accounts for the presence of such toxic interactions, enabling it to provide accurate predictions even for complex ion mixtures where single chemical data may be misleading.

It must be emphasized that when the STR is applied to produced water, its predictions of toxicity are expected to be accurate only if major ions are the primary toxicant. Yet, therein lies another powerful application of the STR. If the observed toxicity of a sample is considerably higher than STR predictions for one or more species, then the presence of an additional toxicant can be inferred. The STR has already been used to detect produced water toxicity caused by hydrogen sulfide (Gulley et al., 1992). Future research plans call for additional applications of the STR to produced water samples containing both elevated TDS and potentially toxic concentrations of trace constituents.

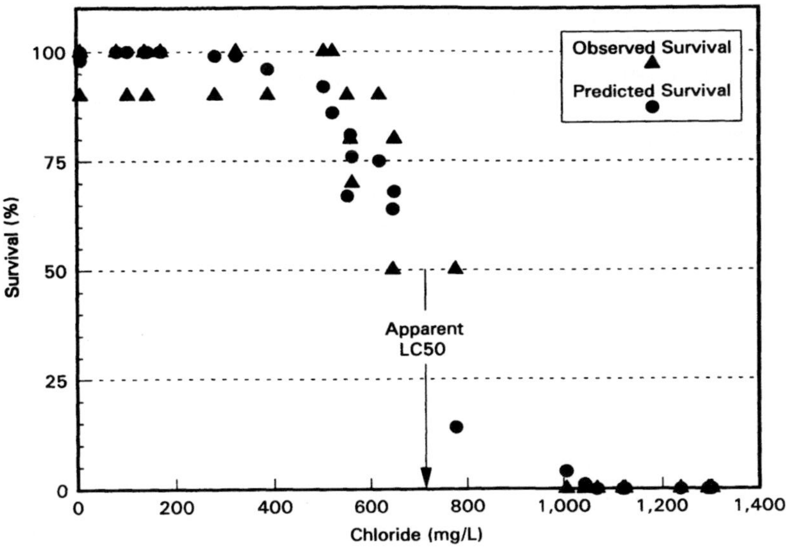

Figure 3. Relationship between chloride concentration and *Ceriodaphnia* survival for ambient water samples from the Salt Creek area of Wyoming (Boelter et al., 1992). LC_{50} estimated visually.

SUMMARY

Research conducted at Cedar Cove suggests that laboratory toxicity tests can be used to predict instream effects of produce water discharge, concurring with earlier USEPA conclusions regarding other types of industrial and/or municipal discharges (USEPA, 1984; 1985a, b; 1986a, c, d, e). While the utility of laboratory tests is clear, some freshwater organisms are sensitive to the elevated TDS concentrations typical of produced waters, in addition to traditional "toxics." Hence, properly interpreting the results of laboratory toxicity tests conducted on produced waters is dependent in part on understanding the relationships between major ion concentrations

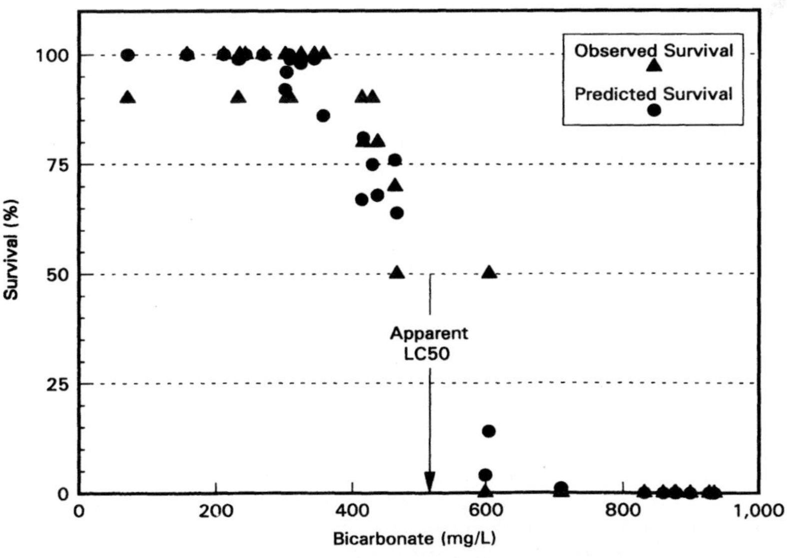

Figure 4. Relationship between bicarbonate concentration and *Ceriodaphnia* survival for ambient water samples from the Salt Creek area of Wyoming (Boelter et al., 1992). LC_{50} estimated visually.

and any observed toxicity. Although these relationships are obviously complex, studies completed thus far indicate that the STR approach will provide a valuable research and management tool for evaluating the efficacy of produced water discharge to surface waters.

ACKNOWLEDGEMENTS

This research was supported by the Gas Research Institute under Contract Nos. 5084-253-1019, 5084-253-1117, and 5091-253-2160. We thank the GRI project manager, J.M. Evans, for his advice, encouragement, and helpful comments on the manuscript. Comments from an anonymous reviewer also improved the manuscript. Finally, we thank K. Barten, T.D. Garrison, S.C. Harris, J.R. Hockett, S.W. McGregor, M.F. Mettee, S.J. Patti, and T.E. Shepard for their respective contributions to the research.

REFERENCES

Boelter, A.M., F.N. Lamming, A.M. Farag, and Bergman, H.L., 1992, Environmental effects of saline oil-field discharges on surface waters. Environ. Toxicol. Chem. (in press)

Gulley, D.D., D.R. Mount, J.R. Hockett, and Bergman, H.L., 1992, A statistical model to predict toxicity of saline produced waters to freshwater organisms. in "Produced Water: Technological/Environmental Issues and Solutions," J.P. Ray and F.R. Engelhardt, eds., Plenum Publishing Corp., New York, New York, USA.

Hamilton, M.A., R.C. Russo, and R.V. Thurston, 1977, Trimmed Spearman-Karber method for estimating median lethal concentrations in toxicity bioassays. Environ. Sci. Technol. 11:714-719 (correction [1978] 12:417).

Hester, F.E., and Dendy, J.S., 1962, A multiple-plate sampler for aquatic macroinvertebrates. Trans. Am. Fish. Soc. 91:420-421.

O'Neil, P.E., S.C. Harris, M.F. Mettee, K.R. Drottar, D.R. Mount, and Fillo, J.P. 1989, Biomonitoring of a Produced Water Discharge from the Cedar Cove Degasification Field, Alabama. Final report. Gas Research Institute, Chicago, Illinois, GRI-89/0073.

O'Neil, P.E., S.C. Harris, M.F. Mettee, S.W. McGregor, and Shepard, T.E., 1991, Long-term Biomonitoring of a Produced Water Discharge from the Cedar Cove dEgasification Field, Alabama. Geological Survey of Alabama, Tuscaloosa, Alabama, Bulletin 141.

O'Neil, P.E., S.C. Harris, M.F. Mettee, H.R. Isaacson, and Evans, J.M., 1992a, Biological Fate and Effect of Coalbed Methane Produced Waters Discharged into Streams of the Warrior Basin, Alabama. Proceedings of the 1992 Produced Water Symposium. Plenum Publishing Corporation: New York, New York.

O'Neil, P.E., H.R. Isaacson, and Evans, J.M., 1992b, Surface discharge of coalbed methane produce waters into the Warrior Basin of Alabama, the Cedar Cove model. in: "Produced Water: Technological/Environmental Issues and Solutions," J.P. Ray and F.R. Engelhardt, eds., Plenum Publishing Corp., New York, New York, USA.

U.S. Environmental Protection Agency (USEPA), 1984, Effluent and ambient toxicity testing and instream community response on the Ottawa River, Lima, Ohio. USEPA, Washington, D.C., EPA/600/3-84/080.

USEPA, 1985a, Validity of Effluent and Ambient Toxicity Tests for Predicting Biological Impact, Scippo Creek, Circleville, Ohio. USEPA, Washington, D.C., EPA/600/3-85/044.

USEPA, 1985b, Validity of Effluent and Ambient Toxicity Tests for Predicting Biological Impact on Five Mile Creek, Birmingham, Alabama. USEPA, Washington, D.C., EPA/600/8-85/015.

USEPA, 1985c, Methods for Measuring the Acute Toxicity of Effluents to Freshwater and Marine Organisms. USEPA, Cincinnati, Ohio, EPA/600/4-85/013.

USEPA, 1985d, Short-term methods for Estimating the Chronic Toxicity of Effluents and Receiving Waters to Freshwater Organisms. USEPA, Cincinnati, Ohio, EPA/600/4-85/014.

USEPA, 1986a, Validity of Effluent and Ambient Toxicity Tests for Predicting Biological Impact, Ohio River, near Wheeling, West Virginia. USEPA, Washington, D.C., EPA/600/3-85/071.

USEPA, 1986b, Validity of Effluent and Ambient Toxicity Tests for Predicting Biological Impact, Kanawha River, Charleston, West Virginia. USEPA, Washington, D.C., EPA/600/3-86/006.

USEPA, 1986c, Validity of Effluent and Ambient Toxicity Tests for Predicting Biological Impact, Skeleton Creek, Enid, Oklahoma. USEPA, Washington, D.C., EPA/600/8-86/002.

USEPA, 1986d, Validity of Effluent and Ambient Toxicity Tests for Predicting Biological Impact, Naugatuck River, Waterbury, Connecticut. USEPA, Washington, D.C., EPA/600/8-86/005.

USEPA, 1986e, Validity of Effluent and Ambient Toxicity Tests for Predicting Biological Impact, Back River, Baltimore Harbor, Maryland. USEPA, Washington, D.C., EPA/600/8-86/001.

USEPA, 1988a, Methods for Acute Toxicity Identification Evaluations: Phase I Toxicity Characterization procedures. USEPA, Duluth, Minnesota, EPA/600/3-88/034.
USEPA, 1988b, An Interpolation Estimate for Chronic Toxicity: the ICp Approach. USEPA, Duluth, Minnesota, National Effluent Toxicity Assessment Center Technical Report 05-88.
USEPA, 1989, Short-term Methods for Estimating the Chronic Toxicity of Effluents and Receiving Waters to Freshwater Organisms. Second edition. USEPA, Cincinnati, Ohio, EPA/600/4-89/001A.

TOXICITY IDENTIFICATION AND CHARACTERISTICS OF PRODUCED WATER DISCHARGES FROM COLORADO AND WYOMING

K. W. Fucik

T.H.E. Laboratories
Broomfield, Colorado
USA

INTRODUCTION

In 1987, the Region VIII office of the Environmental Protection Agency began requiring dischargers holding National Pollutant Discharge Elimination System (NPDES) permits to monitor effluents for biological toxicity. This affected numerous municipal and industrial dischargers including oil and gas producers discharging produced waters to surface waters.

The biomonitoring procedures measure toxicity to an invertebrate, *Ceriodaphnia dubia*, and the fathead minnow, *Pimephales promelas*. One advantage of toxicity tests is their ability to directly measure potential aquatic effects compared to chemical analyses which are difficult to extrapolate. This is particularly true in the case of complex effluents where a broad range of toxicants can be present in low levels.

In the western U.S., produced waters from oil and gas operations are frequently put to beneficial use. The state of Wyoming, for example, recognizes the value of produced waters as a water source for cattle and other range animals. Many of these waters are discharged to stock ponds and provide much if not all the water in these arid regions. However, other produced waters are discharged to streams and rivers that can be sources of drinking waters or are of important wildlife and recreational value. The regulations are designed to protect these resources.

The present results are a compilation of biomonitoring data collected for oil and gas produced water discharges over a four year period. The tested discharges came from operating wells throughout Colorado and Wyoming. Many of the samples have been monitored over periods of three years or more as part of normal permit requirements. In other cases, single samples were collected as part of an operator's data gathering procedures for operational decisions. Toxicity reduction evaluations were performed on various samples to isolate the causes of toxicity to the aquatic test species. These investigations generally involved multiple samples over short periods of time and were used by operators to establish permit requirements as well as discharge procedures.

MATERIALS AND METHODS

Effluent samples generally arrived at the lab within 24-48 hours after collection. The water was refrigerated at 4° C until used. Upon receipt, samples were analyzed for alkalinity, hardness, ammonia, and chlorine. Tests procedures followed the guidelines in EPA 600/4-85/013 (Peltier

Produced Water, Edited by J.P. Ray
and F.R. Engelhart, Plenum Press, New York, 1992

and Weber, 1985). Ceriodaphs exposed to the effluent were less than 24 hours old at the start of the test while the fathead minnows were seven days old or less. The organisms were obtained from in-house cultures or from commercial sources and were monitored through reference toxicant tests using sodium dodecyl sulfate.

The invertebrate tests were conducted in 30 ml disposable plastic cups, each containing 15 ml of the exposure medium. Fish were exposed in 260 ml plastic disposable cups with 200 ml of medium. Exposures were conducted over 48 hours for invertebrates and 96 hours for fish and followed a 16 hr light/8 hr dark cycle in an environmental chamber at 20° C or 25° C. All waters were changed daily and the temperature, dissolved oxygen, and temperature in each concentration measured daily. Conductivity of the effluent was measured at the beginning of the test. Concentrations included 100%, 75%, 50%, 25%, 12%, and 6% dilutions. A minimum of four replicates of five organisms each were exposed in each exposure concentration. Nearby receiving stream water was used as the control and dilution water source. When unavailable, moderately hard laboratory reconstituted water was used for dilutions.

In cases where toxicity reduction evaluations (TRE's) were initiated to isolate causes of toxicity, procedures generally followed the guidelines described in Mount and Anderson-Carnahan (1988) for Phase I toxicity identification evaluations (TIE). Additional testing was initiated as required to confirm suspected causes of toxicity. More detailed discussions about these tests are included in the following sections. Statistical analyses were performed on individual samples as required using the TOXSTAT software (Gulley et al., 1989). LC50's were calculated using either the binomial, moving average, or probit methods.

RESULTS

Produced Water Characterization

Since 1987, thirty-six different produced waters from locations in Wyoming and Colorado have been tested using bioassay techniques. The various produced waters showed a wide range of toxicities as well as possible causes of toxicity (Table 1). These waters could be characterized as belonging to one of four groups based on toxicity (Figure 1). Group A included those waters in which the *Ceriodaphnia* were more sensitive to the discharge than the fathead minnows. For the majority of these waters, the discharge was not toxic to the fathead minnows. Approximately 36% of the tested discharges fell into this category.

In Group B, the two test organisms showed fairly equivalent sensitivities to the produced water exposures. This included 33% of the tested discharges. Fathead minnows were more

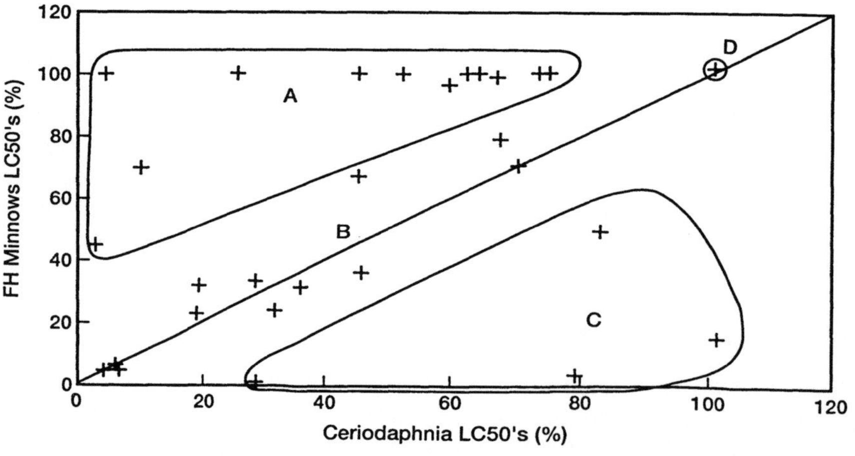

Figure 1. *Ceriodaphnia* LC_{50}'s measured in % of effluent plotted against the fathead minnow toxicity.

Table 1. Produced waters physicochemical and toxicological characteristics

Sample	pH	Alkalinity (mg/l)	Hardness (mg/l)	Conductivity (μmhos/cm)	LC_{50} (%) *Cerios*	FH Minnows	Comments/ Suspected Causes
			Group A - *Ceriodaphnia* Sensitivity				
2	8.1	3540	135	5080	67.6	>100	TDS/hydrocarbons
5	7.7	729	96	1390	26.4	>100	1-time toxicity
6	8.7	1170	260	5500	75.5	>100	TDS
7	8.5	1050	28	2300	62.9	>100	hydrocarbons
8	8.4	670	26	1250	26.6	>100	hydrocarbons
11	8.1	390	38	7300	3.2	45.5	TDS
14	8.1	811	133	6500	45.9	>100	hydrocarbons
17	8.9	850	6	2030	64.8	>100	unknown
18	7.1	230	1264	7500	52.9	>100	TDS; high sulfate
20	8.5	943	23	1330	10.7	70.9	semivolatiles
21	8.3	693	15	1400	74	>100	semivolatiles
22	7.9	1027	55	1800	5.4	>100	semivolatiles
27	8.5	1200	11	4400	60.2	97.2	low D.O.
			Group B - Equal Toxicity				
3	8.0	1201	115	4000	46.2	36.6	0.20 mg/l zinc; high alkalinity
9	8.1	633	31	3000	36.4	31.9	hydrocarbons
10	7.5	844	44	3000	29.2	33.8	hydrocarbons
12	8.1	740	60	6300	45.9	67.6	unknown
13	8.5	784	28	4300	70.7	70.7	unknown
15	9.5	>2500	126	-	<6	<6	hydrogen sulfide
16	8.5	392	50	-	<6	<6	2 mg/l H_2S;
19	8.4	303	2480	3000	32.4	24.7	H_2S
23	8.0	315	1568	4000	6.4	4.5	H_2S; hydrocarbons
24	8.4	300	1736	4400	19.8	32.7	H_2S
25	9.0	550	1680	5000	4.2	<6	H_2S
26	8.3	1075	45	6000	19.5	23.1	semivolatiles
			Group C - Fathead Minnow Sensitivity				
28	8.2	335	770	3800	79.1	3.2	semivolatiles; H_2S
29	8.4	259	880	4000	>100	15.3	H_2S
30	8.5	465	330	6100	29.0	1.0	5 mg/l H_2S; semivolatiles
31	8.0	585	330	6300	83.0	50.0	oily odor
			Group D - No Toxicity				
A	9.0	690	9	1574	>100	>100	
B	8.3	380	357	1050	>100	>100	
C	8.8	680	9	1320	>100	>100	
D	7.7	764	92	1370	>100	>100	
E	8.3	763	65	3200	>100	>100	
F	8.2	125	214	1470	>100	>100	
G	8.4	1128	102	2000	>100	>100	

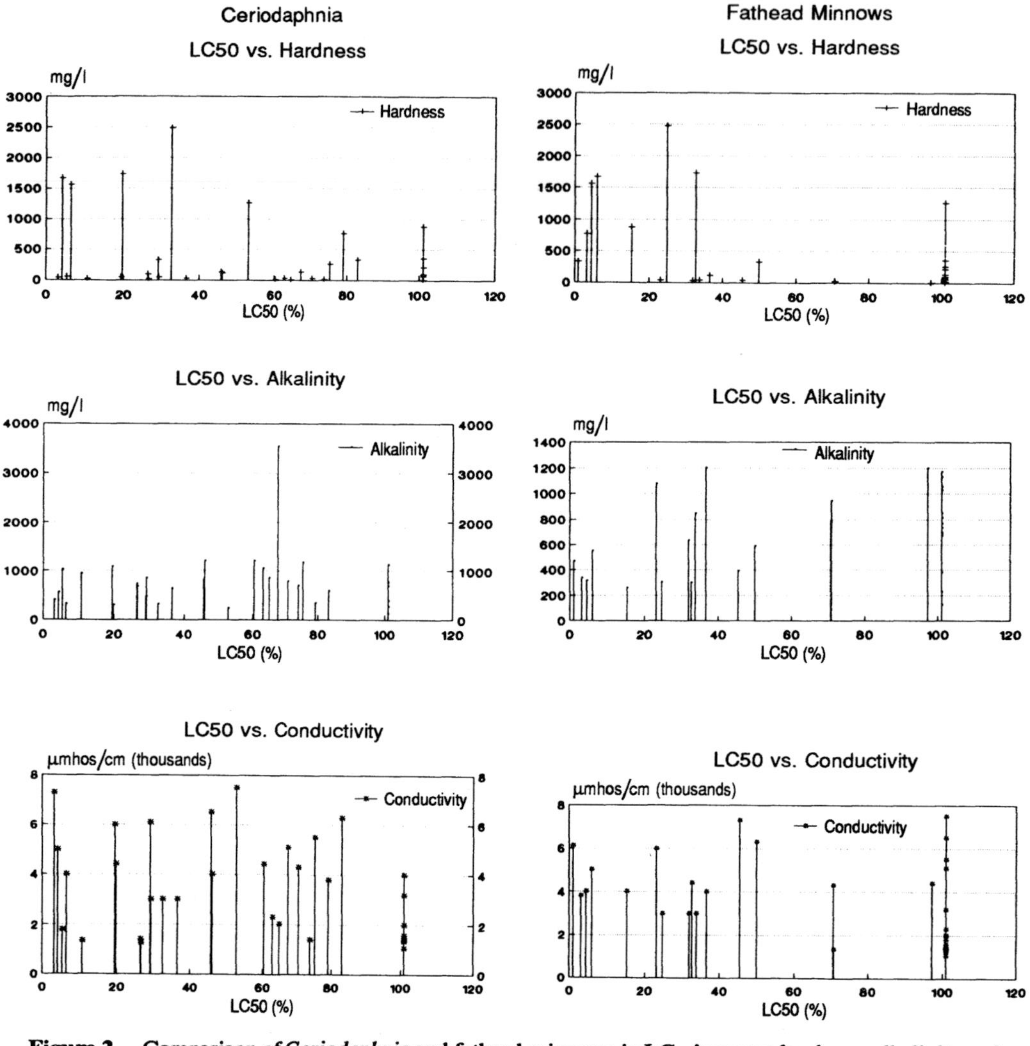

Figure 2. **Comparison of *Ceriodaphnia* and fathead minnows in LC_{50}'s versus hardness, alkalinity and conductivity in produced water samples.**

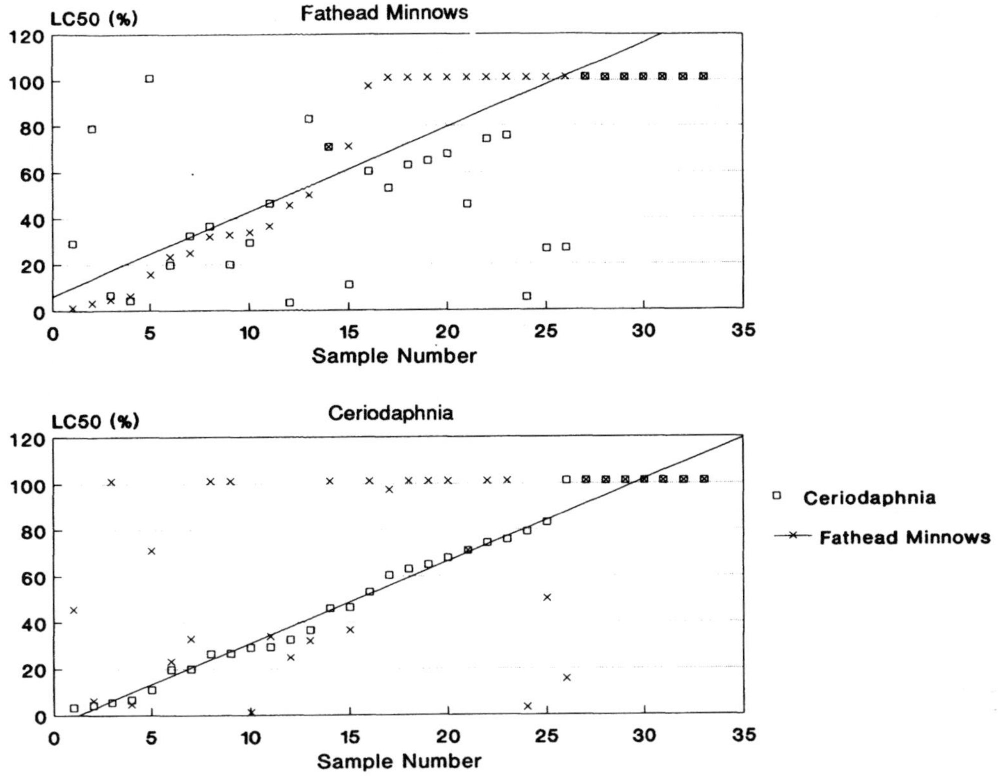

Figure 3. Rank ordering of *Ceriodaphnia* and fathead minnow LC_{50}'s (measured as % of effluent) in 33 different produced water samples. In each case, toxicity is compared for the two species. A trend line for the rank ordered species is also shown.

sensitive in exposures to waters making up Group C. This was the smallest of the groups as it contained only 11% of the discharges. Group D included those discharges that were not toxic to either of the test species. This group was made up of approximately 20% of the samples.

Overall, the samples showed a broad range of pH's from the low 7's to the low 9's in a few cases. There was no apparent correlation between pH and a particular group. It was noted, however, that in most tests, pH generally rose during the period of exposure and in some cases reached as high as pH 9.5. The Group A, B and D waters were characterized by high alkalinities and low hardnesses. For those exceptions where hardness tended to be higher, the difference in concentration between hardness and alkalinity was large. By comparison, the Group C waters were better balanced between these parameters.

Conductivities tended to be high in most of the samples. However, the conductivities of the Group D waters were generally lower than those in the other three groups. For example, the conductivities of the Group B and C waters were uniformly above 3,000 µmhos/cm. The Group A waters had conductivities that incorporated characteristics of both the Group D and the Group B and C waters with a range of values from 1,250 µmhos/cm to as high as 7,500 µmhos/cm. As shown in Figure 2, no apparent correlations could be found between toxicity and hardness, alkalinity, or conductivity.

TOXICITY IDENTIFICATION

A similar range of toxicities was found when compared for the two species against one another (Figure 3). The *Ceriodaphnia* showed a range of sensitivities to the various effluents up to LC50's of approximately 80%. In the case of the fathead minnows, toxicity was generally measured at LC_{50}'s of about 70% or less. Few of the fathead minnow LC_{50}'s were clustered around the *Ceriodaphnia* regression line. The opposite was true with the fathead minnow trend line where a majority of the *Ceriodaphnia* LC_{50}'s fell on or about the minnow trend line. This may indicate that the contaminants affecting fathead minnows probably also affected the ceriodaphs. However, the fathead minnows were not as readily affected by those toxicants producing effects on the ceriodaphs. The graphs also suggest the variety of toxic effects on the test organisms.

Toxicity identification was performed on a number of the produced waters to isolate the sources of contamination. The most common identified or suspected causes of toxicity were either individually or a combination of hydrogen sulfides, total dissolved solids (TDS), and petroleum hydrocarbons. In no instance were trace metals identified as a potential source of toxicity. Toxicity in the Group A samples was generally attributed to volatile and semivolatile hydrocarbons and TDS. The Group B and C toxicities seemed to be due mostly to hydrogen sulfide and semivolatile hydrocarbons. The characteristics of each of these are discussed below.

Total Dissolved Solids

Mount and Anderson-Carnahan (1988) report a 48 hr LC_{50} for *Ceriodaphnia* of 2.3 g/l sodium chloride compared to a 96 hr LC_{50} of 7.7 g/l for fathead minnows. Our experience has shown that waters which have a conductivity that exceeds 3,500 μmhos/cm can probably be suspected of having an acutely toxic effect on the ceriodaphs. Almost half (i.e., 17) of the samples were proven or suspected of having acute toxicity due to high TDS levels. The majority of these discharges were found in the Group A and B categories.

In addition to a high conductivity, a typical pattern for effluents that have toxicity associated with high TDS concentrations is a greater sensitivity of *Ceriodaphnia* compared to the fathead minnows. When TDS alone was involved in producing toxicity, no increase in survival rates was observed with any of the published Phase I TIE procedures. The test history for one sample which had demonstrated symptoms of TDS toxicity is shown in Table 2. The sample had shown low level effects to ceriodaphs and had a conductivity that was typically greater than 7,000 μmhos/cm.

Table 2. LC_{50}'s (%) of a produced water with suspected TDS contamination

Date of Test	*Ceriodaphnia*	Fathead Minnows (μmhos/cm)	Conductivity
07/17/89	52.9	>100	-
09/05/91	>100	>100	6880
09/10/91	95.5		6910
09/12/91	92.4		7180
10/26/91	57.1	>100	6780

A Phase I Toxicity Identification Evaluation (TIE) (Mount and Anderson-Carnahan 1988) was initiated with this discharge but none of the procedures were successful in reducing toxicity. However, it was noted that several treatments that included pH adjustment increased toxicity (Table 3). This probably occurred because of a borderline toxicity in the original sample which was exacerbated by the additions of acid and base to adjust pH and which resulted in higher TDS concentrations.

No methods for isolating TDS as a causative agent for toxicity have been prescribed in the Mount and Anderson-Carnahan (1988) protocols. However, McCulloch and Goodfellow (1989) have suggested a method which has been successfully applied on a wide range of effluent samples. Table 4 demonstrates the results of one such test where this methodology was used with a produced water sample. The sample was diluted to produce exposure concentrations of 100%, 75%, 50%, and 25%. The conductivity of each of the diluted samples (i.e., 75%, 50%, and 25% concentrations) was then adjusted upward with sodium chloride to match the conductivity of the original (i.e., 100% concentration) sample. Each of these adjusted samples was then diluted to produce four dilution series with 100%, 75%, 50%, and 25% exposures.

Table 3. Summary of Phase I TIE tests showing manipulations which resulted in reductions in toxicity (+), increases in toxicity (-), and no change ( ). The LC_{50} is included in parentheses.

TEST	pH 3	pH	pH 11
EDTA Chelation		()	
Oxidant Reduction		()	
pH Adjustment	- (71.2%)	()	— (86.6%)
Aeration	- (82.2%)	()	()
Filtration	- (93.3%)	()	()
SPE Filtration	()	()	—

Table 4. Number of surviving organisms at each concentration (out of five exposed) and LC_{50}'s of various dilution series after adjusting conductivity with NaCl to maintain TDS levels constant

Dilution	100%	75%	50%	25%
Control	5	5	5	4
100%	0	0	1	0
75%	3	3	2	0
50%	5	3	5	2
25%	4	5	3	5
LC_{50} (%)	71.8	63.6	65.1	45.7

Table 5. Cationic and anionic composition of a produced water discharge suspected of producing toxicity due to TDS concentrations.

Analyte	Concentration (mg/l)
Dissolved Solids	4,140
Calcium	370
Magnesium	54
Potassium	140
Sodium	970
Chloride	1,500
Fluoride	4.4
Nitrate + Nitrite (as N)	0.06
Sulfate	2,500
Alkalinity	222

The effect of these procedures is to maintain constant conductivities while diluting other contaminants which may be present. Under such a procedure when TDS is the cause of toxicity, the LC_{50}'s of each dilution series will be similar. LC_{50}'s will increase in the series (i.e. toxicity will decrease) if another toxicant is a key contributor to toxicity. This is due to the fact that other contaminants will be diluted. In the present situation, LC_{50}'s increased with dilution.

These results demonstrate one of the problems associated with identifying TDS toxicity when compounds other than chloride comprise the majority of the TDS. The present sample had a total dissolved solids content of 4,140 mg/l comprised mainly of sulfate with a lesser contribution from chloride (Table 5). However, in the dilution series, a greater concentration of chloride was added in the more dilute series than what was present in the original discharge. This resulted in the increasing toxicity with dilution. This indicated that toxicity cannot be estimated solely from a measure of TDS but must take into account the components of TDS.

One additional set of data for a discharge with a high TDS due to alkalinity can be used to demonstrate this point (Table 6). The initial LC_{50} for the sample was 67.6%. Note that the toxicities decreased with increasing dilution when just the sodium chloride was added whereas toxicities increased when only the sodium bicarbonate was used. Toxicities were equivalent when a combination of the two compounds was used.

Table 6. *Ceriodaphnia* test results to confirm the role of TDS in producing toxicity. Conductivity in each dilution series was adjusted with sodium chloride, sodium bicarbonate, or a combination of the two compounds. Numbers represent survivors in each exposure out of five exposed

Dilution	100%	75%	50%	25%
		Sodium Chloride		
100%	0	0	0	0
75%	0	0	0	0
50%	2	1	4	4
25%	3	5	4	4
Control	5	5	4	5
LC50 (%)	32.4	40.1	46.4	61.2
		Sodium Bicarbonate		
100%	0	0	0	0
75%	0	1	1	0
50%	1	1	1	2
25%	5	4	3	3
Control	5	4	4	4
LC50 (%)	40.1	37.6	30.1	32.4
		Sodium Chloride + Sodium Bicarbonate		
100%	0	0	0	0
75%	0	0	0	0
50%	4	4	5	4
25%	5	5	4	5
Control	4	4	4	4
LC50 (%)	58	58	61.2	58

Petroleum Hydrocarbons

Table 7 compares hydrocarbon levels in several different produced waters and their respective toxicities to the ceriodaphs and fathead minnows. The toxicity in all of these samples was suspected of being caused by the petroleum hydrocarbons. The volatiles concentrations were higher than semivolatile concentrations in all of the samples. The semivolatile hydrocarbons were comprised mostly of naphthalene and 2-methylnaphthalene in all samples. Sample

B which had the highest volatiles concentration but the lowest semivolatile levels had the lowest toxicity. This probably indicates a minor contribution of volatile compounds to the observed toxicity.

Table 7. Comparison of toxicity (LC_{50}-%) to *Ceriodaphnia* and fathead minnows with petroleum hydrocarbon concentrations (μg/l) in different produced water samples.

Analyte	Sample			
	A	B	C	D
VOLATILES				
Acetone	430	620	27	20
2-Butanone	1.5			
Toluene	15	26	40	110
Benzene			19	40
Ethylbenzene	8.9	6.0	3.7	8
Xylenes	100	36	41	130
SEMIVOLATILES				
Phenol			3.4	9
2-Methylphenol			7.8	6
4-Methylphenol	9.0		31	4
2,4-Dimethyl phenol			6.6	6
Naphthalene	21	12	21	19
2-Methylnaphthalene	100	34	37	34
Acenaphthylene	2.1			
Acenaphthene	3.3			
Dibenzofuran	3.6			
Phenanthrene	10	1.6	1.9	2
Pyrene			1.9	
Fluorene				2
TOTAL VOLATILES	555.4	688	130.7	308
TOTAL SEMIVOLATILES	149	47.6	110.6	82
VOLATILES + SEMI-VOLATILES	704.4	735.6	241.3	390
LC_{50}				
Cerios	10.7	74	5.4	44.1
FH Minnows	70.9	>100	>100	-

A Phase I TIE was performed on the B sample and non-polar organics were implicated as the source of the toxicity. Toxicity was removed from sample D when the water was filtered through activated carbon. The A and C samples, which had been collected at the well, lost all toxicity when the samples were collected from a holding pond. No hydrocarbons were detected in the holding pond samples which suggested that the toxicity was due to the non-polar organics. The loss of the hydrocarbons in the holding pond was due to natural degradation. The data suggested that *Ceriodaphnia* were more sensitive to the hydrocarbons than were the fathead minnows.

Reduction in toxicity from produced waters due to weathering and natural degradation has been demonstrated in other samples with suspected hydrocarbon contamination. Table 8 provides the results of toxicity tests with *Ceriodaphnia* in which volatile compounds were suspected of producing significant mortalities due to a rapid loss of toxicity both in the field and in the laboratory. The effluent at the NPDES discharge point was highly toxic compared to a ditch sample approximately one-half mile below the discharge point. Toxicity was also lost rapidly in the lab over time and with aeration.

Table 8. Comparison of toxicity (LC_{50}-%) to *Ceriodaphnia* in a sample following natural weathering or aeration in the laboratory

Treatment	LC_{50} (%)
Original Effluent	5.4
	3.2
After 24 hours	4.9
After 48 hours	62.4
After 90 min aeration	>100
0.5 mile down drainage ditch	>100

The significance of natural degradation in toxicity reduction is demonstrated in Figure 4. Samples A and B were collected from discharge pits which had an ice cover. This probably cut down on volatilization and degradation of the hydrocarbons. Sample C flowed through a discharge canal for approximately three miles before it was collected just prior to entering the receiving stream. High levels of toxicity characterized the A and B samples which also had significant levels of hydrogen sulfide. The effect of the hydrogen sulfide was noted in sample C which had lost most of the hydrocarbons by the point of collection. Toxicity to the ceriodaphs had been eliminated but a residual effect continued to be measured on the fathead minnows, probably as a result of the hydrogen sulfide. In tests conducted in the lab on another sample with hydrogen sulfide contamination, a low level toxicity remained even after two hours of vigorous aeration.

Hydrogen Sulfide

Fathead minnows have a greater sensitivity to hydrogen sulfide than do daphnids. A search of Chemical Information Systems AQUIRE database found 62 references of hydrogen sulfide toxicity to fathead minnows. LC_{50}'s generally ranged from approximately 6 µg/l to 70 µg/l total H_2S. One reported value of 1,000 µg/l was found for *Daphnia magna* (none were reported for *Ceriodaphnia*). Hence, a greater fathead minnow sensitivity would be expected in waters with hydrogen sulfide as a primary contaminant.

Eight of the samples (15, 16, 19, 23, 24, 25, 28, and 30) had proven or suspected toxicity due mainly to hydrogen sulfide contamination. These waters were generally characterized by a high level of toxicity (i.e. low LC_{50}'s). All were in the Group B and C categories. Three others (10, 11 and 27) were characterized by low dissolved oxygen concentrations that may have been influenced by the presence of hydrogen sulfide. However, no measurements were made of hydrogen sulfide in these samples and there was not a noticeable odor of the gas present. Two of these three were in Group A in which the ceriodaphs showed a greater sensitivity. This suggested low levels of hydrogen sulfide if indeed present and the possible presence of another contaminant to affect the ceriodaphs.

DISCUSSION

The results demonstrate the variability that exists between different produced waters and their associated toxicities. This is not totally unexpected given the range of toxicities to aquatic organisms that have been demonstrated by investigators over the years from water soluble fractions of various crude oils. Of perhaps greater significance in these tests was the fact that toxicity could often be attributed to more than one contaminant.

Taraldsen et al., (1987) investigated the toxicity from 11 produced waters in Wyoming. Their results basically agree with the present data and included a broad set of responses ranging

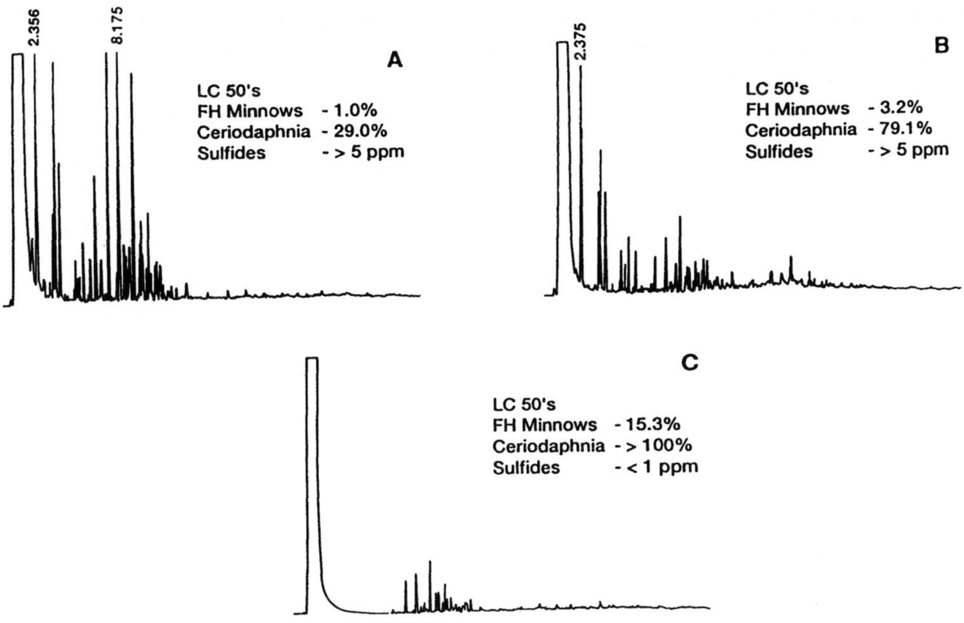

Figure 4. Gas chromatograms of produced water samples. A and B were samples taken at discharge points. C had travelled approximately three miles down a discharge canal before collection.

from no toxicity to equal sensitivities and fathead minnow and *Ceriodaphnia* sensitivities. Toxicity reduction evaluations were conducted on some of the effluents and identified non-polar organics and hydrogen sulfide as the main causes of toxicity. TDS was a minor contributor to toxicity in spite of salinity values as high as 4.0 g/l.

Our data differ from those of the above authors in the fact that TDS could be definitively implicated as the cause of toxicity in some samples. Also, combinations of contaminants were responsible for varying levels of toxicity in a number of samples. This combination of toxicants complicates efforts to isolate causes of toxicity in produced waters.

The sensitivity of *Ceriodaphnia* to TDS makes it particularly difficult to isolate other causes of toxicity which might be present. TDS cannot be easily removed from a sample without eliminating other contaminants as well. However, we have used an amphipod, *Hyalella azteca*, in comparative tests with daphnid species when TDS was a suspected cause of toxicity. Table 9 shows the relative sensitivity of this species and the daphnids to TDS and trace metals. The amphipod has a lower sensitivity to TDS compared to *Ceriodaphnia* and *Daphnia magna* while being equally sensitive to trace metals. Although the amphipod has not yet been tested in produced waters with a combination of TDS and hydrocarbon contamination, the results suggest a useful procedure for evaluating produced water toxicity.

The combination of produced water physicochemical parameters and the relative sensitivities of the test species to produced waters can be important first clues in isolating the causes of toxicity. A suggestion has been made that a classification scheme be developed based on produced water characteristics. The original purpose of this paper was to determine whether some generalizations could be made about produced waters on the basis of toxicity data. However, the present data indicate that such categorization will be difficult on the basis of toxicity. Many factors contribute to the aquatic toxicity of produced waters and these are not always easily separated. While a broad approach to classification on the basis of toxicity was made in this study, it is obvious from the data that differential sensitivities on the part of the test

Table 9. Number of surviving organisms out of 20 exposed for three invertebrate species in various effluents producing toxicity due to Total Dissolved Solids (TDS) and trace metals contamination

Suspected Contaminant	*Hyalella*	*Ceriodaphnia*	*Daphnia*
Trace metals			
#1	0	0	0
#2	8	11	8
#3	1	0	1
#4	6	5	4
TDS			
#5	8	0	—
#6	17	0	11
#7	14	0	0
#8	19	—	8
#9	12	0	0
#10	18	0	4
#11	18	1	2

species did not always provide sufficient information on which to adequately predict the water quality characteristics of the discharges.

REFERENCES

McCulloch, W.L. and Goodfellow, W.L., 1989, The influence of total dissolved solids (TDS) on the interpretation of TIE fractionation results, Paper presented at the 1989 SETAC Conference, EA Engineering Science and Technology, Inc., Sparks, Maryland.

Mount, D.I. and Anderson-Carnahan, L., 1988, "Methods for Aquatic Toxicity Identification Evaluations, Phase I Toxicity Characterization Procedures," EPA 600/3-88/034, Environmental Research Laboratory, Duluth, Minnesota.

Peltier, W.H. and Weber, C.I., 1985, "Methods for Measuring the Acute Toxicity of Effluents to Freshwater and Marine Organisms (Third Edition)," EPA 600/4-85/012, Environmental Protection Agency, Washington, D.C.

Taraldsen, J.E., J.R. Amato, and Mount, D.I., 1987, "Toxicity Testing and Characterization of Toxicants from Effluents of the Powder River Basin, Wyoming," unpublished report, U.S. Environmental Protection Agency, Duluth, Minnesota.

A COMPARATIVE EVALUATION OF PRODUCED WATER TOXICITY

K.C. Schiff[1], D.J. Reish[2], J.W. Anderson[3], and S.M. Bay[4]

[1] Kinnetic Laboratories Incorporated
Carlsbad, California
USA

[2] California State University Long Beach
Long Beach, California
USA

[3] EMCON Marine Sciences Incorporated
Carlsbad, California
USA

[4] Southern California Coastal Water Research Project
Long Beach, California
USA

INTRODUCTION

Concern has increased in recent years regarding the potential effects of produced water discharged to marine and estuarine environments. Considerable work has already been accomplished using toxicity tests with produced water (Montgomery et al., 1987; Montgomery et al., 1988; Parrish et al., 1988). This study responded to the need for more information on toxicity tests with produced water, and the range of sensitivities for different test organisms. The objectives were threefold: (1) evaluate produced water sample handling procedures and sample changes during storage, (2) compare different toxicity testing procedures using the mysid, *Mysidopsis bahia*, with produced water, and (3) test the suitability of other species in toxicity tests with produced water.

The first task, evaluating sample handling and storage procedures for produced water, was designed to measure sample integrity for toxicity testing and chemical analyses. It was assumed that volatile aromatic hydrocarbons such as benzene, ethylbenzene, toluene and *o,m,p*-xylenes (BETX) represent readily degradable components of produced water. Therefore, repeated source sampling for BETX over short-term intervals (minutes to hours) to evaluate effluent variability was followed by long-term sampling (days) of stored samples to determine stability.

Mysid toxicity tests are currently one of the standard tests for evaluating the toxicity of produced water to marine and estuarine environments. The second task was designed to measure differences in mysid tests resulting from the use of various procedures and exposures. By using three different test systems, the effects of both pH/salinity adjustment and volatilization of toxic compounds could be investigated. One procedure used to test mysids followed Peltier and Weber (1985), which routinely adjusts salinity and pH for each test dilution. A second procedure did not adjust salinity or pH, but merely added 20 ppt seawater to produced water to prepare dilutions. Therefore, the use of a hypersaline produced water sample will result in varying salinities at different test dilutions.

To determine if mysid toxicity is correlated with the concentrations of volatile aromatic hydrocarbons (BETX), open and closed containers were employed. In addition to the open, aerated system using the Peltier and Weber (1985) method, a closed-flask system described by Reish and Richards (1957) was used. Data on toxicity and BETX concentrations were used to determine the median lethal concentration (LC_{50}) and the "Toxicity Index" for mysids in ppm-hours of BETX (Anderson et al., 1980).

The third task was designed to compare the toxic responses of five different species to three produced water samples. This comparison provided a wide range of information since these five procedures utilized various exposure times and endpoints. Two of the toxicity tests were "rapid" (1 hour), two were acute tests (96 hours), and one was a 7-day chronic estimator test, with daily replenishment of produced water dilutions.

The five species used were: (1) *Mysidopsis bahia*, estuarine mysids currently being used for biological testing of produced water in the Gulf of Mexico, (2) *Strongylocentrotus purpuratus*, purple sea urchins, (3) *Menidia beryllina*, inland silversides, (4) *Neanthes arenaceodentata*, a polychaetous annelid, and (5) *Photobacterium phosphoreum*, a luminescent marine bacterium.

All five toxicity tests were performed on three separate produced water samples collected from different platforms in the Gulf of Mexico. A suite of chemical analyses characterized each produced water sample used in this task. These analyses were used to indicate the magnitude of differences between samples and reflect the type of variation observed in biological testing.

MATERIALS AND METHODS

Samples for the storage and handling study were obtained from an onshore processing and treatment facility in California that received produced water from offshore platforms. Because of high microbial activity in this particular produced water, it was not used in later toxicity tests. The samples for toxicity testing originated from offshore platforms located in the Mississippi Delta, Gulf of Mexico. All work was completed between August 1988 and September 1989.

Sample Storage and Handling

Sampling generally followed the methods described by Peltier and Weber (1985) for the collection of grab samples and storage of effluents. Produced water samples were obtained by oil company personnel in 20 l polycarbonate carboys with teflon-lined screwcaps. Under normal platform and treatment operating conditions the container was rinsed twice with sample, filled without headspace (i.e. bubbles), and capped securely. Samples were then delivered either by

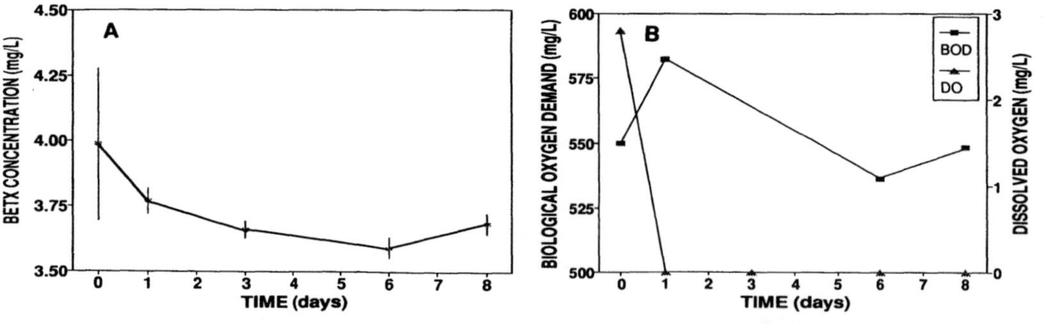

Figure 1. Concentrations of benzene, ethyl benzene, toluene, and *o,m,p*-xylenes (BETX) versus time during the storage of produced water (1A). Biological oxygen demand and dissolved oxygen concentration were also measured (1B).

hand or by overnight delivery in specially constructed, insulated (without refrigeration) shipping crates. Biological testing always began less than 24 hours after arrival of produced water sample.

Samples of produced water were captured every 36 minutes for three hours at the treatment facility and analyzed for volatile aromatic hydrocarbons (BETX) to determine short-term effluent variability. Storage of produced water samples was evaluated by subsampling large carboys into 4 l amber glass jugs with no headspace, securely capped, and kept at 4°C. Water samples for analysis of BETX and total organic carbon (TOC) were taken from separate jugs sampled at 1, 3, 6, and 8 days.

Toxicity Tests

The mysids used were originally purchased from the University of Tampa, Florida, and cultured in a temperature-controlled laboratory (19 ± 1°C). Seven to 10 day old individuals were randomly selected for testing and fed *Artemia* nauplii daily. Mortality was the endpoint, observed daily, for 96 hours.

Mysid tests designed to measure the effect of pH and salinity adjustment were conducted in 60 l aquaria (19 l of test solution) with aeration. Twenty mysids were placed in rolled Nytex™ "cages" (363-m mesh) fastened to 150-mm diameter petri dishes. There were three cages per aquarium. Each aquarium represented one treatment and there were five concentrations, plus a control, per test. Salinity-adjusted (20 ± 1 ppt) and pH-adjusted (8.0 ± 0.2) treatments were tested side-by-side with non-adjusted treatments, whose water quality varied with concentration of produced water added. Produced waters used in these experiments were hypersaline (188 to 196 ppt) and acidic (5.7 to 6.1 pH) relative to dilution water.

Table 1. A summary of the toxicology for mysids, *Mysidopsis bahia*, exposed to produced water (PW) using aerated aquaria and stopped flask test systems

		10% Produced Water	
Test System	LC_{50} as %PW (95% to C.I.)	Time (hrs) to 50% Mortality	Toxicity Index [a] (ppm-hr BETX)
Salinity-adjusted flask	3.74 (3.01 - 4.65)	14.0	3.25
Salinity-adjusted aquaria	5.50 (4.34 - 6.98)	13.5	0.31
Salinity-adjusted flask	4.49 (3.45 - 5.85)	12.0	_[b]
Salinity-adjusted aquaria	5.27 (4.69 - 5.92)	13.0	0.91
Non-adjusted aquaria	8.15 (6.49 - 10.22)	46.0	1.06
Hypersaline aquaria[c] (no produced water)	41.6 ppt		

[a] Integration of BETX decay over time to 50% mortality.
[b] Decay curve not calculable, therefore no toxicity index estimated.
[c] This test was for salinity tolerance; endpoint is no-effect concentration.

Mysid tests designed to measure toxicity relative to changes in BETX concentration employed the above aquaria side-by-side with sealed 500 ml Erlenmeyer flasks (100 ml of test solution). There were 10 mysids per flask, three flasks per treatment, and five concentrations,

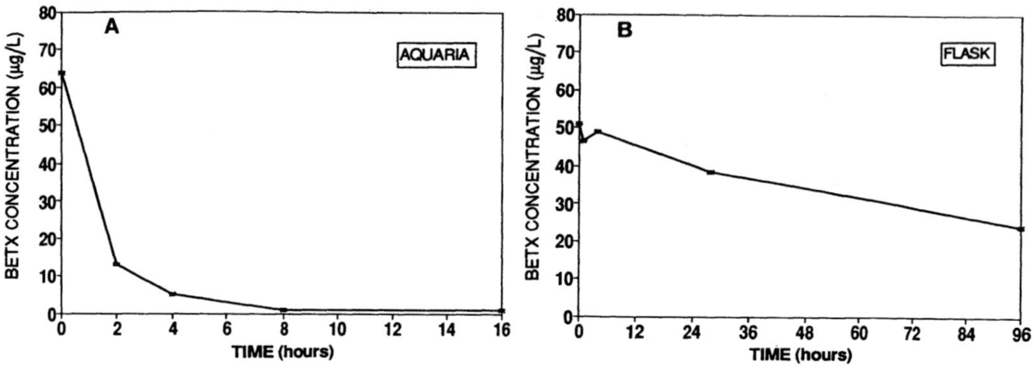

Figure 2. Concentrations of benzene, ethyl benzene, toluene, and xylenes (BETX) versus time during toxicity testing. Five percent produced water was measured from aerated aquaria (2A) and sealed flasks.

plus a control, per test. Additional flasks without mysids were used to measure BETX while aquaria were sampled directly.

Mysid testing during the comparative toxicity testing phase employed 2 l finger bowls (1 l of test solution), with 10 organisms each. There were three replicate bowls per treatment and five concentrations, plus a control, per test. Chromium (as potassium dichromate) was used as a reference toxicant. Salinity- and pH-adjusted as well as non-adjusted treatments were tested.

Toxicity testing with annelids generally followed the procedure described by Reish (1980). Ten organisms, two to three weeks old, were randomly placed in 2 l bowls (1 l of test solution). There were three replicate bowls per treatment and five concentrations, plus a control, per test. Chromium (as potassium dichromate) was used as a reference toxicant. Only salinity-adjusted (34 ppt) and pH-adjusted (8.0) treatments were tested. Mortality was the endpoint, observed daily for 96 hours.

Toxicity testing with silversides generally followed the procedure described by Weber et al., (1988). This was the only test that used solution renewals (75% daily). Ten larvae, five days old, were used in Norberg chambers (0.5 l of test solution). There were four replicate chambers per treatment and six concentrations, plus a control, per test. Organisms were fed *Artemia* nauplii daily. Chromium (as potassium dichromate) was used as a reference toxicant. Only salinity-adjusted (25 ppt) and pH-adjusted (8.0) treatments were tested. Mortality was the endpoint, observed daily, for 156 hours.

Toxicity testing with urchins generally followed the procedure described by Dinnel et al., (1987). Field-collected specimens were induced to spawn in the laboratory by the addition of 0.5 ml of 0.5 M KCl. Approximate sperm-to-egg ratios of 200:1 were used to achieve 80% fertilization, which was verified prior to testing. Male gametes were exposed for 60 minutes in 15-ml test tubes (10 ml of test solution) before adding 2,000 female gametes. Twenty minutes was allowed for fertilization, after which samples were preserved with formalin. There were three replicate tubes per treatment and six concentrations, plus a control, per test. Zinc (as zinc sulfate) was used as a reference toxicant. Only salinity-adjusted (34 ppt) and pH-adjusted (8.0) treatments were tested. Fertilization success, as percentage of control, was the endpoint.

Toxicity testing with luminescent bacteria (referred to as Microtox), generally followed the procedure described by Microbics Corporation (1988). Approximately 1×10^6 bacteria were exposed for 30 minutes in 2-ml cuvettes (1 ml of test solution). There were three replicate cuvettes per treatment and five concentrations, plus a control, per test. Zinc (as zinc sulfate) was used as reference toxicant. Only salinity-adjusted (34 ppt) and pH-adjusted (8.0) treatments were tested. Light production as percent of control was the endpoint.

Median effect concentrations (LC_{50} or EC_{50}) were calculated by probit analysis, using software provided by the U.S. EPA (1988). When data violated the assumptions required for this procedure, the binomial distribution method (Peltier and Weber 1985) was employed. The Microtox EC_{50} was calculated according to Microbics Corporation (1988). The "Toxicity Index" of BETX to mysids in different exposure vessels was concentration-time dependent and calculated as ppm-hours. Based upon Anderson et al. (1980, 1985) and Anderson (1985), ppm-hours was calculated as the integration of the BETX decay rate to the time of 50% mortality for a single concentration.

Chemical Analysis

Analytical Methods are detailed in Anderson et al. (1990), but abbreviated methods were as follows. Volatile hydrocarbons- benzene, ethylbenzene, toluene, and *o,m,p*-xylenes- were measured by purge-and-trap capillary gas chromatography. Saturated and aromatic hydrocarbons, carboxylic acids, and polar substances were liquid-liquid extracted with dichloromethane. Adsorption chromatography (silica gel:alumina, 2:1) was used to isolate hydrocarbon fractions. Capillary gas chromatography/mass spectrometry was used for analysis of compounds. Samples of produced water for metals analyses were preserved with concentrated sulfuric acid. Mercury was measured by cold-vapor atomic absorption technique. Barium, zinc, and chromium were measured by direct-injection atomic absorption spectroscopy. Copper, nickel, silver, lead, and cadmium were measured by graphite furnace Zeeman Atomic Absorption Spectroscopy. Calcium, magnesium, and potassium were measured by using Inductively Coupled Plasma/Mass Spectrometry.

RESULTS

Results for this study are divided into three task objectives: sample storage and handling; response of *Mysidopsis* under different test conditions and changes in BETX concentration; and comparative toxicities of three produced water samples to five different species.

Sample Storage and Handling

Samples of produced water, refrigerated at 4°C, and in sealed containers without headspace, showed minimal changes in total organic carbon and volatile aromatic hydrocarbons (benzene, ethylbenzene, toluene, and xylenes; BETX) over an eight-day storage period (Figure 1A). Total organic carbon ranged from 240 to 260 mg/l with no decreasing trends. Mean concentrations of BETX ranged from 3.99 to 3.59 mg/l, decreasing a maximum of 10% over the eight-day period. Variations in BETX concentration in sequential effluent samples over a three hour period differed by 6.8% (5.47 to 5.84 mg/l). High biological oxygen demand (5-day BOD 536 mg/l) in the sample from a California facility depleted dissolved oxygen in less than one day (Figure 1B).

Response of *Mysidopsis* Under Different Test Conditions and Changes in BETX Concentration

Mysids showed similar responses to produced water in sealed flasks and aerated aquaria, despite distinct differences in BETX exposure. Based upon nominal concentrations of produced water (PW), the 96-hour LC_{50} for sealed flasks was 3.74 to 4.49% PW while the aerated aquaria 96-hour LC_{50} ranged from 5.50 to 5.27% PW (Table 1). However, aerated aquaria showed exponential losses of BETX while sealed flasks showed a more gradual, linear decay of BETX (Figure 2). Although 10% PW concentrations in both exposure systems resulted in similar times to reach 50% mortality (12 to 14 hours), the toxicity index for flask exposures was 3.25 ppm-hours BETX compared to 0.31 ppm-hours BETX for aerated aquaria exposures (Table 1).

The results from mysid tests with produced water using salinity- and pH-adjusted aquaria versus non-adjusted aquaria were comparable (Table 1). The 96-hour LC_{50} for adjusted aquaria was 5.27% PW while non-adjusted aquaria was 8.15% PW. Although the time to reach 50%

mortality in 10% PW concentrations was different between the adjusted and non-adjusted aquaria (13 versus 46 hours), the toxicity index was similar (0.91 versus 1.06 ppm-hrs BETX).

Water quality measurements from adjusted aquaria showed relatively consistent concentrations of salinity (20.5 ± 2 ppt). Salinity measured from non-adjusted aquaria increased up to 35.8 ppt, covarying with increased concentration of PW. The hypersaline no-effect concentration for *Mysidopsis bahia* was 41.6 ppt (Table 1). Changes in pH were less noticeable, perhaps due to the buffering capacity of the dilution seawater; pH ranged from 7.6 to 8.1 for both systems. Dissolved oxygen concentrations were 77% of saturation or higher.

Comparative Toxicity of Three Produced Water Samples to Five Species

Results of the median effect concentrations (EC_{50} or LC_{50}) for five different species exposed to three produced water are presented in Table 2. Chemical characterization of the produced water samples are given in Table 3.

The rank of sensitivity for each test organism to produced water was similar from sample to sample (Table 2). The sea urchin sperm test with *S. purpuratus* was consistently the most sensitive for all three produced water samples tested. The EC_{50} for sea urchins ranged from 0.74% to approximately 1.2% produced water. Survival of polychaetes, *N. arenaceodentata*, was the least sensitive to produced water of all test methods. The LC_{50} for polychaetes ranged from 18.1% to 28.6% produced water. The mysid test, with *M. bahia*, resulted in the median or central tendency of toxic responses. The LC_{50} for mysids ranged from 4.9% to 11.4% produced water. The relative sensitivity among the different test organisms remained nearly the same from sample to sample.

Produced water toxicity varied between sources. The first produced water sample (PW#1) was approximately twice as toxic as the second and third produced water samples (PW#2 and

Table 2. A comparison of the median effect concentrations (EC_{50} or LC_{50}) and 95% confidence limits for five different organisms exposed to three different produced waters (PW) and three different reference toxicants (benzene, chromium, and zinc)

	EC_{50} or LC_{50} (% PW) 95% confidence limits			EC_{50} or LC_{50} (mg/l) 95% confidence limits		
Organism	PW#1	PW#2	PW#3	Benzene	Cr^{6+}	Zn^{2+}
Strongylocentrotus purpuratus (EC_{50})	0.74 (0.56-0.90)	1.73 (1.36-2.07)	* (0.32-1.80)	9.6 (1.0-20.3)	>100	0.09 (0.05-0.20)
Menidia beryllina (LC_{50})	>1.12	5.8 (4.8-6.7)	5.5 (3.1-7.8)	17.2 (9.7-26.7)	9.6 (7.7-11.2)	
Mysidposis bahia salinity adjusted (LC_{50})	4.9 (4.5-5.8)	11.8 (10.6-12.7)	11.4 (10.0-13.3)	27.4 (18.0-32.0)	7.0 (6.5-7.7)	
salinity non-adjusted (LC_{50})	5.2 (4.5-5.8)	12.4 (10.8-13.4)	14.5 (13.3-17.8)			
Microtox (EC_{50})	13.4 (12.6-14.3)	8.9 (7.6-10.3)	16.6 (15.5-17.7)	27.5 (23.1-32.8)	960 (906-1,017)	3.60 (3.37-3.85)
Neanthes arenaceodentata (LC_{50})	18.1 (16.0-20.0)	28.6 (23.6-31.5)	28.6 (26.7-31.4)	>100	3.9 (3.6-4.2)	

* Could not calculate an EC_{50} for this test by probit analysis, 95% confidence limits calculated by the binomial method.

PW#3). The PW#2 and PW#3 elicited similar toxic responses from the five organisms tested (Table 2). As an example, the LC_{50} for the first sample to mysids was 4.9% PW, while the LC_{50} of the second and third samples were 11.8% and 11.4% PW.

No single sample was consistently higher for all the chemical constituents measured (Table 3). Results also showed that none of the samples were strikingly similar to each other. All three samples were characterized as being hypersaline and slightly acidic, with high BOD and correspondingly low dissolved oxygen concentrations. Ammonia measurements were also very high. Total BETX comprised between 9% and 24% of the total hydrocarbons measured. Total hydrocarbons comprised between 0.4% and 2.1% of the total organic carbon measured. High concentrations of barium, calcium, and zinc were also measured.

Table 3. Chemical characteristics of three produced waters

	Concentration (mg/l) except where noted		
Parameter	PW#1	PW#2	PW#3
Salinity (ppt)	172	88	94
pH	6.28	6.94	6.83
Dissolved Oxygen	0.80	0.05	<0.01
B.O.D. (5-day)	750	1,220	750
Total Ammonia	161	244	187
Total Hydrocarbons	24.0	19.0	20.0
Total Petroleum Hydrocarbons	14.0	6.0	6.0
Total BETX	1.216	4.510	2.935
Total Phenols	1.360	3.290	0.759
Total PAH (Low Molecular Weight)	0.252	0.186	0.146
Total Alkanes	0.720	0.272	0.232
Total Organic Carbon	1,150	1,300	4,500
Barium	8.5	94.8	20.1
Calcium	5,300	1,900	2,000
Chromium	0.001	<0.001	<0.001
Copper	0.002	<0.001	<0.001
Potassium	850	340	170
Magnesium	1,100	200	730
Nickel	<0.001	0.002	0.002
Lead	0.009	<0.001	<0.001
Zinc	0.524	0.017	0.017

DISCUSSION

On the basis of the storage study results, a produced water sample could be transported and stored with minimal losses in volatile components and total organic carbon. Results showed less than 10% decreases in BETX stored over eight days at 4° C while BETX at the point of discharge varied by nearly 7% over three hours. If BETX is a conservative indicator of the toxic constituents in produced water, then with appropriate sampling to reduce degradation (i.e. insulation, minimal headspace, rapid delivery), sample integrity could be sustained long enough for biological testing and chemical analysis. Also, subsamples could be stored at 4°C in appropriate containers and used in toxicity tests that call for solution renewals. Results from Montgomery et al., (1988) showed similar toxic responses when exposing mysids to initial produced water samples and samples aged 4- and 8-weeks.

Changes in BETX concentration had little effect on results from the mysid tests. Losses of BETX were exponential from aerated aquaria with concentrations dropping three orders of magnitude within eight hours. In sealed flasks however, linear losses of BETX were slower, with more than half their original concentration retained after 96 hours. Regardless of BETX concentration, mysids showed similar toxic responses to produced water samples in either container. Median lethal concentrations (LC_{50}) and the time to 50% mortality were not significantly different between the two types of test systems. The toxicity index (ppm-hours of BETX), was different by an order of magnitude between flasks and aquaria exposure systems. One likely conclusion is that BETX was not the primary toxicity agent.

Salinity- and pH-adjustment of treatments for mysid testing appeared to have little effect on the results. In fact, non-adjusted treatments (dilutions made with 20 ppt seawater) fared slightly better than salinity-adjusted (to 20 ppt) and pH-adjusted (to 8.0) treatments in identical containers (aerated aquaria). Salinity tolerance of *M. bahia* was relatively wide, as would be expected from an estuarine organism, but interpreting salinity effects as a non-problem should be cautioned. If higher concentrations of produced water are required in future testing, then salinity could rise above the no-effect concentration for salinity alone.

The task on comparative toxicity showed the relationships among different organisms and test methods to produced water samples. Although the rank of sensitivity was similar between produced water samples for the five organisms tested, there was an order of magnitude difference in median effect concentrations between the most sensitive and least sensitive species. Sea urchins and the silversides showed the greatest sensitivity. Test methods used for the urchins are rapid (60-minute exposure), but very sensitive life-history stages (i.e. gametes) are exposed. Silversides tested larval life-history stages, and also used longer exposures (7-days) with daily replenishment of produced water. The polychaete test was a static, 96-hour exposure of juvenile organisms. The polychaetes were consistently less sensitive to produced water than were urchins, silversides, or mysids. Typical concentration-response curves for the polychaete showed an "all-or-nothing" response with 100% mortality and 100% survival separated by as little as 4% produced water. Sublethal endpoints now being employed with *N. arenaceodentata* (Johns and Ginn, 1990) may result in greater sensitivity. The Microtox test also showed reduced sensitivity compared to urchins, silversides, or mysids. However, it has the advantages of being rapid (30 minutes) and portable (for on-site testing).

No single chemical constituent could be strongly correlated with the toxicity observed during the multi-species comparative task. The concentration of as many as 11 of the 19 constituents measured in 100% produced water samples covaried with toxicity depending upon the species tested. For example, higher total hydrocarbon and total petroleum hydrocarbon concentrations covaried with the increased toxicity observed for silversides, mysids, and polychaetes to PW#1. Zinc concentrations also covaried with toxicity observed for these three species. However, there was no correlation of zinc with sea urchin fertilization success although concentrations of zinc in the dilutions of produced water approached those used during reference toxicant assays. The high calcium concentrations recorded for these samples has also been documented as being toxic to *M. bahia*, but toxicity was variable depending upon concentrations of other constituents in those effluents tested (Dorn and Rodgers, 1989).

It was not possible to identify the primary constituent responsible for the observed toxicity in these produced water samples. An adequate multivariate statistical approach for evaluating toxicity would require more produced water samples (i.e. larger sample size). A second approach to correlating toxicity could be a toxicity identification evaluation (TIE). However, some technique refinements are required since brine effluents may present obstacles to sample manipulation (Higashi et al., 1992). More data on other produced water samples and the likely toxic constituents will be required to determine whether or not a single species will be adequate to describe the responses of an entire ecosystem to the discharge of a complex effluent such as produced water.

ACKNOWLEDGEMENTS

This project was funded by the American Petroleum Institute. It was accomplished with the assistance of many very capable and qualified people that should be acknowledged at this time. Personnel that participated in this project were Robert Eganhouse, Darrin Greenstein, Richard Gossett, JoEllen Hose, Andrew Jirik, Valerie Raco, Laura Targgart and Skip Westcott. Analytical assistance was provided by Battelle Northwest Laboratories and West Coast Analytical Service.

REFERENCES

Anderson, J.W., Kiesser, S.L. and Blaylock, J.W., 1980, The cumulative effect of petroleum hydrocarbons on marine crustaceans during constant exposure. Rapp. P.-v. Reun. Cons. Int. Explor. Mer 179:62-70.

Anderson, J.W., Kiesser, S.L., McQuerry, D.L., Riley, R.G. and Fleischmann, M.L., 1984, Toxicity testing with constant or decreasing concentrations of chemically dispersed oil, pp. 14-22 In: "Oil Spill Chemical Dispersants: Research, Experience and Recommendations", STP 840, Tom Allen, Ed., American Society for Testing and Materials, Philadelphia.

Anderson, J.W., 1985, Toxicity of Dispersed and Undispersed Prudhoe Bay Crude Oil Fractions to Shrimp, Fish and Their Larvae. API Publication No. 4441, Health and Environmental Sciences Department, American Petroleum Institute, Washington, D.C., 70 pp.

Anderson, J.W., Schiff, K.C., Bay, S.M., Eganhouse, R.P. and Reish, D.J., 1990, An Evaluation of Produced Water Toxicity Methods. Southern California Coastal Water Research Project Final Report to the American Petroleum Institute. Southern California Coastal Water Research Project, Long Beach, CA. 61 pp.

Dinnel, P.A., Link, J.M. and Stober, Q.J., 1987, Improved methodology for a sea urchin sperm cell bioassay for marine water. Arch. Environ. Contam. Toxicol. 16:23-32.

Dorn, P.B. and J.H. Rodgers, Jr., 1989, Variability associated with identification of toxics in national pollutant discharge elimination systems (NPDES) effluent toxicity tests. Environ. Tox. Chem. 8:893-902

Higashi, R.M., Cherr, G.N., Bergens, C.A., Fan, T.W-M., and Crosby, D.G., 1992, Toxicant isolation from a produced water source in the Santa Barbara Channel. In: "Produced Water: Technological/Environmental Issues and Solutions," J.P. Ray and F.R. Engelhardt, eds., Plenum Publishing Corp., New York, New York, USA.

Johns, D.M. and Ginn, T.C., 1990, Development of a *Neanthes* sediment bioassay for use in Puget Sound. PTI Environmental Sciences, Bellevue, WA. 57 pp.

Microbics Corporation, 1988, Microtox Manual-How to Run Toxicity Tests Using the Microtox 500. Carlsbad, CA.

Montgomery, R.M., Parrish, P.R., and Friedman, S.D., 1987, Effects of produced (formation) water on mysids (*Mysidopsis bahia*). 8th Annual Meeting, Society of Toxicology and Chemistry, Pensacola, FL, 9-12 November 1987.

Montgomery, R.M., Parrish, P.R., and Friedman, S.D., 1988, Produced (formation) water: Methods development and preliminary toxicity test results. In: Proceedings of the 10th Annual EPA Conference on Analysis of Pollutants in the Environment, 13-14 May 1987, U.S. EPA Office of Water Regulations and Standards, Washington, DC, pp. 663-682.

Parrish, P.R., Montgomery, R.M., Friedman, S.D., and Hemmer, M.J., 1988, Effects of oil and gas production Wastes on Saltwater Organisms. 12th Symposium on Aquatic Toxicology and Hazard Assessment, American Society for Testing and Materials, Reno, NV, 24-26 April 1988.

Peltier, W.H. and Weber, C.I. 1985, Methods for Measuring the Acute Toxicity of Effluents to Freshwater and marine organisms, 3rd ed. United States Environmental Protection Agency, Cincinnati, OH. EPA/600/4-85/013.

Reish, D.J. and Richards, T.L., 1957, A technique for studying the effect of varying concentrations of dissolved oxygen on aquatic organisms. Int. J. Air Water Pollut. 10: 69 - 71

Reish, D.J., 1980, The Effect of Different Pollutants on Ecologically Important Polychaete Worms. United States Environmental Protection Agency, Cincinnati, OH. EPA-600/3-80-053.

US Environmental Protection Agency, 1988, EPA Probit Analysis Program Used for Calculating EC values. Version 1.4. Biological Methods Branch, Environmental Monitoring and Support Laboratory, U.S. Environmental Protection Agency. Cincinnati, OH.

Weber, C.I., Horning, W.B. II, Klemm, D.J., Neiheisel, T.W., Lewis, P.A., and Robinson,E.L., 1988, Short-term Methods for Estimating the Chronic Toxicity of Effluents and Receiving Water to Marine and Estuarine Organisms. Environmental Protection Agency, Cincinnati, OH. EPA-600/4-87/028.

IDENTIFICATION OF TOXICITY IN LOW-TDS PRODUCED WATERS

T.C. Sauer, Jr.[1], T.J. Ward[2], J.S. Brown[1], S. O'Neil[1], and M.J. Wade[1]

[1] Arthur D. Little, Inc.
Cambridge, Massachusetts
USA

[2] T.R. Wilbury Laboratories, Inc.
Marblehead, Massachusetts
USA

ABSTRACT

A toxicity identification evaluation (TIE) was performed on four low total dissolved solids (<3 parts per thousand salinity, low-TDS) produced water samples collected from Wyoming, Texas, and Louisiana production facilities. The objective was to identify those compound groups or properties that contribute to produced water toxicity. Information about the characteristics of the toxicants was to be determined through a series of sample manipulations or fractionations that isolate the toxicity of the produced water. Ten fractionation procedures were used to remove or isolate groups of compounds from the produced water. Each fraction was used in 24-hour acute toxicity tests with two freshwater species *Pimephales promelas* (fathead minnow) and *Ceriodaphnia dubia* (daphnid).

Toxicity in low-TDS produced water samples was found to be associated with no more than two of four major fractions or compound classes: volatile compounds, neutral semivolatile organic compounds, particulate matter precipitated at basic pH, and suspended solids. Although not confirmed, the toxicants identified in these fractions appeared to be hydrogen sulfide and hydrocarbons. Toxicants associated with particulates and suspended solids were not identified.

INTRODUCTION

In response to the Clean Water Act (CWA) mandate that "the discharge of toxic pollutants in toxic amounts be prohibited," EPA issued a national policy, "Policy for the Development of Water Quality-Based Permit Limitations for Toxic Pollutants" (Federal Register, Vol. 49, No. 48, 1984). In this policy, EPA recommends the use of whole produced water toxicity testing to complement chemical analyses in controlling toxic constituents. Wastewater discharge permittees, regulated by the National Pollutant Discharge Elimination System (NPDES), are required to perform produced water toxicity identification and reduction evaluations. The objective of the evaluation is to determine the measures necessary to maintain the produced water's toxicity at acceptable levels (Burkhard and Ankley, 1989; Mount and Anderson-Carnahan, 1988, 1989).

In response to the EPA policy, the petroleum industry has taken the initiative to identify the causes of aquatic toxicity in produced water. Results of toxicological analysis on a number of different produced waters over the years have indicated some acute and chronic toxicity in

Produced Water, Edited by J.P. Ray
and F.R. Engelhart, Plenum Press, New York, 1992

produced water (Neff et al., 1987a,b); however, the specific cause(s) have not been completely identified. The primary toxicants need to be identified so that appropriate treatment technologies may be selected to remove or reduce their concentration in produced water.

Because produced waters contain mixtures of organic and inorganic constituents (Stephenson, 1992; Tibbetts et al., 1992), an understanding of their toxicity must begin with an identification of the various toxicants present. Traditionally, identifications have relied upon a chemical inventory of all of the components of the mixture and a separate evaluation of the toxicity of the individual components (usually accomplished through a literature review). The major shortcomings associated with this approach include 1) the limitations of analytical methods often limit the accuracy and completeness of the chemical characterization, 2) the potential additive, synergistic, and antagonistic effects of multiple toxicants in a mixture are not considered, 3) the bioavailability of potential toxicants is not determined, and 4) the toxicity of all components may not be well understood.

The TIE approach to understanding the toxicity of mixed wastes like produced waters has been developed by the U.S. EPA as summarized in Burkhard and Ankley (1989). This toxicity-based approach consists of subdividing the mixture into various fractions and experimentally determining the toxicity of each fraction. A series of fractionation procedures, each designed to remove a class of compounds, subdivides the mixture into a number of samples that can be tested individually. By relating any reduction of toxicity to the class of chemicals removed from the sample, it is theoretically possible to determine a cause-and-effect relationship. An important consideration in a TIE is the variability of material from a single source over time and variability from location to location within an industry. This can only be evaluated by characterizing a number of samples. Produced waters are well suited for this toxicity identification approach.

To assess the causes of toxicity in produced water, TIEs were performed on a number of samples collected from different discharge locations in the U.S. Physical/chemical fractionations and corresponding acute toxicity tests were conducted on low ($<$ 3 parts per thousand [ppt]) TDS produced water samples to 1) identify those broad classes of compounds that are responsible for the toxicity, and 2) examine the variability associated with produced waters from a variety of sources. This type of evaluation is the first step (Phase I) of a TIE (Mount and Anderson-Carnahan, 1988). It is often followed by a second phase of testing (Mount and Anderson-Carnahan, 1989) intended to identify individual toxic compounds, and usually by a third confirmatory phase (Mount, 1989) intended to prove the source of toxicity rather than inferring toxicity by correlating a toxicity reduction with the removal of a specific compound or class of compounds. The second and third phases of a TIE were not part of this study.

METHODS

TIEs were conducted on a total of four low-TDS produced water samples (Table 1); two samples (WY1 and WY2) from separate locations in Wyoming, one sample (TX1) from Texas, and one sample (LA1) from Louisiana. Two freshwater species, the fathead minnow *Pimephales* promelas and daphnid *Ceriodaphnia dubia*, were used in these TIEs. Up to ten major physical/chemical fractionation procedures, each with specific objectives, were conducted on the produced water samples (Table 2).

Most fractionations involved performing a manipulation at different pHs (pH = i [initial pH of the produced water], 3, or 11). The state of organic and inorganic compounds, whether they are in an ionic or un-ionized form (e.g., NH_4+ or NH_3), are influenced by the acidity/basicity of the solution and the equilibrium constant (K_a) of the compound. When the pH of the solution is equal to the pK_a ($-\log_{10} K_a$) of the compound, there are equal amounts of ionic and un-ionized forms (for monoprotic compounds). For acidic compounds, if the pH > pK_a, then the ionic form predominates; if the pH < pK_a, then the un-ionized form is predominate. When the pH is one unit below the pK_a of a compound, ninety percent of the chemical species is in the un-ionized from, the remaining is in the ionized form. For basic compounds, the opposite relationship of ionization is observed. For produced waters, the adjustment of pH will cause a change in the speciation of a selected target compound group(s) such that this group(s) can be removed from the produced water by a fractionation procedure involving pH adjustment.

Table 1. Low-TDS produced water (PWF) samples selected for TIE

Sample ID	Date Received	Origin	Total Dissolved Solids Content	Type of Production	Production Additives
PWF-WY1	10/10/90	Wyoming	low	oil	EB, CI, REB SI, GTC
PWF-WY2	3/27/91	Wyoming	low	oil	EB
PWF-TX1	3/26/91	Texas	low	gas	CI
PWF-LA1	8/7/91	Louisiana	low	oil	CI

EB = Emulsion breaker
CI = Corrosion Inhibitor (oil soluble)
REB = Reverse emulsion breaker
SI = Scale inhibitor
GTC = Gas treating chemical

Table 2. Fractionation manipulations performed on the low-TDS produced water samples

Fractionation Type	Fractionation Objectives
pH Adjustment at pHs 3 and 11	Determination of the effect of pH adjustment in fractionations; provides a "blank" for pH-adjusted fractionations addresses issue of increased ionic strength and chemical reversibility.
Aeration with N_2 at pHs i, 3, and 11	Removal of volatile compounds under neutral, acidic, and basic conditions without presence of oxygen
Aeration with air at pHs i, 3, and 11	Removal of volatile and oxidizable compounds under neutral, acidic, and basic conditions
Filtration at pHs i, 3, and 11	Removal of particulates (with a filter) under neutral, acidic, and basic conditions
Solid-Phase Extraction (SPE) at pHs i, 3, and 9	Removal of semivolatile non-polar organic compounds and metal chelates removal of some acidic and basic organic compounds
Acidic Organic Compound Extraction (ACE) at pH < 2	Isolation of acidic organic compounds
Basic Organic Compound Extraction (BCE) at pH 12	Isolation of basic organic compounds
Neutral Organic Compound Extraction (NCE)	Isolation of neutral organic compounds
Graduated pH Test at pH 6, 7, and 8	Determination of the possible presence of ammonia
EDTA Chelation	Removal of cationic metals

Except for the ACE/BCE/NCE manipulations, the fractionation procedures followed the EPA methods presented in Mount and Anderson-Carnahan (1988). The ACE/BCE/NCE fractionation isolates the acidic, basic, and neutral semivolatile organic compounds in the produced water. Although the EPA solid-phase extraction (SPE) fractionation method is used to evaluate the toxicity of the semivolatile organic compounds, the ACE/BCE/NCE fractionation extends the limited pH range of the SPE fractionation to less than 3 and greater than 9 for the acidic and basic fractions, respectively. The ACE/BCE/NCE fractionation is not an EPA TIE method, and unlike the EPA methods in which toxicity is determined on "reduced" fractions, allows for direct toxicity determination of the target compound fractions without potential effects from salinity or individual cations and anions.

Sampling

Sixteen (16) liters of produced water were sampled from the discharge points of the selected production facilities. Samples were taken at a point immediately downstream from the last conventional water-treatment system. The line was flushed thoroughly for several minutes before sampling. The four 4 liter bottles were filled slowly to the top from the bottom to minimize turbulence and loss of volatile compounds. The samples were sent from the field in coolers packed with ice by one-day delivery service to the laboratory. On the day of receipt (Day 1) a whole sample was tested for toxicity and fractionated (Table 2) for further toxicity testing the next day (Day 2). Field information on each of the samples is presented in Table 1.

Analysis

On Day 1, for each produced water sample, the contents of the four 4-l bottles were combined in a 25-l polycarbonate container and gently mixed for 30 sec. The container was tightly capped and the sample aliquoted from the container by positive pressure decanting to minimize agitation. After mixing, sample aliquots were subjected to initial toxicity testing and preliminary chemical analyses.

The initial (or range finding) toxicity tests were conducted with two freshwater species, the fathead minnow *Pimephales promelas* and the daphnid *Ceriodaphnia dubia*. Toxicity test conditions are summarized in Table 3. Chemical analyses of the produced water sample included pH, alkalinity, conductivity, dissolved oxygen (DO), ammonia, iron, and oil/grease (with quantification by gravimetry and infrared spectrophotometry) determinations.

Table 3. Summary of test conditions for acute toxicity tests using the freshwater species fathead minnow, *Pimephales promelas*, and daphnid, *Ceriodaphnia dubia*

	Species	
Condition	*Pimephales promelas* (Fathead Minnow)	*Ceriodaphnia dubia* (Daphnid)
Reference:	U.S. EPA, 1985	U.S. EPA, 1985
Test Type:	Static	Static
Test Duration:	24 hours	24 hours
Dilution Water:	Well or surface water	Pond water
Age/Size of Test Organisms:	2-24 hours old	2-24 hours old
Temperature:	20 ± 1°C	20 ± 1° C
Photoperiod:	16 hours light/8 hours dark	16 hours light/8 hours dark
Light Intensity:	Ambient laboratory lighting	Ambient laboratory lighting
Number of Concentrations:	5 plus a control	5 plus a control
Number of Replicates:	1	1
Number of Organisms/Replicate:	5	5
Test Vessel Water Volume:	100 ml	20 ml
Loading Rate:	5 animals per 100 ml	5 animals per 20 ml
Feeding:	Live newly-hatched brine shrimp	Yeast, trout chow, cerophyl and/or fresh water algae
Aeration:	None	None
Effects Measured:	Death	Death

With the remaining produced water, the fractionation manipulations (Table 2) were performed on Day 1. The EPA procedures in Mount and Anderson-Carnahan (1988) were used in preparing all fractions except the ACE/BCE/NCE fractions. The ACE/BCE/NCE fractionation procedures are summarized:

Isolation of the Neutral Compound Fraction. The produced water (200 ml) was acidified to pH <2 with 6 N HCL and then extracted three times with Freon. (The acidic aqueous phase was saved for basic compound extraction.) The extracts were combined and back-extracted three times with 1.0 N NaOH. (The basic aqueous phase was saved for acidic compound extraction.) The Freon extract was dried with sodium sulfate and concentrated to 100 l under a stream of clean nitrogen gas. The Freon was displaced with acetone and the extract, after concentration, was added to 200 ml of dilution water. This extract was called the Neutral Compound Extraction (NCE) fraction.

Isolation of the Basic Compound Fraction. The acidic aqueous phase from the NCE isolation was adjusted to a pH > 12 with 1.0 N NaOH and then extracted three times with Freon. The Freon extract was dried and concentrated. The Freon was displaced by acetone and the extract concentrated to 100 ml before addition to 200 ml of dilution water. This extract was called the Basic Compound Extraction (BCE) fraction.

Isolation of the Acidic Compound Fraction. The basic aqueous phase from the NCE isolation was diluted to a volume of 200 ml with dilution water. The pH of the diluted extract was adjusted to the pH of the dilution water. This extract was called the Acidic Compound Extraction (ACE) fraction.

All fractions of the produced waters and archive volumes of the whole produced waters were stored overnight in a refrigerator at 2 to 4° C for toxicity testing on Day 2. On Day 2, after pH adjustment, initial, baseline, and fraction toxicity tests were performed on a single replicate at five concentrations (except for the Graduated pH Test and EDTA Chelation Test) with the *Pimephales* and the *Ceriodaphnia* (Table 3). The five concentrations for each species were determined based upon the results of the initial toxicity tests. Sample blanks were prepared and processed in conjunction with each fractionation procedure, using dilution water adjusted to the appropriate pH. For the sample blank, toxicity tests were conducted at the highest concentration of the fraction solution. Control sample toxicity tests were performed with each set of fractions. A 24-hour LC_{50} was calculated by binomial/nonlinear interpolation, moving average or the probit method from the number of animals that died at each concentration. Based on the LC_{50} of the initial toxicity test and the hardness of the produced water, a stock solution of EDTA was prepared for addition of different amounts to a calculated concentration of produced water.

For the Graduated pH and EDTA Chelation tests, one preselected concentration of whole produced water was tested in which either the pH was changed (Graduated pH Test) or different amounts of EDTA were added (EDTA Chelation Test). The number of living organisms compared to original number of test organisms was recorded at each pH level or EDTA concentration.

RESULTS

During sample processing and fractionation, visual and olfactory observations were recorded on changes to the produced water or fractions. These observations (Table 4) provided some insight on the behavior of the produced water during fractionation and complemented the measured chemical and toxicological data collected as part of the TIE.

Water quality parameters (e.g., hardness, pH) and one of the NPDES required measurements, oil and grease analysis, were determined for each produced water (Table 5). The oil and grease concentrations were calculated with two types of analytical end points, gravimetry and IR, to indicate the possible variability in measurement technique. Because of the uncertainty of the Graduated pH Test in identifying the presence of ammonia in samples, a simple ammonia determination was performed to support the toxicity test result. Iron concentration was also measured in this TIE to possibly explain the observations made during pH adjustments on precipitate formation, and to indicate the gross amount of other trace metals in the produced water.

Table 4. Summary of general observations of the behavior of produced water samples during sample processing and fractionation

Produced Water Sample	Initial Sample Observations	Fractionation Observations		Day 2 Fraction Observations
PWF-WY1	Strong "rotten egg" odor Light surface sheen	pH Adjust: Aeration/Air:	pH3 - cloudy pHi - cloudy	Whole effluent: cloudy, color change to green
PWF-WY2	Strong "rotten egg" odor Light surface sheen	pH Adjust:	pH11 - cloudy	Aeration: pH11 - both fractions exhibit "oily" surface ring Adjustment: pH11 - "rotten egg" odor
PWF-TX1	Sweet "oily" odor Slight "rotten egg" odor	pH Adjust:	pH11-cloudy, yellow to green to rust, rust ppt.	pH3 all fractions: yellowish color pH11 all fractions: light brown color
	Light surface sheen	Aeration:	pH11-ppt settled leaving clear effluent pH11, pHi-color change	
		BCE:	Extract was slightly green, oily smell	
PWF-LAI	Sweet "oily" odor Murky yellow/brown	pH Adjust:	pH11- green ppt pH3 - yellow/orange color change	All sample fractions: orange ppt Whole effluent: color change to rust

Initial Sample Observations - observations of the produced water before sample fractionation
Fractionation Observations - observations of the produced water and fractions during fractionation
Day-2 Fraction Observations - observations of the fractions during Day-2 pH adjustment

Toxicity test results for the whole produced water and fractions are presented in Table 6 for those toxicity tests from which LC_{50}s could be calculated. Details on the toxicity test results, such as the number of organisms alive at the start and end of the tests and supporting water quality results, are provided in Sauer et al. (1992). The Graduated pH and EDTA Chelation test results are summarized in the text.

Table 5. Summary of chemical properties of produced water samples

Property	PWF-WY1	PWF-WY2	PWF-TX1	PWF-LA1
Hardness (mg/l as $CaCO_3$)	1,580	2,300	1,440	647
Alkalinity (mg/l)	400	1,834	456	167
pH	7.4	6.4	6.4	6.5
Conductivity (μmhos/cm)	2,870	3,640	9,930	2,770
Dissolved oxygen (mg/l)	3.0	2.4	8.3	3.3
Oil and grease (IR-mg/l)	—	8.4	100	18
Oil and grease (gravimetric-mg/l)	19	8.7	66	20
Total ammonia (mg/l as N)	5.0	0.5	41	24.8
Iron (mg/l)	—	0.081	21	50

(—) - Not analyzed

Sample PWF-WY1

Over a one-day period, the toxicity of this Wyoming produced water was observed to decrease. The *Ceriodaphnia* LC_{50} of the produced water was initially (Day 1) 5.8%. By the time the baseline toxicity test was conducted (Day 2), the LC_{50} for the *Ceriodaphnia* had increased to 26%. This same trend was observed for the *Pimephales*, but the toxicity was an order of magnitude higher for the *Pimephales* than the *Ceriodaphnia*. The initial (Day 1) *Pimephales* LC_{50} was 0.59% whereas the baseline (Day 2) LC_{50} was 2%.

The results of the toxicity tests on the produced water fractions were generally similar for both the *Pimephales* and the *Ceriodaphnia*; however, the *Pimephales* was more sensitive to the produced water toxicity than the *Ceriodaphnia*.

- During processing, a very strong "rotten egg" odor emanated from the produced water. This odor is a possible indicator of the presence of H_2S gas.
- In almost every fractionation type, the toxicity of the produced water was retained in fractions that were manipulated at a basic pH (pH 11).
- The original produced water toxicity was completely lost in those fractions that had acidic and neutral pH adjustments.
- There was a noticeable change in toxicity when the produced water (pH 11) was aerated with air.
- None of the produced water toxicity was attributed to the acidic organic compounds as determined by the ACE fractionation.

The Graduated pH Test was not performed for this produced water. The EDTA Chelation test for the two species gave mixed results. For the *Ceriodaphnia*, survival was less than 30% for all EDTA additions, whereas for the *Pimephales*, survival decreased with increasing EDTA additions. The control and the low-level EDTA additions had 100% survival, the mid-level addition survival decreased to 60%, and at the higher additions, total mortality was observed. The results are inconclusive.

Sample PWF-WY2

This low-TDS Wyoming produced water exhibited an initial LC_{50} of 0.32% for *Pimephales* and an initial LC_{50} of 22% for *Ceriodaphnia*. The baseline tests performed on Day 2 showed a reduction in toxicity in both species. In the baseline test, toxicity was essentially completely lost for the *Pimephales* (LC_{50} > 80%) and the test concentration range for *Ceriodaphnia* was too narrow to detect LC_{50}s above 10%.

- During processing, a slight "rotten egg" odor emanated from the sample.
- All initial toxicity was reduced in all fractions for both species.
- No toxicity was observed in the ACE, BCE, and NCE fractions, except for the *Ceriodaphnia* ACE fraction where there is slight toxicity.

The Graduated pH Test showed complete survival at all pH levels for both species and therefore no trend to indicate ammonia toxicity. For the EDTA Chelation test, control, blank, and low-level additions of EDTA showed no mortality in both species. There was a slight mortality at mid-level additions, and complete mortality at the higher additions. If metals are contributing to toxicity, the toxicity will decrease at the mid-level EDTA additions. The pHs in the mid and higher EDTA solutions were in the 4.0 to 5.0 pH range.

Sample PWF-TX1

The toxicity of this low-TDS Texas produced water resulted in an initial LC_{50} of 17% for the *Pimephales* and an initial LC_{50} of > 8% for *Ceriodaphnia*. Considering that the baseline test for the fathead *Pimephales* was conducted at a narrow concentration range, the baseline LC_{50} of 35% essentially exhibits little change from Day 1 to Day 2 for the whole produced water.

Table 6. Summary of toxicity test results (LC_{50} with 95 percent confidence interval) of the low-TDS produced water samples

	PWF-WY1		PWF-WY2		PWF-TX1		PWF-LA1	
Fractionation	Minnow LC_{50}	Daphnid LC_{50}	Minnow LC_{50}	Daphnid LC_{50}	Minnow LC_{50}	Daphnid LC_{50}	Minnow LC_{50}	Daphnid LC_{50}
Initial	0.59% (0.1-5%)	5.8% (1-10%)	0.32% (0.1-1.0%)	22% (0.1-50%)	17% (5.0-50%)	8% (5-50%)	22% (10-50%)	8.0% (5-50%)
Baseline	2% (1-5%)	26% (15-50%)	> 10%	> 80%	35% (25-50%)	11% (10-25%)	22% (10-50%)	-
pH Adjustment pH 3	> 50%	> 50%	> 10%	> 80%	> 50%	9% (8-10%)	NT	30% (5- > 50%)
pH Adjustment pH 11	2% (1-4%)	> 50%	> 10%	> 80%	> 50%	8% (4-12%)	NT	22% (10-50%)
N_2 Aeration pH i	> 50%	> 50%	> 10%	> 80%	> 50%	11% (5-25%)	NT	19% (5- > 50%)
N_2 Aeration pH 3	> 50%	> 50%	> 10%	> 80%	35% (25-50%)	15% (12-22%)	NT	27% (10-50%)
N_2 Aeration pH 11	2% (1-5%)	17% (10-38%)	> 10%	> 80%	> 50%	32% (25-50%)	NT	27% (10-50%)
Air Aeration pH i	> 50%	> 50%	> 10%	> 80%	> 50%	13% (10-19%)	NT	16% (7.5-50%)
Air Aeration pH 3	> 50%	> 50%	> 10%	> 80%	> 50%	46% (25- > 50%)	NT	30% (25-50%)
Air Aeration pH 11	22% (10-50%)	27% (16-50%)	> 10%	> 80%	40% (7.5- >50%)	16% (13-25%)	NT	20% (7.5-50%)
Filtration pH i	> 50%	> 50%	> 10%	> 80%	> 50%	> 50%	NT	30% (25-50%)
Filtration pH 3	> 50%	50%	> 10%	> 80%	40% (25- >50%)	46% (25- > 50%)	NT	27% (10-50%)
Filtration pH 11	7% (5-10%)	12% (5-50%)	> 10%	> 80%	> 50%	46% (25- > 50%)	NT	46% (25- > 50%)
SPE (25-55) pH i	> 50%	35% (>0.1-50%)	> 10%	> 60%	> 25%	> 25%	NT	NT
SPE (25-55) pH 3	> 50%	> 50%	> 10%	> 40%	> 25%	> 25%	NT	NT
SPE (25-55) pH 9	2% (1-5%)	6% (1-10%)	> 10%	>60%	> 25%	> 25%	NT	NT
SPE (150-180) pH i	> 50%	> 50%	> 10%	> 60%	> 10%	> 10%	NT	NT
SPE (150-180) pH 3	> 50%	> 50%	> 10%	> 60%	> 25%	> 25%	NT	NT
SPE (150-180) pH 9	3% (1-10%)	6% (2-8%)	> 10%	> 60%	> 10%	> 10%	NT	NT
ACE	> 50%	> 50%	100%	100%	100%	61% (50-75%)	NT	100%
BCE	NT	NT	100%	100%	100%	60% (46-73%)	NT	100%
NCE	NT	NT	100%	100%	100%	19% (10-25%)	NT	100%

NT - Test not conducted

The same trend was observed for the *Ceriodaphnia,* as the baseline LC_{50} was 11%, even less of a change from the initial test result.

The LC_{50}s of the *Pimephales* for all the pH i fractions were > 50% and reflected a decrease in the toxicity of the whole produced water (17 to 35%). This change, from Day 1 to Day 2, in toxicity may be due to loss of volatile compounds over time. The LC_{50} values of the *Pimephales* for all pH 3 fractions with the exception of the SPE and air aeration were higher, ranging from 35 to > 50%. The SPE fractions would likely follow the same pattern as the filtered fractions (> 50% LC_{50}), had the SPE tests been performed at higher concentration levels.

The results of the toxicity tests on the produced water fractions with *Pimephales* and *Ceriodaphnia* show that toxicity was reduced by many of the fractionation procedures. The *Ceriodaphnia* was more sensitive to the produced water toxicity than the *Pimephales.*

- Filtration and SPE fractionations at all pH levels effectively removed toxicity from the samples.
- Simple pH adjustment did not reduce the toxicity of the sample to the *Ceriodaphnia.*
- Aeration (N_2 and air) was more effective in reducing toxicity to the *Pimephales.*
- The toxicity of the isolated NCE fraction was similar to that of the whole produced water.
- Some toxicity was observed in the isolated ACE and BCE fraction.
- Elevated concentration of oil and grease, 60 to 100 mg/l, was determined in the produced water.
- Ammonia concentration was relatively high, approximately 40 mg/l as N (Table 5).

In the Graduated pH Test, the survival rate for the *Pimephales* did not change. There was a survival rate of 80 to 100% at all three pH levels. Eighty percent of the organisms were alive at pH 6 and pH 8, and 100% were alive at pH 7. The survival rate for the *Ceriodaphnia* was different: at pH 6.0, there was 20% survival, at pH 7.0, 40%, and at pH 8.0, 100%. These results indicate that ammonia was not present in the produced water at toxic levels. When ammonia is present in the produced water at toxic levels, the pH 6 test solution should be less toxic than the pH 7 solution, which in turn, should be less toxic than the pH 8 solution.

In the EDTA Chelation test, both test species, *Ceriodaphnia* and *Pimephales,* had a 0% survival rate at all concentrations including blanks. If metals are contributing to the toxicity of the produced water, the mortality will not change from that observed in the control to those resulting from the low EDTA additions. In the EDTA mid-range, toxicity will be reduced, but at high EDTA additions, toxicity will be as high or higher due to unbound EDTA. If toxicity is higher at all EDTA levels, as was the case in this produced water, then usually excess EDTA was added (Mount and Anderson-Carnahan, 1988).

Sample PWF-LA1

The initial LC_{50}s for this low-TDS Louisiana produced water were 22% for *Pimephales* and 8.0% for *Ceriodaphnia* (Table 6). The baseline tests on Day 2 showed identical toxicity for the *Pimephales,* but greatly reduced toxicity for *Ceriodaphnia* (LC_{50} > 50%). The baseline toxicity for the *Ceriodaphnia* was also reduced in comparison to all the fraction toxicities, indicating that the *Ceriodaphnia* baseline test may be inaccurate. Because of the low toxicity with the *Pimephales,* toxicity tests were conducted using *Ceriodaphnia* only. The LC_{50}s of the fractions were generally three to five times greater than the initial LC_{50}. The following trends were observed:

- Toxicities of almost all pH-adjusted fractions were similar to those of the pH Adjustment fractions (pH 3 and pH 11).
- Both the pH i aerated fractions exhibited toxicity closest to the initial toxicity of the whole produced water.
- Toxicity of the produced water was most reduced in the pH 11 fraction.

- The organic compound ACE, BCE, and NCE fractions showed no toxicity.

The Graduated pH Test resulted in 80 to 100% survival at all three pH levels.

The EDTA Chelation test showed a 60% survival in the low EDTA addition range tests (control to 100% survival), and complete mortality in the mid- and high-EDTA addition range tests. The pHs in the mid and high-EDTA addition solutions were approximately 4.0.

DISCUSSION

The compilation of the results of the fractionation observations, chemical property analysis, and toxicity tests (Tables 4, 5, and 6) provided insight into the types of compounds contributing to the toxicity of the four produced water samples. For the most part, the interpretation of the study data was based on the relative differences in toxicities between whole produced water and fractions, and among fractions which were sometimes subtle. Usually more than one manipulation affected toxicity, the information from which was used to either substantiate a potential toxicant or indicate an additional toxicant in the sample. Although not definitive, primary, and in some cases, secondary classes of toxicants were hypothesized for each sample.

Sample PWF-WY1

The toxicity test results of both the *Pimephales* and *Ceriodaphnia* species indicate that the major cause of the toxicity was a volatile compoumd(s). Except for the pH 11 fractions in which whole produced water toxicity was retained, toxicity was lost in every fraction. In the case of the *Pimephales,* the pH 11 (or 9) solutions had LC_{50}s similar to or less than the baseline produced water for all fractionation types. The same trend was observed in the fraction toxicity tests with *Ceriodaphnia,* except that for the pH Adjustment fractionation, where the pH 11 fraction had lost its toxicity.

With some indication from the observed odor, the toxicity of this produced water can be attributed to, at least in part, the presence of hydrogen sulfide (or mercaptans) in solution (Table 4). The presence of hydrogen sulfide in solution is highly dependent on the pH of the solution. Hydrogen sulfide can be present in solution in the form of hydrogen sulfide gas or the ion HS^- (Stumm and Morgan, 1981). At acidic pH, hydrogen sulfide in solution is present as hydrogen sulfide gas, which is volatile; at higher pH, the HS^- ion forms. At the pH of this produced water (pH = 7.4), hydrogen sulfide would evolve from solution.

For the pH Adjustment fractionations, the pH of the produced water is adjusted to the assigned acidic or basic pH, and then after the fractionation, is returned to the original produced water pH. At high pH, the manipulation of the produced water would be ineffective in removing hydrogen sulfide from solution since the hydrogen sulfide is in the ionic form and not available for release to the atmosphere. When the pH is lowered to the original produced water pH, the sulfide ions would then be converted back to hydrogen sulfide, and become available as a toxicant. The behavior of this produced water upon fractionation suggests that this mechanism of retaining toxicity occurred throughout the fractionations and toxicity tests.

The pH 9 SPE fractions of both the 25- to 55-ml and 150- to 180-ml samples for both species exhibited toxicity similar to all the other fractions. For the *Ceriodaphnia,* the initial toxicity (LC_{50} of 5.8%) of the whole produced water was retained during the SPE fractionation. The cause of the toxicity is related to the presence of hydrogen sulfide in the fraction. The complete mortality of the test organisms in an EDTA Chelation Test at the higher levels of EDTA addition may be explained by the increased toxicity of hydrogen sulfide due to decreased pH of the EDTA test solutions. Solution pHs in a EDTA Chelation Test will commonly decrease with increasing amounts of EDTA which, for this sample, would cause increased toxicity of hydrogen sulfide at the lower pH. End-of-test pHs decreased from 6.0 to approximately 4.0 with increasing EDTA addition for both organisms.

Besides the general retention of toxicity at basic pH, there was a subtle change in toxicity when the produced water was aerated. At the basic pH, the LC_{50}s of the aerated (with air) fraction were higher than LC_{50}s of the other fractions. The reason for this may still be related to the presence of hydrogen sulfide in the produced water. If hydrogen sulfide is the major toxicant, the sulfur in the sulfide ($S^=$) form can be oxidized to SO_x in the presence of oxygen. This oxidized form of sulfur will not convert to hydrogen sulfide gas when the pH is decreased to the pH of the original produced water. As a result, some of the toxicity of the produced water is lost.

Other potential volatile toxicants that were considered included volatile hydrocarbons and ammonia. Because toxicity was retained at basic pH during aeration, the presence of toxic quantities of volatile hydrocarbons is doubtful. Volatile hydrocarbons would have been lost at all pH levels during aeration. From the chemical analysis, ammonia at a concentration of 5 mg/l does not seem to be sufficient to cause acute toxicity.

Sample PWF-WY2

Similar to the other Wyoming produced water (WY1) sample, the toxicity of the low-TDS produced water can be attributed to a volatile compound. The initial toxicity of the whole produced water was lost in the Day 2 baseline test, and in all subsequent sample fractions.

The "rotten egg" odor indicates that the toxicity of produced water is due to the presence of hydrogen sulfide/mercaptans in solution. Unlike the PWF-WY1 produced water, however, the retention of toxicity in the basic fractions, indicating the behavior of hydrogen sulfide in pH adjusted solutions, was not observed in this produced water. Like the PWF-WY1 produced water, the *Pimephales* was considerably more sensitive to the toxicant than the *Ceriodaphnia*. At the pH of this produced water (pH = 7.4), hydrogen sulfide would evolve from solution, resulting in the observed decrease in toxicity during sample storage and fraction manipulation.

Sample PWF-TX1

The toxicity in the NCE fraction and the loss of toxicity in all filtered fractions suggest that the major toxicants reside in the neutral semivolatile organic extract and suspended solids of the produced water. The *Ceriodaphnia* was more sensitive to the toxicants than the *Pimephales* and provided the distinguishing toxicological information for interpretation.

Hydrocarbons are the probable toxicants in this sample which are demonstrated by the presence of an oil sheen on the produced water sample surface, a sweet oily odor, a relatively high concentration of oil and grease (60 to 100 mg/l), and the toxicity of the NCE fraction. The toxicity of the NCE fraction, which isolates neutral semivolatile organic type compounds including hydrocarbons, was approximately the same as the whole baseline produced water toxicity. There is evidence from the filtration fractionation results that the hydrocarbons are in the form of emulsified oil, oil droplets, or oil sorbed on particulates. In those fractions that were filtered, toxicity was reduced or eliminated.

Because of its slight rotten-egg smell, the toxicity of this produced water can be attributed to, in part, the presence of hydrogen sulfide in solution. As explained previously, the presence of hydrogen sulfide in solution is highly dependent on the pH of the solution. At higher pHs, the HS^- ion forms. At the pH of this produced water (pH = 6.1), hydrogen sulfide would evolve from solution, resulting in the decrease in toxicity observed in the *Pimephales* baseline test, the species most sensitive of the two test species to hydrogen sulfide.

Although the ammonia concentration was approximately 40 mg/l, the Graduated pH Test results indicated that ammonia was probably not a major cause of toxicity in the produced water.

Sample PWF-LA1

Although initial and baseline tests with *Pimephales* showed no reduction in toxicity from Day 1 to Day 2, the loss of toxicity in all fractions compared to the initial whole produced water in the tests with *Ceriodaphnia* indicated that at least some of the toxic components of this produced water were volatile or easily degradable. Alternatively, the lack of complete loss of toxicity in the aeration fractionations indicated that not all the toxicants were volatile compounds.

If some of the toxicants were volatile compounds, hydrogen sulfide is an unprobable candidate. In this sample, the "rotten egg" smell was not obvious, and when fractionated, the toxicity was not retained at the basic pH during aeration. Ammonia is probably not a toxicant either since the results of the Graduated pH Test showed no trend in survival for the three pHs and ammonia concentrations were only 25 mg/l as N (Table 5).

The best evidence of a possible toxicant in this produced water is provided in the filtration fractionation. The pH 11 filtered fraction showed reduced toxicity that was not similarly observed at the other two pHs (i.e., pH 3 and pH i) or in any other fraction. As a result, potential toxicants for the Louisiana sample may be those materials that precipitate at a basic pH. The loss of toxicity in the basic filtered fraction and the relatively high concentration of iron (50 mg/l) suggest that transition and B-type metal elements are toxicants in this produced water. Possibly, a produced water additive (e.g., corrosion inhibitor) may precipitate at high pH. The iron in the produced water may act to coprecipitate with transition metals or a pH-sensitive additive at the basic pH, thereby removing the toxicant from solution when filtered.

Although the result of the EDTA Chelation Test may suggest the presence of chelatable metals, the survival pattern is probably due to the effect of lower pHs in the high-EDTA test solutions. EDTA solutions are typically acidic and it is common for the pH to decrease with increasing levels of EDTA. The pHs measured at the end of this test were approximately 4.0 for the mid and higher-EDTA solutions. If metals are contributing to the toxicity of the produced water, the mortality will not change from that observed in the control to those produced by low EDTA addition solutions. In the EDTA mid-range, toxicity will be reduced, but at high-EDTA additions, toxicity will be as high or higher due to unbound EDTA. Time to mortality is used to detect partial toxicity removal. Toxicity may be removed at all exposures if the lowest EDTA additions remove metal toxicity and the highest addition does not cause EDTA toxicity. If toxicity is not reduced at any addition, either the produced water has low amounts of chelatable metals or insufficient EDTA was added.

CONCLUSION

Five produced water samples—two Wyoming samples (PWF-WY1 and PWF-WY2), one coastal Texas sample (PWF-TX1), and one coastal Louisiana sample (PWF-LA1)—from an oil production facility (the Texas sample was from a gas facility) were subjected to the TIE procedures. These procedures involved up to 10 fractionation manipulations and subsequent toxicity testing of the fractions (and whole produced water) with two freshwater species (*Pimephales* and *Ceriodaphnia*).

The Phase I TIE procedures applied to these produced waters provided valuable preliminary information on the causes of toxicity in low-TDS produced waters. Major potential toxicant classes were identified (summarized in Table 7) based on the toxicity test and chemical property results, and knowledge of potential types of compounds in produced water.

Volatile compounds were considered as potential toxicants in all the low-TDS samples. In the PWF-WY1 sample the volatile compound suspected as primarily responsible for the observed toxicity was hydrogen sulfide (or mercaptans). This preliminary identification was based on the subtle changes in toxicity during aeration at different pH levels and the pervasive "rotten egg" odor that emanated from the sample. Although not as conclusive in the sample fractionation results as with the PWF-WY1 sample, the major potential toxicant in the PWF-WY2 sample also appeared to be hydrogen sulfide. This produced water possessed a strong rotten-egg odor. Some of the toxicity in the PWF-TX1 and PWF-LA1 samples was also attributed to volatile compounds based on the slight rotten-egg odor and/or decrease in toxicity from the initial toxicity test to the subsequent baseline test. For the PWF-LA1, there was no discernible rotten-egg odor, indicating that volatile components other than hydrogen sulfide were likely responsible for the observed toxicity.

Table 7. Summary of the compound classes associated with toxicity for the four low-TDS produced water samples

Samples	Possible Causes for Toxicity
PWF-WY1	Volatile compounds (probably hydrogen sulfide)
PWF-WY2	Volatile compounds (probably hydrogen sulfide)
PWF-TX1	Neutral semivolatile organic compounds (probably hydrocarbons)
	Suspended solids - filterable material at any pH
	Volatile compounds (probably hydrogen sulfide)
PWF-LA1	Material precipitated at basic pH
	Volatile compounds (not hydrogen sulfide)

Neutral semivolatile organic compounds (probably hydrocarbons) were found to be the primary toxicant class in the Texas PWF-TX1 sample. This is supported by evidence of an oil sheen on the produced water surface, a relatively high oil-and-grease concentration (approximately 100 mg/l), and toxicity associated with the isolated NCE fraction. Material that was filterable (at any pH), probably emulsified oil or oil adsorbed on particles, was also found to contribute to the toxicity of this produced water sample.

Material that precipitated at basic pH was determined to be a possible toxicant for the Louisiana PWF-LA1 sample. The type of compounds associated with precipitate formed at basic pH is not easily identifiable without further chemical analysis, but with the iron analysis (50 mg/l) suggest that chelatable transition metals may be potential toxicants.

ACKNOWLEDGEMENTS

The authors express their appreciation to Cherie A. Howell and Robert L. Boeri for their assistance in the laboratory and Jerry M. Neff and Helder J. Costa for reviewing the document. This work was funded by the American Petroleum Institute.

REFERENCES

Burkhard, L.P., and Ankley, G.T., 1989, Identifying toxicants: NETAC's toxicity-based approach, Environ. Sci. & Technol., 23 (12): 1438-1443.

Mount, D.I., and Anderson-Carnahan, L., 1988, "Methods for Aquatic Toxicity Identification Evaluations: Phase I Toxicity Characterization Procedures," U.S. EPA Environmental Research Laboratory, Duluth, MN, EPA/600/388/034.

Mount, D. I., and Anderson-Carnahan, L., 1989, "Methods for Aquatic Toxicity Identification Evaluations: Phase II Toxicity Identification Procedures," U.S. EPA Environmental Research Laboratory, Duluth, MN, EPA/600/388/035.

Mount, D. I., 1989, "Methods for Aquatic Toxicity Identification Evaluations: Phase III Toxicity Confirmation Procedures," U.S. EPA Environmental Research Laboratory, Duluth, MN, EPA/600/388/036.

Neff, J. M., Rabalais, N.N., and Boesch, D.F., 1987a, Offshore oil and gas development activities potential causing long-term environmental effects. In: "Long-Term Environmental Effects of Offshore Oil and Gas Development," D.F. Boesch and N.N. Rabalais (Eds.). Elsevier, New York. pp. 149-173.

Neff, J.M., Sauer, T.C., Jr., and Maciolek, N., 1987b, Fate and Effects of Produced Water Discharges in Nearshore Marine Waters. Final Report, Submitted to American Petroleum Institute, Washington, D.C.

Sauer, T.C., Brown, J.S., O'Neil, S., and Ward, T.J., 1992, "Identification of Produced Water Toxicity - Final Report Submitted to American Petroleum Institute," Washington DC. 143 p.

Stephenson, M.T., 1992, A survey of produced water studies. In: "Produced Water: Technological/ Environmental Issues and Solutions," J.P. Ray and F.R. Engelhardt, eds., Plenum Publishing Corp., New York, New York, USA.

Stumm, W., and Morgan, J.J., 1981, "Aquatic Chemistry: An introduction emphasizing chemical equilibria in natural waters," John Wiley & Sons, New York. 776 p.

Tibbetts, P.J.C., Buchanan, I.T., Gawell, L.J., and Large, R, 1992, A comprehensive determination of produced water composition. In: "Produced Water: Technological/Environmental Issues and Solutions," J.P. Ray and F.R. Engelhardt, eds., Plenum Publishing Corp., New York, New York, USA.

U.S. EPA, 1985, Methods for measuring the acute toxicity of effluents to freshwater and marine organisms. U.S. EPA Environmental Monitoring and Support Laboratory, Cincinnatti, OH. EPA/600/4-85/013.

AN APPROACH TO TOXICANT ISOLATION FROM A PRODUCED WATER SOURCE IN THE SANTA BARBARA CHANNEL

R.M. Higashi[1], G.N. Cherr[1], C.A. Bergens[2], and T.W-M. Fan[2]

[1] University of California Bodega Marine Laboratory
Bodega Bay, California
USA

[2] Dept. of Land, Air and Water Resources
University of California
Davis, California
USA

INTRODUCTION

Produced water (PW) composition can be highly variable among formations, but in all cases appears to be very complex, consisting of non-polar and polar organic compounds, as well as inorganic cations and anions, and combinations of these diverse chemical catagories (National Research Council, 1985; Boesch and Rabalais, 1987). Little is known about the composition of even major chemical catagories of PW such as polar organic compounds and inorganic anions, and only the transition and B-type (Stumm and Morgan, 1981) metals and hydrocarbons have received attention with respect to effects, fate and transport (Boesch and Rabalais, 1987). The complexity of PW and lack of information on major chemical catagories directly leads to formidable obstacles to understanding of the distribution and effects of PW as:

(a) it is logistically not feasible to analyze for even a small portion of the known constituents present, not to say of the unknown constituents;
(b) just the known constituents represent vastly different physical and chemical properties, and therefore can have different transport characteristics;
(c) the known constituents range from very stable to those that are readily transformed, and therefore can have different fates; and,
(d) because of all the above, if unknown constituents are playing a major role in environmental effects of PW, their presence may continue to go chemically undetected, and therefore may wreak havoc with interpretation and relevance of future modeling studies.

A logical strategy for tackling these obstacles, of course, is to drastically reduce the analytical complexity by first defining the properties of interest in PW, as well as selecting a source and receiving water characteristics. This concept of bioactivity-driven chemical fractionation to investigate unknown, highly complex mixtures is not new; it is historically the basis of major scientific fields such as biochemistry. In the present case, the PW property chosen was toxicity to early life stages of invertebrates, and the source and site was a nearshore marine discharge of PW in Southern California, USA. This paper describes such an isolation effort of

Produced Water, Edited by J.P. Ray
and F.R. Engelhart, Plenum Press, New York, 1992

the constituents from a Santa Barbara Channel PW source that account for chronic toxicity to mussel embryos.

MATERIALS AND METHODS

Materials and Toxicity Tests

PW samples were obtained from an onshore plant in Carpenteria, CA, USA. They were sampled from a tap in the discharge pipe and shipped chilled with headspace eliminated in teflon-capped borosilicate glass bottles. Samples were received within 24 h, immediately filtered on 0.8 μm cellulosic filters, and processed as described below. There was no significant loss of toxicity upon filtration (see Results below).

All common reagents were obtained from Fisher Scientific, Inc., as were all common solvents, which were of HPLC grade. N-methyl pyrrolidinone of HPLC grade was obtained from Aldrich Chemical Co.

Purple sea urchins, *Strongylocentrotus purpuratus*, were selected from a continuous culture maintained at the University of California's Bodega Marine Laboratory. Sea urchins were spawned by the injection of KCl into their coelemic cavity. Toxicity assays were conducted on the gametes following a literature procedure (Cherr et al., 1987). This test consists of incubating sperm in various concentrations of the fractions, and quantitating their ability to initiate fertilization of the eggs at 15° C.

California mussels, *Mytilus californianus*, were collected from Bodega Bay, CA. Mussels were spawned by immersing them in high concentrations of an algal monoculture, *Isochrysis galbana*. Toxicity assays were conducted on the embryos using a slightly modified procedure of Cherr et al. (1990). This consisted of scoring mussel embryos for the normality of their shell formation after 48 hours of culture at 15° C in various concentrations of the fractions.

For both toxicity tests, various fraction-dependent considerations had to be met. The fractions generated are shown in Figure 1 and the various fractions are referred to by the roman numerals from this figure. For the organic solvent-based (III, IV, and VIb) and HPLC fractions (IVa and IVb), a dispersant was used because the constituents were expected to have low solublity in water. We have found that a solvent of exceptionally broad polarity and solvation range, N-methyl pyrrolidinone (NMP), exhibits low toxicity (NOEC > 100 ppm) in these marine tests (G.N. Cherr, unpublished data). Thus, for these fractions, the dried samples (see sample generation procedures below) were first dissolved in small volumes of NMP and then dispersed into seawater. The highest concentration of NMP used in a given test (highest among all tests = 50 ppm) served as control and blank HPLC runs were included as additional controls for IVa and IVb. The water soluble dry-residue fractions (II and V) were first dissolved in distilled water, then these and aqueous fractions (volatile fraction, I, VIa, and VII) were salinity-adjusted with appropriate volumes of 2x seawater for toxicity testing. Seawater served as controls for measurements of these fractions. The exception was VIa which was highly saline after neutralization, which precluded salinity adjustment by addition of seawater, so 1N HCl was used as control to detect any false-positive "toxicity" due to Ca^{2+} deficiency from the lack of seawater. All seawater used in this study was obtained from Bodega Bay, CA and was 0.45 μm filtered.

Sample Fractionation

The scheme for fractionation of PW samples is shown in Figure 1. The major procedures in the method are:

Lyophilization or Freeze-Drying. This procedure, used widely in biochemistry and food technology to stabilize chemical structure and composition, was found to preserve the toxicity of PW while at the same time highly concentrating the effluent and rigorously removing volatile compounds. The filtered PW samples were divided into 500 ml aliquots in Teflon beakers and frozen at 77 K in liquid nitrogen. Because of its low eutectic - or "freezing" - point, PW samples were difficult to lyophilize and had to be lyophilized on glass apparatus with a low dead volume vacuum system and employing two -70° C cold fingers in tandem, one glass and the other steel. The volatile fraction was trapped on the first cold finger (glass) while the second cold finger (steel) was necessary to isolate the first finger from vacuum pump back-vapors, as well as to provide an additional condensation pump to prevent PW sample meltback. The ice-bound volatile sample trapped on the glass cold finger was slipped off as a single piece by allowing the finger to warm to about 0° C, then stored in solid form at -70 ° C until use. Pieces of the ice were thawed in airtight vials with minimal headspace to generate the aqueous sample for toxicity testing.

Sequential Solvent Extraction. The dry, white, powdery residue resulting from lyophilization (II) was subjected to three sequential nonpolar extractions by placing the sample in a borosilicate flask containing methylene chloride ($MeCl_2$) at 50 ml/g residue, and micro-agitation of the flask in a 100-watt sonic bath for 30 min. The extracted residue in each case was retained on 0.45 μm teflon filters, and after the final filtration, traces of solvent was removed using a stream of N_2 flowed through the filter.

The dried residue from $MeCl_2$ extraction was then subjected to acetonitrile (MeCN) extraction in a manner identical to the $MeCl_2$ extraction. The left-over residue (V) was found to be soluble only in water.

The $MeCl_2$ (III) and MeCN (IV) soluble extracts were concentrated to near dryness on a rotary evaporator, transferred to small glass vials with the aid of the respective solvents, and concentrated to dryness under a N_2 stream and stored dessicated at -20° C until use.

Preparative Chromatography. Semi-preparative high-pressure liquid chromatography (HPLC) was used to further fractionate IV. The procedure employed an ISCO (Lincoln, NB) ternary gradient HPLC system with a 10 mm x 250 mm 5 μm spherical particle C_{18} column and ISCO V_4 variable-wavelength UV-visible detection. The chromatographic conditions that generated IVa and IVb are described in detail in the figure legend. Fractions were collected with the detector turned off to avoid photodegradation, concentrated by rotary evaporation with excess MeCN to azeotrope off the water, transferred to a small vial for final concentration by N_2 stream to dryness, and stored dessicated at -20° C until use.

For fractionation of the water-only soluble fraction (V), Chelex 100 polyvalent cation exchange resin (Bio-Rad Laboratories, Richmond, CA), mesh 100-200, was used. The resin was prepared by rinsing in pH 7.2 unbuffered HPLC-grade (17.8 megaOhm) water until the pH of the eluate was close to pH 7, and this material was stored hydrated until use. For fractionation, 1 ml of hydrated Chelex was used per "10 ml" of V (i.e. the amount of V that was generated from 10 ml of I). A typical sample preparation constituted "20 ml" of V. Using volumes for this sample size, 2 ml hydrated Chelex was gravity packed into a polypropylene microcolumn (0.8 cm i.d. x 6 cm) fitted with a polyethylene 10 μm frit. The column was rinsed with three bed volumes of distilled water (3 x 2 ml) and allowed to flow by gravity until the meniscus of the water reached the top of the Chelex bed ("eluted"). The approximate flow rate was 1 ml/min. V was dissolved in 20 ml of distilled water, added to the column, and eluted; constituents retained on the column was VI and the resulting eluant was VII.

VI was further separated into two fractions by the following procedure. VIa was generated by eluting the column with 3 x 2 ml of 1N HCl, which was then neutralized with NaOH to generate the aqueous sample for toxicity testing. VIb was generated by eluting the VIa-devoid Chelex column with 3 x 2 ml of 1:1 MeOH:1N HCl, neutralized with NaOH, rotary-evaporated with excess MeOH to azeotrope off the water, transferred using MeOH to a small vial to eliminate excess salt, concentrated by N_2 stream to dryness, and stored dessicated at –20° C until use.

Chemical Analysis

Analytical methods used in various phases of this study included gas-liquid chromatography ("GC") using a Varian (Varian Instruments, Walnut Creek, CA) 3300 GC outfitted with a

0.25 mm x 40 m 0.4 µm DB-5 coated open-tubular column and flame ionization detector (FID), or Hewlett-Packard (Hewlett-Packard Analytical Products, Palo Alto, CA) 5890 GC with the same column interfaced to a VG Instruments (Altrincham, UK) Trio-2 quadrupole mass spectrometer ("GC-MS"); direct MS using a VG Zab-HS double-focusing mass spectrometer with Xe fast-atom bombardment; atomic spectroscopy using either ARL (Applied Research Laboratories, Inc., Dearborn, MI) 3510 inductively-coupled Ar plasma emission ("ICP-AES") or Varian AA-1475 absorption with Varian VGA-76 hydride-generator; HPLC (system described above); and UV-visible spectrophotometry (Hewlett-Packard 8452A diode-array spectrophotometer).

For analysis by GC and GC-MS, III and IV were dissolved in $MeCl_2$ and MeCN, respectively, and injected into the instrument. GC conditions are given in the legend for Figure 3. Electron-impact MS conditions were: transfer line = 250° C, source = 200° C, electron energy = 70 eV, mass range = 40-700 m/z, positive ion detection. For direct MS, II and V were dissolved in glycerol to provide the matrix, the fast-atom Xe gun operated at 9 kV, accelerating potential = 8 kV, spectrum swept in magnet scan mode, scan time = 20 s, mass range = 100-2000 m/z, negative ion detection. HPLC analysis details are described in Figure 2.

Samples for atomic spectroscopy (I, II, IV, IVa, V, VIa, VII) were analyzed by a digestion procedure using 20% $HClO_4$ + 50% HNO_3 (Agemian and Thomson, 1980) and measurement by ICP-AES (Buchanan and Hannaker, 1984). For analysis of As, the same digestion procedure was coupled with measurement by hydride generation-AA (Brumbaugh and Walther, 1989). NIST-traceable standards were used to determine the response curves for each element.

Samples for UV-visible spectrophotometry (all fractions) were dissolved in the respective solvents used for their isolation, which also served as reference solvent. All spectra were obtained using 1 cm path length quartz cuvettes.

RESULTS AND DISCUSSION

The toxicant fractionation approach used is described above and summarized as follows, outlined by the overall scheme shown in Figure 1. PW samples were collected into teflon-capped glass bottles from a tap in the discharge pipe and shipped refrigerated to our laboratory with no headspace in the samples. PW samples (I) were separated into volatile and nonvolatile (II) fractions by lyophilization (freeze-drying), then II was subjected to sequential organic solvent extraction that generated nonpolar (III, containing hydrocarbons), intermediate polar (IV, containing many other organic compounds such as phenolics), and water-only soluble (V, containing metal ions, polar organic and anionic constituents) fractions. Further fractionation of IV was achieved using HPLC, while V was extracted by an ion-exchange resin (Chelex resin) designed for trapping polyvalent cations (fraction VI), finally leaving a polar fraction (VII) that contained polar organic, inorganic anion and other (non-polyvalent) cationic substances. VI was further separated into two fractions by elution with acid (VIa) and acidic MeOH (VIb).

This fractionation procedure was developed using the more rapid sea urchin sperm cell test, but the mussel embryo test was used for the remainder of the study for greater interpretive relevance to other projects studying this PW source (Fan et al., 1992; Osenberg et al., 1992). To the extent practical, the fractionation was performed in a fashion that made as few presumptions as possible about which classes of constituents were the sources of toxicity. Chemical stability and the structures relevant to the receiving water conditions were also considered. For example, the pH of PW was set to 8 (seawater pH) and was not manipulated further to "optimize" organic solvent extraction; large swings of pH could alter structures, and in any case, it was not known *a priori* for what types of structures to optimize. On the other hand, drastic changes in pH occurred at elution from the Chelex column simply because it was unavoidable. However, as Chelex traps free cations, the elution by strong acid was unlikely to have altered the composition of that fraction.

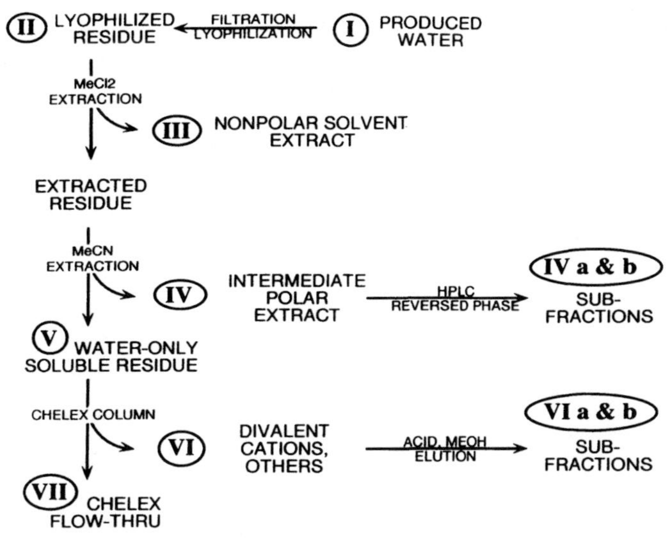

Figure 1. Fractionation scheme.

The salient findings are determined by the relative toxicity of each fraction. These data for mussel embryo toxicity are summarized in Table 1, which shows that the lyophilization step removed a small portion of toxicity, but the majority remained with the residue (II). For illustration of the trend, the data in Table 1 are taken from two to three replicates of a single PW sample. As lyophilization represents rigorous removal of volatile constituents (even water is removed), it follows from this data that relatively non-volatile constituents accounted for the majority of toxicity. This finding also demonstrated that lyophilization is an outstanding method of archiving both the volatile and non-volatile portions of PW samples for investigations into variability as well as for future chemical and biological measurements.

Extensive organic solvent extraction of II likewise removed only a small part of the toxicity to mussel embryos (Table 1); the fractions generated, III and IV, contained little or no measureable toxicity (Table 1). However, as preliminary evidence suggested that significant toxicity to sea urchin sperm existed in IV (data not shown), further fractionation of IV by HPLC was undertaken. These fractions were collected as shown in Figure 2. Not surprisingly, due to the low toxicity of IV towards mussel embryos, its subfractions IVa and IVb also exhibited no measurable toxicity (Table 1). GC-MS analyses of fractions III and IV (for clarity, the corresponding GC-FID tracings are shown in Figure 3) revealed relatively low (< ppb) levels

Table 1. Toxicity of PW fractions to mussel embryos

	Fraction	EC_{50} (%)
I	Whole PW	2.12
II	Lyophilized Residue of I	2.86
III	$MeCl_2$ Extract of II	~15
IV	MeCN Extract of II	»20
IVa	Polar HPLC Fraction of IV	»20
IVb	Nonpolar HPLC Fraction of IV	»20
V	Water-only Soluble Residue of II	2.65
VIa	Chelex-Retained V, acid eluted	2.87
VIb	Chelex-Retained V, acidic MeOH eluted	4.63
VII	Chelex-Flow Thru of V	»20

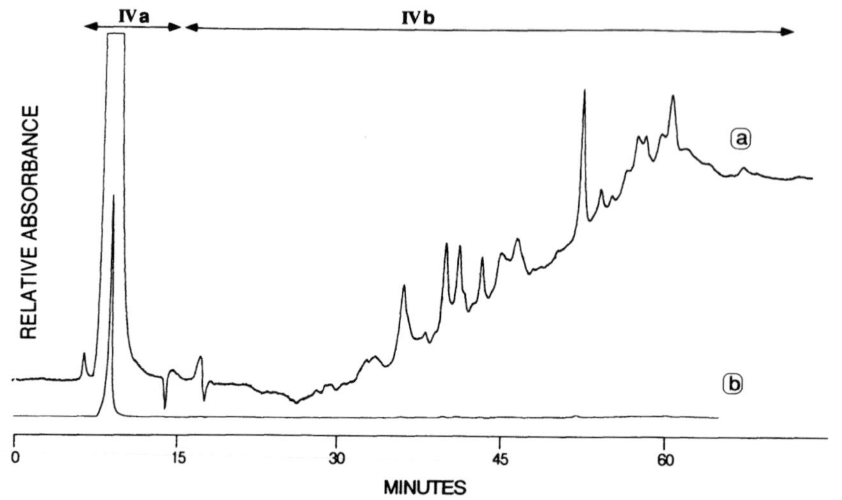

Figure 2. HPLC separation of IV into IVa and IVb. Tracing (a) is the same as (b), except that the ordinate is expanded to show detail. The arrows at the top of the diagram indicate the collection range for IVa and IVb. These fractions were nontoxic to mussel embryos, but IVa exhibited toxicity to sea urchin sperm. The HPLC system employing a semi-preparative C_{18} column is described in the text. Chromatographic conditions were (water:MeCN:isopropanol) 95:5:0 held for 10 min, then a linear gradient to 5:90:5 over the next 40 min, and held at 5:90:5 for the next 30 min. The flow was held constant over the entire run at 2 ml/min, and the void time was 5 min. The UV-visible detector was set to 220 nm to obtain this tracing, but was turned off during sample collection to avoid photodegradation of sample.

of non-aromatic and polynuclear aromatic hydrocarbons, and appeared to contain similar levels of phenolic compounds and possibly structures related to indenes and indolines. The final MeCN extract of the residue contained no detectable GC peaks (data not shown) and almost no absorbance down to 200 nm (data not shown), indicating that further solvent extraction would not have been fruitful.

The remaining residue (V) proved to be soluble in only water (methanol, ethanol, isopropanol, acetone, tetrahydrofuran, N-methyl pyrrolidinone, dimethylsulfoxide, dimethylformamide, ethyl ether and hexane were tested) and still contained the majority of the toxicity (Table 1).

At this point, the likelihood of organic compounds contributing to toxicity had diminished, so we turned to atomic spectroscopy to examine selected element composition of the fractions. The analyte list consisted mostly of transition and B-type metal elements, as many of them are aquatic toxicants and some of these have been reported to exist at significant levels in PW (Boesch and Rabalais, 1987). The results are shown in Table 2. For the particular sample shown, V did not appear to have much of any element of known toxicity, although there were higher levels of Ba and Sr. Other PW samples periodically showed slightly more elevated levels of toxic As or Cu, but these elements do not presently appear to correlate with PW toxicity. Note that some of the As was lost upon lyophilization, which was a consistent trend across many samples. This suggests that volatile forms of As may be a routine contributor to the moderate toxicity exhibited by the volatile fraction.

In order to test if divalent cations in general may be involved in toxicity, V was passed through a bed of hydrated Chelex resin and toxicity of the flow-thru (VII) was tested. Chelex resin is a divinylbenzene/styrene resin with imidodiacetate functionality. The resulting dramatic

Table 2. Elemental analysis of a sample of PW and its fractions (Sampled: July 30, 1991)

	Concentration (ppm)						
Element	I	II	IV	IVa	V	VIa	VII
Ag	<0.01						
Al	0.07						
As*	0.0067	0.0045	<0.001	<0.001	<0.001		
Ba	13		1.6	<0.002	18	12	0.004
Cd	<0.01						
Co	0.05		<0.01	<0.01	0.04		
Cu	0.0044		<0.005	<0.005	<0.005	0.012	<0.005
Cr	0.10		<0.075	<0.075	0.08		
Fe	0.46						
Mn	0.45		0.03	0.03	0.05		
Mo	0.02		<0.02	0.02	0.08		
Ni	<0.02						
Pb	<0.10		<0.10	<0.10	<0.10		
Sn	0.01						
Sr	13		<0.01	<0.01	10	7.5	<0.010
Zn	0.05		0.06	0.06	0.01	0.015	<0.005
B	55						
Ca	189						
K	44						
Mg	35						
Na	5445						
P	20						
S	21						

Blank spaces = not determined, usually due to toxicologically insignificant levels in the parent fraction
* By hydride generation atomic absorption spectrometry
Fractions III, IVb, and VIb are "organic" fractions and were not analyzed.

reduction in toxicity is shown at the bottom of Table 1. However, it could not be immediately concluded that simple divalent cations were the source of toxicity in V, as a pale yellow material was strongly adsorbed in a narrow band at the top of the resin bed. This observation made it clear that at least two more major fractions, the colorless divalent cations and the yellow material, needed to be resolved for toxicity testing.

The divalent cations were first eluted from the Chelex column using 1N HCl, which generated VIa. Table 2 shows that even the strongly-adsorbed Zn^{2+} ion (Bio-Rad, 1988) was eluted with good recovery, and Table 1 shows that the majority of toxicity resided in this fraction. For the pale yellow band, it was found that acidic MeOH was necessary for elution. Acidic MeOH, in fact, eluted the band very rapidly, while the band remained at the top of the column with either aqueous acid or MeOH alone. The resulting eluate, VIb, harbored significant toxicity as well, as seen from Table 1. VIb, when titrated to basic condition, evolved an ammoniacal odor, consistent with our initial suspicion that the yellow band was composed of organic material.

To gain a qualitative sense of the relative toxicity among fractions, we converted the data from Table 1 to relative "toxicity units", and these data are shown graphically in Figure 4. Clearly, the majority of toxicity to mussel embryos lay in II, which persisted in V upon further fractionation, and the final fractionation found equivalent toxicity in the "divalent cations" sub-fraction VIa. There was also substantial toxicity expressed in VIb. In contrast, our initial

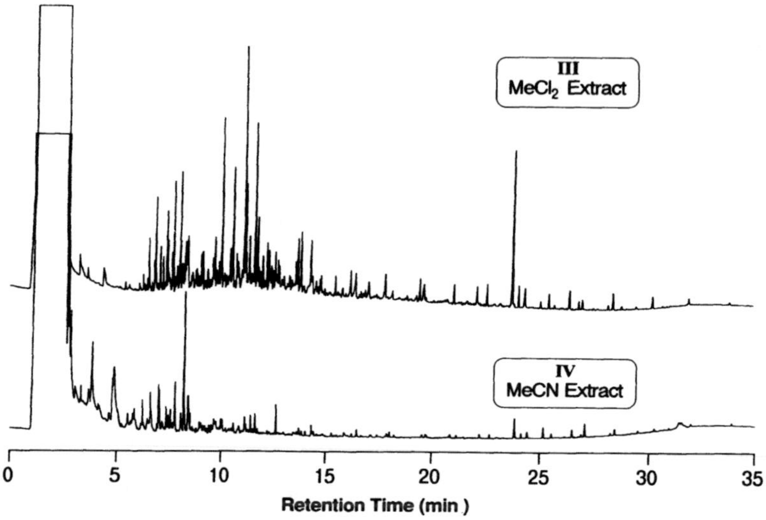

Figure 3. GC-FID analyses of III and IV. Note the large difference in composition of the fractions. III was rated slightly toxic to mussel embryos but not to sea urchin sperm, while IV was nontoxic to mussel embryos but toxic to sea urchin sperm. The instrument setup is described in the text. Analyses conditions were: injector = 250° C, splitless valve held for 1.0 min, column = 60° C held for 1.5 min, ramped to 150° C at 20° C/min, then ramped to 280° C at 6° C/min, and held there for the duration of the run; H_2 carrier gas = 60 cm/s, FID = 320° C.

survey using the sea urchin sperm cell test indicated that fraction IV, and more specifically, fraction IVa (polar HPLC fraction), appeared to harbor the largest portion of the toxicity (data not shown).

It should be noted that Figure 4, while useful qualitatively, illustrates some problems inherent in this type of representation in conjunction with toxicity testing. For instance, based on the darkest bars, the conclusion can be drawn (as we did above) that "all" of the toxicity of II, expressed through V, is retained in VIa. On the other hand, there is substantial additional toxicity in VIb. Thus it appears that the total toxicity units has "grown" past the original level of toxicity (VIa + VIb > II or V). One explanation for this discrepancy is that the non-linear shapes of the toxicity curves invalidate such simple additivity of toxicity units, emphasizing the limitations of representing toxicity data as single numbers such as EC_{50}. A related explanation is that the toxicity of VIa was masking the action of VIb when they were together. On the other hand, it's possible that there was actual antagonism of toxicity between VIa and VIb. Finally, it is possible that the fractionation process has altered some previously nontoxic constituent, which is now expressing toxicity. In all of these cases, further experimentation can resolve the discrepancies, but the chemicals causing toxicity must be determined first.

For the acidic eluate from Chelex, VIa, the fraction seems unlikely to contain constituents other than polyvalent cation elements, considering the properties of the fraction and how it was isolated. Temporarily assuming that a single, polyvalent cationic toxicant was involved, in order to arrive at the observed level of chronic toxicity in VIa (EC_{50} = 2.5%), even the more toxic metals such as Cu would have to be present in PW at the low ppm to high ppb range. However, among the metals, only the alkali earth elements Ba and Sr in our analyte list were routinely present at or above that concentration range (e.g. Table 2).

It appears that very little is known about aquatic toxicity of Ba, but it can exert acute toxic effects to amphipods (Vincent, et al., 1986) and algae (Jernelov, et al., 1976) at the levels found in our PW samples. Strontium was reported to have low acute toxicity to an aquatic nematode

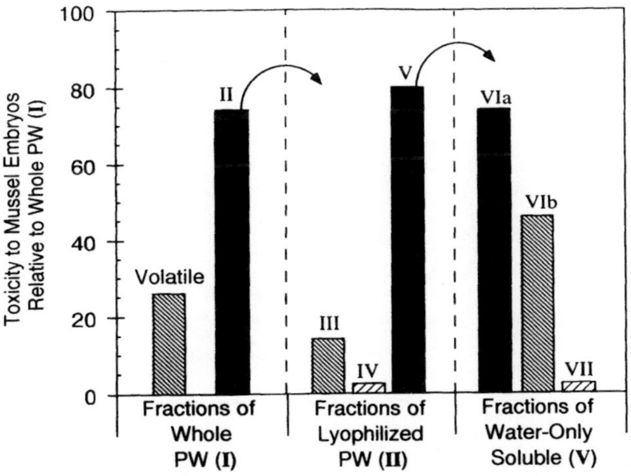

Figure 4. Graphical representation of toxicity data from Table 1, converted to relative "toxicity units". The full scale of the ordinate represents toxicity of whole PW. The bars in each box represent toxicity from sub-fractions of the indicated fraction, and the arrows denote the fraction that was used to generate the sub-fractions in the next box. Note that the toxicity of VIa is equivalent to that of the nonvolatile portion of PW (II).

in comparison with other metals (Williams and Dusenbery, 1990). To the best of our knowledge, there is no previous report of chronic toxicity levels of Ba or Sr to marine life.

The Sr levels found in our PW samples were slightly elevated from the range normally found in seawater as well as formation water (Collins, 1975), but the Ba levels we found were about three orders of magnitude higher than in seawater, and was in the range typical of formation water (Collins, 1975). In addition, marine phytoplankton (Martin and Knauer, 1973) and clams (Sadiq, et al., 1990) were reported to accumulate Ba. Thus, the best candidate in our analyte list for the toxicant in VIa is currently Ba.

Of course, it is just as possible that several of the detected toxicants, each at well below toxic levels, can collectively express the observed toxicity. Related to that issue is synergism. In this study, the differences in toxicity between fractions was large, suggesting that synergism between fractions was not likely, but that still leaves the possibility of synergism within a fraction such as VIa. Finally, some other polyvalent element not on our analyte list for Table 2 could account for the toxicity. For example, Hg was occasionally found in PW at levels that could cause the observed toxicity (Collins, 1975; Middleditch, 1984), but our analyte list otherwise covers most metals typically found in PW (Collins, 1975; Boesch and Rabalais, 1987). Interestingly, some correlation to toxicity in VIa was revealed by UV-visible spectrophotometry, shown in Figure 5. The feature of note is the peak centered around 326 nm, the peak area of which appears to correlate qualtitatively with toxicity of the fractions. The identity of this peak is currently being pursued. We plan further fractionation of VIa and expansion of the analyte list in order to address all of the above possiblities.

For the acidic MeOH eluate from Chelex, VIb, further fractionation by ion-exchange HPLC is planned. Figure 5 shows that this fraction does not contain significant chromophores, and as mentioned previously, possibly contains a polar amine compound. At the current time, its behavior on the Chelex column is inadequately but best explained if VIb contained an organic quarternary amine. There is very little information in the literature concerning polar organic compounds in PW, but various phenols appear to be present (Lysyj, 1982). Amino acids are polar nitrogenous compounds found in petroleum associated waters (Degens, et al., 1964), but these would be components of IV or VII. One study reports that PW from North Sea operations

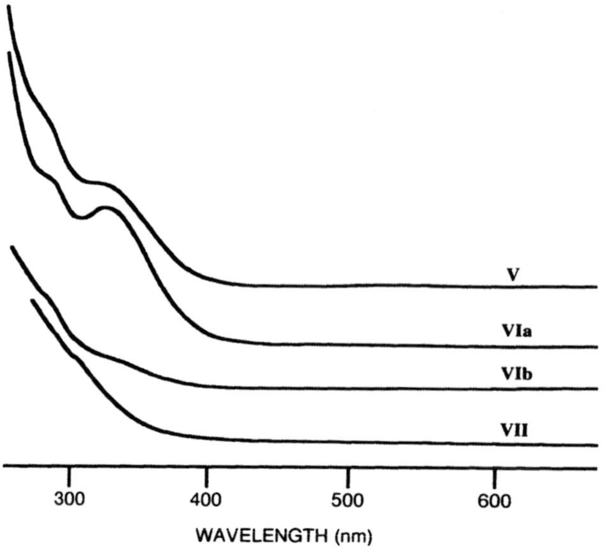

Figure 5. UV-visible spectra of selected fractions. The spectra are offset vertically for clarity, but all of the spectra are presented on the same ordinate scale to allow direct comparison of absorbances. Note the peaks at ca.300 and 326 nm in V, which is more pronounced in VIa, probably in part due to removal of the interferences from VIb and VII. No other peaks were present in these fractions.

consists of largely phenolate, propionate, and butyrate salts, with 20-30 ppm of nitrogen as NH_4+ salts (Somerville, et al., 1987). Another report indicates that NH_3 ("unionized") accounted for the majority of toxicity to *Ceriodaphnia* and fathead minnows in PW from a simulated *in situ* oil shale retort (Hill, 1986). In our case, barring any unusual ligand chemistry, it is very unlikely that ammonium would be present at high levels after adjustment to pH 8, lyophilization, and extraction by MeCN, followed by an even more unlikely strong retention on Chelex column, so strong that it resists elution by 1N HCl.

In summary, we have found that the majority of toxicity to mussel embryos are found in the "divalent cation" fraction, as summarized in Figure 4. Future work entails further fractionation of VIa and VIb, expanding the analysis into elements less typical of PW, and testing the hypothesis that the peak in the UV-visible spectrum represents the source of toxicity. The toxicity pattern according to the sea urchin sperm toxicity survey, however, appears to be very different, and that investigation appears to be heading for work with more polar organic substances (IVa). Thus, in our case, fractionation of whole produced water, coupled with chemical detection by toxicity, do not appear to be leading to either the transition and B-type metals or the hydrocarbons. Yet, these compounds represent the bulk of toxicity and chemistry work reported in the literature.

ACKNOWLEDGEMENTS

Supported by US Department of the Interior grant #14-35-0001-30-471. The authors wish to thank the assistance of Dr. Muralidharan Pillai and Tammy Shields in the toxicity tests, Paul Krause and Dr. Craig Osenberg for PW sampling, Dr. Donald Crosby for advice, Chevron Corporation for their cooperation in supplying PW samples, and two anonymous reviewers for their comments.

REFERENCES

Agemian, H. and Thomson, R., 1980, Simple semi-automated atomic absorption spectrometric method for the determination of arsenic and selenium in fish tissue, Analyst 105: 902-907.

Bio-Rad Laboratories, 1988, "Chelex 100 and Chelex 20 Chelating Ion Exchange Resin Instruction Manual," Bio-Rad Laboratories, Richmond, CA.

Boesch, D.F. and Rabalais, N.N., eds., 1987, "Long-term Environmental Effects of Offshore Oil and Gas Development, Elsevier Applied Science," New York, 708pp.

Brumbaugh, W.G. and Walther, M.J., 1989, Determination of arsenic and selenium in whole fish by continuous-flow hydride generation atomic absorption spectrophotometry, J. Assoc. Off. Anal. Chem. 72: 484-486.

Buchanan, A.S. and Hannaker, P., 1984, Inductively coupled plasma spectrometric determination of minor elements in concentrated brines following precipitation, Anal. Chem. 56: 1379-1382.

Cherr, G.N., Shenker, J.M., Lundmark, C., and Turner, K.O.,1987, Toxic effects of selected bleached kraft mill effluent constituents on the sea urchin sperm cell, Environmental Toxicology and Chemistry 6: 561-569.

Cherr, G.N., Shoffner-McGee, J., and Shenker, J.M., 1990, Methods for assessing fertilization and embryonic/larval development in toxicity tests using the California mussel (Mytilus californianus), Environmental Toxicology and Chemistry 9: 1137-1145.

Collins, A.G., 1975, Geochemistry of Oilfield Waters, Elsevier Scientific Publishing, Amsterdam, 496 pp.

Degens, E.T., Hunt, J.M., Reuter, J.H., and Reed, W.E., 1964, Data on the distribution of amino acids and oxygen isotopes in petroleum brine waters of various geologic ages, Sedimentology 3: 199-225.

Fan, T.W-M., Higashi, R.M., Cherr, G.N., and Pillai, M.C., 1992, Use of noninvasive NMR spectroscopy and imaging for assessing produced water effects on mussel reproduction, *in*: "Produced Water: Technological/Environmental Issues and Solutions," J.P. Ray and F.R. Engelhardt, eds., Plenum Publishing Corp., New York, New York, USA.

Hill, S. L., 1986, Toxicological characterization of water produced during simulated in situ processing of Green River Formation oil shale, U.S. Dept. of Energy Report # DOE/FE/60177-2390, 29 pp.

Jernelov, A., Landahl, C. C., Larsson, T., Martin, A. L., Westling, O., 1976, Barium ecology in aquatic systems, Inst. Vatten- Luftvardsforsk., [Publ.] B, B 277, 25 pp.

Lysyj, I., 1982, Chemical composition of produced water at some offshore oil platforms, U.S.E.P.A. Report # EPA-600/2-82-034.

Martin, J.H. and Knauer, G.A., 1973, The elemental composition of plankton. Geochimica et Cosmochimica Acta 37:1639-1653.

Middleditch, B.S., 1984, Ecological effects of produced water discharges from offshore oil and gas production platforms, Final Report on API Project No. 248, American Petroleum Institute, Washington, D.C., 160 pp.

National Research Council, 1985, "Oil in the Sea: Inputs, Fates, and Effects, National Academy Press," Washington, D.C., 601pp.

Neff, J.M., Rabalais, N.N., and Boesch, D.F., 1987, Offshore oil and gas development activities potentially causing long-term environmental effects, *in:* "Long-term Environmental Effects of Offshore Oil and Gas Development", Boesch, D.F. and Rabalais, N.N., eds., Elsevier Applied Science, New York, 708pp.

Osenberg, C.W., and Schmitt, R.J., and Holbrook, S.J., 1992, Spatial scale of ecological effects associated with an open coast discharge of produced water, *in*: "Produced Water: Technological/Environmental Issues and Solutions" J.P. Ray and F.R. Englhardt, eds., Plenum Publishing Corp., New York, New York, USA.

Sadiq, M., Zaidi, T.H., and Al-Mohana, H., 1990, Barium bioaccumulation in clams collected from different salinity regimes along the Saudi coast of the Arabian Gulf, Bull. Environ. Contam. Toxicol. 45:329-335.

Somerville, H. J., Bennett, D., Davenport, J. N., Holt, M. S., Lynes, A., Mahieu, A., McCourt, B., Parker, J. G., and Stephenson, R.R., 1987, Environmental effect of produced water from North Sea oil operations, Mar. Pollut. Bull. 18: 549-558.

Stumm, W. and Morgan, J.J., 1981, "Aquatic Chemistry," John Wiley & Sons, New York, p.323-417.

Vincent, M., Debord, J., and Penicaut, B., 1986, Comparative toxicity of metal chlorides and of a synthetic organic molluscicide, N-trityl-morpholine, upon two aquatic amphipod crustaceans, *Gammarus pulex* and *Echinogammarus berilloni*, Ann. Rech. Vet. 17(4): 441-6.

Williams, Phillip L. and Dusenbery, D. B., 1990, Aquatic toxicity testing using the nematode, *Caenorhabditis elegans*, Environ. Toxicol. Chem. 9: 1285-90.

SHORT-TERM CHRONIC TOXICITY OF PRODUCED WATER AND ITS VARIABILITY AS A FUNCTION OF SAMPLE TIME AND DISCHARGE RATE

C.M. Moffitt[1], M.R. Rhea[1], P.B. Dorn[2] J.F. Hall[3], J.M. Bruney[4], and S.H. Evans[2]

[1] SeaCrest Environmental Services Inc.
Houston, Texas
USA

[2] Shell Development Company
Houston, Texas
USA

[3] Texaco Inc.
Port Arthur, Texas
USA

[4] Exxon Production Research Company
Houston, Texas
USA

INTRODUCTION

On April 18, 1991, the Environmental Protection Agency (U.S. EPA, 1991a) published a notice of proposed NPDES general permit reissuance for the Western Portion of the Outer Continental Shelf of the Gulf of Mexico. In this notice, Region VI EPA proposed adding aquatic toxicity limits for produced water discharges. Permittees would be required to test produced water samples for toxicity in accordance with the most recent revision of the Agency's short-term methodology assessing the chronic toxicity of effluents and receiving waters to marine and estuarine organisms.

This policy was consistent with EPA's water quality-based approach to toxics control, where the chronic No Observable Effect Concentration (NOEC) is measured for an effluent after a mixing zone is determined within which the effluent mixes with receiving water (U.S. EPA, 1991b). From the permit guidance it was estimated that at least a 100:1 conservative estimate of dilution of the produced water discharge would be achieved within a specific boundary, although industry modelling efforts suggested that closer to 1,000:1 would be more realistic. The target for the program, would be for the NOEC concentration to be >0.001% effluent, or >1% using a very conservative estimate. Previous studies using acute data showed a range in toxicity for 108 samples showing LC_{50} >10% in 51.9% of the samples, LC_{50} >1< 10% in 37% of the samples, and <1% in 11.1% of the samples (Neff, 1987). Similar results were observed by Girling (1989) who studied toxicity of North Sea produced waters to the calanoid copepod *Acartia tonsa*.

In order to help assess the validity of these test methods in their application to discharged produced water, a series of toxicity tests were conducted on samples collected from eight offshore sites in the Gulf of Mexico. Four of the eight platforms were identified as "low" discharge rate sites (<10,000 barrels/day) and four as "high" discharge rate sites (>10,000 barrels/day). To assess potential temporal variability in toxicity, three samples were collected from each platform and sampling times were separated by approximately two weeks.

This report summarizes the results of this produced water toxicity study.

METHODS AND MATERIALS

A total of twenty-four produced water samples were collected and delivered to SeaCrest Environmental Services Inc.'s laboratory in Houston, Texas during the period of April 15, 1991 to June 6, 1991. Because of the difficulties in obtaining 24-hour composite samples from offshore facilities, single grab samples were collected three times from each platform, with sampling times separated by approximately two weeks. Detailed written sampling instructions were provided to all offshore personnel[1]. Compliance with the instructions was confirmed by pre- and post-sampling interviews with offshore personnel by the testing laboratory. Samples were shipped by sealed and locked 48-quart Igloo^R coolers, accompanied by signed chain-of-custody forms. Upon receipt at the laboratory, the produced water samples were stored unopened in a chiller at 4°C until test initiation, and kept cool for the duration of the test to provide fresh media for daily renewals. Laboratory-prepared artificial seawater (Hawaiian Marine Mix^R) was used as the dilution water and reference control for the tests.

Seven-day toxicity tests utilizing *Mysidopsis bahia* (mysid shrimp) and *Cyprinodon variegatus* (sheepshead minnow) were conducted on each of the produced water samples. Testing was conducted with serial dilutions of the produced water samples according to the procedures specified by the U.S. EPA (1988) for short-term chronic toxicity tests using marine species. In conformance with the protocol, the salinity of the dilution water was 25.0 - 26.0 parts per thousand (ppt).

In each of the mysid shrimp tests, eight replicates of five 7-day old mysids were exposed to five concentrations of produced water (0.75%, 1.5%, 3.0%, 7.0%, and 14.0%)[2] and the reference control for seven days with daily renewal of test media. In each of the sheepshead minnow tests, four replicates of ten larvae (less than 24 hours old) were exposed to six concentrations of produced water (same dilutions as above with the addition of 28.0%) and the reference control. Effluent sample aliquots were removed from the sample chillers, warmed, and used to renew the test solutions daily. Aeration was utilized if the dissolved oxygen level fell below 60% of saturation during the testing period. Temperature, salinity, dissolved oxygen, and pH of test concentrations were monitored daily at renewal. A photoperiod of 16 hours light: 8 hours dark was maintained throughout the testing period for the mysids and 14 hours light: 10 hours dark for the sheepshead minnows. Test organisms were obtained from SeaCrest Environmental's in-house culturing facilities. In the mysid shrimp tests, a twice-a-day feeding regime of approximately 75 *Artemia* nauplii per mysid was utilized during the testing. In the sheepshead minnow tests, a once-a-day feeding regime of 0.10 g wet weight of *Artemia* per replicate on Days 0-2 and 0.15 g wet weight of *Artemia* per replicate on Days 3-6 was used.

For each test, No Observable Effect Concentrations (NOEC's) for produced water toxicity were determined for survival and growth in both test species, and for fecundity in the mysid shrimp. The U.S. EPA defines the NOEC as, "the highest concentration of toxicant to which

---

[1] The two-page written sampling instructions are not included in this paper for sake of brevity, but were included as an appendix to the original report on this study submitted to the Offshore Operators Committee (OOC) in October, 1991.

[2] The first thirteen mysid shrimp tests were conducted with six concentrations, which included 28.0%. The 28.0% dilution water was later excluded because of consistently high mortality at 48 hours.

organisms are exposed in a full lifecycle or partial lifecycle test, that causes no observable adverse effects on the test organisms — i.e., the highest concentration of toxicant in which the values from the observed parameters are not statistically significantly different from the control." (U.S. EPA, 1988). Both 7-day and 96-hour LC_{50}'s were calculated using probit analyses. The LC_{50} is that concentration of produced water that causes 50% mortality in the test organisms.

RESULTS

Toxicity Results

The results of the toxicity tests for the 24 produced water samples are contained in Tables 1 and 2. Table 1 shows the lethal and sublethal 7-day NOEC's and LC_{50} values for each sample collected from the "low" discharge rate platforms. The means for each toxicity endpoint examined are included along with their associated standard deviations. Table 2 contains the same information for the "high" discharge rate platforms. U.S. EPA test acceptance criteria detailed by the U.S. EPA (1988) for control organism survival and weight were met in every case. Three Standard Reference Toxicant (SRT) tests were conducted during the time period of the produced water testing. Cupric sulfate (as copper {Cu^{II}}, J.T. Baker Inc.) was used as the reference toxicant for the sheepshead minnows, and 95% sodium dodecyl sulfate (SDS, J.T.Baker Inc.) was used in the mysid tests. The 7-day LC_{50} values calculated for these tests compare favorably with historic LC_{50}'s obtained by SeaCrest Environmental's laboratory. The mean in-house 7-day LC_{50} for seven sheepshead minnow tests conducted in the previous six-month period was 265.0 parts per billion (ppb) Cu^{II} compared to an average value of 250.4 ppb Cu^{II} for the three SRT tests conducted in this study. A mean 7-day LC_{50} of 15.9 parts per million (ppm) SDS was obtained for the three mysid SRT tests conducted in this study compared to a mean of 17.5 ppm SDS obtained in eight previous mysid SRT tests conducted over the previous six-month period.

Table 1. Toxicity of platforms with a "low" discharge rate (<10,000 bbl/day) (Data Expressed as Percent Produced Water)

Platform (Discharge Rate)	Sample #	96-Hr. LC_{50} Mb[1]	96-Hr. LC_{50} Cv[1]	7-Day LC_{50} Mb[1]	7-Day LC_{50} Cv[1]	Survival NOEC Mb[1]	Survival NOEC Cv[1]	Growth NOEC Mb[1]	Growth NOEC Cv[1]	Fecundity NOEC Mb[1]
A (est. 3,800 bbl/day)	1	8.27	27.30	7.29	>28.00	3.00	14.00	3.00	7.00	3.00
	2	4.66	25.70	4.48	21.70	1.50	14.00	1.50	7.00	1.50
	3	7.20	15.10	7.47	18.10	3.00	14.00	1.50	1.50	3.00
B (est. 4,000 bbl/day)	1	9.87	>28.00	8.72	>28.00	3.00	28.00	1.50	1.50	<0.75
	2	3.95	>28.00	2.80	>28.00	1.50	14.00	1.50	3.00	1.50
	3	5.65	*	4.38	*	3.00	*	0.75	*	1.50
C (est. 1,700 bbl/day)	1	5.58	9.61	5.92	9.61	3.00	7.00	<0.75	<0.75	3.00
	2	2.90	18.14	2.55	17.56	1.50	14.00	0.75	1.50	1.50
	3	2.78	5.44	2.81	4.49	1.50	1.50	<0.75	1.50	1.50
D (est. 4,000 bbl/day)	1	6.18	>28.00	4.45	7.72	3.00	7.00	3.00	3.00	3.00
	2	6.34	>28.00	7.37	>28.00	3.00	28.00	1.50	<0.75	3.00
	3	3.15	>28.00	2.76	>28.00	1.50	7.00	<0.75	1.50	1.50
LOW RATE MEAN	1	7.47	23.21	6.60	18.33	3.00	14.00	2.06	3.06	2.44
	2	4.46	24.97	4.30	23.82	1.88	17.50	1.31	3.06	1.88
	3	4.70	16.19	4.36	16.85	2.25	7.50	0.94	1.50	1.88
STD. DEVIATION	1	1.97	9.08	1.83	11.19	0.00	9.90	1.12	2.79	1.12
	2	1.44	4.68	2.22	5.12	0.75	7.00	0.38	2.79	0.75
	3	2.10	11.32	2.21	11.80	0.87	6.26	0.38	0.00	0.75

[1]Mb (*Mysidopsis bahia*) Cv (*Cyprinodon variegatus*)
*Test not conducted due to project budgetary constraints

Table 2. Toxicity of platforms with a "high" discharge rate (>10,000 bbl/day) (Data Expressed as Percent Produced Water)

Platform (DIscharge Rate)	Sample #	96-Hr. LC50		7-Day LC50		Survival NOEC		Growth NOEC		Fecundity NOEC
		Mb[1]	Cv[1]	Mb[1]	Cv[1]	Mb[1]	Cv[1]	Mb[1]	Cv[1]	Mb[1]
E (est. 16,000 bbl/day)	1	4.76	11.30	6.21	3.68	1.50	1.50	1.50	0.75	1.50
	2	7.21	25.50	4.49	19.43	3.00	7.00	<0.75	<0.75	3.00
	3	3.88	>28.00	3.00	24.77	1.50	14.00	<0.75	3.00	0.75
F (est. 22,900 bbl/day)	1	9.90	16.60	4.82	14.58	7.00	7.00	<0.75	3.00	7.00
	2	6.15	13.30	6.97	15.95	3.00	7.00	<0.75	1.50	0.75
	3	9.61	20.60	9.61	17.46	7.00	14.00	1.50	1.50	1.50
G (est. 18,000 bbl/day)	1	7.47	19.80	6.78	19.80	3.00	14.00	3.00	7.00	0.75
	2	14.60	23.00	10.05	21.84	3.00	7.00	0.75	3.00	3.00
	3	16.00	>28.00	5.22	>28.00	3.00	14.00	3.00	7.00	3.00
H (est. 11,000 bbl/day)	1	7.65	17.60	9.03	18.88	7.00	14.00	7.00	<0.75	3.00
	2	>14.00	>28.00	9.16	>28.00	7.00	7.00	0.75	3.00	3.00
	3	2.30	23.00	2.05	21.96	0.75	14.00	0.75	3.00	0.75
HIGH RATE MEAN	1	7.44	16.31	6.71	14.24	4.62	9.12	3.06	2.88	3.06
	2	10.49	22.45	7.67	21.30	4.00	7.00	0.75	2.06	2.44
	3	7.94	24.91	4.97	23.05	3.06	14.00	1.50	3.62	1.50
STD. DEVIATION	1	2.10	3.62	1.75	7.40	2.81	6.06	2.79	2.95	2.79
	2	4.43	6.43	2.48	5.08	2.00	0.00	0.00	1.12	1.12
	3	6.20	3.69	3.37	4.47	2.79	0.00	1.06	2.36	1.06

[1]Mb (*Mysidopsis bahia*) Cv (*Cyprinodon variegatus*)

To determine whether produced water toxicity results differed significantly as a function of the two factors, sampling time and discharge rate, statistical comparisons were made between group values utilizing Two-Factor Analysis of Variance (ANOVA) procedures (with replication) and Tukey's Method of Multiple Comparisons (Neter, et al., 1985; Zar, 1984). The interaction between the two factors was also assessed. All analyses were conducted at $p \leq 0.05$ level of confidence. No significant differences were observed between results from samples collected at different time periods or from offshore platforms with varied discharge rates or between any combinations. All toxicity endpoints (i.e., LC_{50}'s and lethal and sublethal NOEC's) were included in the statistical analyses.

Table 3 represents the summarized data for all toxicity endpoints examined for both test species. Corresponding mean values from an unpublished produced water toxicity database obtained by the Louisiana Department of Environmental Quality (LADEQ) are included in Table 3 for visual comparison. The data from the two studies are in general agreement. We have no information whether any of the approximately 400 produced water samples tested in the LADEQ program contained biocides or other water treatment chemicals.

Figure 1 contains the overall mean concentration-response curves for 7-day mortality for each species. The produced water concentration-response function is clearly steeper for the mysid shrimp (compare Figure 1a to Figure 1b), evidencing the greater potency to the invertebrate species. Only one sample elicited a 7-day LC_{50} value which was lower for the sheepshead minnow.

Typical seven-day concentration-response curves for three separate produced water samples obtained from an offshore platform can be seen in Figures 2a, 2b, and 2c.

Table 3. Summarized toxicity data (Data Expressed as Percent Produced Water)

Produced Water Study Results	96-Hr. LC_{50}		7-Day LC_{50}		Survival NOEC		Growth NOEC		Fecundity NOEC
	Mb[1]	Cv[1]	Mb[1]	Cv[1]	Mb[1]	Cv[1]	Mb[1]	Cv[1]	Mb[1]
Overall Mean	7.08 n=24	21.55 n=23	5.77 n=24	19.72 n=23	3.14 n=24	11.70 n=23	1.60 n=24	2.75 n=23	2.20 n=24
STD. Deviation	3.73	6.99	2.48	7.71	1.92	6.66	1.41	2.17	1.38
Minimum	2.30	5.44	2.05	3.68	0.75	1.50	<0.75	<0.75	<0.75
Maximum	15.96	>28.00	10.05	>28.00	7.00	28.00	7.00	7.00	7.00
Louisiana DEQ Database									
Overall Mean (n= approx. 400)	10.05	19.21	Data not available for this endpoint		3.39	6.28	2.44	5.23	2.67
STD. Deviation	10.36	14.82			5.03	8.96	3.64	8.12	3.24
Minimum	0.00	0.00			0.00	0.00	0.00	0.00	0.00
Maximum	86.20	100.00			50.00	100.00	42.90	100.00	25.00

[1]Mb (*Mysidopsis bahia*) Cv (*Cyprinodon variegatus*)

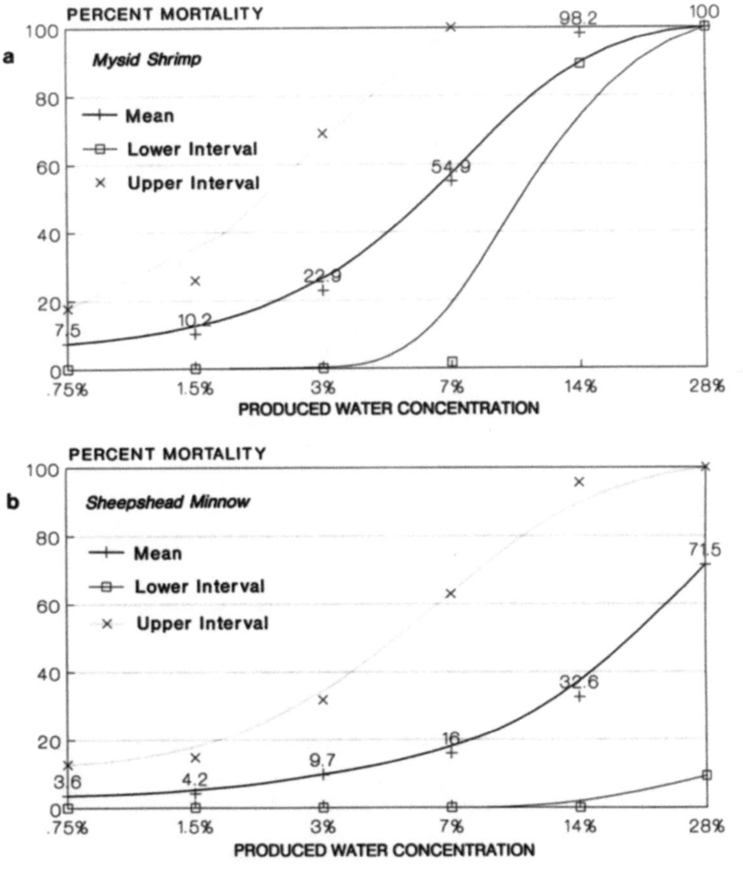

Figure 1. Mean seven-day mortality with 95% confidence intervals for a) mysid shrimp tests (n=24); and b) sheepshead minnow tests (n=23).

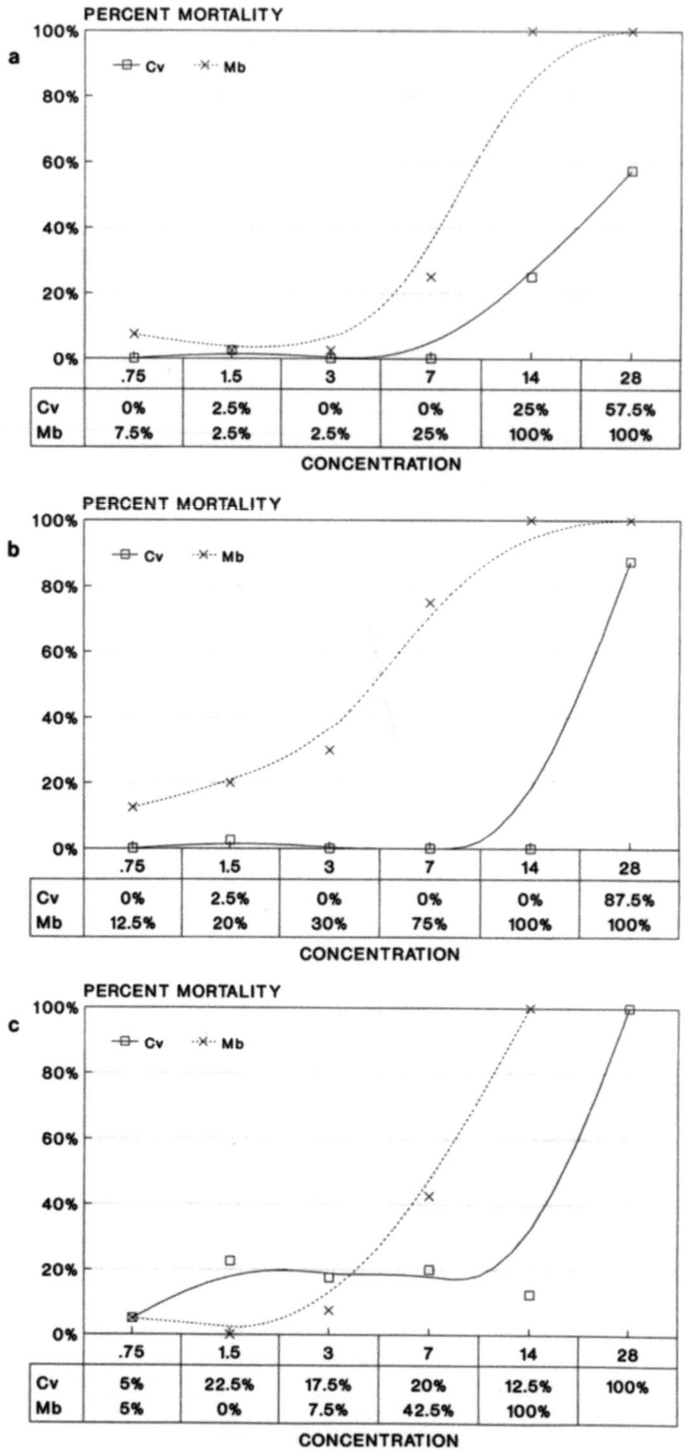

Figure 2. Typical seven-day concentration-response curves for individual produced water samples.

Water Quality Parameters and Correlational Analyses

Initial chemical characterizations of the produced water samples indicated high levels of ammonia and salinity. The ranges for various water quality parameters for the 24 samples prior to and during testing are shown in Tables 4 and 5.

To determine whether the observed toxicity to the two species could be attributed to the high initial salinity of the produced water samples, correlation coefficients (Pearson's rho) (Snedecor and Cochran, 1967) were calculated for 7-day and 96-hour LC_{50} values. None of the correlation coefficients were significantly different from zero (i.e., no significant relationship between survival toxicity and salinity at ≤ 0.05).

The same statistical procedures were applied to investigate the relationship between observed ammonia levels and survival toxicity. Levels of the toxic form of ammonia (NH_3, un-ionized), at the 7.0% test concentrations were determined from total ammonia measurements of each produced water sample. The un-ionized ammonia levels were calculated using the methods found in Whitfield (1974) and U.S. EPA (1989). Since pH of the test media is a major determining factor in the amount of un-ionized ammonia present, the highest daily pH recorded at the 7.0% concentration for each 7-day test was utilized in each calculation. The resulting un-ionized ammonia level was compared to the percent 7-day mortality value at the 7.0% concentration in each test.

Table 4. *Mysidopsis bahia* seven-day range of water quality parameters

Parameter	Produced Water	Synthetic Seawater
Temperature °C		
Initial:	7-25	26
Test:	26*	25-26
Dissolved Oxygen (ppm)		
Initial:	<0.1 - 4.2	6.2-6.6
Test:	3.4 - 6.4*	4.0-6.6
pH		
Initial:	5.9 - 7.2	8.0-8.1
Test:	7.7 - 8.1*	7.8-8.1
Salinity (ppt)		
Initial:	58 - 150	25
Test:	26 - 36*	25-26
Conductivity (μmhos)		
Initial:	135,000 - 245,000	—
Test:	—	—
Hardness (ppm $CaCO_3$)		
Initial:	2,100 - 29,000	—
Test:	—	—
Alkalinity (ppm $CaCO_3$)		
Initial:	200 - 4,400	—
Test:	—	—
Ammonia-N (ppm)		
Initial:	60 - 240	—
Test:	—	—
Residual Chlorine (ppm)		
Initial:	<0.1	—
Test:	—	—

*Ranges at the midpoint 7% dilution for representational purposes.

Table 5. *Cyprinodon variegatus* seven-day range of water quality parameters

Parameter	Produced Water	Synthetic Seawater
Temperature °C		
Initial:	7 - 25	26
Test:	25 - 26*	26
Dissolved Oxygen (ppm)		
Initial:	<0.1 - 4.2	6.0 - 6.6
Test:	2.6 - 6.6*	2.5 - 6.7
pH		
Initial:	5.9 - 7.2	7.9 - 8.1
Test:	7.5 - 8.1*	7.6 - 8.1
Salinity (ppt)		
Initial:	58 - 150	25
Test:	26 - 36*	25
Conductivity (μmhos)		
Initial:	135,000 - 245,000	—
Test:	—	—
Hardness (ppm $CaCO_3$)		
Initial:	2,100 - 29,000	—
Test:	—	—
Alkalinity (ppm $CaCO_3$)		
Initial:	200 - 4,400	—
Test:	—	—
Ammonia-N (ppm)		
Initial:	60 - 240	—
Test:	—	—
Residual Chlorine (ppm)		
Initial:	<0.1	—
Test:	—	—

* Ranges at the midpoint 7% dilution for representational purposes.

Table 6 contains the associated values for each produced water sample toxicity test. The correlation between sheepshead minnow 7-day mortality and un-ionized ammonia levels across tests was not significant ($r=0.16$, $df=21$). However, the correlation between 7-day mortality and un-ionized ammonia for the mysid shrimp was small but significant ($r=0.36$, $df=22$, $p<0.05$), indicating that ammonia present in the produced water samples was a possible contributing factor in the observed toxicity.

DISCUSSION

The results of this study show that the chronic toxicity of produced water to mysid shrimp and sheepshead minnows is not dependent on the rate of the offshore discharge, nor does toxicity vary significantly with time of sampling. The temporal variability in toxicity appears no greater than that associated with this laboratory's results from SRT tests conducted during the same time period. The mysid shrimp proved the more sensitive species of the two to the produced water samples, and mysid growth was the most sensitive endpoint examined. Although mysid survival endpoints appeared to be lower, by a factor of three or four, than sheepshead minnow survival endpoints, mean growth NOEC's for the two species were quite similar (Table 3). The overall observed toxicity of the produced water samples does not seem attributable solely to

Table 6. Seven-day mortality values and associated un-ionized ammonia levels

Platform	Sample Date	Time Period	7-Day Mortality (%) @7.0% Concentration Mb	Cv	Un-Ionized Ammonia (ppm) @7.0% Concentration
	4/16/91	1	25.0	0.0	0.32
Platform A	5/01/91	2	75.0	0.0	0.32
	5/15/91	3	42.5	20.0	0.32
	4/26/91	1	27.5	0.0	0.27
Platform B	5/13/91	2	95.0	15.0	0.68
	5/30/91	3	55.0	*	0.21
	4/29/91	1	70.0	2.5	0.57
Platform C	5/13/91	2	90.0	27.5	0.52
	5/28/91	3	85.0	87.5	0.35
	4/15/91	1	67.5	10.0	0.48
Platform D	4/29/91	2	47.5	0.0	0.32
	5/13/91	3	95.0	15.0	0.95
	4/22/91	1	55.0	87.5	0.22
Platform E	5/06/91	2	82.5	20.0	0.53
	5/20/91	3	97.5	12.5	0.35
	4/19/91	1	47.5	0.0	0.48
Platform F	5/05/91	2	57.5	10.0	0.56
	6/03/91	3	2.5	10.0	0.54
	4/26/91	1	50.0	0.0	0.24
Platform G	5/07/91	2	22.5	7.5	0.27
	5/21/91	3	90.0	10.0	0.48
	5/01/91	1	12.5	5.0	0.39
Platform H	5/13/91	2	10.0	7.5	0.55
	5/24/91	3	80.0	7.5	0.55

* Test not conducted due to project budgetary constraints.

high salinity and ammonia levels, although un-ionized ammonia levels did appear to slightly contribute to mysid shrimp mortality effects at the 7.0% test concentration. Finally, results of this study agree well with toxicity data obtained from a program of produced water testing conducted recently in the State of Louisiana.

Regarding methodology, it does appear that the marine species chronic testing protocol detailed in EPA/600/4-87/028 (U.S.EPA, 1988) is appropriate for produced water testing with the following exceptions:

- Although only two of twenty-four samples were late in arriving at the laboratory, the EPA protocol's 36-hour "sample collection to test initiation" requirement is very difficult for offshore facilities to meet. Weather conditions and sample transportation availability are major factors involved in compliance success. This laboratory experienced numerous delays in sample receipt due to inclement offshore weather, forcing testing schedules to be altered accordingly.
- The protocol's requirement for flow weighted, 24-hour composite samples is impractical for offshore facilities. This study utilized grab sampling and, given the relatively low overall temporal variability observed in the toxicity data, this technique appears to produce a "representative" sample. However, it should be noted that the sample collection instructions for this study prohibited the use of batch biocide or chemical treatment 72 hours prior to sample collection. The presence of varying amounts of biocides in the produced water samples could have introduced more variability into the toxicity results.

The above factors should certainly be considered in the development by EPA of any final requirements for produced water toxicity testing for offshore facilities.

Acknowledgements - This study was funded by the Offshore Operators Committee (OOC), whose Produced Water Ad Hoc Group provided the design and excellent guidance throughout the project. We are also grateful to SeaCrest Environmental's Martin Blum and Aaron Arredondo for their diligent technical assistance during the testing phase of the project, and to Shahriar Laghaeian, Mark Johnston, and Pam Bicknell for their valuable assistance in text preparation and editing.

REFERENCES

Girling, A.E., 1989, Acute and chronic toxicity of produced water from a North Sea oil production platform to the calanoid copepod Acartia tonsa, Bull. Environ. Contamin. Toxicol. 8: 231-245.

Neff, J.M., 1987, Biological effects of drilling fluids, drill cuttings, and produced waters, *in*: "Long Term Environmental Effects of Offshore Oil and Gas Development," D.F. Boesch and N.N. Rabalais, eds., Elsevier Applied Science, New York, N.Y.

Neter, J., Wasserman, W., and Kutner, M.H., 1985, "Applied Linear Statistical Models," Richard D. Irwin Inc., Illinois.

Snedecor, G.W. and Cochran, W.G., 1967 "Statistical Methods," Sixth Edition, Iowa State Press, Ames, Iowa.

U.S. EPA, 1988, "Short-Term Methods for Estimating the Chronic Toxicity of Effluents and Receiving Waters to Marine and Estuarine Organisms," Environmental Monitoring and Support Laboratory, U.S. Environmental Protection Agency, Cincinnati, Ohio, EPA/600/4-87/028.

U.S. EPA, 1989, "Ambient Water Quality Criteria for Ammonia (Saltwater)," Office of Water Regulations and Standards, EPA/440/5-88/004.

U.S. EPA, 1991a, Proposed Reissuance of the NPDES General Permit for the Western Portion of the Outer Continental Shelf (OCS) of the Gulf of Mexico (GMG 290000), Federal Register, Volume 56, Number 73, Pages 15349-15372.

U.S. EPA, 1991b, "Technical Support Document for Water Quality-based Toxics Control," Office of Water, EPA 505/2-90/001, Washington, D.C.

Whitfield, M., 1974, The hydrolysis of ammonium ions in seawater - A theoretical study, *J.* Mar. Biol. Assoc. (UK) 54:565-580.

Zar, J.H., 1984, "Biostatistical Analysis," Prentice-Hall, Inc., Englewood Cliffs, N.J.

TOXICITY TESTING AND CHEMICAL CHARACTERIZATION OF PRODUCED WATER - A PRELIMINARY STUDY

J. Brendehaug,[1], S.Johnsen[2], K. H.Bryne [3], A. L. Gjøse[1], T.H. Eide,[4] and E. Aamot,[2]

[1] STATOIL, Biosentrum
Stavanger, Norway

[2] STATOIL, Research Centre
Trondheim, Norway

[3] STATOIL
Stavanger Operation
Stavanger, Norway

[4] STATOIL
Bergen Operation
Bergen, Norway

INTRODUCTION

The discharge of produced water in the North Sea has been given much closer attention in the last few years due to the increasing volumes of produced water from the different fields. The volumes of produced water discharged in the Norwegian sector of the North Sea in 1990 was close to 16 million m^3 (100.6 x 10^6 bbl), and is expected to increase in the years to come. In the year 2000 the total amount of produced water from the Norwegian offshore fields is expected to reach 90 million m^3 (Olf, 1991). Due to this prognosis, Statoil and other oil companies operating in the North Sea are currently focusing on the discharge of produced water and its possible effects on the marine environment.

At the end of 1990 a total of 22 fields were on stream in the Norwegian sector of the North Sea. The total amount of oil and gas produced in 1990 was 79.0 and 25.4 million m^3 respectively. The largest oil field in the North Sea is the Statfjord field which in 1990 produced 39.6 million m^3 oil and discharged 10.5 million m^3 produced water. During the same period, Gullfaks, Statoil's second largest producing oil field, accounted for 13.3 million m^3 oil and 3.1 million m^3 produced water.

Up to now, the maximum allowed concentration of oil in produced water has been 40 mg/l. The expected increase in the discharged volumes may lead to further restrictions in this level in the future, because the total amount of oil discharged into the environment will increase. The oil companies are working intensively to develop improved separation techniques to minimize the amount of suspended oil in the produced water. In addition to dissolved hydrocarbons, the water soluble fraction of produced water consists of a large variety of polar organic compounds originating from the oil itself, formation water in the reservoir, and the addition of chemicals in

Produced Water, Edited by J.P. Ray
and F.R. Engelhart, Plenum Press, New York, 1992

the production process. The composition of produced water has been described by several authors over the past two decades, and some of these studies have been reviewed by Middleditch (1984) and Stephenson (1991). Stephenson (1991) reported average values for fatty acids, phenols, heavy metals, radionuclides and hydrocarbons in produced water from several platforms in the Gulf of Mexico. The dissolved hydrocarbons are dominated by the volatile aromatic fraction of the oil and the most abundant compounds are benzene, toluene, xylenes, ethylbenzene and naphthalenes. The concentration of these compounds are in the same range or lower compared to the amount of suspended oil in the water (Middleditch 1984; Sommerville et al. 1987; Tibbets, 1988; Stephenson, 1991). The concentration of polar organic compounds has been reported at much higher levels than the hydrocarbons, up to more than 1000 mg/l (Tibbets, 1988; Sommerville et al., 1987).

Heavy metals have also been detected in produced water, and Middleditch (1984) and Stephenson (1991) have given concentrations for some of the more interesting elements from an environmental point of view. Karstensen and Enger (1991) have analyzed produced water from the Statfjord field and found heavy metal concentrations in the low μg/l range. For the elements barium, strontium and iron, concentration levels in the mg/l range were found.

Other chemical compounds which may contribute to the potential environmental effect of produced water are those added during the oil/water separation process. Chemicals like corrosion inhibitors, scale inhibitors, emulsion breakers, flocculants, antifoams and biocides are all used during offshore oil production. Their influence on the produced water quality can affect several factors, including oil/water partition coefficient, acute and chronic toxicity, bioavailability and biodegradability. It is an overall goal for Statoil to avoid the use of production chemicals representing potential environmental hazards, and the parameters listed above are therefore all carefully evaluated for chemicals used in our offshore activities.

To predict the toxic potential of trace amounts of numerous chemicals is difficult and has stimulated the use of biological assays to help define the integrated effects of complex chemical mixtures on environmental systems (Miller et al., 1985). Laboratory toxicity tests have been reported to give good predictions of impacts on the ecosystem when two or more species are used. Biomonitoring can be undertaken either on effluents and/or on the receiving water (Thomas, 1987).

The purpose of the toxicity testing in the present preliminary study is to obtain data on acute toxicity of produced water from the different platforms, and to use the data together with dilution/dispersion models to predict potential acute impacts on the environment. Results of chemical characterization for one produced water sample from each platform will be reported and evaluated against the toxicity data.

MATERIALS AND METHODS

Toxicity Testing

Preparation of Produced Water Samples. A total of 13 samples were collected, four from Gullfaks A and B respectively (flotation cell), and five from Statfjord B and C (degas tank and flotation cell). Samples were not preserved, but quickly transported from the platform, and cooled immediately when they arrived at the laboratory. Samples for toxicity testing were analyzed within 3-5 days after sampling. The samples were filtered through a Whatman No.1 filter before analyses. Toxicity of unfiltered samples was also tested by the Microtox method for comparison. The different concentrations of produced water were made using sea water as the diluent. The sea water used in the experiments was fresh and collected at 40 m depth. The sea water was filtered before use through a Millipore type SS 3.0 μm filter.

Microtox. Luminescent bacterial tests were performed using the Beckman Model 2055 Microtox Toxicity Analyzer System (Beckman, 1988). Bacteria and other necessities for the reconstitution solution, a diluent and a solution for osmotic adjustment, were supplied from

ABK tek A/S, Oslo. The incubator block and chamber in which light production was measured were maintained at a temperature of 15° C. The reconstituted *Photobacterium phosphoreum* reagent was kept for a maximum of about 4 h at a temperature of approximately 5° C during the test. Four replicates were performed for each concentration of the produced water samples. The results were reported as EC_{50} values.

Artemia salina: *Artemia salina* nauplii were hatched from desiccated cysts in natural sea water at 20°C with constant illumination and aeration (Foster and Tullis, 1984). After 48 hours, 50-80 *Artemia* nauplii were transferred to each Erlenmeyer flasks containing produced water samples and controls. Test volume was 100 ml and minimum three replicates were run for each sample. The larvae were exposed to produced water for 24 hours at 20°C with constant illumination. The number of dead and live larvae were counted after 24 hours.

Skeletonema costatum. Toxicity of produced water against *Skeletonema costatum* was determined according to ISO-method DP 10253, 1988. Growth inhibition was measured by microscopic cell counting of the inoculum in each sample after 24 and 48 hours. Inhibition was measured as reduction in growth rate relative to control cultures grown under identical conditions, and reported as an EC_{50} value.

Test for Biodegradability

"Ready Biodegradability" of the produced water was determined according to "Shake flask sea water variant of OECD Method 301 E, 1990" with small adjustments. Two different dilutions of produced water in sea water were used. In addition to analyses of dissolved organic carbon, acute toxicity was measured during the experiment by the Microtox method and chemical analyses were also performed.

Chemical Characterization

A total of six samples were collected, four from Statfjord (degas tank and flotation cell), and two from Gullfaks (flotation cell). The samples were collected in 1 liter glass flasks, precleaned with methylene chloride, and transported to the laboratory. The Statfjord samples arrived at the laboratory within 24 hours after sampling, while the Gullfaks samples were stored in a refrigerator and arrived at the laboratory five days after sampling. All samples were filtered through Whatman No.1 filters in order to remove particulate matter and dispersed oil. This procedure was followed to enable direct comparison with the results from the toxicity tests. After filtration the samples were acidified (to pH 2) by addition of hydrochloric acid (1.0 M).

The samples (2 liters each) were transferred to separation funnels and extracted with methylene chloride (200 ml x 1 h + 100 ml x 1 h) by vigorously shaking. Internal standards, trichlorophenol, deuterated decanoic acid, deuterated benzene, deuterated naphthalene, deuterated phenanthrene, deuterated pyrene and deuterated perylene were dissolved in the extraction solvent prior to extraction. The two extracts from each sample were combined, and the volume reduced to approx. 5 ml by rotavapor (ambient temperature), leaving a total of six samples. Each sample was divided into three aliquots of equal volume.

Aliquot 1. The extract was transferred to methanol solution and eluted through a solid phase extraction column (Bond Elut C_{18}, 500 mg) by 2 x 1 ml methanol, leaving the nonpolar phase adsorbed to the column material. The eluate (polar phase) was analyzed by a HP 5890/VG Trio-1 GC/MS system operating in full scan mode. The total polar fraction was determined by comparing the total area of the chromatogram to the internal standard (trichlorophenol) area. The amounts of phenol and it's C_1 - C_7 alkyl homologues were determined quantitatively by selected ion monitoring (SIM) GC/MS, focusing on the molecular ions of the different phenols.

Aliquot 2. The second aliquot was analyzed quantitatively for fatty acids content. A part of the original water sample was shipped to the University of Bergen for analysis of C_1 -C_6 fatty acids by isotachophoresis (Barth 1987). A solution of boron trifluoride in methanol (10%) was added to the remaining extracts (aliquot 2), and the mixture was flushed with nitrogen (grade 4.0, highly purified), sealed, shaken vigorously for 1 min and heated to 110° C for 10 minutes. After cooling the sample was extracted with hexane (2 x 1 ml), and the extract was eluted through a solid phase column (Bond Elut silica, 500 mg). The column was eluted with hexane (4 x 1 ml), methylene chloride in hexane (1%, 2 x 1 ml) and the fatty acid methyl esters were finally retrieved from the column by eluting with methylene chloride (2 x 1 ml). The content of C_8 - C_{18} fatty acids was determined by SIM GC/MS, focusing on m/z 74 and 87 and using deuterated decanoic acid as internal standard.

Aliquot 3. The last part of each sample was transferred to hexane solution by evaporation of the methylene chloride under a gently stream of nitrogen (grade 4.0, ambient temperature) after addition of hexane (2 ml). The samples were applicated on combined silica (3.6 g, 70-230 mesh) and alumina (6.0 g, Fluka type 501 6A) columns (50 cm x 1 cm) and eluted with pentane (100 ml) giving the aliphatic fractions, and methylene chloride and pentane (100 ml, ratio 1:2) giving the aromatic fractions. The total amounts of aliphatic and aromatic compounds and the content of benzene, toluene, xylenes and higher alkylated benzenes (BTX) were determined by GC/FID analysis (HP 5890A) using crude oils from the actual oilfields as external standards. The content of naphthalenes, phenanthrenes, dibenzothiophenes (NPD) and their C_1 - C_3 alkyl homologues was determined by internal standard SIM GC/MS. In addition the content of 16 priority pollutants (PAH) (EPA test method 610 - PAH) was determined, also by internal standard SIM GC/MS.

Biodegradation Samples. From one of the biodegradation experiments (Statfjord C produced water) samples were collected for chemical characterization after 0, 7, 21 and 28 days. The samples (100 ml water each) were directly extracted by methylene chloride (100 ml x 1 h + 50 ml x 1 h) containing trichlorophenol as internal standard. The volumes of the extracts were reduced by rotavapor evaporation (ambient temperature) to 1 ml and the extractable organic material in the samples were determined quantitatively by external standard GC/FID. The content of phenols was also determined following the procedure described above.

Quality Control. A comprehensive program was followed in order to control the accuracy and precision of the analytical results. The program included blank analysis, spiking and analysis of seawater samples with authentic standards of some selected compounds and repeated analysis of samples and standards during the instrumental analysis.

For the toxicity tests a minimum of three replicates were analyzed of each sample. Average values are reported. The standard deviation was $< 10\%$.

RESULTS AND DISCUSSION

Toxicity testing

The toxicity was measured with three different test organisms: *Skeletonema costatum, Artemia salina* and *Photobacterium phosphoreum*. The produced water samples showed different toxicity values towards the different test species (Table 1 and 2).

The lethality to *Artemia salina* was below 20% in undiluted produced water for all the samples except one. The toxicity was considerably higher with Microtox than with the algae and the herbivore. Microtox EC_{50} values for produced water from Statfjord varied between 4.0-5.5% (g produced water/100 g sea water). Toxicity in produced water from North Sea has been measured for *Photobacterium* and five other organisms by Somerville et al. (1987). Their results on Microtox are consistent with ours. EC50 values between 5 - 10% were reported.

The produced water from Gullfaks was a little less toxic to *Photobacterium*; 5.7-19.2% (Table 1 and 2), than produced water from Statfjord with the *Skeletonema* and *Artemia* tests the situation was opposite. The Gullfaks samples were somewhat more toxic to these species than the Statfjord samples.

Table 1. Toxicity of produced water from Gullfaks

Platform		Toxicity		
	Sampling Date	*Skeletonema* EC50 (%)	*Artemia* Lethality(%)*	Microtox EC50 (%)
Gullfaks A	09/15/91	67.6	13.0	8.7
Gullfaks A	10/01/91	16.5	16.4	7.0
Gullfaks A	10/23/91	4.5	35.7	5.7
Gullfaks A	11/03/91	9.9	5.6	5.7
Gullfaks B	09/16/91	38.3	5.6	9.6
Gullfaks B	09/30/91	39.0	1.4	11.3
Gullfaks B	10/15/91	20.0	0	9.4
Gullfaks B	11/03/91	25.0	0	19.2

* Percent lethality of *Artemia salina* in undiluted produced water.

Table 2. Toxicity of produced water from Statfjord

Platform		Troxicity		
	Sampling date	*Skeletonema* EC50 (%)	*Artemia* Lethality(%)*	Microtox EC50 (%)
Statfjord C Degastank	06/05/91	53.5	5.0	4.0
Statfjord B Flotation cell	09/04/91	11.7	10.0	4.3
Statfjord B Degastank	09/04/91	43.0	0	4.7
Statfjord B Flotation cell	09/18/91	16.0	3.0	4.8
Statfjord b Dagastank	09/18/91	50.0	1.9	5.5

* Percent lethality of *Artemia salina* in undiluted produced water.

The average EC_{50} value to *Skeletonema* for the Gullfaks samples was 28% produced water, while an average EC_{50} of 35% produced water was measured for the Statfjord samples with the *Skeletonema* test. Also toxicity measured towards *Skeletonema* varied more between the different samples than the Microtox results. For the Statfjord samples the EC_{50} values were between 11.7-53.5% and for the Gullfaks samples the toxicity varied between 4.5-67.6%.

Produced water did not give any LD_{50} effect on *Artemia salina*. Lethality differed between 0-10% for undiluted produced water from Statfjord, and between 0-35.7% for the Gullfaks samples (Table 1 and 2). Toxicity measured by Microtox and *Skeletonema* seem to be correlated if we look at Gullfaks and Statfjord samples separately. Also samples from the flotation cell and from the degas tank on Statfjord have to be looked at separately. Samples from degas tank on Statfjord seem to be less toxic to the three test organisms than samples from the flotation cell on Statfjord. The samples from degas tank have passed through the hydrocyclone and the produced water oil content was normally higher from the flotation cell than from the hydrocyclones at Statfjord. In this study we have filtered the water before analysis, but some (variable) amount of oil could slip through the filter and influence the toxicity and chemical composition (see below).

The produced water is treated with a flocculent in the flotation cells. The flocculent used at Statfjord has been analyzed for toxicity to *Skeletonema*, Microtox and *Artemia*. EC_{50} values for the chemical were > 1000 ppm (Table 6), which excludes the toxicity from chemical treatment with flocculents as an explanation for higher toxicity against *Skeletonema* in produced water from flotation cell at Statfjord.

On the average, the produced water from Gullfaks had somewhat higher *Skeletonema* and *Artemia* toxicity than the Statfjord samples. The content of aliphatic hydrocarbons was higher in produced water from Gullfaks than from Statfjord, aromatic hydrocarbons were similar, polar components were lower and the content of fatty acids was drastically lower in the Gullfaks samples than in the Statfjord samples (Table 3). At Gullfaks we have measured 8 ppm of corrosion inhibitor in the produced water. EC_{50} values of the corrosion inhibitor for *Skeletonema*, Microtox and *Artemia* was respectively 0.22, 10.8 and 35.5 ppm. From these data one would suggest that corrosion inhibitor may contribute to the overall toxicity of produced water, at least for *Skeletonema*. This will be examined further. The flocculants used at Gullfaks have EC_{50} values between 10-180 ppm for *Skeletonema*. Toxicity for the flocculants is lower than for the corrosion inhibitors, but has to be taken into consideration in further studies of produced water.

To remove particulate solids the samples were filtered before toxicity testing. Toxicity measured by Microtox in samples that were not filtered, was found to be a little higher.

A matter of particular concern among ecologists has been the aromatic hydrocarbon content of produced water (Middleditch, 1984). Benzo(a)pyrene, a known carcinogen, was not detected in this study (see below). Attention has also focused on the two-ring aromatic hydrocarbons, naphthalene and alkylnaphthalenes. According to Table 4 the Gullfaks B sample had higher naphthalenes content than the others. Phenol content was higher in the Statfjord samples than in the Gullfaks samples. Acute toxicity of phenol for the most sensitive marine organisms was reported to be 5.8 ppm (EC_{50}) according to Middleditch, (1984). EC_{50} values for Microtox are reported to be 10-14 ppm (Kamlet et al. 1986).

The heavy metals were not analyzed in detail during the present study. However, previous results from elemental composition analysis of Statfjord and Gullfaks water showed that only a few elements are present at concentrations higher than seawater background levels (Karstensen and Enger, 1991; Olsen, 1991). While mercury, zinc, cadmium and lead all have been found at concentrations corresponding to marine background levels, chromium and copper in Statfjord and Gullfaks produced water were found at higher concentrations than the background concentration in sea water (Olsen,1991). Chromium varied between 7-16 ppb and copper between 4-15 ppb. Acute toxicity of heavy metals was reported by Miller et al. (1985). Algae is reported to be more sensitive to heavy metals than Microtox. Mean EC_{50} values were respectively 0.042 and 4.98 ppm. The possible variation in heavy metals in produced water will be analyzed in further studies.

Chemical Characterization

The total amount of the aliphatic, aromatic and polar fractions (determined by GC/FID) are given in Table 3. The results show that the total amount of organic matter (OM) in the produced water was much higher for the Statfjord samples than for the Gullfaks samples. The dominate compounds in all samples were the fatty acids, which accounted for more than 90% of the OM in the Statfjord samples and 65-75% in the Gullfaks samples. Within the fatty acid fraction, the C_1-C_6 acids accounted for more than 95% of the total for all samples. Among these, acetic acid was recognized as the most abundant single compound, yielding more than 75% of the total fatty acids for the Statfjord samples. However, since the fatty acids are regarded as relatively non-hazardous compounds to the environment, their presence and fate are not discussed further in the present paper.

The total content of the aliphatic and aromatic fractions are also given in Table 3. For the Statfjord samples (STB) the ratios of aliphatic to aromatic hydrocarbons show a relatively higher aromatic content in the water phase than in the parent oil. This is explained by the higher water solubility of the aromatics. The ratios of the Gullfaks samples (GFA and GFB) indicate relatively high aliphatic content in the water soluble fraction. In addition, the total hydrocarbon content (sum aliphatics and aromatics) was higher for the Gullfaks samples. The reason for this was not studied further, but a breakthrough of dispersed oil in the filtration of the water samples could cause such an effect, and may be one explanation.

Table 3. Chemical characterization of produced water, main fractions (mg/l)

Sample Fraction	GFA Flotcell 9/15/91	GFB Flotcell 9/16/91	STB Flotcell 9/18/91	STB Degas 9/18/91	STB Flotcell 9/4/91	STB Degas 9/4/91
Aliphatic	2.45	5.81	0.0254	1.65	0.316	0.888
Aromatic	1.04	3.99	1.50	1.63	1.05	2.12
Polar	6.77	15.7	72.4	31.8	27.5	65.3
Fatty acids	36.7	53.5	690	790	543	*
Total OM	47.0	79.0	763	825	575	*

* Sample was lost prior to analysis

Another interesting observation was that the concentration of dissolved hydrocarbons (aliphatics and aromatics) was higher in the Statfjord degas tank samples than in the flotation cell samples. Normally, the amount of suspended oil is higher in the water discharged from the flotation cell than the produced water from the degas tank. The oil-in-water concentration is routinely measured every day at all Statoil's production platforms by freon extraction of the water followed by quantitative infrared spectroscopy analysis. The results from the actual days of sampling indeed showed higher oil concentration in the flotation cell water, 11.0 vs. 4.3 mg/l and 66.0 vs. 3.5 mg/l respectively. In future studies, the significance of this observation will be studied further. At this point, the number of samples is far to low to draw any conclusions from the results. Poorly performed filtration of the samples will cause such effects, and may be the reason. A more detailed study of the results in Table 4 (see below) supports this theory. The concentration of volatile aromatics was approximately the same for all Statfjord samples, while the concentration of aliphatics was more variable. The presence of suspended oil in the samples after filtration will influence the aliphatic fraction, since the abundance of aromatic compounds in the crude oil is far lower than the abundance of aliphatics.

The content of the different aromatic compounds analyzed is given in Table 4. However, the volatile aromatics, benzene, toluene and xylenes (BTX), were retrieved quantitatively in the silica/alumina column of the aliphatic fraction, showing that the separation of the two fractions was not complete. An optimization of the stationary and mobile phase is necessary to obtain sufficient separation. As a consequence, these compounds had to be analyzed in the aliphatic rather than the aromatic fraction. To control the situation, both the aliphatic and the aromatic fraction were analyzed for all aromatic compounds. The results of these analyses and the NPD and PAH analysis are given in Table 4. The results show that the BTX and naphthalenes were the dominating aromatic compounds in the nonpolar fraction, but PAH, up to 5-ring structure (benzfluoranthenes) was detected at very low levels in the samples.

The polar fraction of the water soluble organic matter in the produced water was regarded as the most interesting for the present study. Previous reports (Smith, 1986; Sommerville et al., 1987) have suggested that this fraction may yield levels of organic compounds far higher than the dispersed and water dissolved oil in the system. The results presented in Table 3 strongly support such findings. However, the major part of the polar fraction consisted of low molecular weight fatty acids, with acetic acid as the dominate compound. These compounds are not regarded as hazardous to the environment, and are therefore of minor interest when discussing possible effects of produced water discharge. A more interesting class of produced water compounds are the phenols, which have been detected in previous studies of produced water (Tibbets 1988; Neff et al., 1988). In addition to the quantitative analysis of phenol and cresoles, the full scan GC/MS traces from the samples were searched for higher alkylated phenols by studying the mass spectra of the dominate peaks in the chromatograms.

Even though the total ion current chromatograms of the polar fraction proved to be rather complex, good mass spectra were obtained for several of the dominate peaks. Some of these were by library search, and manual interpretation found to represent higher alkylated phenols than cresol. By SIM analysis focusing on the molecular ions of phenol - C_7-phenol these compounds were quantified. Unfortunately, authentic standards of the higher alkylated phenols were not available for this study, meaning that exact response factors could not be determined. A conservative approach, assuming equal response factors for the higher alkylated phenols, e.g., C_2-phenol, were therefore employed. A more detailed study is presently being performed in our laboratory, and the results from this will be published in the near future.

Table 4. Aromatic compounds in produced water

Compound	Concentration (μg/l)					
Sample	GFA Flotcell 9/15/91	GFB Flotcell 9/16/91	STB Flotcell 9/18/91	STB Degas 9/18/91	STB Flotcell 9/4/91	STB Degas 9/4/91
Fraction						
BTX	582	533	168	130	145	159
Naphthalenes	49.3	2160	622	942	708	845
Phenanthrenes	12.5	90.0	23.0	41.7	36.8	21-3
Dibenzothio-phenes	3.17	22.7	8.16	10.8	10.2	5.07
Acenaphthene	<0.001	<0.001	<0.001	<0.001	<0.001	<0.001
Acenaphtylene	0.047	1.53	1.11	1.04	0.666	0.787
Fluorene	0.254	11.3	12.0	12.0	8.89	9.50
Anthracene	<0.001	<0.001	0.0320	0.0460	0.0341	0.0396
Fluoranthene	0.0269	0.195	0.0415	0.0892	0.0854	0.0458
Pyrene	0.0291	0.194	0.0559	0.121	0.118	0.0669
Benzo(a)-anthracene	<0.001	0.0311	0.0045	0.0193	0.01510	.0120
Chrysene/triphenylene	0.329	0.398	0.0403	0.203	0.2260	.112
Benzo(bjk)-fluoranthene	<0.001	0.0528	<0.001	0.0197	<0.001	<0.001
Benzo(a)pyrene	<0.001	<0.001	<0.001	<0.001	<0.001	<0.001
Benzo(ghi)-perylene	<0.001	<0.001	<0.001	<0.001	<0.001	<0.001

In Table 5 the concentrations of phenols in the six samples are presented. As seen from the results, the phenols account for approximately 10% of the non-fatty acid part of the polar fraction. The results indicated higher phenol concentrations in the Statfjord samples than in the Gullfaks samples. Tibbets (1988) analyzed water from the Gullfaks A platform, and found concentrations of phenol and cresols which are in good agreement with the Statfjord levels obtained in the present study. These differences may be explained by the possible biodegradation of phenols in the Gullfaks samples during the relatively long period (5 days) between sampling and extraction. The biodegradation results discussed below support this theory.

Tibbets (1988) also indicated that higher alkylated phenols might be present in the water. Neff et al. (1988) found concentration levels in the low μg/l range of phenol (C_4-phenols) in produced water from Eugene Island and Lake Pelto off Louisiana. The presence of higher alkylated phenols in produced water may be a potential environmental hazard since they most likely have higher fat affinity and are less biodegradable than their parent compound. The extremely low concentration of these compounds together with the rapid dilution of the produced water after discharge, will probably reduce their environmental hazard potential substantially.

No attempt was made to identify the remaining part of the polar fraction in this portions of the study. Previous investigations have shown that organic acids other than the fatty acids may account for a major part of this fraction (Tibbets 1988).

The final class of compounds present in produced water are the production chemicals, added for different purposes through the process or separation line. Table 6 summarizes the

Table 5. Concentration of phenols (mg/l)

Sample Compound	GFA 9/15/91	GFB 9/16/91	STB Flotcell 9/18/91	STB Degas 9/18/91	STB Flotcell 9/4/91	STB Degas 9/4/91
Phenol	0.262	0.152	4.86	2.59	2.84	2.30
C_1-phenol	0.209	0.375	2.84	3.09	2.92	2.51
C_2-phenol	0.0603	0.107	1.26	0.720	0.771	0.884
C_3-phenol	0.0540	0.725	0.634	0.253	0.322	0.603
C_4-phenol	0.0117	0.237	0.0658	0.0217	0.0279	0.0805
C_5-phenol	0.0124	0.0946	0.0235	0.0048	0.0081	0.0352
C_6-phenol	0.0060	0.0243	0.0073	<0.001	0.0016	0.0127
C_7-phenol	0.0026	0.0062	<0.001	<0.001	<0.001	0.0021
TOTAL	0.618	1.72	9.69	6.69	6.89	6.43

Table 6. Chemicals added in process/separation line

Chemical Type	Acute Toxicity EC_{50} (mg/l)			Concentration (mg/l)
	Algae	*Artemia*	Microtox	
Corrosion Inhibitor	0.2-2	>20-25	15-50	1-4
Scale Inhibitor	60	1000	>1000	20-30
Emulsion Breaker	20	30	20	n.a.*
Flocculent	>1000	>15000	<15000	n.a
Anti-foam	120	150	9	n.a.

* n.a. = not analyzed

different classes of chemicals used at Statfjord, their acute toxicity, and indicates the concentration of some of these in the produced water.

Biodegradation

Produced water from Statfjord (flotation cell) is readily degradable according to Figure 1 and more than 95% of dissolved organic carbon is degraded during 28 days. The degradation was most rapid the first weeks (80 - 90%), but continued also for the rest of the period. Biodegradation was examined in 50% and 25% produced water, diluted with sea water, with similar result. Addition of inoculum did not influence the biodegradation. Toxicity was decreased gradually, with less overall reduction at a slower rate than one would expect from the removal of dissolved organic carbon. Toxicity was reduced approximately 7-10 times during biodegradation (Fig. 1 and 2).

In Table 7 the results from the chemical characterization of the biodegradation samples are presented. The results show a rapid decrease in EOM from the samples, with only 6% of the initial EOM extracted from the samples after 28 days. These findings appear to correlate with the development of total dissolved organic carbon in the samples, presented in Figure 1 and 2. The removal of phenol was even more rapid than for EOM. After only seven days of exposure, more than 99.9% of the initial phenol had disappeared. For the alkylated phenols, the results presented in Table 7 clearly demonstrates that the biodegradability decreases with increasing number of carbon atoms bound to the aromatic structure. For C_7-phenol the concentration in the water remained almost at a constant level during the biodegradation period. Sommerville et al. (1987) also found that the organic matter in produced water from North Sea oil operations was easily degradable. When grown in continuous culture the bacterial activity almost completely removed the lower aromatics and their substituted analogues as well as the carboxylic acids. They also reported degradation of phenanthrene.

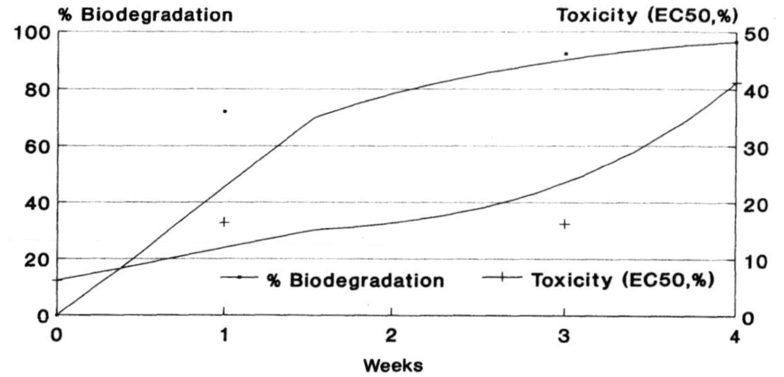

Figure 1. Biodegradation and changes in toxicity over four weeks interval. Statfjord flotation cell 04.09.1991.

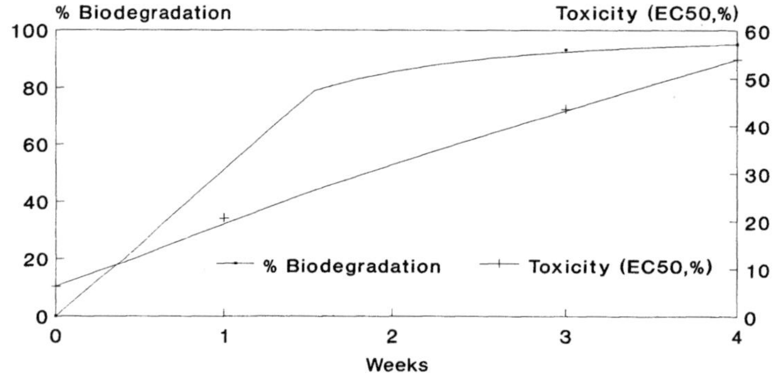

Figure 2. Biodegradation and changes in toxicity.

Table 7 Biodegradation samples

Compound	Initial	Day 7	Day 21	Day 29
EOM (mg/l)	25.6	7.18	4.52	1.526
Phenol (μg/l)	5120	3.84	1.65	1.08
C_1-phenol (cresol)	2450	1.78	0.27	0.33
C_2-phenol	396	11.1	1.90	1.50
C_3-phenol	130	36.4	7.88	6.19
C_4-phenol	26.2	16.6	21.0	3.39
D_5-phenol	16.8	12.2	4.60	2.96
C_6-phenol	13.3	12.2	9.64	4.41
C_7-phenol	5.49	4.61	4.36	3.34
Total phenol	8158	91.0	51.3	23.2

Quality control

The results from the blank analysis were used to determine the limits of detection and quantification for the different chemical parameters. From the spiking experiments the level of recovery of the different chemical parameters from the water phase was determined. In general, the recovery was quantitative within the analytical precision for all parameters (95 - 100%) with exception of phenol and the cresols which only were recovered to a degree of 30 and 75% respectively. The results presented above have been corrected for this loss. It is anticipated that the loss takes place during the evaporation of the extraction solvent. Since the internal standard, trichlorophenol, has a higher boiling point than phenol, its ability to compensate for the loss is limited. A deuterated phenol would be a better choice for internal standard in future analysis. The repeatability of the instrumental systems (GC and GC/MS) was found to be within 5% standard deviation of five replicates.

Dilution and dispersion of produced water

For Statfjord and Gullfaks the primary dilution of produced water will take 40-70 seconds and give a dilution between 50-400 times depending on the platform and the actual discharge. Further dilution will vary , but generally we expect an extra 10 fold dilution at a distance of 800-1000 m from the platform (Rye, 1989 and 1991).

CONCLUSION

In this study acute toxicity of produced water has been measured with three single species laboratory toxicity tests. Microtox was most sensitive and EC_{50} values of produced water varied between 4 - 20 g produced water/100 g sea water for the different samples. Produced water was easily degraded, 95% removal of dissolved organic carbon was measured in analysis of biodegradability. Toxicity was reduced up to 10 fold during the biodegradation. The results from chemical characterization of produced water support earlier findings. In this study also phenol and alkylated phenols up to C_7-phenol are also quantified and reported. Toxicity data of chemicals added continuously in the process system are also discussed.

The results indicate that acute toxicity towards marine organisms is low for produced water. It might have some effect in the mixing zone, but according to our distribution models, acute toxicity effect is negligible outside the mixing zone. The limited amount of data in the present study does not allow unambiguous conclusions to be drawn. Possible long term effects of discharge of produced water will be investigated in future studies.

ACKNOWLEDGMENTS

The authors wish to thank B. Thorvaldsen and T. Berg at the Continental Shelf and Petroleum Technology Research Institute, Trondheim, Norway for their assistance with the chemical analysis. We also wish to thank T. Barth, University of Bergen, Norway for performing the isotachophoresis analysis of volatile fatty acids.

REFERENCES

Barth, T., 1987, Quantitative determination of volatile carboxylic acids in formation waters by isotachophoresis. Anal. Chem. 59:2232.

Beckman, 1988, "Microtox Manual. How to Run a Standard Microtox Test." Microbics Corporation, Part Number 555 880, 45 pp.

Draft OECD guideline for testing chemicals 301, Ready biodegradability, 1990, 301E Modified OECD screening test. Room document No 16, Annex 3: 45-68.

Foster, G.D. and Tullis, R.E., 1984, A quantitative structure-activity relationship between partition coefficients and the acute toxicity of naphthalene derivatives in *Artemia salina* nauplii. Aquatic Toxicology, 5: 245-254.

ISO Method DP 10253, 1988, 4th Working Draft "Water Quality - Marine Algae Growth Inhibition Test with Skeletonema costatum and Phaeodactylum triconutum".

Kamlet, M.J., Doherty, R.M., Vetth, G.D., Taft, R.W. and Abraham, M.H., 1986, Solubility properties in polymers and biological media. 7. an analysis of toxicant properties that influence inhibition of bioluminescence in *Photobacterium phosphoreum* (The Microtox Test). Environ. Sci. Technol. 20: 690-695.

Karstensen, K.H., Enger, B., 1991, Analyze av metaller i 4 vannprøver. SI Oslo Norway. Report no. 91 05 12. 1pp.

Middleditch, B.S. 1984, Ecological Effects of produced water effluents from offshore oil and gas production. Ocean Management, 9: 191-316.

Miller, W.E., Peterson, S.A., Greene, J.C. and Callahan, C.A., 1985, Comparative toxicology of laboratory organisms for assessing hazardous waste sites. J. Environ. Qual., Vol. 14, No. 4: 569-574.

Neff, J.M., Sauer, T.C., and Maciolek, N., 1988, "Fate and Effects of Produced Water Discharges in Nearshore Marine Waters." Final report to the American Petroleum Institute, Washington DC, Battelle Ocean Sciences, Duxbury, MA. 304 pp.

OLF, 1991, "Environmental Programme. Report Phase 1, Part B. Discharges to Sea."

Olsen. S., 1991, Analysis of heavy metals in produced water. Internal report. Rogaland Research.

Rye, H., 1989, Spredningsberegninger for Statfjord platformene. Internal report. Veritas Miljø plan.

Rye, H., 1991, Spredningsberegninger for Gullfaks og Veslefrikk B plattformene. Internal report. Veritas Miljø Plan.

Smith, M.G., 1986, "Measuring Methods for Petroleum Compounds in Discharged Waters from Offshore Installations." M-scan ltd. Ascot UK, Report 8611/1526 78 pp.

Sommerville, H.J., Bennett, D., Davenport, J.N., Holt, M.S., Lynes, A., Mahieau, A., McCourt, B., Parker, J.G., Stephenson, R.R., Watkinson, R.J., and Wilkinson, T.G., 1987, Environmental effect of produced water from North Sea oil operations. Marine Poll. Bull. 16.10:549.

Stephenson, M.T., 1991, Components of produced water: A compilation of results from several industry studies. Paper presented at The first International Conference on Health, Safety and Environment, The Hague 10-14 November 1991. SPE 23313, 14 pp.

Tibbets, P.J.C., 1988, "Examination of the Organic Compounds and Heavy Metals in Formation Water Discharge from the Gullfaks A Platform in January 1988." M-Scan ltd. Ascot UK. Report no. 8803/3605/1 48 pp.

Thomas, N.A., 1987, Use of Biomonitoring to Control Toxics in the United States. U.S.Environmental Research Laboratory - Duluth, 6201 Congdon Boulevard, MN. 55804, 11 pp.

RADIOCHEMICAL DETERMINATION OF NORM IN PRODUCED WATER UTILIZING WET CHEMISTRY SEPARATION FOLLOWED BY RADIOCHEMICAL ANALYSIS

D.L. Demorest and E.S. Wallace

Core Laboratories
Casper, Wyoming

ABSTRACT

The issue of radium contamination in produced water has become one of the most significant environmental problems presently facing the oil and gas industry. Several states have addressed this issue by placing additional requirements on National Pollution Discharge Elimination System (NPDES) permits that limit the amount of radium that can be discharged (60 pCi/l [2.22 Bq/l] in Louisiana and Wyoming). There are also regulations and/or proposed regulations concerned with the buildup of natural occurring radioactive material (NORM) and technologically enhanced natural radioactive (TENR) material, resulting from the discharge of production waters containing these materials (in Texas and Louisiana). These regulatory requirements and the liability issues associated with NORM make it essential that accurate analytical methods be employed to determine the activity levels present in a discharge stream.

The chemical constituents contained in many briny production waters often present significant interference when utilizing conventional analytical methods. If these samples are not prepared properly, the radium components will be lost during the initial separation. The outcome is a false negative result for samples that may actually contain significant amounts of NORM.

The method outlined here provides a quick and accurate means of determining Radium-226 and Radium-228 in produced waters utilizing alpha-scintillation and low background proportional counting techniques.

INTRODUCTION

The ramifications of inaccurate or unverifiable analytical data are immense in situations where regulatory compliance and civil or financial penalties may depend on the outcome of the analytical results. It is essential that analytical laboratories involved in providing data in these situations utilize effective and provable analytical technologies. This paper discusses the methodology utilized in analysis of NORM isotopes in produced water. We will outline and demonstrate how we qualify our data and demonstrate the precision and accuracy of the methods utilized.

The technology for separation of radium by co-precipitation with barium or lead sulfate is very well documented and is the general approach for most EPA (1980) and Standard Methods (American Public Health Association, 1985) methodologies. However, the use of these technologies is limited when dealing with production fluids and produced waters which often contain high chloride and calcium concentrations. High levels of chloride (130,000 mg/l) will interfere

Produced Water, Edited by J.P. Ray
and F.R. Engelhart, Plenum Press, New York, 1992

with the precipitation of lead sulfate carrier (Percival and Martin, 1994), and hence limit the effectiveness of this carrier for isolation of radium from such solutions. High concentrations of calcium (4,000 mg/l), again associated with many produced waters, tend to form insoluble precipitates during the initial precipitation making the isolation of radium very difficult if not impossible in subsequent preparation steps.

One method that has been reliable and effective in dealing with these interferences was developed by Percival and Martin (1974) at the U.S. Department of Energy Health Services Laboratory in Idaho Falls, Idaho. This method was developed to deal specifically with sequential determination of the radium and thorium isotopes in environmental and process waste samples, and is very applicable for use with production water.

In the Percival and Martin method, the isotopes are co-precipitated on lead sulfate. The precipitate is dissolved in diethylenetriaminepentaacetic acid (DTPA) and radium is isolated by preferential precipitation on barium sulfate. Radium-226 is then counted by alpha-scintillation. Radium-228 is determined by isolation of the Radium-228 decay product and beta counted on a low background proportional counting system.

The method described in this paper alters the original Percival and Martin method by diluting the sample twenty fold prior to co-precipitation on lead sulfate, eliminating the potential interference from high chloride and calcium content. This precludes the use of large volumes (1 to 2 liters) usually required to achieve detection limit requirements for drinking waters. However, the sensitivity requirements for production waters are not as stringent as for drinking water. Also, many production waters often contain significant activity eliminating the need for ultra-low detection limits.

Since ^{210}Pb is often required along with radium analysis, the method described here allows ^{210}Pb to be salvaged from the original DTPA solution. The lead sulfate is converted to the carbonate form, dissolved in a scintillation cocktail and the ^{210}Pb beta counted directly by low level liquid scintillation counting as demonstrated earlier in a paper by Blais and Marshall (1988).

COUNTING TECHNIQUES

Radium-226 - determined by counting barium sulfate precipitate directly on a Ludlum scintillation counter.

Radium-228 - ^{228}Ac, the decay product from ^{228}Ra, was counted on the cerium oxalate precipitates using a Berthold 770 10-channel low background proportional counter.

Lead -210 - determined by lead acetate suspended in 10 ml of Packard Ultima Gold-XL scintillation cocktail. The samples were counted in a Packard 2260XL Tri-carb Low Background Liquid Scintillation Counter with the MCA window set at 7 to 30 Kev and 30 to 250 Kev.

REAGENTS

All chemicals utilized were ACS Reagent Grade or better. Radioactive purity was such that the measured radioactivity of blank samples did not exceed 1 pCi/gm (0.037 Bq/gm) in all reagents. Reagent water was specified to be less than 2 μmhos and with an alpha activity of less than 1 pCi/l (0.037 Bq/l).

Reagents for Radium -226 and Radium-228: Reagents for the initial precipitation and isolation of ^{226}Ra and ^{228}Ra for counting were provided in the paper by Percival and Martin (1974).

Reagents for lead 210 - analysis: Concentrated acetic acid; ammonium acetate solution, saturated; sodium carbonate, anhydrous, dissolved 80 g sodium carbonate (Na_2CO_3) in 400 ml reagent water. Diluted to 500 ml with reagent water.

Liquid Scintillation Solution: Packard Ultima Gold XL

CALIBRATION STANDARDS

All calibrations standards were NIST (National Institute of Standards and Technology formerly NBS , National Bureau of Standards) traceable. Strict records of receipt, storage, preparation, and documentation of the Radioactive Material Standards were maintained in order to preserve NIST traceability. The techniques for preparation of these standards were derived from NCRP Report #58.

INSTRUMENT CALIBRATION

Ludlum 148-2 5-channel alpha counter - Calibrated daily by analysis of a 950 dpm Thorium 230 alpha source prepared by Eberline Instruments in Albuquerque, NM.

LB770 proportional counter - Calibrated daily with NIST traceable Americium-241 and Amerium-137 sources.

Liquid Scintillation Counting

The Packard 2260 XL system performance was monitored automatically by performing normalization and calibration(SNC) and instrument performance assessment (IPA) protocols daily. Hard copies along with the computer files were maintained as records of daily performance. The system was also calibrated by analysis of a NIST traceable ^{210}Pb standard solution in Ultima Gold cocktail with each analytical series.

METHODOLOGY

Sample Preparation

Prior to the initial separation on lead sulfate, samples that contained high concentrations of chloride and/or calcium were diluted 20 fold (100 ml to 2 liters) with deionized makeup water. These samples were then brought to a boil and 30 ml of 1:1 H_2SO_4 and 80 g of anhydrous potassium sulfate were added.

^{226}Ra, ^{228}Ra and ^{210}Pb were isolated by separation on lead sulfate by multiple additions of lead nitrate as described by Percival and Martin (1974).

Separation of Radium

The lead sulfate precipitate was dissolved in 0.17 M DTPA and the radium isolated by co-precipitation on barium sulfate per the Percival and Martin method. The remaining DTPA supernatant was utilized for isolation of ^{210}Pb. This process is described below.

Radium-228 Analysis

The resulting barium sulfate precipitate was redissolved in DTPA. The ^{228}Ac decay product was allowed to equilibrate by setting the sample aside for a minimum of 36 hours. The radium was reprecipitated and the DTPA solution extracted with 15% bis(2ethyl-hexyl)phosphoric acid (HDEHP) per Percival and Martin (1974). The ^{228}Ac was then isolated as an oxalate and counted in a low background proportional counter.

Radium-226 Analysis

The barium sulfate precipitate resulting from the ^{228}Ra analysis was analyzed by gross alpha-counting.

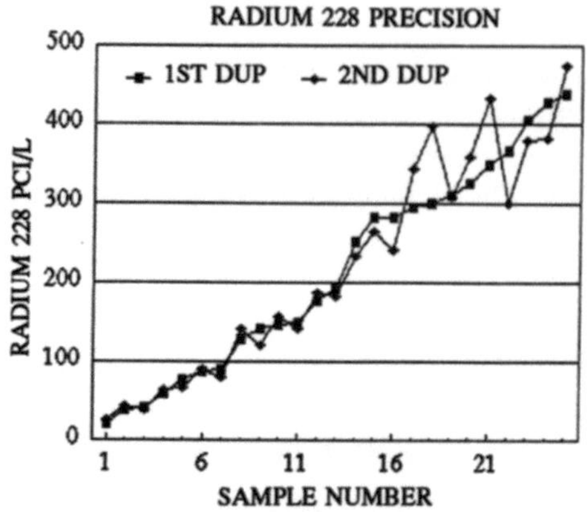

Figure 1. Radium-226 precision.

Lead-210 analysis

The initial DTPA solution set aside after the first separation of radium by barium sulfate precipitation described above was heated and 5 g of anhydrous potassium sulfate added and dissolved. The lead was then reprecipitated as a sulfate by addition of 5 ml of 1:1 H_2SO_4 while mixing. The sulfate precipitate was then dissolved in 25 ml of ammonium acetate. The lead was reprecipitated as a carbonate by addition of 15 ml of sodium carbonate. This process was repeated two times in order to remove all sulfate ions. The carbonate precipitate was dissolved in 1 ml of acetic acid and 4 ml of deionized water and transferred to a 20 ml glass scintillation vial containing 15 mls of Ultima Gold XL scintillation cocktail. The samples were then analyzed by liquid scintillation counting as described by Blais and Marshall (1988).

RESULTS AND DISCUSSION

In qualification of analytical data generated in a commercial analytical laboratory, it is necessary to assess data generated according to specific acceptance/rejection criteria. As a reference, the following quality control acceptance criteria are utilized as the standard limits for

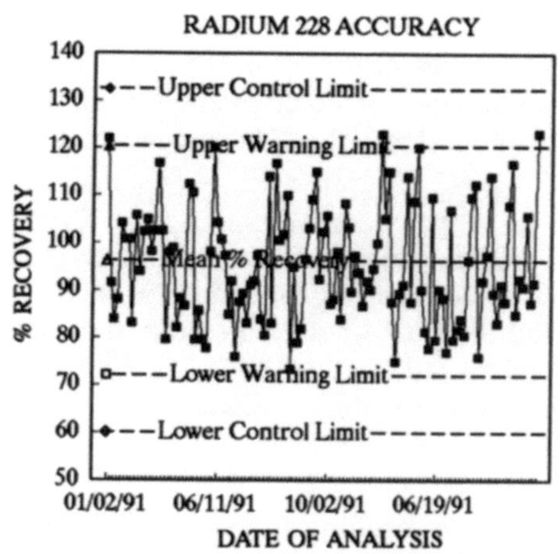

Figure 2. Radium-226 accuracy.

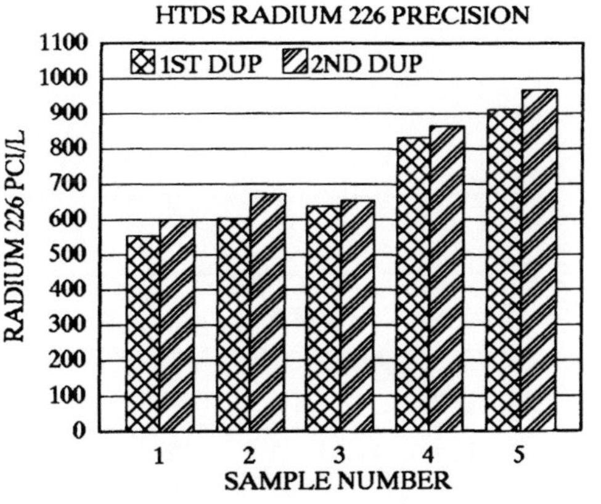

Figure 3. HTDS Radium-226 precision.

data considered acceptable in the EPA CLP program. If any of these criteria are not met the analytical group is automatically reanalyzed.

Method Blank	Analyzed value less than or equal to the Detection Limit
Blank Spike	100 +/- 10% recovery
Duplicates	20% relative percent difference, or +/- the Detection Limit
Matrix Spikes	100 +/- 25% recovery

In addition, method precision and accuracy are tracked and documented on a continuing basis. Precision is determined by calculating the relative percent difference between duplicates and accuracy by examining the percent recovery of the matrix spikes. Upper and lower warning limits (UWL and LWL) and control limits (UCL and LCL) are maintained for use in the lab as a means of examining trends and as a mechanism to identify non-compliance. This is accomplished by evaluating a minimum of 25 sets of knowns and duplicates for accuracy and precision.

In the study described here, we have provided control charts for precision and accuracy measurements on ^{226}Ra, ^{228}Ra and ^{210}Pb analyses performed utilizing data generated from

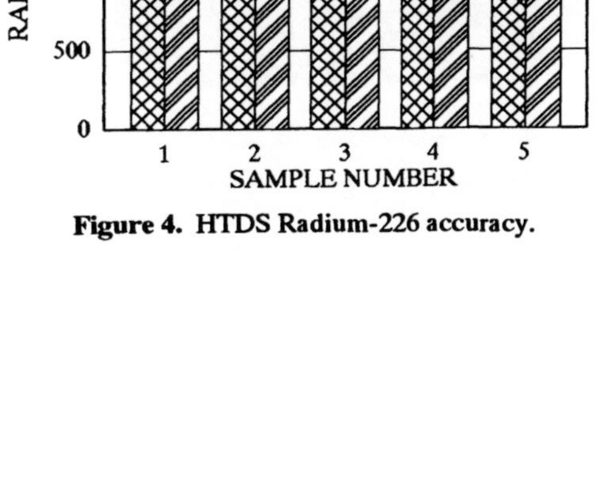

Figure 4. HTDS Radium-226 accuracy.

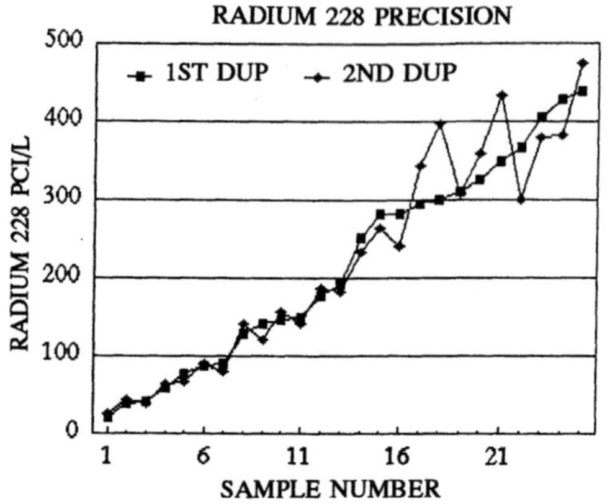

Figure 5. Radium-228 precision.

measurements of produced water or environmental water samples containing large amounts of dissolved solids.

Radium-226 Analysis Precision and Accuracy

Precision measurements demonstrated in Figure 1 represent the results of replicate data over a range of activity from 10 pCi/l (0.37 Bq/l) to 400 pCi/l (14.8 Bq/l) of ^{226}Ra in produced water samples. The standard deviation between duplicate pairs was 5.9% utilizing 30 pairs. The 95% warning limit was 17% and the 99% control limit was 25% (Table 1).

Accuracy measurements shown in Figure 2 and Table 2 represent the measurements for ^{226}Ra in high total dissolved solid content water and produced water from January to August of 1991. The average recovery of a known analytical spike was 95.7% with a standard deviation of 10.1%. This is well within the allowable range of +/- 25 percent. The upper and lower warning and control limits were set at 2 and 3 sigma respectively. The total number of spike analyses was 86, with a maximum of 122% and a minimum of 75% of the expected recovery.

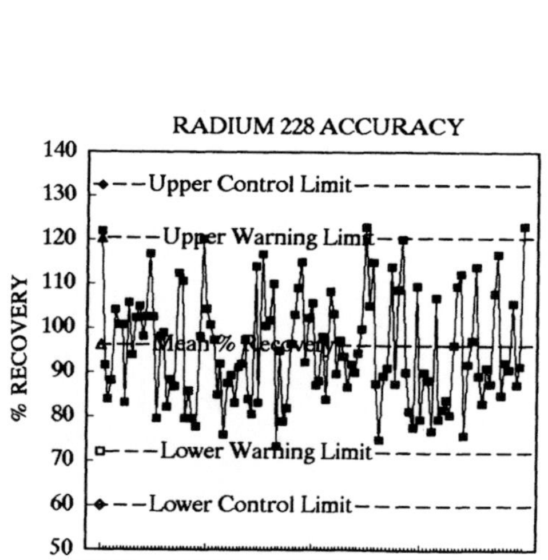

Figure 6. Radium-228 accuracy.

Table 1. Precision Measurements

Parameter	Standard Deviation	95% WL	99% CL	# n	Max Value
^{226}Ra	5.94	16.8	25.2	30	28
Ra226 HTDS	3.5	9.9	14.8	5	11
^{228}Ra	6.5	18.3	27.4	25	28
Ra228 HTDS	2.3	7.3	10.9	5	7.3
^{210}Pb	6.6	18.7	28	25	25

WL - Warning Limit #n - Number of tests in group
CL - Control Limit HTDS - High Total Dissolved Solids

In order to clarify the methods' effectiveness at high activity and high total dissolved solids content, five sets of samples with chloride content of 119,000 mg/l and 4,500 mg/l calcium were analyzed in duplicate and also spiked for a recovery determination. Figure 3 represents the results from duplicate pairs. Even in these high TDS and high activity samples the replicates demonstrate very good precision with a standard deviation between pairs of 3.5 %, warning limit of 9.8% and control limit of 14 % (Table 1).

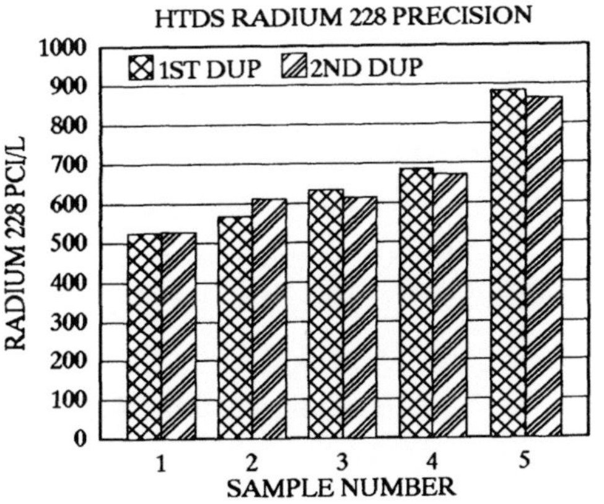

Figure 7. HTDS Radium-228 precision.

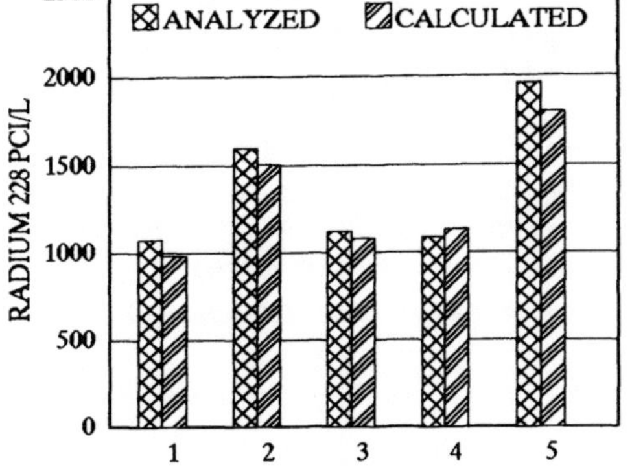

Figure 8. HTDS Radium-228 accuracy.

Table 2. Accuracy Measurements

Parameter	Avg % Rec	Std Dev	99% UCL	95% UWL	95% LWL	99% LCL
^{226}Ra	95.7	10.1	126	116	75.4	65
Ra226-HTDS	94.4	8.0	118	110	78.0	70
^{228}Ra	96.2	12.1	133	120	72	60
Ra228-HTDS	104	5.8	122	116	93	87
^{210}Pb	97.3	6.4	117	110	85	78

UCL - Upper Control Limit UWL - Upper Warning Limit
LWL - Lower Warning Limit LCL - Lower Control Limit
Avg%Rec - Average % Recovery HTDS - High Total Dissolved Solids

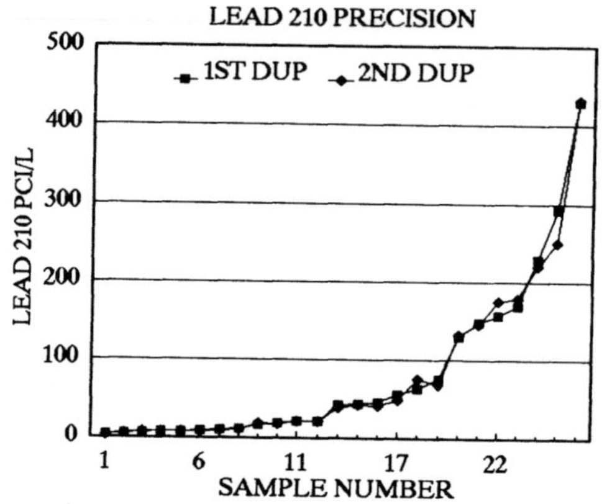

Figure 9. Lead-210 precision.

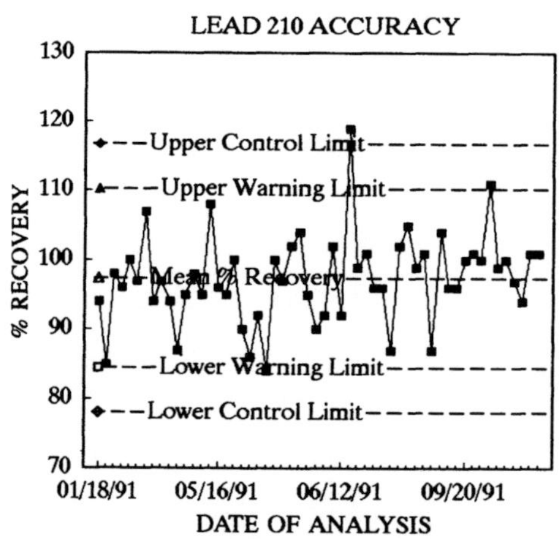

Figure 10. Lead-210 accuracy.

The accuracy is demonstrated in Figure 4 in which the spike sample analysis is compared to the calculated activity. The average recovery is 94.4% with a standard deviation of 8% which is very similar to the accuracy measurements for ^{226}Ra demonstrated in Table 2.

Radium-228 Analysis Precision and Accuracy

Figure 5 represents the results of replicate ^{228}Ra analyses over a range of activity from 10 pCi/l (0.37 Bq/l) to 500 pCi/l (18.5 Bq/l) of ^{228}Ra in produced water samples. The standard deviation between duplicate pairs is 6.5% utilizing 30 pairs. The 95% warning limit is 18% and the 99% control limit is 27 % (Table 1) which are very similar to those already demonstrated for ^{226}Ra.

The data presented in Figure 6 and Table 1 represent the accuracy measurements for ^{228}Ra in high total dissolved content water and produced water already examined for ^{226}Ra. The average recovery of a known analytical spike is 96.2% with a standard deviation of 12.1%, which is very similar to that observed for ^{226}Ra and is well within the allowable range of +/- 25 percent. The upper and lower warning and control limits are set at 2 and 3 sigma, respectively. The total number of spike analyses is also 86 with a maximum value of 123% and a minimum value of 7% of the expected recovery.

As already described for ^{226}Ra, the same five sets of samples with chloride content of 119,000 mg/l and 4500 mg/l calcium were analyzed in duplicate and also spiked for a recovery determination for ^{228}Ra. Figure 7 represents the results from duplicate pairs. Again in these high TDS and high activity samples, the replicates demonstrate very good precision with a standard deviation between pairs of 2.6%, warning limit of 7.3% and control limit of 11%.

The accuracy of the high level ^{228}Ra analysis is demonstrated in Figure 8 in which the spike sample analysis is compared to the calculated activity. The average recovery is 104% with a standard deviation of 6% which is very similar to that observed from ^{226}Ra and for ^{228}Ra.

Lead-210 Analysis Precision and Accuracy

As in the ^{226}Ra and -228 analyses, Figure 9 represents the results of replicate ^{210}Pb liquid scintillation analyses over a range of activity from 10 pCi/l (0.37 Bq/l) to 450 pCi/l (16.7 Bq/l) of ^{210}Pb in produced water samples. The standard deviation between duplicate pairs is 6.6% utilizing 30 pairs very similar to the 5.9% for ^{226}Ra and 6.5 % for ^{228}Ra. The 95% warning limit is 18.7% and the 99% control limit is 28% (Table 1) which are very similar to those already demonstrated for ^{226}Ra and ^{228}Ra.

The data presented in Figure 10 and Table 1 represent the accuracy measurements for ^{210}Pb in high total dissolved content water and produced water already examined for ^{226}Ra and ^{228}Ra . The average recovery of a known analytical spike is 97.3% with a standard deviation of 6.4% which is very similar to that observed for ^{226}Ra and ^{228}Ra and is well within the allowable range of +/- 25 percent. The upper and lower warning and control limits are set at 2 and 3 sigma, respectively. The total number of spike analyses is also 86 with a maximum value of 119% and a minimum value of 84% of the expected recovery.

Results demonstrated in Table 1 and 2 confirm that the methods utilized in this study provide results that are well within the precision and accuracy requirements of the EPA CLP program. These results confirm that this methodology is capable of providing more than adequate precision and accuracy for measurement of these nuclides in produced water samples. Finally, this study affirms that if a proper quality assurance/quality control program is employed, these methodologies are capable of generating defensible data.

CONCLUSIONS

The results of this study of accuracy and precision demonstrate that the methodology employed for the analysis of ^{226}Ra, ^{228}Ra and ^{210}Pb are effective in providing precise and accurate data for measurement of these nuclides in produced water samples.

By maintaining this level of quality assurance, the analytical data produced can be evaluated at the laboratory level and non-conformances identified and corrected before the data is reported.

This information can be provided as documentation of the effectiveness of these methodologies and is the basis for providing defensible data that will survive the strictest validation scrutiny.

These methods are not the only methods available and there may be other approaches with equal merit. The main issue is, that whatever technologies are employed for such measurements, it is essential that a proper quality assurance/quality control program be employed that will demonstrate the accuracy and precision of the measurements beyond all scrutiny. This is essential if a commercial or in-house laboratory is to provide defendable data.

REFERENCES

American Public Health Association, 1985, Standard Methods for Measurement of Radioactivity in Drinking Water. American Public Health Association, 1015 15th Street NW, Washington, D.C. 20005.

Blais, J.S., and Marshall, W.D., 1988, Determination of ^{210}Pb in admixture and bismuth-210 and polonium-210 in quenched samples by liquid scintillation counting.

EPA, 1980, Prescribed Methods for Measurement of Radioactivity in Drinking Water. EPA-600/4-80-032, pp. 31 and 38.

NCRP Report No. 58. A Handbook of Radioactivity Measurements Procedures. National Council on Radiation Protection and Procedures. 7910 Woodmont Avenue, Bethesda, MD 20814.

Percival, D.R., and Martin, D.B., 1974, In: Analytical Chemistry 46, 1742. Ed. by H.L. Krieger and E. Whittaker, EMSL, August 1980, EPA.

RADIUM FATE AND OIL REMOVAL FOR DISCHARGED PRODUCED SAND

T.M. Randolph[1], R.C. Ayers, Jr.[2], R.A. Shaul[3], A.D. Hart [3], W.T. Shebs[4], J.P. Ray[5], S.A. Savant-Malhiet[6], and R.V. Rivera[6]

[1] Shell Offshore Inc.
New Orleans, Louisanna
USA

[2] Exxon Production Research Co.
Houston, Texas
USA

[3] Continental Shelf Associates
Jupiter, Florida
USA

[4] Consultant
Houston, Texas
USA

[5] Shell Oil Co.
Houston, Texas
USA

[6] Shell Development Co.
Houston, Texas
USA

INTRODUCTION

Study Objectives

The Minerals Management Service (MMS) and the Environmental Protection Agency (EPA) have regulatory responsibilities for waste disposal and discharge in U.S. waters and are developing policies concerning radioactivity. These federal agencies need additional data on oil and radium content and their fate from produced sand discharges from offshore platforms for policy and regulatory development. To meet both industry and regulatory agency information needs, a monitored sand washing and discharge was conducted with three primary objectives:

1. Determine if produced sand with low levels of associated radium increased sediment radioactivity after discharge at the West Delta 103B oil production platform in the Gulf of Mexico.
2. Determine the performance of the sand washing system in reducing oil content and eliminating free oil from the discharged sand.
3. Examine existing data on Gulf of Mexico sediment radionuclide levels and data from this study to assess influence of produced sand and produced water discharges.

Produced Sand Characteristics

Produced sand is defined by the EPA as "sands and other solids removed from the produced waters" (EPA Region 6, 1986). Produced sand is generated in highly variable amounts during hydrocarbon production. In the unconsolidated sands typical of much of the Gulf of Mexico, sand control is accomplished by downhole placement of gravel packs which filter the liquid flow to retain the sand. Failures of sand control devices occur and may cause large volumes of

Produced Water, Edited by J.P. Ray
and F.R. Engelhart, Plenum Press, New York, 1992

sand to enter surface separation equipment. Most of the sand reaching the surface settles out in equipment and must be periodically removed during shutdowns or by blowdown from low point drains while operating. Small amounts of sand may remain in suspension in discharged produced water and other small quantities may remain with the crude oil as basic sediment and water. The Offshore Operators Committee (OOC)[1] determined that about 28,000 barrels[2] of produced sand (68% of the total generated) was discharged to the ocean annually with the remainder hauled to shore. An extrapolation of these results indicated that about 170,000 barrels of produced sand could be generated annually in the Gulf of Mexico and adjacent coastal production areas (OOC Letter, August 1991).

Produced sand may contain Naturally Occurring Radioactive Material (NORM). In the Gulf of Mexico, NORM is usually associated with barium sulfate scale. The radionuclides are radium-226 ($t_{1/2}$ = 1600y), radium-228 ($t_{1/2}$ = 5.75y), and their daughters, such as radioactive lead: lead-210 ($t_{1/2}$=22.3y). Radium is soluble under downhole conditions and may be brought to the surface in the water phase (Snavely, 1989) where precipitation is possible. Radium may also reach the surface associated with formation solids not retained by the gravel pack. Produced sand that settles in offshore equipment is oily and salty but does not always contain NORM.

Regulatory Background

Current permits authorize discharge of produced sand subject to the no free oil limit (i.e., no sheen allowed on the receiving waters as a result of the discharge). The EPA is considering changes (EPA Office of Water, 1991) and has requested information on the characteristics and volumes of sand generated during offshore activities to make a final decision on the discharge limit for produced sand. The MMS is responsible for leasing mineral rights underlying the Federal waters and thus has broad authority regarding pollutants. MMS advised lessees in November 1990 that produced sand discharges would not continue to receive approval pending NORM study (MMS, 1990). Several operators were forced to terminate discharges in progress when this policy was adopted.

In response to the EPA request for additional data on offshore discharges and the MMS need for NORM data, Shell Offshore Incorporated (SOI) obtained MMS approval with concurrence from EPA to conduct a monitored discharge of produced sand at West Delta 103B (WD 103B).

PROJECT DESCRIPTION

Sand washing and discharge took place for six days in February 1991 at WD 103B. SOI contracted with Mud Control Equipment (MCE) for sand washing and with Continental Shelf Associates (CSA) for monitoring. Sample materials from other projects were analyzed and compiled with published radionuclide results to provide a description of regional radionuclide levels in seafloor surface sediments in the Northern Gulf of Mexico. Analytical work was done by Shell Development Company, Bellaire Research Center (BRC) and through contracts with Core Laboratories in Casper, Wyoming, and Lafayette, Louisiana, and NUS Corporation in Clear Lake City, Texas. Exxon Production Research Company used the OOC dispersion computer model to predict sand deposition on the seafloor.

Study Site

The monitored discharge occurred on the B platform in West Delta 103 lease block located in 69 m of water about 21 km from the Southwest Pass of the Mississippi River (Fig. 1). WD 103B is one of five platforms in the West Delta 105 oil field which began production in 1965 and currently has 27 producing wells. WD 103B is unmanned and has no continuous discharges. The platform is an eight pile structure with twelve 50 cm diameter well conductors and three

---

[1] A 95 member organization which provides assistance to regulatory agencies in offshore oil and gas activities.

[2] 42 U.S. gallon = 0.159m^3

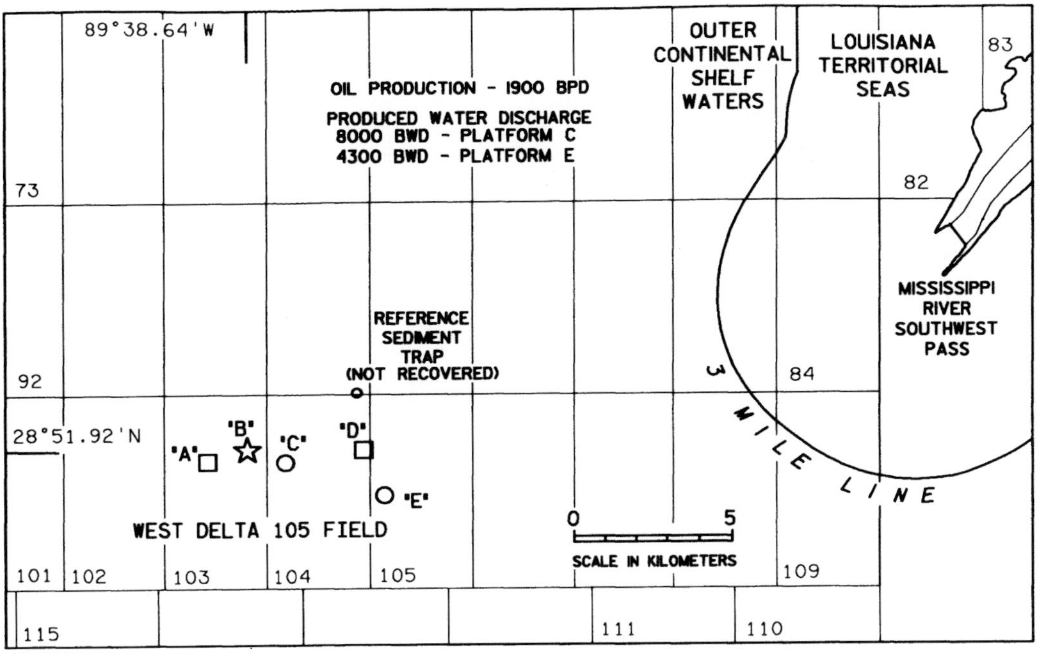

Figure 1. West Delta 103B sand discharge location showing platforms A-E in the West Delta 105 field and the surrounding lease blocks in the West Delta area of the Gulf of Mexico OCS.

pipeline risers. No sand washing discharges are known to have occurred in this field prior to the test discharge.

During the discharge period, tidally driven water movements with greenish coastal water were observed in the mornings with browner Mississippi River water present after tidal front passage. Sea state was calm (0.6-1.2m) and intermittent fog occurred during the discharge period. Two cold fronts generated wave heights approaching 3.6 m after discharge was complete but before recovery of all sediment samples.

Near-surface currents (9.8 m depth) ranged up to 49 cm/s (day 1) during the discharge period and the mid-depth currents (21.3 m and 42.7 m) ranged up to 22 cm/s (day 3). Hourly readings of near-bottom currents (67.1 m) were a maximum of 12.2 cm/s (day 3) (CSA, 1991).

Sand Washing System

The sand discharged at WD 103B was removed from facilities at another platform, Mississippi Canyon 194 (MC 194), by using the MCE system to wash the sand with sea water from the MC 194 equipment into ten 15-bbl cutting boxes. This reduced the volume from an estimated 400-600 barrels to 144 barrels and also removed much of the *in situ* oil. The two step procedure allowed discharge washing to be conducted away from the congested work area of the generating platform and recovered some of the oil contained in the sand during the removal step. The MCE Solids Decontamination Unit consists of solids control equipment components used to condition drilling fluids and was larger than can be accommodated at some offshore platforms. The heaviest component of the MCE system at WD 103B was 9,090 kg, which was within the capacity of the platform crane.

The washing system flow scheme as used for discharge at WD 103B is given in Figure 2 with sample location codes that are used in Tables 4 and 5. The system used a fresh water and biodegradable surfactant solution which was circulated to each cuttings box to wash the solids to a pump suction hose for transfer into the Feed Tank. Solids were rejected from the system at

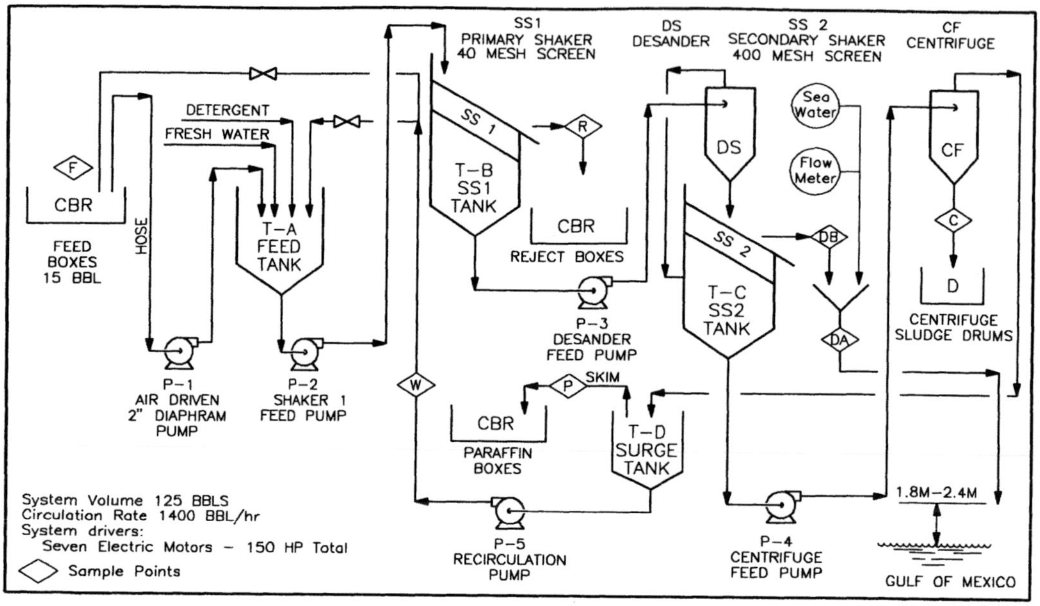

Figure 2. Flow schematic Mud Control Equipment Solids Decontamination Unit at WD 103B.

each of the two vibrating screen shakers and at the centrifuge. The sand from the secondary shaker was the only material discharged after being suspended in sea water. The paraffin, which was skimmed from the wash water, the solids from the primary shaker and the sludge from the centrifuge were sent onshore for storage until acceptable NORM disposal facilities become available. The wash water (115 bbl) containing some of the removed oil was hauled to shore for disposal in a permitted injection well.

Modeling Approach and WD 103B Sample Station Selection

The Offshore Operators Committee (OOC) Dispersion Model was developed by Brandsma Engineering and Exxon Production Research Company (EPR) with funding provided by the OOC and EPR (Brandsma et al, 1980; Brandsma and Sauer, 1983a). The Model has been verified for drilling mud discharges (O'Reilly et al, 1989; and Brandsma and Sauer 1983b). The Model was used in this study to predict the pattern of sand deposition in order to increase the possibility of sampling representative portions of the sand that was actually discharged.

The OOC Model requires information on ocean currents and ambient water column density stratification, discharge rate, material compositions, particle settling velocity data, as well as the diameter, depth, and orientation of the discharge pipe. Results from simulations using current and water column density data obtained at the site prior to discharge were used to locate pre-discharge box core sites and the sediment traps. Results of the simulations with input data from the six day discharge were used as a guide for the post-discharge discretionary box core sampling.

A simulation was carried out for each discharge day with a duration four hours longer than that of the discharge time to estimate the deposition of washed sand on the seafloor. The additional four hours allowed settling time for the sand remaining in the water column after cessation of discharge. (Refer to Table 1 for Model input.) In each simulation, currents were updated every hour. The water column density profiles were relatively constant over the test period and one density-depth profile was used for all the simulations. The simulation results

Table 1. WD 103B daily discharge model input data. The remainder of the discharged material was seawater used to convey the washed sand

1991 Date	Discharge Rate (bbl/hr)	Discharge Duration (h)	Sand weight (kg)	Slurry Density (g/cc)	Weight % Sand
Feb. 7	53.7	8.5	3,008	1.05	3.9
Feb. 8	54.0	9.4	4,051	1.06	4.8
Feb. 9	53.7	9.9	3,823	1.06	4.4
Feb. 10	61.4	9.4	5,281	1.07	5.5
Feb. 11	61.0	7.8	3,235	1.06	4.1
Feb. 12	60.2	6.0	455	1.04	0.8

from the Model were processed using a SAS 5.18 program to calculate weight of sand deposited on a seafloor grid.

Field Sampling

Significant effort was made to obtain representative samples and data at WD 103B. Environmental conditions were monitored every hour during discharge including water column parameters using a Sea Bird Conductivity, Temperature and Depth System (CTD). A primary current meter array was deployed with four InterOceans Model S-4 meters to provide a continuous record of current speed and direction.

Locations of the 10.2 cm diameter discharge pipe, the box core sites, and the sediment traps are shown in Figure 3. The discharge point was 19.8 m from the southwest corner of the platform. This location extended beyond the platform footprint so that box core sampling could be used to avoid the cost of divers for sampling within the subsurface steel work, well conductors, and pipeline risers.

Surface sediment samples were taken prior to discharge at 20 sites and after discharge at 45 sites. Collection was done using a Gray-O'Hara 0.0625 m^2 box core. Each acceptable box core was subsampled using a 10.2 cm internal diameter (ID) teflon tube to obtain three subsamples of the top 2 cm. These samples were used for radionuclide and oil and grease analysis. Other subsamples were removed from each box core and were arhived.

Five sediment traps were deployed near the platform and a sixth, which could not be recovered, was located 3.7 km to the northeast (Fig. 1). Deployment was within 48 hours before beginning discharge and recovery was completed within 24 hours of ending discharge. Each trap was designed to "over trap", ensuring that particles were not resuspended after entering the trap. The trap was designed with an aspect ratio (height/opening diameter) that was adequate for particle retention for the near-bottom current speeds anticipated in the area. The mouth of the trap was about 2 m above the seafloor and the trap was constructed of cylindrical schedule-40 polyvinylchloride (PVC) pipe with 15 cm diameter and 120 cm height (aspect ratio = 8.0). Based on Lau (1979) near bottom currents exceeding 12.5 cm/s would be necessary to resuspend trapped particles. As an additional measure to prevent resuspension out of a trap, a baffle of similar material, consisting of 1 cm spaced grids, was placed in the mouth of the trap. Previous studies indicate that these traps will perform efficiently in terms of particle collection (Hargrave and Burns, 1979; Lau; 1979, Gardner, 1980a,b; Lorenzen et al., 1981).

In order to determine the efficiency of the washing system, unwashed sand sampling was accomplished by coring the full sand depth at five places in each cuttings box with a 3.1 cm ID steel pipe and then compositing this material. The weight of unwashed sand or feed material was determined by lifting each box using a strain gage assembly in the rigging from the WD 103B platform crane. One empty box was weighed to provide a tare weight. The rejected and

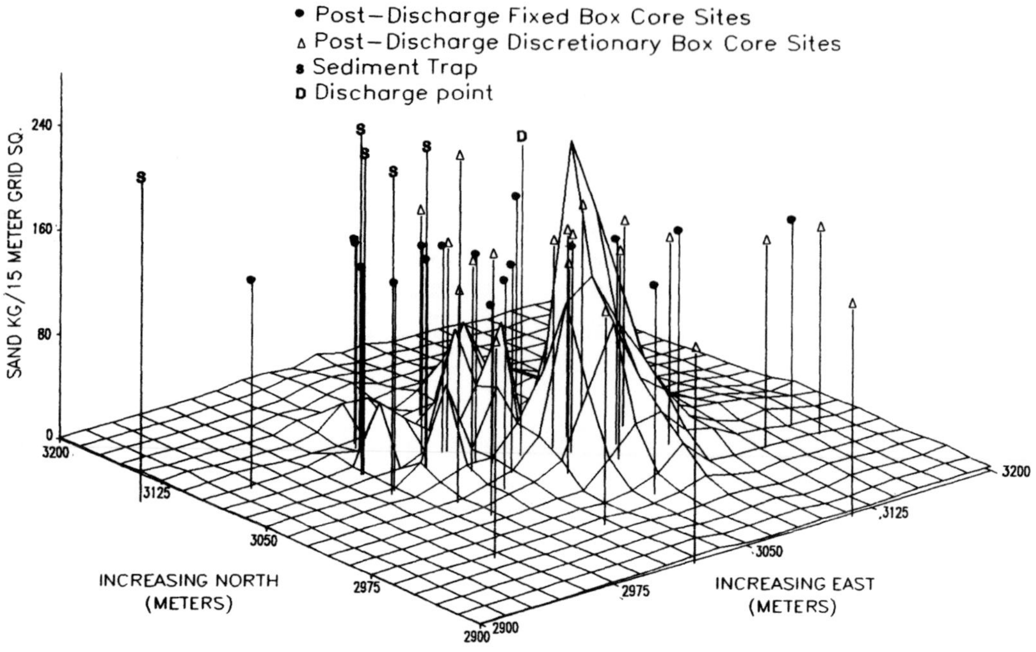

Figure 3. Model hindcast sand distribution on the seafloor at West Delta 103B.

the discharged solids and the wash water were frequently sampled to obtain a representative set of composite samples for analysis.

Regional Sediment Radionuclide Data Set

Some produced water discharges may have very low concentrations of ^{226}Ra, ^{228}Ra, and ^{210}Pb (Snavely, 1989). These radionuclides may also be found in produced sand as discussed previously. Concerns have been expressed about possible elevated NORM levels in sediments near these discharges and this study investigated the influence of local discharges on regional trends. A compilation of surface sediment results was made for the three radionuclides. The corresponding sample sites were added to the MMS 1988 Gulf of Mexico platform map maintained by SOI to determine distances between sites and platforms. A site was classified as remote where it was greater than 1830 m from the nearest production platform or otherwise it was called "nearby". Published data for ^{210}Pb from two studies were included (Rabalais et al, 1991; Boothe and Presley, 1985), as were ten sets of radium data from one study (Rotter, 1985). The geographical extent was from the North Padre Island (NPI) area near Corpus Christi, Texas to the Mississippi Canyon area south of the Mississippi River delta. Included in the data were sample site water depth, analytical methods employed, and grain size analysis where available.

Most of the input to the data set came from LUMCON[3], which collected box core samples of the surface sediments in the Gulf of Mexico during two series of cruises in 1990 and July 1991 (Rabalais, 1991). LUMCON analyzed twelve of the 1990 samples for ^{210}Pb and BRC analyzed 25 of the samples from both years for the radium isotopes. These samples were taken mainly in less than 20 m water depth.

The operators of platforms within 1830 m of the sample sites furnished produced water discharge rates and recent produced sand discharge history in response to a November 1991

[3] Louisiana University Marine Consortium; Chauvin, LA 70314

telephone survey. The produced water data were compared to MMS records. No sand washing discharges had recently occurred at these platforms. At one site, NPI 975, a platform was installed about 5 years after the sediment was sampled and a second platform, South Timbalier 52B, was removed after sampling.

Analytical Procedures

Sample control was accomplished by Bellaire Research Center (BRC). Over 600 samples were taken at WD 103B and sent to BRC, plus over 70 sediment samples provided by LUMCON from the other project sites. Appropriate samples were composited for subsequent analysis. Samples were shipped to NUS Corporation for oil and grease determination, Core Laboratories, Casper, Wyoming, for radionuclide analysis, and Core Laboratories, Lafayette, Louisiana for discharge particle size determination. BRC radiochemistry lab performed gamma spectroscopy, quality control, and archived samples.

Determination of radium isotopes by gamma ray spectroscopy (GRS) was made by comparison with standard samples of radium in a sand matrix. Sample material weighing 20-40 grams was dried, placed in a tared plastic Petri dish, and weighed. The dish was then placed in a 4 oz metal soil can for counting by a high purity germanium crystal connected to an ORTEC multichannel analyzer. Each sample was counted for six hours or more. Gamma rays absorbed within the detector were converted to a signal which was proportional to the energy absorbed. The sample gamma peaks were displayed as an energy versus frequency histogram superimposed on a background and Compton continuum. Calibration was accomplished by mixing standard solutions of each isotope in a sand of negligible radium content. The ^{226}Ra standard was traceable to NIST standards and the ^{228}Ra standard was made from thorium (^{232}Th) nitrate at greater than 97% secular equilibrium with daughter isotopes ^{228}Th and ^{228}Ra. Each standard contains 30-50 nanocuries of the parent and was counted for about two hours for calibration.

Activity was calculated with correction for background gamma from the number of counts in three energy windows or regions of interest (ROI) set up for each isotope to minimize cross interference. The equation giving activity in pCi/g is below and the reported values for each isotope are the averages from the three peaks which are the most sensitive as determined from the three largest values of $F_{c,r}$.

$$Activity = \frac{C_{s,r} \times 1000}{t_s \times F_{c,r} \times W_s}$$

$C_{s,r}$ = sample counts accumulated in ROI r
t_s = sample counting time in hours,

$F_{c,r}$ = calibration factor, ROI r, counts/nCi/hr,

1000 = conversion factor nanocuries to picocuries,
W_s = weight of sample in grams.

Statistical determination of lower detection limit (LLD) and 95% confidence limits followed Sumerlin and Darby (1981). Since radioactive decay is a random process, each decay takes place independently and follows Poisson statistics. The standard deviation was taken as the square root of the number of counts accumulated corrected for the continuum underlying the spectrum. The LLD for ^{226}Ra in low-activity samples, such as the sediments, is 0.1 to 0.3 pCi/g, and for ^{228}Ra, 0.2 to 0.4 pCi/g with the 95% confidence limits ranging from 0.1 to 0.6 pCi/g.

Gamma ray spectroscopy (GRS) was used for determining ^{226}Ra and ^{228}Ra for all samples which contained significant amounts of barium. Radiochemical separation requires less costly instrumentation and permits many samples to be processed simultaneously, but gives lower radium estimates in the presence of barium because the radium is incorporated in the barium scale. This can be quantitatively assayed only by dissolving the $BaSO_4$, requiring fusion of $BaSO_4$, a difficult process. All gamma rays present will be detected by GRS. This is no disadvantage for samples originating from oilfield NORM, since the only activity present is from ^{226}Ra, ^{228}Ra and their daughters (Snavely, 1989). The germanium detector was used to obtain the resolution needed to quantify the radium isotopes. Scintillation detectors, such as NaI crystal detectors have insufficient resolution to separate the peaks in the rich and complex spectrum that the NORM samples generate.

Determination of ^{210}Pb was carried out by chemical separation of lead from the sample, followed by beta particle counting of the ^{210}Bi daughter to quantitate the ^{210}Pb (EPA, 1984). Radiochemistry was also used to determine radium in the sediment traps (Percival, 1974). Oil and grease was determined by Method 503 D for solid samples and by Method 503 C for water samples (APHA, 1985).

Quality practices were followed in the laboratory analysis. Quality assurance samples were prepared by mixing small amounts of previously analyzed barium sulfate scales with produced sand, and submitting duplicate samples of these materials as blind samples. In addition, aliquots of a standard solution of ^{226}Ra were added to sand and analyzed. The ratio of Radium Found/Radium Added for ^{226}Ra ranged from 1.07 to 1.31. For ^{228}Ra, the ratio ranged from 1.30 to 1.49. The cause of these high ratios is not known. Quality control was carried out by submission of blind replicates for approximately 10% of the samples. The coefficient of variation ($100 \cdot 2\sigma$/mean value) for these replicates ranged from about 4% to 33%, with the average being about 15% for ^{226}Ra. Comparable procedures were used for the oil and grease analysis, with the coefficient of variation ranging from 9% to 220% while the average was 80%.

RESULTS AND DISCUSSION

OOC Dispersion Model Output

Figure 3 shows that most of the post-discharge sampling sites are within the depositional area hindcast by the Model runs. The final Model results indicated that significant sand accumulated under the platform where sediment sampling was not done since divers were not used. There was no correlation between ^{226}Ra, ^{228}Ra, or ^{210}Pb activity in sediment samples taken from a grid block and the calculated quantity of sand deposited in that grid block (Fig. 4). This lack of correlation is not surprising since there was no statistically significant difference between results of the pre-discharge sediment samples and results of the sediment samples acquired after discharge (Table 2).

Analytical Results

Gamma ray spectroscopy analysis requires high resolution detectors and calibration by standards in a matrix comparable to the sample at the low radium activity levels found in produced sand. Density and weight differences between the sample and standard have been found to cause errors in the calculated amount of radium by as much as 30%, due to self-absorption of gamma rays within the sample. Therefore, standards should be made up in a matrix that matches the samples.

The results of the sediment analyses are summarized in Table 2. The maximum values pre-discharge were 0.6 pCi/g for ^{226}Ra and 0.9 pCi/g for ^{228}Ra compared to 2.6 pCi/g and 2.2 pCi/g after discharge. All the after discharge results for ^{226}Ra are shown on Figure 4.

The five sediment traps deployed near the discharge point were recovered but the reference trap could not be located. The trap nearest the discharge point caught the largest accumulation of material with an activity of 12.5 pCi/g ^{226}Ra, compared to 37 pCi/g in the discharged sand (refer to Tables 3 and 5.). The maximum hourly average near bottom current was 12.2 cm/s which was less than the 12.5 cm/s velocity needed to resuspend (Lau, 1979). Approximately 98% of the near bottom currents observed were less than 10 cm/s indicating that the traps retained the discharged sand.

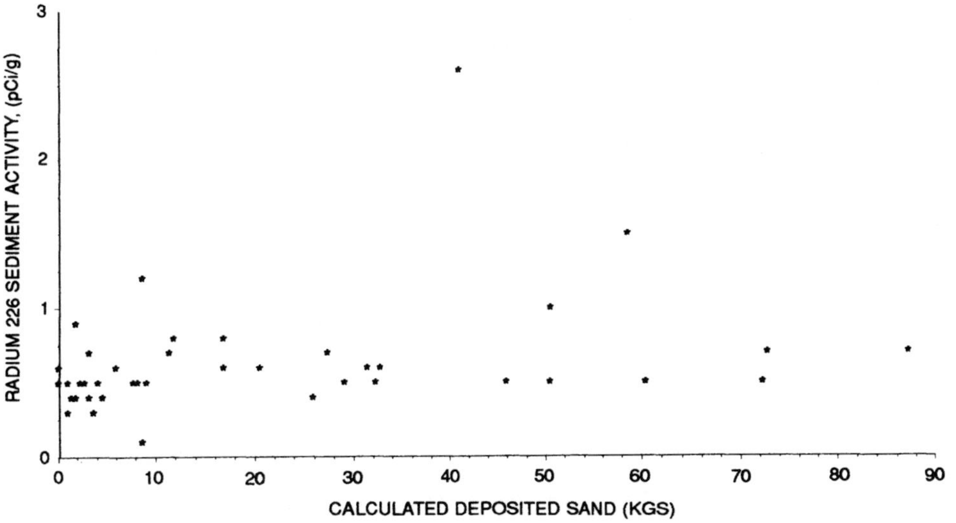

Figure 4. Radium 226 sediment activity vs calculated sand seposition.

Analysis of the wash water and detergent averaged 0.1 pCi/g dissolved ^{226}Ra and 120 mg/l oil and grease in the initial solution. The oil and grease increased to 2020 mg/l at the end of the test, but the dissolved radionuclides were essentially unchanged indicating that the radionuclides were not leached from the sand during the washing. The suspended solids activity was in the range of 1.1 to 19 pCi/l for the two radium isotopes. Assuming these fines are represented by the reject solids (see Solids R in Table 5), these values represent 0.01 to 0.2 g of suspended fines per liter of wash water showing that the buildup of suspended fines was small. The total mass of the suspended fines in the 115 barrels of wash water amounted to less than 4 Kg.

Material Balance

Using the sample codes in Figure 2, the quantities and concentrations for the various streams are given in Tables 4 and 5, respectively. The total activity of the radium isotopes was 1.818 mCi (specific activity: 40 pCi/g ^{226}Ra and 38 pCi/g ^{228}Ra) in the feed and 1.385 mCi (specific activity: 37 pCi/g ^{226}Ra and 35 pCi/g ^{228}Ra) in the discharge. The discharged radium was associated with 19,170 kg of washed sand that contained 1.6 wt% oil and grease (O&G) indicating that the sand washing removed about 60% of the oil and grease associated with the feed material. The material not suitable for discharge included the reject solids (total activity of

Table 2. Average analytical results for sediments at WD 103B platform. Twenty stations were sampled pre-discharge and post-discharge. Discretionary samples were acquired only during post-discharge sampling. SD = 2 σ

	Sediment Activity pCi/g						Oil & Gas	
	226Ra		^{228}Ra		^{210}Pb		% wt.	
Sample Type	**Av**	**SD**	**Av**	**SD**	**Av**	**SD**	**Av**	**SD**
Pre-Discharge (N=20)	0.5	0.15	0.6	0.30	3.9	2.5	0.06	0.07
Post-Discharge (N=20)	0.7	1.01	0.8	0.79	3.0	1.5	0.09	0.13
Discretionary (N=25)	0.6	0.48	0.7	0.44	4.3	1.9	0.04	0.03

0.189 mCi radium and 7.0 wt% O&G), paraffin (22.9 wt% O&G) and centrifuge sludge (total activity of 0.245 mCi radium and 5.4 wt% O&G). The wash water contained 35 kg of oil and grease which is tabulated in the paraffin total in Table 4.

Table 3. Sediment trap sample results at WD 103B. Trap location is given by the compass bearing and the distance in meters from the discharge point

			Activity, pCi/g		
Trap	Location	Weight, g	^{226}Ra	^{228}Ra	^{210}Pb
A	290° - 46	11.217	12.5	5.1	6.8
B	310° - 76	9.973	0.2	0.0	6.0
C	290° - 76	8.004	1.7	0.0	5.5
D	270° - 76	7.513	1.4	0.0	4.6
E	290° - 183	7.904	1.8	3.0	4.

Table 4. Material balance for WD 103B sand washing (before sea water addition)

	Feed F	Solids R	Paraffin P	Centrifuge C	Discharge DB
^{226}Ra mCi	0.931	0.099	0.00	0.126	0.707
^{228}Ra mCi	0.887	0.090	0.00	0.119	0.678
O&G Kg	970	95	284	221	370
Solids Kg	23201	889	0	3142	19170
Water Kg	5870	377	840	723	3930

Table 5. Material concentrations for WD 103B sand washing

	Feed F	Solids R	Centrifuge C	Discharge DB
^{226}Ra pCi/g	40	111	40	37
^{228}Ra pCi/g	38	101	38	35
O&G wt %	3.2	7.0	5.4	1.6
Solids wt %	77.3	65.3	76.9	81.7
Water wt %	19.5	27.7	17.7	16.7

Regional Sediment Radionuclide Examination

The results of the compilation of the 35 observation Gulf-wide ^{226}Ra data were classified in two ways. The first classification examined the 14 sites that were at a distance less than 1830 m from five platforms with produced water discharges at rates ranging from 300 to 20,000 barrels of water per day (BWD). The mean ^{226}Ra value of this group of 14 sites was compared to the value for the 21 remote sites not potentially influenced by nearby produced water discharge. These results are shown in Table 6 with the pre-discharge sediment data from WD 103B.

A statistical, nonparametric test of these data found no significant difference between mean ^{226}Ra measurements for the two groups, but cannot say that the results are identical. The Mann-Whitney test obtained a p-value of 0.360; a value greater than 0.05 indicates that the radium mean of the nearby group is not significantly different than that of the remote group. This test does not rely on assumption or evidence of data distribution.

Produced water was discharged from WD 103C at 8,000 BWD. This platform is located 1220 m from WD 103B. Results of produced water samples taken from WD 103C for this project are 363 pCi/l ^{226}Ra and 367 pCi/l ^{228}Ra, which is typical of the Gulf of Mexico produced water analysis reported by Stephenson and Supernaw (1990). This results in a discharge of total radium of 0.93 mCi/d. Produced water radionuclide analyses were not available for the five platforms discharging nearby the 14 sample sites. With produced water activity data, a more comprehensive regional model could be undertaken.

Table 6. Gulf of Mexico sediment ^{226}Ra activity (not influenced by sand discharge) relative to the proximity of produced water (PW) discharges. D = distance between sample site and PW discharge pipe

	Activity pCi/g			
Site Location	**Avg**	**Max**	**Variance**	**SD**
WD 103B Pre-Discharge (N=20)	0.5	0.6	0.005	0.15
D < 1830 m from PW Discharges (N=14)	0.76	1.8	0.129	0.36
D > 1830 m from Platforms (N=21)	0.79	1.6	0.086	0.29

Produced sand discharges are not known to have occurred from any nearby platforms for any of these sites except WD 103B, but definitive information is very difficult to obtain on sand discharges that may have occurred prior to recent approval requirements. A second classification and Mann-Whitney test of the 35 observations also found no significant difference (p-value = 0.480) comparing the ^{226}Ra mean for the 16 sites nearby to any type production structure to the mean for the 19 remote sites. If produced sand discharges have occurred nearby these sites, there has not been a noticeable effect on sediment activity.

SUMMARY AND CONCLUSIONS

A monitored sand washing and discharge was conducted at WD 103B to provide data on the fate of radium isotopes and oil and grease associated with produced sand discharges. The field effort was successful except for the lost reference sediment trap which precluded drawing conclusions from the trap data. Representatives from the MMS and the State of Louisiana Department of Environmental Quality made field visits and were satisfied that the discharge operation met regulatory requirements. The washing system was able to remove a majority of the oil and grease associated with the sand.

Analysis of samples from sediment traps, box cores and process streams, as well as data from the Gulf-wide compilation showed that the radium isotopes associated with the sea floor samples after the produced sand discharge were comparable to natural occurring background radium. No correlation was found between sediment radionuclide activity and calculated sand deposition at the WD 103B site.

The conclusions of the monitored sand washing and discharge are as follows:

- Radium levels did not significantly increase in the WD 103B sediments after a discharge of 1.4 millicuries total radium contained in 19,170 kilograms of washed produced sand, probably due to the dispersed nature of the discharge.
- Sand washing achieved a discharge concentration of 1.6 weight percent oil and grease. The removal efficiency was approximately 60% from a prewashed sand of reduced oil content. No visible sheen occurred on the receiving waters during this discharge, and the discharge was in compliance with all regulatory requirements. The discharge did not measurably increase oil and grease in the WD 103B sediments.
- Sediment radium levels do not appear elevated near oil and gas production platforms in the Gulf of Mexico based on 35 samples divided into nearby and remote groups. Fourteen of the sample sites were within 1830 meters of five platforms with produced water discharge.

ACKNOWLEDGMENTS

The authors would like to thank the Minerals Management Service Gulf of Mexico OCS Region and the Environmental Protection Agency for authorizing this monitored discharge and the Offshore Operators Committee for reviewing the study.

In addition, the authors would like to acknowledge the following people for their help: Peter Cook of Exxon Production Research Company, who carried out the OOC model and SAS calculations used in this paper; Joe Raia, Shell Development Company for oil and grease analytical consultation; the SOI West Delta 105 Field Operations Crews, Phil Hale, Production Foreman; Lynn LeBleu and his three man crew who operated the Solids Decontamination Unit furnished by Mud Control Equipment, Shelby Roussell, President, Youngsville, LA; Nancy Rabalais of LUMCON who graciously provided sediment samples from two projects: MMS Cooperative Agreement No. 14-35-001-30470 with LUMCON, and shelf-wide NOAA NECOP MAR program award to N. Rabalais, et al; Martha Scott of Texas A&M who furnished the coordinates for the samples reported by Rotter; Rodney Rester of SOI CAD mapping who mapped the sediment samples; Mike Orlando of SOI who surveyed the operators of platforms nearby sediment sample sites; Joyce Wellman of Shell Development for statistical analysis of the regional data; Mike P. Winzig and Carol Garcia who helped in the sample processing and analysis at BRC; and special thanks to Zeke Austin of SOI who coordinated the field effort and the contractors. We also deeply appreciate the support of Shell and Exxon management.

REFERENCES

American Public Health Association, 1985, "Standard Methods for the Examination of Water and Wastewater", 16th Edition.

Boothe, P. N. and Presley, B. J., 1985, Distribution and Behavior of Drilling Fluids and Cuttings Around Gulf of Mexico Drilling Sites, Published by Environmental Affairs Department, American Petroleum Institute, 105 pp plus appendices, Washington D.C.

Brandsma, M. G., Davis, L. R., Ayers, R. C., Jr., and Sauer, T. C., Jr., 1980, A computer model to predict the short-term fate of drilling discharges in the marine environment, in "Proceedings of Symposium, Research on Environmental Fate and Effects of Drilling Fluids and Cuttings," January 21-24, 1980, Lake Buena Vista, Florida, Volume I, pp. 588-608.

Brandsma, M. G., and Sauer, T. C., Jr., 1983a, Mud Discharge Model Report and User's Guide, Available from Exxon Production Research Co., POB 2189, Houston, Texas 77252-2189.

Brandsma, M. G., and Sauer, T. C., Jr., 1983b, The OOC model: prediction of short-term fate of drilling mud in the ocean. Part 2: model results. Proceedings of MMS Workshop on an Evaluation of Effluent Dispersion and Fate Models for OCS Platforms, February 7-10, 1983, Santa Barbara, California, Minerals Management Service, Contract No. 14-21-0001-29122, pp. 86-106.

Continental Shelf Associates, 1991, Produced Sand Discharge Monitoring Study, West Delta Area Block 103 Platform B Survey Report, CSA, Jupiter, FL.

EPA Eastern Environmental Radiation Facility, 1984, Radiochemical determination of lead-210 in water and solid samples, in: "Radiochemistry Procedures Manual".

EPA Region 6, 1986, Federal Register, July 9, 1986, Gulf of Mexico NPDES Permit GMG-280000, page 24927.

EPA Office of Water, 1991, Federal Register, March 13, 1991, Proposed Rules, Offshore Subcategory Effluent Limitations Guidelines, page 10696.

Gardner, W. D., 1980a, Field assessment of sediment traps, J. Mar. Res., 3:41-52.

Gardner, W. D., 1980b, Sediment trap dynamics and calibration: a laboratory evaluation, J. Mar. Res., 38:17-39.

Hargrave, B. T., and Burns, N. M., 1979, Assessment of sediment trap collection efficiency, Limnol. Oceanogr., 24:1124-1136.

Lau, Y. L., 1979, Laboratory study of cylindrical sedimentation traps, J. Fish. Res. Bd. Can., 36:1288-1291.

Lorenzen, C. J., Shuman,F.R., and Bennett, J. T., 1981, *in situ* calibration of a sediment trap, Limnol. Oceanogr., 26:580-585.

MMS Gulf of Mexico Region Letter to Gulf of Mexico Lessees, Nov. 20, 1990.

OOC letter to Ron Jordan, US EPA Office of Water, August 30, 1991.

O'Reilly, J. E., Sauer, T. C., Jr., Ayers, R. C., Jr., Brandsma, M. G., and Meek, R. P., 1989, Field verification of the OOC mud discharge model, *in* "Drilling Wastes", F. R. Engelhardt, J. P. Ray, and A.H. Gillam, eds., Elsevier Applied Science, 867 pp.

Percival, D. R. and Martin, D. B., 1974, Sequential determination of radium-226, radium-228, actinium-227, and thorium isotopes in environmental and process waste samples, Analytical Chemistry, Vol. 46, No. 12, October 1974, pp 1742-1749.

Rabalais, N. N., McKee, B. A., Reed, D. J., and Means, J. C., 1991, Fate and Effects of Nearshore Discharges of OCS Produced Waters, MMS 91 0004, New Orleans, LA.

Rabalais, N. N., 1991, Personal communications June and August transmitting Gulf of Mexico sediment samples and data from work in progress.

Rotter, R. J., 1985, Pb-210 and Pu-239, 240 in Nearshore Gulf of Mexico Sediments, A Thesis Submitted to Texas A&M Oceanography, College Station, TX.

Snavely, E. Jr., 1989, Radionuclides in Produced Waters, a Literature Review, American Petroleum Institute, API Publication, Washington, DC.

Stephenson, M. T., and Supernaw, I. R., 1990, OOC 44 Platform Study, Radionuclide Analysis Results, Available from OOC, PO Box 50751, New Orleans, LA 70150.

Sumerlin, T. J. and Darby, S. C., 1981, Statistical Aspects of the Interpretation of Counting Experiments Designed to Detect Low Levels of Radioactivity, National Radiological Protection Board, NRPB-R113, Her Majesty's Stationery Office.

PRODUCED WATER RADIONUCLIDES FATE AND EFFECTS

M.M. Mulino and M. F. Rayle

Steimle and Associates, Inc.
Metairie, Louisana
USA

INTRODUCTION

The Louisiana Department of Environmental Quality (LDEQ) has recently proposed and received approval for stringent produced water (PW) effluent criteria. To define ambient produced water radionuclide levels, LDEQ performed a study of four PW discharges in shallow, brackish habitats in southeastern Louisiana (St. Pé, 1990). Three of the four sites sampled showed elevated sediment radium 226 (^{226}Ra) activities as far as 150 m from the discharge. These levels were determined by gamma spectroscopy (GS) rather than by the radon (Rn) emanation method which had been used in other studies of sediments near PW discharges (Hanan, 1981; Steimle & Associates, Inc., 1991).

Hanan (1981) sampled sediments in the vicinity of a single PW discharge located near the mouth of the Mississippi River. His findings showed modest but discernible increases in sediment ^{226}Ra activities within 50 m of the discharge.

Steimle & Associates, Inc. (1991) collected sediment ^{226}Ra and 228 radium (^{228}Ra) samples near six of 36 PW discharges sampled. These samples showed ^{226}Ra activities one to two orders of magnitude lower than the values reported by St. Pé. The sediment ^{226}Ra activities found by Hanan (1981) were similar to those reported by Steimle and Associates, Inc.

The Louisiana Division of the Mid-Continent Oil & Gas Association (LAMOGA) contracted Steimle & Associates, Inc. to repeat the sediment radionuclide sampling of the LDEQ Lirette Field (LF) site. This site was chosen because it had higher sediment ^{226}Ra activities than the rest of the sites sampled by St. Pé (1990). The purpose of this study was to compare ^{226}Ra values obtained by GS and Rn emanation analytical methods.

Subsequently, the API contracted Steimle and Associates, Inc. to collect water, sediment and tissue radionuclide data near two inshore and one offshore PW discharges. The purpose of these studies was to provide sediment radionuclide data in close proximity to the discharges and to provide data for radionuclide risk assessment studies.

METHODS

An inshore sampling site, in addition to the LF site, for the API study representative of a canal environment was Texaco Golden Meadow Tank Battery #3 (GM) with PW discharge ^{226}Ra + ^{228}Ra activities greater than 500 pCi/l. Stations were arranged so that samples could be taken 500 m to 1,000 m away from the discharge point without interference with land or other discharges.

Produced Water, Edited by J.P. Ray
and F.R. Engelhart, Plenum Press, New York, 1992

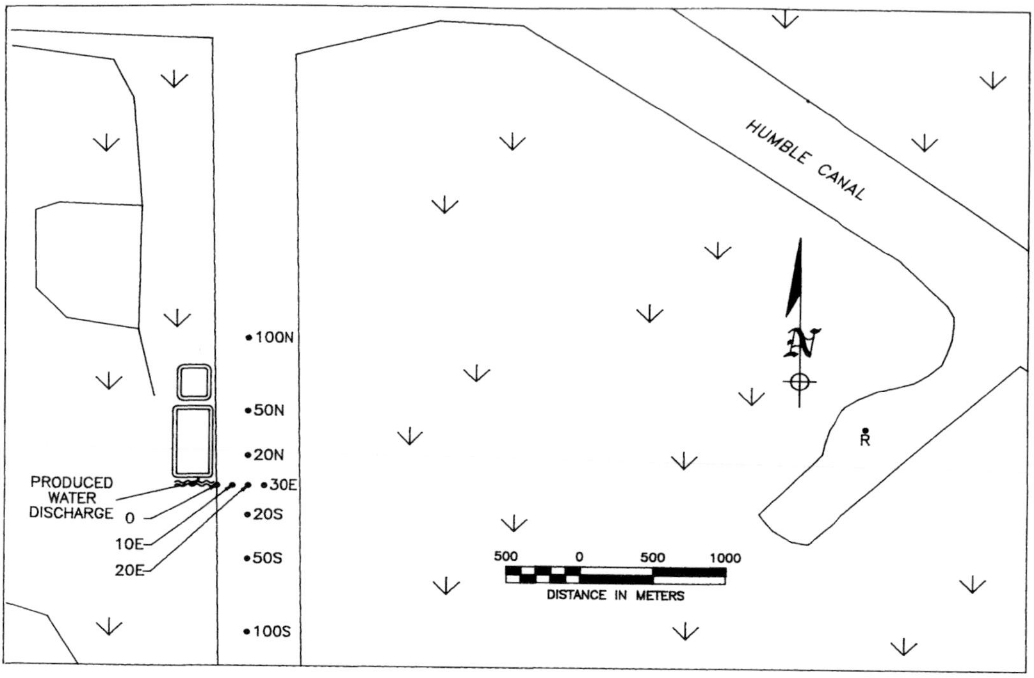

Figure 1. Sediment sample locations at Lirette Field.

The offshore sampling site, Chevron South Timbalier Block 52 Platform C (ST), was chosen because of high ^{226}Ra + ^{228}Ra activity in the PW discharge (greater than 500 pCi/l) and a water depth at the platform was less than 20 m. The platform was located at least 1,000 m away from other PW discharges.

Locations for various samples at the three study sites are shown in Figures 1-3. In addition, bottom water samples were collected offshore at 20 m, 30 m, 50 m north, south, east and west of the ST platform; 100 m north, east and west; and 1,000 m north, east and northwest.

Sediment samples inshore were collected using a pole mounted Wildco liner type corer with a 5 cm diameter and a 50 cm liner length. Offshore sediment samples were taken by subsampling a 0.09 m^2 box corer.

Separate samples of the 0-10 cm depth and 10-20 cm depth core sections were taken at the LF and GM sites and placed in labeled plastic jars. At the @0, 20 N and 20 S Stations at the LF site, a 20-30 cm depth core section was collected from the same core. A 20-30 cm sample was collected at all stations at the GM site. All samples were placed on ice and subsequently frozen.

Sediment radionuclide analyses on LF and GM samples were done by GS performed on a EG&G ADCAM System multi-channel analyzer using EG&G ORTEC MINIGAM software. The samples were counted for at least 6,000 seconds on a 20% high purity, high resolution germanium detector. Sediment samples were calibrated to a NIST-traceable ^{226}Ra standard in sand in a sealed tuna can. Lead 210 (^{210}Pb) was determined on ST sediment samples using a liquid scintillation method developed by Core Laboratories.

Method 903.1 (EPA, 1980) was also used on samples to analyze sediment radionuclide samples for Total ^{226}Ra by radon emanation. During the sample analysis an insoluble barium precipitate formed in some of the samples. Duplicate QA/QC samples produced inconsistent results, indicating that some of the Ra was being trapped in the precipitate as it formed. The

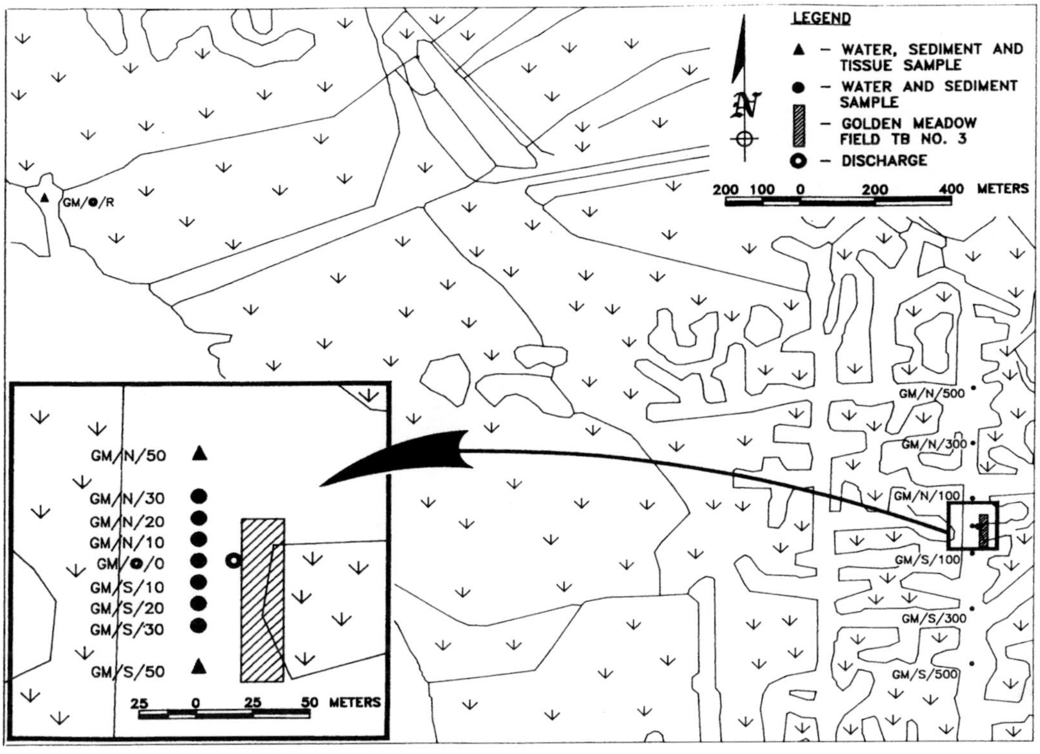

Figure 2. Sediment, water and tissue sample locations at Golden Meadow.

precipitate and solution were redigested, treated with {[(Carboxymethyl) imino] bis(ethylenenitrilo)} tetraacetic acid (DTPA) and centrifuged in a Beckman Model 2000 centrifuge at 2,000 RPM for ten minutes. The supernate was used as the sample for analysis by EPA Method 900.1 (radium, alpha count) (EPA, 1980) to get ^{226}Ra.

All water samples for radionuclide analysis were collected as grab samples in plastic 1 liter containers containing HNO_3 preservative. A separate, unacidified plastic 1 liter sample bottle was used to collect a PW sample for chloride analysis. Bottom water samples were collected with a 3 l Van Dorn bottle.

Biological tissue was retrieved from the offshore location by divers who scraped epifauna from the legs of the platform into a net bag. The bag was returned to the surface and the contents sorted by species. Tissue samples of the blue crab from inshore location were collected using commercial crab traps. Hard parts were separated from soft parts where possible and all tissue was frozen.

The PW end of pipe sample was analyzed for chloride using Method 325.3 (EPA, 1983). All radionuclide water and tissue samples from all locations were analyzed for ^{226}Ra and ^{228}Ra using the Rn emanation method. ^{210}Pb was determined on tissue samples from ST by the liquid scintillation method developed by Core Laboratories. Tissue samples were first digested multiple times with nitric acid. The water and tissue samples were calibrated to National Institute of Standards and Technology (NIST, formerly National Bureau of Standards or NBS) traceable standards and counted in a 500 ml Marinelli beaker.

Laboratory QA/QC procedures follow EPA SW-846 (EPA, 1986).

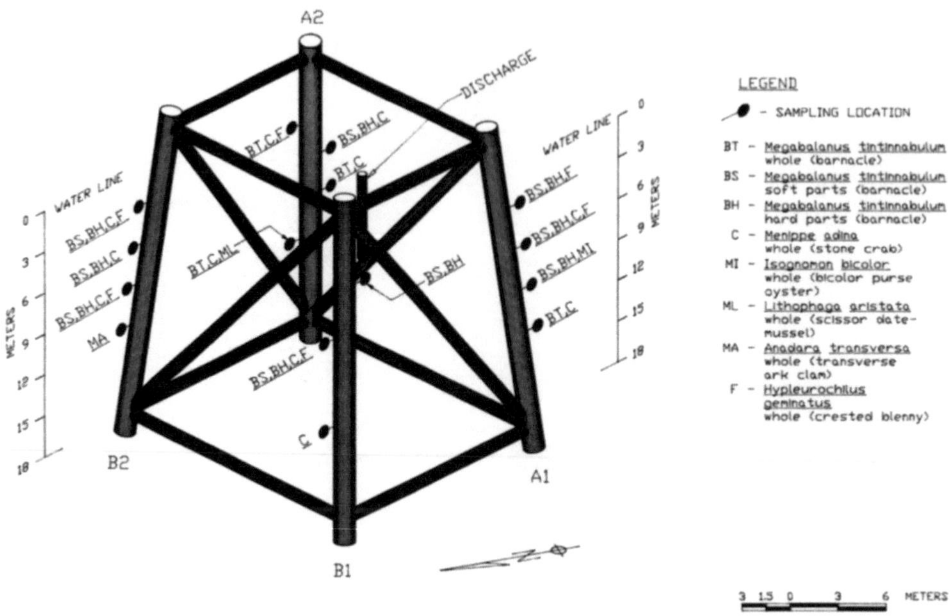

Figure 3. Tissue sample locations at South Timbalier.

RESULTS

Produced Water Samples

Results of analyses of the end of pipe water samples for each of the locations are given in Table 1.

Table 1. Analyses of end of pipe discharge samples

Location	^{226}Ra (pCi/1)	^{228}Ra (pCi/1)	Salinity (ppt)	Volume (bpd)
Lirette Tank Battery #1	477 ± 30	—	134	1,320
Golden Meadow Tank Battery #3	143 ± 6.5	149 æ 5.6	137	3,000
South Timbalier Block 52 Platform C	190 ± 7	198 æ 13	134	16,200

bpd = barrels per day

Sediment Samples

The comparison of methods for analysis of ^{226}Ra and ^{228}Ra in sediments from the LF showed that the Ra emanation method was subject to interference from barium, a common constituent of drill mud and thus the sediments in oil fields. Additionally, the percent recovery of spiked samples was low, 77% and 83%. In contrast, the readings of reference standards by

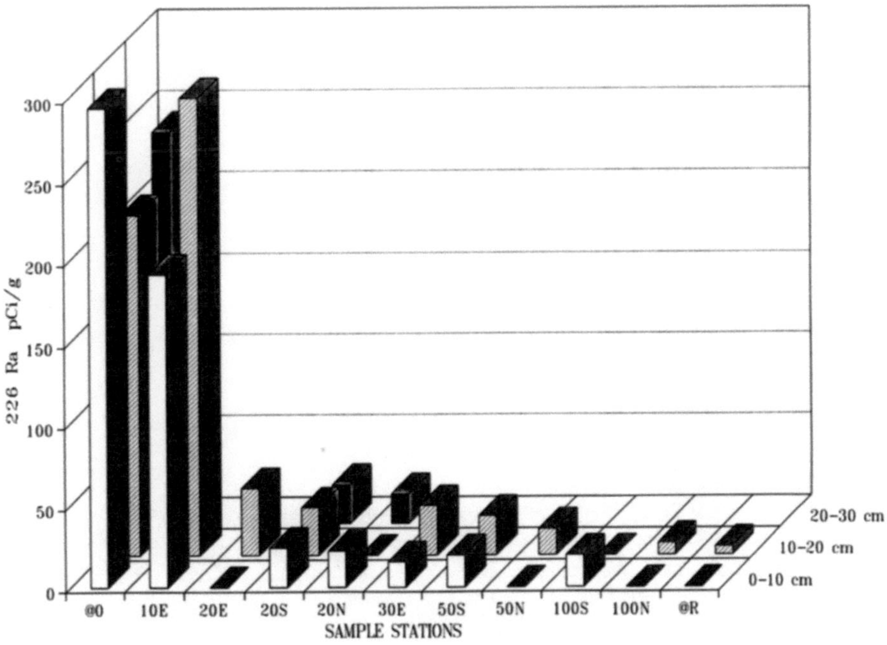

Figure 4. ^{226}Ra **activities in sediment with depth at Lirette Field.**

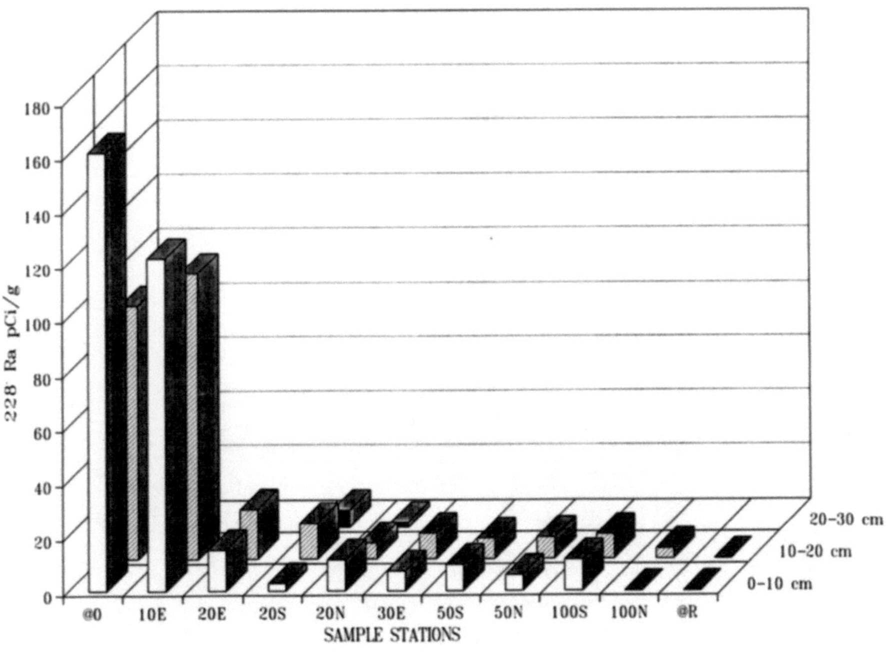

Figure 5. ^{228}Ra **activities in sediment with depth at Lirette Field.**

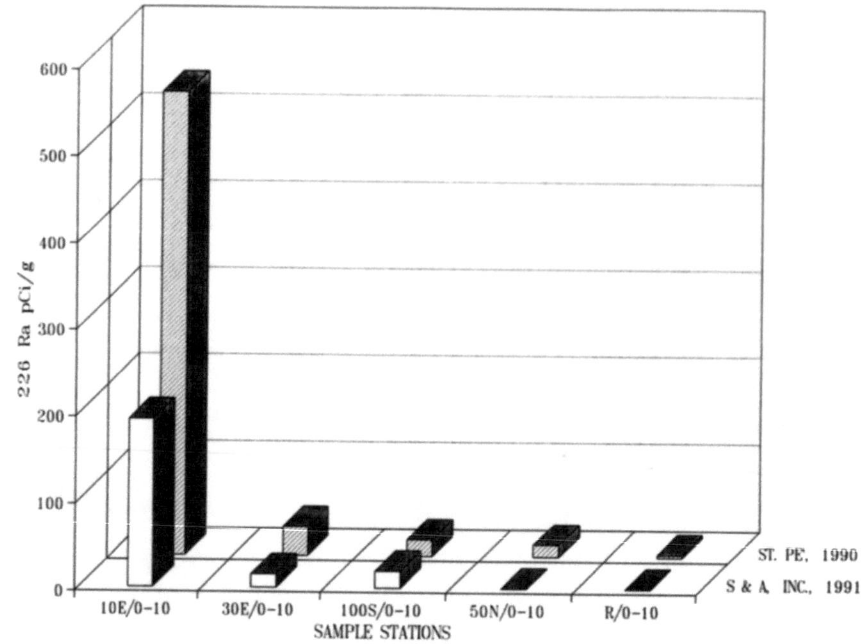

Figure 6. Comparison of ^{226}Ra values from comparable stations in St. Pé (1990) and this study.

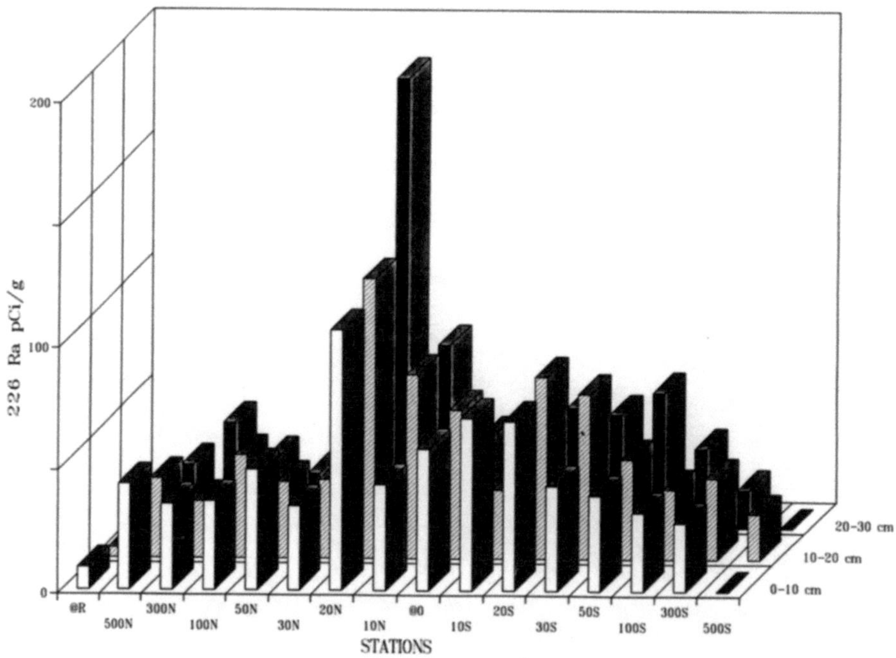

Figure 7. ^{226}Ra activities in sediment with depth at Golden Meadow.

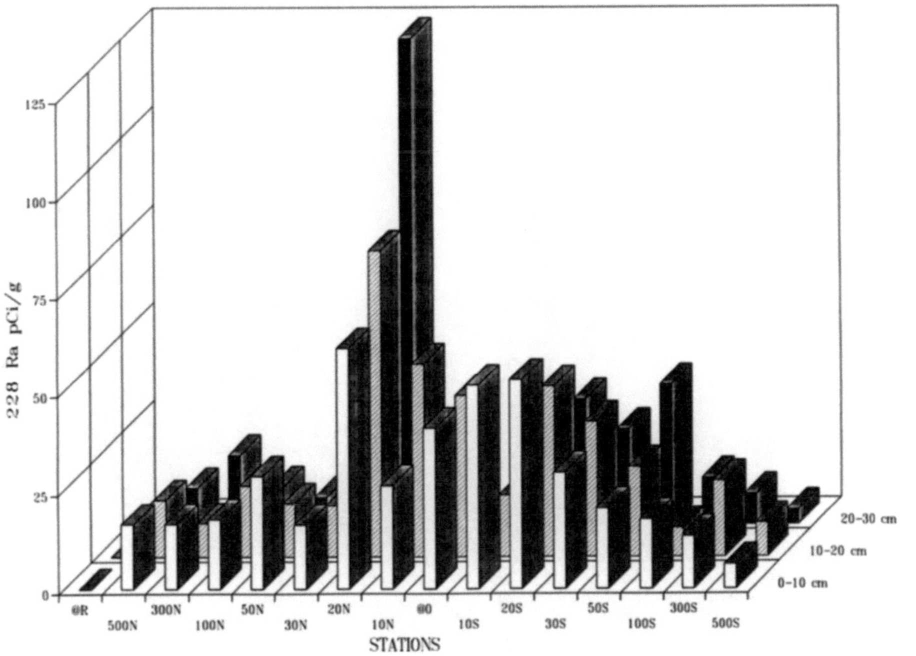

Figure 8. ^{228}Ra **activities in sediment with depth at Golden Meadow.**

GS were 97%, 93%, and 92% of the true value. Consequently, only sediment samples analyzed by GS are reported.

Results of ^{226}Ra and ^{228}Ra analysis on sediment samples from LF Tank Battery #1 with depth are given in Figures 4 and 5, respectively. A comparison of ^{226}Ra values with those of St. Pé (1990) at comparable stations to those in the St. Pé study is given in Figure 6.

^{226}Ra and ^{228}Ra sediment values from the GM site are shown by station and by depth in Figures 7 and 8, respectively. ^{210}Pb analysis was done on the same samples; however, values for all stations and all depths were below 10 pCi/g and are not reported.

Water Samples

Results of ^{226}Ra and ^{228}Ra analysis in surface and bottom water samples at the GM site are given in Figure 9.

Radionuclide analyses of bottom water samples taken at ST showed only two samples which were above the lower limit of detection (LLD). One of these was ^{228}Ra of 2.5 ± 1.1 pCi/l at 100 m west of the platform (LLD = 1.6 pCi/l). The other was a value of ^{226}Ra at 20 S of 0.9 ± 0.5 pCi/l (LLD = 0.5 pCi/l). The LLD of the other samples ranged from 0.3 to 0.4 pCi/l for ^{226}Ra samples and from 1.6 to 5.1 pCi/l for ^{228}Ra samples.

Tissue

Results of *Callinectes sapidus* (blue crab) tissue analyzed for ^{226}Ra, ^{228}Ra and ^{210}Pb from GM are given in Table 2.

Values of ^{226}Ra, ^{228}Ra, and ^{210}Pb for tissue samples from *Megabalanus tintinnabulum* (barnacle), *Menippe adina* (stone crab), *Isognomon bicolor* (bicolor purse oyster), *Lithophaga aristata* (scissor datemussel), *Anadara transversa* (transverse ark clam), and *Hypleurochilus geminatus* (crested blenny) collected offshore at ST are given in Table 3.

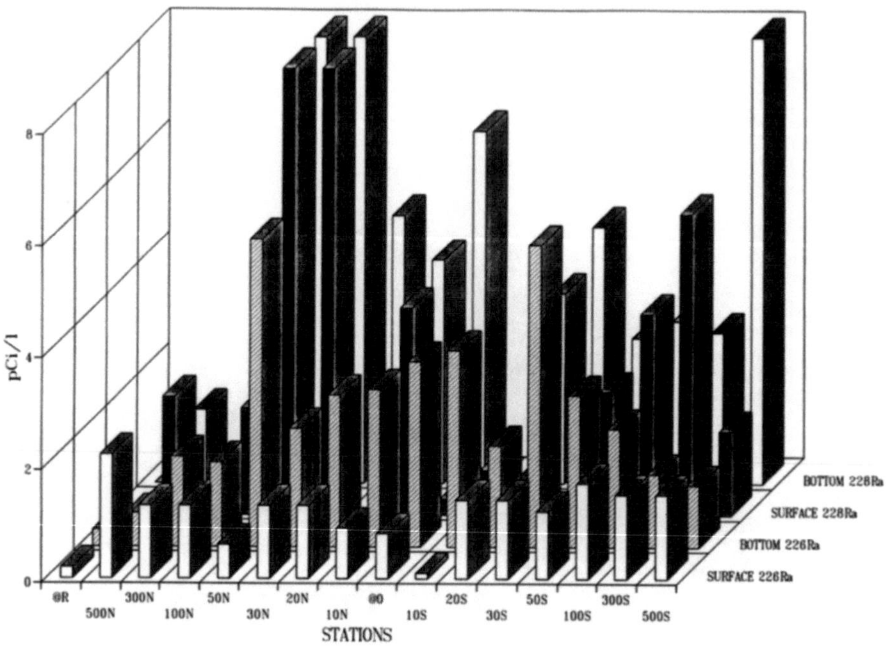

Figure 9. ^{226}Ra and ^{228}Ra values from surface and bottom water samples at Golden Meadow.

Table 2. Radionuclide levels in *Callinectes sapidus* (blue crab) tissue collected at GM

Station	Tissue*	Total ^{226}Ra (pCi/g)	LLD** ^{226}Ra (pCi/g)	Total ^{228}Ra (pCi/g)	LLD** ^{228}Ra (pCi/g)	Total ^{210}Pb (pCi/g)	LLD ^{210}Pb (pCi.g)	Total ^{226}Ra (pCi/g)	Total ^{210}Pb (LLD) pCi/g
Ref.	Soft	<0.02	0.02	ND ***	0.1	<0.06	0.06	<0.06	0.06
Ref.	Hard	2.08 ± 0.23	0.06	0.9 ± 0.2	0.2	<0.22	0.22	<0.22	0.22
50N	Soft	<0.03	0.03	ND	0.1	<0.11	0.11	<0.11	0.11
50N	Hard	1.26 ± 0.19	0.07	1.1 ± 0.2	0.2	<0.24	0.24	<0.24	0.24
50S	Soft	0.03 ± 0.01	0.01	ND	0.1	<0.04	0.04	ND	0.04
50S	Hard	0.81 ± 0.13	0.05	0.3 ± 0.1	0.2	<0.19	0.19	<0.19	0.19

* *Callinectes sapidus*
** Lower Limit of Detection
*** None Detected

DISCUSSION

The field water quality data taken at the LF site showed salinity stratification at all stations except the reference. Bottom salinities were highest at the discharge and decreased with increasing distance from the discharge, indicating that the PW was the source of the stratification. The high salinity (134 ppt) of the discharge relative to the ambient water salinities (0.0 ppt at the surface and bottom at Station R), the shallow water depths of the canal (less than 2 m) and the low current velocities found during the sampling all favor the formation of salinity stratification. St. Pé (1990) also found salinity stratification near the discharge. The St. Pé (1990) study reported a ^{226}Ra by GS activity of 537 ± 77.6 pCi/l from the PW sample. This activity is comparable to the 477 ± 30 pCi/l value found in our sampling.

Table 3. Radionuclide levels in tissues collected offshore (ST)

Sample Location/ Depth (m)	Sample Type/1	^{226}Ra (pCi/g)	LLD (pCi/g)	^{228}Ra (pCi/g)	LLD (pCi/g)	Total ^{210}Pb (pCi/g)	LLD (pCi/g)
B-1/3	BH	0.3 ± 0.1	0.1	< 0.9	0.9	< 0.8	0.8
B-1/3	BS	ND	0.1	ND	3.0	< 0.2	0.2
B-1/6	BH	0.1 ± 0.1	0.1	<1.1	1.1	< 0.7	0.7
B-1/6	BS	ND	0.1	< 0.3	0.3	0.2 ± 0.1	0.2
B-1/6	C	0.1 ± 0.1	0.1	ND	1.3	< 0.8	0.8
B-1/6	F	ND	0.1	ND	1.0	< 0.7	0.7
B-1/12	C	0.1 ± 0.1	0.1	ND	0.7	ND	0.4
A-1/3	BH	0.2 ± 0.1	0.1	1.6 ± 0.8	1.1	< 0.8	0.8
A-1/3	BS	ND	0.1	1.3 ± 0.6	0.8	< 0.6	0.6
A-1/3	F	ND	0.2	2.7 ± 1.6	2.2	< 1.06	1.6
A-1/6	BH	0.2 ± 0.1	0.1	2.3 ± 1.0	1.4	< 1.0	1.0
A-1/6	BS	ND	0.1	0.2 ± 0.1	0.2	< 0.2	0.2
A-1/6	C	0.1 ± 0.1	0.1	< 0.2	0.2	0.2 ± 0.1	0.1
A-1/6	F	ND	0.2	2.1 ± 1.4	2.1	< 1.5	1.5
A-1/9	BH	ND	0.1	<1.0	1.0	<0.8	0.8
A-1/9	BS	0.1 ± 0.1	0.1	< 0.2	0.2	< 0.2	0.2
A-1/9	MI	0.1 ± 0.1	0.1	<0.8	0.8	<0.6	0.6
A-1/12	BT	0.2 ± 0.1	0.1	< 1.0	1.0	< 0.8	0.8
A-1/12	C	ND	0.1	<0.2	0.2	<0.2	0.2
B-2/3	BH	0.4 ± 0.2	0.1	3.7 ± 1.2	1.6	<1.3	1.3
B-2/3	BS	0.1 ± 0.1	0.1	ND	1.2	< 0.9	0.9
B-2/3	C	ND	0.1	ND	1.4	< 1.2	1.2
B-2/3	F	ND	0.1	ND	1.5	< 1.2	1.2
B-2/6	BH	0.3 ± 0.1	0.1	ND	1.1	ND	0.8
B-2/6	BS	ND	0.1	ND	1.2	< 0.8	0.8
B-2/6	C	0.2 ± 0.1	0.1	1.5 ± 0.8	1.2	< 0.9	0.9
B-2/9	BH	0.4 ± 0.2	0.1	ND	1.3	< 0.9	0.9
B-2/9	BS	ND	0.1	ND	1.5	< 1.1	1.1
B-2/9	C	ND	0.1	ND	1.1	< 0.8	0.8
B-2/9	F	ND	0.1	ND	1.4	< 1.0	1.0
B-2/12	MA	0.1 ± 0.1	0.1	1.9 ± 1.1	1.6	< 1.1	1.1
A-2/3	BT	0.2 ± 0.1	0.1	< 0.9	0.9	< 0.8	0.8
A-2/3	C	0.1 ± 0.1	0.1	<0.8	0.8	< 0.7	0.7
A-2/3	F	ND	0.1	ND	1.8	< 1.3	1.3
A-2/6	BH	0.7 ± 0.2	0.1	<0.9	0.9	< 0.8	0.8
A-2/6	BS	0.4 ± 0.2	0.1	< 1.0	1.0	< 0.9	0.9
A-2/6	C	0.1 ± 0.1	0.1	< 0.9	0.9	<0.8	0.8

Table 3. Radionuclide levels in tissues collected offshore at South Timbalier (Continued)

Sample Location/ Depth (m)	Sample Type/1	^{226}Ra (pCi/g)	LLD (pCi/g)	^{228}Ra (pCi/g)	LLD (pCi/g)	Total ^{210}Pb (pCi/g)	LLD (pCi/g)
A-2/9	BT	0.8 ± 0.2	0.1	<0.9	0.9	< 0.8	0.8
A-2/9	C	1.3 ± 0.3	0.1	< 0.9	0.9	< 0.8	0.8
A-2/12	BT	0.2 ± 0.1	0.1	1.3 ± 0.7	0.9	<0.7	0.7
A-2/12	C	ND	0.1	2.0 ± 0.8	1.0	<0.7	0.7
A-2/12	ML	0.2 ± 0.1	0.1	<0.8	-0.8	<0.8	0.8

BT = *Megabalanus tintinnabulum*, barnacle (whole)
BH = *Megabalanus tintinnabulum*, barnacle (hard parts)
BS = *Megabalanus tintinnabulum*, barnacle (soft parts)
C = *Menippe adina*, stone crab (whole)
MI = *Isognomon biocolor*, biocolor purse oyster (whole)
ML = *Lithophaga artistata*, scissor datemussel (whole)
MA = *Anadara transversa*, transverse ark clam (whole)
F = *Hypleurochilus geminatus*, crested blenny (whole)

High sediment ^{226}Ra and ^{228}Ra activities were found in the samples from Stations @0 and 10 E, both located in shallow (less than 1.0 m) water depths immediately adjacent to where the discharge entered the canal. All the other stations located in slightly deeper water (more than 1.0 m) had lower sediment radionuclide levels by one to two orders of magnitude. Although the single measurements of surface current velocities at each station were low (± 0.05 m/sec), it is possible that under other conditions current velocities are higher and that dilution of the PW reduces the amount of radionuclides available to be absorbed into the sediment at the deeper stations.

Sediment data collected by St. Pé (1990) at the LF site show a similar pattern to those in our study. Both data sets show high values in close proximity to the discharge which decrease rapidly by an order of magnitude within about 20 m. The major difference in the data sets is the value of ^{226}Ra at the discharge site. The St. Pé (533 ± 25.9 pCi/g) surface sediment value obtained at the location of our 10 E site is much higher than that obtained in this study (192 ± 15 pCi/g).

Sediment data collected near the GM site show a general decrease of both ^{226}Ra and ^{228}Ra away from the discharge. The highest value for both parameters (Figures 7 and 8), however, was at the station located 20 m north of the discharge. This particular sediment sample was unusual and appeared to have a consistency similar to that of drill mud although the depositional environment was similar to other stations at this site.

Concentrations of sediment radionuclides did not show the sharp decrease 20 m away from the discharge observed in the LF data. Radionuclide concentrations at the discharge were not nearly as high in sediment near the GM discharge as in the sediment near the LF discharge. Concentrations in the sediment near the GM discharge decreased over a larger area to the near background levels at the 500 m stations. The area is a major access canal in the field, and boat traffic as well as current may play a role in diffusing the discharge over the area. Within 500 m south of the GM discharge is a compressor station, and approximately 1,000 m south of the discharge is a tank battery which ceased discharging within the past few years.

At neither inshore site did radionuclide levels show a consistent pattern of increasing or decreasing with depth in the sediment. St. Pé (1990) suggested that activity may increase with depth at some stations based on observations made on sediment Ra levels at the Lake Washington Field site.

Data from sediment ^{226}Ra and ^{228}Ra sampling locations near six PW discharges reported by Steimle and Associates, Inc. (1991) were generally from water deeper than 1.0 m. Those results, based on Total ^{226}Ra (Rn emanation method) analysis, ranged from below the LLD (0.1 pCi/g) to 6.3 ± 0.6 pCi/g for ^{226}Ra activity and below the LLD (0.7 pCi/g) to 3.9 ± 0.8 pCi/g for ^{228}Ra activity for samples taken at the 0-10 cm depth. These data were collected from stations 15 m upstream and downstream of the discharge.

Hanan (1981) reported sediment ^{226}Ra activities for samples from 0-10 cm depths within 20 m of a PW discharge and collected in "waist" to "chest" deep water. His data showed activities ranging from 0.98 ± 0.04 pCi/g to 7.43 ± 0.24 pCi/g.

Radionuclide concentrations reported from surface and bottom water column samples near the GM discharge were all below 8 pCi/l. Bottom water concentrations were slightly higher than surface water concentrations although concentrations at the discharge were not the highest recovered. The canal was salinity stratified when the samples were taken as far as 300 m north and 100 m south of the discharge, and the halocline was very near (0.4 m) the bottom. The stratification was related to the discharge.

Bottom water samples taken at ST had very low ^{226}Ra and ^{228}Ra activities. Temperature and salinity stratification extended 1,000 m north, south and east of the platform. The stratification was an area wide phenomenon, with dissolved oxygen concentrations near 1.0 ppm in the bottom water samples. Oxygen depleted bottom waters is an annual summer phenomenon which is widespread off the coast of Louisiana (Rabalais, 1986).

Epifaunal tissue was recovered from the ST platform legs primarily between the surface and 10 m where the epifauna was healthy. From about 12 m to about 15 m, enough tissue for a sample could sometimes be retrieved. These samples always, however, smelled of rotting organisms. Below 15 m, no live epifauna could be retrieved. No benthos was found within the area sampled around the platform. The sediments retrieved were black and anoxic.

Levels of ^{226}Ra, ^{228}Ra, and ^{210}Pb measured in tissue soft and hard parts were low. Most values were near the LLD even in samples taken in close proximity to the discharge. Live epifauna was found to occur as close as about 3 m to the discharge, indicating that the discharge was rapidly dissipated. Sampling was accomplished across trophic levels in the epifaunal and associated community. High radionuclide concentrations did not occur in filter feeders (barnacles and molluscs) or micro-predators/opportunists (blennies).

Samples of blue crab tissue (*Callinectes sapidus*) from GM showed low concentrations of ^{210}Pb (<0.2 pCi/l). Most of the concentrations of ^{226}Ra and ^{228}Ra which were above the LLD were in the hard tissue (shell). The highest concentration of ^{226}Ra found in crab shell was from the reference station. The number of samples in this study was very limited, and crab populations are highly mobile. The levels do represent, however, those in seafood available for commercial harvest.

Comparable data for tissue concentrations of radionuclides are not available for inshore crab and offshore epifaunal populations. St. Pé (1990) exposed oysters to PW discharges at Lirette, Bully Camp and Lake Washington. He found that only the group exposed to the discharge at LF accumulated ^{226}Ra above the LLD. He concluded that radiological data demonstrated the potential for oysters growing near PW discharges to accumulate ^{226}Ra. Data collected in this study show that epifaunal and associated organisms are not accumulating ^{226}Ra, ^{228}Ra or ^{210}Pb to high levels near the ST discharge.

LITERATURE CITED

EPA, 1980, Prescribed procedures for the measurement of radioactivity in drinking water. EPA 600/4-80-032. Office of Research and Development. Cincinnati, OH.

EPA, 1983, Methods for chemical analysis of water and wastes. EPA 600/4-79-020. Office of Research and Development. Cincinnati, OH.

EPA, 1986, Test methods for evaluating solid waste. Third Edition. Laboratory Manual Vol. 1 Section A. EPA SW-Office of Solid Waste and Emergency Response. Wash., D.C.

Hanan, M. A., 1981, Geochemistry and Mobility in Sediments of Radium from Oilfield Brines - Grand Bay, Plaquemines Parish, Louisiana. M.S. thesis, University of New Orleans, 89 p.

Rabalais, N., 1986, Oxygen depleted waters on the Louisiana continental shelf. Pages 314-316 in "Proceedings Seventh Annual Gulf of Mexico Information Transfer Meeting". U.S. Department of the Interior. Minerals Management Service. MMS 87-0058 381 p.

St. Pé, K., M. Ed., 1990, An assessment of produced water impacts to low energy, brackish water systems in Southeast Louisiana. Louisiana Dept. of Env. Qual., Water Poll. Contr. Div., Baton Rouge, LA. 204 p.

Steimle & Associates, Inc., 1991, Produced Water Impacts on Louisiana Wetlands. American Petroleum Institute Publ. 4517. American Petroleum Institute, Washington, D.C. 132 p.

RADIUM CONCENTRATION FACTORS AND THEIR USE IN HEALTH AND ENVIRONMENTAL RISK ASSESSMENT

A.F. Meinhold and L.D. Hamilton

Brookhaven National Laboratory
Upton, New York
USA

INTRODUCTION

Produced waters discharged to coastal and offshore waters can contain high levels of radium (Snavely, 1989). Radium is known to be taken up by aquatic animals, and tends to accumulate in bone, shell and exoskeleton. The biological uptake of radium by aquatic animals is of concern because of the potential for: (1) dose to aquatic animals from accumulated radium and the effects on individual animals, populations and ecosystems; and (2) increased cancer risk to people from ingestion of radium in food.

The most common approach to estimating the uptake of a radionuclide by aquatic animals in health and environmental risk assessments is the concentration factor method. This method relates the concentration of a contaminant in an organism to the concentration in the surrounding water. Site specific data are not usually available; generic, default values are often used.

This paper describes the concentration factor method, summarizes some of the variables which may influence the concentration factor for radium, reviews reported concentration factors measured in marine environments and presents concentration factors derived from data collected in a study in coastal Louisiana. Use of generic default values for the concentration factor is also discussed.

THE CONCENTRATION FACTOR METHOD

Concentration factors are commonly used in dose assessments to estimate the levels of radionuclides in aquatic organisms. The concentration factor (CF) is a function of the concentration in the water or sediment (C) and the equilibrium concentration (on a wet weight basis) in the organism (Q).

$$CF = Q / C \tag{1}$$

It is usually assumed that there is a linear relationship between C and Q and that the concentration factor is independent of the concentration in the environment. This assumption, along with the assumption that groups of similar organisms have similar concentration factors for a specific radionuclide, allow the use of generic concentration factors in dose assessment models. These assumptions may not always be justified. Generic factors can only be used in a preliminary assessment as a first order estimate of bioaccumulation. Site and organism specific factors are desirable for such a study, but obtaining them can be difficult and expensive.

Produced Water, Edited by J.P. Ray
and F.R. Engelhart, Plenum Press, New York, 1992

VARIABLES AFFECTING THE CONCENTRATION FACTOR FOR RADIUM

Values for the concentration factor in the literature have a considerable range. There are several factors which contribute to this variation; they should be considered when using generic factors in dose-assessment modeling. Important influences on the concentration factor include inter-specific differences, differences between organs, the effects of environmental factors, the extent to which the organism is in equilibrium with its surroundings, the relationship between the concentration factor and the concentration of the radionuclide in the water, and the presence of chemically similar elements. These factors contribute to the variation in reported concentration factors and are discussed below.

Species and Trophic Level Differences

Radium uptake varies among groups of organisms (fish, molluscs, crustaceans) and among species in a group. Differences in habitat, food preference and position in the food web account for some of the differences in concentration factors noted between species. In general, the concentration factors of radionuclides decrease at higher trophic levels.

Swanson (1983) studied the levels of radionuclides in fish from lakes affected by effluents from a uranium mine and mill. Concentration factors for radium varied among species in the same lake (Table 1). Small bottom-feeding forage fish (trout-perch, nine spine stickleback, spottail shiner and lake chub) generally had greater radium levels than the larger species. Cisco, a planktivorous species, accumulated smaller amounts of radium than the bottom-feeding foragers. Among the larger fish (white sucker, lake whitefish and lake trout), radium levels were highest in the bottom feeding white suckers, moderate in the omnivorous lake whitefish, and lowest in the lake trout feeding mainly on cisco. The inter-specific differences in concentration factors suggest that use of a single concentration factor to represent all species in a group may not be always appropriate.

Table 1. Concentration factors for whole fish from Beaverlodge Lake [1]

Species	Concentration Factor
Nine spine stickleback	387
Trout-perch	1473
Cisco	53
Spottail shiner	986
Lake chub	187
White sucker	287
Lake whitefish	33
Lake trout	13

[1]from Swanson, (1983)

Differences Between Organs

Certain radionuclides preferentially concentrate in particular organs of aquatic organisms. Radium is chemically similar to calcium and concentrates in bone, shell and exoskeleton. Table 2 shows the differences in the concentration factor for radium in various parts of three species of fish from a lake contaminated by effluent from a uranium mine and mill (Swanson, 1983). In all three species, bone accumulated the most radium, flesh the least.

Table 3 gives concentration factors reported for radium in marine molluscs (Iyengar, 1984). The shell invariably accumulated more radium than the soft edible parts of the animal. These studies suggest that concentration factors based on radium content of the whole organism can overestimate the level of radium in the edible portion.

Table 2. Concentration factors for various parts of fish from Beaverlodge Lake [1]

Species	Portion	Concentration Factor
White sucker	flesh	12
	skin	31
	bone	1793
	whole	287
Lake whitefish	flesh	3
	skin	93
	bone	360
	whole	33
Lake trout	flesh	1
	skin	20
	bone	10
	whole	13

[1]from Swanson, (1983)

Table 3. Concentration factors for radium in marine molluscs [1]

Species	Portion	Concentration Factor
Oyster	soft parts	50
(*Ostrea* sp.)	shell	500
Green mussel	soft parts	46
(*Perna viridis*)	shell	419
Snail	soft parts	63
(*Thais* sp.)	shell	156
Snail	soft parts	44
(*Petalla radiate*)	shell	256

[1]from Iyengar, (1984)

Equilibrium With Surroundings

The concentration factor method assumes that the organism is in chemical equilibrium with its surroundings. The time required to reach equilibrium depends on the radionuclide half-life and the biological half-life of the element in the organism (Till and Meyer, 1983). For fresh water mussels, Jeffree observed that the radium concentration in flesh continued to increase with age in mature animals from a single location (Jeffree, cited in Williams, 1984). Twenty-eight day experiments in which radium uptake was induced also failed to reach equilibrium (Jeffree, cited in Williams, 1984). These studies suggest that for situations in which the environmental concentrations change over time, organisms will not reach complete equilibrium with their surroundings. Mobile organisms such as fish may also fail to reach equilibrium.

Major Ion Effects

The presence of major ions in solution influences the biological uptake of non-nutrient trace substances because chemically analogous substances compete for uptake and retention by organisms. Radium and calcium are chemically similar; the amount of calcium available can affect the rate of radium uptake.

De Bortoli and Gaglione (1972) found that for three of four lakes studied in Italy, the concentration of calcium in the water was inversely related to the concentration of radium in fish. Jeffree found that uptake of radium by the fresh water mussel *(Velesunio angasi)* was suppressed in water containing 5 mg/l calcium, but not in water containing 0.5 mg/l calcium (cited in Williams, 1984). Barium is also chemically similar to radium, and was found to reduce the uptake of radium by unicellular algae (Sebesta et al., cited in Williams, 1984).

Other Environmental Factors

Environmental factors, including temperature, salinity and pH affect the growth and metabolism of organisms, and consequently the uptake of radium and other radionuclides. Small increases in temperature tend to increase biological activity and the uptake and excretion of radionuclides (Till and Meyer, 1983). In a study of ^{65}Zn uptake by shellfish, Duke et al. (1969) found that the primary factors affecting the concentration factor were salinity and temperature.

Relationship Between Concentration in Water and in Organism

Most applications of the concentration factor method in dose assessment assume a linear relationship between C (the concentration of the radionuclide in water) and Q (the concentration of the radionuclide in the organism). This assumption allows use of a single concentration factor (CF = Q/C) over a range of environmental concentrations. Available data suggest that this assumption is not valid for all groups or species of organisms.

Williams (1984) investigated the relationship between C and Q for several groups of organisms. He found a linear relationship for insects and fresh water mussels. The data for fresh water mussels were for a single species *(Velesunio angasi)*, but the animals had not reached equilibrium with the surrounding water (Jeffree, cited in Williams, 1984).

Williams (1984) pooled data from a number of studies to test the linearity of the relationship between C and Q for fish. This analysis resulted in a highly nonlinear model:

$$CF = 13\ C^{-0.32} \tag{2}$$

Because of the importance of the factors described in this section (e.g., inter-specific differences), it may not be appropriate to pool data for different species of fish. Other studies also suggest that use of a single concentration factor over a range of water concentrations is not appropriate. In the studies by Swanson (1983, 1985), the concentration factor for radium in several species of fish generally decreased with an increase in water concentration.

CONCENTRATION FACTORS REPORTED FOR MARINE ORGANISMS

Radium concentration factors for marine organisms were reported in a only a few published studies. These studies are reviewed here.

Fish

Table 4 presents concentration factors for Radium-226 and Radium-228 in marine fish taken off the coast of India (Iyengar et al., 1980). The concentration factor for ^{226}Ra in muscle had a range of 55-130, and ranged 65-610 in bone (water concentration: 0.0015 Bq/l; 0.04 pCi/l). Concentration factors for ^{228}Ra had a range of 21-63 in muscle and 98-700 in bone (water concentration: 0.0119 Bq/l; 0.32 pCi/l). The concentration factors were generally higher for ^{226}Ra than ^{228}Ra, which had a higher concentration in water.

Molluscs and Crustaceans

Concentration factors for ^{226}Ra in marine molluscs were published by Iyengar (1984). The range of concentration factors in these animals was small (Table 5). The concentration factor for ^{226}Ra in soft parts of molluscs ranged 44-50, and in shell 156-500 (water concentration: 0.0016 - 0.0056 Bq/l; 0.04 - 0.15 p Ci/l).

Table 6 presents concentration factors for both ^{226}Ra and ^{228}Ra in marine molluscs (Iyengar et al., 1980). The concentration factor for ^{226}Ra in the soft parts of the shellfish ranged 73-140. The concentration factor for ^{228}Ra ranged from 130-170.

Table 4. Concentration factors for ^{226}Ra and ^{228}Ra in marine fish (India) [1]

Species	Portion	^{226}Ra Water Bq/l (pCi/l)	CF	^{228}Ra Water Bq/l (pCi/l)	CF
Sole	muscle	0.0015 (0.04)	55	0.0119 (0.32)	21
	bone	0.0015 (0.04)	370	0.0119 (0.32)	160
Mackerel	muscle	0.0015 (0.04)	120	0.0119 (0.32)	63
	bone	0.0015 (0.04)	-	0.0119 (0.32)	700
Ray	muscle	0.0015 (0.04)	60	0.0019 (0.32)	-
	bone	0.0015 (0.04)	65	0.0019 (0.32)	98
	liver	0.0015 (0.04)	-	0.0119 (0.32)	72
Oil	muscle	0.0015 (0.04)	130	0.0019 (0.32)	44
Sardine	bone	0.0015 (0.04)	610	0.0019 (0.32)	180
RANGE	muscle		55 - 130		21-63
	bone		65 - 610		98 - 700

[1]from Iyengar et al., (1980)

Table 5. Concentration factors for ^{226}Ra in marine molluscs (India) [1]

Species	Organ/body part	Conc in water Bq/l	(pCi/l)	CF
Oyster	soft parts	0.0016	(0.043)	50
(*Ostrea* sp.)	shell	0.0016	(0.043)	500
Green mussel	soft parts	0.0016	(0.043)	46
(*Perna viridis*)	shell	0.0016	(0.043)	419
Clam	soft parts	0.0016	(0.043)	-
(*Meretrix* sp.)	shell	0.0016	(0.043)	200
Snail	soft parts	0.0056	(0.15)	63
(*Thais* sp.)	shell	0.0056	(0.15)	156
Snail	soft parts	0.0016	(0.043)	44
(*Petalla radiate*)	shell	0.0016	(0.043)	256

[1]from Iyengar, (1984)

Marine crustaceans were also studied by Iyengar et al. (1980). For ^{226}Ra the concentration factor in crustacean muscle ranged from 80 to 210, and in exoskeleton from 220 to 830 (Table 7). The range for ^{228}Ra was 35-360 in muscle and 230-800 in exoskeleton. Table 8 summarizes the range of concentration factors for marine molluscs and crustaceans in the reviewed literature.

Table 6. Concentration factors for ^{226}Ra and ^{228}Ra in marine molluscs (India) [1]

Species	Portion	^{226}Ra Water Bq/l (pCi/l)	CF	^{228}Ra Water Bq/l (pCi/l)	CF
Clam	soft parts	0.0015 (0.04)	140	0.0119 (0.32)	170
Oyster	soft parts	0.0015 (0.04)	73	0.0119 (0.32)	130
	shell	0.0015 (0.04)	—	0.0119 (0.32)	100

[1]from Iyengar, (1980)

GENERIC CONCENTRATION FACTORS

Generic or average values for concentration factors were suggested based on surveys of published data (Thompson et al. 1972, IAEA 1982). Generic factors are meant for use in radiological assessment models for estimating the dose to man from a number of pathways. Commonly used models contain default concentration factors for a number of radionuclides and groups of organisms. The generic factors suggested by IAEA (1982) are used by many authors and models and are given in Table 9. IAEA (1985) revised these estimates upward for molluscs and fish (Table 9), based on the highest values reported in the literature.

Table 7. Concentration factors for ^{226}Ra and ^{228}Ra in marine crustaceans (India) [1]

Species	Portion	^{226}Ra Water Bq/l (pCi/l)	CF	^{228}Ra Water Bq/l (pCi/l)	CF
Crab	muscle	0.0015 (0.04)	110	0.0119 (0.32)	180
(*Scylla serrata*)	gills	0.0015 (0.04)	340	0.0119 (0.32)	160
Crab	muscle	0.0015 (0.04)	—	0.0019 (.32)	35
(*Neptanus* sp.)	gills	0.0015 (0.04)	11000	0.0019 (0.32)	12000
	exoskel	0.0015 (0.04)	830	0.0019 (0.32)	800
Mole crab	muscle	0.0015 (0.04)	210	0.0019 (0.32)	68
	exoskel	0.0015 (0.04)	220	0.019 (0.32)	230
Prawn	muscle	0.0015 (0.04)	80	0.0019 (0.32)	360
	muscle	0.0015 (0.04)	-	0.0019 (0.32)	56
	exoskel	0.0015 (0.04)	360	0.0019 (0.32)	300

[1]from Iyengar, (1980)

Site and species specific concentration factors are not usually available, and developing them is expensive and time consuming. The variation within and between species, and the many other variables which can affect the concentration factor make using generic factors problematic, but also, in most cases, necessary. In a preliminary risk assessment study the use of generic factors is appropriate. If the resulting estimates are high compared to background or to an acceptable level of risk, detailed site specific data will be needed.

Table 8. Molluscs and crustaceans - ranges of concentration factors in the reviewed literature

	Water Bq/l (pCi/l)	Soft Parts/ Muscle	Shell/ Exoskeleton
Molluscs - marine[1] (^{226}Ra)	0.0016 (0.04)	44 - 63	156 - 500
Molluscs - marine[2] (^{226}Ra)	0.0015 (0.04)	73 - 140	—
Molluscs - marine[2] (^{228}Ra)	0.0119 (0.32)	130 - 170	100
Crustaceans - marine[2] (^{226}Ra)	0.0015 (0.04)	80 - 210	220 - 830
Crustaceans - marine[2] (^{228}Ra)	0.0119 (0.32)	35 - 360	230 - 800
Molluscs - overall range	0.0015 - 0.016 (0.04-0.43)	44 - 170	100 - 500
Crustaceans - overall range	0.0015 - 0.0119 (0.04-0.32)	35 - 360	220 -830

[1] Iyengar, (1984)
[2] Iyengar et al., (1980)

Table 9. IAEA generic concentration factors for radium

	IAEA (1982)	IAEA (1985)
Marine fish	100	500
Marine molluscs	100	1000
Marine crustaceans	100	100

CONCENTRATION FACTORS DERIVED FROM A FIELD STUDY IN COASTAL LOUISIANA

Data collected included ^{226}Ra and ^{228}Ra concentrations in water, sediment, fish, shellfish and crustaceans at three produced water outfalls and three background stations in Louisiana (CSA, 1991). The three outfalls were located in canals along the Louisiana coast. Concentrations of ^{226}Ra and ^{228}Ra in the discharges were 228.9 and 383.0 pCi/l for outfall 1, 110.6 and 244.4 pCi/l for outfall 2, and 251.9 and 254.9 pCi/l for outfall 3.

Radium concentrations were measured in water and sediment at stations located 25 and 50 feet from the outfalls, in three directions (the outfalls were located along one side of a canal). Biological samples were taken within 50 feet of the discharge point. All mollusc samples were oyster (*Crassostrea virginica*), and all fish samples were seatrout (*Cynoscion* sp.). Crustaceans sampled at Site 2 were crabs (*Callinectes* sp.) and at Sites 1 and 3 were shrimp (*Penaeus* sp.). Oysters were removed from the shell before analysis. Fish and crustacean samples were whole-body samples, including bone, skin and exoskeleton.

Concentration factors were calculated for each site/organism combination as follows:

$$CF = [Q\ (pCi/g) \times 1000\ (g/l)]/C\ (pCi/l) \qquad (3)$$

where:

CF= concentration factor
Q= concentration of radium in organism (pCi/g)
C= concentration of radium in water (pCi/l)

Table 10 gives the concentration data and concentration factors for ^{226}Ra and ^{228}Ra at the three sites and the six background reference stations. The concentration factor was calculated for the water concentration at 50 feet from the outfall. Because the biological samples were taken within (not at) 50 feet of the discharge, these values represent maximum concentration factors. The values reported here are also uncertain because concentrations at or below the detection limit for the analytical system used were reported as zeros, underestimating the concentration of radium in water, and overestimating the concentration factor. Because water samples were not taken exactly at the locations where biological samples were harvested, the data set does not provide enough data to develop a model for the concentration factor of radium based on water concentrations.

Table 10. Calculated radium concentration factors[1] for organisms sampled in the CSA (1991) study

Station	Conc. at[1] 50 feet (pCi/l) $^{226}Ra/^{228}Ra$	Concentration Factors Fish $^{226}Ra/^{228}Ra$	Oysters $^{226}Ra/^{228}Ra$	Crustaceans $^{226}Ra/^{228}Ra$
Station 1	0.43/0.93	33/5	16/3	170/101
Station 2	0.52/0.22	NC/100	15/50	130/114
Station 3	0.92/4.0	46/3	ND/ND	139/61
Reference 1a	0.25/0.15	ND/ND	4/240	56/27
Reference 1b	0.2/0	25/NC	20/NC	120/NC
Reference 2a	0.7/0	3/NC	6/NC	ND/ND
Reference 2b	0.5/0	62/NC	2/NC	138/NC
Reference 3a	0.7/3.3	28/18	ND/ND	23/18
Reference 3b	0.7/10.3	43/2	ND/ND	110/4

ND: no data
NC: concentration factor not calculated because water or animal concentration was reported in CSA (1991) as 0.
[1] average of measured concentration at three stations, each 50 feet from discharge.

The calculated concentration factors based on this data set cover a wide range. Figure 1 shows the ranges of concentration factors for whole fish (seatrout), the soft parts of molluscs (oysters), and whole crustaceans (crabs and shrimp). Concentration factors ranged from 2 to 100 for whole fish, from 2 to 240 for the soft parts of molluscs and from 4 to 170 for whole crustaceans.

The concentration factors calculated from the CSA data set (1991) were used to derive a distribution of concentration factors, including a mean or expected value and a 95% confidence value for whole fish, the soft parts of molluscs and whole crustaceans. A latin hypercube sampling process was used to generate a distribution of concentration factors, by sampling the available data 5,000 times.

Table 11 gives the expected value and 95% confidence limit for fish, molluscs and crustaceans generated in this analysis.

Table 11. Expected value and 95% confidence interval for simulated distributions of the concentration factor

Organism Type	Expected value	95% Confidence Interval
Whole fish	30	100
Molluscs soft parts	40	240
Whole crustaceans	86	170

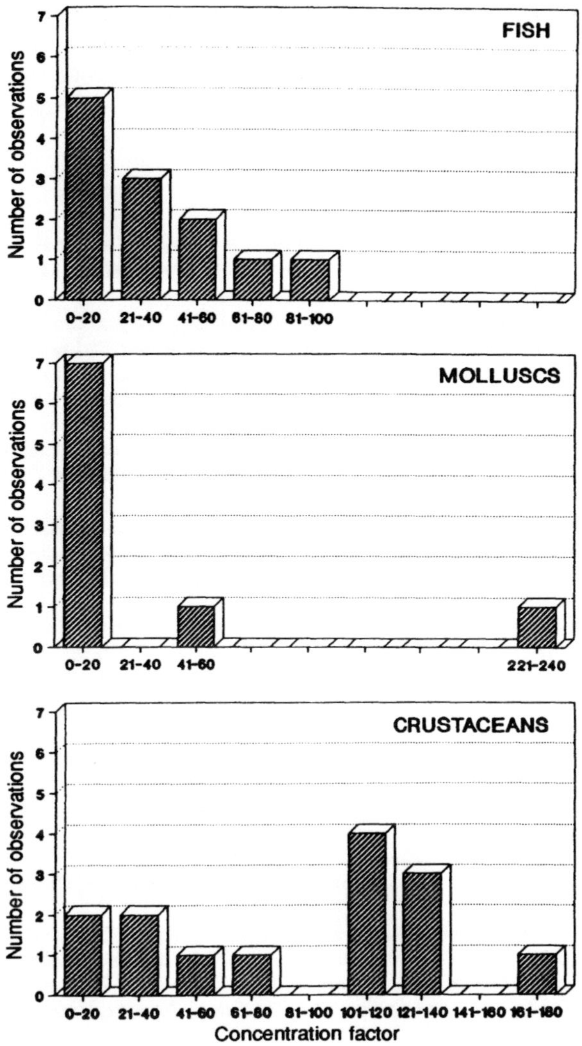

Figure 1. Distribution of calculated concentration factors for whole fish, the soft parts of molluscs and whole crustaceans sampled in coastal Louisiana.

SUMMARY AND DISCUSSION

The generic concentration factors suggested by IAEA (1982) (100 for marine fish, molluscs and crustaceans) are consistent with the values reported in the literature for the edible parts of fish, molluscs and crustaceans (Table 4, Table 8). The higher values suggested in IAEA (1985) (Table 9) are not supported by these data.

An analysis of data collected in coastal Louisiana suggests that an average or expected value of 30 for whole fish, 40 for the edible parts of molluscs and 86.4 for whole crustaceans would be applicable concentration factors for radium discharged in produced waters to the Gulf Coast. The concentration factors calculated from this data set are uncertain, because of a problem with values near the detection limit, and because animals were not sampled exactly at stations sampled for water radium concentrations. These data also overestimate the concentration factors for the edible parts of fish and crustaceans because they were based on whole animal samples.

The concentration factor for radium is a critical value in a risk assessment study of the health and environmental effects of radium in coastal and offshore environments. Additional data are needed to describe better the distribution of radium concentration factors in the edible portions of animals used for food. Generic concentration factors are appropriate for use only in a screening-level assessment.

REFERENCES

CSA, 1991, "Measurements of Naturally Occurring Radioactive Materials (NORM) at Three Produced Water Outfalls", Continental Shelf Associates, Inc., Jupiter, Florida, prepared for Mid-Continent Oil and Gas Association, Baton Rouge, Louisiana.

de Bortoli, M., and Gaglione, P., 1972, Ra-226 in environmental materials and foods. Health Physics. 22:43.

Duke, T., Willis, J., Price T., and Fischler, K., 1969, Influence of environmental factors on the concentration of Zn-65 by an experimental community, *in*: "Symposium on Radioecology", Proc. Second National Symposium, Ann Arbor Michigan, May 1967, D.J. Nelson and F.C. Evans, eds, CONF-670503, March 1969, United States Atomic Energy Agency, Washington, D.C.

IAEA, 1982, "Generic Models and Parameters for Assessing the Environmental Transfer of Radionuclides from Routine Releases", Safety Series No. 57. Procedures and Data. International Atomic Energy Agency, Vienna.

IAEA, 1985, "Sediment K_ds and Concentration Factors for Radionuclides in the Marine Environment", Technical Report Series No. 247, International Atomic Energy Agency, Vienna.

Iyengar, M.A.R., 1984, Distribution in nature, *in*: "The Behavior of Radium in Waterways and Aquifers", IAEA-TEC DOC 301. International Atomic Energy Agency, Vienna.

Iyengar, M.A.R., Rajan, M.P., Ganapathy, S. and Kamath, P.R., 1980, Sources of radiation exposure in a low monazite environment, *in*: " The Natural Radiation Environment III", Symp. Proc. Houston, 1978, CONF-780422. Technical Information Center, Springfield pp. 1090-1106.

Snavely, E.S., Jr., 1989, "Radionuclides in Produced Water", Production Department, American Petroleum Institute, Washington, D.C.

Swanson, S.M., 1985, Food chain transfer of U-series radionuclides in a northern Saskatchewan aquatic system, Health Physics. 49:747.

Swanson, S.M., 1983, Levels of Ra226, Pb210 and total uranium in fish near a Saskatchewan uranium mine and mill, Health Physics. 45:67.

Thompson, S.E., Burton, C.A., Quinn, D.J., and Ng, Y.C., 1972. "Concentration Factors of Chemical Elements in Edible Aquatic Organisms", Lawrence Livermore National Laboratory, University of California, Report UCRL-50564, Revision 1.

Till, J.E. and Meyer, H.R., 1983, "Radiological Assessment: A Textbook on Environmental Risk Analysis", NUREG/CR-3332, ORNL-5968, U.S. Nuclear Regulatory Commission, Washington, D.C.

Williams, A.R., 1984, Biological uptake and transport, *in*: "The Behavior of Radium in Waterways and Aquifers", IAEA-TEC DOC 301, International Atomic Energy Agency, Vienna.

HEALTH RISK ASSESSMENT FOR RADIUM DISCHARGED IN PRODUCED WATERS

L.D. Hamilton[1], A.F. Meinhold[1] and J. Nagy[1]

[1] Brookhaven National Laboratory
Upton, New York
USA

INTRODUCTION

Produced water generated during the production of oil and gas can contain enhanced levels of radium. This naturally occurring radioactive material (NORM) is discharged into freshwater streams, estuarine, coastal and outer continental shelf waters. Large volumes of produced waters are discharged to coastal waters along the Gulf Coast of Louisiana. The Gulf of Mexico is an important producer of fish and shellfish, and there is concern that radium discharged to coastal Louisiana could contaminate fish and shellfish used by people for food, and present a significant increase in cancer risk.

This paper describes a screening-level assessment of the potential cancer risks posed by radium discharged to coastal Louisiana in oil-field produced waters. This screening analysis was performed to determine if a more comprehensive and realistic assessment is necessary, and because of the conservative assumptions embedded in the analysis overestimates the risk associated with the discharge of radium in produced waters.

RADIUM IN COASTAL LOUISIANA

Two isotopes of radium (^{226}Ra and ^{228}Ra) are the radionuclides of most concern in produced water in terms of potential human health effects. Radium isotopes probably make up the bulk of nuclide activity in produced water (Snavely, 1989). Radium is important because of its high solubility in produced water and tendency to bioaccumulate in food organisms. ^{226}Ra and ^{226}Ra arise during the radioactive decay of their naturally occurring parent radionuclides. The isotopes of radium give rise to a series of radioactive daughters that lead to a stable isotope of lead. ^{226}Ra, created is a member of the decay series. ^{226}Ra has a half-life of 1602 years and decays to ^{222}Rn through the emission of an alpha particle. ^{228}Ra is part of the thorium decay series, and its parent radionuclide is ^{232}Th. ^{228}Ra has a half-life of 6.7 years and decays to ^{228}Ac with the emission of a beta particle.

Radium-226 and ^{228}Ra are naturally occurring radionuclides, and are present in the earth's crust, in freshwater rivers and lakes, and in coastal waters, oceans and sediments. Radium-226 concentrations in coastal waters have been reported to be between 0.01 and 0.59 pCi/l (Cherry and Shannon, 1974). A study of three produced water outfalls and six background stations in coastal Louisiana reported a range of background concentrations of 0.2 to 0.7 pCi/l for ^{226}Ra and 0 to 10.3 pCi/l for ^{228}Ra (CSA, 1991).

Produced Water, Edited by J.P. Ray
and F.R. Engelhart, Plenum Press, New York, 1992

The range of radium concentrations in produced waters discharged to coastal Louisiana was estimated from a data base compiled by the State of Louisiana. Radium-226 concentrations varied from 0 to 930 pCi/l, with an average concentration of 159.2 pCi/l. Radium-226 concentrations varied from 0 to 928 pCi/l, with an average of 164.5 pCi/l.

Radium in produced water undergoes rapid mixing and dilution when discharged to surface water. A study of three low energy, coastal produced water outfalls in Louisiana (CSA, 1991) demonstrated dilution factors of 64 to 1,111 within 50 feet of the outfalls. These stations were considered worst case examples in terms of environmental impact because they are located in estuarine, low energy canal environments.

BIOACCUMULATION OF RADIUM BY FISH AND SHELLFISH

Radium is known to be accumulated by freshwater and marine organisms. Radium is a chemical analogue of calcium and tends to accumulate in bone, shell and exoskeleton. To describe the risk associated with the ingestion of radium in fish and shellfish harvested in coastal Louisiana, the amount of radium in organisms used for food must be estimated.

Concentration factors are commonly used in risk assessment studies to estimate the levels of radionuclides in aquatic organisms. The concentration factor (CF) is a function of the concentration in the water or sediment (C), and the equilibrium concentration (on a wet weight basis) in the organism (Q).

This method assumes there is a linear relationship between C and Q and that the concentration factor is independent of the concentration in the environment. In reality, however, concentration factors are affected by many variables, including the species of animal and the concentration of radium in the water.

The concentration factors commonly used in dose assessment studies (IAEA, 1982, 1985) are appropriate for screening-level assessments, but may overestimate the concentrations in edible portions of marine organisms. There are few studies available which describe concentration factors for radium in the edible portions of marine animals, but concentration factors of 100 (IAEA, 1982) appear to be conservative estimates (Table 1). A more recent recommendation in IAEA (1985) suggested larger concentration factors for radium in marine fish and molluscs, based on the largest values reported in the literature. Table 1 summarizes the International Atomic Energy Agency (IAEA) recommended values, and values from the literature summarizing the concentration factors found in marine organisms used for food. The assessment described here used the values of 100 for fish, molluscs and crustaceans recommended in IAEA (1982), because they are more representative of radium concentration factors reported in the literature than the values recommended in IAEA (1985).

HUMAN HEALTH EFFECTS

The health effects of radium can be attributed to the radioactive emissions of the isotopes and their daughters. The alpha, beta and gamma radiation released by the decay of radium and its daughters cause ionization of cellular components followed by the mutation or death of affected cells.

Most of the information concerning the health effects of radium comes from studies of two groups of people: radium dial painters who ingested radium paint and patients who were injected with ^{224}Ra for treatment of spinal arthritis and tuberculosis of the bone (NAS, 1988). Health effects associated with the ingestion of ^{226}Ra include bone sarcoma and cancer of the paranasal sinuses and mastoid air cells (NAS, 1988).

Current practice in radiation protection and risk assessment is to assume there is some cancer risk associated with even very low doses of radiation. Risk factors are derived from epidemiological data and extrapolated down to low doses to describe the cancer risk associated with small exposures. Using a model called RADRISK, and assuming a person drinks two liters of water each day, the United States Environmental Protection Agency (USEPA) has estimated the risk factor associated with the ingestion of ^{226}Ra in drinking water to be 4.4 x 10^{-6} per pCi/l lifetime, and the risk factor for ^{228}Ra to be 3.8 x 10^{-6} per pCi/l lifetime (Federal Register, 1991, USEPA, 1991). These values are equivalent to 2.2 x 10^{-6} per pCi/day lifetime and 1.9 x 10^{-6} per pCi/day lifetime.

Table 1. Recommended concentration factors and reported values

		Literature	IAEA(1982)	IAEA(1985)
Fish	fresh	21 - 130[1]	100	500
	whole	2 - 100[3]		
Molluscs	soft parts	44 - 170[1,2]	100	1000
	soft parts	2 - 240[3]		
Crustaceans	soft parts	35 - 360[1]	100	100
	whole	4 - 170[3]		

[1]Iyvengar et al., 1980
[2]Iyvengar, 1984
[3]Calculated from data in CSA (1991)

The RADRISK model predicts the dose of ionizing radiation delivered to radiosensitive organs by radium and calculates cancer risk based on a synthesis of human epidemiologic data (USEPA, 1991). The RADRISK model results were adjusted for the overprediction of leukemias and lack of prediction of head carcinomas (USEPA, 1991). There is a large amount of uncertainty associated with these estimates. The analysis performed by the USEPA assumes a linear dose-response relationship for bone sarcoma, although the best fit for bone sarcoma in the radium dial painters is quadratic (USEPA, 1991). If the true relationship is quadratic, the USEPA risk factors will be overestimates. There may also exist a practical threshold for bone sarcoma (USEPA, 1991). Additional uncertainties and assumptions in the USEPA analysis are described in USEPA (1991). The assessment reported here used the current USEPA risk factors for radium.

RISK ASSESSMENT FOR RADIUM IN PRODUCED WATER

A health risk assessment for an environmental pollutant describes the discharge of the contaminant, its transport and fate in the environment and the associated human exposure. Human-health risks are then calculated based on data and models that relate exposures to risk (Till and Meyer, 1983). Two measures are commonly used to describe the probability that harm will result from exposure to a risk agent (Cohrssen and Covello, 1989): 1) Individual lifetime risk — the estimated increase in probability that an individual will experience a specific adverse health effect as a result of exposure to a risk agent over a lifetime; and 2) Population risk — the estimated number of deaths in the exposed population.

The results of a risk assessment are used by risk managers, such as regulatory agencies, to determine the need for regulation or remediation, and to set discharge limits. For the purpose of setting standards, the U.S. Environmental Protection Agency (USEPA) currently considers excess individual lifetime cancer mortality risks less than 1 x 10^{-6} (one-in-a-million) to 1 x 10^{-4} (one-in-ten-thousand) to be acceptable (Federal Register, 1991).

A common approach to the assessment of risk associated with the discharge of a contaminant to the environment is to perform an assessment in two steps. The first step is a screening assessment using simple models and conservative assumptions which results in an overestimate of risk. For discharges identified as presenting potentially significant risks by a screening analysis, more realistic and complex assessments are performed in a second step. In this second step, conservative assumptions and simple models are replaced by best estimates and more sophisticated models.

This paper reports on the first phase of a risk assessment for radium discharged to coastal Louisiana in produced waters. An initial screening-level risk assessment was performed to determine if a more comprehensive and realistic analysis is required. The analysis presented here does not use state-of-the art models or analytical techniques, but should be viewed as a screening exercise to determine if there is a potential for significant risk.

The first step in a risk assessment for radium discharged in produced waters is the identification of the major exposure routes and receptors of concern. The ingestion of contaminated fish and shellfish is expected to be the most significant exposure route, because people spend only a small amount of time in contact with coastal water, and because radium is known to accumulate in fish and shellfish. The important receptors for radium discharged in produced water are people ingesting a large amount of fish and shellfish harvested near a produced water outfall. People subsisting on locally harvested fish and shellfish and workers fishing from the platforms are represented by the maximally exposed individual — the hypothetical person ingesting a large amount of seafood harvested near an outfall.

In the screening-level analysis described here, exposures to man from ^{226}Ra and ^{228}Ra were estimated for water concentrations of radium representative of produced water outfalls and for the levels measured in organisms in a study performed by Continental Shelf Associates (CSA, 1991). For comparison, risks associated with background radium concentrations were also calculated. This screening-level assessment represents a conservative analysis, involving a number of simplistic assumptions that necessarily result in an over-estimate of the risk associated with the discharge of radium in produced water.

ESTIMATION OF INDIVIDUAL LIFETIME RISK

Data Sets Used in the Assessment

The assessment of individual risk was performed using two sets of data to describe radium concentrations in water and in seafood resulting from produced water discharges. The first data set is from a field study of three produced water outfalls and six background stations in coastal Louisiana (CSA, 1991). The outfalls included in this field study did not cover the entire range of potential discharges to coastal Louisiana, and five potential (or modeled) discharge scenarios were included in this assessment. These two data sets are described in more detail below.

CSA Study. Three produced water discharge platforms and six reference stations were sampled in this study. ^{226}Ra and ^{228}Ra concentrations were measured in the discharged produced water and at stations located 25 and 50 feet from the outfalls in three directions. Radium concentrations were also measured in sediment and biological samples from each station. The three outfalls were located in canals along the Louisiana coast. At these stations, there were produced water discharges for several years before sampling, with discharge rates typical of produced water discharges in coastal Louisiana (3,000 to 5,000 bbl/day).

Potential Discharge Scenarios. For all of these modeled scenarios, it was assumed that the concentration of ^{226}Ra (pCi/l) was equal to the concentration of ^{228}Ra (pCi/l) in each discharge. To represent background conditions, two zero discharge scenarios were included, with assumed $^{226}Ra/^{228}Ra$ concentrations in water of 0.1 and 1.0 pCi/l. The three discharge scenarios included in the assessment were 30 pCi/l, 500 pCi/l and 2,000 pCi/l. The 2,000 pCi/l discharge represents the very few high discharge concentration outfalls in coastal Louisiana, while the 500 and 30 pCi/l discharges are more typical.

Steps in the Assessment

The calculation of the individual lifetime cancer risk associated with the discharge of radium in produced water involved the following steps: 1) estimation of the concentration of

^{226}Ra and ^{228}Ra in coastal water; 2) estimation of the concentration of ^{226}Ra and ^{228}Ra in fish and shellfish; 3) estimation of the amount of fish and shellfish that could be consumed by individuals; 4) calculation of the amount of ^{226}Ra and ^{228}Ra consumed by individuals in seafood and 5) calculation of the individual lifetime risk associated with the estimated radium intake in fish and shellfish. These steps in the risk assessment are described in detail below.

Radium Concentrations in Water

Field Study. For the data collected by CSA (1991), radium concentrations in water are not needed for a risk assessment study because the concentration of radium in fish and shellfish was measured. The radium concentrations measured in water, fish, oysters and crustaceans at the sites studied by CSA (1991) are given in Table 2.

Potential Discharge Scenarios. The five potential (or modeled) scenarios used in the analysis require the estimation of water concentrations to calculate concentrations in aquatic organisms used for food. The concentrations in water resulting from the three potential discharge concentrations (30, 500 and 2,000 pCi/l) were calculated assuming a dilution factor of 100 within 50-100 feet of the outfall. The use of a dilution factor of 100 is a reasonable, but conservative assumption, because most fish and shellfish will be harvested further than 100 feet from a discharge point and because dilution factors greater than 100 are expected within 100 feet (dilution factors calculated for the three sites studied in CSA (1991) ranged from 64 to 1,111 at 50 feet). A dilution factor of 100 applied to the three discharge scenarios described above (30, 500 and 2,000 pCi/l) results in water concentrations (of both ^{226}Ra and ^{228}Ra) of 0.3, 5 and 20 pCi/l (Table 3). Assumed concentrations in water for the two zero discharge, background scenarios were 0.1 and 1.0 pCi/l.

Concentration of Radium in Aquatic Organisms Used For Food

Field Study. The data set extracted from CSA (1991) includes measurements of ^{226}Ra and ^{228}Ra concentrations in fish, oysters and crustaceans sampled within 50 feet of the outfalls and at the six background stations. These data are given in Table 2.

Potential Discharge Scenarios. The concentrations of ^{226}Ra and ^{228}Ra in fish, molluscs and crustaceans were calculated for the five potential scenarios described below (water concentrations of 0.1, 1, 0.3, 5 and 20 pCi/l) using the concentration factor method. For the purpose of this screening-level assessment, the conservative IAEA Concentration Factors (100 for fish, molluscs and crustaceans) were used. The estimated concentrations of ^{226}Ra and ^{228}Ra in fish, molluscs and crustaceans for the five potential discharge/water concentration scenarios are given in Table 3.

Amount of Radium Ingested in Seafood

To calculate risk from ingestion of radium in seafood, the amount of seafood consumed by an individual per year is needed. Food consumption patterns vary among individuals and with age and geographic location. This study used intake rates from a survey commissioned by the National Marine Fisheries Service in 1973-1974 and described in Rupp et al. (1980). Intake rates for people living in the West South Central Region of the United States (Arkansas, Louisiana, Oklahoma and Texas) were used in the assessment presented here.

Rupp et al. (1980) report consumption rates for three age groups (1-11 years, 12-18 years, and 19-98 years) and for three types of seafood (freshwater fish, saltwater fish and shellfish). A regional distribution of shellfish consumed by species (i.e. percent of the total of each species that is consumed by each U.S. region) is presented in Rupp et al. (1980). These data were used to estimate the percent of shellfish consumption represented by molluscs (23.8%) and by crustaceans (76.2%) for the West South Central Region. Table 4 gives the intake rates (kg/year) for the three age groups, calculated from the data in Rupp et al. (1980).

Intake levels for adults (19-98 years) were used in this analysis. Three intake rates were used in this preliminary assessment: the maximum intake rate, the average intake rate for fish/shellfish consumers, and the average intake rate for the whole population (Table 4). In all cases it was conservatively assumed that 50% of a person's entire seafood intake was harvested near a produced water outfall.

Table 2. Values used in calculation of individual risk for field data from the CSA (1991) study. Concentrations of ^{226}Radium and ^{228}Ra in discharge, water and aquatic organisms

Station	Discharge $^{226}Ra/^{228}Ra$ (pCi/1)	Water Conc.[1] $^{226}Ra/^{228}Ra$ (pCi/1)	Fish[2] $^{226}Ra/^{228}Ra$ (pCi/g)	Oysters[3] $^{226}Ra/^{228}Ra$ (pCi/g)	Crustaceans[4] $^{226}Ra/^{228}Ra$ (pCi/g)
Site 1	228.9/383.0	0.4/0.93	0.014/0.005	0.007/0.003	0.073/0.094
Site2	110.6/244.4	0.5/0.22	0.0.022	0.008/0.011	0.068/0.025
Site3	251.9/254.9	0.9/4.0	0.041/0.012	NS/NS	0.125/0.243
Background 1a	0/0	0.3/0.2	NS/NS	0.001/0.036	0.014/0.004
Background 1b	0/0	0.2/0	0.005/0.021	0.004/0.011	0.024/0.025
Background 2a	0/0	0.7/0	0.002/0.030	0.004/0.003	NS/NS
Background 2b	0/0	0.5/0	0.031/0.127	0.001/0.01	0.069/0.033
Background 3a	0/0	0.7/3.3	0.018/0.058	NS/NS	0.016/0.058
Background 3b	0/0	0.7/10.3	0.030/0.017	NS/NS	0.077/0.045

NS: no sample
[1] Average of measured concentration at three stations, each 50 feet from discharge; for reference samples value is average of two samples taken.
[2] All fish samples were *Cynoscion* spp. (seatrout).
[3] All oysters were *Crassostrea virginica*.
[4] Crustacean samples at stations 1 and 3 were *Penaeus* spp. (shrimp) crustaceans at station 2 were *Callinectes* spp. (crab).

Table 3. Values used in calculation of individual risk for five potential discharge/water concentration scenarios. Concentrations ^{226}Radium and ^{228}Ra in discharge, water and aquatic organisms

Scenario	Discharge (pCi/1)	Water Conc.[1] (pCi/1)	Fish[2] (pCi/g)	Molluscs[2] (pCi/g)	Crustaceans[2] (pCi/g)
Low Background	0	0.1	0.01	0.01	0.01
High Background	0	1	0.1	0.1	0.1
Low Discharge	30	0.3	0.03	0.03	0.03
High Average Discharge	500	5	0.5	0.5	0.5
High Discharge	2000	20	2.0	2.0	2.0

[1] Assumes dilution factor of 100 within 50-100 feet of outfall.
[2] Calculated using concentration factor method, CF=100

Table 4. Marine fish and shellfish consumption in Arkansas, Louisiana, Oklahoma and Texas (kg/year) [1,2]

Age Group (years)	Type	Maximum Consumer	Consumer Average	Population Average
1-11 (n=510)	fish	15.1	1.8	1.5
	molluscs	2.7	0.5	0.1
	crustaceans	8.8	1.5	0.4
12-18 (n=194)	fish	28.6	2.8	2.4
	molluscs	3.4	0.6	0.2
	crustaceans	10.8	1.8	0.5
19-98 (n=1424)	fish	31.2	4.0	3.6
	molluscs	5.9	0.8	0.3
	crustaceans	19.0	2.7	1.1

[1] Data derived from Rupp et al. (1980)
[2] Shellfish divided into molluscs and crustaceans by percentage derived from data in Rupp et al. (1980) for entire United States (23.8% molluscs, 76.2% crustaceans).

The amount of radium consumed in seafood per day was calculated as follows:

$$\underset{\text{(pCi/day)}}{\text{Consumption}} = \underset{\text{(pCi/g)}}{\text{Food Conc.}} \times \underset{\text{(g/kg)}}{1000} \times \underset{\text{(kg/year)}}{\text{Intake}} \times \underset{\text{(year/days)}}{1/365} \quad (1)$$

Using the intake rates described above, ^{226}Ra and ^{228}Ra consumption rates in units of pCi/day were calculated for the concentrations of radium measured in fish and shellfish in the CSA (1991) study, and for the estimated levels of radium in seafood for the five potential discharge scenarios.

Calculation of Individual Lifetime Risk

Conservative, upper bound estimates of individual lifetime risk were calculated using the USEPA risk factors (^{226}Ra: 2.2×10^{-6} per pCi/day lifetime intake, ^{228}Ra: 1.9×10^{-6} per pCi/day lifetime intake). Individual lifetime risk estimates were calculated separately for ^{226}Ra and ^{228}Ra and then summed for each exposure scenario as follows:

$$\text{Individual Lifetime Risk} = [^{226}\text{Ra Consumption (pCi/day)}] \times [^{226}\text{Ra Risk Factor}] + [^{228}\text{Ra Consumption (pCi/day)}] \times [^{228}\text{Ra Risk Factor}] \quad (2)$$

Table 5 gives the estimates of individual lifetime risk for ingestion of ^{226}Ra and ^{228}Ra in seafood for the three produced water discharges and six background stations sampled in the CSA (1991) study. No individual lifetime risk estimates based on the CSA (1991) data exceeded a level of 1×10^{-4}. Levels of 1×10^{-5} were exceeded for Sites 1 and 3 for the maximum consumer (Table 5). Levels of 1×10^{-5} were also exceeded for the hypothetical maximum consumer at three of the six reference sites.

Table 6 presents individual lifetime risk estimates for the five potential discharge scenarios. Individual lifetime risk estimates based on the potential discharge scenarios exceeded 1×10^{-4} for the maximum consumer at the two higher discharge sites (500 pCi/l and 2,000 pCi/l). A risk of 1×10^{-5} was exceeded for the maximum consumer/high background scenario, and for all intake rates at the two high discharge outfalls (500 and 2,000 pCi/l).

Table 5. Individual lifetime risk estimates for consumption of fish, molluscs and crustaceans harvested from waters with enhanced levels of radium - estimates based on data from a field study in coastal Louisiana (CSA, 1991)

	Discharge $^{226}Ra/^{228}Ra$ (pCi/l)	Maximum Consumer	Lifetime Risk Consumer Average	Population Average
Site 1	228.9/383.0	1.1E-5	1.5E-6	7.1E-7
Site 2	110.6/244.4	7.2E-6	1.0E-6	5.1E-7
Site 3 [1]	251.9/254.9	2.4E-5	3.4E-6	1.6E-6
Background 1a	0/0	1.6E-6	2.2E-7	8.8E-8
Background 1b	0/0	5.0E-6	6.9E-7	4.1E-7
Background 2a	0/0	2.7E-6	3.6E-7	3.1E-7
Background 2b	0/0	1.9E-5	2.5E-6	1.8E-6
Background 3a	0/0	1.0E-5	1.4E-6	9.5E-7
Background 3b	0/0	1.1E-5	1.5E-6	8.6E-7

[1] No mollusc samples at station 3, estimates based on concentrations of radium in fish and crustaceans only.

Table 6. Individual lifetime risk estimates for consumption of fish, molluscs and crustaceans harvested from waters with enhanced levels of radium - estimates based on analysis of five potential discharge/water concentration scenarios

Scenerio	Discharge (pCi/1)	Water Conc.[1] (pCi/1)	Maximum Consumer	Lifetime Risk Consumer Average	Population Average
Low Background	0	0.1	3.2E-6	4.3E-7	2.8E-7
High Background	0	1	3.2E-5	4.3E-6	2.8E-6
Low Discharge	30	0.3	9.5E-6	1.3E-6	8.4E-7
High Average Discharge	500	5	1.6E-4	2.1E-5	1.4E-5
High Discarge	2000	20	6.3E-4	8.5E-5	5.6E-5

[1]Water concentration at 50 to 100 feet from outfall.

These risk estimates are based on the following assumptions: (1) radium discharged in produced water is reduced by a factor of 100 before coming in contact with edible fish and shellfish; (2) an individual gets one-half of their yearly fish and shellfish from near (i.e., where the dilution factor reaches 100, probably within 100 feet) a produced water outfall for their entire lifetime; (3) a concentration factor of 100 can be used to estimate the concentration of radium in fish and shellfish; and (4) the USEPA risk factors for radium accurately describe the risk associated with the ingestion of small amounts of radium over a lifetime.

ESTIMATION OF POPULATION RISK

A conservative, upper-bound estimate of the population risk associated with the ingestion of radium from produced waters was calculated using simple models and assumptions. To estimate total population risk several calculations were needed — the "average" increase in radium concentrations in coastal Louisiana; the resulting "average" concentration of radium in fish and shellfish; and the amount of fish and shellfish ingested from the area.

Radium Concentrations in Water

A simple steady state box model (IAEA, 1982) was used to estimate the increase in ^{226}Ra and ^{228}Ra concentrations in the coastal zone of Louisiana resulting from the discharge of radium in produced water. This analysis was needed to assess the total impact of produced water outfalls on coastal Louisiana. All radium discharged into the coastal zone was assumed to remain in the water column. The model assumes a constant input of radium and complete mixing within a given volume. The steady state concentration of radium in water can be estimated as:

$$C = Q / K_e V \quad (3)$$

where:

C = the change in radium water concentration (pCi/l)
Q = the radium discharge rate (pCi/day)
V = the volume of water in the Louisiana coastal zone (l)
K_e =the fractional loss rate of water from the mixed volume (day^{-1})

The removal constant K_e is estimated as:

$$K_e = q + r/V \quad (4)$$

where:

q = the decay constant
r/V = the fractional loss rate of water from the mixed volume.

In tidally influenced coastal seas, V may be described by the tidal excursion, and r/V by the net water movement (IAEA, 1982).

A conservative estimate (i.e. small) of the volume of water in the Louisiana coastal zone was developed based on the area of surface water in the coastal zone of Louisiana (2631 km^2, Department of Commerce, 1990), and an assumed depth of 3 meters. These assumptions result in a calculated volume of 7.9×10^{12} liters. The fractional loss rate used was 0.015 per day. This value is based on studies of the Barataria and Terrebonne-Timbalier Basins in Louisiana reported in Boesch and Rabalais (1989).

The rate of ^{226}Ra and ^{228}Ra discharged into the coastal zone of Louisiana (Q) was estimated by summing the discharges (pCi/day) for all of the stations in the database compiled by the State of Louisiana. These totals were 4.39×10^{10} pCi/day for ^{226}Ra and 4.83×10^{10} pCi/day for ^{228}Ra. The increase in ^{226}Ra and ^{228}Ra concentrations due to produced water discharges (C) was calculated using the box model and parameters described above. The resulting increases in radium concentrations were 0.37 pCi/l for ^{226}Ra and 0.41 pCi/l for ^{228}Ra.

Concentration of Radium in Aquatic Organisms Used For Food

The "average" concentration of ^{226}Ra and ^{228}Ra in fish and shellfish harvested in coastal Louisiana was estimated using the concentration factor method. The IAEA values (IAEA, 1982) (100 for fish, molluscs and crustaceans) were used in this analysis. The "average" radium concentrations in fish, molluscs and crustaceans were calculated for 1) the estimated increase in water concentration due to produced water discharges (^{226}Ra: 0.37 pCi/l, ^{228}Ra: 0.41 pCi/l); 2) a low estimate of background radium levels (0.1 pCi/l for both ^{226}Ra and ^{228}Ra) and; 3) a high estimate of background radium concentrations (1.0 pCi/l for both ^{226}Ra and ^{228}Ra). The estimated average concentrations in fish and shellfish for these water concentrations are given in Table 7.

Amount of Radium Ingested in Seafood

To estimate the number of excess cancers resulting from ingestion of radium, the amount of fish and shellfish harvested and ingested per year from coastal Louisiana is needed. The catch in millions of pounds live weight for Louisiana in 1989 was 1.23×10^9 pounds (Department of Commerce, 1990). Some of this catch was from offshore — this analysis assumed that 50% of the reported catch for Louisiana was harvested in coastal waters. In 1989, 73.3% of the fish and shellfish harvested in the United States was used for human food (Department of Commerce, 1990). Not all of the fish and shellfish catch is consumed — the value for Louisiana reported above includes fish bones and skin but not clam or oyster shells. In this analysis, it was assumed that 75% of the fish and shellfish used for human food was actually consumed. The amount of fish and shellfish consumed per day was calculated as follows:

$$(1.23 \times 10^9 \text{ lbs/year}) \times (0.5 \text{ inshore}) \times (0.73 \text{ for food}) \times (0.75 \text{ eaten}) \times (453.6 \text{ g/lb}) \times (1/365 \text{ year/days}) \approx 4.2 \times 10^8 \text{ g/day} \quad (5)$$

The amount of radium ingested in fish and shellfish (pCi/day) was calculated by multiplying the amount ingested in the region over a year (4.2×10^8 grams/day) times the "average" concentration of ^{226}Ra and ^{228}Ra in fish and shellfish (pCi/g) (Table 7).

Table 7. Calculated average radium concentrations in seafood and amount ingested

	^{226}Ra			^{228}Ra		
	Water (pCi/l)	Food[1] (pCi/g)	Ingested (pCi/d)	Water (pCi/l)	Food[1] (pCi/g)	Ingested (pCi/d)
Produced Water	0.37	0.037	1.6E7	0.41	0.041	1.7E7
Low Background	0.1	0.01	4.2E6	0.1	0.01	4.2E6
High Background	1.0	0.1	4.2E7	1.0	0.1	4.2E7

[1] Calculated using the concentration factor method, CF=100.

Calculation of Population Risk

Conservative, upper bound estimates of population risk were calculated using the USEPA risk factors (Federal Register, 1991) of 2.2×10^{-6} per pCi/day lifetime intake for ^{226}Ra and 1.9×10^{-6} per pCi/day lifetime intake for ^{228}Ra. The number of excess cancers expected over a 70 year lifetime was calculated as:

$$\text{Population Risk} = \text{Consumption (pCi/day)} \times \text{Risk Factor} \quad (6)$$

The resulting estimates of excess cancers per year of operation are given in Table 8. The risk to the individual associated with the increase in radium in the water predicted by this analysis (0.37 pCi/l ^{226}Ra; 0.41 pCi/l ^{228}Ra) is 2.5×10^{-5} assuming the maximum consumption rate (Table 4) and 2.2×10^{-6} assuming the average population consumption rate.

Table 8. Estimated excess cancers per lifetime from radium in coastal Louisiana waters*

	^{226}Ra (pCi/l)	Cancers	^{226}Ra (pCi/l)	Cancers	Total cancers
Produced Water	0.37	35	0.41	32	67
Low Background	0.1	9	0.1	8	17
High Background	1.0	92	1.0	80	172

* At current levels of produced water discharge.

DISCUSSION

Using the conservative assumptions outlined above, a level of risk which could be considered significant (greater than 1 x 10^{-4}) was estimated for the person consuming the most seafood under the two highest discharge scenarios (500 and 2000 pCi/l). No risk estimates for samples taken at three coastal discharge sites exceeded 1 x 10^{-4}.

Based on a simple, conservative analysis, the total number of excess cancers expected over a 70 year lifetime from ^{226}Ra and ^{228}Ra discharged to coastal Louisiana waters is 67. A similar analysis, using a range of possible background concentrations (0.1 - 1.0 pCi/l) predicted from 17 to 172 excess cancers from background radium in the region. The major uncertainties and conservatisms in this screening-level analysis are:

1. The concentration of radium in water and the geographic distribution of contaminated shellfish. The analysis of individual risk used concentrations likely to be measured at the outfall of a produced water discharge when, in fact, considerable dilution occurs with distance from the discharge point. A reduction in the radium concentrations in fish and shellfish also occurs with increasing distance. The concentration of radium in the water in which fish and shellfish are harvested is critical to the estimation of risk. For the calculation of population risk, a simple box model was used that assumed complete mixing of all discharged radium, and a resultant "average" concentration of radium in fish and shellfish harvested from the region. In fact, radium concentrations in water and in fish and shellfish are variable over the area.
2. The concentration factor used to calculate the concentration of radium in fish, molluscs and crustaceans. Commonly used concentration factors are higher than those derived from the CSA (1991) data set. The concentration factor used in the analysis has a large effect on the resulting risk estimates. Conservative IAEA concentration factors were used in this assessment, which probably resulted in an overestimate of the concentration of radium in food.
3. The distribution of intake rates and the percent of consumed fish and shellfish that is radium contaminated. This analysis assumed that the maximally exposed individual harvested one-half of his seafood from near a produced water outfall. These estimates of intake probably overestimate the amount of fish and shellfish harvested near an outfall even for individuals subsisting on locally harvested seafood.
4. The risk factors for radium. This screening-level analysis used the USEPA risk factors for ^{226}Ra and ^{228}Ra (Federal Register, 1991). These risk factors overestimate the health risks associated with exposure to small amounts of radium.

Because of the many conservative assumptions incorporated into this screening-level analysis, it can be concluded that the risks associated with the discharge of produced water to coastal Louisiana are small. The results of this study do, however, suggest the need for a more detailed analysis of the potential risks to individuals subsisting on locally harvested fish and shellfish which may be harvested near a produced water outfall. The extent to which such a maximally exposed population exists will be investigated and a more detailed, comprehensive assessment of the risks to this population will be performed.

REFERENCES

Boesch, D.F., and Rabalais, N.M., 1989, Environmental Impact of Produced Water Discharges in Coastal Louisiana, Louisiana Universities Marine Consortium, Louisiana State University, A&M College, and Aero-Data Corporation.

Cherry, R.D. and Shannon, L.V., 1974, The alpha radioactivity of marine organisms. Atomic Energy Review. 12:3.

Cohrssen, J.J., and Covello, V.T., 1989, Risk Analysis: A Guide to Principles and Methods for Analyzing Health and Environmental Risks, United States Council on Environmental Quality, Executive Office of the President.

CSA, 1991, Measurements of Naturally Occurring Radioactive Materials (NORM) at Three Produced Water Outfalls, Continental Shelf Associates, Inc., Jupiter, Florida. prepared for Mid-Continent Oil and Gas Association, Baton Rouge, Louisiana.

Department of Commerce, 1990, Fisheries of the United States 1989, Current Fisheries Statistics No. 8900. U.S. Department of Commerce, Bureau of the Census, Washington, D.C.

Federal Register, 1991, Environmental Protection Agency, National Primary Drinking Water Regulations; Radionuclides, 40CFR Parts 141,142. 56:138:33050.

IAEA, 1982, Generic Models and Parameters for Assessing the Environmental Transfer of Radionuclides from Routine Releases, Safety Series No. 57, Procedures and Data, International Atomic Energy Agency, Vienna.

IAEA, 1985, Sediment Kds and Concentration Factors for Radionuclides in the Marine Environment, Technical Report Series No. 247, International Atomic Energy Agency, Vienna.

Iyengar, M.A.R., 1984, Distribution in nature, *in*: "The Behavior of Radium in Waterways and Aquifers", IAEA-TEC DOC 301, International Atomic Energy Agency, Vienna.

Iyengar, M.A.R., Rajan, M.P., Ganapathy, S., and Kamath, P.R., 1980, Sources of radiation exposure in a low monazite environment, *in*: "The Natural Radiation Environment III", Symp. Proc. Houston, 1978, CONF-780422, Technical Information Center, Springfield, pp. 1090-1106.

NAS, 1988, Health Risks of Radon and Other Internally Deposited Alpha Emitters, National Academy of Sciences, National Research Council, Committee on the Biological Effects of Ionizing Radiation, (BEIR IV), National Academy Press, Washington, 1988.

Rupp, E.M., Miller, F.L., and Baes III, C.F., 1980, Some results of recent surveys of fish and shellfish consumption by age and region of U.S. residents. Health Physics 39:165.

Snavely, E.S., Jr., 1989, Radionuclides in Produced Water, Production Department, American Petroleum Institute, Washington, D.C.

Till, J.E. and Meyer, H.R., 1983, "Radiological Assessment: A Textbook on Environmental Risk Analysis", NUREG/CR-3332, ORNL-5968, U.S. Nuclear Regulatory Commission, Washington, D.C.

USEPA, 1991, Final Draft For the Drinking Water Criteria Document on Radium, Prepared for the Drinking Water Standards Division, Office of Ground Water and Drinking Water, Office of Water, United States Environmental Protection Agency, Washington, D.C.

BIOLOGICAL FATE AND EFFECT OF COALBED METHANE PRODUCED WATERS DISCHARGED INTO STREAMS OF THE WARRIOR BASIN, ALABAMA

P. E. O'Neil[1], S. C. Harris[1], M. F. Mettee[1], H. R. Isaacson[2], and J. M. Evans[2]

[1] Geological Survey of Alabama
Tuscaloosa, Alabama
USA

[2] Gas Research Institute
Chicago, Illinois
USA

INTRODUCTION

Disposal of natural gas produced waters is a significant concern from both the economic view of environmental control costs and actual pollutant control. Produced waters are typically disposed by injection underground into permitted geologic horizons. Some produced waters, however, are permitted for surface disposal by either land application or stream discharge. This practice is limited because many produced waters are of such high salinity that surface discharge could cause significant degradation of receiving stream water quality and the biological environment.

Development of methane from coal seams in Alabama has provided a unique opportunity to critically examine many of the issues associated with disposal of produced waters. Because coalbed methane produced waters are typically less saline than conventional natural gas waters, Alabama environmental regulatory agencies permit the surface discharge of most of these produced waters. Consequently, research has been directed at examining instream biological effects, water-quality changes, and the aquatic toxicology of these waters. Alabama experienced significant drilling activity for methane from 1989 through the end of 1990. Produced water generated in 1990 was 32 million barrels (bbls) and in 1991 was 95 million bbls. The majority of produced water generated from development of coalbed methane has been surface disposed, either land application or stream discharge.

This paper will first review studies which contain relevant information to the biological fate and effect of saline waters discharged to stream environments; second, present data relative to coalbed methane studies which illustrate the range of effects encountered; and third discuss the relationship of these studies to regulation of produced water in Alabama.

PREVIOUS INVESTIGATIONS

Instream biological fate and effect studies dealing with produced water discharges have primarily examined effluents associated with oil production and supply (Table 1). Mathis (1965) studied benthic invertebrate community structure in an intermittent stream in Oklahoma receiving brine effluents composed principally of calcium and sodium chloride with some high concentrations

Produced Water, Edited by J.P. Ray
and F.R. Engelhart, Plenum Press, New York, 1992

of sulfate. Mathis found a 34% decrease in species richness immediately below the brine outfall compared to upstream stations. Mayflies were almost totally eliminated, decreasing in density by approximately 92%. Algal blooms in zones of lower turbidity downstream of the discharge generally increased the productivity of benthic invertebrates. Reduction in turbidity due to the high salt concentrations was thought to be responsible for the increase in biological productivity.

Table 1. Summary of instream biological effects in relation to chlorde discharges

Source	Type of effluent	Place of study	Range of chloride (mg/l)	Biological Effects
Mathis (1965)	Oil brines	Oklahoma	29,100 -174,000	Reduced species richness; mayflies almost eliminated; algal blooms in areas of reduced rubidity.
O'Neil et al. (1984)	Oil brines	Alabama	Peak Cl of 4,500 - 6,500	Rich and abundant fauna at stations 5 to 6 miles downstream of salt water disposal wells.
Birge et al. (1985)	Oil brines	Kentucky	70 - 100	No impact
			1,100 - 3,200	Substantial impact
			5,800 - 17,400	Extremely high impact
O'Neil et al. (1987)	Coalbed methane waters	Alabama	Up to 73 downstream of POD; up to 170 in one tributary draining field	No significant changes in stream fauna due to porduced water discharge; surface mine runoff dominant factor impacting many streams receiving produced water discharge.
Cosgrove et al. (1988)	Oil brines	Kentucky	1,900 - 5,900	Significant reduction in species richness of benthic invertebrate community.
			85 - 1,000	Significant recovery following elimination of brine discharges.
O'Neil et al. (1989)	Coalbed methane waters	Alabama	Ambient <200 with short-term peaks up to 500.	No significant effects to either benthic invertebrate or fish community structure.
			565	Instream chronic value.
			>600	Decline in benthic invertebrate density up to 55 percent and species richness up to 95 percent.
O'Neil et al. (1991)	Coalbed methane waters	Alabama	Ambient <200 with short-term peaks up to 650.	No significant effects to either benthic invertebrate or fish community structure over 3 years of continuous discharge.

Cosgrove et al. (1988) examined biological communities of streams draining the Martha oil field in Kentucky. Benthic invertebrate communities recovered significantly following the elimination of produced water discharges from stripper oil wells. Instream chlorides ranged from 1,900 to 5,900 mg/l before and from 85 to 1,000 mg/l after discharge ceased. During discharge of brines in 1986, the species richness of dipterans was reduced by 40%, mayflies 100%, and dragonflies 63%.

A study relating aquatic toxicity testing and actual instream effects was conducted in Kentucky by Birge et al. (1985). This study suggested that no instream impact occurred at chloride concentrations of 70 to 100 mg/l, substantial changes occurred at 1,100 to 3,200 mg/l, and extremely high biological impact occurred at 5,800 to 17,400 mg/l. Interestingly, no mayflies were reported from immediately below the outfall to about 6 miles downstream.

In a study of benthic invertebrates and fishes in streams draining the Citronelle oil field in southern Alabama, O'Neil et al. (1984) found that both faunas were rich and abundant despite having a history of brine and oil spills since 1955. Water-quality records from the region indicated that these past brine spills resulted in peak chlorides ranging from 4,500 to 6,500 mg/l for unspecified periods of time (Powell et al. 1973).

These studies illustrate that saline discharges have variable effects depending on the biology of the receiving stream and the instream concentration of produced water. Biological community structure will typically return to predischarge conditions after discharge ceases. Many of the studies placed emphasis on the element chloride as intimately involved in biological toxicity and the concentration of chloride as a direct correlate of toxicity or impact to biological communities. This paper will continue with this precedent while acknowledging that other substances may contribute to the toxicity of coalbed methane produced waters.

COALBED METHANE PRODUCED WATER QUALITY AND TREATMENT

The chemical quality of discharged produced waters depends on a variety of factors including well completion technique, stimulation methods, volume of water produced, and treatment in holding lagoons. Composite waters from the Cedar Cove field, Alabama are highly buffered with pH ranging from 7.8 to 8.8 and bicarbonate ranging from 450 to 1,000 mg/l. High iron and manganese concentrations cause significant operational problems to most producers in maintaining discharge permit requirements. Cedar Cove waters, after lagoon treatment, rarely exceeded 1.0 mg/l dissolved iron and were typically <0.5 mg/l from 1987 to 1989. Upsets of lagoon chemistry can occur when stimulation acids are mixed with produced waters. At such times metals are generally mobilized from bottom sludge and dissolved concentrations of iron exceeding 20 mg/l can result. Sulfate was usually less than 100 mg/l. Trace metals including arsenic, lead, mercury, strontium, and zinc have been detected in individual well samples (O'Neil et al., 1989). Chloride is a major ionic constituent of these produced waters and the significant controlling parameter relative to discharge permits. Average daily chloride of Cedar Cove lagoon waters ranged from 300 to near 5,400 mg/l, averaging near 1,700 mg/l from 1987-89. Salinity of produced waters will depend upon which wells are contributing to holding lagoons at the time of sampling.

Prior to 1990 produced waters received remedial treatment in holding lagoons before stream discharge. Functionally, the lagoons provided passive aeration to produced waters (waters entering lagoons have dissolved oxygen concentrations usually no greater than 3.0 to 4.0 mg/l), thereby promoting oxidation processes and subsequent precipitation of metals such as iron and manganese. Also, limited biological treatment for nonrefractory organics occurred because algal, benthic invertebrate, and bacterial populations colonize most lagoons. Storage capacity is also created for holding waters during low streamflows. With today's significant gas and water production, produced water treatment has become more important and technology intensive. Forced aeration, larger lagoons, separate aeration-storage lagoon systems, chemical coagulation of some waters, and extensive water-quality monitoring systems are currently used throughout the industry.

CEDAR COVE BIOMONITORING STUDIES

A series of studies was initiated in 1983 to define instream effects associated with development of methane from both the perspective of nonpoint effects resulting from methane field development and that of discharged produced water effects on water-quality regimes and resident biological communities. These studies included a general reconnaissance of the water quality and biological characteristics of streams draining an active methane degasification field and receiving permitted discharge of produced waters; a series of experimental studies undertaken to develop a model relating the chloride concentration of produced water to benthic invertebrate community effects; and finally a long-term monitoring study of water quality and biological communities in a stream receiving controlled produced water discharge over several years.

Biological investigations were conducted in a portion of the Cedar Cove field in Tuscaloosa County, Alabama (Fig. 1). The number of producing wells, over the span of these studies (1983-89), which discharged water into either Hurricane or Little Hurricane Creeks, ranged from 9 to 18. By May of 1991, after biological investigations were completed, there were 349 producing wells in the Cedar Cove field. The Cedar Cove field in the study area is drained by Kepple Creek to the west and Little Hurricane Creek to the east. These streams flow north and are tributaries to Hurricane Creek. Land use in the Hurricane Creek watershed is dominated by surface coal mining, silviculture, coalbed methane development, and limited agriculture. Surface mining in the Hurricane Creek watershed upstream of the point of produced water discharge covered approximately 7.2% of the 158-square kilometer (km^2) watershed. Approximately 7.0% of the Kepple Creek watershed (21 km^2) has been disturbed by mining whereas <0.5% of the Little Hurricane Creek watershed (51 km^2) has been disturbed by mining. Some drainage from former underground mines in the community of Cedar Cove enters Little Hurricane Creek. The Cedar Cove degasification field is partially developed on an old surface mine. Because of the close association between early degasification field development and coal surface mines, environmental factors associated with surface mine runoff were considered important for assessing and monitoring coalbed methane produced water discharges.

Sampling stations were established at eight locations throughout the study area (Fig. 1). Two stations were located on Kepple Creek, three on Little Hurricane Creek, and three on Hurricane Creek. The primary point of discharge during the period of study was into Hurricane Creek between stations HC2 and HC3. Produced water was diverted to Little Hurricane Creek in late 1986 to a controlled experimental facility for conducting concentration-response testing and for the long-term studies. The study area is described in detail by O'Neil et al. (1987, 1989).

Biological and Water Quality Assessment of the Cedar Cove Field

Comparison of water quality between sampling stations (Table 2) revealed that Kepple Creek, particularly the lower reaches, was affected by mine runoff as observed in low pH values and elevated concentrations of total dissolved solids (TDS), sulfate, and manganese. Lower Kepple Creek (KC1) was also affected by two spills of produced water (instream chloride >150 mg/l) resulting from broken water lines. Little Hurricane Creek was relatively undisturbed from mine and gas development during the study as observed in the low TDS, chloride, and manganese values. In comparison, Hurricane Creek was severely affected by mine runoff as observed in the low pH values and elevated concentrations of TDS, sulfate, and manganese. Hurricane Creek additionally received permitted discharge of produced waters just upstream of station HC2 at the mouth of Kepple Creek resulting in elevated chloride concentrations ranging from 1.6 to 73 mg/l and averaging 12.4 mg/l.

Benthic invertebrates were sampled with a modified Hester-Dendy type of artificial substrate. Replicate samplers were collected every six weeks from all stations. Biological community structure within these streams was directly related to ambient water-quality conditions. Mean taxa and density of benthic invertebrates (Fig. 2) was lowest at those stations with the highest degree of mine impact (KC1, KC2, HC2, and HC3) and highest at those stations more reflective of natural water-quality conditions for streams in this part of Alabama (LHC2, LHC3, and HC4). No statistically significant differences ($P>0.05$) were found in benthic invertebrate diversity and density between stations upstream (HC3) and downstream (HC2) of the point of coalbed methane produced water discharge. Consequently, no invertebrate community effects could be attributed to produced water discharge. In contrast, significantly more individuals and taxa were collected at the control station (HC4) than at either stations HC2 or HC3. This effect is due to impacts of mine runoff to benthic invertebrate communities.

The types and proportions of taxa collected in the three Hurricane Creek stations (Table 3) reflect changes in stream habitat due to water-quality effects and physical habitat disruption due

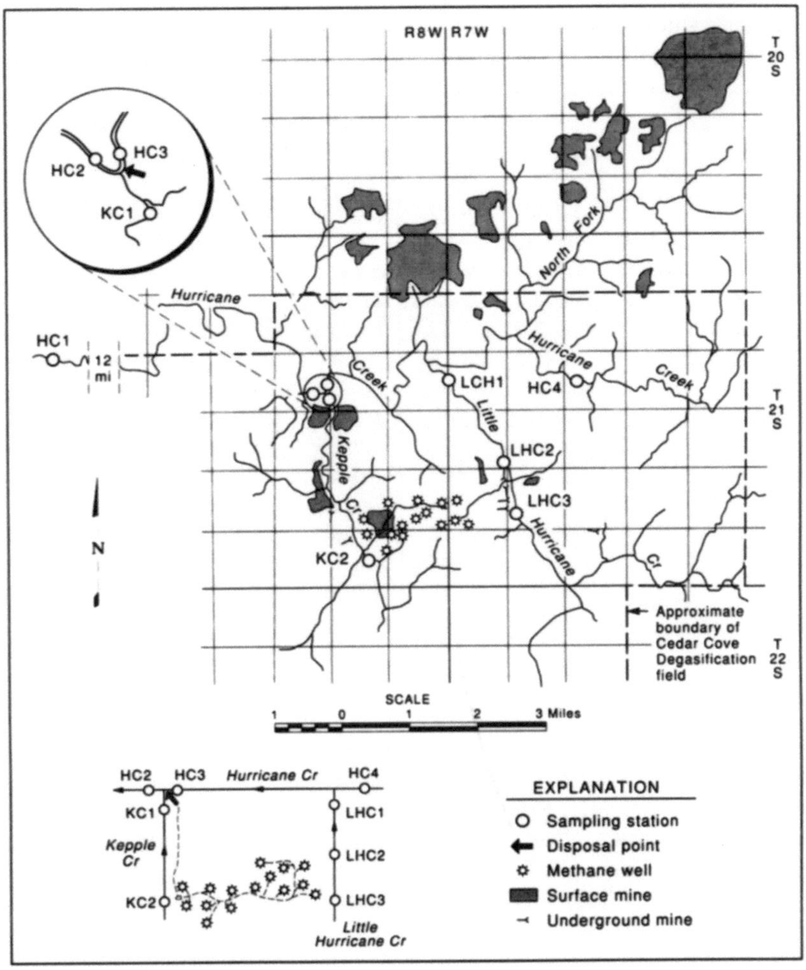

Figure 1. Index map of the Hurricane Creek watershed showing locations of water-quality and biological monitoring stations, methane wells, and coal mines.

to mine runoff. For example, the midge *Microtendipes* and mayfly *Stenacron*, though abundant at the control (HC4), were virtually absent at stations affected by mine runoff. In contrast, taxa such as the alderfly *Sialis* and the caddisfly *Polycentropus* comprised a greater proportion of the total fauna at mine-impacted stations. Some effect appeared likely to the midge *Tanytarsus* due to produced water because of reduced abundance downstream of the discharge point.

Results of this study demonstrate that stream biological communities are influenced substantially by activities which change natural water-quality regimes within watersheds. In particular, runoff from coal surface mines shapes benthic invertebrate communities in a way that results in a less diverse and productive community. Very similar effects to benthic invertebrate community structure, as observed in Hurricane Creek, were reported by Harris et al. (1985) for the Tyro Creek watershed in the Warrior basin. Consequently, within similar sized streams of the Warrior basin similar magnitudes of impact from mining should be predictable. Preliminary evidence from the Hurricane Creek study suggests that produced waters, discharged into streams impacted by mine runoff, contribute no additional impact to previously stressed communities if instream chloride concentrations remain low. Of particular note is the fact that measured chloride downstream of the discharge point did not exceed 100 mg/l during the study (Table 2). A more thorough presentation of study results can be found in O'Neil et al. (1987, 1988).

Table 2. Summary water-quality data collected from seven stations in the Hurricane Creek watershed, October 1983 through January 1984 (O'Neil et al., 1987)

Parameter	Kepple Creek		Little Hurricane Creek		Hurricane Creek		
	KC1	KC2	LHC2	LHC3	HC2	HC3	HC4
Discharge (cfs)	1.1-45	0-17	1.2-106	0.69-108	9.0-274	7.3-230	0.47-86
Specific conductance (μmhos/cm)	42-730	31-310	33-120	30-83	87-675	77-600	17-134
pH (units)	5.0-7.1	5.1-6.7	5.1-7.1	5.1-6.8	3.8-6.3	3.7-6.2	5.1-7.4
Temperature (°C)	2.0-28.0	3.5-25.0	4.0-26.0	1.0-25.5	6.0-30.0	6.0-30.5	3.0-26.0
Dissolved oxygen (mg/l)	6.7-13.9	1.9-13.1	7.1-13.0	6.5-13.2	6.2-12.5	6.2-12.5	6.4-13.1
Total dissolved solids (mg/l)	13-430	20-156	14-90	23-90	44-368	58-388	14-91
Calcium (mg/l)	2.7-15	2.1-9.0	.5-9.2	2.2-4.9	4.2-36	5.1-38	.9-8.3
Magnesium (mg/l)	1.7-12	0.8-4.7	1.0-4.7	0.6-2.3	4.5-29	4.7-30	1.0-7.90
Sodium (mg/l)	2.2-120	1.4-31	1.0-5.3	0.8-5.1	2.5-22	2.0-13	.8-2.2
Potassium (mg/l)	1-3	1-3	0-2	1-2	1-3	1-3	<-1-1
Bicarbonate (mg/l)	1-77	4-30	4-16	6-21	0-19	0-11	2-24
Sulfate (mg/l)	6.4-54	.3-12	3.4-37	0.6-13	24-260	12-300	0.7-57
Chloride (mg/l)	2.2-170	1.2-49	0.6-6.1	0.9-14	1.6-73	0.2-6.2	0-4.8
Iron (μg/l)	90-680	80-750	50-800	40-1900	50-1100	50-1100	30-710
Manganese (μg/l)	200-1900	70-2800	100-600	80-370	30-8600	1000-9000	30-130

Concentration-Response Modelling

A series of instream tests were completed in 1987 and 1988 to define the limiting concentration of produced waters (via chloride) which causes significant changes in benthic invertebrate community structure. Produced water from wells in the Cedar Cove field (Fig. 1) was pumped to Little Hurricane Creek and discharged in a controlled fashion to manipulate the instream concentration of chloride. Benthic invertebrate communities colonized on artificial substrates were exposed to different instream concentrations of produced water, and the effects to community structure quantified. A community response-chloride concentration relationship was subsequently developed using a simple logistic relationship. This relationship has foundation in toxicity testing theory and proved to be appropriate for this data set. The relationship also provided a predictive capability for estimating percent effect to community parameters at particular chloride concentrations.

Seven concentration-response tests were completed during this study (Table 4) ranging from a mean instream chloride of 268 mg/l to 944 mg/l. An instream chronic value of 565 mg/l chloride was calculated by determining the geometric mean of the LOEC (lowest observable effect concentration - 615 mg/l) and the NOEC (no observable effect concentration - 519 mg/l). This chloride concentration provides a reference point below which no significant changes to benthic invertebrate community structure is expected while chloride concentrations exceeding this value will most likely result in significant changes.

Two models were developed (Fig. 3) that predict change in the community structure of benthic macroinvertebrates. The abundance model predicts percent decline in log-density of benthic invertebrates with increasing chloride concentrations. The taxa model predicts percent decline in number of benthic invertebrate taxa with increasing chloride concentrations. Each model consists of four curves with the area between curves B and C representing percent effects most likely to occur. The area between curves A and B and curves C and D represent zones of least likely effect. The overall effect portrayed in each model is a zone where the indicated T percent effect is most likely to occur bordered by areas where effects

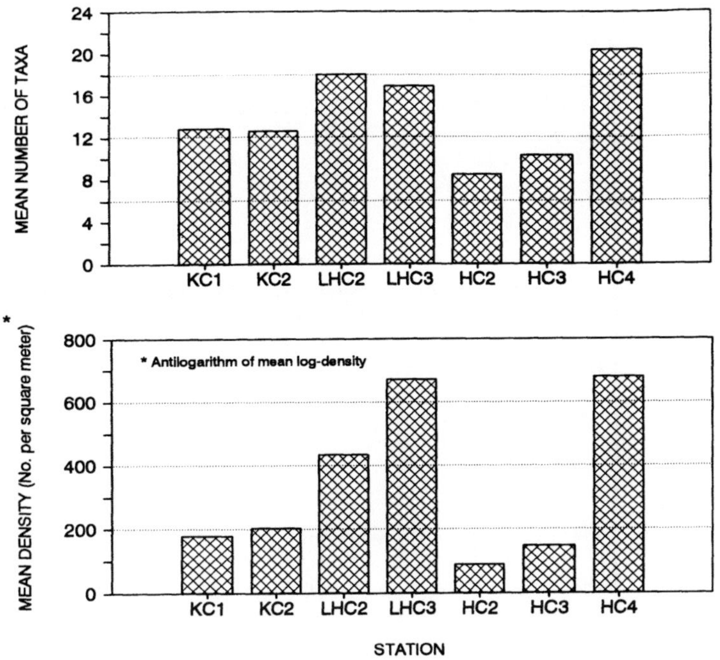

Figure 2. Mean taxa and density of benthic macroinvertebrates collected at different stations in the Hurricane Creek watershed, Warrior basin, Alabama (O'Neil et al., 1987).

may occur but with a much lower probability. The number of invertebrate taxa is predicted to be reduced 62 to 88 percent at chloride concentrations > 650 mg/l whereas log-density is predicted to decline between 23 to 40 percent at chloride concentrations > 650 mg/l. Detailed presentation of concentration-response data and development of the models can be found in O'Neil et al. (1989).

Table 3. Dominant benthic invertebrate taxa from Hurricane Creek, Alabama, September 1984 through January 1986.

	Station HC2 Downstream of POD[1]		Station HC3 Upstream of POD		Station HC4 Control	
Taxa	N[2]	%[3]	N	%	N	%
Oligochaeta (worm)	8	5.30	12	4.12	105	9.65
Ancyronyx (beetle)	2	1.32	3	1.05	19	1.75
Ablabesmyia (midge)	5	3.31	6	2.10	40	3.68
Microtendipes (midge)	2	1.32	2	.70	169	15.53
Tanytarsus (midge)	7	4.64	61	21.40	118	10.85
Stenacron (mayfly)	0	0	0	0	162	14.89
Sialis (alderfly)	10	6.62	11	3.86	11	1.01
Polycentropus (caddisfly)	19	12.58	53	18.60	4	.37

[1]POD - Point of discharge.
[2]N - Number of organisms collected during sampling period (September 1984-January 1986).
[3]% - Percent relative abundance (number collected in a taxon/total collected).

Table 4. Summary results of concentration-response testing and calculation of instream chronic value (O'Neil et al., 1989)

	Test 1	Test 2	Test 3	Test 4	Test 5	Test 6	Test 7
Testing dates	June 2-22, 1987	July 14-Aug. 3, 1987	Aug. 19- Sept 8, 1987	Oct. 6-26, 1987	June 8- 28, 1988	Aug. 3-23, 1988	Sept 22 Oct. 12, 1988
Percent change for tests with statistically significantly difference[1]							
Number of taxa	NS[2]	-53.2	-64.0	NS	-54.5	-87.9	NS
Log-density	NS	-15.1	-23.0	NS	-22.6	-40.6	+ 13.0
Diversity	NS	-40.1	-30.5	NS	-30.0	-84.4	NS
Net result of test	No Effect	Effect (NOEC)	Effect	No effect (NOEC)	Effect	Effect	No effect
Backgound water quality (experimental)							
Mean chloride (mg/l) calculated from conductivity data	268	615	944	519	641	883	490
Chloride (mg/l) range of laboratory values	180 - 410	490 - 740	660 - 980	440 - 640	390 - 840	740 - 910	220 - 800
Iron (mg/l)	0.24 - 0.72	0.17 - 0.58	0.03 - 0.43	0.15 - 0.57	0.06 - 1.4	0.03 - 0.11	0.03 - 0.44
Bicarbonate (mg/l)	90 - 180	240 - 330	360 - 470	290 - 400	510 - 660	550 - 740	120 - 330
pH	6.7 - 7.7	8.1 - 8.3	8.2 - 8.3	8.0 - 8.4	7.2 - 8.4	7.5 - 8.0	7.3 - 7.6

[1] Percent change is the percent decline (or increase) in community parameter, from the control to the experimental, for samples with statistically different means at day 20 of each test.
[2] NS - means not statistically different (P>0.05).
ChV = $(\text{NOEC} \times \text{LOEC})^{1/2} = (519 \times 615)^{1/2} = 565$ mg/l chloride

Birge et al. (1985) recommended an instream chloride value of 600 mg/l to protect Kentucky streams, and our studies confirm their recommendations. It should be emphasized that the instream chronic value of 565 mg/l is directly related only to instream biological community effects. Laboratory aquatic toxicity tests completed with the same produced waters (Drottar et al. 1989) using U.S. Environmental Protection Agency (USEPA) standardized testing procedures resulted in chronic values of 600 to 900 mg/l chloride to *Ceriodaphnia* sp. These data were therefore in agreement with the tests conducted instream and support the conclusion that laboratory toxicity tests can adequately predict instream ecological effects to the benthic fauna. An ambient water-quality criterion for chloride was recently established by the USEPA (1988). This criterion states that the four-day average concentration of dissolved chloride, when associated with sodium, should not exceed 230 mg/l more than once every three years on the average.

The USEPA chloride criterion is approximately 2.5 times less than chronic values determined from instream sampling and laboratory aquatic toxicity tests. Typical application factors for toxicity data are 3 times or greater. The experimental data supports the conclusion that the ambient water-quality criterion for chloride (230 mg/l) was protective of stream life examined in these studies. The criterion can be applied to coalbed methane produced waters found in Alabama for limiting produced water effluent discharges.

Long-term Biomonitoring of a Produced Water Discharge

The final study examined the long-term effects of a continuous produced water discharge to aquatic biological communities in a small stream where the limiting instream values of chloride

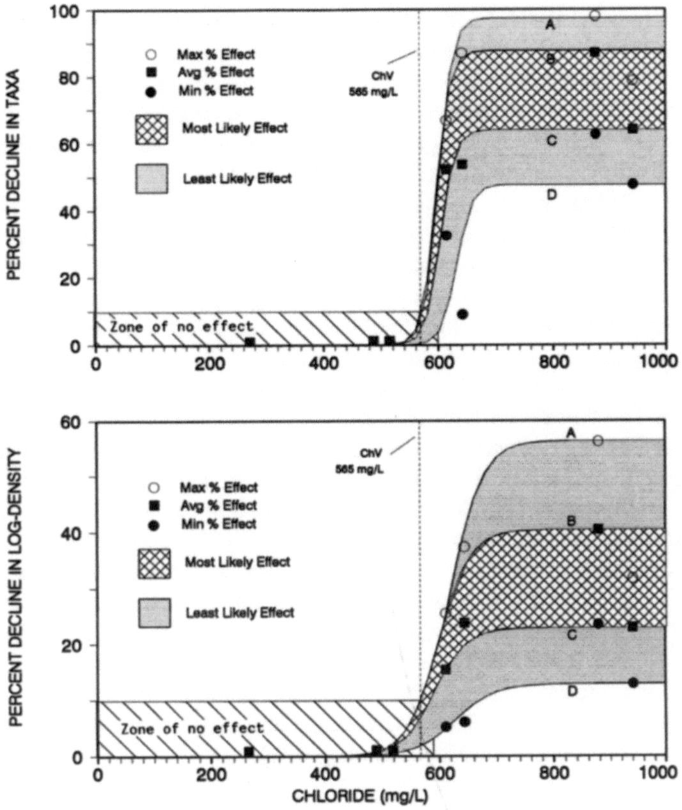

Figure 3. Concentration-effect models describing change in benthic invertebrate community structure with increasing chloride (O'Neil et al., 1989).

discussed previously were intentionally approached and occasionally exceeded. The concept behind this study was to test chloride limiting values in a true worst-case discharge scenario.

Little Hurricane Creek was selected as the study area for several reasons. A significant biological and water-quality data base had been established for the watershed and a mechanism was in place to control the discharge of produced waters and monitor instream concentrations of chloride. Also, the hydrologic regime of Little Hurricane Creek was such that streamflows low enough to allow significant increases in chloride with the given volumes and quality of produced water occurred during summer months. Discharge of produced waters into Little Hurricane Creek began in late 1986 and continued until January 1990. The point of discharge was located at the experimental test site between stations LHC2 and LHC3.

Water quality at the control site, LHC3, was typical for a stream draining Pottsville rocks. The pH was slightly acidic, ranging from 5.6 to 7.0, with bicarbonate typically <5.0 mg/l resulting in a poorly buffered water. Iron concentrations were consistently < 1.0 mg/l. The overall chemical quality at station LHC3 and the significant influence of ground water on water-quality regimes was best illustrated by the concentration of TDS. During periods of average streamflow and greater, TDS normally ranged from 20 to around 50 mg/l. During periods of summer drought TDS increased with peak concentrations ranging from 100 to 200 mg/l (Fig. 4). The major inorganic constituent contributing to TDS during summer streamflows was sulfate originating from solution of Pottsville sandstones containing pyrite. Consequently, increasing ground-water contributions of sulfate during drought periods increased the minerality of Little Hurricane Creek substantially.

The water quality of Little Hurricane Creek downstream of the produced water discharge point depended upon the relative flow contributed by both the stream and discharged produced waters and seasonally related streamflow phenomena such as drought and flood. The pH regime at station LHC2 was distinctly different from the control with a majority of measurements ranging from 6.4 to 8.0. Bicarbonate ranged from 10 mg/l or less to near 500 mg/l during times of low streamflow. Iron concentrations of the mixed water were variable but the majority of measurements were <1.0 mg/l. The average daily chloride concentration at station LHC2 was also variable ranging from 5.0 mg/l or less to in excess of 650 mg/l. Average daily chloride exceeded the USEPA ambient water-quality criterion for chloride (230 mg/l) several times during 1987 and for extended periods during 1988 (Fig. 4). The criterion was not exceeded during 1989. The instream chronic value discussed earlier (565 mg/l chloride) was exceeded once in 1987, several times during 1988, and not at all in 1989.

Prior to discharge in 1984 and 1985, the number of benthic invertebrate taxa at both stations LHC2 and LHC3 tracked closely with more taxa generally collected at LHC3 (Fig. 5). In 1987, fewer taxa were consistently collected at LHC2 during the first two-thirds of the year than at LHC3. However, during the last third of 1987 and into 1988, an equal or greater number of taxa were collected at LHC2 than at LHC3. Patterns of taxa richness at stations LHC2 and LHC3 were similar and tracked closely during the last half of 1988 and through 1989. Following produced water discharge, there was no dramatic decrease in density of benthic invertebrates at LHC2. Aside from isolated peaks in density due to the genera *Tanytarsus, Parametriocnemus,* and *Stenacron,* populations upstream and downstream of the discharge point were similar in seasonal variation in 1987, 1988, and 1989. Analysis of the types of benthic invertebrate taxa collected revealed no significant shifts in the faunal composition downstream of the discharge point. Water-quality and biological results pertaining to long-term biomonitoring in the Cedar Cove field are discussed at length in O'Neil et al. (1991).

CONCLUSIONS

The series of biological and water-quality studies briefly discussed in this paper represent a comprehensive attempt to analyze the instream biological effects associated with the discharge of a coalbed methane produced water in Alabama. Results of these studies provide a broad basis for evaluating likely instream biological effects of a coalbed methane produced water discharge and for developing biological monitoring strategies for future permit compliance requirements. Specific conclusions include the following:

- Instream biological communities including benthic macroinvertebrates and fishes (the fish data was not included in this paper but can be found in the detailed reports in the bibliography) are reflective of stream water-quality conditions resulting from development of a methane gas field, associated surface mining activity, and a combination of both. As such they are good tools for monitoring the biological integrity of streams receiving produced water discharges.
- Discharge of coalbed methane produced waters resulting in instream chloride concentrations <100 mg/l yielded no additional biological impact to a stream fauna experiencing severe stress from surface coal mine runoff.
- A threshold value of 565 mg/l chloride was experimentally determined through concentration-response testing of entire benthic invertebrate communities. Below this value no significant changes to the structure of benthic communities was observed while above this value significant changes were documented.
- Long-term biomonitoring of benthic invertebrate and fish communities in a stream receiving continuous discharge of produced water resulted in no detectable toxic effects to either group of organisms. Chloride during these studies ranged from 5.0 to 650 mg/l.

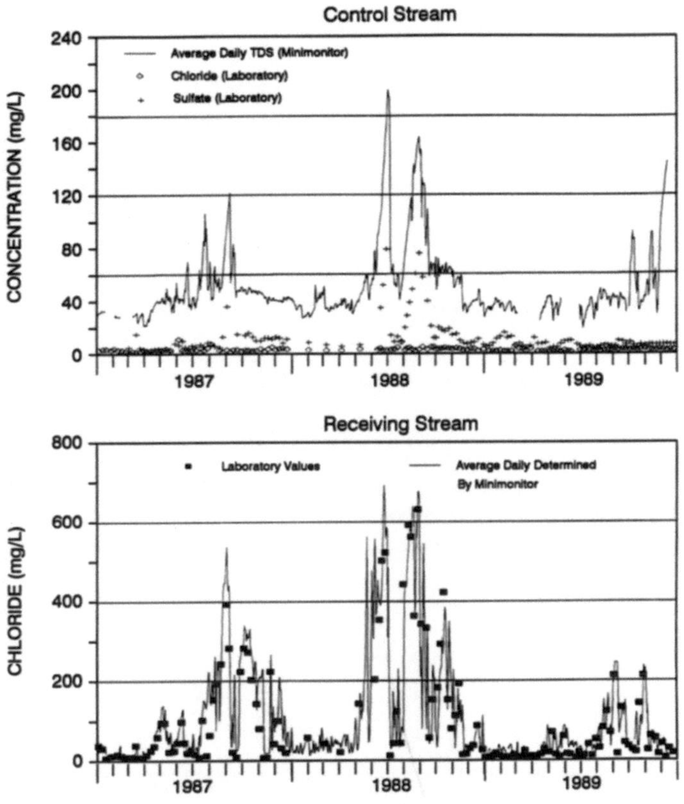

Figure 4. Water-quality characteristics of a control stream and a stream receiving coalbed methane produced water effluent, Warrior basin, Alabama (O'Neil et al., 1991).

While these studies were being conducted, the regulatory process for permitting surface water discharges made a significant change from the traditional technology-based approach for permitting to the water quality-based approach now implemented by the USEPA and the states. The water quality-based approach applies water-quality standards and criteria to mixing zones of receiving streams requiring that chemical-specific and aquatic toxicity limits be met instream. Aquatic toxicity testing has added a biological component to the permitting process and, in the future, new initiatives such as biocriteria will no doubt be included in the water quality-based approach for permitting. Biocriteria will require that biological integrity (as measured using techniques and parameters illustrated in this paper) be maintained downstream of a discharge point. As these initiatives are implemented throughout the oil and gas industry, more information will be needed on the biological fate and effect of all discharges.

Application of biomonitoring techniques to produced waters as demonstrated provides a reasonably accurate and sensitive method for determining acceptable levels of discharge into surface waters. Complacent regulation of produced water discharges by chloride alone, however, is not recommended and is not in the best interest of industry or the regulating community. Toxic responses of both standard test organisms and whole aquatic communities are complex, dependent on individual physiological tolerances, ecological interactions among populations of species and multispecies communities, and chemical reactions of effluents introduced into the environment. Consequently, although a significant data base now exists on the toxicity of

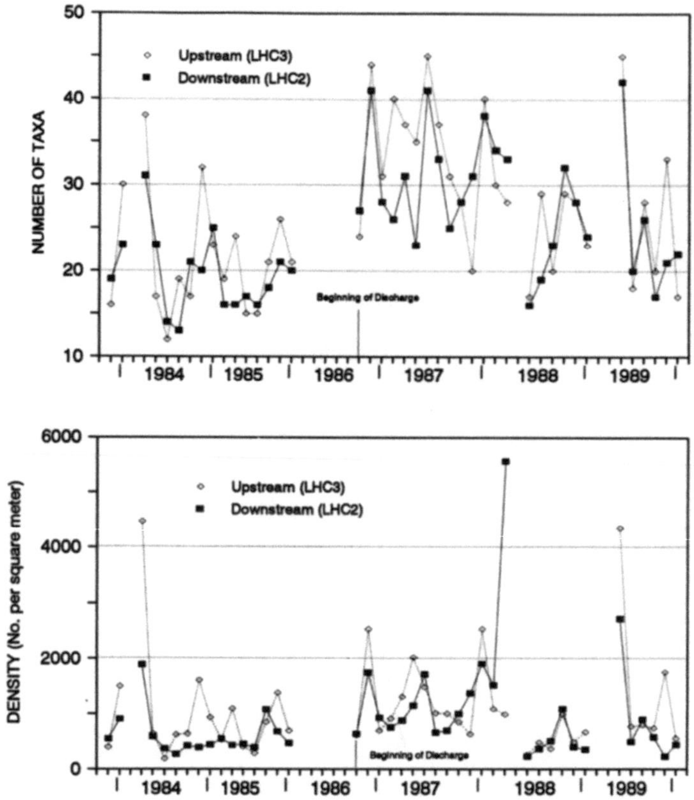

Figure 5. Monthly number of taxa and density of benthic macroinvertebrates collected in a stream receiving coalbed methane produced water effluent, Warrior Basin, Alabama (O'Neil et al., 1991).

chloride, undue reliance on this one parameter may lead to serious underestimates of toxic effect and result in significant deterioration of the stream environment. Through the judicious use of both applicable chemical-specific water-quality criteria and appropriate biological evaluation techniques, practical and environmentally protective programs of produced water discharge can be developed. The Gas Research Institute is currently funding studies to develop salinity-toxicity models to assist producers in estimating the aquatic toxicity of a produced water effluent based on its chemical quality, and studies detailing an instream biomonitoring methodology applicable to produced waters similar in chemistry to coalbed waters.

ACKNOWLEDGEMENTS

Research outlined in this paper was supported by the Gas Research Institute under contract No. 5084-253-1019, and by the Geological Survey of Alabama. We also thank the following individuals for their field efforts and other contributions to the research: Paul Moser, Stuart McGregor, Tom Shepard, Matthew Rhode, and Marek Kaliszewski.

REFERENCES

Birge, W.J., J.A. Black, A.G. Westerman, T.M. Short, S.B. Taylor, D.M. Bruser, and Wallingford, E.D., 1985, Recommendations on numerical values for regulating iron and chloride concentrations for the purpose of protecting warm water species of aquatic life in the commonwealth of Kentucky. Kentucky Natural Resources and Environmental Protection Cabinet Memorandum of Agreement No. 5429, Frankfort, Kentucky. 73 pp.

Cosgrove, W.M., H.S. Howard, and Schultz, D.A., 1988, Water chemistry and benthic invertebrate improvements one year following the discontinuation of oil brine discharges in the Blain Creek watershed, Martha, Kentucky. U.S. Environmental Protection Agency open-file report, Athens, Georgia, unnumbered.

Drottar, K.R., D.R. Mount, and Patti, S.J., 1989, Biomonitoring of coalbed methane produced water from the Cedar Cove degasification field, Alabama. Pages 363-368 *in* "Proceedings of the 1989 Coalbed Methane Symposium", April 17-20, 1989. University of Alabama, College of Continuing Studies, Tuscaloosa, Alabama.

Harris, S.C., P.E. O'Neil, Mettee, M.F. and Chandler, R.V. 1985, Impacts of surface mining on the biology and hydrology of a small watershed in west-central Alabama. Alabama Geological Survey Bulletin 125. 124pp.

Mathis, B.J., 1965 Community Structure of Benthic Macroinvertebrates in an Intermittent Stream Receiving Oil Field Brines. Ph. D. Thesis. Oklahoma State University, Stillwater, Oklahoma. 52 pp.

O'Neil, P.E., S.C. Harris, K.R. Drottar, D.R. Mount, J.P. Fillo, and Mettee, M.F., 1989, Biomonitoring of a produced water discharge from the Cedar Cove degasification field, Alabama. Gas Research Institute Final Report, June 1989, GRI-89/0073. 195 pp.

O'Neil, P.E., S.C. Harris, and Mettee, M.F., 1984, The distribution and abundance of aquatic organisms inhabiting streams draining the Citronelle oil field. Alabama Geological Survey Circular 99. 26 pp.

O'Neil, P.E., S.C. Harris, M.F. Mettee, S.W. McGregor, and Shepard, T.E., 1991, Long-term biomonitoring of a produced water discharge from the Cedar Cove degasification field, Alabama. Gas Research Institute Final Report, January 1991, GRI-90/0233. 117 pp.

O'Neil, P.E., M.F. Mettee, and Harris, S.C., 1987, Coalbed methane development in Alabama: biological and hydrological conditions of streams draining the Cedar Cove degasification field. Gas Research Institute Topical Report, January 1987, GRI-87/0038. 113 pp.

O'Neil, P.E., M.F. Mettee, and Harris, S.C., 1988, Stream monitoring of produced water discharge from methane wells in the Warrior basin, Alabama. Gas Research Institute Quarterly Review of Methane from Coal Seams Technology 6(1):50-54.

Powell, W.J., M.E. Davis, B.L. Bailey, and German, E.R., 1973, Water resources monitoring and evaluation: a key to environmental protection in Alabama oil fields. Alabama Geological Survey Information Series 44. 82 pp.

United States Environmental Protection Agency, 1988, Ambient Water Quality Criteria For Chloride. U.S. Environmental Protection Agency Office of Regulations and Standards, Washington, D.C., EPA 440/5-88-001. 39 pp

SURFACE DISCHARGE OF COALBED METHANE PRODUCED WATERS IN THE WARRIOR BASIN OF ALABAMA, THE CEDAR COVE MODEL

P. E. O'Neil[1], H. R. Isaacson[2], and J. M. Evans[2]

[1] Geological Survey of Alabama
Tuscaloosa, Alabama
USA

[2] Gas Research Institute
Chicago, Illinois
USA

INTRODUCTION

Recent implementation of water quality-based control of effluents, with toxicity testing requirements and possible biocriteria requirements, is evidence of a more comprehensive regulatory environment in the future for discharge of wastewaters. For the gas industry to successfully respond to these changes, research efforts should be directed at fully defining the characteristics of their wastewaters relative to new regulatory initiatives and in context with toxicity testing and biological sampling methodologies now proposed. The Gas Research Institute in conjunction with the Geological Survey of Alabama and ENSR Consulting and Engineering have completed a series of studies that analyzed coalbed methane produced waters from the perspective of effluent and receiving stream chemical characteristics, aquatic toxicity, and instream biological effects. Information from these studies has been incorporated into the regulatory process by defining how coalbed methane produced water effluents can be controlled and monitored when surface discharged. These studies successfully demonstrated that, when properly monitored and controlled, certain produced waters can be discharged to surface waters in a biologically and toxicologically acceptable manner.

The conceptual basis of specific monitoring and control components comprise what we term the Cedar Cove model, which is a systematic approach for characterizing, monitoring, and controlling produced water effluents discharged into surface waters. This paper will develop and illustrate the Cedar Cove model within the framework of existing discharge permit requirements, future regulatory initiatives, and studies completed in the coalbed methane fields of Alabama.

COALBED METHANE DEVELOPMENT IN ALABAMA

Development of methane from coal seams in Alabama has generated a significant new industry in addition to demonstrating that development of unconventional gas resources can be accomplished in a profitable manner. Since the beginning of the industry in the late 1970's through the end of 1991, approximately 4,400 permits to drill had been issued by the State Oil

Produced Water, Edited by J.P. Ray
and F.R. Engelhart, Plenum Press, New York, 1992

and Gas Board with near 2,800 producing wells in the Warrior basin. During this same period, approximately 160 million barrels of produced water were generated and disposed either by land application or stream discharge. Production of water increased significantly beginning in 1989 (Fig. 1) as the tax credit deadline (December 31, 1990) for producing fuel from a nonconventional source approached.

Methane is recovered from a coal seam by drilling through either single or multiple coal zones, fracturing these zones hydraulically to provide a drainage network for the gas, and finally removing formation waters associated with the coal seam(s) to lower hydraulic pressures thereby promoting gas flow to the well bore. Regulations affecting the disposal of coalbed methane produced waters in Alabama have changed significantly since 1980. Prior to 1980 all produced waters generated in Alabama were from conventional deep-well sources and were disposed by underground injection under control of the state through the Underground Injection Control program. The first NPDES (National Pollutant Discharge Elimination System) permits issued for disposal of coalbed produced waters allowed surface discharge via land application or stream discharge because the salinity of these waters was much lower compared to conventional produced waters. Discharge permit limits were derived primarily from coal surface mine criteria and regulated instream concentrations of total dissolved solids, effluent concentrations of iron and manganese, and pH. These discharge limits were implemented by the Alabama Department of Environmental Management (ADEM) at the same time the Environmental Protection Agency (EPA) was introducing a new policy direction for control of surface discharged effluents.

The new policy stipulated that regulation of surface discharged effluents was to be water quality-based in addition to technology-based. The water quality-based approach emphasized protection of water quality through water-quality criteria within strictly defined mixing zones and through evaluation of effluent toxicity using standardized aquatic toxicity testing methods. Implementation of the water quality-based approach compelled ADEM to revise their original coalbed methane NPDES permit limitations to update applicable water-quality criteria, implement effluent toxicity testing, perform extensive instream chemical monitoring, and establish systems to continuously monitor stream water-quality conditions.

The latest edition of the Technical Support Document (TSD) for water quality-based control of effluents (USEPA, 1991) recommends three levels for controlling effluents discharged to surface waters. These are chemical specific (water-quality criteria) controls, aquatic

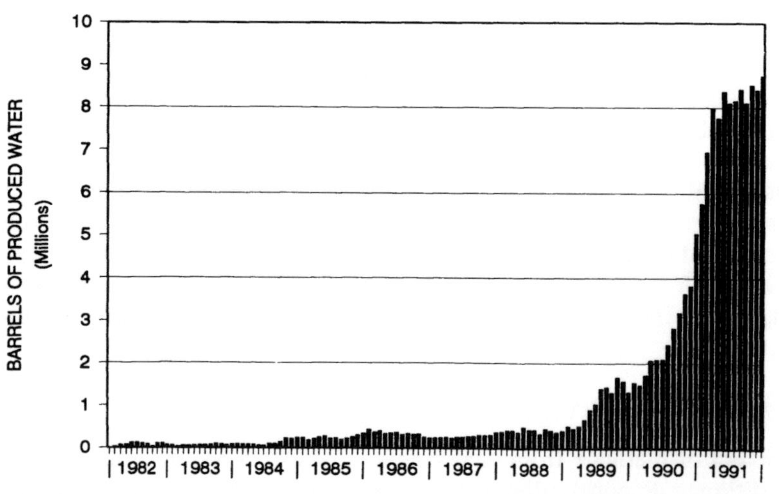

Figure 1. Monthly volume of produced water generated from development of coalbed methane in Alabama, 1982-1991.

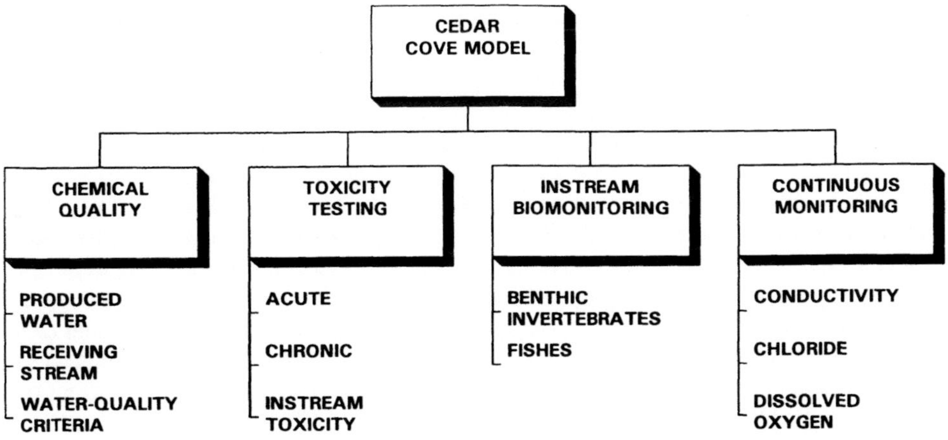

Figure 2. The Cedar Cove model illustrating levels of effluent discharge control.

toxicity testing, and instream monitoring of biological populations (biomonitoring). These three levels comprise, in EPA's judgement, a complete program for controlling and monitoring the discharge of effluents to surface waters. Each level of control is considered equal in that a negative response from any level (either exceeding water-quality criteria, demonstrating toxicity, or impacting stream populations) would require action from the permittee to regain compliance of their effluent with permit limitations.

THE CEDAR COVE MODEL

A series of studies (for summary see O'Neil et al., 1992) were completed in the Warrior basin of Alabama that addressed all phases of the recommended water quality-based approach for controlling effluents as outlined in the TSD. Elements of these studies can be organized into levels of control parallel to those recommended in the TSD and regulatory requirements of discharge permits. These levels of control include chemical characterization, both receiving stream and produced water effluent; aquatic toxicity testing of produced waters; biomonitoring of instream biological communities; and continuous evaluation of instream water-quality conditions. The Cedar Cove model (Fig. 2) was organized around these four levels of effluent control. When all levels of the model were fully implemented, surface discharge of produced waters was possible within limitations currently written into NPDES permits by ADEM.

At a minimum, the Cedar Cove model is nothing more than judicious application of recommended techniques for effluent control outlined in the TSD combined with regulatory requirements for continuous monitoring to allow surface discharge of saline effluents. Development of the model with the Cedar Cove data set, however, has broad application industry-wide beyond discharge permit compliance. Water-quality criteria, as specifically related to coalbed methane produced waters, were critically evaluated for both toxicological and instream effects; instream effect models were developed quantifying degree of biological community impact in relation to chloride concentration; and long-term effect studies to both biological communities and water quality were conducted. Two significant conclusions from the model were that the national water-quality criterion for chloride, 230 mg/l, (USEPA, 1988) was protective of biological communities examined in streams draining the Cedar Cove field; and that stream discharge of low to moderately saline produced waters can be accomplished in an acceptable manner relative to current regulatory requirements. The Cedar Cove model will be illustrated in this paper by examining field data derived within each of the four levels of effluent control illustrated in Figure 2.

Chemical Quality

Produced Waters. Discharged coalbed produced waters range in concentration from slightly to very saline (Table 1). Produced water effluents from the middle and northeastern parts of the degasification region in Alabama range from approximately 1,500 to 5,000 mg/l total dissolved solids (TDS). Hem (1989) ranks waters of this concentration as slightly to moderately saline. Produced water effluents from the southwestern part of the degasification region are more concentrated ranging from approximately 15,000 to near 22,000 mg/l TDS. These waters are ranked as very saline and approach the concentration of sea water (>35,000 TDS). Sodium and chloride are the major contributors to TDS in coalbed produced waters. Chloride in waters from the southwestern region can range from 8,000 to 15,000 mg/l while waters from the middle regions range from around 500 to 2,000 mg/l. Coalbed waters in the very northeastern part of the Warrior basin have a very low salinity and some approach the TDS concentration of receiving streams in the area.

Table 1. Water-quality Characteristics of Produced Waters and Receiving Streams in the Warrior basin, Alabama

Parameter	Units	Produced waters Warrior basin		Unmined		Mined	
		SW[1]	NE[2]	Big Sandy Creek[3]	Little Hurricane Creek[4]	Mud Creek[5]	Hurricane Creek[6]
Specific conductance	µS/cm	27,600-33,800	3,000-6,900	45 - 179	46 - 195	1,220 - 1,300	77 - 600
Turbidity	JTU	2-10	0 - 20	3 - 40	0 - 20	1 - 6	—
Dissolved oxygen	mg/l	4.2 - 9.8	1.4 - 8.2	7.5-11.2	4.1-9.3	7.0 - 7.6	6.2 - 12.5
Hardness	mg/l	1,700 - 2,700	35 - 130	13 - 89	15-130	370 - 430	—
Nitrate as N	mg/l	0 - 8.6	0 - 1.2	0.07 - 0.23	0 - 0.46	0.1 - 0.9	—
Ammonia as N	mg/l	4.4-7.4	0 -0.61	0 - 0.1	0 - 0.04	<0.01	—
Phosphate as PO_4	mg/l	0 - 4.6	—	ND	—	ND	—
TDS	mg/l	16,700-21,200	1,900-4,300	31 - 100	20 - 141	855 - 1,000	58 - 388
BOD	mg/l	<1.0 - 61	—	—	—	<1.0	—
pH	s.u.	7.7 - 8.0	7.3 - 8.5	6.6-8.0	5.7 - 7.0	7.7 - 7.9	3.7 - 6.2
Bicarbonate	mg/l	170 - 380	670 - 810	9.0 - 100	5.6 - 58	82 - 100	0 - 11
Carbonate	mg/l	0	0 - 74	0	0	0	0
Calcium	mg/l	460 - 740	7.8 - 41	3.0 - 23	3.3 - 24	68 - 78	5.1 - 38
Magnesium	mg/l	130 - 200	3.4 - 15	1.3-7.8	1.6-20	48-60	4.7-30
Sodium	mg/l	5,900 - 7,200	540 - 1,800	1.4 - 5.3	0.4 - 69	120 -150	2.0-13
Potassium	mg/l	6.3 - 20	1.9 - 3.3	1.0 - 1.0	0.6 - 27	3.0 - 3.0	1.0 - 3.0
Sulfate	mg/l	ND	0 - 110	3.7 - 10	0 - 79	550 - 630	12 - 300
Chloride	mg/l	9,900 - 13,000	480 - 2,000	2.3 - 8.4	1.3 - 74	14 - 21	0.2 - 6.2
Silica	mg/l	1.9 - 7.2	—	7.2 - 11	—	8.4 - 9.3	—
Iron	µg/l	0 - 30	10 - 410	70 - 340	250 - 3,200	20 - 50	50 - 1,100
Manganese	µg/l	160 - 440	0 - 130	120 - 500	50 - 1,600	150 - 170	1,000 - 9,000

ND, not detected; —not measured.
[1]Southwest part of basin, 11 samples February-July 1991 (unpublished data).
[2]Northeast part of basin, 43 samples June 1987 - August 1988 (O'Neil et al., 1989).
[3]17 samples August 1990 - May 1991 (unpublished data).
[4]43 samples June 1987 - August 1988 (O'Neil et al., 1989).
[5]5 samples August 1991 (unpublished data).
[6]41 samples June 1984 - January 1986 (O'Neil et at. 1981).

Waters directly from wells can have high biochemical oxygen demand (BOD) resulting from both reduced inorganic elements, such as iron and manganese, and from fluids used to drill and complete wells. If drilling fluids contain significant quantities of organic material, such as the guar bean used in some gels, then very high BOD's can result. Water of this nature must be substantially oxidized to prevent discharge of high BOD materials into receiving streams. High suspended solids are typically not a disposal problem with most produced waters.

The pH of coalbed produced waters will typically range between 7.5 and 9.0 with most waters between 8.0 and 8.5. These waters are highly buffered with bicarbonate ranging from approximately 200 to 1,000 mg/l and carbonate ranging up to 100 mg/l on occasion. High concentrations of calcium and magnesium, particularly in the southwestern part of the region, result in very hard waters (Table 1). Produced water directly from wells is frequently devoid of oxygen while the oxygen concentration of waters stabilized in lagoons is variable ranging from <1.0 mg/l to saturation.

Coalbed methane produced waters in Alabama also contain detectable concentrations of some trace elements including arsenic, barium, lead, strontium, and zinc (O'Neil et al., 1989). The distribution of trace metals within coalbed produced waters has not been studied in sufficient detail to generalize concentrations throughout the Warrior basin degasification region.

Receiving Streams. The water-quality characteristics of streams receiving produced waters should be thoroughly examined prior to discharge of effluents and as discharge is proceeding. Chemical contaminants exceeding permit limits may be present in receiving waters upstream of a discharge point. Consequently, it is in the interest of the producer to know all sources of runoff and discharge within a basin and to characterize receiving streams appropriately relative to these sources. This process should continue as discharge proceeds. Three classes of streams can generally be recognized in the Warrior basin that receive produced waters: those streams that are relatively natural and do not have altered chemical regimes; those affected by coal surface mine runoff, either mildly or severely; and those receiving produced waters upstream of a permitted discharge point. Most receiving streams in the Warrior basin have characteristics of all three types to varying degrees. Another important consideration is stream or watershed size which affects the volume of water flowing past a permitted discharge point. In the Warrior basin, produced waters are discharged into dam regulated waters, such as the Black Warrior River (10,927 square kilometers [km^2] at Holt Lock and Dam), and free-flowing streams such as Big Sandy Creek (461 km^2 at mouth) and Hurricane Creek (300 km^2 at mouth).

Typical water-quality characteristics for two free-flowing streams not significantly affected by nonpoint runoff are presented in Table 1 (under the unmined heading). Warrior basin streams of this type are typically low in dissolved solids (30 to 100 mg/l), neutral to slightly acidic pH (5.5 to 7.0), and poorly buffered (<50 mg/l bicarbonate). Iron concentrations generally range between 0.5 to 3.0 mg/l and may be higher when significant ground water contributions to flow occur during drier months.

Receiving streams contaminated with mine runoff have a significantly different chemical regime compared to unmined receiving streams. TDS increases substantially due to sulfate, iron, manganese, calcium, and magnesium from mine spoil. Acidity also increases due to increased quantities of hydrogen ions when pyritic rock material is oxidized. The pH of some streams may drop to 3.5 or lower during extremely low streamflows. Produced waters discharged into streams of this chemical nature can have significant effects to resident biota if not controlled and monitored properly.

Water-Quality Criteria. Produced water discharge permits in Alabama are written on the basis of either chemical or parameter-specific criteria not to be exceeded or on the basis of mass quantities of material that may be discharged over specific time periods. Permits written for streams will have criteria-specific requirements (Tier I) while permits written for larger rivers are usually on a mass discharge basis (Tier II). Chemical-specific criteria are usually written as follows: dissolved oxygen (>5.0 mg/l); pH (6.0-9.0); total dissolved iron (3.0 mg/l monthly or daily average, 6.0 mg/l daily maximum); total dissolved manganese (2.0 mg/l monthly or daily average, 4.0 mg/l daily maximum); biochemical oxygen demand (30 mg/l monthly or daily average, 45 mg/l daily maximum); and total dissolved chloride (230 mg/l monthly or daily average, 860 mg/l daily maximum). Response levels for chloride are lower than those just indicated, usually in the 150 to 200 mg/l range. Calculation for mass discharge (lbs/day) is based on the 7-day 10-year low flow (7Q10) and the average chloride concentration reported for wastewaters. All permits are written specifically for a produced water effluent under application and most contain the above criteria as a minimum. Some permits may require additional chemical-specific criteria for selected trace metals if data for the wastewater and receiving stream support such a stipulation.

Aquatic Toxicity Testing

Requirements for conducting aquatic toxicity tests are now written into most NPDES permits issued. This requirement introduced a biological component to the permitting process that, in the opinion of many researchers and permit writers, was a much needed addition. Methodology manuals for conducting both acute (USEPA, 1985a) and chronic (USEPA, 1985b, 1989a) toxicity tests have been prepared for laboratories conducting effluent monitoring evaluations. Acute toxicity testing requirements are generally written for effluents discharged into receiving waters with relatively large average annual flows, such as the Black Warrior River, while chronic toxicity testing requirements are typically written for smaller receiving waters. The chronic test is generally a more sensitive indicator than the acute test for determining toxicity.

The aquatic toxicity of whole-effluent produced waters is variable as reported in O'Neil et al. (1989). *Ceriodaphnia dubia* was the most sensitive of two species tested in 1986-88 with acute LC_{50} values to produced waters ranging from 715 to 1,385 mg/l chloride (Table 2). A synthetic sodium-chloride solution, similar in salinity to produced waters tested in 1988, had an acute LC_{50} of 860 mg/l chloride. The acute toxicity of two different produced water effluents in the Warrior basin, tested in 1991, ranged from approximately 960 to 1,588 mg/l chloride (unpublished data). Produced waters examined in 1987 were not toxic to fathead minnows (Table 2) or to rough shiners (*Notropis baileyi* a species collected from the receiving stream) (O'Neil et al., 1989).

The chronic toxicity of whole-effluent produced waters in 1986-88 ranged from 210 to 742 mg/l as reported in Drottar et al. (1989) and O'Neil et al. (1989). Synthetic NaCl solutions with different formulations of dilution water ranged in chronic toxicity from 215 to 858 mg/l chloride in 1988. Produced waters tested in 1991 ranged in chronic toxicity from 480 to 731 mg/l chloride (Table 2). The chronic toxicity of laboratory prepared sodium-chloride solutions is very predictable to standard test organisms. In fact, sodium-chloride solutions are often used as reference toxicants to determine species sensitivity ranges. The reported historical range of chronic toxicity to *Ceriodaphnia dubia* is 600 to 900 mg/l chloride. Although several of our earlier *C. dubia* chronic test values fall outside of this range, the majority of tests conducted since 1988 are within this range. The low values of chronic toxicity observed for earlier tests was attributed to unknown factors associated with different qualities of produced waters examined over the years.

A series of tests were completed in 1987 and 1988 (O'Neil et al., 1989, 1990) that quantified the effects of produced water discharges on the structure of benthic invertebrate communities. Samples of instream fauna were exposed to controlled concentrations of produced water and subsequently compared to samples of fauna in control regions of the receiving stream. Seven tests were completed during these studies with each test lasting 21-days. An instream limiting value for chloride of >565 mg/l was calculated, below which no significant effects to the community structure of benthic invertebrates was observed. Concentrations of chloride >565 mg/l always resulted in a degraded faunal community, statistically different ($P<0.05$) relative to benthic invertebrate community structure, compared to the control. Quantitative models relating chloride concentration to benthic invertebrate community effects were developed for estimating the degree of instream biological effect to produced water concentrations. These models are presented in O'Neil et al. (1989).

Table 2. Summary of Aquatic Toxicity Tests for Warrior Basin Produced Waters (O'Neil et al., 1989 and unpublished data)

	Test solutions		48 hr. Acute LC_{50} (USEPA, 1985a) (as dissolved chloride mg/l))		
Year	Effluent[1]	Dilution[2]	*Ceriodaphnia dubia*	Fathead minnow	Notes
1986	PW	LHC	1,385	>2,160	Pit treated
1987	PW	LHC	994	>1,655	Pit treated
1988	PW	LHC	1,048; 1,253	—	Pit treated
1988	PW	LHC	715; 736; 1,304	—	Untreated[3]
1988	NaCl	RECON	860	—	Lab solution
1991	PW	MC	960	>2,700	Untreated
1991	PW	BSC	1,588	2,612	Pit treated

	Test solutions		7 Day Chronic ChV (USEPA 1985b, 1985a) (as dissolved chloride mg/l)		
Year	Effuent[1]	Dilution[2]	*Ceriodaphnia dubia*	Fathead minnow	Notes
1986	PW	LHC	210	841	Pit treated
1987	PW	LHC	260; 287; 336	1,103	Pit treated[4]
1988	PW	LHC	742	—	Pit treated
1988	PW	RECON	742	—	Pit treated
1988	NaCl	RECON	858	—	Lab solution
1988	NaCl	LHC	215	—	Lab solution
1991	PW	MC	480	960	Untreated
1991	PW	BSC	731	1,483	Pit treated

—, test not conducted.
[1]Effluent sources:
PW - Produced water from production operation either treated in a stabilization lagoon (pit treated) or untreated before water enters lagoon.
NaCl - Laboratory prepared sodium chloride solution to match salinity of produced water.
[2]Dilution water sources:
LHC - Little Hurricane Creek
MC - Mud Creek
BSC - Big Sandy Creek
RECON - Reconstituted dilution water with similar hardness as receiving stream
[3]Untreated effluent received treatment as follows: aerated, filtered at pH 11, filtered, respectively.
[4]All tests were identical with respect to effluent and dilution water types.

Instream Biomonitoring

Instream plant and animal communities are the ultimate targets for protection in any effluent permitting program. Inclusion of some type of instream biomonitoring into effluent control programs can assist in determining if effluents have instream toxicity not predicted by water-quality criteria and aquatic toxicity tests. Although not a part of NPDES permit requirements, instream biomonitoring will most likely become more widespread and incorporated into NPDES permits with time.

Problems arise in examining the aquatic system at the community level, however, because of the complexity in defining and sampling the system, and subsequent assignment of effluent fate and effect to specific faunal or floral components in the aquatic system. The EPA has taken steps to formalize sampling methodologies and analysis techniques to provide a standard basis for conducting instream biomonitoring evaluations. Policy statements incorporating instream biomonitoring into the permitting process may be expected within the next few years. The general approach for implementing instream biomonitoring criteria, or more appropriately called biocriteria, is to sample below a discharge point and compare results with data collected from a reference station, either upstream or in an adjacent watershed, that has not experienced the effluent in question. Program guidance for implementation of the biological criteria concept by states has been published (USEPA, 1990).

Biological studies completed in the Warrior basin of Alabama have demonstrated the utility of sampling aquatic faunal communities to monitor the biological condition of streams receiving coalbed produced waters (see summary in O'Neil et al., 1992). One of these studies (O'Neil et al., 1987) presented data showing that communities of benthic invertebrates and fishes are reflective of stream water-quality conditions resulting from methane development, coal surface mine activities, and a combination of both. The biological integrity of the Hurricane Creek watershed can be illustrated by relating the total number of benthic invertebrate taxa collected at a station to specific water-quality parameters which were not necessarily causal but were symptomatic of water-quality problems at the respective stations. In Figure 3 for example, at stations where the highest chloride concentrations were measured (KC1, KC2, and HC2), which relates to coalbed methane discharges, and at stations where the highest sulfate concentrations were measured (HC2 and HC3), which relates to runoff from coal surface mines, the least number of invertebrate taxa were collected.

Another study (O'Neil et al., 1991) examined the effects that long-term discharge of a coalbed methane produced water may have on communities of benthic invertebrates and fishes occurring in a small receiving stream. Produced waters were experimentally discharged into Little Hurricane Creek beginning in late 1986 and continued uninterrupted until the end of 1989. Average annual instream chloride concentrations during the time of active discharge were 75 mg/l (1987), 209 mg/l (1988), and 35 mg/l (1989). Biological sampling in Little Hurricane Creek was initiated in 1984 and continued for two years prior to the initiation of discharge in 1986. Biological sampling also continued as produced waters were actively discharged. Examining benthic invertebrate community parameters of taxa diversity and density (Fig. 4), it was concluded that chloride concentrations well above ambient background had no affect on the diversity of the benthic invertebrate fauna. Density downstream of the discharge point appeared

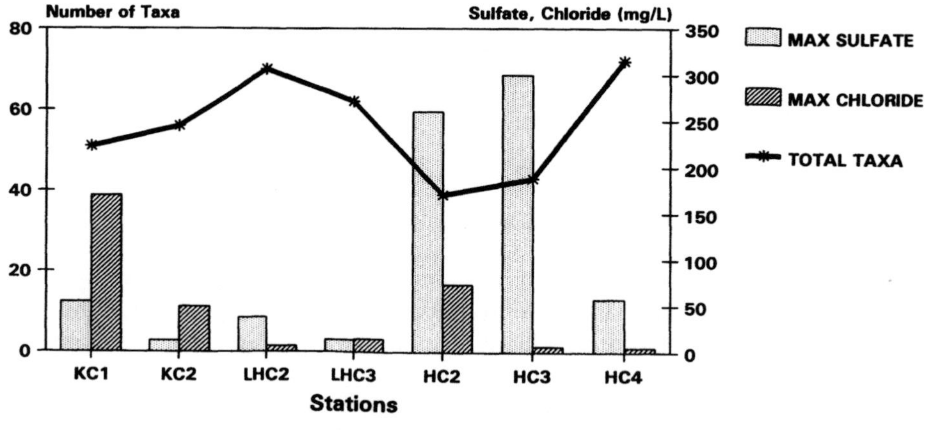

Figure 3. Water-quality and biological parameters at seven stations in the Hurricane Creek watershed, Alabama.

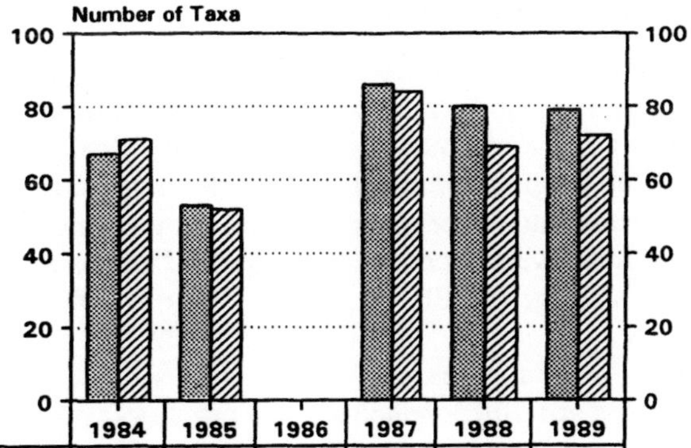

	1984	1985	1986	1987	1988	1989
Downstream	67	53		86	80	79
Upstream	71	52		84	69	72
Cl Down (mg/L)	2.5	3		75	209	35
Cl Up (mg/L)	1.9	3		4	3.4	5.6

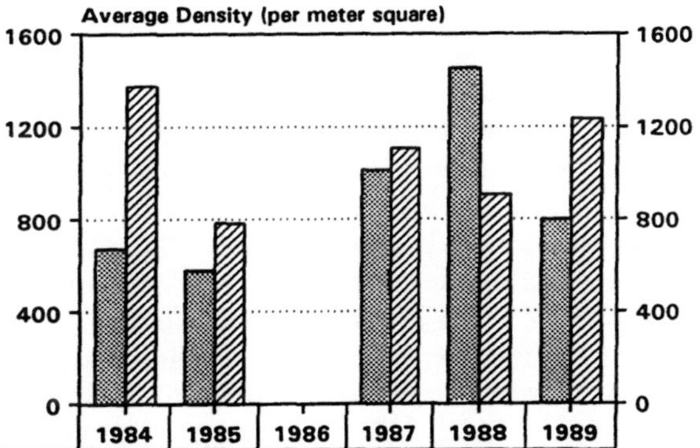

	1984	1985	1986	1987	1988	1989
Downstream	672	578		1015	1453	797
Upstream	1375	781		1109	906	1234
Cl Down (mg/L)	2.5	3		75	209	35
Cl Up (mg/L)	1.9	3		4	3.4	5.6

Figure 4. Total number of taxa and density of benthic invertebrates collected in Little Hurricane Creek, Alabama. Numbers in tables are yearly taxa totals and annual average density. Chloride values are yearly averages.

to increase suggesting increased secondary productivity due to produced waters. Collectors were generally more abundant downstream due principally to increased numbers in the midge genera *Ablabesmyia, Microtendipes, Tanytarsus,* and *Parametriocnemus.* Mayfly diversity and abundance did not change compared to the control. Patterns of fish diversity and abundance were similar to the benthic invertebrate results.

Results of these two studies effectively demonstrate that biocriteria is an achievable objective in effluent control and that produced waters can be controlled in such a way that little, if any, effects to instream faunal communities occur. A methodology for conducting instream bioassessments for benthic invertebrates and fishes has been developed and published by the EPA (USEPA, 1989b). A methodology using a different approach is being developed by the Gas Research Institute in conjunction with the Geological Survey of Alabama. The methodology is being developed for produced water effluents specifically and will provide a more critical determination and statistically based definition of significant effect to biological community parameters. The methodologies differ primarily in the way faunal communities are sampled and defined, and in the way that data are analyzed to determine changes.

Continuous Water-Quality Monitoring

The final component of the Cedar Cove model is a system for continuously evaluating water quality of the receiving stream and/or effluent. Continuous water-quality monitoring is important for real-time assessment of instream chemical conditions to comply with water quality-based NPDES permit limits. This is particularly critical given the time and exposure concentrations written into water-quality criteria. For example, one part of the chloride criteria states that the one-hour average concentration should not exceed 860 mg/l more than once every three years on the average (USEPA, 1988). Some form of continuous water-quality measurement coupled with a control feedback loop is an ideal way to monitor compliance and achieve water-quality criteria stipulations such as this.

Measurement of specific conductance is a practical approach for continuously monitoring chloride concentrations because when produced waters are mixed with receiving waters, the increased dissolved solids translates directly into increased specific conductance. Ratios of chloride:conductivity vary with produced water salinity but can be expected to range from 0.25 to 0.40 if the effluent has a low or high salinity, respectively. Commercially available water-quality monitoring systems have a variety of options for sampling frequency, data storage and processing, communications interface, parameter selection, and operational control. Several producers have telecommunications access to their instrumentation and are developing automated discharge controls if chloride exceeds a certain limiting value. Several parameters are currently available for most automated water-quality systems. The standard parameters include specific conductance, temperature, dissolved oxygen, and pH. Only specific conductance (chloride) is required to be continuously monitored in Alabama. Other parameters may be appropriate in other combinations of produced water and receiving stream quality in other states.

Continuous water-quality measurement is very effective for monitoring instream permit compliance as illustrated in Fig. 5. Hourly measurements of specific conductance revealed that significant quantities of chloride were being discharge with peak instream concentrations approaching, and occasionally exceeding NPDES permit requirements. Produced waters during this time were disposed in a batch fashion by pumping large volumes in short periods of time. The operator has since spread the volume discharged over a longer period of time thereby lowering and smoothing the peak instream chloride concentrations.

Data in Fig. 6a illustrate discharge into a typical flowing stream that is not affected by other sources of runoff or permitted discharge. Instream concentrations of chloride were easily

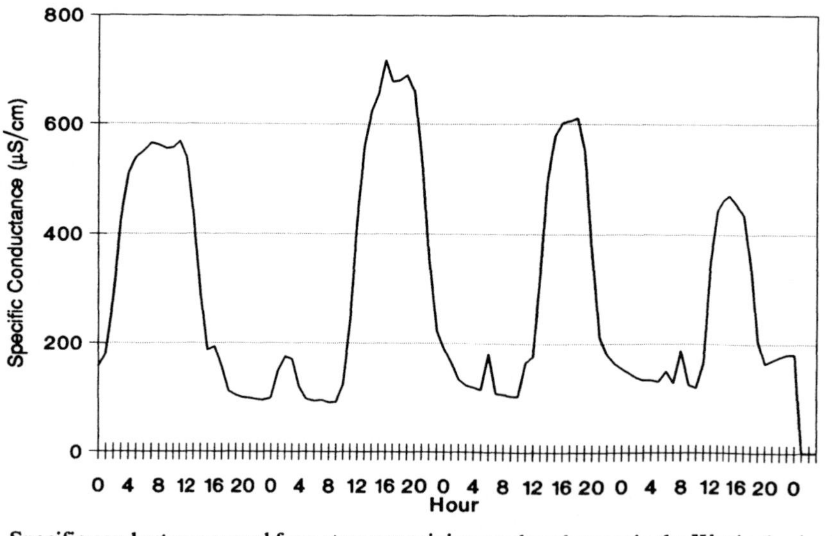

Figure 5. Specific conductance record for a stream receiving produced water in the Warrior basin, Alabama.

managed even during low streamflows in September. Fig. 6b illustrates how continuous water-quality measurement can be used as a management tool for produced waters. Settling lagoons are important for not only chemical stabilization of produced waters but also to equalize the effluent concentration. Produced waters not receiving lagoon treatment resulted in greater instream specific conductance variation (Fig. 6b) compared to the same effluent that was equalized in lagoons prior to stream discharge. Discharged effluent with a smaller variation in salinity allows for greater control over volumes of produced water discharged. The produced waters in Fig. 6b were discharged into a stream with significant runoff from coal surface mines (conductance exceeding 1,000 µS/cm upstream of discharge).

DISCUSSION

The Cedar Cove model as outlined in this paper demonstrates that produced waters from the Warrior basin in Alabama can be discharged to receiving streams without causing significant biological effects to the instream faunal communities. Water-quality criteria can be monitored

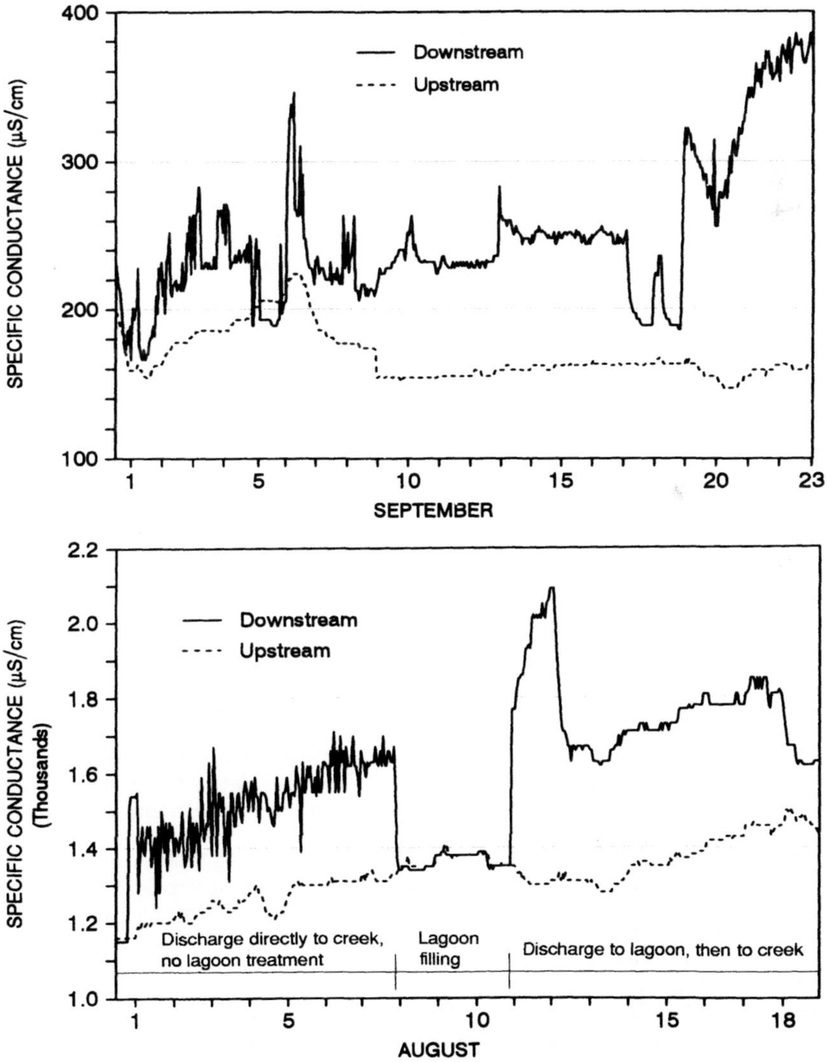

Figure 6. Specific conductance plots for two streams receiving produced waters in the Warrior basin, Alabama.

by judicious water sampling and subsequent chemical analyses, and by operation of a system to continuously evaluate the concentration of chloride instream. Other effluent limited water-quality criteria such as iron, manganese, and biochemical oxygen demand can be controlled in managed lagoons through aeration and chemical treatment of produced waters prior to stream discharge.

One problem encountered by producers is that at certain times of the year streamflows are insufficient to effectively dilute produced water discharges to permitted limits within allowed mixing zones. Some operators who have NPDES permits to discharge into smaller streams have constructed large storage/treatment lagoons to provide adequate volume for water storage until streamflows increase. Other operators have opted for transporting produced waters significant distances to large rivers which can dilute produced waters to acceptable limits most, if not all, of the time. Application of the Cedar Cove model provides for complete environmental monitoring (chemical, toxicological, biomonitoring) of a produced water discharge within the current philosophy of water quality-based effluent control. The Cedar Cove model also has anticipated future regulatory initiatives such as instream biomonitoring and biocriteria by providing critical fate and effect data for instream biological systems. Data generated during the Cedar Cove studies can help producers by providing a quantitative reference data base concerning the chemical and toxicological nature of produced waters in Alabama and the instream biological community effects of such.

ACKNOWLEDGEMENTS

Research outlined in this paper was supported by the Gas Research Institute under contract Nos. 5084-253-1019, 5084-253-1117, and 5091-253-2100. Support was also given by the Geological Survey of Alabama. Many individuals have been involved in research outlined in this paper including S.C. Harris, M.F. Mettee, S.W. McGregor, T.E. Shepard, K.R. Drottar, D.R. Mount, J.P. Fillo, and S.J. Patti. We truly appreciate their contributions.

LITERATURE CITED

Drottar, K.R., D.R. Mount, and Patti, S.J., 1989, Biomonitoring of coalbed methane produced water from the Cedar Cove degasification field, Alabama. Pages 363-368 in Proceedings of the 1989 coalbed methane symposium, April 17-20, 1989. University of Alabama, College of Continuing Studies, Tuscaloosa, Alabama.

Hamaker, T., N. Santarella, D.F. Keefe, and Fillo, J.P., 1987, Static renewal toxicity tests with *Ceriodaphnia affinis/dubia* and fathead minnows (*Pimephales promelas*) exposed to coalbed methane degasification waters. Gas Research Institute, unpublished topical report (March-June 1986).

Hem, J.D, 1989, Study and interpretation of the chemical characteristics of natural water. United States Geological Survey Water-Supply Paper 2254. 263 pp.

O'Neil, P.E., S.C. Harris, K.R. Drottar, D.R. Mount, J.P. Fillo, and Mettee, M.F., 1989. Biomonitoring of a produced water discharge from the Cedar Cove degasification field, Alabama. Gas Research Institute Final Report, June 1989, GRI-89/0073. 195 pp.

O'Neil, P.E., S.C. Harris, and Mettee, M.F., 1990, Experimental stream mesocosms as applied in the assessment of produced water effluents associated with the development of coalbed methane. North American Benthological Society 1990 Technical Information Workshop Proceedings. Virginia Polytechnic Institute, Blacksburg, Virginia. 14pp.

O'Neil, P.E., S.C. Harris, M.F. Mettee, H.R. Isaacson, and Evans, J.M., 1992, Biological fate and effect of coalbed methane produced waters surface discharged into streams of the Warrior basin, Alabama. in: "Produced Water: Technological/Environmental Issues and Solutions," J.P. Ray and F.R. Engelhardt, eds., Plenum Publishing Corp., New York, New York, USA.

O'Neil, P.E., S.C. Harris, M.F. Mettee, S.W. McGregor, and Shepard, T.E., 1991, Long-term biomonitoring of a produced water discharge from the Cedar Cove degasification field, Alabama. Gas Research Institute Final Report, January 1991, GRI-90/0233. 117 pp.

O'Neil, P.E., M.F. Mettee, and Harris, S.C., 1987, Coalbed methane development in Alabama: biological and hydrological conditions of streams draining the Cedar Cove degasification field. Gas Research Institute Topical Report, January 1987, GRI-87/0038. 113 pp.

O'Neil, P.E., M.F. Mettee, and Harris, S.C., 1988, Stream monitoring of produced-water discharge from methane wells in the Warrior basin, Alabama. Gas Research Institute Quarterly Review of Methane from Coal Seams Technology 6(1):50-54.

U.S. Environmental Protection Agency, 1985a, Methods for measuring the acute toxicity of effluents to freshwater and marine organisms. Third edition. United States Environmental Protection Agency, Environmental Monitoring and Support Laboratory, Cincinnati, Ohio. EPA/600/4-85/013. 216 pp.

U.S. Environmental Protection Agency, 1985b, Short-term methods for estimating the chronic toxicity of effluents and receiving waters to freshwater organisms. United States Environmental Protection Agency, Environmental Monitoring and Support Laboratory, Cincinnati, Ohio. EPA/600/4-85/014. 162 pp.

U.S. Environmental Protection Agency, 1988, Ambient water quality criteria for chloride. U.S. Environmental Protection Agency Office of Regulations and Standards, Washington, D.C., EPA 440/5-88-001. 39 pp.

U.S. Environmental Protection Agency, 1989a, Short-term methods for estimating the chronic toxicity of effluents and receiving waters to freshwater organisms. Second edition. United States Environmental Protection Agency, Environmental Monitoring Systems Laboratory, Cincinnati, Ohio. EPA/600/4-89/001. 249 pp.

U.S. Environmental Protection Agency, 1989b, Rapid bioassessment protocols for use in streams and rivers. United States Environmental Protection Agency, Office of Water, Washington D.C., EPA/444/4-89-001.

U.S. Environmental Protection Agency, 1990, Biological criteria, national program guidance for surface waters. United States Environmental Protection Agency, Criteria and Standards Division, Office of Water Regulations and Standards, Washington D.C. EPA-440/5-90-004. 57pp.

U.S. Environmental Protection Agency, 1991, Technical support document for water quality-based toxics control. Second edition. United States Environmental Protection Agency, Office of Water, Washington D.C., EPA/505/2-90-001.

PRODUCED WATER IMPACTS IN LOUISIANA COASTAL WATERS

M.F. Rayle and M.M. Mulino

Steimle and Associates, Inc.
Metairie, Louisiana

INTRODUCTION

The fate and possible effects of discharged saline formation waters from oil and gas operations on shallow water coastal wetlands has recently become an issue of significant environmental concern. Hydrocarbon concentrations in produced water (PW) are known to impact species diversity in benthic communities close to shallow water discharges (Middleditch, 1984; Neff, 1987). Neff et al. (1989) found elevated total petroleum hydrocarbon (TPH) concentrations less than 300 m from a shallow water discharge and less than 100 m from a deep water discharge. Neff et al. (1989) concluded that benthic communities within 20 m of both discharges were influenced by sediment contamination, probably derived from produced water discharges.

Boesch and Rabalais (1989), in a recent study of the effects of relatively low volume non-Outer Continental Shelf (OCS) origin PW discharges, found no significant increases in mean salinity related to PW discharge as a result of a modeling study of salinity gradients in Barataria Bay with actual PW discharge volumes as an input variable. They also found no salinity related impacts on marsh vegetation loss rates although elevated bottom water salinities were found in canals and bayous near discharges.

The significance of naturally occurring radionuclide concentrations in PW discharged to shallow coastal habitats is largely unknown. Hanan (1981) documented a doubling of Radium-226 concentration in bottom sediments near a PW discharge when compared to a nearby control site. The uptake of this radionuclide was found in the study to be inhibited by high salinity brines accumulating on the bottom. Clay-sized particles were also found to provide the highest number of sorption surfaces for ^{226}Ra. Landa and Reid (1983) further found that the ^{226}Ra in the sediments is tied almost exclusively to the mineral fraction.

As part of an ongoing data collection program, the American Petroleum Institute (API) and the Louisiana Division of the Mid-Continent Oil & Gas Association (LAMOGA) commissioned Steimle and Associates, Inc. to collect physical and chemical data near 38 PW discharges in Louisiana coastal wetlands to determine their potential area of impact. The study goals were to 1) determine if and to what extent salinity or hydrocarbon stratification occurs near PW discharges 2) determine sediment hydrocarbon concentrations within 300 m of discharges, and 3) determine if and to what extent naturally occurring radionuclides accumulate in sediments close to discharges. This paper is a summary of the project report which is published by API (Steimle and Associates, Inc., 1991).

Produced Water, Edited by J.P. Ray
and F.R. Engelhart, Plenum Press, New York, 1992

MATERIALS AND METHODS

The individual PW sites were chosen from the API-LAMOGA PW data base. Site selection criteria were based on fresh/intermediate, brackish and saline marsh habitat types from Chabreck and Linscombe (1978), site configuration (open water, canal, dead end canal), three categories of discharge flow (Low (1) = 200-600 bbl/day; Moderate (2) = 1,000-3,000 bbl/day; and High (3) = 3,000-6,000 bbl/day); and three categories of Ra activity: Low (1) $^{226+228}Ra < 200$ pCi/l, Moderate (2) $^{226+228}Ra = 200\text{-}600$ pCi/l and High (3) $^{226+228}Ra > 600$ pCi/l.

Figure 1 is a map of coastal Louisiana showing the appropriate locations of the sites sampled. Table 1 is a list of stations showing the result of sections based on the above criteria.

One sample of the undiluted PW from the discharge was collected at each discharge station for TPH analysis. A seperate water sample was collected for chloride analysis.

The salinity and current profiling locations around the discharge were typically on transects 15, 90 and 300 m from the discharge. At sites where salinity stratification was present, additional profiling locations were placed on the transect at 45 and 150 m from the discharge. For this study, salinity stratification is defined as a greater than 1.0 part per thousand (ppt) difference between the surface and bottom salinities. Also for purposes of this study, stations located north of the discharge are denoted in the figures and tables as negative distances. Those located south of the discharge are denoted as positive distances. Basic sampling schemes for different site configurations are given in Figure 2.

At each profile location, salinity, temperature and conductivity were measured at 0.3 m intervals from the surface down to 0.7 m from the bottom using a YSI Model 33 S-C-T meter. In the lowest 0.7 m of the water column, salinity, temperature and conductivity were measured at 15 cm intervals. Current velocity and direction data were collected at the surface, mid-depth and 15 cm above the bottom using a Marsh-McBirney Model 201 flow meter.

Water column TPH samples were collected 15, 90 and 300 m from the discharge, both at the surface and below any salinity stratification layer, if present. PW and water column samples were collected in washed and solvent rinsed one quart borosilicate glass containers with foil lined caps as per ASTM D 3325-85.

At six of the discharge locations, single bottom sediment samples were collected for TPH at the discharge and 15, 45, 90, 150 and 300 m upstream and downstream of the discharge where structures, shallow water or land did not interfere. Samples were collected using a Wildco liner type 5.0 cm diameter corer with a new, washed and solvent rinsed acrylic liner for each sample. The samples consisted of the top 10 cm of each core sample.

Single surface and bottom water column and sediment samples were collected at six stations for radionuclide analysis. These samples were collected 15 m upstream and downstream of the discharge pipe. The water samples were acidified and both water samples and the top 10 cm of the core samples were iced immediately after collection.

PW, ambient water column and bottom sediment samples were collected and preserved in accordance with ASTM Procedure D 3326-85. Water samples were analyzed for TPH using liquid extraction and infrared spectrophotometry according to U.S. Environmental Protection Agency (EPA) Method 418.1. (EPA, 1983) PW samples were analyzed for chlorides to calculate equivalent salinity (American Public Health Association, 1985). Bottom sediments were analyzed for TPH using a Soxhlet extraction procedure according to EPA Methods 9071 (EPA, 1986) and 418.1 (EPA, 1983).

Water samples for radionuclide analysis were analyzed by Core Laboratories for ^{226}Ra and ^{228}Ra according to EPA Methods 903.1 and 904.0, respectively (EPA, 1980). Errors reported are those provided by Core Laboratories and reflect counting error.

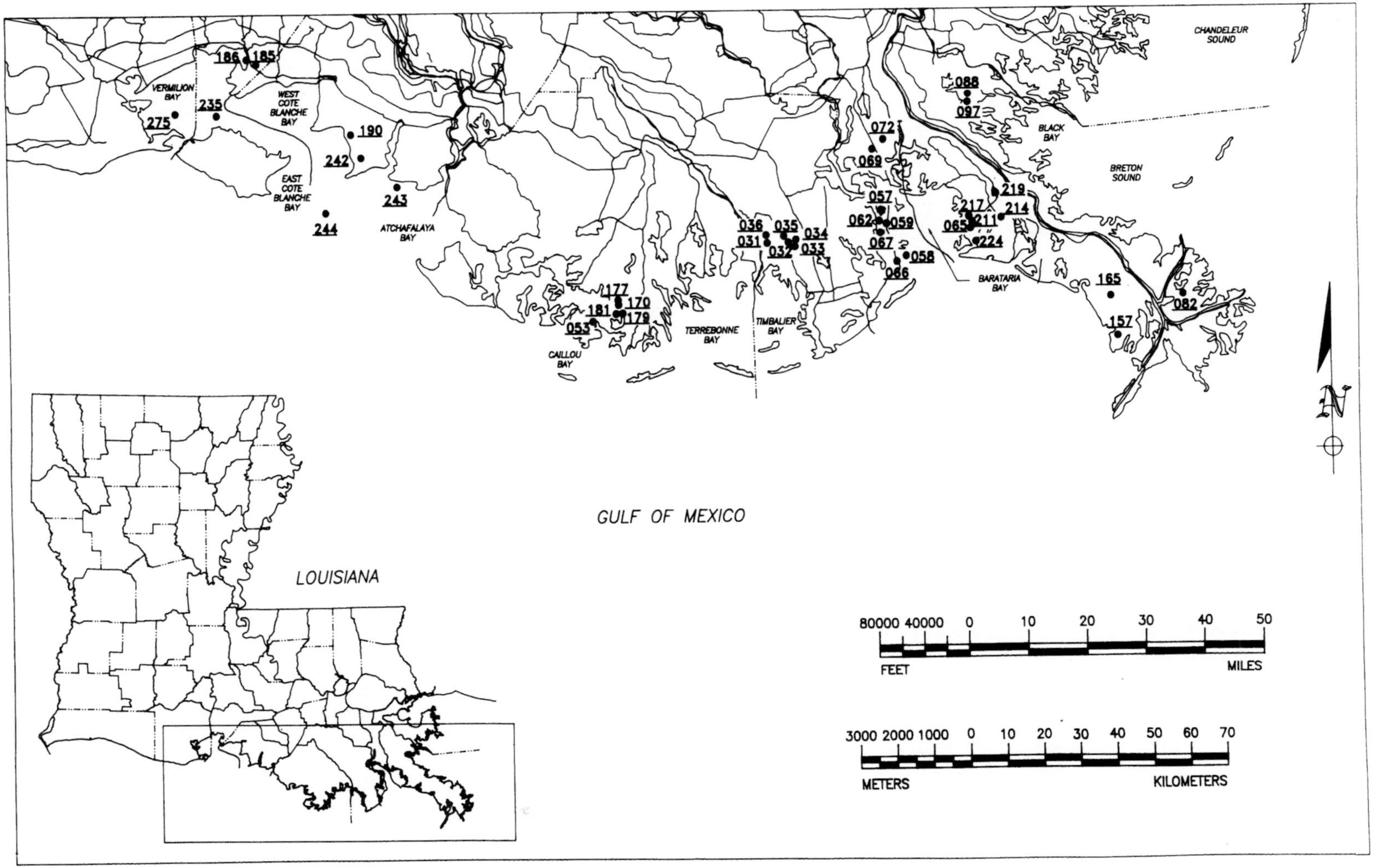

Figure 1. Sampling station locations in coastal Louisiana.

Table 1. Stratification and bottom water TPH summary

Station Number	Field	Facility Name	Marsh Habitat Type	Station Config.	Sample Types	Prod. Water Disch. Criteria	Ra 226 +228 Criteria	Measured PW Discharge TPM (mg/l)	Measured PW Discharge Salinity (ppt)
31	Golden Meadow	TB#12	B	C2	W	2	3	27	147
32	Golden Meadow	TB#13	B	OW4	W	1	1	14	115
33	Golden Meadow	TB#5	B	C2	W	2	3	55	128
34	Golden Meadow	CF#3TB#3	B	C2	W	2	3	26	138
35	Golden Meadow	CF06TB#6	B	OW4	W	2	1	13	121
36	Golden Meadow	TB#10	B	C2	W	2	2	31	152
53	Dog Lake	T8#7CF#6	S	C3	W	1	2	18	153
57	Bay De Chene	CF#2TB#5	S	OW4	W,S,R	3	2	30	161
58	Queen Bess	CF#4TB#3	S	OW4	W	1		1	10
59	Bay De Chene	CF#ITB#6	S	OW4	W	3	1	62	130
62	Bay De Chene	CF#4TB#4	S	OW4	W	2	2	24	137
65	Lake Washington	S/L 212A CF#l	S	OW4	W	2	2	7	196
66	Queen Bess	CF#3TB#4	S	OW4	W	1	3	13	178
67	Bay De Chene	CF#5TB#7	S	OW4	W	3	2	43	115
69	Lafitte	CF#IOTB#8	B	C2	W	2	2	29	135
72	Lafitte	CF#4TB#3	B	C2	W	2	2	30	138
82	Delta Duck Club	CF#4	F-I	C2	W	1	1	8	63
88	Delacroix Island	TB#l	B	C1-OW3	W,S,R	2	3	2	154
97	Delacroix Island	TB#2	B	C2	W	2	2	24	158
157	West Bay	FAC. #12	F-I	C2	W,S,R	3	3	23	111
165	W/D117-41&S.E.B.B.	Del.Gtr.Sta.	F-I	C2	W,S,R	3	2	2	158
170	Four Isle Dome	LL&E 12-14	S	C2	W	1	3	5	185
177	Four Isle Dome	Tank Battery	S	C2	W	3	2	14	185
179	Four Isle Dome	LL&E 22-7	S	*	*	2	2	-	-
181	Four Isle Dome	LL&E 22-2D	S	C1	W,S,R	2	3	20	162
185	Weeks Island	CTB#1	F-I	C2	W	2	1	26	173
186	Weeks Island	R.H.Goodrich	F-I	C2	W	3	2	34	198
190	Bayou Sale	SWD#1	F-I	*	*	1	1	-	-
211	Lake Washington	Central TB	S	OW3	W	1	3	8	201
214	Lake Washington	Homeplace TB	S	C1	W,S,R	1	1	63	34
217	Lake Washington	Cockrell 2 TB	S	OW3	W	2	3	40	191
219	West Potash	CTB#4	S	OW2	W	2	2	40	135
224	Lake Washngton	CM-3	S	OW4	W	2	1	5	45
235	Vermillion Bay	TB#l	B	OW4	W	2	2	80	139
242	Bayou Sale	CF#2TB#2	F-I	C2	W	1	1	119	79
243	Belle Isle	TB"A"	F-I	C3	W	1	2	27	167
244	Rabbit Island	TB"A"	B	OW4	W	2	3	28	148
275	Redfish Point	S/L3306	B	OW4	W	1	2	12	227

F-I= Fresh-Intermediate Marsh

B= Brackish Marsh

S= Saline Marsh

C#- Canal Configuration & Number of Transects

Ow# = Open Water Configuration & Number of Transects

* Locations Selected But Not Sampled

Sample Types: W-water, S-sediment, R-radionuclide

Disch. Criteria: 1-200-600 Bpd, 2-1000-3000 Bpd, 3-3001-6000 Bpd

Ra 226+228 Criteria: 1-≤ p Ci/l, 2=200-600 p Ci/l, 3>600 p Ci/l

Note: Marsh Types Based on Chabreck and Linscombe (1978)

RESULTS

Produced Water

Data from the PW discharge samples are presented in Table 1. The mean TPH concentration was 28.2 mg/l, with a standard deviation (S.D.) of 23.5 mg/l. Ten of the 36 samples contained TPH concentrations greater than 30 mg/l.

The salinity of the PW discharges is also given in Table 1. The mean salinity of all samples was 141 ppt (S.D.= 47).

Salinity Stratification

Table 2 is a summary of the salinity stratification and bottom water TPH data. Stratification was found at 20 of the 36 locations sampled. Of these, three stations (170,177, and 244) were stratified at all locations. Stations 53, 62, and 157 were stratified only at locations away from the discharge.

The remaining 14 stations showed salinity stratification at varying distances from the discharges. Six stations (32, 34, 35, 59, 181 and 219) were stratified only at 15 m from the discharge. Station 275 was stratified at a distance of 45 m from the discharge.

Four stations (31, 33, 67 and 165) were stratified at a distance of 90 m from the discharge. Stations 72 and 185 were stratified at a distance of 150 m from the discharge and only Station 186 exhibited stratification at the 300 m distance. No data from beyond the stratified 150 m

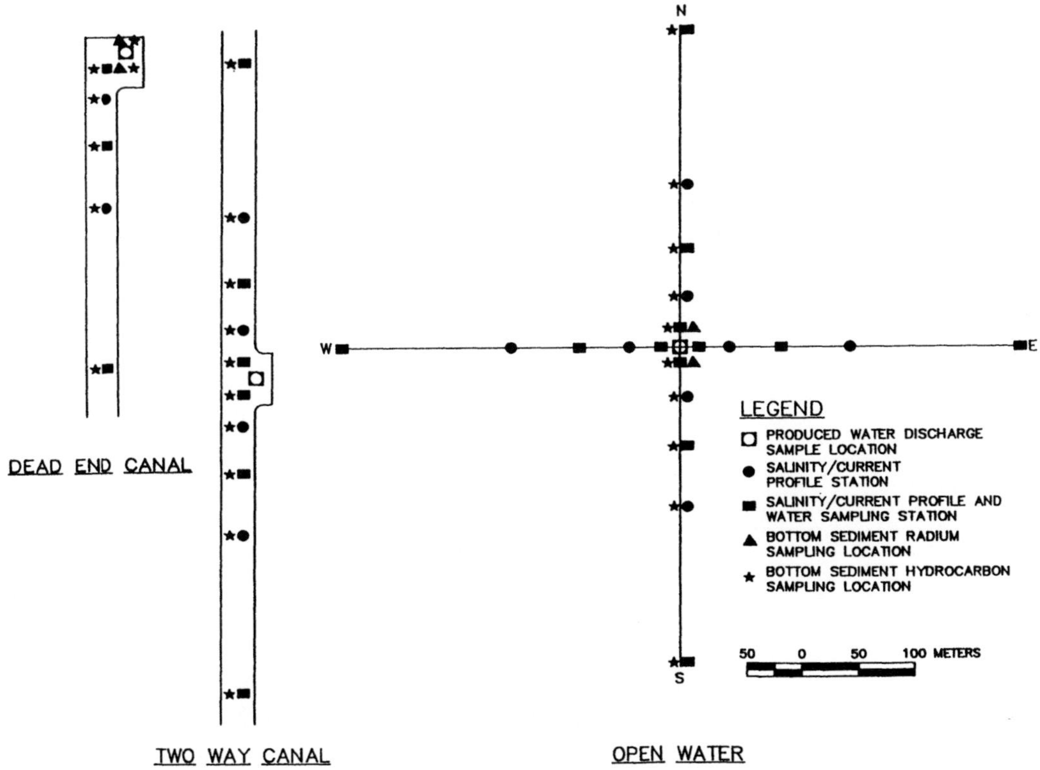

Figure 2. Typical sampling locations.

locations were collected at Stations 72 and 185 due to equipment malfunction and shallow water access problems, respectively.

Water and Sediment Total Petroleum Hydrocarbons

The surface water samples contained no detectable (0.5 mg/l or greater) concentrations of TPH. Data from the bottom water concentrations of the 36 stations sampled are given in Table 2. Of these, 25 stations had maximum bottom water TPH concentrations of < 0.5 mg/l.

The bottom sediment sample data from the top 10 cm of cores taken at Stations 57, 88, 157, 165, 181 and 214 are given in Figure 3 which presents the average TPH data for interstation comparison.

Water and Sediment Radium-226 and Radium-228

Radionuclide samples were collected at Stations 57, 88, 157, 165, 181 and 214. Three radionuclide water samples (Stations 57, 88, and 157) were lost in transit to the laboratory. Figures 4 and 5, respectively, present the ^{226}Ra data from water and sediment samples. Radium-228 concentrations from water and sediment samples are shown in Figures 6 and 7, respectively.

DISCUSSION

Produced Water Hydrocarbons

The PW TPH and salinity data presented do not provide a long-term data set to determine "average" inputs to coastal Louisiana. The data do provide a "snapshot" of the discharges to be used in conjunction with other data to determine how far the discharge remained detectable in the water column at the time of sampling.

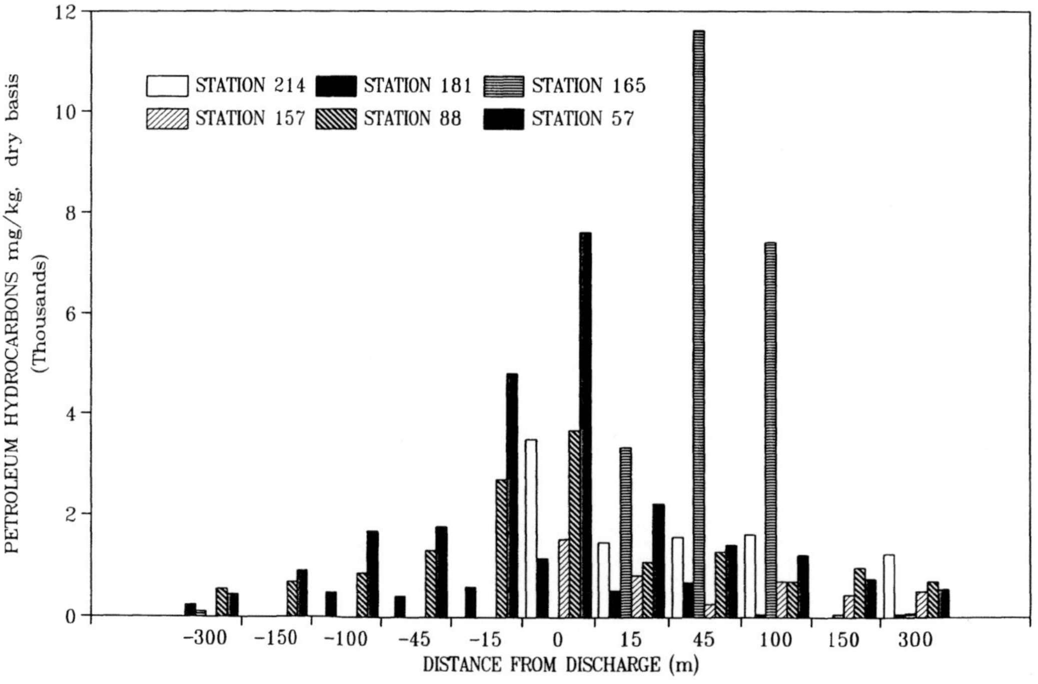

Figure 3. Average sediment TPH.

Table 2. Louisana PW discharge sampling sites and discharge data

Station Number	Disch. Criteria	Stratification Present	Strat. Max. Dist. from Discharge (m)		Maximum Water TPH Conc. (mg/kg)	Detectable TPH Max. Dist. (m) from Disharge
31	2	Y	90		2.4	90
32	1	Y	15		8.1	15
33	2	Y	90		15	90
34	2	Y	15		*179	15
35	2	Y	15		8.6	15
36	2	N	0		<0.5	—
53	1	Y	—	1/	<0.5	—
57	3	N	0		<0.5	—
58	1	N	0		4.6	15
59	3	Y	15		<0.5	—
62	2	Y	300	2/	<0.5	—
65	2	N	0		<0.5	—
66	1	N	0		3.2	90
67	3	Y	90		3.0	15
69	2	N	0		<0.5	—
72	2	Y	—		2.5	15
82	1	N	0		<0.5	—
88	2	N	0		<0.5	—
97	2	N	0		<0.5	—
157	3	Y	300	1/	<0.5	—
165	3	Y	90		<0.5	—
170	1	Y	300	3/	<0.5	—
177	3	Y	300	3/	<0.5	—
181	2	Y	15		<0.5	—
185	2	Y	150+		*11	90
186	3	Y	300		31	15
211	1	N	0		<0.5	—
214	1	N	0		<0.5	—
217	2	N	0		<0.5	—
219	2	Y	15		<0.5	—
224	2	N	0		<0.5	—
235	2	N	0		<0.5	—
242	1	N	0		<0.5	—
243	1	N	0		<0.5	—
244	2	Y	300	3/	<0.5	—
275	1	Y	45		<0.5	—

1/ Only @ 300 m location, not stratified near discharge
2/ Only @ 90 m & 300 m locations, not stratified near discharge
3/ All stations stratified
* Emulsion problems, water contained large amount of suspended solids

The TPH concentrations of the PW discharges in this study ranged from 1.7 (Station 165) to 119 mg/l (Station 242) and averaged 28.2 mg/l æ 23.5 mg/l S.D. (Table 3). Total hydrocarbons in PW samples analyzed by IR or GC and reported by Neff et al. (1989) ranged from 20 to 48 mg/l. Other studies of PW from platforms in the Gulf of Mexico such as Jackson et al. (1981) and Lysyj (1982) reported means from 11 to 106 mg/l total hydro-carbons using IR. The TPH concentrations reported in this study are similar to those cited in spite of differences in analytical methodology.

Dilution of the discharge TPH concentration to below detectable levels (<0.5 mg/l) was observed within 300 m of the discharge. Comparison of water column hydrocarbon data collected in this study with other studies of PW discharges in coastal habitats is difficult due to differences in methodology. Boesch and Rabalais (1989) reported the concentrations of a number of specific hydrocarbons and classes of hydrocarbons analyzed by GC/MS in near surface water samples collected in the vicinity of a PW discharge in the Bayou Sale Field. Their results showed trace to below detection limit concentrations of volatile, polyaromatic and saturated hydrocarbons both at the discharge and 500 m from the discharge. No bottom water hydrocarbon sample data were available from the Bayou Sale site.

Salinity Stratification

Determination of which sampling locations around a station were stratified due to the PW discharge was obvious in many cases. The PW was warmer and much more saline than the rest of the water column. Sampling was done in fall when water temperatures were cool. The PW also formed a sharp pycnocline in the water column.

The bottom water stratification associated with PW discharges observed in this study was localized within 300 m of the discharge in most cases. At only three of 36 sampling stations was PW induced stratification observed beyond 150 m. The non-OCS discharge study sites of Boesch and Rabalais (1989) had similar discharge rates, receiving water body configurations and were located in the same fields as a number in this study. The extent of stratification in the Boesch and Rabalais (1989) study ranged from 0-500 m. At comparable sites in our study stratification effects ranged from 0 to > 150 m.

Sediment Hydrocarbons

The sediment hydrocarbon data collected in this study were limited both in numbers of stations sampled and distance from the discharge. Where sediment TPH values at or within 300 m of the PW discharge approached background concentrations as defined by Boehm and Requejo (1986) for the coastal Gulf of Mexico, 10-50 mg/kg, unique circumstances existed. At both stations, sampling locations with low TPH concentrations were located in waterways

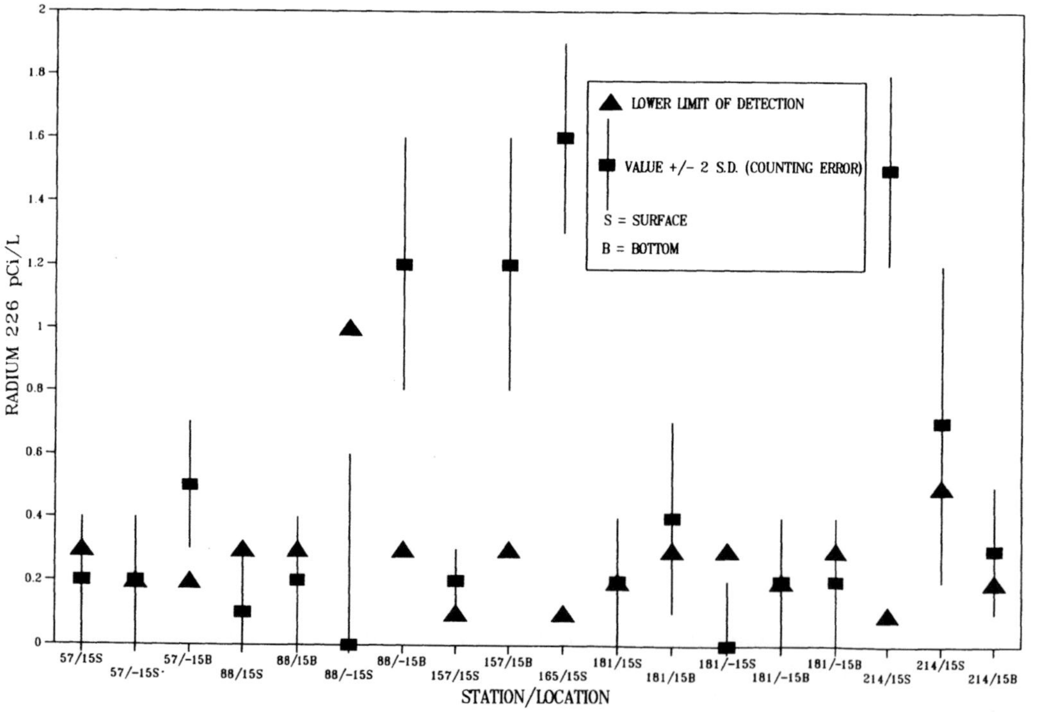

Figure 4. Average sediment TPH.

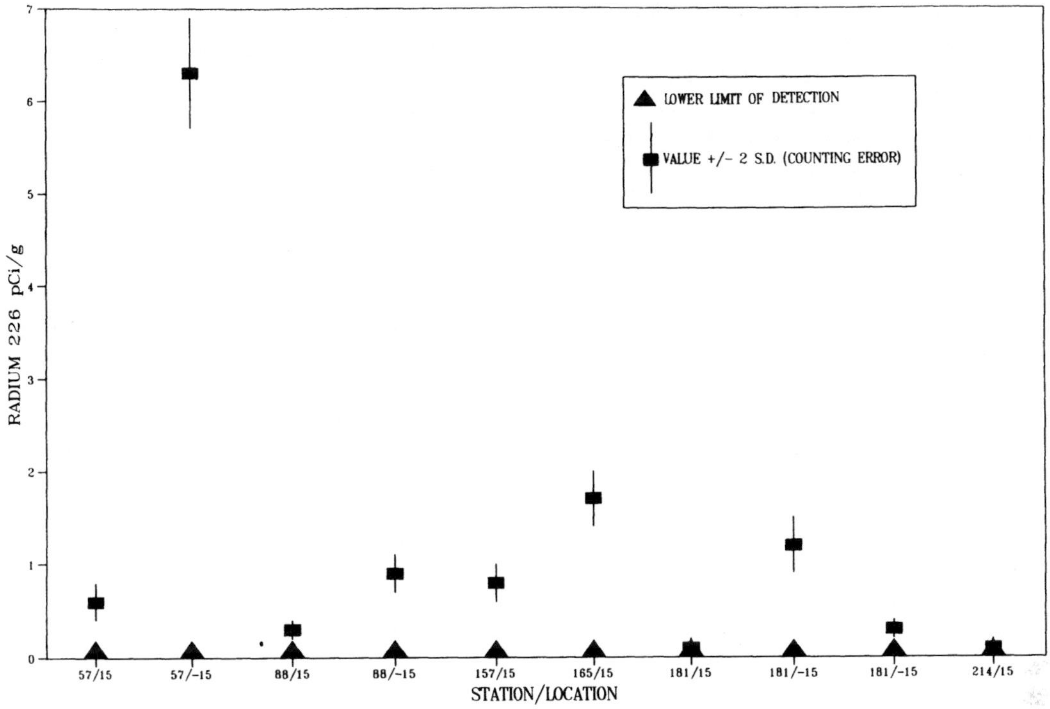

Figure 5. Sediment sample ^{226}Ra activity.

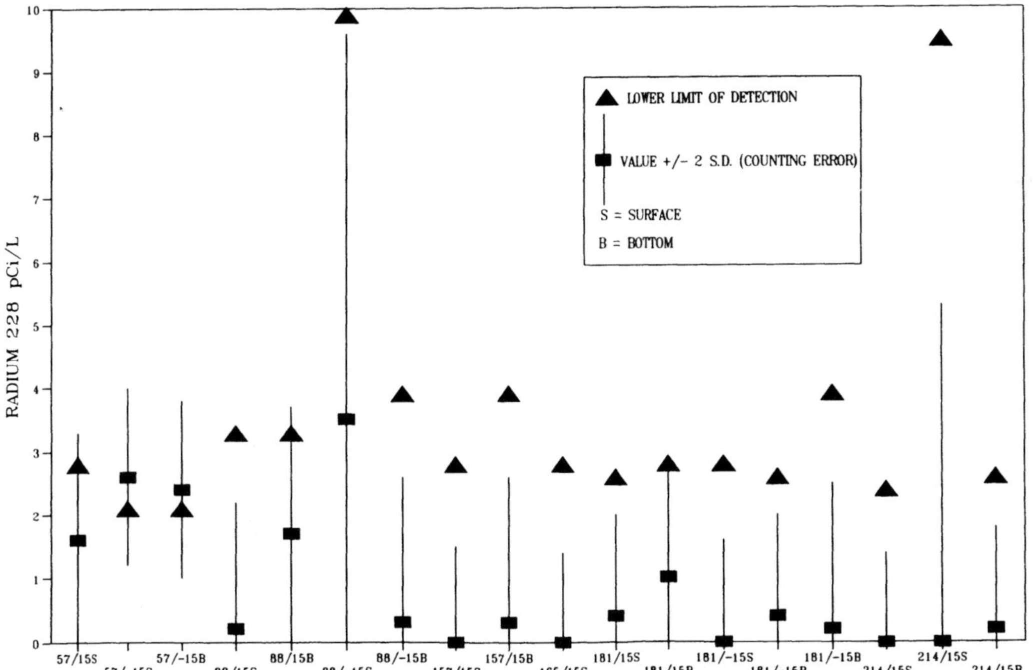

Figure 6. Water sample ^{228}Ra activity.

adjacent to the one in which the discharge was located. In both cases, Stations 165 (Delta Gathering) and 181 (LL&E 22-2D) this nearby waterway was deeper than the canal receiving the discharge and with stronger current.

The PW discharges sampled in this and other studies are never the only source of petroleum hydrocarbons in the vicinity. The pattern of generally decreasing sediment hydrocarbon concentrations with distance from the discharge, however, does indicate that PW from the facilities are the primary source of the hydrocarbons.

Water and Sediment Radionuclides

Water and sediment ^{226}Ra and ^{228}Ra data were collected at the same six stations as the sediment hydrocarbon data. Water column radionuclide activities were elevated above background levels in the vicinity of some discharges. Rabalais et al. (1991) define total Radium activity for natural waters as about 0.9 pCi/l (range .02 to 2.3 pCi/l).

No pattern is apparent as to which stations have higher activities. Some high radionuclide criteria discharges have low water column activities (Station 181). The low radionuclide criteria discharge sampled (Station 214) had the second highest activity of all the samples. The open water stations had both elevated (Station 88) and below background (Station 57) activities in some samples. The canal stations were similar with both below background (Station 181) and elevated (Station 165) radionuclide activities.

Sediment ^{226}Ra activities ranged from below the detection limit to 6.3 ± 0.6 pCi/g in the samples from the six discharge stations. Eight of the ten samples had activities higher than the LLD (Lower Limit of Detection).

Hanan (1981) found sediment ^{226}Ra activities of 1.00 ± 0.06 pCi/g to 5.17 ± 0.15 pCi/g at sampling locations within 20 m of the PW discharge he examined. Although the minimum values in our samples are well below those of his samples, the maximum values are similar.

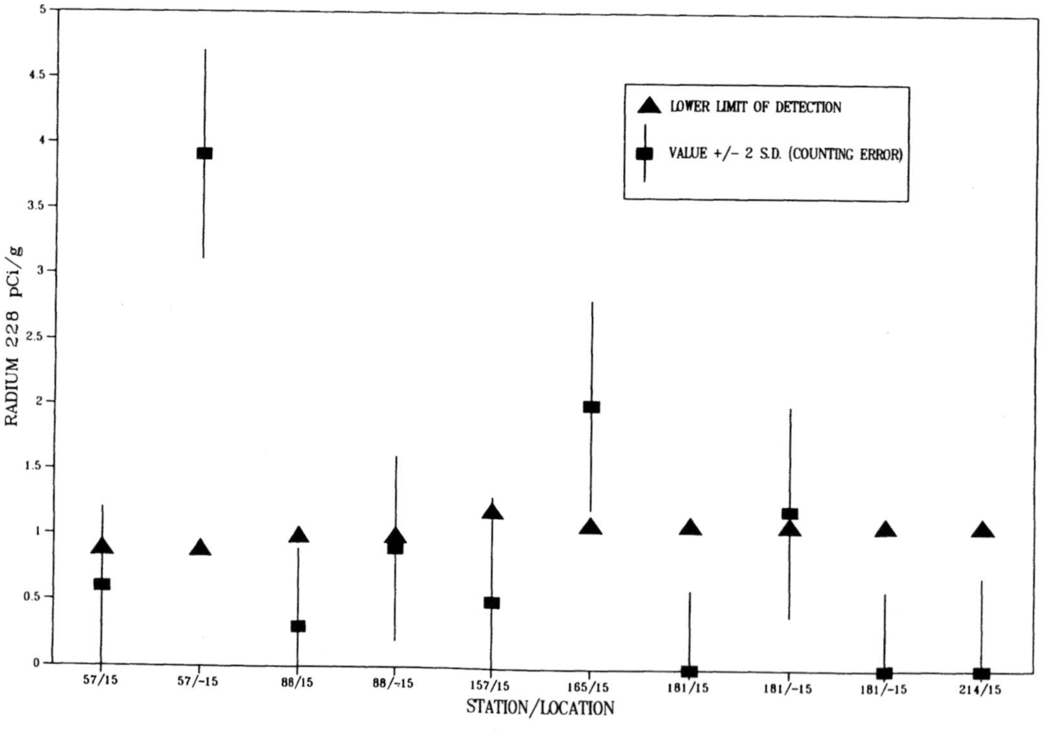

Figure 7. Sediment sample ^{228}Ra activity.

The small number of samples reported here and the limited scope of the sampling preclude broad generalizations about radionuclide accumulation in the sediments near the discharges sampled. From the limited sampling and literature data available for comparison, the stations sampled do not show high levels of sediment radionuclides near the discharges.

CONCLUSIONS

The PW discharges sampled in this study cover a wide range of discharge rates, PW salinities and hydrocarbon concentrations, marsh habitat types and receiving water body configurations. With the exception of the high volume inshore coastal OCS discharges, they are typical of many other facilities located in coastal Louisiana.

No detectable (± 0.5 mg/l) TPH levels were found in the surface water samples collected around the discharges. Dilution of the PW TPH concentrations decreased to below detection limits within 300 m of the discharge. Less than half of the stations sampled had any detectable TPH concentrations in the bottom water samples.

Salinity stratification was found at 20 of 36 PW discharge stations sampled. Three of the 20 stations exhibited area wide stratification which was the result of large fresh water inputs not related to the discharges.

Bottom water salinity stratification associated with PW discharges observed in this study, when present, occurred at sampling locations within 300 m of the production facility. Discharge related stratification extending 150 m was found at only three stations. Current velocity, water depth and bottom topography were the major factors affecting if and to what extent PW related stratification occurs near the discharges.

The limited sediment TPH data collected in this study show high levels (below detection limits to 1,181 mg/kg) of sediment contamination within 300 m of the discharges sampled. Sediment TPH levels approached background concentrations of 10-50 mg/kg for coastal Gulf of Mexico sediments (Boehm and Requejo, 1986) within 300 m of the discharge only in areas in which changes in hydrologic factors such as water depth and current velocities occurred abruptly. The pattern of generally decreasing sediment hydrocarbon concentrations away from the discharge indicates that PW is the primary source of the sediment hydrocarbons.

Water column radionuclide activities which ranged from 0.0 (below the detection limit) to 3.5 pCi/l ^{226}+^{228}Ra were elevated above background levels at a distance of 15m from the discharge at some stations. None of the samples exceeded the primary drinking water standards of 5.0 pCi/l ^{226} + ^{228}Ra activity.

Two of ten sediment samples contained ^{226}Ra of 1.7 and 6.3 pCi/g, both above the control station values (1.05 ± 0.04 pCi/g to 1.46 ± 0.07 pCi/g) reported by Hanan (1981). From the limited data presented here and available for comparison from Hanan (1981), who sampled a single discharge station and a single control station, sediment radionuclide levels measured in this study are generally low and do not appear to be accumulating to high levels near the discharges.

ACKNOWLEDGMENTS

We thank the API staff and the members of the Production Effluent Guidelines Task Force for their contributions to the design and review of the project. We also thank the operators of the various facilities sampled for their permission and assistance in the project.

REFERENCES

American Public Health Association (APHA), 1985, "Standard Methods for the Examination of Water and Wastewater." 16th Edition. American Public Health Association. Washington, D.C. 1268 p.

Boehm, P. D and Requejo, A.C., 1986, Overview of the recent sediment hydrocarbon geochemistry of Atlantic and Gulf Coast outer continental shelf environments. Estuar. Cstl. Shelf Sci. 23:29-58.

Boesch, D. F., and Rabalais, N.N.,(eds.), 1989, Environmental Impact of Produced Water Discharges in Coastal Louisiana. Report to La. Div. Mid-Continent Oil and Gas Association. LUMCON, Chauvin, La. 287 p.

Chabreck, R. H. and Linscombe, G., 1978, Vegetative Type Map of the Coastal Louisiana Marshes. Louis. Dept. Wildl. and Fish., New Orleans, Louisiana.

Environmental Protection Agency, 1980, Prescribed Procedures for the Measurement of Radioactivity in Drinking Water. EPA 600/4-80-032. Office of Research and Development. Cincinnati, OH.

Environmental Protection Agency, 1983, Methods for Chemical Analysis of Water and Wastes. EPA 600/4-79-020. Office of Research and Development. Cincinnati, OH.

Environmental Protection Agency, 1986, Test methods for evaluating solid waste. Third Edition. Laboratory Manual Vol. 1 Section A. EPA SW-846 Office of Solid Waste and Emergency Response. Wash., D.C.

Hanan, M. A., 1981, Geochemistry and Mobility in Sediments of Radium from Oilfield Brines - Grand Bay, Plaquemines Parish, Louisiana. M. S. thesis, Univ. of New Orleans, 89 p.

Jackson, G. F., Hume, E., Wade, M.J., and Kirsch, M., 1981, Oil Content in Produced Brine on Ten Louisiana production platforms. Report to U.S. EPA, Municipal Environmental Research Lab., Cincinnati, OH. EPA 600/2- 81-209.

Landa, E. R., and Reid, D.F., 1983, Sorption of radium-226 from oil-production brine by sediment and soils. Environ. Geol. 5:1-8.

Lysyj, I., 1982, Chemical composition of produced water at some offshore oil platforms. Report to U.S. EPA, Municipal Environmental Research Lab. Cincinnati, OH. EPA 600/2-82-034.

Middleditch, B. S., 1984, Ecological Effects of Produced Water Discharges from Offshore Oil and Gas Production Platforms. Final Report on API Project No. 248. American Petroleum Institute, Washington, D.C. 160 p.

Neff, J. M., 1987, Biological effects of drilling fluids, drill cuttings and produced waters. Pages 469-538 in D. F. Boesch and N. N. Rabalais (eds.), "Long-Term Environmental Effects of Offshore Oil and Gas Development," Elsevier Applied Science Publishers Ltd., London.

Neff, J. M., Sauer, T.C., and Maciolek, N., 1989, Fate and Effects of Produced Water Discharges in Nearshore Marine Waters. American Petroleum Inst. Publ. No. 4472, American Petroleum Institute, Washington, D.C. 300 p.

Rabalais, N. N., McKee, B.A., Reed, D.J., and Means, J.C., 1991, Fate and Effects of Nearshore Discharges of OCS Produced Waters. Volume II, Technical Report. OCS Study/MMS 91-005, U.S. Dept. of the Interior, Minerals Management Service, Gulf of Mexico OCS Regional Office, New Orleans, Louisiana, 337 p.

Steimle and Associates, Inc., 1991, Produced Water Impacts on Louisiana Wetlands. American Petroleum Inst. Publ. No. 4517. American Petroleum Institute, Washington, D.C. 132 p.

FATE AND EFFECTS OF PRODUCED WATER DISCHARGES IN COASTAL LOUISIANA, GULF OF MEXICO, USA

N.N. Rabalais,[1] B.A. McKee,[1] D.J. Reed,[1] and J.C. Means[2]

[1] Louisiana Universities Marine Consortium
Chauvin, Louisiana
USA

[2] School of Veterinary Medicine
Louisiana State University
Baton Rouge, Louisiana
USA

INTRODUCTION

During the production of crude oil, condensates, or natural gas, water that is brought to the surface with the product stream is removed. The separated produced water (or oilfield brine) is reinjected down a well, either for disposal or enhanced recovery of hydrocarbons, or as is the case of the vast majority of production from the northern Gulf of Mexico region, is discharged into surface waters of the ocean or coastal areas.

Approximately 95% of the oil and 98% of the gas produced on the Federal Outer Continental Shelf (OCS) comes from the central and western regions of the Gulf of Mexico. Seventy-five percent (75%) of the water produced in this process is separated and discharged at the offshore platforms. Of the estimated 1 million barrels per day ($bbl \cdot d^{-1}$, 1 barrel = 159 liters) of produced water generated on the Gulf of Mexico Federal OCS (Rabalais et al., 1991), 25% (or 253,994 $bbl \cdot d^{-1}$) is piped ashore with the oil or gas streams, separated, and discharged within the State of Louisiana's territorial sea. While the number of facilities that discharge Federal OCS produced waters into coastal environments of Louisiana are few in number, they account for large volumes, individually and collectively. These OCS-generated discharges represent 13% of all produced waters discharged into Louisiana's State waters (Rabalais et al., 1991). A U. S. Minerals Management Service funded study (Rabalais et al., 1991) provided a detailed resolution of the fate and effects of Federal OCS produced waters discharged into Louisiana State waters. This paper represents a summary of the results from that research.

STUDY AREAS

Of the 15 facilities which were identified as discharging Federal OCS-generated produced water into 11 coastal environments of Louisiana (Rabalais et al., 1991), 10 facilities in seven areas were studied (Figure 1). Except for a few small volume discharges in southwestern and south-central Louisiana, most were located in the Mississippi River deltaic plain from Atchafalaya Bay to the Mississippi River delta proper (Figure 1). The volumes of the OCS-gen-

Produced Water, Edited by J.P. Ray
and F.R. Engelhart, Plenum Press, New York, 1992

erated produced water discharges ranged from approximately 3000 to 106,000 bbl·d^{-1} (Rabalais et al., 1991). While the focus of this study was the discharge of Federal OCS-generated produced waters into Louisiana State waters, the treatment facilities are very often located in close proximity to OCS-generated but OCS-disposed effluents and/or effluents generated in State waters. In addition, some effluents mixed OCS- and State-generated produced waters. Combination of the various volumes for each study area, both OCS- and State-generated, indicated there were large discharges of produced waters in any single receiving environment from one or as many as four facilities (Table 1).

The receiving environments for these effluents included the shallow, nearshore continental shelf; high energy, freshwater distributaries of the Mississippi River delta; marshes adjacent to facilities; and brackish to saline coastal environments with moderately to poorly flushed waters. This study expanded on the initial assessment of Boesch and Rabalais (1989a) with increased temporal and spatial studies at Bayou Rigaud, Port Fourchon and East Timbalier Island, additional study areas at Emeline Pass, Romere Pass and Empire Waterway, and an abandoned discharge at Empire Waterway, and expanded analytical and field observations at all study areas.

Two locations, Bayou Rigaud and Pass Fourchon, were selected for seasonally and spatially intense sampling. These areas were sampled in the Boesch and Rabalais (1989a) study on two occasions in 1987-1988 and displayed considerable sediment contamination. Subsequent sampling for this study was conducted in February, May and October of 1989 and February 1990. These locations were also selected for bioaccumulation studies.

Bayou Rigaud is the location of two large discharges of OCS produced waters: the Conoco Grand Isle Shore Base (105,760 bbl·d^{-1}) and the Exxon Grand Isle Station (40,000 bbl·d^{-1}).

Pass Fourchon at the time of the field studies received two OCS produced water discharges 9,645 bbl·d^{-1} and 4,768 bbl·d^{-1}) and a discharge generated in State waters (33,756 bbl·d^{-1}). Changes in handling of product streams have occurred at all three facilities, but a large combined volume is still discharged at the location (Rabalais et al., 1991).

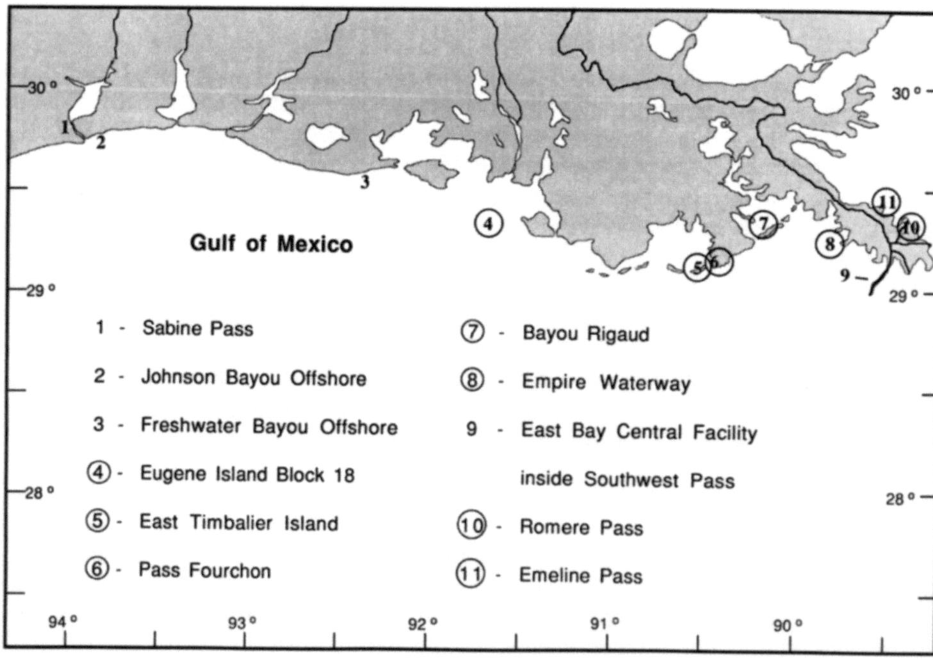

Figure 1. General location of OCS-generated produced water discharges in Louisiana State waters with circled numbers indicating areas studied.

Table 1. Comparison of produced water discharge volumes and loadings by study area. Study areas in order of decreasing total daily discharge of both OCS- and State-generated produced waters. [VH=volatile hydrocarbons, PAH=polynuclear aromatic hydrocarbons, Ba=barium, Pb=lead]

Study Area	Key to Location in Fig. 1	Daily Discharge[a] (bbl·d^{-1})	Number of Current Discharges	Mass Loadings[b] VH (kg·d^{-1})	PAH (kg·d^{-1})	Ba (kg·d^{-1})	^{210}Pb (dpm·d^{-1})
Bayou Rigaud	7	145,760	2	98.8	10.5	1030.1	1.02
Pass Fourchon	6	48,169	3	11.2	3.5	207.2	0.07
East Timbalier Island	5	25,558	4	1.2	3.6	934.6	0.03
Eugene Island Block 18	4	21,000	1	19.4	1.4	634.4	0.01
Romere Pass	10	20,158	2	5.1	3.4	25.7	0.01
Empire Waterway (Current discharge)	8	10,959	1	2.6	0.9	4.1	<0.01
Emeline Pass	11	3,693	1	0.5	1.4	14.4	<0.01

[a]Total for study area (based on monthly averages for period 02/89 - 02/90 provided by industry representatives; Rabalais et al., 1991).
[b]Calculated from average concentration in effluent(s), average daily discharge, and conversion of 1 bbl = 159 l.

Additional locations were selected because they represented the range of coastal environments receiving OCS produced waters and because of the size of the discharges. These areas were sampled on one or two occasions:

East Timbalier Island, sampled once in May 1989, is the location of several discharges. At the time of the field studies, two were discharging OCS-generated produced waters (13,258 bbl·d^{-1} and 5,005 bbl·d^{-1}). [One of these (the greater volume discharge) has since been rerouted to the OCS (Rabalais et al., 1991).] A third discharge, which had been generated on the OCS, was discontinued in 1988. In the general area for field studies, two additional discharges handle State-generated produced waters (4,866 bbl·d^{-1} and 2,429 bbl·d^{-1}).

Eugene Island Block 18 is the location of a platform in the Louisiana territorial sea just offshore of the remnant oyster reefs which demark the lower end of Atchafalaya Bay. The platform, situated in 2 to 3 m of water on the shallow continental shelf, is influenced by alternating deposition of sediments from the Atchafalaya River and wave scour. Platform M, which discharges OCS-generated produced waters (100 bbl·d^{-1}), also discharges produced water from several wells in State waters (20,900 bbl·d^{-1}). This study area was sampled in May 1989 during high river flow and in November 1989 during low river flow.

Emeline Pass is one of the two study areas located in the Mississippi River delta proper. A single facility handles OCS produced waters (3,693 bbl·d^{-1}) and discharges into a marsh area next to the facility. On high river flow, which was the situation in April 1989 during sampling, the discharge may interchange with the waters of Emeline Pass, a main distributary of the river. During low flow periods, as in the collections of October 1989 and as seen in aerial photographs taken in November 1988, the discharge flows into the marsh away from the treatment facility and Emeline Pass.

Romere Pass is the second study area located in the Mississippi River delta proper. Romere Pass connects to Main Pass, a major distributary of the river. There are two discharges of produced waters into Romere Pass. One is generated on the Federally-controlled OCS (3,021 bbl·d^{-1}) and the second is generated within the State's territorial sea (17,137 bbl·d^{-1}).

Empire Waterway is the location of a facility which currently discharges OCS-generated produced waters (10,959 bbl·d^{-1}), as well as a discontinued discharge (see below). The West Delta 30 Terminal is on an access canal which joins Empire Waterway. The final treatment process is a separation pond which empties through a pipe into a slough in the marsh adjacent to the facility and not into the access canal where sample stations accessible to our research vessel were located. Drainage patterns from the slough and the marsh observed during the November 1989 field trip, however, indicated the potential for some of the effluent to flow into the slough and, subsequently, into the channel adjacent to the facility.

The abandoned site, Pelican Island Terminal, was located on a dredged canal with access to Empire Waterway to the south of the West Delta 30 Terminal described above. The facility is no longer in operation, and, at the time of the field studies in November 1989, was being dismantled. When discharging, the volume reported was an estimated 12,000 to 15,000 $bbl \cdot d^{-1}$; operations ceased in 1987. Based on aerial photographs and communications with industry representatives, the point of the produced water discharge was determined to be near a pit along a slough in the marsh, somewhat removed from the dredged access canal. The distance from the old discharge point to the closest station sampled was 250 m.

METHODS

Within each of the study areas, a series of stations was designed for a general assessment of the fate and effects of the produced water discharge(s) in the receiving environment. Gradients away from the discharge followed, as closely as possible within the limits of local physiography, distances of 0, 150, 300, 500, 750 and 1,000 m. Detailed maps of study areas and sample methodology are given in Rabalais et al. (1991). The general characteristics of the water column in the study area were determined from profiles obtained from a Hydrolab Surveyor II for water depth, temperature, salinity, pH and dissolved oxygen. In order to better characterize the receiving environments, current speed and direction were measured through the water column with a Braystroke BFM 108 current meter over a 26 h period (30 min intervals) at a station near the discharge point. Bed shear stress was calculated from current velocity profiles using Karman-Prandtl methodology (Leeder, 1982). Bed shear stress, which is proportional to the slope of the current velocity profile, indicates the potential for resuspension and transport of sediments.

At each study area, produced waters and the near-bottom waters of the receiving water column were examined for contaminants. Produced waters from the discharge points were sampled for salinity, hydrocarbons, trace metals, several radionuclides and sulfides. Near-bottom waters collected in a horizontally-mounted 30 l Niskin bottle were sampled for hydrocarbons, selected radionuclides and sulfides. Surficial sediments (down to 10 cm) collected in a 0.1 m^2 Ekman-type box corer were sampled along a gradient away from the discharge for interstitial salinity, hydrocarbons, trace metals and a single radionuclide lead-210 (^{210}Pb). At some stations, sediment cores were collected for vertical distribution of hydrocarbons, trace metals and ^{210}Pb. In order to properly characterize the receiving environments, to place the level of contaminants in context and to explain possible macroinfaunal distributions, general analyses of sedimentary characteristics were made, including sediment total organic carbon (Control Equipment Elemental Analyzer), sediment grain size (Coulter Multisizer), and sedimentation rate measurements (^{210}Pb vertical profiles).

Salinity of produced waters and interstitial waters was determined by a potentiometric chloride titration on a digital chloridometer (Haak Buchler). The determination of sulfides followed spectrophotometric methods of Parsons et al. (1984). Water samples for volatile organics were analyzed by a purge and trap/gas chromatography (GC) technique similar to EPA Method 624 (Tekmar LS-2 purge and trap interfaced to a Hewlett Packard 5890 GC equipped with a 30-m J&W 624 megabore capillary column and a flame ionization detector). Semivolatile organics in sediments, oysters, near-bottom waters, and the F1/F2 and F3 fractions of the produced water discharges were analyzed using a Hewlett Packard 5890 GC equipped with a 30-m J&W DB-5 capillary column interfaced to a Hewlett Packard 5970B Mass Spectrometer. A Fossil Fuel Pollution Index (FFPI) was modified from Boehm and Farrington (1984) (Rabalais et al., 1991) and indicated the relative percentage of petroleum-derived hydrocarbons in the measured total polynuclear aromatic hydrocarbon (PAH) fraction. Trace metals were analyzed on a VG PlasmaQuad Inductively Coupled Plasma Mass Spectrometer. Radium (Ra) isotopes were counted directly via gamma spectrometry on an intrinsic germanium detector (Michel et al., 1981); thorium (Th) and uranium (U) were isolated via ion exchange methods (Aller and Cochran, 1976; McKee et al., 1986); ^{234}Th was plated onto stainless steel planchets and counted by low level beta spectrometry, ^{210}Pb (via ^{210}Po) and ^{238}U were counted by alpha spectrometry. Standard laboratory quality control measures were conducted for all chemical analyses.

Biological assessments included sampling of bottom sediments for benthic macroinfauna along the same gradients as outlined for sedimentary characteristics. Five replicate 0.025 m^2 Ekman grabs were removed from the larger box cores, sieved through 0.5 mm screens, fixed, and the fauna later enumerated to lowest possible taxon. Standard benthic community parameters of number of individuals, number of species, diversity (Shannon -Wiener H') and evenness (J') were calculated (Levinton, 1982). Basic statistics were computed using procedures of the Statistical Analysis System (SAS). Infaunal data were analyzed using the General Linear Models procedures for analysis of variance (SAS). Data were transformed when necessary to meet the assumptions of homogeneity of variance and normal distribution. Duncan's multiple range test was performed to identify significantly different stations within a group. The level of significance was established at $P < 0.05$.

Another objective of this study was to examine the potential impacts of organic and inorganic contaminants associated with the discharges of produced waters into coastal environments, by addressing the bioaccumulation potential of these contaminants. The approach used was to deploy filter-feeding bivalves (the American oyster, *Crassostrea virginica*) in a gradient of distances from the discharge points for known periods of time. Oyster tissues were extracted as described in MacLeod et al. (1985) with some modifications and analyzed for PAHs, trace metals and total radium as done for water and sediments.

CHARACTERIZATION OF PRODUCED WATERS

Produced waters analyzed in this study ranged between 43 and 192 ppt salinity. There was little variation in the salinity of the produced water discharge through time for effluents sampled two or four times. Sulfide concentrations were high (120 µg-atoms $S \cdot l^{-1}$) in one of the discharges and elevated in several others (3.6 to 48 µg-atoms $S \cdot l^{-1}$). The remaining discharges were found to be trace or not detected. Where elevated levels of sulfide were detected, the findings were consistent across time for discharges sampled more than once.

Produced waters contain many species of organic compounds, primarily petroleum hydrocarbons, but also partially oxidized organics. Similar compounds are found in each of the effluents, but the relative proportions differ with the facility and through time (detailed results in Rabalais et al., 1991). The components representing the highest proportion of the organic load in the produced water effluents in this study were the aliphatic fatty acids and the aromatic acids. The acid-extractable compounds are very water soluble and not readily sorbed onto particulate matter (Boesch and Rabalais, 1987). Therefore, these compounds are less likely to be deposited in sediments, but are more likely to be diluted in the water column. The saturated hydrocarbons were the next highest in concentration, but these compounds are the least toxic fraction of crude oil and very susceptible to microbial degradation when deposited in the environment (National Research Council, 1985; Boesch and Rabalais, 1987). The volatile hydrocarbons (VH) and phenols were the components with the next highest contribution to the total of identified organics. Of the volatiles, benzene and toluene composed 75% to 95%. The volatiles and phenols are highly water soluble and are acutely toxic to organisms exposed to high concentrations (National Research Council, 1985, Boesch and Rabalais, 1987). Their long-term fate, however, is to be volatilized or diluted and dispersed in the water column.

The polynuclear aromatic hydrocarbons (PAH) contributed the lowest proportion of the total identifiable hydrocarbons. This fraction, however, is the heaviest, most toxic and environmentally stable fraction of crude oil. The toxicity of crude oil is a reflection of its aromatic content, primarily the alkyl-substituted naphthalenes and phenanthrenes (National Research Council, 1985; Boesch and Rabalais, 1987). PAH are the most likely components of produced water discharges to be incorporated into the sediments because of low water solubilities and high sorption coefficients (Boesch and Rabalais, 1987). The highest PAH component in all samples was the naphthalenes followed by the phenanthrenes, fluorenes and dibenzothiophenes.

The produced water effluents sampled generally had highest concentrations (of the nine trace metals sampled) of barium, vanadium and nickel. Zinc, copper and chromium were also found in higher than ambient concentrations in most of the discharges. Cadmium, mercury and lead were also detected in various effluents. In most cases, the concentrations of all nine metals listed above exceeded the levels normally found in freshwater and seawater by a factor of 1000 (Forstner and Wittman, 1983).

Water samples were analyzed for a variety of radionuclides: ^{210}Pb, ^{238}U, ^{234}Th and total radium (^{226}Ra plus ^{228}Ra). [For comparison to values of radioactivity reported in the literature as picocuries, the conversion from disintegrations per minute (dpm) to picocurie units (pCi) is dpm ÷ 2.2 = pCi.] In both produced waters and near-bottom waters, ^{238}U and ^{234}Th activities were below detection levels (<0.5 and <0.8 $dpm \cdot l^{-1}$, respectively). This lack of enhanced uranium and thorium activity indicates strongly that the radioactivity associated with produced waters originates with the radium isotopes and is not supported by radioactive parents of radium which are higher in the decay chain. Produced waters from the sites examined had total radium activities of 304 to 2312 $dpm \cdot l^{-1}$. The ratio of ^{226}Ra/total Ra varies from site to site for produced waters, but in most cases the ^{226}Ra activity constitutes more than two-thirds of the total radium activity. The total radium activities observed are approximately 150 to 1,150 times higher than natural waters (Reid, 1983). All discharges sampled had total radium activities which were in excess of the 111 $dpm \cdot l^{-1}$ activity designated by the Environmental Protection Agency (EPA) as hazardous waste (Reid, 1983). Total radium activities of these produced waters correlated positively and strongly ($R^2 = 0.805$) with the salinity of the effluent. Produced waters from the sites examined had ^{210}Pb radioactivities ranging from 0.21 to 25.4 $dpm \cdot l^{-1}$. These values are two to 254 times higher than natural waters (Reid, 1983). Differences in ^{210}Pb activities in produced water may be due to differences in the suspended sediment concentrations of these waters. ^{210}Pb is particle-reactive, quickly sorbing onto particle surfaces; therefore, the settling out of particulate matter during treatment may change the total ^{210}Pb activity of the water sample. Total radium activity was fairly consistent through time for a single effluent, while ^{210}Pb varied.

Several factors determine the fate and effects of produced waters in coastal environments and organisms. These include the volume and composition of the discharge and the hydrologic and physical characteristics of the receiving environment. Comparisons of the produced water discharges can be approached by calculations of loadings of particular contaminants. Examples of selected contaminant daily discharge rates for combined effluents within each study area are listed in Table 1. A lower volume discharge in a study area may be compensated for by a high concentration of a single constituent, e.g. volatile hydrocarbons or barium; or vice versa, a contaminant may not be particularly concentrated in the effluent, but the volume may be large and the total loading may be similarly large. These comparisons provide a context for the relative contribution of potentially traceable contaminants to a receiving environment.

COMPARISON OF RECEIVING ENVIRONMENTS

An evaluation of the hydrological and sedimentological characteristics of each study area was made to determine the dilution potential of the environment. Dilution of water soluble contaminants would be influenced primarily by the volume of the receiving waters, the current velocity, and the potential for resuspension of sediments. With the exception of measurements taken on a high spring tide (Empire Waterway) and during a storm (Romere Pass and Eugene Island Block 18 in May 1989), current velocity measurements were assumed to be representative of similar conditions throughout the year.

Dispersion of sediment-adsorbed contaminants would be influenced by the bed shear stress, sedimentation rates, and the grain size distribution of the surface sediments. None of the ^{210}Pb vertical sediment profiles were acceptable for calculating sedimentation rates, because influx of ^{210}Pb greatly exceeded the natural decay rate for all the study areas. To calculate sedimentation rates, some decay in ^{210}Pb downcore must be observed. Accumulation of sediments was concluded from grain size distributions and/or surface sediment and downcore chemical contamination. Comparison of the receiving environments is shown in Table 2.

Table 2. Comparison of study areas with respect to dilution potential of the receiving environment. Study areas in order of decreasing total daily discharge (see Table 1)

Study	Dilution Potential of Environmnet[a]			Extent of Plume
Area	High	Medium	Low	on Bottom[b]
Bayou Rigaud		Depositional environment Periodic resuspension (storm related)		No plume observed
Pass Fourchon			Depositional environment	800 m on bottom
East Timbalier Island		Depositional environment/ Erosional environment		100 m on bottom
Eugene Island Block 18	Depositional environment Periodic resuspension (wind related)			No plume observed
Romere Pass		Depositional environment Seasonal resuspension (river flow related)		No plume observed
Empire Waterway		Depositional environment Periodic resuspension (storm related)		No plume observed
Emeline Pass	Erosional			No plume observed

[a]Based on current velocity profiles and sedimentary characteristics.
[b]Based on bottom water salinity.

Two environments were characterized as having a high potential for dilution of contaminants: Emeline Pass and Eugene Island Block 18. Mean current velocities reach 0.9 $m{\cdot}s^{-1}$ in Emeline Pass, and the coarse-grained, low organic content sediments indicate active sediment transport close to the bed. Eugene Island Block 18 is located in open, shallow offshore waters adjacent to the Atchafalaya River delta. High freshwater flow influences the area. The high silt content of the surface sediments at this study area indicate the influence of sediment deposition from the Atchafalaya River plume; however, the shallow, inner continental shelf is expected to be subject to wind-induced events causing resuspension and rapid turnover of surface sediments.

On the other end of the continuum is the Pass Fourchon study area. Current velocities are consistently negligible; potential for sediment resuspension and transport is low. The sediments within the dead-end canal system are sandy muds or muddy sands. Abrupt transitions within the study area are apparent in increasing sand content at the station in closest proximity to the back of the sandy barrier shoreline and at those stations in the more tidally active Belle Pass. A notable change in the grain size distribution of the canal stations between two sample periods with a decrease in sand and an increase in silt indicates the potential for sediment accumulation but periodic shifts to sandier sediments were probably related to storm events.

The remainder of the environments were characterized as having a medium potential for dilution of produced water contaminants. Current velocities were strong in Bayou Rigaud and sufficient to resuspend and transport sediments during peak tidal flow. The sedimentary regime, however, indicated periods of high silt content. Major shifts in sediment grain size composition between two sample periods indicate periods of accumulation and periodic resuspension during storms. Temporal variability in sediment-adsorbed hydrocarbon contaminants also support periodic resuspension and accumulation of sediments.

A similar situation to Bayou Rigaud was found in Empire Waterway. Tidal currents were strong when measured (as noted above on a high spring tide) and indicated that the potential for sediment resuspension and transport was high. However, the high silt content of the sediments and the accumulation of contaminants near the abandoned discharge point indicated a depositional environment. Resuspension of sediments and shifts in sediment grain size distribution may be storm related as in Bayou Rigaud.

In Romere Pass sediments were sands, but fine-grained, and graded to silts on the northern end of the transect. Resuspension events in Romere Pass are thought to be seasonal and related to the flow of the Mississippi River rather than storm-related as in Bayou Rigaud and Empire Waterway. The influence of Mississippi River discharge along Romere Pass was seen in a north-south gradient in water column salinities.

Two hydrological and sedimentary environments were distinguished at the East Timbalier Island study area. The East Timbalier Island area was complex with regards to sedimentary, hydrological conditions and numerous produced water discharges (including one discontinued effluent). Current velocities, during the single sample period, indicated moderate flushing in the east-west canal parallel to the back side of the island along the southern end of Timbalier Bay where sediments were generally siltier but sands were coarser-grained. The access dead-end canal which was situated north-south in the study area was shallower, had sandier sediments typical of a back barrier washover area, and more sluggish tidal currents. The north-south canal was less well flushed and contained higher levels of sediment-adsorbed contaminants.

DISPERSION OF BRINE EFFLUENT

Because produced water effluents have salinity levels in excess of that of ambient sea water and may act as a dense plume upon discharge into receiving waters, the salt content of the bottom waters may be used as a conservative tracer of the brine plume. A clear density plume was identifiable only at the Pass Fourchon study area, and extended up to 800 m from the point of discharge (Table 2). Elevated salinity was found in the bottom water at a single station in the East Timbalier Island study area and was related to the produced water effluent (Table 2). Otherwise, no clear salinity signals were found that could be attributed to the produced water effluents (Table 2). Elevated interstitial water salinities of surface sediments identified the same stations at Pass Fourchon and East Timbalier Island as being affected by the salinity plume.

Other tracers of the produced water effluent (volatile hydrocarbons, total radium and sulfide) are water soluble and are not conservative, but their presence in near-bottom waters indicates the extent of the produced water plume to at least the distance they were detected. In the case of volatile hydrocarbons, the extent of contamination of the water overlying the sediments by these chemicals was the same for Pass Fourchon and East Timbalier Island as the salinity signal. At Eugene Island Block 18, elevated volatile hydrocarbons (VH) in near-bottom waters were detected adjacent to the discharge and to some distance from the discharge whereas an elevated bottom water salinity was not. In a somewhat similar situation at Bayou Rigaud, elevated VH were detected in near-bottom waters near a discharge point and at uniformly low levels at several other stations even though a salinity plume was not detected in the bottom waters. Comparison of total loadings among study areas (Table 1) provides an additional context for the observed near-bottom distribution of the brine effluent in the adjacent receiving environment. For example, loadings of VH were large for Bayou Rigaud (primarily the result of the large volume) and Eugene Island Block 18 (primarily the result of high concentrations in the effluent). Detectable levels of VH were found in near-bottom waters at both of these study areas. In spite of a high dilution potential at Eugene Island Block 18 (Table 2), there were detectable levels of several VH at some distance from the discharge; the net loading of these constituents was important in their retention in the water column. Total loadings of VH were also high in the Pass Fourchon study area; this combined with a poorly flushed water column contributed to the retention of these contaminants in a density plume for some distance from the discharge. Loadings of VH were much lower for the remaining study areas. Of the five remaining areas, elevated levels of VH were found only at a single station at East Timbalier Island in a dead-end canal near a discharge point. The lack of flushing in this particular point compared to the moderately- to well-flushed characteristics of the other areas would account for VH contamination in spite of the relatively low loadings.

Sulfide levels were elevated in some of the near-bottom waters at Pass Fourchon; one of the effluents for the study area was high in sulfide. Elevated levels of sulfides in the near-bottom waters of some stations in the Romere Pass study area may be related to an effluent with a high sulfide content. Sulfides were not elevated in the remaining produced water discharges; but where an effluent contained high sulfides, there were indications that this constituent may be traced in the environment.

Total radium activity in the near-bottom water potentially provides additional information to the salinity, VH or sulfide tracers concerning the dispersion of water column contaminants. At the Pass Fourchon study area, total radium activity was elevated above background level to a distance of 1,000 m from the discharge; VH and salinity signals were detectable only within 800 m of the discharge point. Elevated total radium levels were detected at Bayou Rigaud, but there was no clear pattern to the distributions. At Eugene Island Block 18, higher total radium activities paralleled the elevated levels of VH. In Romere Pass, total radium three times above natural levels was detected at 450 m north of the discharge point where there were no detectable levels of VH or elevated salinity. For both the East Timbalier Island and Empire Waterway study areas, total radium activities were elevated well above background levels throughout the study areas. This indicates some level of water column contamination which was not indicated by the other tracers.

SEDIMENT CONTAMINATION

Surface Sediments

Substantial contamination of fine-grained sediments with petroleum hydrocarbons of produced water origin was observed at all study areas with the exception of Emeline Pass and the active discharge facility at Empire Waterway. A comparison was made by study area of the distance from the discharge point where a sediment contaminant signal was identifiable (Table 3). Distance of contaminant effect was determined as the greatest distance where the concentration of the constituent was elevated above background levels or the fossil fuel pollution index (FFPI) indicated petrogenic origin hydrocarbons.

The extent to which sediment contamination was documented in distance from a discharge point is a function of the loadings of the contaminant (Table 1), the nature of the receiving environment (Table 2), and the partitioning of the chemical constituent in the environment. The fate of produced water source constituents were best detected from direct measurements of effluent-source hydrocarbons, particularly the alkylated PAH, rather than other constituents which are also associated with the discharge, e.g. Ba and ^{210}Pb. With the exception of the Bayou Rigaud study area, Ba was detected at least as far as or to greater distances from the discharge point than alkylated PAH and ^{210}Pb (Table 3). Because of inconsistent gradients of Ba with distance from the discharge, characteristics of the geochemistry of Ba (sorption/desorption mechanisms of Ba related to dissolved and particulate phases which are salinity-dependent), or additional inputs to the systems from drilling fluids, Ba is not a particularly good tracer of the dispersion of produced water effluents. With the exception of the Bayou Rigaud and Empire Waterway study areas, ^{210}Pb was found at elevated levels to equal to or less distances from the discharge point than elevated levels of PAH. Lead-210 is, therefore, not a particularly good tracer of produced water effluents either. The concentration ratios of maximal ^{210}Pb concentrations to background levels were less than 10:1 compared to ratios of usually >100:1 for alkylated PAH.

Elevated levels of alkylated PAH were found in the sediments at most of the study areas. Produced water origin PAH were found to the maximal distances of the study transects at both Bayou Rigaud and Pass Fourchon, 1,300 and 1,000 m, respectively. More limited distances of detection were found for alkylated PAH at Romere Pass (450 m), East Timbalier Island (250 m), Eugene Island Block 18 (250 m) and the abandoned discharge at Empire Waterway (250 m).

Table 3. Comparison of study areas with respect to distance of elevated contaminants, benthic community impacts, and effects of bioaccumulation from discharge points (in meters). Study areas in order of decreasing total daily discharge (see Table 1)

	Extent of Sediment Contamination			Benthic Community Effects		Bioaccumulation
Study Area	Ba	Alkylated PAH	^{210}Pb	Species	Individuals	
Bayou Rigaud	1,000	1,300	1,000	500	700	350
Pass Fourchon	1,000	1,000	1,000	800	800	1,000
East Timbalier Island	580	360	<100	100	100	—
Eugene Island Block 18	250	250	<50	300	300	—
Romere Pass	450	450	<250	None	None	—
Empire Waterway						
(Current discharge)	1,000	NA	<300	None	None	—
(Abandoned discharge)	1,000	250	<450	250	<550	—
Emeline Pass	<100	<100	<100	None	None	—

NA - not applicable, because only trace levels detected.
— not studied.

Elevated levels of alkylated PAH were not found in surface sediments in the Emeline Pass study area and near the currently active discharge facility at Empire Waterway. The location of the discharge in both of these areas may account for lack of a contaminant signal in the access canal adjacent to the facility. The discharge point for the active treatment facility near Empire Waterway is next to a holding pond into a slough in the marsh adjacent to the facility. Drainage patterns from the slough and the marsh observed during field sampling indicated the potential for some of the effluent to flow into the access channel adjacent to the facility. The closest a sample station could be located to the discharge pipe was at 550 m. Drainage of the produced water effluent at Emeline Pass was not directly into the Pass but into a marsh area next to the facility. On high river flow, which was the situation in April 1989, the discharge may interchange with the waters of Emeline Pass. During low flow periods, as in the collections of October 1989 and as seen in the aerial photographs taken in November 1988, the discharge flows into the marsh away from the treatment facility and Emeline Pass. Emeline Pass has the added feature of being a high energy, river-dominated environment which would rapidly dilute any effluent which may enter the system.

Vertical Sediment Profiles

The concentration of hydrocarbons, trace metals and ^{210}Pb were analyzed for several vertical sediment cores at several of the study areas and for different time periods at some of them. As stated earlier, ^{210}Pb activities in the vertical sediment profiles were not appropriate for calculation of sedimentation rates. We concluded that the profiles were due to several factors including varying input and disturbance of the sedimentary structure by dredging, boat traffic and/or natural events such as storms. This explanation is consistent with the hydrocarbon contamination seen in some of the vertical cores which indicates maxima in produced water origin hydrocarbons in both subsurface sediments and/or with depth in the core.

Hydrocarbon concentrations in vertical cores indicated 1) a decrease in overall concentrations with increasing distance from the produced water discharge point, 2) maximal concentrations in the upper sections of the cores at those stations closest to the discharge point, 3) subsurface maxima and additional peaks with depth in cores separated by lower concentrations in intervening sections, particularly in those closer to the discharge points, 4) very often a weathering of saturated hydrocarbons with depth in the core, and 5) elevated levels of hydrocarbons in the upper sections of vertical cores collected from the distal ends of study area transects usually indicative of pyrogenic hydrocarbon inputs. Changes in concentration of produced water hydrocarbons with depth in vertical cores are related to changes in mass loadings through time, variations in transport and deposition of contaminated sediments, resuspension and transport of contaminated sediments from the area during storm events, and transport of uncontaminated sediments into the area during storm events. The presence of high concentrations of produced water derived hydrocarbons to depths of 25 to 30 cm in a vertical sediment core in some of the study areas indicates the long-term accumulation of these contaminants and their resistance to degradation.

Abandoned Discharge

The study area at Empire Waterway included a discontinued discharge. Alkylated PAH were detected in sediments at low concentrations at several stations adjacent to the abandoned discharge facility. In contrast, no significant levels of alkylated PAH were detected in any of the sediments along the transect near the active discharge at Empire Waterway. The maximum alkylated PAH value was found at the closest station sampled (250 m from the old discharge point at the Pelican Island facility), and concentrations decreased away from there.

Vertical sediment cores analyzed from this study area showed maximal petrogenic contamination in surface sections, either 0-2 or 2-5 cm. In fact, the 2-5 cm section of the core from the station closest to the abandoned discharge point had the highest concentration of alkylated PAH for the entire Empire Waterway study area. Subsurface elevated levels of alkylated PAH of petrogenic origin were found in 20-25 cm sections of cores closest to the discharge point and to some distance from the discharge point. Despite the categorization of this study area as medium for dilution potential (Table 2), the sediments accumulate but are periodically resuspended and transported from the area (or into the area). The presence of produced water origin hydrocarbons in relatively high concentrations with depth in the vertical sediment cores indicates the long-term accumulation of these constituents and their resistance to degradation.

BIOLOGICAL ASSESSMENTS

Biological assessments of the impact of the produced water effluents were determined as changes in benthic communities along a gradient of distance from the discharge point(s) within each study area and by bioaccumulation studies of filter-feeding molluscs in a similar gradient at Pass Fourchon and Bayou Rigaud. The general results from these studies are shown in Table 3.

Benthic Communities

Benthic community parameters of mean number of species, mean number of individuals, mean diversity and mean evenness were calculated for each station along a transect away from the discharge or adjacent to the discharge. Analysis of variance indicated statistically significant differences ($P < 0.05$) among the station means of the number of species and number of individuals for most study areas. Results for diversity and evenness were ambiguous and were complicated by unequal sample sizes where the number of individuals in a replicate was zero. This discussion will be restricted to the parameters of number of species and number of individuals. Duncan's multiple range tests of the station means within each study area indicated that some groups of stations were distinct from others with regards to impacted fauna, but more often there was a gradient in impacts with distance from the discharge indicated by overlapping station means in the analysis. In three cases, the means for the benthic community parameters differed among stations, but these differences were not related to a gradient of distance or sediment contamination from the discharge point. These are labeled "None" in Table 3. Effects on benthic communities were negative in that they indicated loss of numbers of species and number of individuals compared to communities at the distal ends of the transects. In no instances were there found to be trends for increases in numbers of species or individuals with increasing hydrocarbon content of the sediments. It was not possible to determine from these studies whether ecological effects such as the loss of biomass or specific organisms might adversely impact the structure of the ecosystem.

The most severely depressed benthic macroinfaunal communities were found within 500 to 800 m of the discharge point in Pass Fourchon. Benthic infauna was absent or substantially reduced at 400, 500, 600 and 800 m north of the discharge point for most of the sample periods. Depending upon the positioning of the ship in the dead-end channel, benthic infauna was usually absent or reduced at 600 and/or 800 m north of the discharge. Infauna was reduced in the deeper parts of the channel where the effluent density plume followed the bottom contours, but was present at higher topographic locations not impacted by the bottom-hugging effluent. The benthic fauna 600 m south of the discharge and at distances of greater than 800 m to the north was not impacted by the discharges.

Impacts to the benthic communities were observed as far as 700 m in Bayou Rigaud, 300 m at Eugene Island Block 18, 100 m at East Timbalier Island, and within 550 m of the abandoned discharge at Empire Waterway. The impacts of accumulated and residual produced water contaminants adjacent to the abandoned discharge site indicate the persistent effects of these constituents on the benthic fauna and an unknown length of time for adequate recovery of the environment following cessation of effluents. It was not possible to estimate the time required for the affected benthic community to recover from the impact of the produced water, but evidently it exceeded the three years since the cessation of the discharge.

No effects on the benthic communities for the parameters analyzed were observed in the Emeline Pass, Empire Waterway (current discharge), and Romere Pass study areas. Implications of the flow of the discharge into adjacent marshes rather than the access canals accessible to the research vessel were already discussed for the Empire Waterway and Emeline Pass study areas. The Empire Waterway and Romere Pass environments are normally low in numbers of species and individuals. The depauperate fauna of Emeline Pass is a characteristic of the fine-grained, sandy, low organic sediments and the high energy river flow. The number of species and number of individuals are low in the Romere Pass study area, as expected in a brackish-to-fresh transition environment. Most species and individuals were found at the northernmost two stations in the Romere Pass transect where salinities were higher. The few number of stations in the Romere Pass study area limited the ability to find relationships with environmental parameters. There were statistical differences among stations for three of the benthic community parameters calculated, but the differences were not obviously related to proximity to a discharge point nor any chemical parameter, and there was considerable overlap in the station means.

Impacts on the benthic macroinfauna were consistent across season in the Pass Fourchon and Eugene Island Block 18 study sites (analysis of co-variance, $P < 0.05$). For the Bayou Rigaud and Emeline Pass study sites, there were significant differences among stations with regards to season.

Benthic community impacts were associated with levels of produced water chemical constituents in the sediments which generally decreased with distance from the discharge point(s). The threshold concentrations of contaminants at or above which the benthic community was affected varied among study areas. For example, the threshold level of alkylated PAH for East Timbalier Island was 5,000 ppb; Pass Fourchon 5,900 ppb; Bayou Rigaud, 7,700 ppb; and Eugene Island Block 18, 430 ppb. These differences are a function of the loadings to the environment, the sedimentary characteristics of the environment and the type of benthic community associated naturally with the environment.

Table 4. Comparison of discharge rate, receiving environment, extent of sediment contamination, and extent of benthic community impacts for produced water discharges in the Gulf of Mexico region

Location	Discharge ($bbl \cdot d^{-1}$)	Receiving Water Depth (m)	Environment	Zone of Sediment Contaminants	Extent of Benthic Community Impacts
Bayou Rigaud[1,2]	146,000	4-5	Dredged Bayou	1,300 m	700 m
Pass Fourchon[1,2]	48,000	3-4	Canal-Dredged Bayou	1,000 m	800 m
East Timbalier Island[1,2]	26,000	1.5-2	Canals near Bay	360 m	100 m
Eugene Island Block 18[1,2]	21,000	2	Shallow Shelf	250 m	300 m
Romere Pass[1,2]	20,200	2	Miss. R. Distributary	450 m	None
Empire Waterway[1,2]	11,000	3	Marsh,Dredged Canal	None	None
Trinity Bay[3]	4,000-10,000	3	Open Bay	250-300 m	150m
Emeline Pass[1,2]	3,700	3-6	Marsh,Miss. R. Distrib.	None	None
Lake Pelto[4]	3,700	2	Open Bay (near Pass)	100 m	20 m
Lafitte Field[5]	3,700	2	Dredged Canal	500 m	250 m
Eugene Island 120[4]	3,000	12	Shallow Shelf	100 m	20 m
Golden Meadow Field[5]	2,800	2-3	Dredged Canal,Bayou	100 m	100 m
Bayou Sale Field[5]	2,500	2-3	Dredged Canal	500 m	100 m
Buccaneer Field[6]	120-2,000	20	Shallow Shelf	200 m	NA

[1]Boesch and Rabalais (1989a)
[2]Rabalais et al. (1991)
[3]Armstrong et al. (1979)
[4]Neff et al. (1989)
[5]Boesch and Rabalais (1989b)
[6]Middleditch, 1981

Bioaccumulation

American oysters, *Crassostrea virginica*, were deployed at known distances from the discharge points at the Pass Fourchon and Bayou Rigaud study areas in April and May for 14 days (Deployment I) and 27 days (Deployment II), respectively. Two reference sites where chosen in the Pass Fourchon area, far from oil-related activity. Oysters were obtained the day before deployment from a local seafood supplier. A set of oysters was retained as controls and returned to the laboratory for analysis. Oysters were deployed in cages in water depths of 1-2 m at a height of approximately 0.1 m above the sediment.

A comparison of the extent of accumulation of produced water contaminants in oysters is shown in Table 3. The comparisons are based on Deployment I in April as indicative of a predicted moderate level of accumulation compared to expected higher values in winter and the lower values found a month later when the oysters were presumed to be spawning and mobilizing lipids and associated hydrocarbons. There was higher mortality during the longer deployment period and at stations closer to the discharge for Pass Fourchon. Generally, mortality was higher in the Pass Fourchon area than in the Bayou Rigaud area. During the first deployment, oysters at the 0 m, 200 m and 400 m distances at Pass Fourchon, and a 300 m station at Bayou Rigaud displayed lower weight gain (<50% of reference) than the oysters at the reference stations. The lower weight gain for the oysters from Pass Fourchon was probably the result of the produced water discharges; with the Bayou Rigaud station, this was not the case. The low weight gain may have resulted from stress induced during the deployment from burial of the tray with silts. Oysters from the 0 m and 200 m stations at Pass Fourchon again showed lower weight gain (<25% of reference) during the second deployment than oysters deployed at all other stations.

Oysters at Pass Fourchon and Bayou Rigaud accumulated levels of alkylated PAH and total hydrocarbons above background concentrations during the first deployment (3 to 18 times and 3 to 7 times above background, respectively). The levels of accumulation during the second deployment were about half those of the first. These reductions are attributed to season and reproductive condition. Oysters accumulated levels of produced water origin hydrocarbons in excess of the concentrations of the contaminants found in ambient sediments in the region. Results for the trace metals were inconsistent with distance from the discharge point, and inconsistent with the values found in the sediments of the study areas. Total radium activity above the detection limit of 0.2 $dpm.g^{-1}$ was measured in four of the 14 samples analyzed. These four elevated levels were in oysters deployed adjacent to a discharge, one each at Pass Fourchon and Bayou Rigaud, or in one instance 200 m from a discharge at Pass Fourchon. The bioaccumulation studies show the clear potential for uptake and accumulation of produced water origin contaminants by oysters, both in close proximity to the discharge and to great distances from the discharge (350 m at Bayou Rigaud and 1,000 m at Pass Fourchon).

SUMMARY

Several studies have been conducted on the fate and effects of produced water discharges in coastal environments of the Gulf of Mexico (several reported in these proceedings). While an exhaustive review of these works is beyond the scope of this paper, a summary of similar studies is provided in Table 4. The distance to which sediment contamination related to a produced water discharge has been identified in the receiving environment was found in our present study to be greater than for any studies to date. The zone of sediment contamination extends to greater distances where the volume of discharge is large, the loadings of petrogenic-origin hydrocarbons are high, and the receiving environment is characterized by restricted water exchange. A feature of our study, not provided previously, was the measurement of current velocity profiles and bed shear stress in the receiving environment as an indicator of whether sediment resuspension and transport were likely to occur. Regardless of hydrographic regime in a receiving environment, some level of sediment contamination is evident at most of the locations studied (Table 4); but benthic infaunal community impacts (usually identified as reductions in number of species or individuals) are not always evident, or may be localized. Benthic fauna are severely reduced or absent where the threshold of certain chemical contaminants (such as volatile hydrocarbons and alkylated PAHs) are exceeded. The presence of high concentrations of produced water derived hydrocarbons to depths of 25 to 30 cm in vertical sediment cores at some of the study areas indicates the long-term accumulation of these contaminants and their resistance to degradation. Similar accumulations in surficial sediments, and particularly sediments to depth, in an access canal adjacent to a discontinued produced water discharge identified the long-term (at least three years) persistence of produced water origin hydrocarbons which continued to impact the benthic infaunal community. The bioaccumulation studies show the clear potential for uptake and accumulation of produced water origin contaminants by oysters, both in close proximity to discharges (as shown previously) but also to great distances from discharges (350 m to 1,000 m).

ACKNOWLEDGEMENTS

This research was supported under Cooperative Agreement 14-12-0001-30454 between the U. S. Dept. of Interior Minerals Management Service (MMS) and the Louisiana Universities Marine Consortium (LUMCON). The views and conclusions contained in this manuscript are those of the authors and should not be interpreted as necessarily representing the official policies, either expressed or implied, of the the U.S. Government. The research program could not have proceeded without the cooperation of Chevron U.S.A. Inc., Chevron Pipe Line Company, Exxon Pipeline Company, Conoco Inc., and Shell Offshore, Inc. Dr. James J. Kendall, Dr. Murray S. Brown and Ms. Gail Rainey represented the MMS. We thank the captains and crew of the R/V *Acadiana* for assistance in sampling. Many research and laboratory assistants helped with field work and sample analyses.

REFERENCES

Aller, R. C., and Cochran, J.K., 1976, Th-234/U-238 disequilibrium and diagenetic time scales, Earth and Planetary Letters 29:37-50.

Armstrong, H. W., Fucik, K., Anderson, J.W., and Neff, J.M., 1979, Effects of oil field brine effluents on sediments and benthic organisms in Trinity Bay, Texas, Mar. Environ. Res. 2:55-69.

Boehm, P. D., and Farrington, J.W., 1984, Aspects of polycyclic aromatic hydrocarbon geochemistry of recent sediments in the Georges Bank region, Environ. Sci. Technol. 18:804-845.

Boesch, D. F., and Rabalais, N.N., eds., 1987, "Long-Term Environmental Effects of Offshore Oil and Gas Development," Elsevier Applied Science Publishers Ltd., London.

Boesch, D. F., and Rabalais, N.N., eds., 1989a, Produced Waters in Sensitive Coastal Habitats: An Analysis of Impacts, Central Coastal Gulf of Mexico. OCS Report/MMS 89-0031, U.S. Dept. of Interior, Minerals Management Service, Gulf of Mexico OCS Regional Office, New Orleans, Louisiana.

Boesch, D. F., and Rabalais, N.N., eds., 1989b, Environmental Impact of Produced Water Discharges in Coastal Louisiana. Final Rept. to Louisiana Div. of Mid-Continent Oil and Gas Assoc., Louisiana Universities Marine Consortium, Chauvin, Louisiana.

Forstner, U., and Wittman, G.T.W., 1983, "Metal Pollution in the Aquatic Environment," Springer-Verlag, New York.

Leeder, M. R., 1982, "Sedimentology," George, Allen and Unwin, London.

Levinton, J. S., 1982, "Marine Ecology," Prentice-Hall Inc., Englewood Cliffs, New Jersey.

MacLeod, W. D., Brown, D.W., Friedman, A.J., Burrows, D.G., Maynes, O., Pearce, R.W., Wigren, C.A., and Bogar, R.G., 1985, Standard Analytical Procedures of the NOAA National Analytical Facility, 1985-1986. NOAA Techn. Mem. NMFS F/NWC-92.

McKee, B. A., DeMaster, D.J., and Nittrouer, C.A., 1986, Temporal variability in the partitioning of thorium between dissolved and particulate phases on the Amazon shelf: Implications for the scavenging of particle-reactive species, Continental Shelf Res. 6:87-106.

Michel, J., Moore, W.S., and King, P.T., 1981, Gamma-ray spectrometry for determination of radium-228 and radium-228 in natural waters, Analytical Chem. 53:1885-1889.

Middleditch, B. S. (ed.), 1981, "Environmental Effects of Offshore Oil Production. The Buccaneer Gas and Oil Field Study," Plenum Press, New York.

National Research Council, 1985, "Oil in the Sea. Inputs, Fates, and Effects," National Academy Press, Washington, D.C.

Neff, J. M., Sauer, T.C., and Maciolek, N., 1989, Fate and Effects of Produced Water Discharges in Nearshore Marine Waters, API Publ. No. 4472, American Petroleum Inst., Washington, D.C.

Parsons, T. R., Maita, Y., and Lalli, C.M., 1984, "A Manual of Chemical and Biological Methods for Seawater Analysis," Pergamon Press, New York.

Reid, D. F., 1983, Radium in formation waters: How much and is it of concern?, *in* Proceedings, Fourth Annual Gulf of Mexico Information Transfer Meeting. OCS Report/MMS 84-0026, U.S. Dept. of Interior, Minerals Management Service, Gulf of Mexico OCS Regional Office, New Orleans, Louisiana.

Rabalais, N.N., McKee, B.A., Reed, D.J., and Means, J.C., 1991, Fate and Effects of Nearshore Discharges of OCS Produced Waters. Volume I. Executive Summary. OCS Study/MMS 91-0004; Volume II. Technical Report. OCS Study/MMS 91-0005; Volume III. Appendices. OCS Study/MMS 91-0006. U.S. Dept. of the Interior, Minerals Management Service, Gulf of Mexico OCS Regional Office, New Orleans, Louisiana.

SAS Institute, Inc., 1982, "SAS User's Guide," SAS Institute, Inc., Cary, North Carolina.

COMPOSITION, FATE AND EFFECTS OF PRODUCED WATER DISCHARGES TO NEARSHORE MARINE WATERS

J.M. Neff[1], T.C. Sauer, Jr.[1], and N. Maciolek[2]

[1] Arthur D. Little, Inc.
Cambridge, Massachusetts
USA

[2] Duxbury, Massachusetts

INTRODUCTION

Produced water often is permitted for discharge from offshore oil and gas production platforms or shore-side treatment facilities to the territorial seas of many countries. However, there is a limited amount of information available about the chemical composition of produced water (Neff et al, 1987) and about effects of chronic discharges on shallow-water, nearshore marine environments (Neff, 1987).

Water is frequently produced with oil and gas from a well. Before the crude oil can be refined or the gas processed, this produced water must be removed. The produced water is treated to remove as much oil as possible and then may be discharged to the ocean from the platform or from the on-shore treatment facility. Chemicals may be added to the fossil fuel/water mixture at different steps to improve the efficiency and completeness of oil/water separation (Hudgins, 1989). Some of these chemicals remain in the produced water and are discharged with it to the ocean. The oil/water separation is not 100 percent efficient. Some oil and other natural ingredients of the original mixture remain in the produced water and are discharged with it.

Produced water contains a variety of chemicals that have been dissolved from the geologic formations in which the produced water resided for millions of years. These chemicals include inorganic salts (essentially the same salts that are found in seawater), several metals, and a wide variety of organic chemicals. The most abundant organic chemicals are petroleum hydrocarbons and low molecular weight fatty acids. The natural radioactive elements, radium-226 and radium-228 (^{226}Ra and ^{228}Ra) also may occur at trace concentrations in produced water.

The investigation reported here was designed to determine (1) the chemical composition of produced water typical of that discharged to the US Gulf of Mexico, (2) if sediments located near platforms that have been discharging produced water to shallow coastal waters for many years contain elevated concentrations of organic and metal contaminants, and (3) if communities of benthic animals resident in the sediments are adversely affected by the chronic discharge. The study design was based on the hypothesis that any significant impacts to the marine environment resulting from more or less continuous discharges of produced water for several years would be manifest as gradients of decreasing sediment contamination and alterations of normal benthic community structure with distance from the produced water discharges.

Produced Water, Edited by J.P. Ray
and F.R. Engelhart, Plenum Press, New York, 1992

MATERIALS AND METHODS

Two sites were selected from an initial screening survey of four oil-water separation platforms in less than 20 meters of water off the coast of central Louisiana, USA. One site, Lake Pelto, Tank Battery No. 1, located in Terrebonne Bay, represented a shallow enclosed estuarine environment; the other site, Eugene Island, Block 105, represented a shallow offshore environment. The average rate of discharge of produced water from the separator platform in Eugene island, Block 105, is 1,570 barrels/day (250,000 l/day). The discharge was initiated in May 1981, about five years before this study. The average rate of discharge of produced water from Lake Pelto, Tank Battery No. 1, is 2,750 barrels/day (437,000 l/day). The discharge was started in the mid-1940s. Water depth at Eugene Island is 7.6 to 8.5 meters, whereas that at Lake Pelto is only 1.8 to 3.0 meters.

Two field surveys were performed at each platform, the first in March 1986 and the second in October 1986. On the first survey, a group of 16 stations, arranged in a radial array with four stations equally spaced at distances of 20, 100, 300, and 1,000 meters from each produced water discharge, was sampled. On the second survey, a subset of the stations was resampled to determine if there were seasonal trends in the chemical and biological parameters. Stations were located by Loran C coordinates.

Three separate produced water samples were collected at each production platform during both surveys. The samples were preserved according to standard methods for water samples destined for analysis of semivolatile organic compounds, volatile aromatic analytes, and metals.

On both surveys, three sediment grab samples for chemical analysis and six sediment grab samples for biological analysis were collected from each station with a Kynar-coated 0.04 m^2 Van Veen grab sampler. Subsamples of surficial sediment were collected from all grab samples for sediment grain size analysis. Subsamples of the upper 2 cm of sediment from chemistry grab samples were collected with a Kynar-coated scoop for trace metals and organics analyses. Samples were frozen immediately and transported to the laboratory on dry ice. The contents of each biology grab sample was placed in a 9.5 liter bucket, elutriated into a large diameter, 0.3 mm mesh sieve, and washed with filtered seawater. Animals retained on the 0.3 mm screen were transferred quantitatively to glass jars and preserved in 10% buffered formalin in seawater.

At each station on both surveys, dissolved oxygen concentration and other hydrographic data were recorded 0.3 m below the sea surface and 1 m above the bottom with a Hydrolab 4041 indicator unit and a Type 70 SONDE unit. The oxygen electrode was calibrated regularly by comparison with results of a Winkler titration.

Produced water and sediment samples were analyzed for total organic carbon, total and individual saturated hydrocarbons, 2- through 4-ring polycyclic aromatic hydrocarbons (PAHs) and their C_1 through C_5 alkyl homologues, steranes, triterpanes, phenol and alkyl phenols, barium, cadmium, chromium, copper, lead, mercury, nickel, silver, and zinc. Produced water samples also were analyzed for ^{226}Ra and ^{228}Ra, volatile organic compounds, and naphthenic acids. Water samples for analysis of hydrocarbons, phenols, and naphthenic acids were acidified to pH<4 with 1 N HCl and extracted twice with 100 ml of methylene chloride and a third time with 100 ml 50:50 methylene chloride:ethyl ether. Before extraction, 500 μg each of *o*-terphenyl and androstane were added to each sample as internal standards. The combined extracts were dried over Na_2SO_4 and split into two aliquots, one for analysis of hydrocarbons and naphthenic acids and the other for analysis of phenols.

The aliquot for hydrocarbon analysis was passed through an alumina/silica gel column to separate the extract into f_1(saturates), f_2 (aromatics), and f_3 (polars) fractions. The f_1 fraction was eluted with hexane, the f_2 fraction was eluted with 1:1 hexane:methylene chloride, and the f_3 fraction was eluted with 100% methanol.

The f_1 and f_2 fractions were analyzed with a Hewlett-Packard D5790 gas chromatograph (GC) with a mass-selective detector (MSD) equipped with a 30 m x 0.25-mm (i.d.) DB-5 fused silica capillary column. The temperature of the GC was programmed from 40 to 290° C at 4°

C/min with a one-min. initial hold. The GC data analysis system was used to quantify specific hydrocarbons in the f_1 and f_2 fractions relative to the internal standards. The n-alkanes from C_{10} and C_{34} were determined in the f_1 fraction. The f_2 fraction was analyzed for the homologous series of benzenes containing 2 to 5 alkyl carbons (C_2-C_5), and C_0 through C_5 naphthalenes, biphenyls, fluorenes, phenanthrenes, dibenzothiophenes, pyrenes, chrysenes, and perylenes. Concentrations of total n-alkanes and aromatic hydrocarbons were determined by GC/MSD analysis of the f_1 and f_2 fractions, respectively. Both resolved and unresolved hydrocarbons were quantified. The resolved hydrocarbons were quantified by summing the total resolved area, integrating the valleys, and using an average n-alkane or PAH response factor to calculate an amount relative to the internal standard. The concentrations of unresolved hydrocarbons were calculated by integrating the total area of the chromatogram (both resolved area and unresolved complex mixture [UCM] area), subtracting the resolved area, and determining the amount relative to the internal standard. Concentrations of UCM were not corrected for instrumental response.

Steranes and triterpanes were determined in the f_1 fraction by GC/MSD with selective ion monitoring (SIM). Column conditions were the same as those used to analyze n-alkanes. Mass ions m/e 217 and 318 were used to identify steranes, and mass ions m/e 191 and 192 were used to identify triterpanes. Individual compounds were identified by comparison with authentic standards. Concentrations of total steranes and triterpanes were estimated based on the relative response of the 217 and 191 mass ions and the response of the internal standard, androstane.

Attempts were made to identify and quantify naphthenic acids in the f_3 fraction. The fraction was derivatized with boron trifluoride. Both underivatized and BF_3-derivatized extracts were analyzed by GC/MSD. An attempt was made to identify the major peaks on the chromatograms by comparison of mass spectra with those in the Hewlett Packard NBS computer library.

The extract for phenol analysis was partitioned two times with 50 ml of 10 N NaOH in a separatory funnel. The aqueous phases were combined, acidified to pH $\approx$ 1 with 6 N HCl, and extracted three times with ethyl ether. The extracts were combined and dried over $Na_2 SO_4$. The phenol extract was analyzed by a GC/MSD equipped with a 30 m x 0.25 mm DB 5 capillary column for phenol and C_1 through C_4 phenols. The phenols were quantified relative to the internal standard, d_8p-cresol.

For analysis of volatile hydrocarbons, methane through C_3-benzene, water samples were partitioned according to the head space extraction method of McAuliffe (1974) and quantified by GC with flame ionization detection (FID). Analytes were identified by comparison with relative retention indices of known standard mixtures. Volatile priority pollutants were analyzed by EPA Method No. 624.

Total organic carbon in water samples was quantified by the method of Menzel and Vaccaro (1964) as modified by Mills and Quinn (1979). The concentration of total organic carbon in sediment samples was determined by the method of Froelich (1980).

It was necessary, because of the high concentration of salts in produced water, to coprecipitate metals from produced water (Buchanan and Hannaker, 1984) before analysis. A 5% ammonium 1-pyrrolidine dithiocarbamate (APDC) chelate solution was added to the water sample and the pH of the sample was adjusted to 7-8 with Ultrix NH_4OH. The precipitate was collected on a 0.4 μm Nucleopore filter and redissolved in a heated water bath with a dilute acid solution. Concentrations of cadmium, chromium, copper, lead, nickel, and silver were determined with a Zeeman effect graphite furnace atomic absorption spectrophotometer by the method of standard additions. Barium was determined by neutron activation analysis. Mercury was determined by cold vapor atomic absorption spectrophotometry, and zinc was determined by direct injection of the original water sample onto the atomic absorption spectro photometer without prior coprecipitation. Radium isotopes in produced water were analyzed by standard methods by a contract laboratory.

Sediment grain size was determined by the methods described by Folk (1974). The homogenized sediment was wet-sieved into gravel, sand, and silt/clay fractions. The gravel and sand fractions were fractionated further by dry sieving and the silt/clay fraction was fractionated by wet pipet analysis.

Sediment samples for analysis of organic compounds were thawed and mixed thoroughly with a spatula. A 100 g sample was extracted and cleaned up by the ambient temperature method of Brown et al., (1980). The internal standards, *o*-terphenyl, d_8p-cresol, and 5α-androstane, were added before extraction. The volume of the combined organic extract was measured and it was divided into two equal aliquots, one for hydrocarbon analysis and the other for phenol analysis.

The aliquot for hydrocarbon analysis was concentrated to 1 ml in a Kuderna-Danish flask at 70° C, displaced into hexane, and subjected to column chromatography on an activated copper/partially deactivated alumina column. The fraction eluted from the column in 5% methylene chloride in hexane was analyzed by GC/MS according to the methods described above for produced water extracts. The aliquot for analysis of phenols was cleaned up and analyzed by the methods described above for produced water extracts.

Concentrations of barium in sediments were determined by neutron activation analysis. For the other metals, sediments were digested with aqua regia and analyzed by flame or Zeeman effect atomic absorption spectrophotometer according to the methods of Agemian and Chau (1976), Fisher (1986), and Lamothe et al., (1986).

Samples for biological analysis were sieved through nested 0.5 mm and 0.3 mm screens and transferred to 80% ethanol. Animals were stained with Rose Bengal and sorted under a dissecting microscope to major taxonomic groups. All animals subsequently were identified to lowest possible taxon, usually species, and counted. Juveniles of 35 polychaete species and the bivalve mollusk *Mulinia lateralis* were identified and counted separately from the adults.

The mean and standard deviation of the abundance of several dominant species were calculated. Student's T-test was used to compare mean abundances among stations and between surveys. The percent contribution of juveniles to the total fauna at each station was calculated, as was the percent of juveniles for several species.

Shannon-Wiener diversity (H′) and its associated evenness value (E) were calculated for each sample and Hurlbert's modification (1971) of the rarefaction method (Sanders, 1968) was used to predict the number of species in a random sample with a given population size.

An agglomerative clustering technique was used to determine similarity among benthic faunal samples. The similarity was measured between all pairwise combinations of samples, beginning with the most similar pairs, and subsequently combining samples until they all formed one large group. The similarity measure was NESS, the normalized expected species shared (Grassle and Smith, 1976), in which the comparison of expected species shared is between random samples of a set number of individuals in each replicate. The clustering strategy used was the group average (Boesch, 1977).

RESULTS

Composition of Produced Water

The mean compositions of the produced water from the two platforms were similar to those of produced water from other sources (Table 1). Both produced waters were hypersaline brines with salinities about three times that of coastal water from the northern Gulf of Mexico.

Produced water from Eugene Island contained copper, lead, zinc, and barium at concentrations four to ten times higher than their concentrations in produced water from Lake Pelto. Concentrations of the other metals were about the same in produced water from both platforms. The metals present at highest concentration in the produced water samples were barium, lead, and zinc. The other metals had concentrations similar to or only slightly higher than their concentrations in clean natural seawater. Elevated concentrations of barium, lead, and zinc have been reported in other samples of produced water from coastal Louisiana and elsewhere (Neff et al., 1987; Boesch and Rabalais, 1989). High concentrations of barium apparently can exist in solution in produced water if the produced water contains low concentrations of sulfate and carbonate (Barth, 1991). This barium probably precipitates as barium sulfate (barite) when the produced water mixes with seawater (high in sulfate).

Concentrations of lead and zinc in the produced water were more variable than concentrations of other analytes and may have been derived from corrosion or chipping of galvanized structures on the platform or in the oil/water separator system.

Total radium ($^{226}Ra + ^{228}Ra$) was measured in produced water from four platforms as part of the initial screening study. Concentrations ranged between 605 and 1,215 pCi/l. There was a strong positive correlation (r= 0.90) between the concentrations of radium and total organic carbon in the produced water samples. However, because of the small number of samples analyzed, it is uncertain whether this is a general relationship. Earlier, Kraemer and Reid (1984) reported a positive correlation between radium concentration and salinity of a larger number of produced water samples. This relationship was not evident for the four samples of produced water analyzed in this investigation.

The concentration of total organic carbon in the samples was in excess of 200 mg/l. About 7 to 14% of this organic carbon is accounted for by total resolved and unresolved hydrocarbons (as measured by GC). The rest probably consisted of partly-characterized, nonvolatile, soluble organic matter, including organic acids (Lysyj, 1982; Somerville et al., 1987; Means and Hubbard, 1987; Barth, 1991). Phenols and ketones contributed 0.7 to 2% to the total organic carbon of the samples. Steranes and triterpanes were only trace components of the organic carbon in produced water. Naphthenic acids were detected in the produced water samples but could not be characterized or quantified because of interferences.

The most abundant ketones in the produced water samples were 2-butanone, 3-methyl-2-butanone, 2-hexanone, and 3-hexanone. Ketones of these types are used as solvents in paints and for a variety of purposes on a platform and have been identified in produced water samples from other sources (Sauer, 1981; Middleditch, 1984). They may enter the oil/water separator system with platform runoff water. A small amount of tetrahydrothiophene was present in produced water from Lake Pelto and may be a natural component of the crude oil from this platform. Small amounts of phenol and C_2 through C_4 phenols were present in the produced water samples. The concentration of phenols was about four times higher in produced water from Eugene Island than in produced water from Lake Pelto. The most abundant phenols in the produced water samples were the m- and p-cresols.

Steranes and triterpanes are polycyclic alkanes that are derived from diagenesis of plant materials. The composition of the sterane/triterpane assemblage varies from one crude oil to another and these compounds are quite resistant to degradation (Seifert and Moldowan, 1978). Therefore, they can serve as indicators "biomarkers" of the presence of crude oil in environmental samples. The C_{27} steranes were most abundant in the Lake Pelto produced water and the C_{29} steranes were most abundant in the Eugene Island produced water. The triterpane composition of the two produced water samples was similar, and indicated that the crude oils from the two fields are very mature (Seifert and Moldowan, 1978).

The mean concentrations of total petroleum hydrocarbons measured by GC were 30.9 mg/l and 20.4 mg/l in produced water from Eugene Island and Lake Pelto, respectively. Nearly 10% of the total hydrocarbons in produced water from Eugene Island were semivolatile aromatics; the hydrocarbon assemblage in produced water from Lake Pelto contained about 20% aromatics.

The pattern of concentrations of n-alkanes in the two produced water samples was typical of that from a petroleum source (Figure 1). In both samples, the n-alkanes above C_9 present at highest concentration were the C_{14} or C_{15} alkanes and concentrations gradually decreased with increasing chain length to C_{34}. Both produced water samples contained higher concentrations of methane, ethane, and propane (190 to 666 μg/l) than of higher molecular weight liquid or solid alkanes. Only 25% of the n-alkanes quantified in either sample of produced water were the higher molecular weight C_{21} through C_{34} n-alkanes. This alkane composition of produced water was undoubtedly due to the fact that the solubility on n-alkanes decreases logarithmically as chain length increases (McAuliffe, 1966), and most crude oils contain a predominance of low over high molecular weight n-alkanes.

The most abundant aromatic hydrocarbon in the produced water samples was benzene. The C_1 through C_4 benzenes also were abundant. The most abundant PAH were naphthalene and its C_1 through C_4 alkyl homologues (Figure 2). The C_2-naphthalenes were the most abundant PAH in both produced waters. The Lake Pelto produced water also contained small amounts of phenanthrenes (22.4 μg/l), fluorenes (11.3 μg/l), and dibenzothiophenes (4.8 μg/l). The Eugene Island produced water contained only traces of these PAH. Neither produced water sample contained more than traces of higher molecular weight PAH (pyrene through perylene).

Table 1. Mean concentrations of several classes of organic compounds and metals in produced water from production platforms at Eugene Island, Block 105, and Lake Pelto, Tank Battery No. 1 (values in μg/l unless indicated otherwise)

Compound Class	Eugene Island	Lake Pelto
Total Organic Carbon	220,000	298,000
Total Saturates (GC)	28,400	17,100
Total Aromatics (GC)	2,500	3,300
Total Hydrocarbons (GC)	30,900	20,400
Total Targeted n-Alkanes (C_{10}-C_{34})	2,677	606
Total Targeted PAH (Naph.-Peryl.)	80	142
Total Volatile Hydrocarbons (Methane - C_3-Benzenes)	11,308	5,515
Steranes	63	92
Triterpanes	76	80
Ketones	2,230	1,320
Phenols (C_0 - C_4)	2,720	675
Salinity (parts per thousand)	183	150
Metals		
Barium	37,400	11,500
Cadmium	0.32	0.12
Chromium	1.11	0.71
Copper	6.36	0.40
Lead	17.9	1.50
Mercury	0.06	0.07
Nickel	0.40	1.27
Silver	<0.20	<0.20
Zinc	1,200	125

The relative abundance of different aromatic hydrocarbons in produced water was related in part to the inverse relationship between the aqueous solubility of aromatic hydrocarbons in water and hydrocarbon molecular weight (Neff, 1979). Thus, light aromatics will be enriched in produced water compared to their concentrations in the oil. The traces of higher molecular weight alkanes (including steranes and triterpanes) and aromatics probably were present in the produced water as particulate oil.

Produced Water Chemicals in Sediments

Patterns of sediment grain-size distribution were quite different at stations around the two platforms. At Eugene Island, sediment texture was finest near the platform and became coarser with distance from the platform. At Lake Pelto, the coarsest sediments were located 20 m from the platform and mean sediment grain size decreased with distance from the platform. It is possible that some of the coarse sediments near the tank battery at Lake Pelto were placed there

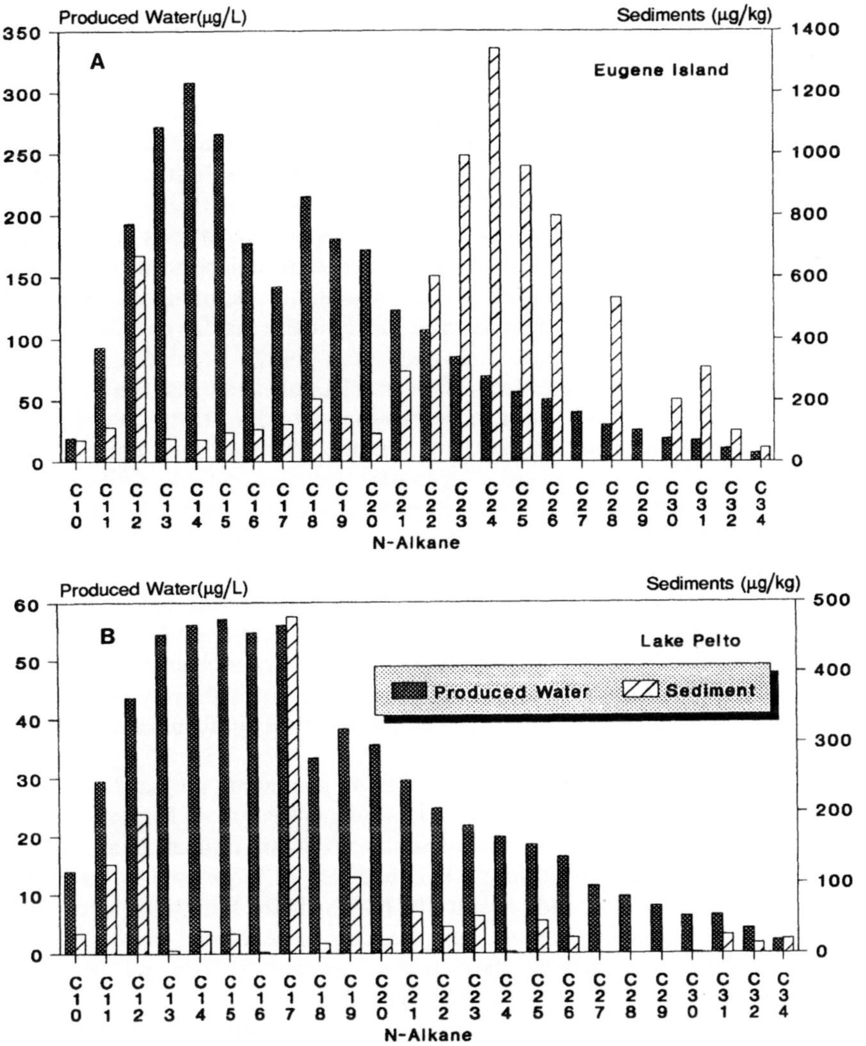

Figure 1. Average concentrations in produced water and in the most heavily contaminated samples of surficial sediments from Eugene Island, Block 105, and Lake Pelto, Tank Battery No. 1, of individual n-alkanes form C_{10} through C_{34}.

to stabilize the bottom near the platform or were produced sand. Grain size has important effects on the distribution of benthic metals, petroleum hydrocarbons, and fauna. Concentrations of dissolved oxygen were near saturation in surface and bottom water at all stations around both platforms at both sampling times. Seasonal episodes of hypoxic ($\leq$ 2.0 mg/l O_2) or anoxic (<0.5 mg/l) bottom water have been reported frequently over large areas of the inner continental shelf off Louisiana (Rabalais et al, 1985). Hypoxic or anoxic bottom water have not been reported from the immediate vicinity of either Eugene Island, Block 105 or Lake Pelto, Tank Battery No. 1. However, the possibility of transitory hypoxic events must be kept in mind when interpreting data on the benthic community structure around the platforms.

Of the nine metals analyzed in sediments from the vicinity of both platforms, only barium and to a much lesser extent mercury were present in sediments from some stations at concentrations higher than would be expected for clean natural nearshore sediments (Table 2). Boothe and Presley (1989) reported background concentrations of barium in sediments from the northwest Gulf of Mexico from less than 200 to about 800 μg/g dry weight. Barium concentrations were higher in sediments from Lake Pelto than from Eugene Island. However, no gradients of decreasing barium concentration were detected in sediments with distance from either platform. At Lake Pelto, the highest mean concentration of barium (3,047 μg/g) was observed in sediments collected 100 m northeast of the platform, and the highest concentration in a single sediment sample (5,040 μg/g) was from 20 m southwest of the platform. At Eugene Island, the highest mean concentrations of barium in sediments were from stations located 300 and 1,000 m south of the platform. These results suggest that elevated concentrations of barium in sediments occur throughout the areas around the two platforms, particularly Lake Pelto. This barium could be derived from drilling mud or produced water discharges (both high in barium) or from the Mississippi River outflow. Chan and Hanor (1982) detected higher concentrations of dissolved barium in coastal water from the central Louisiana coast than in water from the open Gulf of Mexico, and attributed it to desorption of barium from river-borne suspended material during natural processes of estuarine mixing. This desorbed barium would tend to precipitate as barium sulfate as it mixed with coastal seawater high in sulfate.

Mean concentrations of mercury were slightly elevated at only a few stations near Lake Pelto. The highest mean concentration (0.76 μg/g) was from 100 m northeast of the platform. Although this was the same station that had the highest mean concentration of sediment barium, the correlation between concentrations of barium and mercury in sediments near Lake Pelto was poor ($p \geq 0.2$). Nevertheless, the excess mercury in those sediments that contained slightly elevated concentrations of mercury could have been derived in part from metal sulfide impurities in drilling mud barite. Boothe and Presley (1989) reported a correlation between elevated barium and mercury concentrations in sediments near two production platforms and concluded that the mercury was derived from impurities in barite.

Concentrations of total organic carbon (TOC) in sediments ranged from 0.17 to 0.69% around Eugene Island and from 0.37 to 0.83% around Lake Pelto. At Eugene Island, there was a tendency for the concentration of TOC in sediments to be highest near the platform and decrease with distance from the platform. This was due in large part to the strong positive correlation between percent TOC and percent silt plus clay in the sediments around the platform, and the gradients of decreasing percent silt plus clay in sediments with distance from the platform. At Lake Pelto, the coarsest sediments were near the platform and there were no clear gradients of sediment TOC with distance from the platform.

The patterns of hydrocarbon distributions in sediments around the platforms tended to parallel those of sediment TOC, with a few exceptions. At Eugene Island, concentrations of total hydrocarbons and PAH were highest in sediments 20 m from the platform along all four transects (Figure 3). Concentrations of total hydrocarbons dropped to background at 100 or 300 m from the platform; however, PAH concentrations were at apparent background in sediments 100 m or farther from the platform along all four transects.

At Lake Pelto, concentrations of total hydrocarbons were elevated at 20 m from the platform in all four directions and were at or near background at all distances greater than 20 m from the platform (Figure 4). However, concentrations of PAH in sediments were elevated along all four transects out to 100 m from the platform.

The relative compositions of n-alkanes in sediments near Eugene Island and Lake Pelto were different from those in the produced water discharged from the two facilities (Figure 1). At Eugene Island, the dominant n-alkanes in sediments were the higher molecular weight C_{22} through C_{31} alkanes, and the lower molecular weight alkanes, except C_{12}, were depleted relative to their concentrations in the produced water. At Lake Pelto, all alkanes except C_{12} and C_{17} were present at low concentrations. The C_{17} and possibly C_{12} n-alkanes probably were biogenic.

The PAH had similar distribution patterns to those of the n-alkanes in sediments at Eugene Island and Lake Pelto (Figure 2). Relative to the PAH distribution in the produced water, concentrations in sediments of lower molecular weight PAH were depleted and concentrations of higher molecular weight PAH were enhanced. The most abundant PAH in sediments at Eugene Island was perylene (probably biogenic [Venkatessan, 1988]). In sediments at Lake Pelto, the most abundant PAH was C_3-naphthalene. Phenanthrenes were relatively abundant in sediments at both platform sites and pyrene was abundant in sediments from Lake Pelto. The distribution of hydrocarbon assemblages in sediments relative to their concentrations in produced water can be attributed in part to selective partitioning of the higher molecular weight, less soluble hydrocarbons onto the solid substrate. The more soluble lower molecular weight hydrocarbons remained in solution in the water column and did not accumulate in sediments. This phenomenon was also observed in sediments near the produced water discharge from a treatment facility in Trinity Bay, Texas (Armstrong et al., 1979).

Table 2. Comparison of typical concentrations (µg/g) or concentration ranges of nine metals in clean nearshore marine sediments and in sediments collected at stations around the platforms at Eugene Island, Block 105 and Lake Pelto, Tank Battery No. 1 during the spring and fall surveys

Metal	Marine Sediments[1]	Eugene Island	Lake Pelto
Barium	80-8,100	528-2,270	864-5,040
Cadmium	0.1-0.6	0.015-0.118	0.068-0.480
Chromium	10-100	4.32-17.93	4.96-13.60
Copper	10-50	1.66-8.17	4.34-14.22
Lead	5-30	4.44-96.16[2]	5.37-26.6
Mercury	0.1-0.3	0.010-0.234	0.010-1.02
Nickel	30-70	6.71-18.6	5.63-15.1
Silver	0.06	<0.005-0.034	<0.005-0.140
Zinc	2-70	24.3-61.3	32.4-114.9

[1]From Forstner and Wittmann (1983), Salomons and Forstner (1984), and Neff (1990).
[2]Value for one replicate. Other replicates contained less than 5.0 µg/g lead.

Steranes and triterpanes were present in sediments at stations close to the two platforms. However, there was no clear concentration gradient of steranes and triterpanes with distance from either platform. At Eugene Island, the steranes and triterpanes in sediments did not resemble those in the produced water; in sediments 20 m from the discharge at Lake Pelto, they did. Concentrations of phenols were high around both platforms. However, the phenol assemblages in the sediments did not resemble those in the produced water. The dominant phenol in sediments was the C_3-phenol; the dominant phenol in produced water was the C_1-phenol. This difference may be due to selective sorption to sediments of the more highly alkylated phenol or to the existence of other sources than produced water of phenols in sediments.

The fossil fuel pollution index (Boehm and Farrington, 1984) and other diagnostic indices of petroleum contamination indicated that sediments 20 m from the platform at Eugene Island contained some petroleum contamination. Sediments greater than 20 m from the platform contained only traces of petroleum. At Lake Pelto, the diagnostic indices indicated that sediments from 20 and 100 m from the platform, but not those from 300 and 1,000 m away contained primarily petroleum hydrocarbons. Thus, the area of sediments around the two platforms that was contaminated with petroleum hydrocarbons from multi-year discharges of produced water was very small: about 1,250 m^2 at Eugene Island and 31,000 m^2 at Lake Pelto.

Benthic Fauna

There were many unusual features of the benthic communities around the platforms at Eugene Island and Lake Pelto. At both locations, the fauna were dominated by one species, the polychaete, *Mediomastus ambiseta*, at all stations. Along two of the four transects at Lake Pelto, the abundance of *Mediomastus* increased with distance from the platform to a high of more than

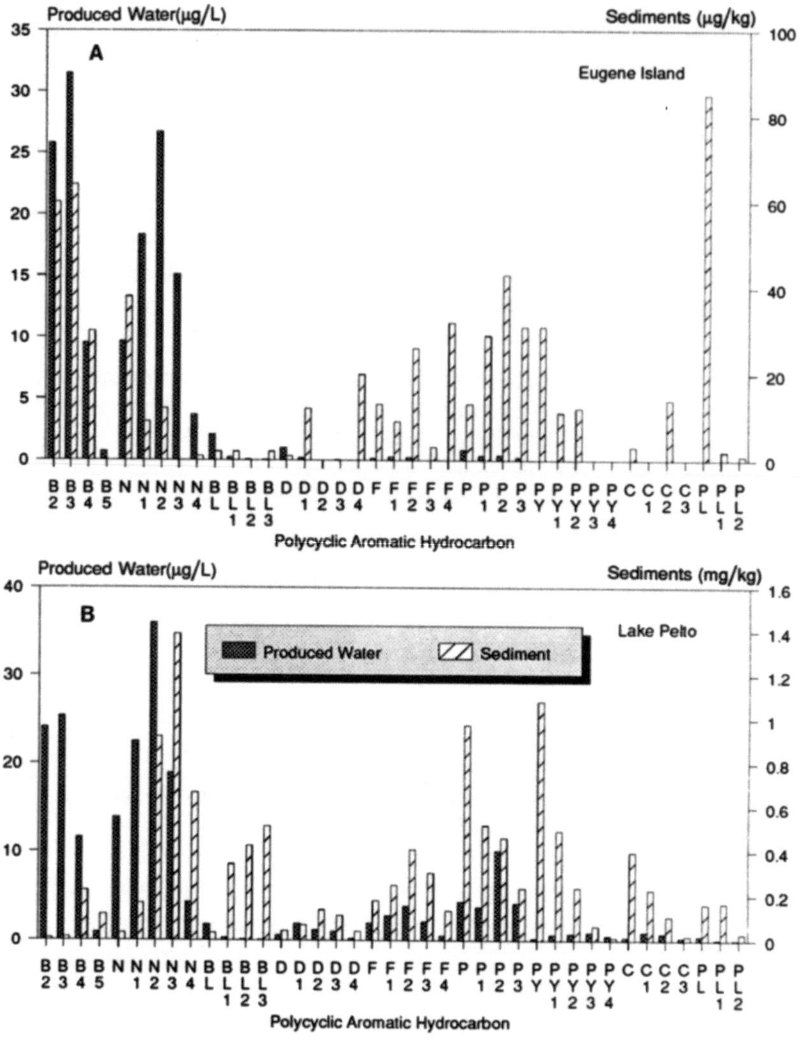

Figure 2. Average concentrations in produced water and in the most heavily contaminated samples of surficial sediments form Eugene Island, Block 105, and Lake Pelto, Tank Battery No.1, of individual parent polycyclic aromatic hydrocarbons and the alkyl homologues. B, benzene; N, naphthalene; BL, biphenyl; D, dibenzothiophene; F, fluorene; P, phenanthrene; Py, pyrene; C, chrysene; PL, perlyene.

70% of the total benthic fauna at the two 1,000 m stations. There was no clear gradient of abundance of *Mediomastus* with distance from the platform at Eugene Island.

The dominant fauna at all stations were short-lived, opportunistic species such as the polychaetes *M. ambiseta* and *Streblospio benedicti* and the bivalve mollusc *Mulinia lateralis*. Juvenile forms were very abundant, representing more than 50% of the total number of individuals of several species at many stations. At Lake Pelto, but not at Eugene Island, the relative abundance of juveniles in the benthic fauna decreased with distance from the platform.

There were substantial differences in benthic community structure at both study sites between the spring and fall surveys. Large seasonal fluctuations are typical of shallow coastal and estuarine polyhaline benthic communities (Boesch et al., 1976). These communities are dominated by opportunistic species that are adapted by high reproductive potential and rapid growth to exploit variable, unpredictable habitats.

Benthic communities with many of the same characteristics were described by Gaston (1985) and Gaston et al. (1985) off Cameron, LA, by Shaw and Vittor (1985) in Chandeleur and Breton Sounds east of the Mississippi River delta, and by Armstrong et al. (1979) in Trinity bay, TX. The reasons for a benthic fauna dominated by opportunistic species and juvenile life stages at many locations along the Gulf coast, including the two study sites for the present investigation, are unclear. The most likely explanation is recurrent disturbance that prevents the development of a more persistent and long-lived benthic community. This disturbance could be periodic hypoxia or low salinity of bottom waters in nearshore waters along the northwest Gulf coast. Although it is unlikely that the disturbance influencing benthic community structure throughout the area is produced water or drilling mud discharges from offshore platforms, it is possible that such discharges may be causing effects near platforms that are superimposed on the area-wide disturbance. Such gradients of effect should be discernable above the wide natural variability.

At Eugene Island, the total fauna were most abundant near the platform and abundance decreased with distance from the platform. However, diversity was lower near the platform than at stations farther away, due mainly to the dominance of the fauna near the platform by *Mediomastus*. Similarity analysis revealed that the benthic fauna of the four stations 20 m from the platform were similar to one another and different from the fauna at stations further away. Many of the juveniles present in the spring did not survive until fall, but were replaced by new juveniles of the same and different species.

The distributions of benthic communities around the platform can be explained in part by the influence of sediment grain size on benthic community structure. There was a strong correlation ($p<0.01$) between faunal density and percent silt plus clay. There also was a strong correlation ($p<0.01$) between the concentration of total petroleum hydrocarbons and percent silt plus clay in sediments around the platform at Eugene Island. However, there was no correlation between faunal density and the concentration of total hydrocarbons in sediments. These results suggest that, at Eugene Island, the distribution and composition of benthic communities around the platform was influenced more by gradients of sediment grain size than by any concentration gradients in sediments of contaminants derived from produced water discharges. If produced water discharges were influencing infaunal community structure, this effect was minor and extended out to 20 m but not to 100 m from the platform.

At Lake Pelto, the gradient of decreasing sediment grain size with distance from the platform obscured any gradients that might have existed in the benthic fauna because of effects of the produced water discharge. As a general rule, benthic communities were more diverse at Lake Pelto than at Eugene Island. Highest diversities were at stations that had the coarsest sediments near the platform. There was no clear relationship between distance from the platform and abundance of adults or juveniles or diversity of the total benthic fauna. Cluster analysis revealed that the faunas at two of the stations at 20 m were similar to one another. Two stations at 1,000 m were quite different from all other stations. However, there was not a statistically significant correlation between percent silt plus clay in sediments and species abundance or diversity ($p>0.2$). All samples from the fall survey clustered separately from spring samples, reflecting seasonal changes in benthic community structure. Although stations 20 m from the platform contained high concentrations of hydrocarbons and PAH identified as probably derived from petroleum, there was no significant correlation between any biological diversity or abundance parameter and concentrations of total hydrocarbons or PAH in sediments at Lake Pelto. Nevertheless, the distinct community structure at stations nearest the platform could have been caused in part to sediment contamination with hydrocarbons from the produced water discharge.

CONCLUSIONS

Produced water from two platforms in Louisiana coastal waters were hypersaline brines containing complex mixtures of low concentrations of several metals, petroleum hydrocarbons, and other organic compounds. However, metals concentrations in sediments near both produced water discharges were not elevated compared to concentrations in sediments further away.

At Eugene Island, Block 105, sediments from stations 100 m or further from the produced water discharge did not contain greater than trace or background quantities of petroleum hydrocarbons. Sediments from stations 20 m from the platform contained low concentrations of petroleum hydrocarbons. Thus, chronic discharges of about 1,570 bbl/day of produced water to inner continental shelf waters with a depth of about 8 m has resulted in a low to moderate level of contamination of surficial sediments with petroleum hydrocarbons out at least to 20 m

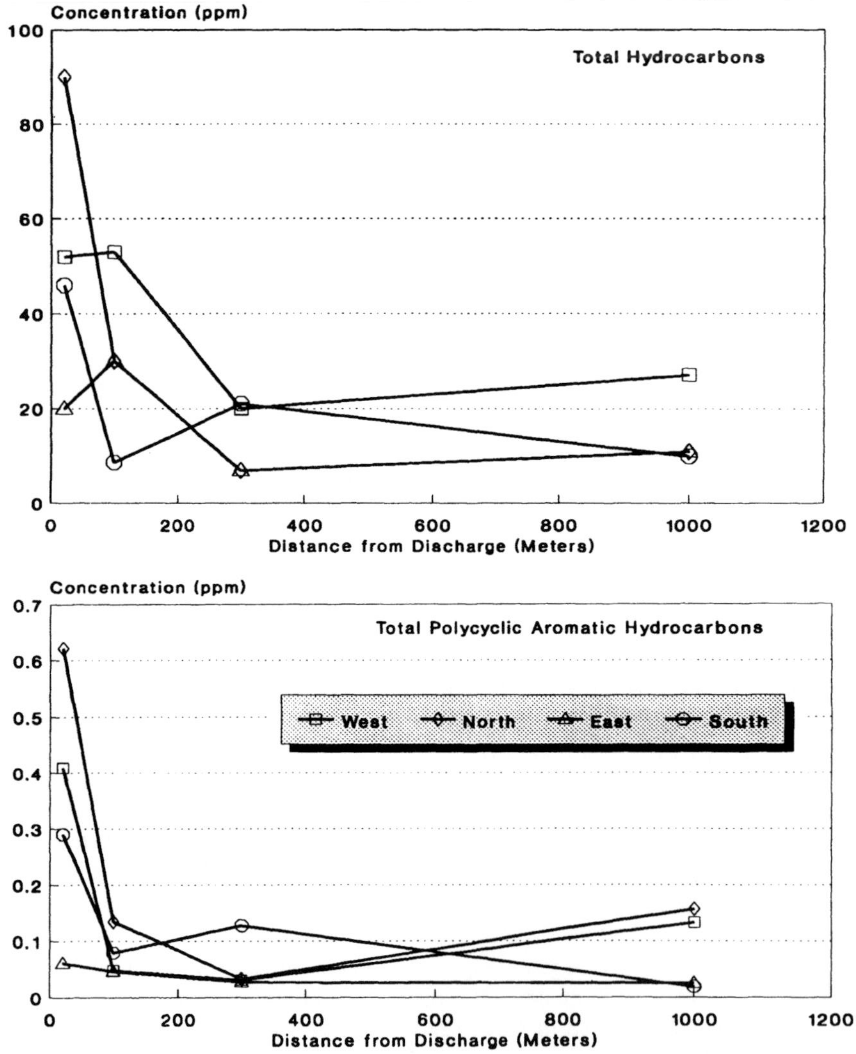

Figure 3. Average concentrations in surficial sediments at different distances along four transects around the platform at Eugene Island, Block 105, of total hydrocarbons and total PAH.

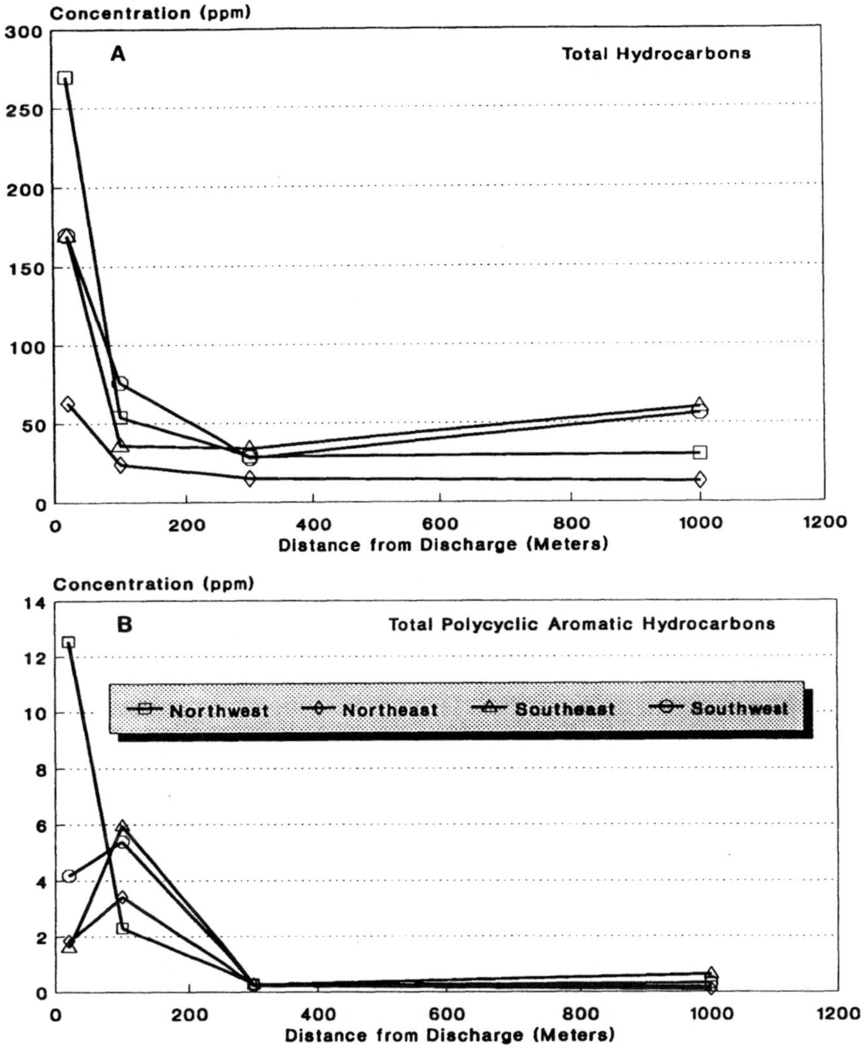

Figure 4. Average concentrations in surficial sediments at different distances along four transects around the platform at Lake Pelto, Tank Battery No.1, of total hydrocarbons and total PAH.

but not to 100 m from the platform. Benthic communities may have been affected by the chronic discharges out to 20 m from the platform but not at distances of 100 m or more.

At Lake Pelto, Tank Battery No. 1, sediments out to at least 100 m but not as far as 300 m from the platform contained petroleum hydrocarbons, probably derived from more than 40 years of produced water discharges at a rate of about 2,750 bbl/day to the shallow (2 m) protected waters of a coastal bay. Benthic communities may have been affected at distances of 20 m but not at distances of 100 m or more from the platform by the produced water discharges.

ACKNOWLEDGEMENTS

This research was performed under a contract between the American Petroleum Institute and Battelle Memorial Institute. We wish to thank Dr. Jack Gould, API technical representative, and the API Drilling Fluid and Produced Water Task Force, in particular Drs. B.A. Cox, C.B. Koons, C.D. McAuliffe, and J.P. Ray, for their guidance and critical review of the technical progress of the program. We also wish to thank the petroleum industry for cooperation in providing access to offshore platform sites and historic information about the platforms chosen for this study.

LITERATURE CITED

Agemian, H. and Chau, A.S.Y., 1976, Evaluation of extraction techniques for the determination of metals in aquatic sediments. Analyst 101:761-767.

Armstrong, H.W., K. Fucik, J.W. Anderson, and Neff, J.M., 1979, Effects of oilfield brine effluent on sediments and benthic organisms in Trinity Bay, Texas. Mar. Environ. Res. 2: 55-69.

Barth, T., 1991, Organic acids and inorganic ions in waters from petroleum reservoirs, Norwegian continental shelf: a multi-variate statistical analysis and comparison with American reservoir formation waters. Appl. Geochem. 6: 1-15.

Boehm. P.D. and Farrington, J.W., 1984, Aspects of the polycyclic aromatic hydrocarbon geochemistry of recent sediments in the Georges Bank region. Environ. Sci. Technol. 18: 840-845.

Boesch, D.F., 1977, "Application of numerical classification in ecological investigations of water pollution," EPA-600/3-77-033. National Technical Information System, PB-269 604. 114 pp.

Boesch, D.F. and Rabalais, N.N. (eds.), 1989, "An Analysis of Impacts. Produced Waters in Sensitive Coastal Habitats." Central Gulf of Mexico" OCS Study MMS 89-0031. U.S. Department of the Interior, Minerals Management Service, Gulf of Mexico OCS Regional Office, New Orleans, LA. 157 pp.

Boesch, D.F., Wass, M.L. and Virnstein, R.W., 1976, Dynamics of estuarine benthic communities. Pages 177-196, *in* M. Wiley (ed.), "Estuarine Processes. Volume 1. Uses, Stresses, and Adaptation to the Estuary". Academic Press, New York.

Boothe, P.N. and Presley, B.J., 1989, Trends in sediment trace element concentrations around six petroleum drilling platforms in the northwestern Gulf of Mexico. Pages 3-22 In: F.R. Engelhardt, J.P. Ray, and A.H. Gillam (eds.), "Drilling Wastes". Elsevier Applied Science Publishers, London.

Brown, D.W., L.S. Ramas, M.Y. Uyeda, A.J. Friedman, and MacLeod Jr., W.D., 1980, Ambient temperature extraction of hydrocarbons from marine sediment - comparison with boiling solvent extraction. Pages 313-326 In: L. Petrakis and F.T. Weiss, eds., "Petroleum in the Marine Environment." Advances in Chemistry Series No. 185. American Chemical Society, Washington, DC.

Buchanan, A.S. and Hannaker, P., 1984, Inductively coupled plasma spectrometric determination of minor elements in concentrated brines following precipitation. Anal. Chem. 56: 1379-1382.

Chan, L.H. and Hanor, J.S., 1982, Dissolved barium in some Louisiana offshore waters: problems in establishing baseline values. Contr. Mar. Sci. 25: 149-159.

Fisher, L.B., 1986, Microwave dissolution of geologic material: application to isotope dilution analysis. Anal. Chem. 58: 261-263.

Folk, R.L., 1974, "Petrology of Sedimentary Rocks". Hemphill Publ., Austin, TX 182 pp.

Forstner, U. and Wittmann, G.T.W. ,1983, "Metal Pollution of the Aquatic Environment. Springer," Verlag, New York. 486 pp.

Froelich, P.N., 1980, Analysis of organic carbon in marine sediments. Limnol. Oceanog. 25: 564-572.

Gaston, G.R., 1985, Effects of hypoxia on macrobenthos of the inner shelf off Cameron, Louisiana. Estuar. Cstl. Shelf Sci. 20: 603-613.

Gaston, G.R., P.A. Rutledge, and Walther, M.L., 1985, The effects of hypoxia and brine on recolonization by macrobenthos off Cameron, Louisiana (USA). Contrib. Mar. Sci. 28: 79-93.

Grassle, J.F. and Smith, W.L., 1976, A similarity measure sensitive to the contribution of rare species and its use in investigation of variation in marine benthic communities. Oecolog. 25: 13-22.

Hudgins, C.M. Jr., 1989, "Chemical Treatments and Usage in Offshore Oil and Gas Production Systems." Report to the American Petroleum Institute, Washington, DC. 52 pp.

Hurlbert, S.H., 1971, The nonconcept of species diversity: a critique and alternative parameters. Ecol. 52: 577-586.

Kraemer, T.F. and Reid, D.F., 1984, The occurrence and behavior of radium in saline formation water of the U.S. Gulf coast region. Isotope Geosci. 2: 153-174.

Lamothe, P.J., T.L. Fries, and Consul, J.J., 1986, Evaluation of a microwave oven system for the dissolution of geologic samples. Anal. Chem. 58: 1881-1886.

Lysyj, I., 1982, "Chemical Composition of Produced Water at some Offshore Oil Platforms." Report to U.S. EPA, Municipal Environmental Research Laboratory, Cincinnati, OH. EPA-600/2-82-034.

McAuliffe, C.D., 1966, Solubility in water of paraffin, cycloparaffin, olefin, acetylene, cycloolefin and aromatic hydrocarbons. J. Phys. Chem. Wash. 70: 1267-1275.

McAuliffe, C.D., 1974, "Determination of C_1 Through C_{10} Hydrocarbons in Water". NBS Spec. Publ. 409. Marine Pollution Monitoring Symposium.
Means, J.L. and Hubbard, N., 1987, Short-chain aliphatic acid anions in deep subsurface brines: a review of their origin, occurrence, properties, and importance and new data on their distribution and geochemical implications in the Palo Duro Basin, Texas. Org. Geochem. 11: 177-191.
Menzel, D.W. and Vaccaro, R.F., 1964, The measurement of dissolved organic and particulate carbon in seawater. Limnol. Oceanog. 9: 138-142.
Middleditch, B.S., 1984, Ecological effects of produced water effluents from offshore oil and gas production platforms. Ocean Manage. 9: 191-316.
Mills, G.L. and Quinn, S.G., 1979, Determination of organic carbon in marine sediments by persulfate oxidation. Chem. Geol. 25: 155-162.
Neff, J.M., 1979, "Polycyclic Aromatic Hydrocarbons in the Aquatic Environment. Sources, Fates, and Biological Effects". Applied Science Publishers, Barking Essex, England. 262 pp.
Neff, J.M., 1987, Biological effects of drilling fluids, drill cuttings and produced waters. Pages 469-538. *in* D.F. Boesch and N.N. Rabalais (eds.), "Long-Term Environmental Effects of Offshore Oil and Gas Development". Elsevier Applied Science, London.
Neff, J.M., 1990, "Forms and Reactivity of Contaminants in Marine Sediments." Report to U.S Dept. of Commerce, National Oceanic and Atmospheric Administration, National Ocean Pollution Program Office, Rockville, MD. 141 pp.
Neff, J.M., N.N. Rabalais, and Boesch, D.F. 1987, Offshore oil and gas development activities potentially causing long-term environmental effects. Pages 149-174 In: D.F. Boesch and N.N. Rabalais (eds.), "Long-Term Environmental Effects of Offshore Oil and Gas Development". Elsevier Applied Science Publishers, London.
Rabalais, N.N., M.J. Dagg, and Boesch, D.F. 1985, "Nationwide Review of Oxygen Depletion and Eutrophication in Estuarine and Coastal Waters: Gulf of Mexico (Alabama, Louisiana, Texas)." Final report to U.S.D.C., National Oceanic and Atmospheric Administration, National Ocean Survey, Ocean Assessments Division, Rockville, MD 60 pp.
Salomons, W. and Forstner, U., 1984, "Metals in the Hydrocycle". Springer Verlag, New York. 349 pp.
Sanders, H.L., 1968, Marine benthic diversity: a comprehensive study. Amer. Nat. 102: 243-282.
Sauer, T.C., Jr., 1981, Volatile liquid hydrocarbon characterization of underwater hydrocarbon vents and formation waters from offshore production operations. Environ. Sci. Technol. 15: 917-923.
Seifert, W.K. and Muldowan, J.M., 1978, Application of steranes, triterpanes and monoaromatics to the maturation, migration and source of crude oils. Geochim. Cosmochim. Acta 42: 77-95.
Shaw, J.K. and Vittor, B., 1985, Benthic macrofauna from Breton-Chandeleur Sound area. In: "Tuscaloosa Trend Regional Data Search and Synthesis Study." Final Report. Volume II. Supplemental Reports, Appendix A. 29 pp. Report to the U.S. Dept. of the Interior, Minerals Management Service, Gulf of Mexico OCS Regional Office, New Orleans, LA.
Somerville, H.J., D. Bennett, J.N. Davenport, M.S. Holt, A. Lynes, A. Mahieu, B. McCourt, J.G. Parker, R.R. Stephenson, R.J. Watkinson, and Wilkinson, T.G., 1987, Environmental effect of produced water from North Sea oil operations. Mar. Pollut. Bull. 18: 549-558.
Venkatesan, M.I., 1988, Occurrence and possible sources of perylene in marine sediments - a review. Mar. Chem. 25: 1-27.

SPATIAL SCALE OF ECOLOGICAL EFFECTS ASSOCIATED WITH AN OPEN COAST DISCHARGE OF PRODUCED WATER

C.W. Osenberg[1,3], R.J. Schmitt[1], S.J. Holbrook[1,2], and D. Canestro[1]

[1] Coastal Research Center, Marine Science Institute

[2] Department of Biological Sciences
University of California
Santa Barbara, California
USA

INTRODUCTION

The biological effects of produced water are poorly understood (Neff, 1987). Not only are laboratory studies of toxicity impeded because of the complex and variable chemical composition of produced water (Middleditch, 1984; Higashi et al., 1992), but field assessments are complicated by other types of production activities that confound effects of produced water (Carney, 1987; Spies, 1987). For example, most field studies of biological effects of "produced water" have compared biological samples collected at various distances from produced water outfalls associated with production platforms. Spatial patterns that are detected in such studies are difficult to interpret, because biological responses can result not only from effects of produced water, but also from discharges of other substances from the platform, or from physical effects of the platform itself. Field studies that examine effects of only produced water discharge are relatively uncommon.

Studies of Trinity Bay (Armstrong et al., 1977), a shallow embayment within Galveston Bay, have provided some of the strongest evidence for effects of produced water on benthic fauna. Depressed infaunal densities were observed out to a distance of approximately one kilometer from the separator platform discharge. Despite other production activities in the area, this study convincingly implicated the discharge of produced water as the causative agent (Armstrong et al., 1977). However, the authors urged caution in extrapolating their results to other situations, because the study was conducted in very shallow (2 - 3 m), turbid water in a sheltered environment. In more exposed and/or deeper areas, dilution rates should be greater and biological effects could be much more localized (Armstrong et al., 1977, Middleditch, 1984). Because other studies of the effects of produced water have also been conducted in poorly mixed coastal or estuarine waters (e.g., Boesch and Rabalais, 1987) or have been confounded with other sources of impact (e.g., Bedinger et al., 1981; Spies, 1987), there exists little information regarding environmental effects of discharge of produced water to high energy coastal habitats.

---

[3] Present Address: Department of Integrative Biology, University of California, Berkley, California 94720 USA

Produced Water, Edited by J.P. Ray
and F.R. Engelhart, Plenum Press, New York, 1992

Although impacts from produced water are likely to occur in the benthos because of accumulation of contaminants in the sediments (e.g., Armstrong et al., 1977), such impacts may be difficult to detect due to natural spatial and temporal variability in population density of infaunal organisms. Examination of individual-based parameters (such as growth) might provide more sensitive indicators of environmental impacts than assessment of population density (Carney, 1987; Osenberg et al., 1993). This approach might reveal a different spatial extent of biological effects than documented by variation in population densities. However, biological effects other than those on densities, such as changes in growth or other measures of individual performance, rarely have been addressed in field studies of impacts of produced water. Even in the absence of natural variability, examination of individual-based parameters might lead to the detection of impacts that would not be identified by focusing only on population density. For example, changes in the production of a population are related not only to the density of the population, but also the amount of growth or reproduction attributed to each individual. Although density might be unaffected by discharge of produced water, production (and therefore the population's ability to persist in face of other sources of stress) might be affected adversely. This can be assessed by direct examination of, for example, individual growth or reproductive output, in addition to density.

Here we examine the spatial scale of biological impacts associated with discharge of produced water to a high energy coastal environmental in southern California. Discharge comes from an onshore facility and occurs through a single pipeline a few hundred meters offshore. Therefore, we were able to study effects of produced water discharge in the absence of localized effects from other production activities and structures (e.g., the nearest platform is at least several km from the outfall). The three principal goals in this study were to: (1) determine if produced water discharge had discernible biological effects in this high energy environment; (2) document the nature of observed effects; and (3) compare the spatial extent of impacts on infaunal densities with effects on growth and body condition of "indicator" species (outplanted mussels).

METHODS

Study Site

The study site is located near Carpinteria, CA, USA (34°23'N, 119°30'W), where the subtidal environment consists of a sand bottom with little or no physical relief. This area is a high energy, open coast environment, although it is somewhat sheltered from the swell due to the presence of the northern Channel Islands and the east-west orientation of the coast. The surf typically ranges from 1 - 2 m, but varies seasonally. The surf is greatest during the winter due to large amplitude, long period swells generated from Pacific storms; however the winter in which infauna were sampled (see below) was a relatively mild winter with few major storms. Current flow primarily is east-west (parallel to shore), with an offshore flow component probably due to tidal flux. Based on depth profiles from 30 sampling dates, currents flowed to the west, east, south (to offshore) and north, 42%, 27%, 25% and 6% of the time, respectively.

Discharge of produced water occurs approximately 200 - 300 m from shore along the last 25 m section of the pipe at bottom depths of 10 - 12 m. The effluent is discharged through 10 T-ports that are oriented perpendicular to the pipe about 0.75 m above the bottom, resulting in a calculated minimum initial dilution of 125:1. The volume discharged is extremely consistent from month to month, averaging 2.64 million liters/day (16,600 bbls/d) from February 1989 - October 1990. Produced water has been discharged from this diffuser array since 1978.

The effluent is comprised primarily of produced water, with minor amounts of wastewater added from natural gas processing vessels and the cleaning of equipment. An unspecified amount of storm water is discharged with the produced water. However, there was no storm run-off during the period of this study. Wastewaters are treated by gravity settling and induced gas flotation units. Chemical characteristics of the effluent are summarized in Higashi et al. (1992). The effluent is approximately 6° C warmer than the receiving waters and has a lower salinity (~20 ppt). However, there is no discernible thermal gradient beyond 0.5 m of the discharge ports.

Infaunal Depth Distribution

In a preliminary study, the depth distribution of infauna was examined to determine the appropriate depth for our infaunal sampling survey. On 31 October 1989, divers sampled infauna at four sites that were within 1 km of the Carpinteria outfall. At each site, 10 cores (20.3 cm^2/core) were taken to a depth of 25 cm. Each core was sectioned into 5 cm depth intervals, and depth fractions from five cores (all from the same site) were pooled into a single sample, yielding two samples/site, each with five depth fractions. Buffered formalin was added to each sample to bring the total solution to approximately 10% formalin, preserved samples then were washed through a series of 2 mm, 1 mm and 0.5 mm sieves, and each fraction subsequently was stored in alcohol. Organisms were picked from the sediments in each fraction, counted and identified to broad taxonomic categories.

Benthic Survey

An intensive spatial survey of infauna was conducted along the 11 m isobath at the Carpinteria study site. A total of 20 sites was sampled along a gradient upcoast (west) and downcoast (east) of the diffusers. Scuba divers collected samples on 10 February 1991 from 2, 3, 5, 10, 25, 50, 100, 250, 500, and 1000 m west of the diffusers and on 11 February 1991 from 5, 7, 10, 15, 30, 55, 100, 250, 500, and 1000 m east of the diffusers. Infaunal densities were sampled by taking 8 coffee can cores (78 cm^2/core to a depth of 10 cm) at each site, and pooling four cores into a single sample. As before, buffered formalin was added to the samples to bring the total solution to approximately 10% formalin, washed through 2 mm, 1 mm and 0.5 mm sieves and stored in alcohol. Organisms were picked from the sediments, counted and identified to broad taxonomic categories. For each site and each taxonomic group a mean based on the two replicates was calculated, and this mean was used in all subsequent analyses.

At the same time that the biological cores were collected, 3 sediment cores also were collected at each site (20.3 cm^2/core, 5 cm deep) and frozen. One core was used to determine percent organic matter in the sediments, one core was used for grain size analysis, and the remaining core was archived. Percent organic matter (POM) was determined by thawing sediments, aspirating the overlying water, and drying the sediments for 48 hours at 45° C. Three subsamples (ca. 5 - 6 g) of sediments from each sample were then dried for 3 hours at 60° C, weighed, combusted for 4 hours at 450° C, and reweighed. POM was calculated as 100% X(ash free dry mass / total mass of subsample), and the mean of the three subsamples was determined for each site.

The percent of sediments comprised by silt and clay was determined by drying each sample at 60° C, breaking up the dried sample with a rubber stopper, and shaking the sample for 10 minutes through a series of standard ASTM soil sieves. The mass of sediments on each sieve was determined, and the percent silt-clay was estimated as the percent of the sample that passed through the 0.063 mm sieve. Surficial sediments also were collected for analyses of metals and organics, and results of their analyses will be reported elsewhere by collaborators; we refer briefly to their findings in the discussion.

Data were analyzed using three related techniques. For all analyses, infaunal densities were log(x+1) transformed and distance was log transformed to better satisfy assumptions of the analyses (e.g., linearity and homoscedasticity of residuals). In the first approach, we examined the correlation (Pearson's r) between infaunal densities and each of three environmental parameters - distance from outfall, POM, and percent silt-clay. Second, we used multiple regression analysis to examine the effects of each environmental variable (e.g., distance) after accounting for variation due to the other two variables (e.g., POM and percent silt-clay). Third, we simplified the infaunal data using Principal Components Analysis, PCA (Dillon and Goldstein, 1984). PCA provided a way to simultaneously examine the patterns for all infaunal taxa by constructing a set of new (and fewer) dependent variables (i.e., PC1, PC2, etc.) from the original variables (i.e., densities of each infaunal group) (see Jolliffe, 1986). Principal components were determined using the SAS Proc Factor procedure, and axes were rotated using the quartimax technique (SAS, 1988). Interpretation of the principal components was achieved by examining the loadings (i.e., correlation) of each original variable (i.e., density of a particular group) on the principal components. Following preliminary analyses, we retained and rotated the first two principal components, as remaining axes yielded little aggregation of the infaunal data.

PCA defined groups of organisms that exhibited similar patterns of spatial variation in abundance: i.e., each of the new variables (e.g., PC1) provided a measure of the aggregated response of several infaunal taxa to environmental variation. We then explored how the principal component scores varied with distance from the outfall and with variation in sediment characteristics. In other words, PCA enabled us to compare the relationship between environmental factors and a much reduced number of biological variables (i.e., the principal component scores), rather than densities of 20 different taxa. In this way, we could more easily infer the spatial extent of impacts due to produced water discharge.

Outplanted Mussels

Mussels (*Mytilus californianus* and *M. edulis*) were transplanted to six of the study sites to determine if proximity to the outfall influenced performance attributes (e.g., growth). At each of the six study sites (1, 5, 10, 50, 100, and 1000 m West of the outfall), we installed buoy arrays, each consisting of a cement anchor, line, and a subsurface buoy. *M. californianus* were collected from the intertidal zone at Montana de Oro State Park, CA, and *M. edulis* were collected from subtidal buoy lines at a location 1.5 km offshore Gaviota, California.

Before being transplanted to the field, individual mussels were measured and the shell margin notched so that initial sizes could be estimated after subsequent growth. Forty mussels from a uniform size distribution (range 20 - 60 mm shell length) of one species were put into a bag of 1.25 mm oyster netting with a vexar mesh skeleton for support. One bag with *Mytilus californianus* and one with *M. edulis* were attached to each buoy line approximately 4.5 m above the sediments on 7 June 1990. A subsample of "control" mussels was frozen without being transplanted to the field to provide an estimate of the initial condition of the transplanted mussels. Transplanted mussels were retrieved on 4 October 1990 and frozen.

Survivorship of outplanted mussels was estimated as the number of marked individuals recovered in each bag divided by the number initially marked. Shell length and initial shell length (as estimated by location of the notch) were measured for every marked mussel of both species recovered alive from each site (distance). For each individual, tissue was removed, separated into gonadal and somatic tissue, dried at 60° C for 24 hours, and weighed. Like most organisms, mussel growth and tissue mass vary with mussel size. To simplify analyses (data were taken over a range of mussel sizes), we obtained size-independent measures of mussel performance using analyses of covariance (ANCOVA) for each parameter for each mussel species, with log(shell length) as the covariate. From these analyses, we obtained adjusted means for each parameter at each of the six sites, and these adjusted means are the relative measures of mussel performance we report. All ANCOVAs were performed on log-transformed data, and adjusted means were either left transformed or back-transformed to original units depending on the specific analysis. Analyses of shell growth (i.e., change in length) used log (initial shell length) as the covariate, while analyses of condition (i.e., gonadal or somatic mass) used log (final shell length) as the covariate.

Principal components analysis was used to summarize the mussel data, which consisted of six different estimates of mussel performance measured at each site (i.e., the log-transformed adjusted means for shell growth, gonadal mass, and somatic mass for each of the two mussel species). We extracted only one principal component because all mussel parameters strongly covaried and therefore the first principal component accounted for most of the variation in the mussel data set.

To estimate tissue production for each of the two mussel species we 1) dissected "control" mussels that were not outplanted, 2) estimated the allometric relationship between tissue mass and shell length for these "control" mussels (for each species the relationship explained over 94% of the variation in tissue mass), and 3) for each outplanted mussel, subtracted its initial tissue mass (estimated by using its initial shell length in the allometric regression derived from control mussels) from its final tissue mass. We used ANCOVA to obtain the adjusted means for tissue production for each site using log(initial shell length) as the covariate, and then examined the relationship between the adjusted means and distance from the outfall. Thus, we determined if mussels that began the transplant period at the same size varied in their production of new tissue as a function of distance from the produced water outfall.

RESULTS

Infaunal Depth Distribution

An average of 73% of all infauna collected from the top 25 cm of sediments occurred in the first 5 cm, with approximately 5 - 10% occurring in each of the deeper 5 cm depth fractions. A total of 82% of the infauna (range among the 4 sites: 80 - 86%) was collected in the top 10 cm. All taxonomic groups shared this general pattern, and similar results were obtained at nearby sites studied by Spies and Davis (1979). Based upon these findings, samples collected during our subsequent infaunal survey were taken to a depth of 10 cm.

Benthic Survey

Analyses revealed a number of strong associations between infaunal densities and the three environmental parameters: distance from outfall, POM and percent silt-clay (Table 1). The environmental parameters were not correlated with one another ($|r| < 0.22$; $P > 0.35$ for each comparison), leading to the similar results from correlation and multiple regression analyses. In general, densities were more strongly associated with distance from the outfall than with either of the sediment characteristics we measured. Among the 20 sites, POM ranged from 1.05 - 1.86 % ($\overline{X} = 1.29$), percent silt-clay ranged from 4.4 - 17.8 % ($\overline{X} = 10.2$), and there was no pattern in variation of either parameter with distance from the diffuser. In particular, sediments near the diffuser were not any finer or coarser (% silt-clay at 7 sites ≤ 10 m from outfall: $\overline{X} \pm$ 1 SE= 9.7 ± 0.8 %) than elsewhere along the gradient, and there was no evidence of elevated POM nearer to the discharge (% POM at 7 sites ≤ 10 m: $\overline{X} \pm$ 1 SE = 1.21 ± 0.06 %). The results of correlation analysis, multiple regression and PCA were similar. We categorized each taxon into one of four groups based on the way density covaried with the environmental variables (Table 1).

One group, consisting of Nematodes only, achieved greatest densities at sites near the outfall, and thus appeared to benefit from the discharge of produced water (Table 1). The second group, consisting of echinoderms, larval crustaceans, nemertean worms and several polychaete families, exhibited the opposite pattern and achieved greatest densities at sites farthest from the outfall. The third group included bivalves, two classes of crustacea (ostracods and cumaceans), and two polychaete families (Capitellidae and Spionidae) and exhibited a mixed pattern in which densities were generally positively associated with POM, and possibly more abundant near the outfall. The fourth group exhibited no discernible pattern with distance from the outfall or POM and included Glycerid polychaetes, gastropods, and all other crustaceans (e.g., gammarid amphipods, copepods).

The PCA produced two principal components that were associated with environmental features (Table 1). PC1 explained 25% of the variation in infaunal densities, and scores on PC1 were highly correlated with distance from the outfall ($r = 0.81$; $P < 0.0001$; $n = 20$). Importantly, the pattern of variation associated with PC1 was symmetrical to the east and west of the diffuser (Fig.1). However, the effect associated with distance was extremely non-linear, suggesting that impacts on the infauna were not present beyond a distance of approximately 50 - 100 m from

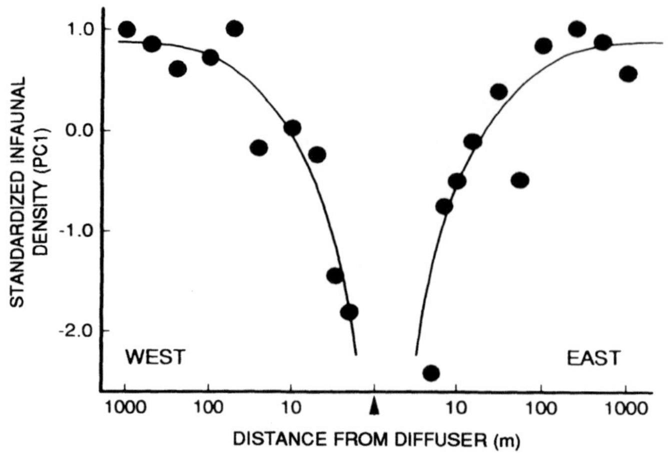

Figure 1. Standardized infaunal density as a function of distance from the produced water outfall. Standardized infaunal density (defined as PC1: Table 1) is an aggregate measure based on the densities of all infauna and is standardized to a mean of 0. The arrow indicates the location of the diffuser.

the outfall. Loadings on PC1 (Table 1) reveal that its variation was primarily associated with variation in the densities of organisms in Groups 1 and 2 (e.g., nematodes, nemerteans and polychaetes (Fig.2)

PC2 explained an additional 19% of the variation in infaunal densities and was primarily associated with organisms in Group 3 (e.g., ostracods, cumaceans and bivalves; Table 1). Scores on PC2 were significantly correlated with POM ($r = 0.48$; $P = 0.03$; $n = 20$). There was a weaker and non-significant relationship between PC2 and distance ($r = -0.36$; $P = 0.12$; $n = 20$). The pattern of variation in PC2 (Fig. 3) was not similar to the East and West of the diffusers. Therefore, it appears that variation in PC2 was driven by variation in POM and was not related to produced water discharge.

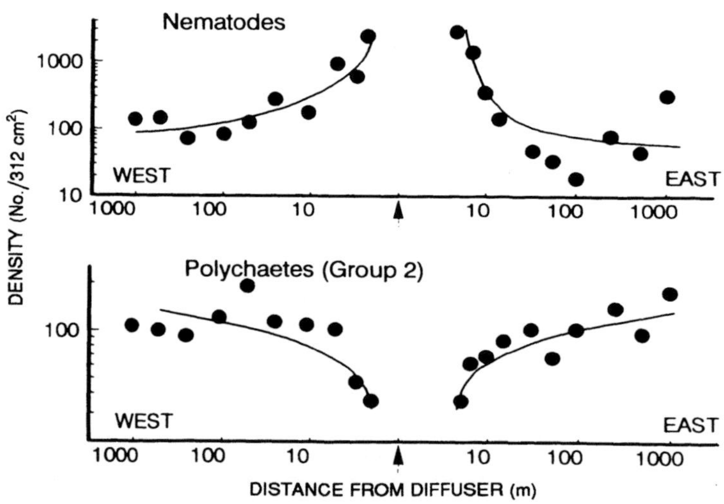

Figure 2. Densities of nematodes and Group 2 polychaetes (see Table 1), as a function of distance from the produced water outfall. Each data point gives the mean of two samples collected at each location. The arrow indicates the location of the diffuser.

Table 1. Summaries of three separate analyses of infaunal densities with environmental variables [log distance from the produced water outfall (DIST), percent organic matter in the sediments (POM), and percent of sediments in the silt-clay fraction (SC)]. Infaunal taxa were placed into one of four groups based upon the way their densities covaried with the environmental variables. Statistically significant ($P < 0.05$, n = 20) results from correlation and regression analyses are given and the sign indicates the direction of the effect. The signs of loadings from principal components analysis are indicated when they exceeded ± 0.50. PC1 and PC2 explained a total of 44% (25% and 19%, respectively) of the variation in infaunal densities. Because site score on PC1 was highly correlated with distance from the outfall (r=0.81, P<0.0001; Figure 1), a positive loading indicates greater densities were achieved far from the outfall. PC2 was positively correlated with POM (r = 0.48, P = 0.03) suggesting that taxa with positive loadings were more abundant at sites with greater POM.

		Multiple	Loadings on	
	Correlation	Regression	PC1	PC2
Group 1: (Greater density near outfall)				
Nematodes	-DIST	-DIST	—	
Group 2: (Greater density far from outfall)				
Polychaetes:				
Cirratulids			+	
Other Sedentariae	+DIST	+DIST	+	
Nephtyds			+	
Nereids	+DIST	+DIST	+	
Syllids	+DIST	+DIST	+	
Other Errantiae				
Nemerteans	+DIST	+DIST	+	
Larval crustaceans	+DIST	+DIST		
Echinoderms	+DIST	+DIST		
Group 3: (Greater density at sites with high POM and/or near outfall)				
Polychaetes:				
Spionids				+
Capitellids				+
Cumaceans	-DIST, +POM	-DIST, +POM		+
Ostracods	-DIST	-DIST, +POM		+
Bivalves				+
Group 4: (No apparent relationship to outfall or POM)				
Polychaetes:				
Glycerids	-SC	-SC		
Copepods		-SC		
Gammarids				
Other Crustaceans				
Gastropods				+

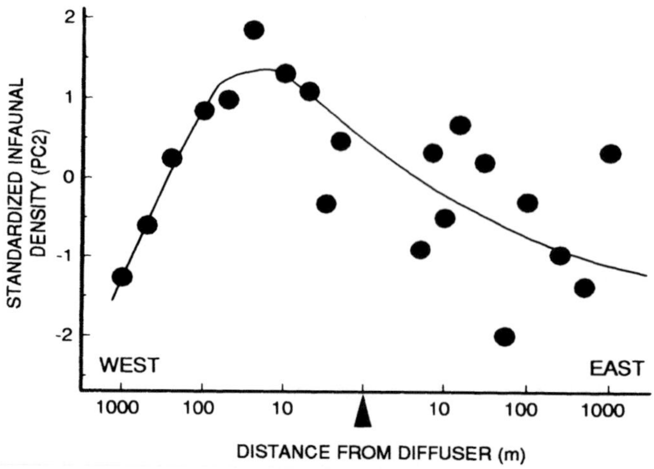

Figure 3. Standardized infaunal density, defined by PC2 (Table 1), as a function of distance from the produced water outfall. The arrow indicates the location of the diffuser. Compare with Figure 1.

Outplanted Mussels

Survivorship of mussels was relatively high (means ± 1 SE among sites: 0.71 ± 0.05 for *M. californianus*; 0.63 ± 0.5 for *M. edulis*), and showed no significant correlation with distance from the outfall ($|r| < 0.4$; $P > 0.4$ for each species). However, mussels did exhibit considerable variation in the intensity of sublethal effects, as measured by shell growth and condition (i.e., size-specific gonadal and somatic tissue masses). Both mussel species exhibited similar patterns of variation among sites in these sublethal components of performance; sites at which *M. californianus* performed well were the same sites at which *M. edulis* performed well (Figure 4).

Performance was highly correlated with distance from the outfall for all three measures of performance for *M. californianus*, and for two of three measures for *M. edulis* (Table 2). Due

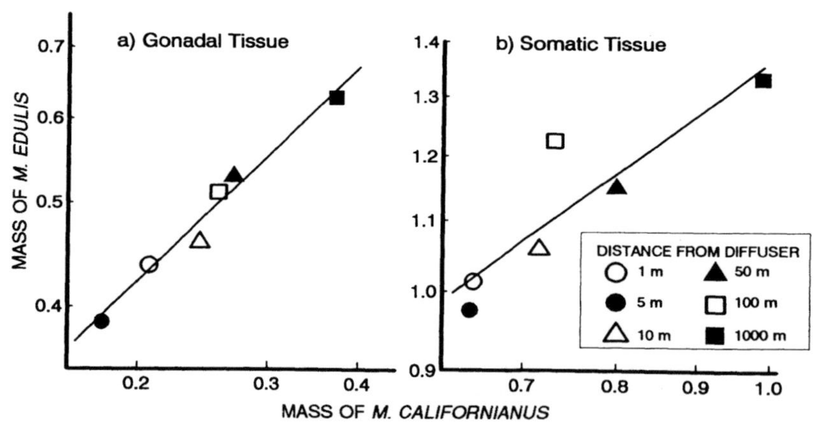

Figure 4. Pattern of covariation in condition (size-specific gonadal mass and somatic mass) of two mussel species (*M. californianus* and *M. edulis*). Each datum gives the response of the two species outplanted to the same site. In general sites close to the outfall yielded low values for each species, whereas sites far from the outfall yielded higher values. Condition of both species was highly correlated ($r = 0.99$, $P < 0.0001$ for gonadal tissue; $r = 0.90$, $P = 0.015$ for somatic tissue).

Table 2. Statistical summaries of analyses of mussel performance. Given are the correlations between adjusted means from analyses of covariance (based on log transformed data) and log (distance), as well as the correlations between the adjusted means and the first principal component (i.e., loadings). Correlations >0.81 are significant ($P<0.05$). PC1 explained 79% of the total variation in mussel performance and was highly correlated with distance from the outfall ($r = 0.94$, $P = 0.005$, $n = 6$).

Species	Parameter	Correlation with Distance	Loading on PC1
M. californianus			
	shell growth	+0.88	+0.92
	somatic mass	+0.92	+0.94
	gonadal mass	+0.90	+0.98
M. edulis			
	shell growth	+0.33	+0.36
	somatic mass	+0.95	+0.98
	gonadal mass	+0.89	+0.99

to the similarity in response of the three measures of performance for both species, the first principal component (PC1) provided a simple description of the aggregate response of mussels to environmental variation. PC1 explained 79% of the total variation in mussel performance, and was highly correlated with distance from the outfall (Fig. 5; Table 2), suggesting a major role of produced water discharge (relative to other factors) in determining variability in mussel performance among sites.

These results demonstrate that mussels near the produced water outfall tended to grow more slowly, and at any given size were in poorer condition (i.e., had less gonadal and somatic tissue) than mussels far from the outfall. These responses provide three separate measures of how produced water affected mussel production. To examine the aggregate effect, we estimated the per capita tissue production at each site. Tissue production was correlated with distance for

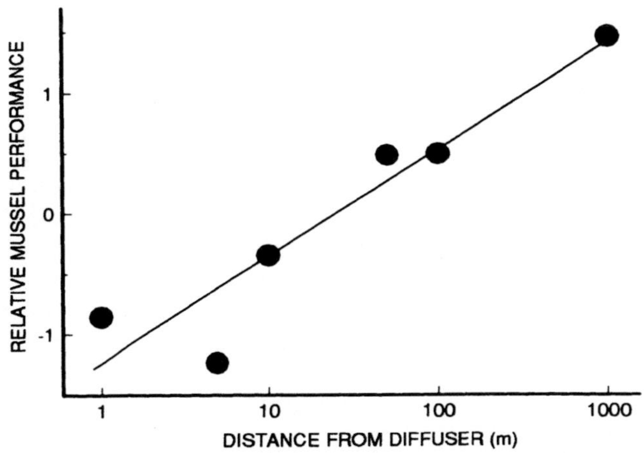

Figure 5. Relative mussel performance as a function of distance from the produced water outfall. Relative mussel performance is defined by the PC1 from analysis of mussel growth and condition (Table 2). Performance and log (distance) were highly correlated ($r = 0.94$, $P = 0.005$).

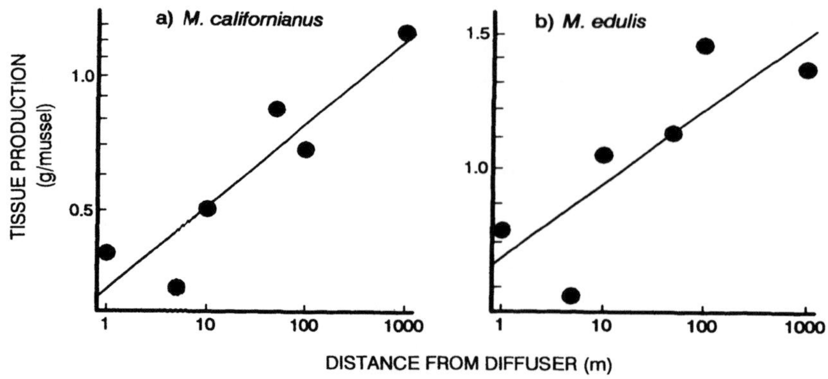

Figure 6. Tissue production for two species of mussel (*M. californianus* and *M. edulis*) as a function of distance from the diffuser. Tissue production and distance were positively correlated for each species ($r = 0.92, P = 0.009$ for *M. californianus*; $r = 0.82, P = 0.044$ for *M. edulis*).

both species (Fig. 6: based on log-transformed data, $r = 0.92$; $P = 0.009$ for *M. californianus* and $r = 0.82$; $P = 0.044$ for *M. edulis*). Sites farthest from the diffusers had production that was two to three times greater than sites near the outfall.

DISCUSSION

Interpretation of Spatial Patterns

Field studies of biological effects arising from point discharges such as produced water are hampered by a number of logistical and design problems. Principal among them are detecting effects amid natural variation and assigning causation for observed patterns (e.g., Carney, 1987). Ideally, this requires knowledge of environmental conditions at "impact" and "control" sites before and after the discharge occurs to determine whether (and by how much) attributes near the outfall change from those expected (e.g., Stewart-Oaten et al., 1986). While such an approach can provide powerful inference (e.g., Stewart-Oaten et al., 1986: Osenberg et al., 1992) , we know of no field assessment of environmental effects of produced water that has been able to collect pre-discharge information (but see ongoing study by Osenberg et al., 1993). Usually by necessity, the alternative approach is to seek spatial correlations between biological variables and environmental characteristics after a long period of produced water discharge. In these situations, natural spatial variability cannot be statistically isolated from the putative impact (Stewart-Oaten et al., 1986).

In addition to design constraints, many previous assessments of produced water discharge have been conducted where other oil and gas production activities occurred in the same immediate area, and/or where large, natural sources of disturbance existed (e.g., Bender et al., 1979; Bedinger et al., 1981; but see Armstrong et al., 1977). In these circumstances, it can be virtually impossible to isolate the contribution of produced water in shaping all but the most extreme environmental pattern (Carney, 1987). By contrast, the produced water outfall we examined is several kilometers from the nearest offshore platform, is the only discharge of any kind in the local vicinity, and, with the exception of small boat traffic, it is the only major industrial activity that occurs in the local area. Further, the closest potential source of a large natural disturbance is Carpinteria Salt Marsh, a small (93 ha) coastal wetland located ~ 3 km west of the outfall (Ferren, 1985). Thus, there are no obvious enterprises that covary in space with the discharge of produced water at Carpinteria that could account for observed patterns. It remains possible that the observed patterns resulted from natural spatial differences in environmental attributes that were unrelated to, but were confounded in space with the discharge of produced water. Without pre-discharge information, it is difficult to reject this possibility. However, several lines of evidence suggest that the observed patterns resulted from the discharge of produced water.

The first supportive evidence for the causal involvement of produced water on mussel performance comes from the correspondence between the observed field effects and responses induced by exposure of mussels in controlled laboratory experiments. For example, based on *in vivo* imaging using nuclear magnetic resonance, Fan et al. (1992) showed that prolonged exposure to produced water from Carpinteria led to a severe reduction in the energetic status of mussels, *M. californianus*. This effect on the ability of mussels to maintain energy stores is a likely cause of our observed reductions in growth, condition, and tissue production.

The second line of evidence is the marked temporal consistency in effects that we have observed in four different outplants conducted after the one reported here (Osenberg et al., *unpublished data*). Although there was seasonal variation in the overall amount of mussel growth at each site, individual performance was consistently lowest at sites near the outfall and always greatest at the 1000 m site. Thus, the results of our first outplant (Figures 4 - 6) were not spurious and clearly represent consistent spatial variation in performance that is correlated with distance from the outfall.

Although suggestive, these results alone cannot isolate effects of the produced water from an unspecified factor that is correlated in space with the outfall. At a minimum, this would require additional evidence demonstrating that mussels actually encounter produced water at the sites with reduced performance. Because bivalves are known to incorporate a variety of trace elements into their shells (e.g., Carriker et al. 1982), and because barium is a marker of the Carpinteria produced water (Higashi et al. 1992), we recently determined the amount of barium in the shells of mussels that were outplanted to our six study sites during the same period as the mussels analyzed for performance (Osenberg, Fan and Collins, *unpublished data*). Barium concentration in shells of both *M. edulis* and *M. californianus* was significantly negatively correlated with distance ($r < -0.91$, $P \leq 0.01$, $n=6$ for each species based on log transformed data). Consequently, barium concentration was negatively correlated with mussel performance ($r < -0.92$, $P < 0.01$, $n = 6$); performance was greatest, and barium content lowest, at the 1000 m site. Thus, it appears that mussel performance varied inversely with relative exposure of mussels to the produced water plume.

The most conclusive evidence for evaluating if produced water exposure caused the observed pattern of mussel performance could be obtained by stopping the discharge of produced water and seeing if the spatial patterns of mussel performance and barium contamination disappear. If so, this would demonstrate the absence of an underlying cause for the patterns that was independent of produced water discharge. However, the mussel outplants require several months in the field, which makes it logistically infeasible for us to conduct such an experiment. Fortunately, operators of the Carpinteria plant plan to stop discharging produced water during the next year or two. At that time a conclusive test might be possible.

In contrast to the mussel outplants, that require several months to complete, shorter term field assessments have been possible. For example, a complementary investigation by Raimondi and Schmitt (1992) clearly demonstrated a cause-effect relationship between the discharge of produced water from the Carpinteria facility and the performance of abalone larvae. They documented marked reductions in survival, settlement and metamorphosis of abalone larvae with increased proximity to the diffuser when produced water was being discharged. The spatial pattern of effects on abalone was qualitatively identical to that we observed for mussels. Because their study relied on short-term exposures ($\leq$ 4 days), they were able to show that effects were not present during a brief (10 day) period when produced water was not discharged.

Interpretation of the infaunal patterns is somewhat more difficult. Unlike parameters obtained from mussels, various infaunal groups did not show the same patterns of density as a function of distance from the outfall: some groups were more abundant, some were less, and others showed no pattern with distance. The sedimentary environment is likely to be more heterogeneous than the water column where mussels were outplanted, and infaunal organisms respond to natural variation in such characteristics as grain size or particulate organic matter (e.g., Pearson and Rosenberg, 1978; Reish, 1979; Ferris and Ferris, 1979: Reish and Barnard, 1979. This heterogeneity can add considerable noise to patterns that arise from human disturbance. Despite this, interpretable relationships between density and proximity to the outfall were apparent, and the collective evidence strongly suggests an involvement of produced water discharge.

As revealed by principal components analysis, the major discernible source of variation in infaunal density was distance from the outfall (Fig. 1). The pattern of variation was symmetrical on both sides of the outfall, with the greatest alteration of densities closest to the diffuser. Symmetry of effects to the west and east of the diffuser would be crudely predicted by our water current data. Further, sediment samples collected concurrently with our infaunal cores revealed localized elevations of barium, a chemical marker of the effluent (Higashi et al., 1992), near to and on both sides of the diffuser (A. Flegal and K. Abu-Saba, *pers. com.*). In contrast to our analyses of barium in mussel shells, the elevation of barium in the sediments was much more localized, suggesting a more limited impact on the benthos (see below).

The particular organisms that varied systematically along the distance gradient also suggest that produced water was involved. Nematodes were more abundant closer to rather than further from the diffuser (Fig. 2a), and they are known to respond positively to organic enrichment of sediments from oil and sewage contamination (Chasse 1978; Spies et al., 1980; Sandulli and De Nicola 1991; Spies and DesMarais 1983) argued that enrichment by petroleum hydrocarbons at natural oil seeps led to higher densities of nematodes by stimulating the production of a food resource, the bacterium, *Beggiatoa sp.* We have observed *Beggiatoa* mats near the produced water outfall and coating buoy lines at our 1, 5, and 10 m sites.

In contrast to Spies' work at oil seeps, we found little evidence of trophic enhancement for infauna other than nematodes. Indeed, most carnivorous groups that we examined (e.g., nemerteans, and several polychaete families, including Nephtyds, Nereids and Syllids) varied inversely with the abundance of nematodes and were least abundant near the outfall. Pearson and Rosenberg (1978) suggested that carnivorous species might decline in abundance under moderate levels of trophic enrichment, and the pattern of variation we observed for this trophic group at Carpinteria was consistent with such a hypothesis.

An alternative explanation is that observed spatial variation in densities of the groups above resulted from current scour caused directly or indirectly by the outfall (e.g., Middleditch, 1981). Several lines of evidence argue against this hypothesis. First, sediment grain size and composition (i.e., percent silt-clay) were not coarser nearer to the pipeline than much further away, as would be expected by scouring. Second, mats of bacteria would not be expected to develop in even modestly scoured areas, yet they occurred on the sediment near the outfall. Third, the infaunal community near the diffuser was not that expected in a scoured area, but was characteristic of areas subject to organic enrichment from oil contamination and sewage discharges (e.g., Gray, 1979).

However, POM did not correlate with distance from the outfall, suggesting either that our estimates of POM were too crude to detect enrichment, or that particulate organic input provided only a minor portion of the source of trophic enrichment near the outfall. Our data did reveal that some groups of deposit and filter-feeding organisms (i.e., those in Group 3, Table 1, such as bivalves, cumaceans, ostracods and Spionid polychaetes) were more abundant in areas with greater POM.

Spatial Extent of Effects

Based on the foregoing discussion, we believe that the major effects of produced water on infaunal densities at Carpinteria are summarized by variation in Principal Component 1 (Figure

1). Effects on densities appeared limited to areas very close to the outfall (≤ 100 m). This spatial extent of effects on the benthos was more limited than reported previously for areas with shallower receiving waters (e.g., Armstrong et al. 1977; Boesch and Rabalais 1987; Neff et al. 1989), and perhaps for deeper sites with lower discharge rates (Middleditch 1981; Neff et al. 1989). Thus, our infaunal survey might lead to the conclusion that biological effects, in general, were extremely localized. However, results from our mussel outplants and from the related study by Raimondi and Schmitt (1992) suggested that important biological effects can occur over larger spatial scales at Carpinteria, despite the discharge to a high energy environment.

The production of mussel tissue was affected by produced water, and, by contrast with the limited spatial effect on infaunal density, mussel performance was reduced out to a distance of at least 100 m and perhaps beyond 1 km (Fig. 4). Unfortunately, our study provides no resolution between the 100 and 1000 m sites and we have no sites more than 1000 m from the outfall. In retrospect, we would be able to make much stronger statements about the spatial extent of these effects if we had also placed additional sites past 100 m. Previous suggestions that dilution was likely to reduce the spatial extent of effects in open coast environments (e.g., Middleditch, 1984; Neff et al., 1989) led us to focus on near-field effects. Since completing our first mussel outplant, we have added a site at 500 m and have completed two subsequent outplants. In each case, performance was greatest at the 1000 m site suggesting that effects on performance might extend past 500 m.

An area of effect of similar size also was observed at Carpinteria for performance of abalone larvae (Raimondi and Schmitt, 1992), and serves to illustrate how study of different biological parameters can reveal different spatial scales of environmental impact. The reasons for the different spatial extents of effects between densities of infauna and performance of mussels and abalone larvae are unknown, but they could be related to the two different types of parameters measured and/or the particular habitats (i.e., benthos vs. water column) sampled. Below, we discuss several possible explanations, which we offer as hypotheses for future field tests.

Mussels were suspended in the water column, and effects on their performance, like those on abalone larvae (Raimondi and Schmitt, 1992), may have arisen from dissolved and suspended fractions of the produced water plume. By contrast, it is possible that effects on infaunal organisms largely resulted from components of the effluent that settled to the benthos (e.g., Reynoldson, 1987). Our data suggest that water-borne contaminants may have caused effects over a much greater spatial scale than did the particulate fraction(s). This is consistent with the finding that the water-soluble fraction of the Carpinteria produced water was responsible for most biological effects observed in laboratory tests (Higashi et al., 1992). If true, this challenges conventional thinking regarding the relatively greater likelihood of produced water effects being mediated by toxicants accumulated in the sediments rather than those dispersed in the water column (e.g., Middleditch, 1984; Neff, 1987; Payne et al., 1987). As noted by Howarth (1989) and Raimondi and Schmitt (1992) this possibility suggests that more attention needs to be given to potential effects mediated through the water column, not only for planktonic organisms but also for benthic and demersal species that are potentially exposed to water soluble contaminants.

Differences in "habitats" notwithstanding, part of the explanation for the greater spatial effects on mussels than infauna probably relates to differences in the biological parameters examined. It has been suggested that individual-based parameters (e.g., growth and reproduction) are more sensitive indicators of pollutants than population-level characteristics (e.g., densities), especially with regard to detectability of actual effects (e.g., Carney, 1987; Osenberg et al., 1992, 1993). For example, the extent to which affected populations are depressed locally will depend on the size and nature of effects on local adults (e.g., survivorship, fecundity) and the supply of new individuals from all sources (e.g., Underwood and Peterson, 1987). Most marine organisms have a planktonic dispersive stage, which tends to decouple local production of propagules from subsequent recruitment into the local adult population. Thus, it is possible for a discharge to have no direct effect on local adult stocks (e.g., when recruitment and survival of subsequent age classes are independent of contaminant levels), but considerable effects on the local production of larvae (e.g., reduced growth and reproductive output of adults). Mussels in the vicinity of the Carpinteria outfall, for instance, have about half the gonadal mass of adults further away, yet because of recruitment of larvae produced elsewhere, local population densities may not be depressed by local reduction in reproductive output. Furthermore, the expression (if any) of the reduced local output of mussel propagules would be spread over a much larger spatial scale, defined by the dispersal distance of the larvae, which would make effects on the "global" population virtually impossible to detect (see Raimondi and Schmitt, 1992; Nisbet et al., 1993).

Because of propagule dispersal, assessment of local population densities can lead to considerable underestimation of both near- and far-field biological impacts of a point source discharge. Conversely, the effect of effluent discharge might be to alter settlement behavior of propagules, resulting in depressed local populations, even in the absence of effects on local adult survivorship or per capita production. In this circumstance, it is possible that the local reduction in density overestimates the ecological impact if larvae that chose not to settle at the discharge site successfully recruited elsewhere. A similar argument can be made for measures of individual-based parameters. For these reasons, we contend that a primary focus of assessment studies should be on the dynamics and persistence of populations. In marine systems especially, this requires information not only on local density, but also production (e.g., assessed by density and changes in demographic rates or individual-based parameters), as well as the connection between local processes and larger scale dynamics. Given the limited information on metapopulation dynamics in marine systems (Roughgarden and Iwasa, 1986), it is unrealistic to expect such thoroughness in assessment studies. However, it is important that field studies begin addressing aspects of production in addition to static analysis of local density. Doing so will provide a more complete picture of the spatial extent of ecological impacts that arise from such local perturbations as the discharge of produced water.

ACKNOWLEDGEMENTS

We thank K. AbuSaba, G. Cherr, T. Fan, A. Flegal, R. Higashi, P. Krause, A. Martinez, F. Piltz, D. Steichen and S. Swarbrick for useful discussions and criticisms, J. Wallace for assisting with coordination of field logistics with the Carpinteria processing facility, staff of the Regional Water Quality Control Board for discharge data, and F. Piltz of MMS for support of the UC/MMS Southern California Educational Initiative. S. Anderson, M. Carr, M. Edwards, T. Herrlinger, B. Hoffman, T. Kaltenberg, Shi-Yong Lin, A. Martinez, M. Perez, K. Phillips, P. Raimondi, D. Reed, K. Sydel, and V. Vredenburg assisted in the field and/or the laboratory, and B. Williamson provided technical assistance. This research was funded by the UC Coastal Toxicology Program, and by the Minerals Management Service, U.S. Department of Interior under MMS Agreement No. 14-35-0001-3071. The views and conclusions in this paper are those of the authors and should not be interpreted as necessarily representing the official policies, either express or implied, of the U.S. Government.

REFERENCES

Armstrong, H.W., Fucik, K., Anderson, J.W., and Neff, J.M., 1977, "Effects of Oilfield Brine Effluent on Benthic Organisms in Trinity Bay, Texas, API Publ. No. 4291," American Petroleum Institute, Washington, D.C.

Bedinger, C.A., Jr., Childers, R.E., Kimball, K.T., Cooper, J.W., and Kwok, A., 1981, Part 1 - Background, program organization and study plan, *in*: "Ecological Investigations of Petroleum Production Platforms in the Central Gulf of Mexico, Volume I. Pollutant Fate and Effects Studies," C.A. Bedinger, Jr., ed., Southwest Research Institute, San Antonio, Texas.

Bender, M.E., Reish, D.J., and Ward, C.H., 1979, Re-examination of the offshore ecology investigation, *Rice Univ. Stud.* 65:35-118.

Boesch, D.F., and Rabalais, N.N., eds., 1987, "Long-term Environmental Effects of Offshore Oil and Gas Development," Elsevier Applied Science, New York.

Carney, R.S., 1987, A review of study designs for the detection of long-term environmental effects of offshore petroleum activities, *in*: "Long-Term Environmental Effects of Offshore Oil and Gas Development," D.F. Boesch and N.N. Rabalais, eds., Elsevier Applied Science, New York.

Carriker, M.R., Swann, C.P., and Ewart, J.W., 1982, An exploratory study with the proton micropobe of the ontogenetic distribution of 16 elements in the shell of living oysters (*Crassostrea virginica*), Marine Biology 69:235-246.

Chasse, C., 1978, The ecological impact on the near shores by the *Amoco Cadiz* oil spill, Mar. Poll. Bull. 9(11):298-301.

Dillon, W.R., Goldstein, M., 1984, "Multivariate Analysis," John Wiley and Sons, New York.

Fan, T., Higashi, R., Cherr, G., and Pillai, M., 1992, Produced water perturbs reproduction in mussels as monitored *in vivo* by NMR spectroscopy and imaging, *in*: "Produced Water: Technological/ Environmental Issues and Solutions," J.P. Ray and F.R. Engelhardt, eds., Plenum Publishing Corp., New York, New York, USA.

Ferren, W.R., Jr., 1985, "Carpinteria salt marsh; environment, history, and botanical resources of a southern California estuary, University of California Santa Barbara Herbarium Publ. 4," University of California, Santa Barbara, California.

Ferris, V.R., and Ferris, J.M., 1979, Thread worms (Nematoda), *in*: "Pollution Ecology of Estuarine Invertebrates," C.W. Hart, Jr. and S.L.H. Fuller, eds., Academic Press, New York.

Gray, J.S., 1979, Pollution-induced changes in populations, Phil. Trans. R. Soc. Lond. B. 286:545-561.

Higashi, R., Cherr, G., Bergens, C., Fan, T., and Crosby, D, 1992, Toxicant isolation from a produced water source in the Santa Barbara Channel, *in*: "Produced Water: Technological/Environmental Issues and Solutions," J.P. Ray and F.R. Engelhardt, eds., Plenum Publishing Corp., New York, New York, USA.

Howarth, R.W., 1989, Determining the ecological effects of oil pollution in marine ecosystems, *in*: "Ecotoxicology: Problems and Approaches," S.A. Levin, M.A. Harwell, J.R. Kelly, K.D. Kimball, eds., Springer-Verlag, New York.

Jolliffe, I.T., 1986, "Principal Component Analysis," Springer-Verlag, New York.

Middleditch, B.S., ed., 1981, "Environmental Effects of Offshore Oil Production. The Buccaneer Gas and Oil Field Study," Plenum Press, New York.

Middleditch, B.S., 1984, "Ecological Effects of Produced Water Discharges from Offshore Oil and Gas Production Platforms. Final Report on API Project No. 248," American Petroleum Institute, Wash., D.C.

Neff, J.M., 1987, Biological effects of drilling fluids, drill cuttings and produced waters, *in*: "Long-Term Environmental Effects of Offshore Oil and Gas Development," D.F. Boesch and N.N. Rabalais, eds., Elsevier Applied Science, New York.

Neff, J.M., Sauer, T.C., and Maciolek, N., 1989, "Fate and Effects of Produced Water Discharges in Nearshore Marine Waters, API Publ. No. 4472," American Petroleum Institute, Washington, D.C.

Nisbet, R.M., Murdoch, W.W., and Stewart-Oaten, A., 1993, Consequences for Adult Fish Stocks of Human-Induced Mortality on Immatures, Ecological Applications, in review.

Osenberg, C.W., Holbrook, S.J., and Schmitt, R.J., 1992, Implications for the design of environmental assessment studies, *in*: "Perspectives on the Marine Environment," P.M. Grifman and S.E. Yoder, eds., USC Sea Grant, Los Angeles, California.

Osenberg, C.W., Schmitt, R.J., Holbrook, S.J., Abu-Saba, K.E., and Flegal, A.R., 1993, Detection of environmental impacts: power analysis and spatial inference, Ecological Applications, in review.

Payne, J.R., Phillips, C.R., and Hom, W., 1987, Transport and transformations: water column processes, *in*: "Long-Term Environmental Effects of Offshore Oil and Gas Development," D.F. Boesch and N.N. Rabalais, eds., Elsevier Applied Science, New York.

Pearson, T.H., and Rosenberg, R., 1978, Macrobenthic succession in relation to organic enrichment and pollution of the marine environment, Oceanogr. Mar. Biol. Ann. Rev. 16:229-311.

Raimondi, P.T., and Schmitt, R.J., 1992, Effects of produced water on settlement of larvae: field tests using red abalone, *in*: "Produced Water: Technological/Environmental Issues and Solutions," J.P. Ray and F.R. Engelhardt, eds., Plenum Publishing Corp., New York, New York, USA.

Reish, D.J., 1979, Bristle Worms (Annelida: Polychaeta), *in*: "Pollution Ecology of Estuarine Invertebrates," C.W. Hart, Jr. and S.L.H. Fuller, eds., Academic Press, New York.

Reish, D.J., and Barnard, J.L., 1979, Amphipods (Arthropoda: Crustacea: Amphipoda), *in*: "Pollution Ecology of Estuarine Invertebrates," C.W. Hart, Jr. and S.L.H. Fuller, eds., Academic Press, New York.

Reynoldson, T.B., 1987, Interactions between sediment contaminants and benthic organisms, *in*: "Ecological Effects of *In Situ* Sediment Contaminants," R. Thomas, R. Evans, A. Hamilton, M. Munawar, T. Reynoldson, and H. Sadar, eds., Dr W. Junk Publishers, Boston.

Roughgarden, J., and Iwasa, Y., 1986, Dynamics of a metapopulation with space-limited subpopulations, Theor. Pop. Biol. 29:235-261.

Sandulli, R., and De Nicola, M., 1991, Responses of meiobenthic communities along a gradient of sewage pollution, Mar. Poll. Bull. 22(9):463-467.

SAS Institute Inc., 1988, "SAS/STAT User's Guide, Release 6.03 Edition." Cary, NC. SAS Institute Inc., 1028 pp.

Spies, R.B., 1987, The biological effects of petroleum hydrocarbons in the sea: assessments from the field and microcosms, *in*: "Long-Term Environmental Effects of Offshore Oil and Gas Development," D.F. Boesch and N.N. Rabalais, eds., Elsevier Applied Science, New York.

Spies, R.B., and Davis, P.H., 1979, The infaunal benthos of a natural oil seep in the Santa Barbara Channel, *Mar. Biol.* 50:227-237.

Spies, R.B., Davis, P.H., and Stuermer, D.H., 1980, Ecology of a submarine petroleum seep off the California coast, *in*: "Marine Environmental Pollution, 1. Hydrocarbons," R.A. Geyer, ed., Elsevier, Amsterdam.

Spies, R.B., and DesMarais, D.J., 1983, Natural isotope study of trophic enrichment of marine benthic communities by petroleum seepage, Mar. Biol. 73:67-71.

Stewart-Oaten, A., Murdoch, W.W., and Parker, K., 1986, Environmental impact assessment: 'pseudoreplication' in time?, Ecology 67:929-940.

Underwood, A.J., and Peterson, C.H., 1987, Towards an ecological framework for investigating pollution, Mar. Ecol. Prog. Ser. 46:227-234.

USE OF NONINVASIVE NMR SPECTROSCOPY AND IMAGING FOR ASSESSING PRODUCED WATER EFFECTS ON MUSSEL REPRODUCTION

T.W.-M. Fan[1], R.M. Higashi[2], G.N. Cherr[2], and M. C. Pillai[2]

[1] Department of Land, Air and Water Resources
University of California
Davis, California
USA

[2] University of California
Bodega Marine Laboratory
Bodega Bay, California
USA

INTRODUCTION

Perturbation of reproductive performance such as gametogenesis, can be a key factor in the decline of a species in a polluted environment since reproductive success is directly related to effects at the population level. The effects of contaminants on reproduction in indigenous marine invertebrates have been documented both in the laboratory (Kluytmans et al., 1988; Hawkins et al., 1989; Langston et al., 1989) and in the field (Lowe, 1988; Moore, 1988; Auffret, 1988). In the Santa Barbara Channel, CA, the discharge of aqueous petroleum waste, primarily the produced water (PW), has been correlated with changes in growth and gonadal-somatic indices (GSI) of mussels outplanted at the discharge site (Osenberg et al., 1992). However, because other complicating environmental factors can exist in field studies, it is difficult to assign a cause-and-effect relationship between the pollution source and reproductive perturbations observed. In fact, this situation was reported for *Mytilus californianus* in an oil seep site study in the Santa Barbara Channel (Straughan, 1976). It has been suggested that the most effective approach would be to relate molecular and cellular alterations to individual-level physiological changes that lead to reproductive perturbations (Spies, 1987).

In order to assess the reproductive success at the biochemical level of indigenous organisms such as the California mussel (*M. californianus*), we have developed noninvasive ^{31}P NMR spectroscopy and ^{1}H NMR imaging methods to acquire unique molecular information on energy budget and detailed morphological and biophysical changes during gametogenesis. These methods were then applied to investigate the chronic effects of PW discharged into the Santa Barbara Channel, on gametogenesis in *M. californianus*.

Mytilus was utilized in this study because it is a worldwide environmental monitoring organism, and can be readily outplanted at polluted sites (Phillips, 1976; George et al., 1979). Noninvasive approaches designed for repeated measurements of individuals were developed for their unique capabilities of observing chronic effects on mussels without interferences from commonly observed individual heterogeneities in reproductive performance. We have focused

Produced Water, Edited by J.P. Ray
and F.R. Engelhart, Plenum Press, New York, 1992

on the use of *in vivo* ^{31}P NMR spectroscopy for monitoring energy metabolism in mussels (Fan et al., 1991) which plays a central role in gametogenesis and organismal health. This technique has rapidly become the method of choice for measuring intracellular pH, energy status, and ATP turnover (see Gadian, 1982) in biomedical research, although its application to environmental pollution research is yet at its infancy. In view of the importance of energy balance in chronically impacted ecosystems, ^{31}P NMR should be a useful tool for assessing the individual reproductive performance.

To assist evaluation of gametogenesis, we also employed the ^{1}H NMR imaging (also known as magnetic resonance imaging, MRI) technique which has been used to provide detailed anatomical features of mammals including human in medical research and diagnostics (see Morris, 1986) as well as in invertebrates such as mussels (Fan et al., 1991). In this report, we demonstrated how the time course of gametogenesis was assessed by MRI in field-collected mussels, and how this information was applied to determine the impact of PW exposure on gametogenesis.

MATERIALS AND METHODS

PW was collected from an onshore plant in southern California, shipped chilled without headspace, divided into 20 ml aliquots, and kept frozen at -70° C until use. Elemental composition of PW was performed on an ARL-3510 Inductively Coupled Plasma/Atomic Emission Spectrometer using filtered (through 0.2 µm filter), acidified (5% HCl) samples. *M. californianus* was collected from Bodega Bay, CA and maintained in flowing sea water at 12-14° C until NMR measurements or dissections were conducted. The temperature of the magnet bore where the mussel was placed for NMR measurements was maintained at 14° C with recirculating cooler (Figure 1). *In vivo* ^{31}P NMR spectroscopy was performed using a 3.0 or 3.5 cm photoetched surface NMR probe (Fan and Higashi, 1989) on a General Electric CSI-7 Tesla or a CSI-2 Tesla imaging NMR spectrometer (Fremont, CA), respectively. The surface probe arrangement allowed selective acquisition of ^{31}P NMR signals originating from the gonadal region of live mussels. Two-dimensional ^{1}H NMR images of live mussels were obtained from the CSI-2T instrument using a 3 inch birdcage imaging probe optimized for marine organisms with high conductance (Fan et al., 1991).

The PW exposure experiment was conducted at 14° C in a 48 hr static/renewal system for 4.5 months @ 1% (w/v) PW in Bodega Head seawater. We have chosen 1% because this is the concentration expected to occur initially at the discharge. For each seawater renewal, a new batch of PW kept at -70° C was thawed and added; this measure was to avoid repeated thawing or refreezing of PW. This method of PW handling did not affect toxicity to sea urchin embryos (Baldwin et al., in press). No attempt was made to control for volatization of PW constituents.

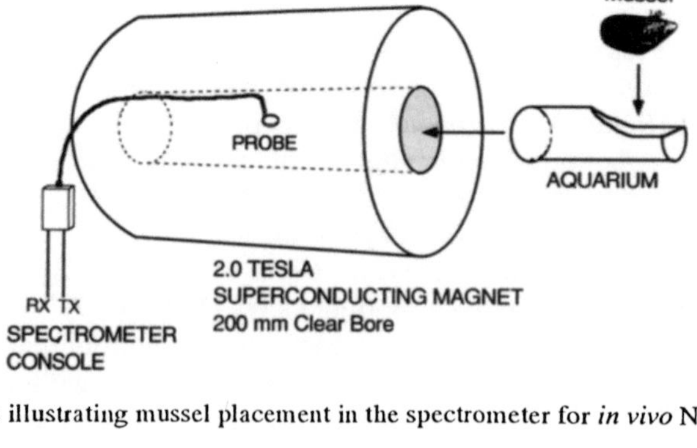

Figure 1. Scheme illustrating mussel placement in the spectrometer for *in vivo* NMR measurements.

The control treatment involved incubation with Bodega Head seawater for 5 months. Five mussels each were used for the exposure and the control. These mussels were fed at the static/renewal interval with a monoculture of *Isochrysis galbana* (Tahitian strain). At a one-month interval, all 10 mussels were subjected to ^{31}P NMR and ^{1}H MRI measurements. No fatality was observed throughout the experiment.

Following *in vivo* NMR spectroscopy and imaging, animals were dissected and their gonads subjected to both perchloric acid (PCA) extraction and correlative histological analysis of gametogenesis. PCA extraction was performed on lyophilized gonads which were pulverized to about 1 μm size particle to maximize extraction efficiency (Fan et al., 1986). Briefly, tissues were extracted with 5% PCA, adjusted to pH 3.5 with K_2CO_3, centrifuged, and lyophilized. The lyophilized extract was dissolved in D_2O and neutralized before analysis by ^{31}P NMR on a Bruker AM-400, a General Electric Omega-300 wide-bore, or an Omega-500 NMR spectrometers. For histological examination, tissues (ovaries and testes) were fixed in phosphate buffered formalin (10%), post-fixed in 0.5% osmium tetroxide, dehydrated in a graded acetone series, and embedded in epoxy resin. Sections (1-2 μm) were stained with borate-buffered toluidine blue, and examined with a light microscopy (Cherr and Clark, 1982).

RESULTS AND DISCUSSION

In Vivo NMR Assessment of Mussel Gametogenesis

***^{31}P* NMR** Using the ^{31}P NMR approach, we found that mussel ovaries and testes had several distinct differences in their phosphate metabolite composition as shown in Figure 2. A phosphodiester (PDE) signal (tentatively assigned as glycerophosphorylcholine) was prominent in both the *in vivo* (Figure 2A) and extract ^{31}P NMR spectra (Figure 2B) of the late-developing ovary but not in those of the late-developing testis. This was the case with at least 50 mussels examined ranging from early-developing to ripe stages. In addition, several phosphate metabolites that contributed to signals in the -27 to -30 ppm region of the extract ^{31}P NMR spectra were present in the ovary but not in the testis (Figure 2B). These metabolites are likely to be NADP(H) and UDP-glucose based on the ^{31}P NMR peak positions (Lee and Ratcliffe, 1983; Pfeffer et al., 1986). We also found that several metabolites were common to both ovaries and testis, including phosphoarginine (PArg), ATP, and orthophosphate (P_i). Furthermore, except for PArg and P_i peak intensities, there was a close resemblance between the *in vivo* and the corresponding extract spectra, which supported that the *in vivo* spectra acquired originated largely from the gonadal tissues. The relatively higher PArg and P_i peak intensities in the extract spectra could be due to two factors: PArg may be hydrolyzed to yield P_i during extraction procedure; alternatively, part of the increase in extract P_i peak intensity may result from mobilization of P_i fraction bound to macromolecules and therefore has become visible in the *in vivo* spectra.

The presence of the PDE signal only in the ovarian spectra provided a noninvasive way of distinguishing the gender of mussels. In addition, time course quantitation of major energy compounds such as PArg and ATP on the same individual could be utilized to detect chronic stress in situation where high variability in energy status existed.

In addition to gender distinction, the relative intensity of the PDE signal increased in ovaries from mid (II) to late-developing (III) stages and decreased from late-developing to spawned stages (IV) as shown in Figure 3 (see below for histological descriptions). The latter was also illustrated in the ^{31}P NMR spectra of the corresponding ovarian extracts (Figure 4). It should be noted that other phosphate metabolites including P_i, ATP, and PArg also decreased in levels after spawning (Figures 3 and 4). These results indicate that the reduction of phosphate metabolite composition in the spawned ovary was through the loss of oocytes; in particular, they suggest that PDE was associated with the oocyte and thus, could be an effective marker for oogenesis.

2-D ¹H NMR Imaging and Histology. Another noninvasive means of assessing mussel gametogenesis that we developed is 2-D ¹H MRI. Figure 5A shows the cross-section ¹H NMR images of the same three female individuals as in Figure 3, where the oogenic stages corresponded to II, III, and IV. The ovary (indicated by arrows) appeared as two crescent shape structures cradled against each half of the shell. It is evident from Figure 5A that the relative image density of the ovary increased from Stage II to III but decreased from III to IV (after spawning). In addition, patches of lower image density areas (*) were seen in the spawned ovary. Thus, the ovarian image density was positively correlated with the extent of oogenesis or the presence of oocytes.

These ovarian stages in Figures 3 and 5 were independently assessed by histological analysis as shown in Figure 5B. We have followed the staging criteria developed for *M. californianus* by Edwards (1985). Briefly, the Stage I ovary represents the early active phase of oogenesis. None of the individuals examined in this study were in this phase of ovarian development. The Stage II ovaries were characterized by the appearance of oocytes with multinucleolated germinal vesicles and randomly-distributed cortical granules in the cytoplasm (Figure 5B). Ripe ovaries (Stage III) contained distended follicles filled with eggs that either lacked nucleoli, or possessed uninucleolate germinal vesicles. In addition, a distinct band of cortical granules distributed along the periphery of the egg was evident (Figure 5B). Spawned ovaries (designated as Stage IV in this study) were characterized by follicles containing few eggs, along with their remnants (Figure 5B).

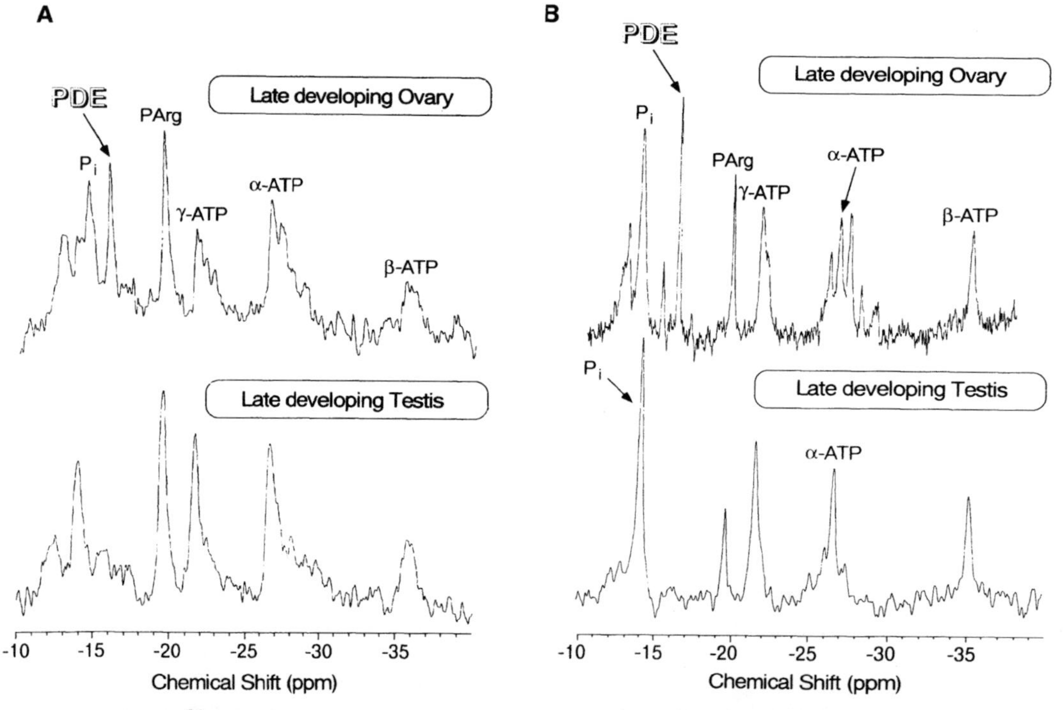

Figure 2. *In Vivo* ³¹P NMR spectra of mussel ovary and testis (A) and ³¹P NMR spectra of corresponding PCA extracts (B). The *in vivo* ³¹P NMR spectra (A) were acquired on the CSI-7T spectrometer using a 30 s pulse width, a 1.13 s repetition rate, a 6500 Hz spectral width, 4096 digital points, 2048 acquisitions, and a 20 Hz line broadening. The extract spectra (B) were obtained on the Omega-500 (ovary) and Omega-300 (testis) spectrometers using a 90° pulse angle, a 1 s repetition rate, a 9000 Hz spectral width, 4096 digital points, 2048 acquisitions, and a 10 Hz line broadening. Peak assignment is as follows: orthophosphate (P_i); phosphodiester (PDE); phosphoarginine (PArg); g-ATP, a-ATP, and b-ATP, g, a, and b-phosphate of adenosine triphosphate (ATP).

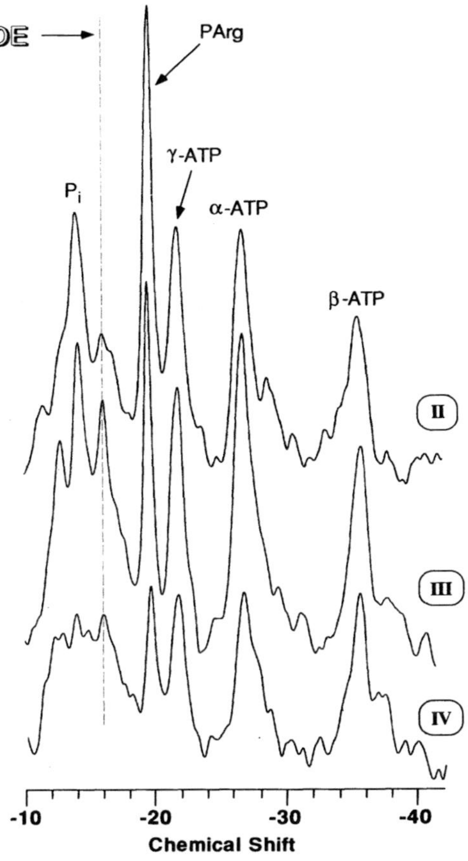

Figure 3. *In vivo* ^{31}P NMR spectra of mussel ovary in three different gametogenic states. These spectra were acquired on the CSI-2T spectrometer using a 45° pulse angle to minimize signal contribution from other parts of the mussel body. Other spectral parameters included a 1 s repetition rate, a 3015 Hz spectral width, 2048 digital points, 2048 acquisitions, and a 20 Hz line broadening. Peak assignment is the same as that in Figure 2. II, III, and IV represent mid-developing, late-developing, and partially spawned ovaries (see text for stage descriptions).

Likewise, the ^{1}H NMR image density of the testis was correlated with the presence of spermatocytes or spermatozoa as illustrated in Figure 6, where a pair of ^{1}H NMR images from a male individual before and right after spawning were compared. Similar to the case of the ovary, the average image density of the testis decreased and patches of low image density (indicated by *) were evident as a result of spawning. These results suggest that the developing gametes themselves were responsible for the increase in the ^{1}H NMR image density and that the patchy appearance of gonads was from spawning. The latter factor led to greatly increased standard deviations of image densities. This, together with parameters such as "patchiness" (hole count and connectivity) measurable by image analysis, could be valuable tools for evaluating gonadal heterogeneity that resulted from partial spawning.

Impact of Produced Water Exposure on Mussel Oogenesis

Applying the *in vivo* methodology described above, we have examined the chronic effect of PW exposure on mussel oogenesis in the laboratory. The PW from this discharge was unusually devoid of hydrocarbons (< 1 ppb; Higashi et al., 1992) and it appeared that the inorganic fraction may be responsible for the chronic bioeffects (Higashi et al., 1992). A selected

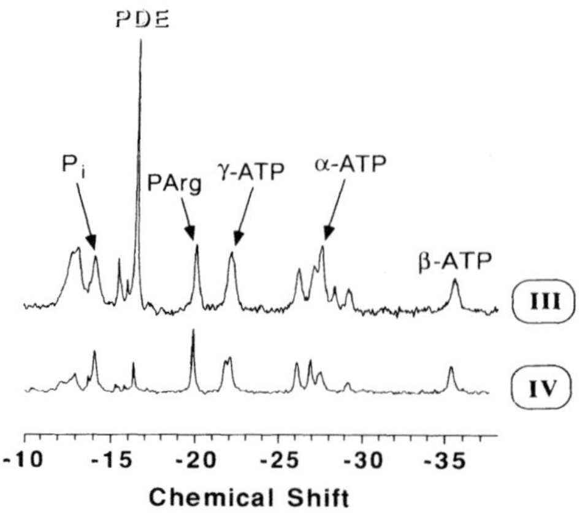

Figure 4. ^{31}P NMR spectra of the PCA extracts of late-developing and partially spawned ovaries. These two spectra were obtained from the PCA extracts of the late-developing (III) and partially spawned (IV) ovaries from Figure 3 on the AM-400 spectrometer. Spectral parameters included a 90° pulse angle, a 1 s repetition rate, a 8064 Hz spectral width, 8192 digital points, and 2048 acquisitions with Waltz-16 ^{1}H broadband decoupling. Peaks were assigned as in Figure 2.

elemental composition of the batch of PW used for exposure is shown in Table 1, where several trace elements including Sn, As, and Cu were present at low ppm level that could impact mussel reproduction (for more detailed composition, see Higashi et al., 1992). Figures 7A and 7B compared the time course changes of the ovarian ^{31}P NMR spectra of PW-exposed (#4) and control (#9) individuals, respectively.

Table 1. Elemental composition of produced water employed for mussel exposure

	Element														
	S	Ca	Na	Sn	As	Zn	Pb	Cu	Ni	Mn	Al	Ba	Cr	Cd	Fe
ppm	10	168	9900	0.2	0.5	0.02	<0.1	0.06	0.04	0.06	0.09	70	0.4	<0.1	0.74

One month after PW exposure, a large decrease in [PArg] occurred along with a large increase in [P_i] and a smaller drop in [ATP] while the control individual exhibited little change. The opposite relationship among [P_i], [PArg], and [ATP] suggest that the breakdown of PArg and ATP may have accounted for the increase in [P_i]. This phenomenon is usually observed when organisms are under stress. Three months after PW exposure, [ATP] and [PArg] continued to drop and [PDE] was hardly detected while a significant portion of PDE, ATP, and PArg was still present in the control individual. Finally, no further major compositional changes in the phosphate metabolites occurred in the PW-exposed individual after 4.5 months. In contrast, a large decrease in [PArg] and [ATP], a moderate decrease in [P_i], and the disappearance of PDE signal were observed for the control individual after 5 months. These changes in the control individual were consistent with the occurrence of spawning where the loss of oocytes was associated with a large decrease in the concentrations of phosphate metabolites (*cf.* Figures 3 and 4).

In addition, according to histological examination, ovaries from the exposed individuals were dominated by connective tissues and dense granular materials with very few normal oocytes within the follicles (Figure 8A). This appearance was more characteristic of degener-

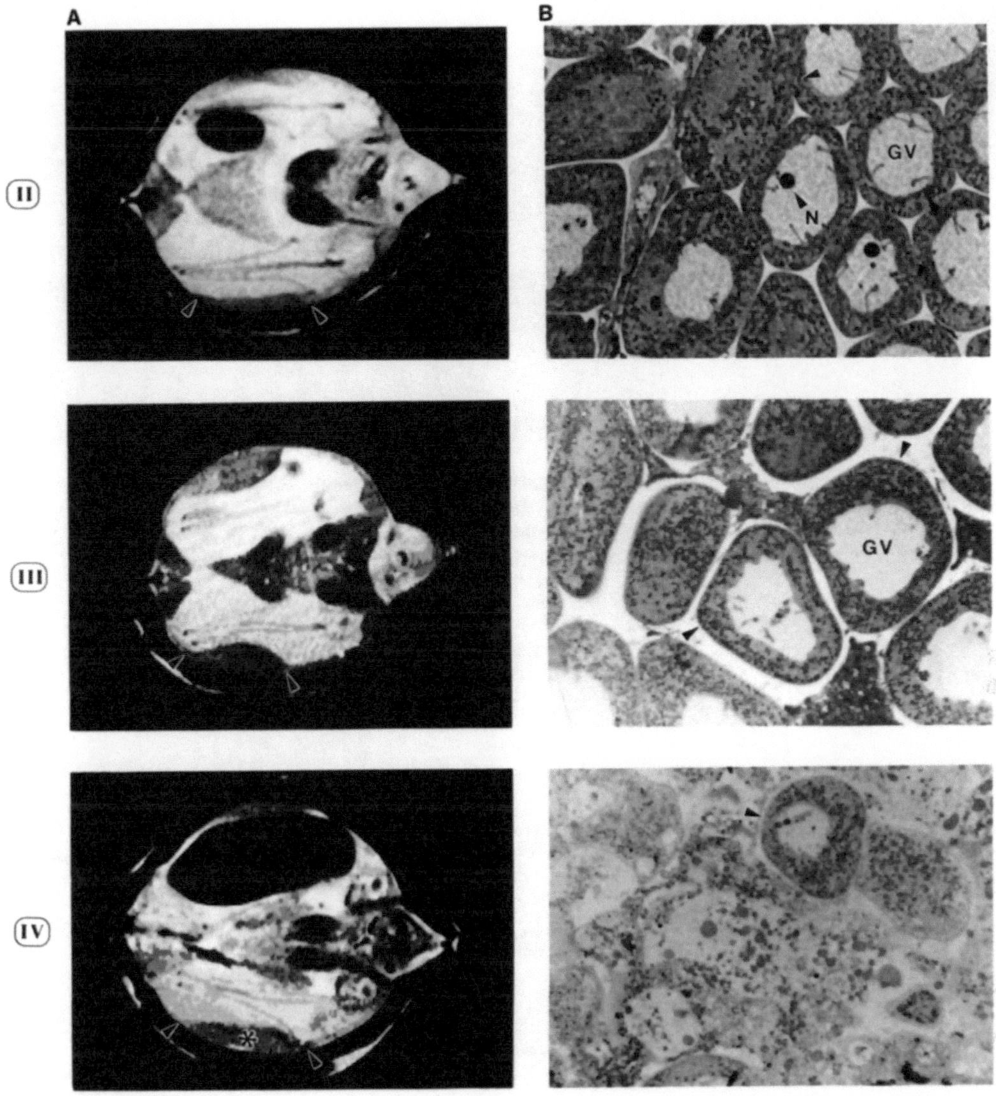

Figure 5. 2-D ^{1}H NMR images (A) and histomicrographs (B) of ovaries in different gametogenic states. The ^{1}H NMR images (A) of the corresponding ovaries, from the same individuals as in Figure 3, were acquired on the CSI-2T spectrometer using a multislice spin-echo pulse sequence that allowed simultaneous acquisition of 8 slices . Each image slice was 2 mm thick and 5 mm apart with a digital resolution of 256x256. The spin-echo delay (TE = 120 ms) was chosen to optimize the image contrast between the ovary and the rest of the mussel tissues. In the examples shown, the ovary (indicated by arrows) appeared as two crescent shape structures lined against each half of the shell. The asterisk indicates patches of lower density areas. Large dark areas in the upper half of *M. californianus* images are air cavities formed by these high-littoral organisms; hence only the lower halves should be compared. The oogenic stages were assessed independently by histological analysis of the corresponding ovaries (B). The three histomicrographs shown are at 450-fold magnification. The Stage II ovary contained oocytes with distinct germinal vesicles (GV), multiple nucleoli (N), and randomly-distributed cortical granules (arrows). The Stage III ovary possessed mature eggs with either one or no nucleolus and a distinct layer of cortical granules along the periphery of the egg (arrows). The Stage IV (partially spawned) ovary was characterized by a few intact eggs (arrow) and remnants of resorbing oocytes.

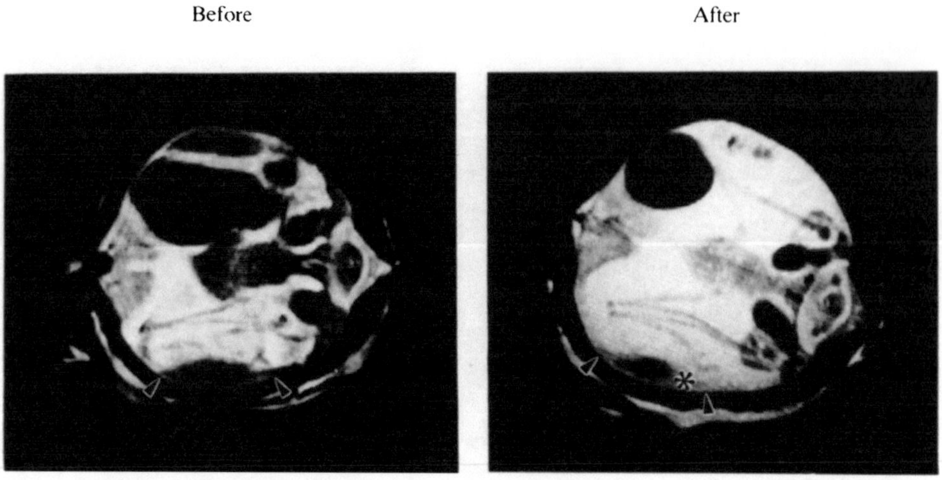

Figure 6. 2-D ^{1}H NMR images of a male mussel before and after spawning. The ^{1}H NMR images were obtained similarly as in Figure 5A. The same individual was scanned both before and right after spawning was induced. Ovarian tissues are indicated by arrows. Patches of low image density areas (indicated by *) were evident in the ovarian image acquired after spawning.

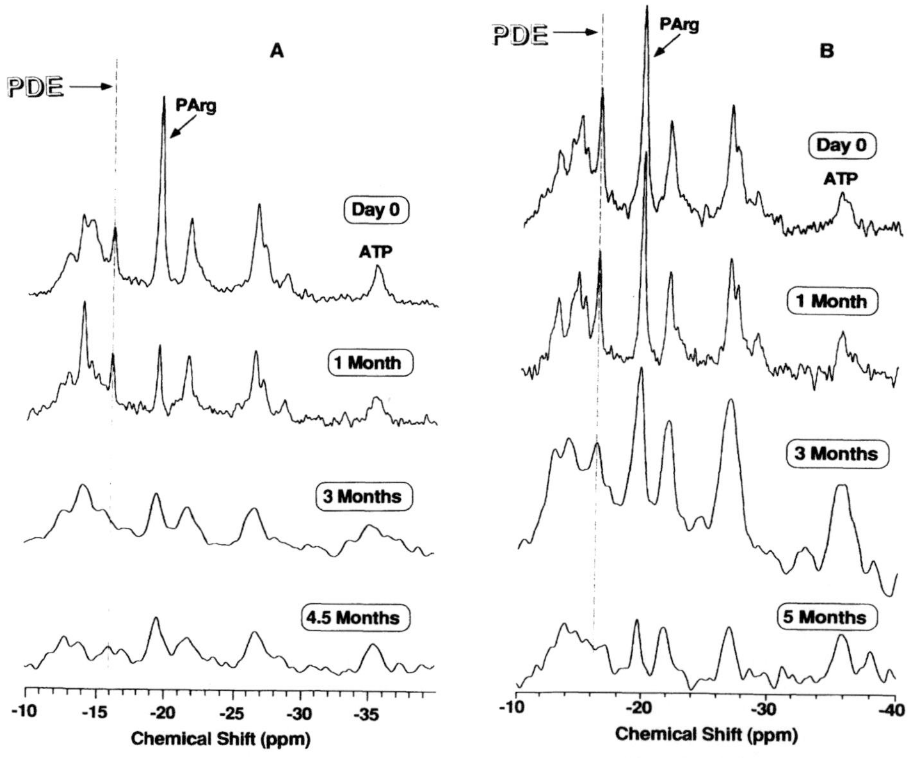

Figure 7. Time course changes in *in vivo* ^{31}P NMR spectra of ovaries from two female mussels under PW (A) and control (B) treatments. The treatment conditions were as described in Materials and Methods. For the first one month treatment, ^{31}P NMR spectra were acquired on the CSI-7T spectrometer using the same parameters as in Figure 2 while those acquired thereafter were from the CSI-2T instrument using the same parameters as in Figure 3. The acquisition parameters used for these two instruments were adjusted to give comparable ^{31}P NMR spectra of the ovaries.

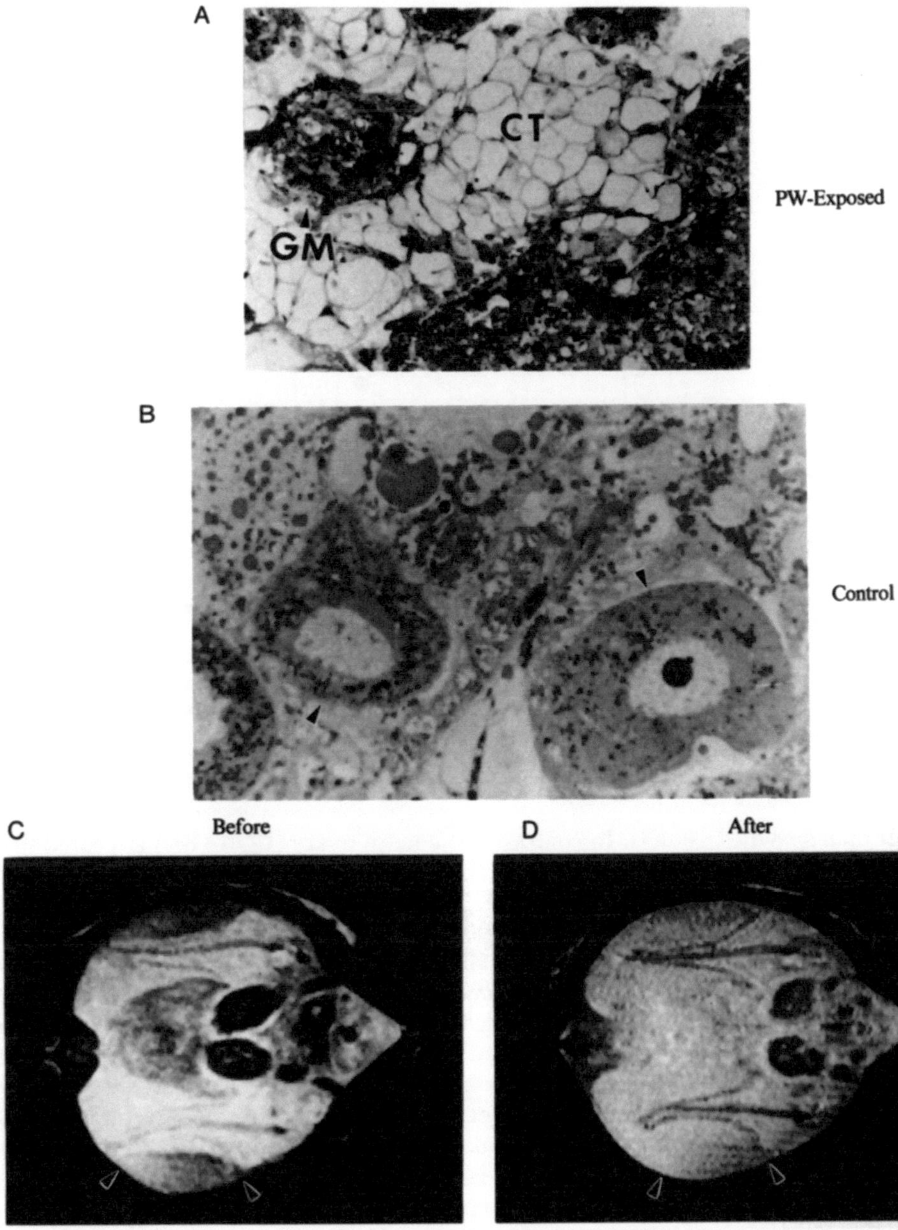

Figure 8. 2-D ^{1}H NMR images of a female before (C) and after (D) exposure to PW, and histomicrographs of ovarian sections from mussels under PW (A) and control (B) conditions. Treatment conditions were as described in Materials and Methods. The two micrographs (A and B) shown are at 450-fold magnification. The PW-exposed (mussel #2) and the control individuals (mussel #9) were the same as those from Figure 9. Degenerating oocytes, connective tissues (CT), and dense granular material (GM) were evident in the PW-exposed ovary while a few intact eggs (arrows) and remnants of resorbing oocytes were seen in the control ovary. The pair of NMR images (C and D) shown were acquired as described in Figure 5 from mussel #2 before and after PW exposure. Ovarian tissues are indicated by arrows.

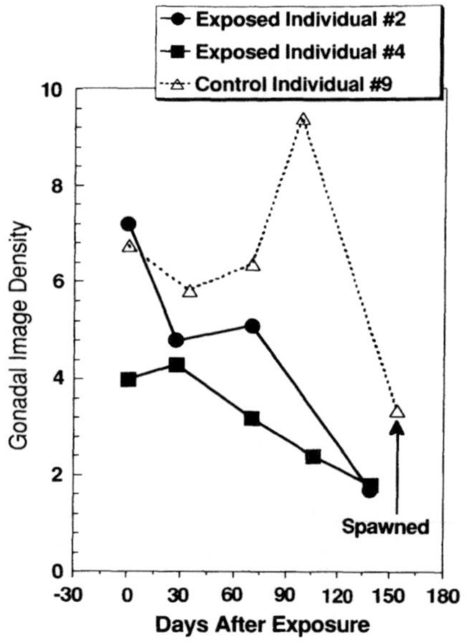

Figure 9. Time course change in the ^1H NMR image density of ovaries from mussels under PW exposure and control conditions. The treatment protocol was as described in Materials and Methods and the NMR image acquisition was the same as in Figure 5. The ovarian image density was obtained by digitizing the NMR images, followed by image analysis using the "Image" program developed by the U.S. National Institutes of Health. The relative image density is defined as the ratio of the image density of the ovary to that of the body water. By taking this ratio, any systematic error associated with image aquisition and digitization was minimized. The closed symbols and open triangles represent PW-exposed and control individuals, respectively.

ating rather than spawned gonads. In contrast, the control ovary (figure 8B) appeared to have undergone spawning prior to the end of the experiment based on the presence of empty follicles and a low count of normal oocytes.

Moreover, impact of PW on oogenesis was evident in the ^1H NMR images of the exposed individuals relative to those of the control. As shown in Figure 8B, a dramatic decrease in the ovarian image density was observed for the individual (#2) after exposure to PW for 4.5 months (Figure 8D. The control individual also exhibited a decline in the ovarian image density but to a lesser extent (result not shown). With repeated image measurements on the same individual, the time course changes in the ovarian image density was established, which is shown in Figure 9. The pattern of change in ovarian image density of two PW-exposed individuals clearly differed from that of the control; the former two declined progressively with time while the latter increased prior to a sudden drop. Together with the evidence acquired from ^{31}P NMR and histology, these patterns of change were consistent with a degeneration of oogenesis in PW-exposed individuals and with occurrence of spawning in the control. Therefore, chronic PW exposure in the laboratory appeared to perturb energy balance in mussel ovary and caused a regression in mussel oogenic activities.

Although preliminary, these results indicated that the noninvasive NMR methodologies that we have developed could be applied to explore the long-term effects of pollutants on reproduction and organismal health in the laboratory. Similar approach could also be extended to investigate effects in field outplant studies. In either case, the added advantage of this approach is that a clear relationship could be discerned without the need for a large number of samples because the same individual is monitored both before and after the exposure and detailed time course changes can be established.

CONCLUSION

This study demonstrated that the noninvasive NMR approaches (^{31}P NMR spectroscopy and ^{1}H NMR imaging), together with correlative histological analysis, provided reliable assessments of gender-specific gametogenesis in mussels. Specifically, *in vivo* ^{31}P NMR revealed a specific phosphodiester indicator for oogenesis, in addition to providing information on gonadal energy budget. In a complementary fashion, ^{1}H NMR imaging enabled the distribution density of developing gametes across the entire gonad to be monitored.

When applied to determine the effect of PW exposure on mussel, both methods revealed the dynamics of chronic ovarian perturbations on the same individual, thus eliminating interferences arising from intrinsic heterogeneities in oogenesis of this species. Consequently, reliable conclusions could be drawn from a much smaller sample size. The changes observed suggest that PW exposure perturbed ovarian energy balance and caused ovarian degeneration in *M. californianus*. To the best of our knowledge, this is the first direct demonstration of perturbations in reproduction of organisms chronically exposed to PW under controlled laboratory conditions. However, it should be kept in mind that as produced water composition may vary from batch to batch and that the results reported here were derived from one batch only, a general conclusion of the impact of produced water cannot be drawn. Nevertheless, such study, together with the fractionation approach employed by Higashi et al. (1992), can lead to the identification of active constituents in PW and help establish a cause-and-effect relationship between PW and observed bioeffects.

More importantly, how such laboratory studies relate to the decreased GSI and growth observed for field-outplanted mussels (Osenberg et al., 1992) await further investigations. To help answer this question, an outplant experiment is underway that utilizes the same noninvasive approaches to explore the relationship between PW discharge and reproductive development in field-exposed mussels.

ACKNOWLEDGEMENTS

We would like to thank Gayle Garman and Tammy Shields for assistance on the histology and John Baldwin for care of the animals. This project is supported in part by the Minerals Management Service of the U.S. Department of Interior, grant #14-35-0001-30-471.

REFERENCES

Auffret, M., 1988, Histopathological changes related to chemical contamination in *Mytilus edulis* from field and experimental conditions, Mar. Ecol. Prog. Ser. 46:101-107.

Baldwin, J.D., M.C. Pillai, and G.N. Cherr. The reponse of sea urchin embryos to aqueous petroleum wastes includes the expression of a high molecular weight glycoprotein. Marine Biology, in press.

Cherr, G.N. and Clark, W.H., Jr., 1982, Fine structure of the envelope and micropyles in the eggs of the white sturgeon, *Acipenser transmontanus* Richardson, Develop. Growth Different. 24:341-352.

Edwards, E.L., 1985, The Reproductive and Percentage Solids Cycles of *Mytilus edulis* and *Mytilus californianus* in Humboldt County, California, Technical Report Series, Humboldt State University, CA., pp. 57.

Fan, T.W.-M. and Higashi, R.M., Lane, A.N., and Jardetzky, O., 1986, Combined use of ^{1}H NMR and GC/MS for metabolite monitoring and *in vivo* ^{1}H NMR assignments, Biochim. Biophys. Acta 882:154-167.

Fan, T.W.-M. and Higashi, R.M., 1989, Reproducible NMR surface coil fabrication by combining computer-aided-design and photoresist process, Analyt. Chem. 61:636-638.

Fan, T.W-.M., Higashi, R.M., and Macdonald, J.M., 1991, Emergence and recovery response of phosphate metabolites and intracellular pH in intact *Mytilus edulis* as examined *in situ* by *in vivo* ^{31}P-NMR, Biochim. Biophys. Acta 1092:39-47.

Higashi, R.M., Cherr, G.N., Bergens, C.A., and Fan, T.W.-M., 1992, An approach to toxicant isolation from a produced water source in the Santa Barbara Channel, *in:* "Produced Water: Technological/Environmental Issues and Solutions," J.P. Ray and F.R. Engelhardt, eds., Plenum Publishing Corp., New York, New York, USA.

Gadian, D.G., 1982, "NMR and its Applications to Living Systems," Oxford University Press, London.

Hawkins, A.J.S., Rusin, J., Bayne, B.L., and Day, A.J., 1989, The metabolic/physiological basis of genotype-dependent mortality during coper exposure in *Mytilus edulis*, Mar. Environ. Res. 28:253-257.

Kluytmans, J.H., Brands, F., and Zandee, D.I., 1988, Interactions of cadmium with the reproductive cycle of *Mytilus edulis* L., Mar. Environ. Res. 24:189-192.

Langston, W.J., Bebianno, M.J., and Mingjiang, Z., 1989, A comparison of metal-binding proteins and cadmium metabolism in the marine molluscs *Littorina littorea* (Gastropoda), *Mytilus edulis* and *Macoma balthica* (Bivalvia), Mar. Environ. Res. 28:195-200.

Lee, R.B. and Ratcliffe, R.G., 1983, Phosphorus nutrition and the intracellular distribution of inorganic phosphate in pea root tips: A quantitative study using ^{31}P-NMR, J. Exp. Bot. 34:1222-1244.

Lowe, D.M., 1988, Alterations in cellular structure of *Mytilus edulis* resulting from exposure to environmental contaminants under field and experimental conditions, Mar.Ecol. Prog. Ser. 46:91-100.

Moore, M.N., 1988, Cytochemical responses of the lysosomal system and NADPH-ferrihemoprotein reductase in molluscan digestive cells to environmental and experimental exposure to xenobiotics, Mar. Ecol. Prog. Ser. 46:81-89.

Morris, P.G., 1986, "Nuclear Magnetic Resonance Imaging in Medicine and Biology", Clarendon, Press, Oxford.

Osenberg, C.W., Schmitt, R.J., and Canestro, D., 1992, Spatial scale of ecological effects associated with an open coast discharge of produced water, *in:* "Produced Water: Technological/Environmental Issues and Solutions," J.P. Ray and F.R. Engelhardt, eds., Plenum Publishing Corp., New York, New York, USA.

Pfeffer, P. E., Tu, S.-I., Gerasimowicz, W.V., and Cavanaugh, J.R., 1986, *In vivo* ^{31}P NMR studies of corn root tissues and its uptake of toxic metals, Plant Physiol. 80:77-84.

Straughan, D., 1976, Sublethal Effects of Natural Chronic Exposure of Petroleum in the Marine Environment, Publ.#4280, Washington, D.C., American Petroleum Institute, p.119.

EFFECTS OF PRODUCED WATER ON SETTLEMENT OF LARVAE: FIELD TESTS USING RED ABALONE

P.T. Raimondi and R.J. Schmitt

Coastal Research Center, Marine Science Institute
University of California - Santa Barbara
Santa Barbara, California
USA

INTRODUCTION

Field studies of biological effects arising from the chronic discharge of produced water into the marine environment have emphasized potential impacts to the benthos, and have not addressed directly possible effects on organisms in the water column (Neff, 1987; Payne et al., 1987; Spies, 1987). In part this arises because harmful components of produced water, such as petroleum hydrocarbons and heavy metals, have relatively short residence times in the water column (e.g. Armstrong et al., 1979; Grahl-Nielsen et al., 1979), but can accumulate in surficial sediments near the point of discharge (e.g. Armstrong et al., 1979; Wheeler et al., 1980; Mittleditch, 1981a). Water soluble contaminants are thought to be diluted rapidly to levels well below those suspected to cause meaningful biological responses based on acute (lethal) laboratory tests (e.g. Montalvo and Brady, 1979; Middleditch, 1981b, 1984; Rose and Ward, 1981; Payne et al., 1987), and produced water typically is only mildly toxic (Neff, 1987). If effects of produced water on planktonic forms occur anywhere, they are thought to be most likely in embayments or other shallow, low-energy environments where the turnover rate of the receiving water is low relative to the discharge volume (e.g. Armstrong et al., 1979; Neff, 1987; Payne et al., 1987; Spies, 1987). However, limited data on petroleum hydrocarbon contamination suggest that effects on planktonic organisms can occur in open waters (e.g. Howarth 1989), although there have not been explicit field evaluations for discharges of produced water (Neff, 1987).

In marine systems, many planktonic forms are larvae and spores of benthic organisms (i.e., meroplankton), and early developmental stages are thought to be more sensitive to petroleum hydrocarbons and other contaminants than older life stages (for review see Capuzzo, 1987; Neff, 1987). Normal development of "wild" larvae that encounter waste plumes can be greatly altered (e.g. Weis et al. 1989; Kingsford, 1993), and there is evidence that brief exposure of early life stages to low concentrations of produced water can result in a developmental response at a later stage (e.g. Krause et al., 1992). Furthermore, effects of toxicants may be most pronounced at transition phases in the life history when biochemical pathways are initiated or repressed. For planktonic larvae, an important transition phase involves settlement and metamorphosis into the adult habitat.

Produced Water, Edited by J.P. Ray
and F.R. Engelhart, Plenum Press, New York, 1992

Planktonically dispersed propagules differ in the extent to which they are susceptible to point source pollution. Principal attributes affecting susceptibility are the duration of the planktonic phase and the behavior of individuals while in the water column. For example, some larvae and spores can be viewed as "passive particles" that drift widely, while others, such as abalone (Prince et al., 1988), complete their planktonic stage in a relatively localized area. These characteristics influence both the likelihood of encountering a waste plume and the duration of exposure once encountered. While the importance of potential toxicity and dilution of produced water plumes have been recognized (e.g. Rose and Ward, 1981; Payne et al., 1987), little attention has been given to the biology of planktonic propagules relative to the spatial extent of waste plumes (e.g. Keough and Black, 1993).

The intent of this paper is to determine whether the discharge of produced water into a high energy, open coast environment can adversely affect planktonic larvae of long-lived benthic organisms. In particular, we address (1) the relationship between proximity to discharge, duration of exposure, and the probability of a larva surviving, settling and metamorphosing, and (2) whether effects differ for larvae of different developmental stages. Laboratory tests are open to the criticism that organisms may not be subjected to realistic exposure regimes, which adds uncertainty in extending results to natural situations. Field tests avoid this problem, and therefore are the approach taken here. However, unlike laboratory tests, assigning causation for observed responses from field approaches can be problematic as other, natural environmental conditions that co-vary with the factor of interest (e.g., produced water plume) could be responsible for observed responses. For that reason, we conducted laboratory tests to verify whether the discharge of produced water was responsible for observed responses.

Study Organism

The organism selected for study was the larva of the red abalone, *Haliotis rufescens* Swainson (Mollusca: Gastropoda). Red abalone is a long-lived (up to 15 years), commercially important species that lives on rocky reefs between the intertidal zone to depths > 150 m on the west coast of the United States (Cox, 1962). The species is common in the Santa Barbara Channel, a region of considerable offshore oil production, and adults can be found on submerged pipelines that link offshore platforms with onshore processing facilities (S. Anderson, personal communication). Abalone have free-swimming lecithotrophic larvae that enter the water column and subsequently settle and metamorphose into the adult body form in the benthic habitat. The planktonic phase can be divided into an initial period of 5 - 7 days when the larva is developmentally incapable of settling, followed by a 2 - 5 week period when it is competent to settle in response to an inducer (e.g. Morse et al., 1979a; Morse, 1990); these are termed the *Pre-competent* and *competent* stages respectively. Because of the short pre-competent period and behavioral mechanisms, the distances that abalone larvae disperse from the point of origin is relatively small (e.g. Prince et al., 1987, 1988); the average distance for larvae of *H. rufescens* is thought to be ~1 - 10 km (M. Tegner, *personal communication*). The best information on dispersal distances is for *H. rubra*, whose larvae disperse ~100 m (Prince et al., 1988).

Competent red abalone larvae can be induced to settle and metamorphose using γ-aminobutyric acid (GABA) (Morse et al., 1979a, b, 1980). GABA is an extremely potent inducer for red abalone larvae. Morse et al. (1979a) found that 10^{-6} molar GABA resulted in 98% settlement in 18 hours. Once settled, metamorphosis to the adult body form typically is completed within 24 hrs (Morse et al., 1980). This potent settlement inducer allows measurement of the effect of different exposure regimes; larvae that are capable of settling will do so in the presence of GABA. This technique has been used to assess interference of settlement of red abalone by pesticides and other contaminants (e.g. Morse et al. 1979b).

METHODS

Study Site and Outfall Characteristics

The diffuser of produced water under study is ~200 - 300 m offshore of Carpinteria, CA (34°23'N, 119°30'W) over an unconsolidated soft bottom at a water depth of approximately 11 m; additional information on the study site and benthic fauna can be found in Osenberg et al. (1992). Carpinteria is at the southern end of the Santa Barbara Channel, and is a high energy, open coast environment. Current speed and direction at the diffuser, measured on 30 dates, indicated that mid-water (3 - 6 m) flow primarily was parallel to shore (i.e., west - east). Flow to the west was more frequent (51%) than to the east (24%), and the average speed was ~12 cm/sec regardless of direction; speeds 5 cm/sec were measured on ~5% of the observation dates. Mean surf zone currents on the mainland coast in the Santa Barbara Channel vary between ~1 - 2 cm/s during relatively calm periods to > 170 cm/s generated by long period swell (Gable, 1981).

Produced water has been discharged from the Carpinteria outfall since at least 1978. The current average effluent volume is approximately 2.3 x 10^6 l/day, and the maximum permitted flow is 4 x 10^6 l/day. The diffuser design is described in Osenberg et al. (1992), and chemical and other characteristics of the produced water discharged at this site are described in Higashi et al. (1992). The effluent is slightly hyposaline and warmer than the receiving water, both of which would tend to increase buoyancy of the plume.

No quantitative physical or chemical measurements of the extent of the plume produced by the discharge of produced water has yet been done. Our experiments were, in part, designed to measure the extent of the plume using measurements of biological response.

Field Experiments of Effects on Abalone Larvae

Parameters measured for larvae were survivorship, ability of survivors to settle after reaching the age of competence, and ability of settlers to metamorphose. Each of these parameters were estimated as a function of proximity to the diffuser and duration of exposure. We also addressed whether gross changes in swimming behavior of larvae occurred. The general procedure to estimate larval performance was to rear abalone larvae in the laboratory, then transplant cohorts of larvae of the appropriate age to cages at a gradient of distances from the diffuser. After a set amount of time, cages were collected and returned to the laboratory where the number of survivors and settlers were scored and any indication of bacterial infection noted; abalone larvae infected by bacteria are quite distinctive (even shortly after death), and we found no evidence of bacterial infection in any cage of any trial. Survivors that had not settled from each cage were then placed in a vial containing 10^{-6} molar of the inducer GABA in fresh seawater. Settlement, defined for this species as plantigrade attachment, in the laboratory was scored after 24 hours; in undamaged settlers, metamorphosis to the adult body form is completed by another 24 hours. Larvae that were exposed during the pre-competent stage were tested for their responsiveness to GABA only after reaching the age at which unexposed siblings became fully competent.

To control for "quality" (viability) of reared larvae, random samples from each batch were kept in the laboratory concurrent with each field trial and subjected to identical laboratory procedures as the outplant organisms. To prevent bacterial infection in the laboratory, antibiotics (2 mg/l rifampicin) were added to laboratory vessels holding field or laboratory (control) larvae; antibiotics do not interfere with normal settlement of red abalone larvae (Morse et al., 1979b). A handling control was done once by transporting two sets of competent larvae to the field and maintaining a third in the laboratory; one of the field sets was placed in the plume and the another was kept in fresh seawater on the boat; after 5.5 hours in the field, both field sets returned to the laboratory where all three groups were subjected to identical procedures to estimate survivorship and settlement. There was no statistical difference in either parameter among the three groups of larvae.

Each field cage was an 8 cm long, 7.6 cm diameter PVC tube capped with 100 μm Nytex screen. Between 500 - 1,000 larvae were placed in each cage and transported to and from the field in seawater. At each field station, one to three replicate cages were attached (by divers) on a buoy line approximately 1 m from the bottom (where abalone larvae are thought to occur). The number of stations in the gradient of distance from the diffuser differed slightly among experiments, but the maximum number included sites at 5, 10, 50, 100, 500, and 1,000 m west of the diffuser, and 5 and 1000 m to the east. Exposure durations included 4 day and 5.5 hour (using fewer stations) exposures. The longer trials were used to represent the likely exposure regime for larvae, such as abalone, that are capable of remaining in a local area, while the shorter exposure was used to represent highly transient larvae that passively drift through the plume. In 5.5 hrs, a passive particle would be transported about 2 kms (i.e., 1 km on each side of the outfall) under average current speed (12 cm/sec); the total "drift" distance in 5.5 hrs would be < 600 m when current speeds were ≤ 3 cm/sec.

To examine if effects differed between developmental stages, pre-competent and competent stage larvae were subjected to separate four day field exposures. Pre-competent larvae were transplanted at age three days and retrieved at age seven days (when siblings were developmentally competent to settle and in our experiments had not yet done so); this was designed to measure the probability that a larva exposed during the pre-competent stage would settle at an age when unexposed siblings were capable of doing so. Stations at 500 m west, 5 m east and 1000 m east were not used. Two replicate cages were placed at the remaining five stations, although one cage was lost at the 1000 m site during the experiment. Daily discharge monitoring records indicated that the average daily volume of produced water released during the course of this experiment was 2.49×10^6 l/day. A similar four day exposure experiment was conducted using competent larvae that were at least seven days old at the time of transplant. All stations were used for this trial, with one replicate cage per station. Many larvae settled and metamorphosed spontaneously in the cages while in the field. Consequently, survivorship, settlement of survivors and metamorphosis of settlers were scored immediately after retrieval from the field. The average volume of produced water discharged during the course of this experiment was 2.48×10^6 l/day.

Data analyses involved linear regressions of the dependent variable (proportion of cohort that (1) survived, (2) that settled given survivorship, and (3) that metamorphosed given settlement) as a function of (log) distance from the diffuser. When available, stations from both east and west of the diffuser were included in the analysis, although it is likely that the exposure regime differed between stations at the same distance on either side of the diffuser. Excluding the two east stations from the analyses did not alter statistical significance. The mean value for each replicate cage at a station was used as the variate, and because data were proportions, the data were arcsine transformed for statistical analysis.

A 5.5 hour exposure trial was conducted to determine (1) whether brief exposure had adverse effects and (2) whether effects differed between water depths. Competent larvae were transplanted to stations at 5 m and 1000 m west of the diffuser; control larvae were maintained in the laboratory as described above. At each station, three replicate cages were placed 1.5 m below the ocean surface, and three additional cages were attached to the buoy line 1.5 m above the bottom at a depth of approximately 9.5 m. The experiment was performed during a period of the year when the entire water column over the diffuser was well mixed (i.e., no stratification). Approximately 1,000 competent larvae were placed in each cage, placed at the assigned station and depth, then retrieved 5.5 hours later. After transport to the laboratory, larvae were exposed to 10^{-6} M GABA, and settlement success was scored after 24 hours. The volume of produced water discharged on the day of this experiment was 2.79×10^6 l.

Gross change in swimming behavior was examined by transplanting competent larvae to all stations along the gradient to the west of the diffuser. Approximately 500 - 1,000 larvae were placed in cages 1.5 m above the bottom, and retrieved 5.5 hours later. The number of replicate cages was either four (5 and 1,000 m west) or two (10, 50, 100, 500 m west). At the time of cage retrieval, seawater from the location of each cage also was collected, and larvae were transported back to the laboratory in the water collected from the specific outplant site. In the laboratory, the proportion of larvae that exhibited normal swimming behavior was scored for each cage. The order in which stations were examined was selected at random. The elapsed time between cage retrieval and scoring of behavior was 2 - 3 hours; hence, the maximum exposure of larvae was 7.5 - 8.5 hours.

Causation for Observed Effects

To test whether observed effects resulted from the discharge of produced water, a four day field test was conducted during a ten day period when the Carpinteria outfall was not releasing effluent. Three day old (pre-competent) larvae were transplanted to all eight stations four days after the discharge of produced water had ceased. At each station, three replicate cages each containing approximately 500 - 1,000 larvae were attached as before to a buoy line 1 m from the bottom. One cage was lost at the 10 m west station. Cages were collected after four days. On the day that cages were collected from the field, the processing facility discharged 3×10^4 l of seawater only through the outfall; commencement of produced water discharge did not occur until the day after larvae were retrieved. All subsequent laboratory analyses and procedures were identical to previous trials; survivorship was scored within 2 - 3 hours of retrieval, survivors were challenged with 10^{-6} *M* GABA, and the number of settlers was scored after 24 hours.

A short-term laboratory bioassay was performed to establish whether exposure to produced water would produce the same class of effects as observed in the field. Undiluted produced water was collected in an airtight glass bottle directly from a sampling spigot on the discharge pipe at the onshore processing facility, and laboratory experiments were initiated within 24 hours of collection. Initial concentrations (treatments) were 10%, 1%, 0.1%, 0.01% and 0% (control) produced water; there were three replicates of each treatment, and antibiotics (2 mg/l rifampicin) were added to each container. Approximately 100 larvae were placed in individual glass vials containing a total of 10 ml of seawater, produced water (at the appropriate concentration), and 10^{-6} *M* GABA. Settlement and survivorship were scored after 24 hours using methods described above.

Effects of produced water on swimming behavior was explored in the laboratory by placing 500 - 1000 larvae in one of three treatments: (1) in clean seawater (no produced water) for 96 hrs (control), (2) in 10% produced water for 96 hrs, and (3) in 10% produced water for 96 hrs followed by 24 hrs in clean seawater. Treatment (3) provides the time scale of behavioral recovery. There were three replicates of each treatment, and antibiotics (2 mg/l rifampicin) were added to each container

RESULTS

Field Experiments of Effects on Abalone Larvae

The pattern of survivorship in relation to the diffuser differed substantially between the two developmental stages (Fig. 1). Survivorship of pre-competent larvae increased with distance from the diffuser (Fig 1a; $r^2 = 0.63$; $F_{1,7} = 11.99$; $P < 0.02$); only about half of the initial cohort survived out to 10 m of the diffuser, and a substantial reduction (~ 30%) in survivorship was apparent up to 100 m from the outfall (Fig. 1a). By contrast, survivorship of competent abalone larvae was unaffected by distance from the diffuser (Fig. 1b; $r^2 = 0.001$; $F_{1,6} = 0.003$; $P > 0.95$). More than 80% of the competent larvae survived at all stations, even at 5 m from the outfall (Fig. 1b).

For individuals that had not settled in the field (no individuals exposed when pre-competent settled in the field in this or subsequent experiments) settlement ability as a function of distance transplanted from the diffusers differed considerably for larvae exposed during the pre-competent and competent developmental stages (Fig. 2). The settlement ability of surviving pre-competent (r^2

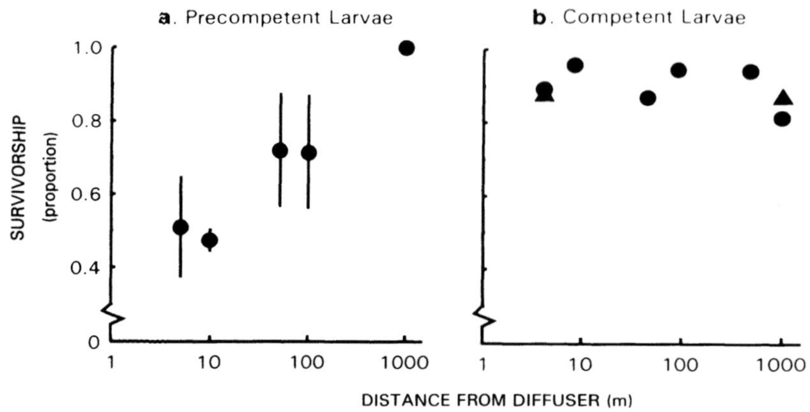

Figure 1. Survivorship of larvae exposed during the (a) pre-competent and (b) competent stage as a function of distance from an active produced water discharge. Data are the proportions of larvae that survived four days of exposure in the field; error bars in (a) indicate 1 SE from replicte cages. Circles are stations to the west of the diffuser, and triangles are stations to the east.

$= 0.79$; $F_{1,7} = 26.44$; $P < 0.001$) larvae was impaired, while that for competent larvae was not (Figs. 2a, 2b). However, for competent larvae, settlement in the field did vary with distance transplanted from the diffusers (Fig. 2c, $r^2 = 0.90$; $F_{1,6} = 51.49$; $P < 0.001$), with the magnitude of the effect decreasing with distance from the outfall. Reductions in settlement success were apparent as far away from the diffuser as 50 - 100 m (Fig. 2a, 2c). To the extent that it was examined, the spatial pattern of effects on settlement was qualitatively symmetrical for both sides of the diffuser (Fig. 2c); the two lowest settlement rates for competent larvae were observed at stations 5 m east and west of the outfall, whereas the highest rates occurred at sites 1 km east and west.

The fraction of settled larvae (in the field) that completed metamorphosis increased with distance from the diffuser (Fig. 3; $r^2 = 0.89$; $F_{1,6} = 46.20$; $P < 0.001$). The depression in successful

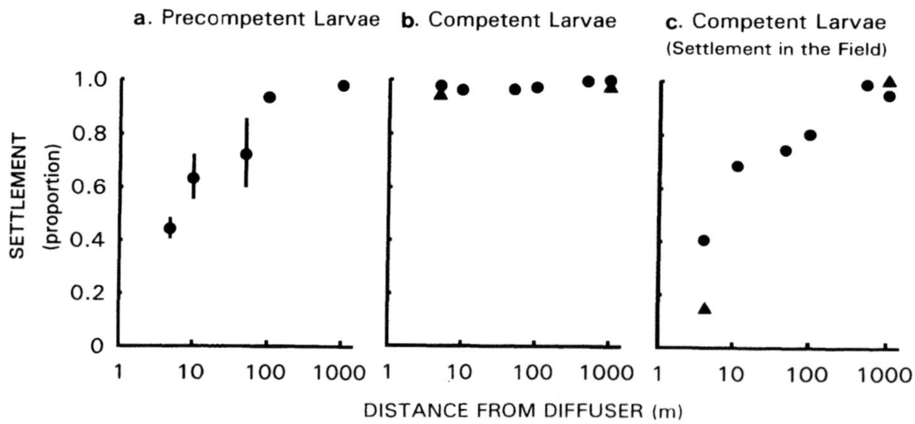

Figure 2. Settlement ability of larvae exposed during the (a) pre-competent or (b) competent stage as a function of distance from an active produced water discharge. The data in Figures 2a and 2b are the number of unsettled survivors that were able to settle in the laboratory in response to GABA. Figure 2c shows the settlement of competent larvae that occurred in the field as a function of distance from an active produced water discharge. See Figure 1 for symbols.

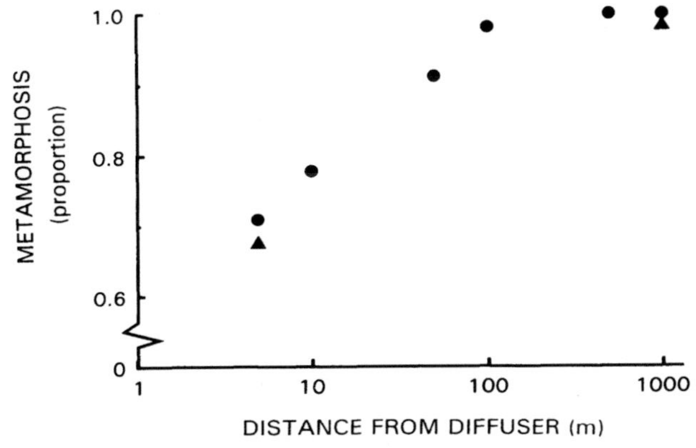

Figure 3. Metamorphosis of competent stage abalone larvae with distance from an active produced water discharge. Data are the proportion of settled larvae that completed metamorphosis after 4 days of exposure in the field. See Figure 1 for symbols.

metamorphosis was not as great as the effect on settlement (Figs. 3 vs 2c); the greatest reduction in metamorphosis, observed for sites 5 m east and west of the outfall, was approximately 35%. As with settlement, the spatial effects on metamorphosis were qualitatively similar on both sides of the diffuser (Fig. 3).

Effects on survivorship, settlement and, for competent larvae, metamorphosis were combined to determine the fraction of the initial cohorts of pre-competent and of competent larvae that ultimately can recruit to the adult habitat. This measure, which we termed viability, represents the proportion of individuals that remained viable following exposure, because those that fail at some point to complete metamorphosis eventually die. As expected, the fraction of initial pre-competent and of competent larvae that could successfully make the transition to the adult body form increased with distance from the diffuser (Fig. 4). For both developmental

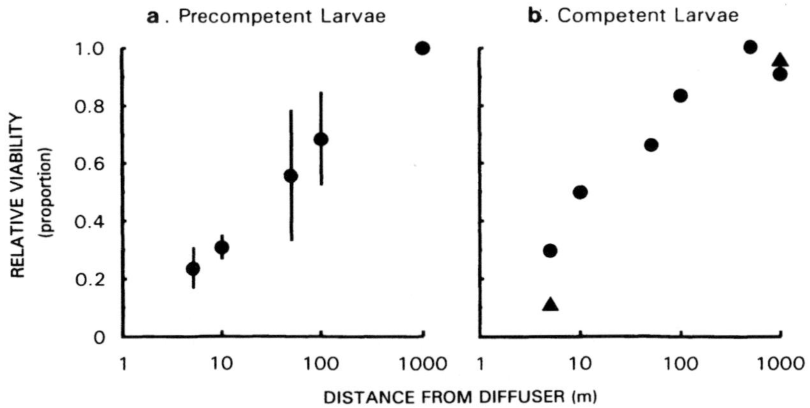

Figure 4. Reduction in viability of abalone larvae when exposed during the (a) pre-competent and (b) competent stage as a function of distance from an active produced water discharge. Data are proportion (standardized) of initial cohorts that eventually completed the planktonic phase. See Figure 1 for symbols.

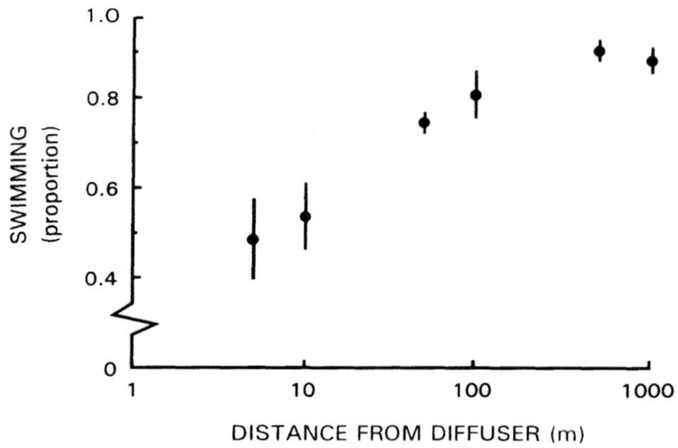

Figure 5. Swimming behavior of competent abalone larvae following 5.5 hours in the field as a function of distance from an active produced water discharge. Data are the mean proportion (± 1 SE) of larvae that were swimming within 3 hours of retrieval from the field.

stages, reductions of 20 - 30% were apparent as far as 100 m from the outfall (Fig. 4); the greatest depressions occurred at the 5 m sites, where nearly 80% of the initial cohorts failed to successfully make the transition to the adult body form (Figs. 4a & b). As before, the depression was qualitatively similar on both sides of the diffuser (Fig. 4b).

A second set of field experiments, involving just a 5.5 hour exposure, was performed to emulate the exposure regime that passively drifting larvae might encounter in two water strata: near the surface or above the bottom. The results indicate that the fraction of competent larvae that were able to settle was markedly reduced at the site near the diffuser (5 m) relative to that far (1000 m) away (Table 1; $F_{1,4} = 11.2$; $P < 0.03$). The brief encounter with the plume reduced the fraction of viable larvae > 15% near the diffuser. No difference in settlement was detected between shallow and deep water strata at either distance (Table 1; $F_{1,4} = 0.001$; $P > 0.95$; interaction between distance and depth: $F_{1,4} = 1.11$; $P > 0.35$). The overall proportion of the initial cohort that settled at the far site (95%) was not statistically different from the proportions observed in field and laboratory controls.

In addition to larval viability, swimming behavior of larvae was influenced by exposure to the produced water plume in the field. After less than nine hours of exposure, the fraction of larvae that were capable of swimming declined markedly with increased proximity to the diffuser (Fig. 5; $F_{5,10} = 6.77$; $P < 0.01$). The greatest decrease, seen at the site closest (5 m) to the diffuser, was > 45%. The data suggest that some reduction in the fraction of swimming larvae occurred to at least 100 m from the outfall (Fig. 5).

Table 1. Settlement success of larvae after 5.5 hr exposure in shallow (1.5 m) and deep (9.5 m) water depths near (5 m) to and far (1000 m) from an active produced water diffuser. Data are the mean (± 1 SE) proportion of larvae that settled in response to GABA

	Near (5 m)	Far (1000 m)
Shallow (1.5 m)	0.82 (0.10)	0.93 (0.02)
Deep (9.5 m)	0.77 (0.01)	0.97 (0.01)

Causation for Observed Effects

A four day field trial when no produced water was discharged revealed that pre-competent larvae were unaffected by proximity to the diffuser (Fig. 6). There was no evidence that survivorship (r^2 = 0.002; $F_{1,18}$ = 0.04; P > 0.85), or settlement after reaching the age of competence (r^2 = 0.006; $F_{1,18}$ = 0.11; P > 0.70) was reduced by proximity to the outfall. Overall, more than 90% of the larvae at all distances survived and settled.

Twenty-four hour exposure of competent stage larvae in the laboratory confirmed that produced water elicits the same responses observed in the field. The fraction of larvae that successfully settled in laboratory trails decreased with increasing concentration of produced water (Fig. 7; r^2=0.918, $F_{1,13}$ = 30.08; P > 0.0001). Depression in settlement was apparent even in the lowest concentration used (0.01%); the greatest reduction, which occurred in the highest concentration (10%) used, was > 30%. By contrast with effects on settlement, there was no mortality of larvae to any concentration of produced water used in the laboratory. These settlement and survivorship patterns of competent stage larvae to direct exposure to produced water paralleled results obtained in the field.

The issue remains as to whether the quantitative responses observed in the laboratory trial correspond to the depression in settlement of competent larvae observed in the field. As a first approximation, we assume that the dilution of produced water seen by larvae placed 5 m from the outfall was ~100:1 (i.e., 1%), which is the roughly theoretical initial dilution achieved by the diffuser design. If it is further assumed that the response to produced water increases linearly with duration of exposure, the observed depressions at the 5 m west site in the 5.5 hr and 96 hr field experiments yields an extrapolated reduction in settlement of 24.6% for a 24 h exposure. This predicted value is remarkably close to the 22.8% depression obtained in the 1% produced water treatment in the 24 hr laboratory trial. While we caution against making strong inference from this quantitative agreement, in general the laboratory results greatly strengthen confidence that effects on settlement observed in the field were caused by produced water.

Laboratory observations of larval behavior revealed that produced water resulted in complete cessation of swimming (Table 2). In the presence of 10% concentration of produced water, essentially all competent stage larvae stopped active swimming, and because of their negative buoyancy, sank to the bottom of the seawater containers. The effect on swimming behavior persisted long after larvae were placed back into clean seawater; only about half of the larvae previously exposed to 10% produced water had commenced swimming after 24 hours in

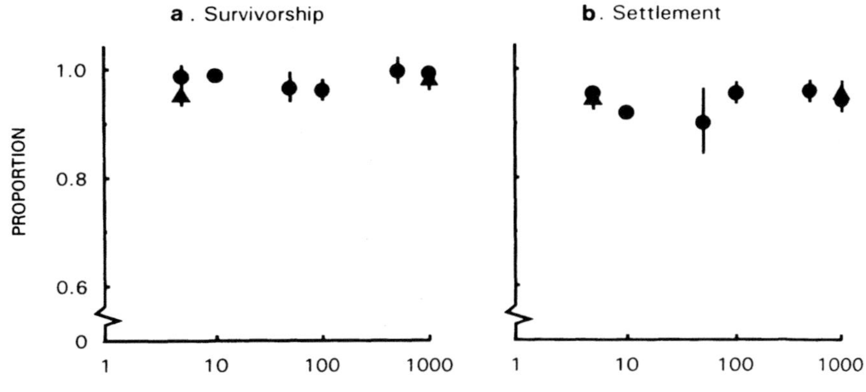

Figure 6. Survivorship (a) and settlement (b) of larvae placed in the field in the pre-competent stage when no produced water was being discharged. Data are the mean proportion (± 1 SE) of larvae that settled in response to the inducer GABA after 4 days in the field. See Figure 1 for symbols.

Table 2. Alteration of swimming behavior in laboratory exposure to produced water. Data are the mean (± 1 SE) proportion of larvae that were actively swimming in the absence of produced water, in 10% produced water after 96 hr exposure, and in clean seawater 24 hours after 96 hour exposure to 10% produced water. All treatment means are statistically different

	Proportion Actively Swimming	
Clean Seawater	10% PW	Clean Seawater after 10% PW
0.910 (0.02)	0.002 (0.002)	0.503 (0.044)

clean seawater (Table 2). The proportion of swimming larvae differed statistically among treatments ($F_{2,11} = 1367.37$; $P < 0.0001$).

DISCUSSION

Data presented here reveal that the chronic discharge of produced water into an open coast environment can have biologically meaningful effects on planktonic larvae. Exposures of red abalone larvae to a plume of produced water in the field reduced the probability that a larva would successfully complete the transition to the adult body form. The magnitude of this effect was directly related to distance from the diffuser and duration of exposure. The effect was most pronounced for longer exposures and for closer proximity to the outfall. Viability of larvae that resided within 10 m of the diffuser for four days was greatly reduced, with ~ 80% of the cohort failing to complete metamorphosis. Even brief exposure resulted in a sizable reduction in settlement success. In addition to direct effects on viability, produced water had an important effect on larval behavior that further reduces the likelihood of successfully completing the planktonic phase. Larvae exposed to produced water stopped swimming, and only return to normal behavior slowly after exposure ceased. Abalone larvae are negatively buoyant and must actively swim to remain in the water column; the consequence to a planktonic larva that prematurely sinks to the benthos is unknown, but it likely greatly reduces the probability of survival. The prevalence of these responses cannot be assessed, primarily because effects of produced water on planktonic organisms virtually have been ignored, and sublethal responses in general have not been thoroughly investigated (Neff, 1987).

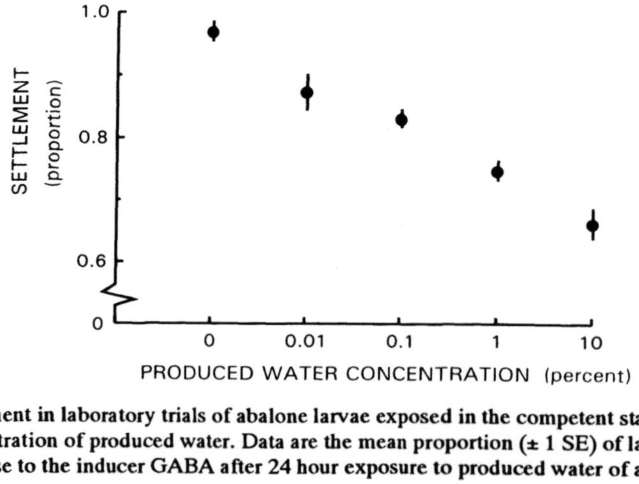

Figure 7. Settlement in laboratory trials of abalone larvae exposed in the competent stage as a function of concentration of produced water. Data are the mean proportion (± 1 SE) of larvae that settled in response to the inducer GABA after 24 hour exposure to produced water of a given dilution.

Developmental stages differed in their response to exposure to produced water. Survivorship for pre-competent larvae decreased with increasing proximity to the diffusers, while for competent individuals distance from the diffusers was unrelated to survivorship (Fig. 1). The most interesting difference in response between stages was for settlement (Fig. 2). Settlement ability decreased for pre-competent but not for competent larvae as a function of increasing proximity to the diffusers. Recall that this "ability" was defined as the probability that a live not settled individual brought back from the field and exposed to clean seawater containing GABA would settle. However, settlement *in the field* did decrease as a function of increasing proximity for competent larvae (Fig. 3c). Here, the larvae were exposed to affected water during the entire settlement process. One interpretation of these results is that for pre-competent larvae exposure to produced water physiologically destroys the ability to settle, and for larvae exposed after attaining the ability to settle such exposure interferes with settlement ability only for the duration of the exposure.

It is possible that some of the reduction in settlement success we observed for abalone larvae exposed during the pre-competent stage could have involved slowing of development, thereby increasing the time needed to become developmentally competent to settle. Krause et al. (1992), found that brief (10 minute) exposure of sea urchin (*Strongylocentrotus purpuratus*) gametes and longer (48 hr) exposure of zygotes to concentrations of Carpinteria produced water as low as 0.0001% significantly slowed embryonic development. Delayed developmental rates of other larvae have been reported from laboratory tests of petroleum hydrocarbon exposure (Wells, 1972; Katz, 1973; Wells and Sprague, 1976; Byrne and Calder, 1977; Caldwell et al., 1977; Mecklenburg et al., 1977; Winters et al., 1977; Laughlin et al., 1978; Cucci and Epifanio, 1979; Laughlin and Neff, 1979). With respect to effects on swimming of abalone larvae, our results suggest that larval ciliary action was reduced by exposure to produced water. It is known that certain types of heavy metals (e.g., strontium, barium) inhibit ciliary beating by altering normal activity of microtubules (e.g. Tamm, 1989, 1990). These classes of heavy metals occur in most produced waters (Neff, 1987), are relatively abundant in the Carpinteria effluent (Higashi et al., 1992), and the fraction including them was identified as responsible for most toxicity in the Carpinteria produced water (Higashi et al., 1992). Such heavy metals affect ciliary activity by competing with calcium. To reach the ciliary microtubules, such metals must compete with calcium for uptake through membranes. However, calcium is typically so much more abundant than any potential competing metal that it may be unlikely that this mechanism operates under natural conditions (e.g. in samples of undiluted produced water, calcium was 10 - 200 times more abundant than barium: P. Krause, personal communication).

With respect to field experiments such as ours, it is important to explore causation since factors that co-vary in space with the agent of interest (e.g., produced water plume) could be responsible for observed responses. Several lines of evidence increase our confidence that the responses we observed in the field for abalone larvae were caused by the plume of produced water. First, the magnitude of the effect increased with proximity to the diffuser, and to the extent that we examined it, the pattern of effect was qualitatively similar on both sides of the outfall. These data indicate that the diffuser location was the area of greatest impact. Second, larvae were unaffected when transplanted to the field during a brief period when no produced water was being discharged, regardless of proximity to the diffuser. These results strongly suggest that observed effects resulted from the operation of the discharge, and that the constituent(s) responsible did not have long residence times in the water column. Third, our short-term laboratory exposures of abalone larvae to graded concentrations of produced water yielded qualitatively similar effects to those obtained from the field. These data directly demonstrate that produced water does cause the same responses in abalone larvae as we observed in the field. Finally, Higashi et al. (1992) have found that the majority of toxicity from the Carpinteria produced water lies in the water-only soluble fraction, especially that containing divalent cations. The precise constituent(s) of produced water from Carpinteria responsible for observed effects on abalone larvae remain to be determined.

Taken together, the data presented here challenge prevailing assumptions (Rose and Ward, 1981; Middleditch, 1981b, 1984; Neff, 1987; Spies, 1987) that the discharge of produced water in coastal marine waters will not adversely affect organisms in the water column, and that the risk to planktonic species is especially unlikely for discharges in high energy, open coast environments. To our knowledge, the present study represents one of the first attempts to directly assess this issue using realistic field experiments. As our data suggest, water soluble constituents of produced water may not be diluted below biologically harmful levels as rapidly as currently believed (e.g. Montalvo and Brady, 1979; Rose and Ward, 1981; Middleditch, 1984). Effects on abalone larvae from 4 day outplants were apparent to somewhere between 100 - 500 m from the diffuser (there was no evidence for effects at 500 m). In longer (4 month) outplants of mussels (*Mytilus californianus* and *M. edulis*) at these stations, Osenberg et al. (1992) found a virtually identical spatial pattern on shell and tissue growth, and on allocation to reproductive tissues. By contrast with individual effects on mussels placed in the water column, Osenberg et al. (1992) found more localized effects of the Carpinteria discharge on the density of infaunal organisms. The study by Osenberg et al. (1992) suggests either that effects in the water column may occur over a larger area than those on the bottom, or that individual responses (e.g., growth) are more sensitive indicators of stress than changes at the population level (Carney, 1987).

It is widely appreciated that the potential spatial scale of biological effects arising from a point source discharge depends on the toxicity of the effluent, other characteristics of the effluent that affect mixing (e.g., specific density relative to receiving waters), the volume discharged, and local oceanographic conditions that influence dilution (e.g. Payne et al., 1987). Relatively little attention has been given to biological attributes of the organisms at risk. It is becoming increasing clear that early developmental stages such as larvae tend to be more sensitive to contaminants than older life stages (Wells and Sprague, 1976; Donahue et al., 1977; Linden 1978; Cucci and Epifanio, 1979; Sharp et al., 1979; Vashchenko, 1980; Capuzzo and Lancaster, 1981; Krause et al., 1992). However, there currently is little recognition that propagules of different species vary in attributes that influence their potential risk to discharge plumes. In particular, the duration of the planktonic stage and the distance dispersed from point of origin vary widely among benthic organisms (for review see Keough, 1988), yet affect both the probability that a larva will encounter a plume and the duration of exposure if a plume is encountered.

The distance a propagule disperses generally is related to the length of the planktonic stage, which varies tremendously among species. The duration and dispersal distance can be as short as ~ 4 minutes and 2 m from the parents, for example, the didemnid ascidian, *Diplosoma similis* (Stoner, 1990) to as long as one year and hundreds of kilometers, for example, the spiny lobster, *Panulirus interruptus* (Serfling and Ford, 1975; Engle, 1979). Jackson and Strathmann (1981) have argued that the longshore distance a larva moves is proportional to the duration of the pre-competent phase; larvae with long pre-competent periods are more likely to be transported offshore, necessitating a long competent period to increase the probability of returning to nearshore habitats suitable for settlement. However, not all larvae with moderately long planktonic periods disperse widely because of behaviors that limit transport. For example, larval dispersal distances of the abalone *Haliotis rubra* could be as small as 10 - 100 m, despite a planktonic stage of several weeks (Prince et al., 1987, 1988; also see Knowlton and Keller, 1986). Considerably more attention needs to be given to the characteristics of larvae in relation to oceanographic features that influence meso-scale transport of larvae (e.g. Kingsford, 1990; Kingsford et al. 1991; Shanks, 1983, 1985, 1986, 1988; Shanks and Wright, 1987) and perhaps pollutants (Shanks, 1987).

Population consequences to benthic populations of mortality of propagules from point source pollution rarely have been addressed (e.g. Capuzzo, 1987), although it often is assumed to be of minor ecological importance because natural mortality rates of these stages typically are prodigious, and the fraction of propagules affected by human disturbances typically is considered small. However, as discussed above, the fraction of propagules for a given population that potentially is affected by an outfall, for example, increases as the dispersal distance of the propagules approaches the spatial area influenced by a plume. Understanding the spatial extent of local (sub) populations (determined by dispersal linkage among sub-populations) and the rate of supply of propagules from other local areas is important in assessing potential ecological effects of point source discharges (Keough and Black, 1993). Effects of propagule mortality on the density and other characteristics of local populations should scale inversely with the supply rate of propagules from unaffected areas, and directly with the fraction of larvae killed.

A central but largely unanswered question is the extent to which mortality of propagules from a human source is in addition to, rather than a substitute for, natural deaths, that is, the degree of compensation (Nisbet et al., 1993). There is growing evidence that the local abundance of many marine populations reflects the supply rate of new propagules (e.g. Connell, 1985; Victor, 1986; Roughgarden et al., 1987; Keough, 1988; Mapstone and Fowler, 1988; Underwood and Peterson, 1988; Underwood and Fairweather, 1989; Hughes, 1990; Jones, 1990; Raimondi, 1990), and that reductions in recruitment from point source pollution can reduce local abundances of benthic population (e.g. Bence et al., 1993). Reductions in densities of benthic populations caused by diminished recruitment are most likely to be detected for species with limited dispersal abilities and which are in the vicinity of an outfall; recruitment to the local population may not be sufficient to balance adult mortality (e.g. Bence et al., 1993), especially if the outfall also increases adult death rates. However, large losses of propagules from a point source can affect abundances of species that disperse widely (Murdoch et al., 1989; Nisbet et al., 1993). In these cases, however, we lack the ability to detect such losses because they are distributed over a large area. For example, Murdoch et al. (1989) and Nisbet et al. (1993) calculated that the death of widely dispersing fish larvae by the cooling system of a single coastal power plant in California reduced adult standing stocks of certain fish species (e.g., queenfish, *Seriphus politus*) in the entire Southern California Bight by as much as 13% (Murdoch et al., 1989; Nisbet et al., 1993). Thus, while not detectable at either the local or "global" scale, local impacts on propagules can have serious, large-scale effects.

There is far too little information to assess whether the discharge of produced water affects populations of benthic organisms by reducing the supply of planktonic propagules. The intent of the present study was to address a key assumption regarding the low likelihood of effects on meroplanktonic forms. Our results indicate that planktonic larvae can be adversely affected by produced water plumes, even from discharges in high energy, open coast environments. As such, these data draw attention to the need for more complete assessments of both the risk to meroplankton and associated population consequences from the discharge of produced water. Such endeavors should include explicit consideration of the biology of propagules.

ACKNOWLEDGMENTS

We thank G. Cherr, S. Holbrook, M. Keough, P. Krause, A. Martinez, D. Morse, F. Piltz, C. Osenberg, D. Reed, D. Steichen and two anonymous reviewers for useful discussions and criticisms. D. Morse graciously provided larvae and access to his laboratory. We thank J. Wallace and his colleagues at the Carpinteria plant for facilitating collection of produced water samples, and F. Piltz of the Minerals Management Service for support of the UC/MMS Southern California Educational Initiative. D. Canestro, D. Reed, A. Martinez, and T. Kaltenberg assisted in the field, and B. Williamson provided technical assistance. This research was funded by the UC Coastal Toxicology Program, and by the Minerals Management Service, U.S. Department of Interior under MMS Agreement No. 14-35-0001-3071. The views and conclusions in this paper are those of the authors and should not be interpreted as necessarily representing the official policies, either express or implied, of the U.S. Government.

REFERENCES

Armstrong, H.W., Fucik, K., Anderson, J., Neff, J.M., 1979, Effects of oilfield brine effluent on sediments and benthic organisms in Trinity Bay, Texas, Mar. Environ. Res. 2:55-69.

Bence, J.R., Stewart-Oaten, A. and Schroeter, S.C., 1993, Estimating the size of an effect from a before-after/control-impact pairs design: The predictive approach applied to a power plant study, *in*: Ecological Applications, submitted.

Byrne, C.J. and Calder, C.J., 1977, Effect of the water-soluble fractions of crude, refined, and waste oils on the embryonic and larval stages of the Quahog clam *Mercenaria* sp. Mar. Biol. 40:225-231.

Caldwell, R.S., Calderone, E.M. and M.H. Mallon, 1977, Effects of seawater-soluble fraction of Cook Inlet crude oil and its major aromatic components on larval stages of the Dungeness crab, *Cancer magister* Dana, *in*: "Fate and Effects of Petroleum Hydrocarbon in Marine Ecosystems and Organisms," D.A. Wolfe, ed., Pergamon Press, New York.

Capuzzo, J.M., 1987, Biological effects of petroleum hydrocarbons: assessments from experimental results, *in*: "Long-Term Environmental Effects of Offshore Oil and Gas Development," D.F. Boesch and N.N. Rabalais, eds., Elsevier Applied Science, NY.

Capuzzo, J.M. and Lancaster, B.A., 1981, Physiological effects of South Louisiana crude oil on larvae of the American lobster (*Homarus americanus*), *in*: "Biological Monitoring of Marine Pollutants," F.J. Vernberg, A. Calabrese, F.P. Thurberg and W.B. Vernberg, eds., Academic Press, NY.

Carney, R.S., 1987, A review of study designs for the detection of long-term environmental effects of offshore petroleum activities, *in*: "Long-Term Environmental Effects of Offshore Oil and Gas Development," D.F. Boesch and N.N. Rabalais, eds., Elsevier Applied Science, New York.

Connell, J.H., 1985, The consequences of variation in initial settlement vs. post-settlement mortality in rocky intertidal communities, J. Exp. Mar. Biol. Ecol. 93:11-45.

Cox, K., 1962, California abalones, family Haliotidae, Calif. Fish & Game Fish. Bull. 118:1-133.

Cucci, T.L. and Epifanio, C.E., 1979, Long-term effects of water-soluble fractions of Kuwait oil on the larval and juvenile development of the mud crab *Eurypanopeus depressus*, Mar. Biol. 55:215-220.

Donahue, W.H., Wang, R.T., Welch, M., and Nicol, J.A.C., 1977, Effects of water-soluble components of petroleum oils and aromatic hydrocarbons on barnacle larvae, Environ. Pollut. 13:187-202.

Engle, J.M., 1979, Ecology and growth of juvenile California spiny lobster, *Panulirus interruptus* (Randall), SeaGrant Dissertation Series, Los Angeles, CA.

Gable, C.G., 1981, "Report on Data from the Nearshore Sediment Transport Study Experiment at Santa Barbara, February 1980," Institute of Marine Research Number 80-5, University of California, San Diego.

Grahl-Nielsen, O., Westerheim, K., and Wilhelmsen, S., 1979, Petroleum hydrocarbons in the North Sea, *in*: "Proceedings 1979 Oil Spill Conference," American Petroleum Institute, Wash,, D.C.

Higashi, R., Cherr, G., Bergens, C., Fan, T., and Crosby, D, 1992, Toxicant isolation from a produced water source in the Santa Barbara Channel, *in*: "Produced Water: Technological/Environmental Issues and Solutions," J.P. Ray and F.R. Engelhardt eds., Plenum Publishing Corp., New York, New York, USA.

Howarth, R.W., 1989, Determining the ecological effects of oil pollution in marine ecosystems, *in*: "Ecotoxicology: Problems and Approaches," S.A. Levin, M.A. Harwell, J.R. Kelly, K.D. Kimball, eds, Springer-Verlag, New York.

Hughes, T.P., 1990, Recruitment limitation, mortality, and population regulation in open systems: a case study, Ecology 71:12-20.

Jackson, G.A. and Strathmann, R.R., 1981, Larval mortality from offshore mixing as a link between precompetent and competent periods of development, Amer. Nat. 118:16-26.

Jones, G.P., 1990, The importance of recruitment to the dynamics of a coral reef fish population, Ecology 71:1691-1698.

Katz, L.M., 1973, The effects of water-soluble fractions of crude oil on larvae of the decapod crustacean *Neopanope texana* (Say), Environ. Pollut. 5:199-204.

Keough, M.J., 1988, Benthic populations: is recruitment limiting or just fashionable?, *in*: "Proceedings of the 6th International Coral Reef Symposium, Australia, Vol. 1: Plenary Addresses and Status Reviews," J.H. Choat, et al., eds, 6th International Coral Reef Symposium, Townsville, Aus.

Keough, M.J. and Black, K.P., 1993, Predicting the scale of marine impacts: understanding planktonic links between populations, *in*: "The Design of Ecological Impact Assessment Studies: Conceptual Issues and Application," R.J. Schmitt and C.W. Osenberg, eds., submitted.

Kingsford, M.J., 1990, Linear oceanographic features: a focus for research on recruitment processes, Austral. J. Ecol. 15:391-401.

Kingsford, M.J., Wolanski, E., Choat, J.H., 1991, Influence of tidally induced fronts and Langmuir circulations on distribution and movements of presettlement fishes around a coral reef, Mar. Biol. 109:167-180.

Kingsford, M.J., 1993, Influence of pollutants and oceanography on abundance and deformities of wild fish larvae, *in*: "The Design of Ecological Impact Assessment Studies: Conceptual Issues and Application," R.J. Schmitt and C.W. Osenberg, eds., submitted.

Knowlton, N. and Keller, B.D., 1986, Larvae which fall short of their potential: highly localized recruitment in a alpheid shrimp with extended larval development, Bull. Mar. Sci. 39:213-223.

Krause, P.R., Osenberg, C.W., and Schmitt, R.J., 1992, Effects of produced water on early life stages of a sea urchin: gender-specific responses and delayed expression," *in*: "Produced Water: Technological/Environmental Issues and Solutions," J.P. Ray and F.R. Engelhardt, eds., Plenum Publishing Corp., New York, New York, USA.

Laughlin, R.B., Jr., and Neff, J.M., 1979, Interactive effects of salinity, temperature and polycyclic aromatic hydrocarbons on the survival and development rate of larvae of the mud crab *Rhithropanopeus harrisii*, Mar. Biol. 53:281-291.

Laughlin, R.B., Jr., Young, L.G.L., and Neff, J.M., 1978, A long-term study of the effect of water-soluble fractions of No. 2 fuel oil on the survival, development rate, and growth of the mud crab *Rhithropanopeus harrisii*, Mar. Biol. 47:87-95.

Linden, O., 1978, Biological effects of oil on early development of the Baltic herring *Clupea harengus membras*, Mar. Biol. 45:273-283.

Mapstone, B.D., and Fowler, A.J., 1988, Recruitment and the structure of assemblages of fish on coral reefs, Tree 3:72-77.

Mecklenburg, T.A., Rice, S.D., and Karinen, J.F., 1977, Molting and survival of king crab (*Paralithoides camtschatica*) and coonstripe shrimp (*Pandalus hypsinotus*) larvae exposed to Cook Inlet crude oil water-soluble fraction, *in*: "Fate and Effects of Petroleum Hydrocarbons in Marine Ecosystems and Organisms," D.A. Wolfe, ed., Pergamon Press, New York.

Middleditch, B.S., ed., 1981a, "Environmental Effects of Offshore Oil Production. The Buccaneer Gas and Oil Field Study," Plenum Press, New York.

Middleditch, B.S., 1981b, Hydrocarbons and sulfur, *in*: "Environmental Effects of Offshore Oil Production. The Buccaneer Gas and Oil Field Study," Mittleditch, B.S., ed., Plenum Press, New York.

Middleditch, B.S., 1984, "Ecological Effects of Produced Water Discharges from Offshore Oil and Gas Production Platforms. Final Report on API Project No. 248," American Petroleum Institute, Wash., D.C.

Montalvo, J.G.,Jr., and Brady, V., 1979, Concentrations of Hg, Pb, Zn, Cd and As in Timbalier Bay and the Louisiana oil patch, *in*: "The Offshore Ecology Investigation: Effects of Oil Drilling and Production in a Coastal Environment," C.H. Ward, M.E. Bender, and D.J. Reish, eds., Rice University Studies, Texas.

Morse, D.E., 1990, Recent progress in larval settlement and metamorphosis: closing the gap between molecular biology and ecology. Bulletin of Marine Science 46:465-483.

Morse, D.E., Hooker, N., Duncan, H., and Jensen, L., 1979a, γ-Aminobutyric acid, a neurotransmitter, induces planktonic abalone larvae to settle and begin metamorphosis, Science 204:407-410.

Morse, D.E., Hooker, N., Jensen, L., and Duncan, H., 1979b, Induction of larval abalone settling and metamorphosis by γ-aminobutyric acid and its congeners from crustose red algae: II. Applications to cultivation, seed production and bioassays: principle causes of mortality and interference, Proc. World Maricult. Soc. 10:91-91.

Morse, D.E., Duncan, H., Hooker, N., Baloun, A., and Young, G., 1980, GABA induces behavioral and developmental metamorphosis in planktonic molluscan larvae. Fed. Proc. 39:3237-3241.

Murdoch, W.W., Fay, R.C., and Mechalas, B.J., 1989, "Final Report of the Marine Review Committee to the California Coastal Commission" Marine Review Committee, Santa Barbara, CA.

Neff, J.M., 1987, Biological effects of drilling fluids, drill cuttings and produced waters, *in*: "Long-Term Environmental Effects of Offshore Oil and Gas Development," D.F. Boesch and N.N. Rabalais, eds., Elsevier Applied Science, New York.

Nisbet, R.M., Murdoch, W.W., and Stewart-Oaten, A., 1993, Consequences for Adult Fish Stock of Human Induced-Mortality Immatures; Ecological Application, submitted.

Osenberg, C.W., Schmitt, R.J., and Holbrook, S.J., 1992, Spatial scale of ecological effects associated with an open coast discharge of produced water, *in*: "Produced Water: Technological/Environmental Issues and Solutions," J.P. Ray and F.R. Engelhardt, eds., Plenum Publishing Corp., New York, New York, USA.

Payne, J.R., Phillips, C.R., and Hom, W., 1987, Transport and transformations: water column processes, *in*: "Long-Term Environmental Effects of Offshore Oil and Gas Development," D.F. Boesch and N.N. Rabalais, eds., Elsevier Applied Science, New York.

Prince, J.D., Sellers, T.L., Ford, W.B., and Talbot, S.R., 1987, Experimental evidence for limited dispersal of haliotid larvae (genus *Haliotis*: Mollusca: Gastropoda), J. Exp. Mar. Biol. Ecol. 106:243-263.

Prince, J.D., Sellers, T.L., Ford, W.B., and Talbot, S.R., 1988, Confirmation of a relationship between the localized abundance of breeding stock and recruitment for *Haliotis rubra* Leach (Mollusca: Gastropoda), J. Exp. Mar. Biol. Ecol. 122:91-104.

Raimondi, P.R., 1990, Patterns, mechanisms, consequences of variability in settlement and recruitment of an intertidal barnacle, Ecol. Monogr. 60:283-309.

Rose, C.D., and Ward, T.J., 1981, Acute toxicity and aquatic hazard associated with discharge formation water, *in*: "Environmental Effects of Offshore Oil Production. The Buccaneer Gas and Oil Field Study," B.S. Middleditch, ed., Plenum Press, New York.

Roughgarden, J., Gaines, S., and Pacala, S.W., 1987, Supply-side ecology: the role of physical transport processes, *in*: "Organization of communities past and present," J.H.R. Gee and P.S. Giller, eds, Blackwell Scientific Press, Oxford, England.

Serfling, S.A., and Ford, R.F., 1975, Ecological studies of the puerulus larval stage of the California spiny lobster *Panulirus interruptus* (Randall), Fish. Bull. Natl. Mar. Fish. Serv. 73:360-377.

Sharp, J.R., Fucik, K.W., and Neff, J.M., 1979, Physiological bases of differential sensitivity of fish embryonic stages to oil pollution, *in*: "Marine Pollution: Functional Responses," W.B. Vernberg, A. Calabrese, F.P. Thurberg, and F.J. Bernberg, eds., Academic Press, New York.

Shanks, A.L., 1983, Surface slicks associated with tidally forced internal waves may transport pelagic larvae of benthic invertebrates and fishes shoreward, Mar. Ecol. Prog. Ser. 13:311-315.

Shanks, A.L., 1985, Behavioural basis of internal-wave-induced shoreward transport of megalopae of the crab *Pachygrapsus crassipses*, Mar. Ecol. Prog. Ser. 24:289-295.

Shanks, A.L., 1986, Vertical migration and cross shelf dispersal of larval *Cancer* spp. and *Randallia ornata* (Crustacea: Brachyura) off the coast of southern California, Mar. Biol. 92:189-199.

Shanks, A.L., 1987, The onshore transport of an oil spill by internal waves, Science 235:1198-1200.

Shanks, A.L., 1988, Further support for the hypothesis that internal waves can cause shoreward transport of larval invertebrates and fish, Fish. Bull. 86:703-714.

Shanks, A.L., and Wright, W.G., 1987, Internal-wave-mediated shoreward transport of cyprids, megalopae, and gammarids and correlated longshore differences in the settling rate of intertidal barnacles, J. Exp. Mar. Biol. Ecol. 114:1-13.

Spies, R.B., 1987, The biological effects of petroleum hydrocarbons in the sea: assessments from the field and microcosms, *in*: "Long-Term Environmental Effects of Offshore Oil and Gas Development," D.F. Boesch and N.N. Rabalais, eds., Elsevier Applied Science, New York.

Stoner, D.S., 1990, Recruitment of a tropical colonial ascidian: relative importance of pre-settlement versus post-settlement processes, Ecology 71:1682-1690.

Tamm, S.L., 1989, Control of reactivation and microtubule sliding by calcium strontium and barium in detergent-extracted macrocilia of *Beore*, Cell. Motil. Cytoskeleton 12:104-112.

Tamm, S.L. and Tamm, S., 1990, Calcium, barium and strontium-induced conformational changes of ciliary axonemes, Cell. Motil. Cytoskeleton 17:187-196.

Underwood, A.J., and Fairweather, P.G., 1989, Supply-side ecology and benthic marine assemblages, Tree 4:16-19.

Underwood, A.J., and Peterson, C.H., 1988, Towards an ecological framework for investigating pollution, Mar. Ecol. Prog. Ser. 46:227-234.

Vashchenko, M.A., 1980, Effects of oil pollution on the development of sex cells in sea urchins, Helgolander Meeresunters. 33:297-300.

Victor, B.C., 1986, Larval settlement and juvenile mortality in a recruitment-limited coral reef fish population, Ecol. Monogr. 56:145-160.

Weis, P., Weis, J.S., Greenburg, A., 1989, Treated municipal wastewaters: effects on development and growth of fishes, Mar. Envir. Res. 28:527-532.

Wells, P.G., 1972, Influence of Venezuelan crude oil on lobster larvae, Mar. Pollut. Bull. 3:105-106.

Wells, P.G., and Sprague, J.B., 1976, Effects of crude oil on American lobster (*Homarus americanus*) larvae in the laboratory, J. Fish. Res. Board Can. 33:1604-1614.

Wheeler, R.B., Anderson, J.B., Schwarzer, R.R., and Hokanson, C.L., 1980, Sedimentary processes and trace metal contaminants in the Buccaneer oil/gas field, northwestern Gulf of Mexico, Environ. Geol. 3:163-175.

Winters, K., VanBaalen, C., and Nicol, J.A.C., 1977, Water soluble extractives from petroleum oil: chemical characterization and effects on microalgae and marine animals, Rapp. P.- V. Reun. Cons. Int. Explor. Mer. 171:166-174.

EFFECTS OF PRODUCED WATER ON EARLY LIFE STAGES OF A SEA URCHIN: STAGE-SPECIFIC RESPONSES AND DELAYED EXPRESSION

P. R. Krause[1,2], C. W. Osenberg[1] and R. J. Schmitt[1]

[1] Coastal Research Center, Marine Science Institute

[2] Department of Biological Sciences
University of California
Santa Barbara, California
USA

INTRODUCTION

Common objectives of many bioassays are to provide a measure of relative toxicity of various effluents (e.g., Allen, 1971; Kobayashi, 1980; National Research Council, 1983), or to identify components in a given effluent that are most responsible for biological effects (e.g., Higashi et al., 1992). There is an indisputable need for such perspectives for produced waters from different coastal formations (Neff, 1987). However, while such a focus can provide a useful toxicity benchmark, it is not particularly well suited for exploring the nature, extent and generality of biological responses. For example, there is an emerging paradigm that earlier developmental stages of a marine organism are more sensitive to a pollutant than later ones (e.g., National Research Council, 1983; Neff et al., 1976), yet no consistent pattern emerges from the relatively few studies that have examined different life stages of the same species (e.g., Neff et al., 1976; Rossi and Anderson, 1978; see Capuzzo, 1987). Even when multiple life stages of a species have been considered, it is difficult to extend laboratory-based results to natural situations, in part because of uncertainty as to whether the concentrations and exposure durations used are realistic (e.g., Castagna et al., 1981; Neff, 1987).

Another feature of bioassays, and one that has not received a great deal of attention, concerns the endpoint (response variable) chosen to measure an "effect." For a number of logical and logistical reasons, the response variable selected often is a biological "milepost" that coincides with the end of an exposure period. For example, fertilization success often is used as the response variable to estimate effects of toxicant exposure on gametes (e.g., Kobayashi, 1971; Dinnel et al., 1987). However, the extent of deleterious consequences of exposure of gametes will be underestimated if additional effects are manifested at a later developmental stage for successfully fertilized zygotes. Delayed expression of effects, while known to occur in marine organisms (e.g., Castagna et al., 1981; Pagano et al., 1982, 1983), has not been explored systematically, and we therefore lack an appreciation for its potential importance. For similar reasons, we know little about the combined effects that might arise from repeated exposure of an individual in different life stages. To facilitate meaningful biological interpretation of laboratory bioassays, the challenge is to match the laboratory protocol with reasonable estimates of which life stages are at risk, the concentrations of contaminant they are likely exposed to, and the duration of the exposure under field conditions.

With respect to general classes of lethal and sublethal effects, one of the most important from an ecological perspective is impaired reproductive success of adults. If early life history stages are relatively more sensitive to toxicants than adults, then major effects on the reproductive success of adults could be mediated through effects on early stages of their progeny. For many benthic marine organisms, gametes are released directly into the water column where they undergo syngamy and form zygotes. Thus, important impacts on reproductive success of adults living near a point source outfall might be mediated through early exposure of their gametes and developing progeny. In the present study, we explore whether and how brief exposure to a range of concentrations of produced water affect gametes and early larval stages of the purple sea urchin, *Strongylocentrotus purpuratus*. In particular, we exposed separately and together eggs, sperm, and zygotes to ascertain (a) the relative sensitivities of these life stages to produced water at durations and concentrations realistic to each stage, (b) the nature of the biological responses, and (c) the potential for delayed expression.

We chose the purple sea urchin as our model organism, in part, because there is an extensive body of toxicological work on various life stages of sea urchins (e.g., Cherr et al., 1987; Dinnel et al., 1981, 1982, 1987, 1989; Hagstrom and Lonning, 1973; Hose, 1985; Kobayashi, 1971; McGibbon and Moldan, 1986; Oshida et al., 1981; Pagano et al., 1982, 1983, 1986). Furthermore, the purple sea urchin is representative of many marine organisms in that benthic adults broadcast eggs and sperm into the water column where fertilization and subsequent larval development occur. Purple sea urchin larvae typically develop into the pluteus stage within 48 hours. Up through the late pluteus (~72 hr), larvae are sustained by egg reserves and do not feed (Strathmann, 1987).

METHODS

Adult purple sea urchins were collected from a pristine reef near Santa Barbara, CA, and maintained in large, continuous flow sea water tanks located outdoors. Animals were supplied *ad libitum* with freshly collected giant kelp (*Macrocystis pyrifera*), a favored food item. For each experiment, gametes were collected from 4 - 6 adults of each sex within 15 minutes following intracoelomic injection of 0.5 - 1.0 ml of 0.5 *M* KCl (Strathmann, 1987). Eggs were cleaned by passing them twice through a 200 µm Nitex mesh and allowing them to settle through 100 ml of filtered sea water. Sperm were collected "dry" (i.e., without sea water, which metabolically activates sperm [Timourian and Watchmaker, 1970]) by pipetting the gametes from the dried surface of a male (Dinnel et al., 1987). Both gamete types were stored on ice until used in a trial, which occurred no later than 3 hours after gamete collection. Separate "control" trials were conducted to determine whether the delay between collection and use (and therefore manner of storage) altered the viability of eggs or sperm; results indicated no difference in viability of eggs or "dry" collected sperm that were stored for three hours compared with freshly released gametes.

Undiluted produced water used for laboratory tests was collected from an oil processing facility in Carpinteria, CA. Samples were collected from an onshore test spigot located on the discharge pipe just after the effluent leaves the final settling tank and just prior to ocean discharge. Produced water was collected without head space in clean amber glass bottles with Teflon lined lids. Samples were maintained on ice and in the dark until experiments were begun within approximately six hours of collection.

The general laboratory protocol involved cross-designed experiments in which produced water concentration (1, 0.01 and 0.0001%) was crossed with a second factor designating the life stage(s) that was exposed (e.g., eggs only, sperm only, zygote only, and/or a combination of these treatments). A control treatment was included and consisted of unexposed life stages kept in filtered sea water but otherwise handled identically to the other treatments. Produced water concentrations were made by serial dilution of the raw effluent with filtered sea water in volumetric labware (Allen, 1971). All sea water was filtered through a 0.45 µm autoclaved filter. In all experiments, there were three replicates of each treatment, and each replicate typically involved approximately 500 eggs. The incubation units were 15 ml plastic Falcon multi-well culture dishes into which the sperm and eggs were introduced. All experiments were conducted in a cold (15° C) room on a continuous shaker table.

Ten minute gamete exposures were accomplished by adding sperm (or eggs) to a glass test tube containing a given concentration of produced water. Unexposed gametes were added to filtered sea water for 10 minutes. Following the 10 minute exposures, gametes were incubated together in 10 ml of filtered sea water. To assess fertilization success, replicates were fixed with 10% formalin 25 minutes after gametes had been added to the incubation units. Fertilization success was estimated by microscopic identification of the fertilization membrane.

In experiments that examined zygote performance, we manipulated zygote exposure by varying the concentration of produced water in which sperm and eggs were incubated (i.e., exposed eggs and sperm were introduced into incubation units that contained a gradient of produced water concentrations). Developing zygotes remained in these units for up to 96 hours. Thus, our "zygote exposures" included brief exposures of gametes during the period between introduction into the incubation chamber and fertilization. After a specified time, replicates were fixed with 10% formalin, and 100 individuals from each replicate were inspected and their developmental stage noted. Modification to these general protocol are noted below.

To facilitate comparisons among experiments and to avoid polyspermy, we standardized fertilization rates by adjusting the ratio of sperm to eggs (Cherr et al., 1987; Dinnel et al., 1987). Prior to each experiment, batches of eggs were flooded with varying amounts of a stock sperm solution to establish the relationship between fertilization success and the ratio of sperm to eggs. Each experiment was then conducted at the sperm:egg ratio that produced 90% fertilization in filtered sea water.

To prevent microbial contamination, 0.05% Penicillin G was added to each replicate in all experiments. An initial experiment was conducted to establish whether the added antibiotic altered normal fertilization or development, and whether there was a synergistic response of penicillin and produced water. The experiment involved two factors: the presence/absence of penicillin (0 or 0.05%) was crossed with the presence/absence of produced water (0 or 10%). Eggs and sperm were separately exposed to produced water (or filtered sea water) for 10 minutes and then incubated together in filtered sea water. Two endpoints were measured: (a) the proportion of successfully fertilized eggs (as measured by the presence of the fertilization membrane) after 25 minutes of adding sperm to the eggs, and (b) the proportion of embryos that had reached the pluteus stage after 48 hours.

The first main experiment explored the effects of produced water on fertilization success. Produced water concentrations used were 1, 0.01, and 0.0001%. Three exposure regimes were crossed with the produced water treatment: sperm exposed but eggs unexposed, eggs exposed but sperm unexposed, and both sperm and egg exposed. A 0% gamete exposure (filtered sea water only) served as the control, and all incubations were done in filtered sea water.

The second main experiment addressed (a) whether effects from the exposure of gametes might be delayed to a later developmental stage, (b) the relative sensitivities of eggs, sperm and zygotes, and (c) the magnitude of effects arising from combined exposure of different life stages. Unlike the first experiment, the incubation period (following the 10 minute exposure of gametes) was increased from 25 minutes to 48 hours. In addition, there were five exposure treatments: 1) sperm (i.e., eggs and zygotes unexposed); 2) eggs; 3) zygotes; 4) sperm and eggs; 5) sperm, eggs and zygotes. After 48 hours, each replicate was fixed in 10% formalin and the proportion of embryos that had reached the pluteus stage was determined microscopically.

In this experiment, the response parameter (proportion of larvae in the pluteus stage) was based on the relative developmental states of only the eggs and larvae that were still intact after 48 hours. Furthermore, the endpoint of this experiment was time dependent (i.e., assessed after only 48 hours). Thus, it was not possible to separate the relative contributions of mortality from delayed development. Consequently, a third experiment was conducted to estimate directly the separate effects of produced water on embryo survivorship and developmental rates. To do this, initial cohorts of eggs were counted and the fates of all initial eggs followed through time. Treatments consisted of exposure to 1% produced water of eggs only, sperm only, and of zygotes. A control treatment consisted of exposure to filtered sea water only. The number of eggs in each replicate was known at the beginning of the experiment, and eighteen "copies" of each treatment were initiated; at specified intervals, three replicates per treatment were fixed for later analysis. The intervals were 25 min, 12 hr, 24 hr, 48 hr, 72 hr and 96 hr. Mortality (1 minus survivorship) was estimated as the number of "dead" eggs or embryos divided by the number of eggs initially present. "Dead" eggs and embryos were defined as those missing or obviously dead (i.e., damaged or broken). The numbers of individuals in the pluteus stage were counted at each sampling time and developmental success expressed as the proportion of the initial cohort that had entered the pluteus stage.

In general, results of each experiment were analyzed by analysis of variance. Comparisons with the control treatment were made using Dunnett's two-tailed *t* test, and *a posteriori* comparisons among treatment main effects were made using the Ryan-Elinot-Gabriel-Welsch multiple *F* test (REGWF: SAS, 1988). All data were arcsine-square root transformed prior to analysis.

RESULTS

Methodological Tests of Penicillin Effects

Before considering effects of produced water, we first present evidence that the addition of penicillin to control bacterial contamination does not obscure interpretation of laboratory experiments. Penicillin G had no detectable effect on the proportion of eggs that were successfully fertilized (Table 1; $F_{1,8} = 1.30$; $P > 0.25$). More importantly, there was no interaction between the antibiotic and the effect of produced water on fertilization (Table 1; $F_{1,8} = 0.43$; $P > 0.50$). Similarly, the proportion of embryos that reached the pluteus developmental stage in 48 hrs was not affected by the antibiotic (Table 1; in sea water, $F_{1,4} = 1.47$; $P > 0.25$). There were, however, marked effects of 10% produced water on fertilization success ($F_{1,8} = 98.4$; $P < 0.0001$) and development (Table 1).

Table 1. The effect of 0.05% Penicillin G on fertilization success and embryo development in the presence and absence of 10% produced water; data are the mean (± 1 SE) proportion of eggs fertilized and of embryos that were in the pluteus developmental stage at 48 hours

Produced Water	Fertilization Success: Penicillin Absent	Fertilization Success: Penicillin Present	Pluteus Stage: Penicillin Absent	Pluteus Stage: Penicillin Present
Absent	0.900 (0.040)	0.840 (0.056)	0.720 (0.034)	0.773 (0.026)
Present	0.434 (0.020)	0.409 (0.020)	0	0

Fertilization Assay for Effects of Produced Water

Ten minute exposure of eggs and/or sperm to produced water resulted in statistically significant depressions in the fraction of successful fertilizations (Fig. 1; $F_{2,18} = 83.23$; $P < 0.0001$). As expected, the magnitude of the depression increased with increasing concentrations of produced water (Fig. 1), although a substantial fraction (>50%) of eggs still were fertilized at the highest concentration used (1%). However, even at the lowest produced water concentration (0.0001%)

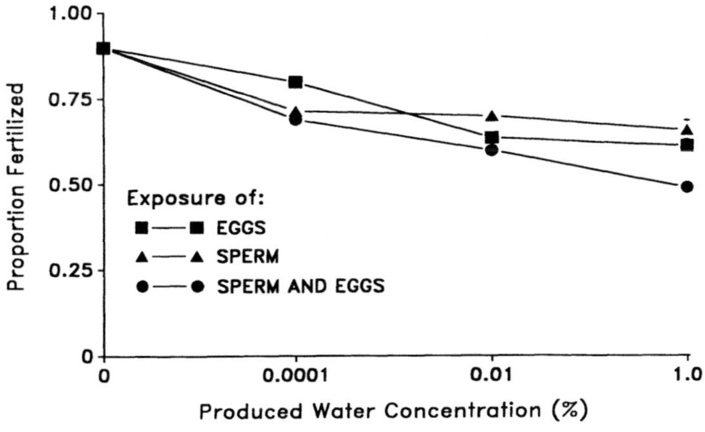

Figure 1. Fertilization success of gametes exposed for 10 minutes to a range in concentrations of produced water. Data are the mean (± 1 SE, n=3) proportion of eggs fertilized after 25 minute incubations of eggs and sperm in filtered sea water.

fertilization success was reduced by as much as ~10 - 20% (Fig. 1); each treatment response in the lowest concentrations differed significantly from the control (Dunnett's test, $P < 0.05$).

There was also a significant difference attributable to which gamete type(s) was exposed (Fig. 1; $F_{2,18} = 42.42$; $P < 0.0001$). *A posteriori* comparison of main effects (using REGWF) shows that exposure of sperm gave similar results as egg exposure, but that exposure of both sperm and eggs yielded a significantly lower fertilization rate. That is, based on fertilization success averaged across produced water concentrations, the sensitivity of sperm was about equal to that of eggs. Further, the greatest impairment occurred when both gamete types were exposed, but the combined effect appeared to be less than the sum of individual effects on each gamete type (Fig. 1). There was, however, an interaction between the gamete type exposed and concentration of produced water ($F_{4,18} = 11.27$; $P < 0.0001$), suggesting a cross-over in relative sensitivity of eggs and sperm to increasing produced water concentrations. It appears that sperm may be relatively more sensitive at low toxicant levels, whereas eggs may be more sensitive at higher concentrations (Fig. 1).

Embryonic Development Assay for Effects of Produced Water

Produced water reduced the fraction of larvae that were in the pluteus developmental stage after 48 hours (Fig. 2). For all gamete and zygote exposure combinations, the occurrence of pluteus larvae decreased with increasing concentration of produced water ($F_{2,30} = 919.37$; $P < 0.0001$). The reductions in pluteus larvae were statistically significant even at the exceedingly low concentration of 0.0001% (Dunnett's test, $P < 0.05$).

Within a concentration of produced water, the smallest reduction in pluteus development occurred when the zygote only was exposed, and the greatest reduction occurred when sperm, egg and the zygote stage all were exposed (Fig. 2; REGWF comparison shows all main effects are significantly different from one another). These data indicate that the brief 10 minute exposure of gametes reduced larval development to a greater degree than did the subsequent 48 hour exposure of zygotes. Thus, gametes were exceedingly more sensitive than zygotes to produced water. Furthermore, most of the effect arising from exposure of gametes was due to sperm, and very little from exposure of eggs (Fig. 2). Based on the pluteus development assay, the rank of sensitivity of each life stage to produced water was:

sperm > > egg > > zygote.

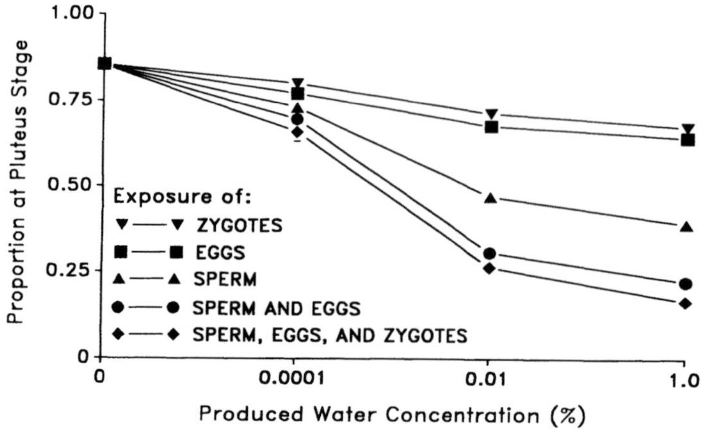

Figure 2. The mean (± 1 SE, n=3) proportion of embryos that are in the pluteus stage of development after 48 hours as a function of life stage exposed and produced water concentration. Gametes were exposed to produced water for 10 minutes prior to fertilization and zygotes were exposed for the entire 48 hour incubation period.

Cohort Analysis for Effects of Produced Water

The effect noted above could be due to a combination of increased mortality of developing embryos and/or a reduction in the development rate of survivors. To distinguish between direct mortality and reduced developmental rates, cohorts of eggs of known number were followed through time and the fraction of the initial number of eggs that eventually reached the pluteus larval stage was estimated. This was done using 1% produced water, which was the effluent concentration that yielded the greatest effects in both our fertilization (Fig. 1) and embryo development (Fig. 2) assays.

The results indicate that 1% produced water had no discernible effect on mortality (Fig. 3). Regardless of the life stage exposed, > 85% of the initial cohort survived to 96 hours, and there was no significant difference among the control or exposure treatments (Fig. 3; $F_{3,8} = 1.23$; $P > 0.35$). It is important to note that zygotes were exposed to produced water for the entire 96 hr period, yet still did not show an increased death rate.

By contrast with no affect on mortality, produced water greatly altered developmental rates of sea urchin embryos (Fig. 4). For example, the fraction of initial eggs that had reached the pluteus stage by 48 hours differed markedly among treatments ($F_{3,8} = 17.65$; $P<0.001$); Once again, zygote exposure yielded the smallest effect, while sperm exposure gave the greatest effect. These patterns at 48 hours (Fig. 4) were both qualitatively and quantitatively similar to those observed in the embryonic development bioassay reported above (Fig. 2). However, the effluent did not affect the number of eggs that ultimately reached the pluteus stage by 96 hours (Fig. 4; $F_{3,8} = 1.02$; $P> 0.40$). More than 70% of the initial cohort of eggs in each treatment had developed into pluteus larvae by 96 hours, and all treatments appeared to be asymptoting near 85%. Thus produced water slowed the developmental rates of sea urchin embryos but did not contribute directly to their mortality. Based on the slowing of developmental rates relative to exposure duration, this experiment indicates that the rank sensitivity of life stages was:

sperm > eggs >> zygotes.

The cohort analysis directly contradicted the finding in our fertilization assay regarding the fraction of eggs that ultimately underwent development: e.g., we observed only 61% of eggs fertilized after exposure of eggs to 1% produced water (Fig. 1), but we subsequently found that

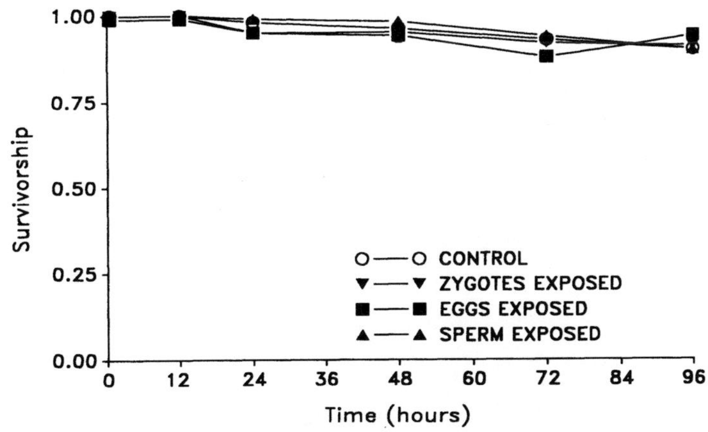

Figure 3. Survivorship of sea urchin eggs and embryos. Data are the mean proportion (± 1 SE, n=3) of cohorts of sea urchins (beginning at the egg stage) alive at each time interval. Gamete treatments involved exposure of eggs or sperm for 10 minutes prior to fertilization (no subsequent exposure of zygotes), and the zygote treatment involved exposure for the entire 96 hour period..

83% of all eggs developed to the pluteus stage (Fig. 4). It is possible that, in the cohort analysis, sperm continued to fertilize eggs long after the 35 minute exposure and incubation period used in the fertilization bioassay. A direct test of this hypothesis indicated that only ~10% of sperm remain viable 35 minutes after activation, and none were capable of fertilizing eggs after 45 minutes (Fig. 5). Exposure of gametes to 1% produced water lowered the fraction of viable sperm of a given age, and did not prolong the length of time that sperm remained capable of fertilization (Fig. 5). The estimate of age-specific sperm viability was independent of the endpoint (i.e., 25 or 120 mins) used to measure successful fertilization (Fig. 5). This experiment also indicated that the effect of produced water on viability of sperm was virtually instantaneous,

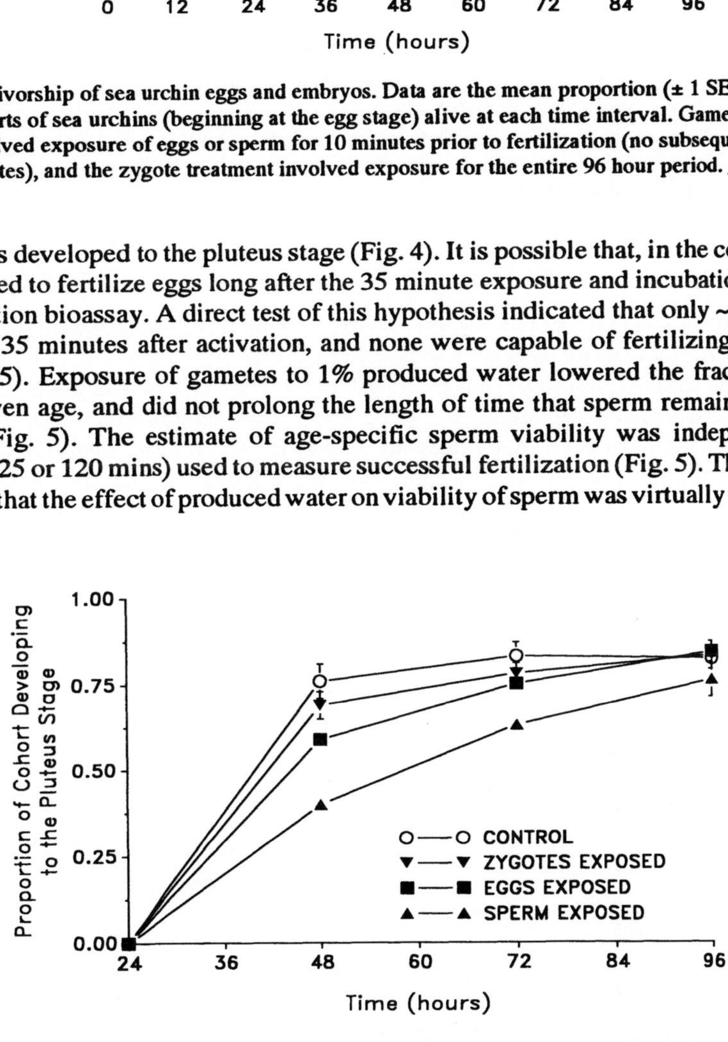

Figure 4. Mean proportion (± 1 SE, n=3) of initial sea urchin eggs that had entered the pluteus larval stage at different times. Gamete treatments involved exposure of eggs or sperm for 10 minutes prior to fertilization (no subsequent exposure of zygotes), and the zygote treatment involved exposure for the entire 96 hour period.

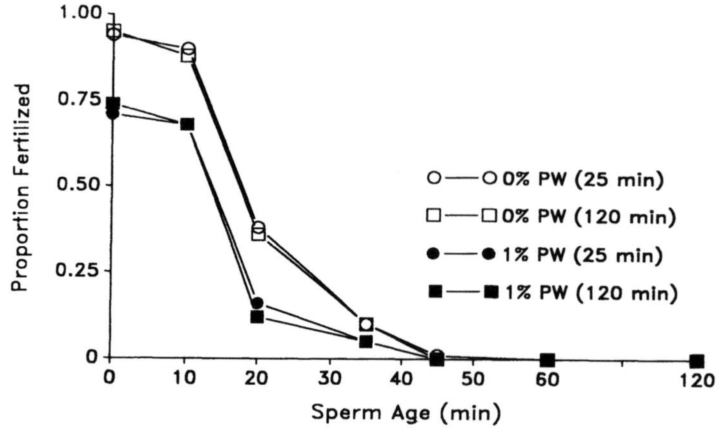

Figure 5. Viability of purple sea urchin sperm as a function of time since activation. Data are the mean (± 1 SE, n=3) proportion of eggs fertilized by sperm in the absence (circles) and presence (squares) of 1% produced attributable water. Only sperm were exposed to produced water: eggs and zygotes were kept in filtered sea water. Fertilization was assayed after 25 (open symbols) and 120 (closed symbols) minute incubations.

and suggests that at least part of the "zygote" sensitivity we have observed might be due to brief exposure of sperm that occurs when we exposure zygotes.

An alternative explanation to prolonged sperm viability is that the presence of a vitelline membrane underestimated successful syngamy when gametes had been exposed to produced water. This was supported by re-examining data from the cohort analysis and expressing fertilization success as the fraction of eggs that either had a fertilization membrane or had undergone cleavage. The initial depression of fertilization in the produced water treatment diminished through time such that by 36 hours there was no difference between treatments in

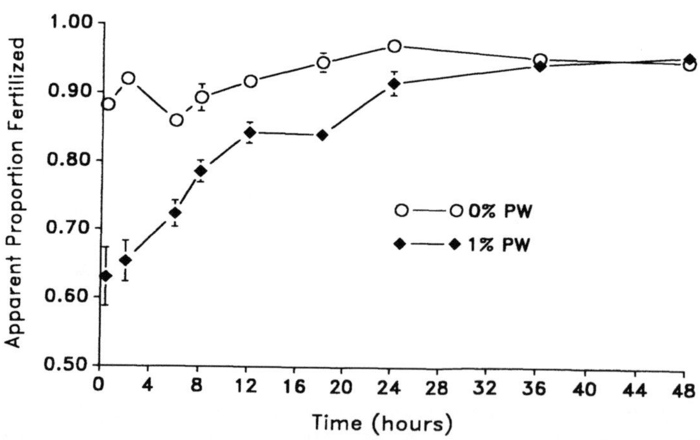

Figure 6. Apparent fertilization success as a function of time. Data are the mean (± 1 SE, n = 3) proportion of egg cohorts known to be fertilized based on the presence of either a vitelline membrane or cleavage at each interval in the absence (open circles) and presence (solid diamonds) of 1% produced water. Sperm, eggs, and zygotes were all exposed to produced water in this example.

the fraction of initial eggs known to have been successfully fertilized (Fig. 6). The results indicate that early indicators of fertilization (e.g., presence of the vitelline membrane) grossly underestimated the fraction of eggs that were actually fertilized and subsequently began embryonic development (Fig. 6).

DISCUSSION

Interpretation of Bioassay Studies

Bioassays typically use a response variable that coincides with the end of the planned exposure period. For example, tests of effects on gametes usually involve estimating changes in fertilization ability immediately after gamete exposure (e.g., Adams and Slaughter-Williams, 1988; Allen, 1971; Dinnel et al., 1981; Hagstrom and Lonning, 1973; Kobayashi, 1971; Pagano et al., 1982, 1983), and fertilization success has become a standard toxicity measure (e.g., Cherr et al., 1987; Dinnel et al., 1982, 1987; Dinnel and Stober, 1987; McGibbon and Moldan, 1986; Pagano et al., 1985, 1986). Post-fertilization assays often measure the fraction of young individuals that reach some developmental "milepost" in a specified time period (e.g., Adams and Slaughter-Williams, 1988; Byrne and Calder, 1977; Dinnel et al., 1989; Hose, 1985; Oshida et al., 1981; Pagano et al., 1985, 1986). However, our application of these event and time-dependent bioassays to produced water revealed several limitations of this general approach.

Our fertilization assay did not capture accurately the effect of produced water on syngamy. The apparent effect on gametes was to reduce successful fertilization (Fig. 1). However, our embryo cohort analyses revealed that the fraction of eggs eventually fertilized was unaffected by produced water (Fig. 4). Failure of the fertilization test to properly convey the lack of an overall affect on syngamy could have been due either to the particular design of the bioassay (i.e., sperm fertilized eggs long after the 35 minute exposure/incubation period used in our fertilization trials), or from the use of an inaccurate measure of syngamy. The first hypothesis can be rejected (Fig. 5). Other workers have also found that, once activated, sea urchin sperm remains viable for a very short period, usually < 25 minutes (e.g., Tyler, 1953; Timourian and Watchmaker, 1970; Pennington, 1985). By contrast, our measure of fertilization appeared to be imprecise. As is typical in these types of assays (e.g., Dinnel et al., 1987; McGibbon and Moldan, 1986; Oshida et al., 1981), we used the presence of a fertilization membrane to score successful fertilization. However, following exposure of gametes to produced water, the fraction of initial eggs that eventually underwent embryonic development was grossly underestimated by the fertilization membrane criterion (Fig. 6). It appears that produced water inhibited the vitelline membrane from lifting off the surface of some eggs, yet did not preclude syngamy. Alternative techniques to estimate fertilization success, such as nuclear staining, may be more reliable (though less practical) measures of syngamy.

With respect to differential sensitivity of gametes, our fertilization assay also incorrectly suggested that effects arising from pre-fertilization exposure on eggs and sperm were roughly equivalent (Fig. 1). However, our embryo development tests demonstrated that effects arising from sperm exposure were far greater than those from egg exposure (Figs. 2 & 4). The inconsistency resulted from an effect from sperm exposure that was manifested much later in development, and which was not evident at the time of fertilization. These results cogently illustrate that a delay in expression of an effect cannot be discovered when the assay endpoint, such as fertilization success, coincides with the end of the exposure period. The scope and magnitude of biological responses from a toxicant can be underestimated by failing to determine whether exposure at one life stage of an individual has an effect that is expressed in a later stage, yet this aspect is largely unexplored (but see Adams and Slaughter-Williams, 1988). In the case of fertilization, this limitation was perhaps presaged by the suggestion of certain authors that fertilization assays may be less sensitive indicators of effects from gamete exposure than exploration of abnormalities later in development (e.g., Allen, 1971; Hose, 1985).

However, the interpretation of bioassays that examine delayed effects can still be problematic. For example, we found that the fraction of sea urchin embryos reaching the pluteus larval stage in 48 hours was greatly impaired by produced water (Fig. 2). A logical interpretation of these results is that produced water exposure killed sperm, eggs, or embryos in some developmental phase prior to the pluteus stage. Our cohort analyses revealed that this interpretation was incorrect. The only effect of produced water was to impede developmental rates of embryos (Fig. 4): it did not directly increase mortality of developing zygotes (Fig. 3) or reduce fertilization of eggs (Fig. 6).

These results underscore a crucial but little appreciated difficulty with standard bioassay approaches; without additional evidence, the existence and contribution of various biological effects (e.g., delayed expression, reduced developmental rates, increased mortality) cannot be inferred safely from bioassays with fixed-time endpoints. While such techniques are useful in ranking the relative toxicity of various effluents, their utility in providing an understanding of the nature of biological responses is more limited, and alternative approaches should be sought. Endpoints and exposure regimes should be selected that enhance biological insight and therefore either help pinpoint mechanisms of toxicity or permit extrapolation to field situations where biological effects arise from exposure of particular life stages (e.g., Somerville et al., 1987; Higashi et al., 1992; Raimondi and Schmitt, 1992).

Biological Effects of Produced Water

Predicting or understanding the potential for long-term environmental impacts arising from the discharge of produced water into coastal waters is impeded because biological effects of whole produced waters have received little attention (Neff, 1987). To the extent that laboratory bioassays have been applied to produced waters, most produced waters do not appear to be acutely toxic (for review see Neff, 1987). Results of our study support this notion for produced water from oil fields in the Santa Barbara Channel; brief (10 minute) exposure of sea urchin gametes and prolonged (96 hr) exposure of zygotes did not result in increased mortality, even at the moderately high concentration of 1% produced water (Fig. 3). We did, however, detect sublethal responses to produced water at substantially lower levels; statistically significant responses of young purple sea urchins were detected at produced water concentrations as low as 1 part per million, with the magnitude of the response(s) increasing with increased concentration of the effluent.

It has been argued that such sublethal or chronic effects may be a better gauge of potential environmental significance than the acute lethality of an effluent (e.g., Capuzzo, 1987; Neff, 1987). Extrapolation from laboratory demonstration of sublethal effects to assessment of ecological consequences under natural conditions is not, however, straightforward. For example, exposure concentrations and durations used in a laboratory study do not necessarily bear any relationship to exposure regimes that arise under field conditions. Two lines of evidence suggest that our laboratory exposures spanned conditions likely to occur in the field. First, our results show that the effect of produced water on gametes, particularly sperm, was virtually instantaneous (Fig. 5). Therefore, adult urchins that spawn in the vicinity of the produced water outfall are likely to produce offspring with delayed developmental rates even though those developing larvae may soon be transported away from the outfall. Second, we have examined effects under field conditions by exposing urchin gametes and larvae to water collected at different distances from the produced water diffuser and compared these results with those obtained from known concentrations of produced water (Krause, unpublished data). These results suggest that although produced water is rapidly diluted (the diffuser is designed for an initial seawater:effluent dilution of 125:1), detectable developmental effects can persist out to 100 - 500 m, where produced water concentrations drop to approximately 1 ppm. Thus, the produced water concentrations we used in our laboratory experiments nicely matched with the concentrations that can occur in the field at sites within 500 m of the outfall.

Another potential difficulty in extrapolating laboratory findings of developmental delays to field consequences is that delayed development can lead to an indirect increase in mortality: e.g., by increasing the length of time that early life stages remain vulnerable to natural sources of mortality, such as size-selective predators (e.g., Werner et al., 1983). Thus, slight sublethal effects, such as delayed development, could indirectly contribute to significant reductions in the number of larvae available to recruit into benthic populations. However, processes linking larval and benthic dynamics are poorly understood and therefore the ecological significance of larval mortality is largely unknown (e.g., Capuzzo, 1987; Nisbet et al., 1993; Raimondi and Schmitt, 1992). Much work remains before impacts on benthic populations can be predicted based on sublethal responses of larval stages.

Because so little is known regarding sublethal effects of produced water, it is not yet possible to assess whether a reduction in developmental rate is a common response of early life stages. Lowered growth or developmental rates have been observed for individuals exposed to certain constituents of produced water. For example, petroleum hydrocarbons can retard individual growth of adult copepod crustaceans (Hay et al., 1988), larval decapod crustaceans (Caldwell, 1977; Cucci and Epifano, 1979; Katz, 1973; Laughlin et al., 1978; Laughlin and Neff, 1979; Wells, 1972; Wells and Sprauge, 1976) and larval bivalve mollusks (Byrne and Calder, 1977). In each of these cases, the life stage exposed was capable of feeding, and hydrocarbons are known to reduce feeding rates of planktotrophic larvae (e.g., Johns and Pechenik, 1980; Wells and Sprauge, 1976). Hence, reduced growth or developmental rates could have resulted primarily from an indirect effect on food intake. This process was not involved in the reduced developmental rates of purple sea urchins; larvae of this species do not feed until the late pluteus stage (Strathmann, 1987), which is not attained until ~72 hours after fertilization in normally developing embryos. The response of young purple sea urchins, especially that arising from exposure of gametes, must have involved direct alteration of normal cellular activity. Delayed development of embryos mediated through exposure of gametes to PCBs has been observed for the sea urchin *Arbacia punctulata* (Adams and Slaughter-Williams, 1988). Like produced water, the actual mechanism producing this effect of PCB exposure is unknown, but like certain petroleum hydrocarbons (Capuzzo, 1987), PCBs are highly lipophilic. Lipid reserves are of primary importance in embryonic development (e.g., Holland, 1978), and it has been suggested that larvae may shunt energy reserves away from differentiation to be used to detoxify such lipophilic compounds as petroleum hydrocarbons (e.g., Sharp et al., 1979).

Produced water, of course, contains numerous potential toxicants in addition to petroleum hydrocarbons (e.g., Neff, 1987), and the particular constituents of the Carpinteria effluent responsible for the observed effect on purple sea urchins remains to be determined. Higashi et al. (1992) explored "toxicity" of various fractions of produced water from Carpinteria, and found that the majority of biological effects arose from the water-only soluble fraction that contained divalent cations. Interestingly, Pagano et al. (1982) found results for other sea urchin species exposed to cadmium that were qualitatively similar to ours. Although cadmium is not common in the Carpinteria produced water, analogous elements, such as barium and strontium, are present (Higashi et al., 1992). Furthermore, barium and strontium are known to impair development (Conrad and Davis, 1980), and recent work suggests that these ions might mediate their effects through modification of microtubule function (Tamm, 1989; Tamm and Tamm, 1990). Because we observed that sperm exposure yielded the greatest effect on development, and because this effect appeared to arise very early in development, we suggest that the toxicological mechanisms most likely involve a feature specific to sperm that involves microtubule function. One likely candidate is the sperm centriole, which is responsible for the transport of the sperm pronucleus to the egg pronucleus (Bestor and Schatten, 1981). If the function of the centriole were impaired, this could lead to an early retardation of development by delaying nuclear fusion. Other studies using produced water from this same source have found responses that might involve microtubule-mediated effects: e.g., swimming and chemoreception of abalone larvae (Raimondi and Schmitt, 1992), swimming of kelp spores (Reed, 1992), migration of kelp nuclei (Pillai et al.,

1990), and growth (possibly feeding) of mussels (Osenberg et al., 1992). If these diverse biological effects of produced water were mediated through a similar toxicological mechanism, it would provide a strong unifying theme to the study of this complex effluent.

ACKNOWLEDGMENTS

We thank G. Cherr, A. Ebeling, T. Fan, R. Higashi, S. Holbrook, A. Martinez, D. Morse, M. Pillai, and D. Steichen for discussions and criticisms, M. Krause for assistance in the laboratory, B. Williamson for technical support, and J. Wallace for facilitating our collections of produced water. This research was funded by the UC Coastal Toxicology Program, and by the Minerals Management Service, U.S. Department of Interior under MMS Agreement No. 14-35-0001-3071. The views and conclusions in this paper are those of the authors and should not be interpreted as necessarily representing the official policies, either express or implied, of the U.S. Government.

REFERENCES

Adams, J.J., and Slaughter-Williams, S., 1988, The effects of PCB's (Aroclors 1254 and 1016) on fertilization and morphology in *Arbacia punctulata*, Water, Air, and Soil Pollut. 38:299-310.

Allen, H., 1971, Effects of petroleum fractions on the early development of a sea urchin, Mar. Poll. Bull. 2:138-140.

Bestor, T.M., and Schatten, G., 1981, Anti-tubilin immunofluorescence microscopy of microtubules present during the pronuclear movements of sea urchin fertilization, Dev. Biol. 88:80-91.

Byrne, C.J., and Calder, J.A., 1977, Effects of the water-soluble fractions of crude, refined, and waste oils on the embryonic and larval stages of the Quahog clam *Mercenaria* sp, Mar. Biol. 40:225-231.

Caldwell, R.S., Calderone, E.M., and Mallon, M.H, 1977, Effects of seawater-soluble fraction of Cook Inlet crude oil and its major aromatic components on larval stages of the Dungeness crab, *Cancer magister* (Dana), *in*: "Fate and Effects of Petroleum Hydrocarbons in Marine Ecosystems and Organisms," D.A. Wolfe, ed., Pergamon Press, New York.

Capuzzo, J.M., 1987, Biological effects of petroleum hydrocarbons: assessments from experimental results, *in*: "Long-Term Environmental Effects of Offshore Oil and Gas Development," D.F. Boesch and N.N. Rabalais, eds., Elsevier Applied Science, NY.

Castagna, A., Sinatra, F., Scalia, M., and Capodicasa, V., 1981, Observations of the effect of zinc on the gametes and various development phases of *Arbacia lixula*, Mar. Biol. 64:285-289.

Cherr, G.N., Shenker, J.M., Lundmark, C., and Turner, K.O., 1987, Toxic effects of selected bleached kraft mill effluent constituents on the sea urchin sperm cell, Env. Toxicol. Chem. 6:561-569.

Conrad, G.W., and Davis, S.E., 1980, Polar lobe formation and cytokinesis in fertilized eggs of *Ilyanassa obsoleta*, Devel. Biol. 74:152-172.

Cucci, T.L. and Epifano, C.E., 1979, Long term effects of water- soluble fractions of Kuwait oil on the larval and juvenile development of the mud crab *Eurypanopeus depressus*, Mar. Biol. 55:215-220.

Dinnel, P.A., Stober, Q.J., and DiJulio, D.H., 1981, Sea urchin sperm bioassay for sewage and chlorinated seawater and its relation to fish bioassays, Mar Env. Res. 5:29-39.

Dinnel, P.A., Stober, Q.J., Crumley, S.C., and Nakatani, R.E., 1982, Development of a sperm cell toxicity test for marine waters, *in*: "Aquatic toxicology and hazard assessment: Fifth Conference, ASTM STP 766," J.G. Pearson, R.B. Foster, and W.E. Bishop, eds., Amer. Soc. Testing and Materials, Philadelphia, PA.

Dinnel, P.A., Link, J.M., and Stober, Q.J., 1987, Improved methodology for sea urchin sperm cell bioassay in marine waters, Arch. Environ. Contam. Toxicol. 16:23-32.

Dinnel, P.A., and Stober, Q.J., 1987, Application of the sea urchin sperm bioassay to sewage treatment efficiency and toxicity in marine waters, Mar. Env. Res. 21:121-133.

Dinnel, P.A., Link, J.M., Stober, Q.J., Letourneau, M.W., and Roberts, W.E., 1989, Comparative sensitivity of sea urchin sperm bioassay to metals and pesticides, Arch. Environ. Contam. Toxicol. 18:748-755.

Hagstrom, B.E., and Lonning, S., 1973, The sea urchin egg as a testing object in toxicology, Acta. Pharm. et Toxicol. 32(Supp. I):1-49.

Hay, S.J., Evans, G.T., and Gamble, J.C., 1988, Birth, growth and death rates for enclosed populations of calanoid copepods, J Plank. Res. 10(3):431-454.

Higashi, R., Cherr, G., Bergens, C., Fan, T., and Crosby, D., 1992, Toxicant isolation from a produced water source in the Santa Barbara Channel, *in*: "Produced Water: Technological/ Environmental Issues and Solutions," J.P. Ray and F.R. Engelhardt, eds., Plenum Publishing Corp., New York, New York, USA.

Holland, D.L., 1978, Lipid reserves and energy metabolism in the larvae of benthic marine invertebrates, *in*: "Biochemical and Biophysical Perspectives in Marine Biology, Volume 4," D.C. Malins and J.R. Sargent, eds., Academic Press, New York.

Hose, J.E, 1985, Potential uses of sea urchin embryos for identifying toxic chemicals: Description of a bioassay incorporating cytologic, cytogenetic and embryonic endpoints, *J.* Appl. Toxicol. 5:245-254.

Johns, D.M., and Pechenik, J.A., 1980, Influence of the water accommodated fraction of No. 2 fuel oil on energetics of *Cancer irroratus* larvae, Mar. Biol. 55:247-254.

Katz, L.M., 1973, The effects of water-soluble fractions of crude oil on larvae of the decapod crustacean *Neopanope texana* (Say), Environ. Pollut. 5:199-204.

Kobayashi, N., 1971, Fertilized sea urchin eggs as an indicatory material for marine pollution bioassay, preliminary experiment, Publs. Seto Mar. Biol. Lab. 18:379-406.

Kobayashi, N., 1980, Comparative sensitivity of various developmental stages of sea urchins to some chemicals, Mar. Biol. 58:163-171.

Laughlin, R.B. Jr., and Neff, J.M., 1979, Interactive effects of salinity, temperature and polycyclic aromatic hydrocarbons on the survival and development rate of larvae of the mud crab *Rhithropanopeus harrisii*, Mar. Biol. 53:281-291.

Laughlin, R.B., Jr., Young, L.G.L., and Neff, J.M., 1978, A long-term study of the effect of water-soluble fractions of No.2 fuel oil on the survival, development rate, and growth of the mud crab *Rhithropanopeus harrisii*, Mar. Biol. 47:87-89.

McGibbon, S., and Moldan, A.G.S., 1986, Routine toxicity testing of toxicants using a sea urchin gamete bioassay, Mar. Pollut. Bull. 17(2):68-72.

National Research Council, 1983, "Drilling Discharges in the Marine Environment," National Academy Press, Washington, D.C.

Neff, J.M., Anderson, J.W., Cox, B.A., Laughlin, R.B., Jr., Rossi, S.S., and Tatum, H.E., 1976, Effects of petroleum on survival, respiration and growth of marine animals, *in*: "Sources, Effects and Sinks of Hydrocarbons in the Aquatic Environment," American Institute of Biological Sciences, Washington, D.C.

Neff, J.M., 1987, Biological effects of drilling fluids, drill cuttings and produced waters, *in*: "Long-Term Environmental Effects of Offshore Oil and Gas Development," D.F. Boesch and N.N. Rabalais, eds., Elsevier Applied Science, New York.

Nisbet, R. M., Murdoch, W.W., and Stewart-Oaten, A., 1993, Consequences for Adult Fish Stocks of Human Induced Mortality on Immatures. Ecol. Appl. in review.

Oshida, P.S., Goochey, T.K., and Mearns, A.J., 1981, Effects of municipal wastewater on fertilization, survival, and development of the sea urchin, *Strongylocentrotus purpuratus*, *in*: "Biological Monitoring of Marine Pollutants," F.J. Vernberg, A. Calabrese, F.P. Thurnberg, and W.B. Vernberg eds., Academic Press, New York.

Osenberg, C.W., Schmitt, R.J., Holbrook, S.J., and Canestro, D. 1992, Spatial scale of ecological effects associated with an open coast discharge of produced water, *in*: "Produced Water: Technological/ Environmental Issues and Solutions," J.P. Ray and F.R. Engelhardt, eds., Plenum Publishing Corp., New York, New York, USA.

Pagano, G., Esposito, A., and Giordano, G.G., 1982, Fertilization and larval development in sea urchins following exposure of gametes and embryos to cadmium, Arch. Env. Contam. Toxicol. *11:47-55.*

Pagano, G., Esposito, A., Bove, P., DeAngelis, M., Rota, A., and Giordano, G.G., 1983, The effects of hexavalent and trivalent chromium on fertilization and development in sea urchins, Env. Res. 30:442-452.

Pagano, G., Cipollaro, M., Corsale, G., Esposito, A., Ragucci, E., Giordano, G.G., and Trieff, N.M., 1985, Comparative toxicities of chlorinated biphenyls on sea urchin egg fertilization and embryogenesis, Mar. Env. Res. 17:240-244.

Pagano, G., Cipollaro, M., Corsale, G., Esposito, A., Ragucci, E., Giordano, G.G., and Trieff, N.M., 1986, The sea urchin: Bioassay for the assessment of damage from environmental contaminants, *in*: "Community toxicity testing, ASTM STP-920," J. Cairns ed., Amer. Soc. Testing and Materials, Philadelphia, PA.

Pennington, J.T., 1985, The ecology of fertilization of echinoid eggs: The consequences of sperm dilution, adult aggregation, and synchronous spawning, Biol. Bull. 169(10):417-430.

Pillai, M.C., Baldwin, J.D., and Cherr, G.N., 1990, Role of cytoskeleton in germtube elongation and nuclear translocation in a marine algal gametophyte, J. Cell Biol. 111:(No.5, Pt. 2) 415a.

Raimondi, P.T., and Schmitt, R.J., 1992, Effects of produced water on settlement of larvae: field tests using red abalone, *in*: "Produced Water: Technological/ Environmental Issues and Solutions," J.P. Ray and F.R. Engelhardt, eds., Plenum Publishing Corp., New York, New York, USA.

Reed, D.C., and Lewis, R.J., 1992, Water column effects of produced water discharges on the production and dispersive potential of giant kelp (*Macrocystis pyrifera*). In preparation.

Rossi, S.S., and Anderson, J.W., 1978, Petroleum hydrocarbon resistance in the marine worm *Neanthes arenaceodentata* (Polychaeta: Annelida) induced by chronic exposure to No. 2 fuel oil, *B*ull. Environ. Contam. Toxicol. 20:515-521.

SAS Institute Inc, 1988, "SAS/STAT User's Guide, Release 6.03 Edition," SAS Institute Inc, Cary, North Carolina.

Sharp, J.R., Fucik, K.W., and Neff, J.M., 1979, Physiological bases of differential sensitivity of fish embryonic stages to oil pollution, *in*: "Biological Monitoring of Marine Pollutants," F.J. Vernberg, A. Calabrese, F.P. Thurnberg, and W.B. Vernberg, eds., Academic Press, New York.

Somerville, H.J., Bennett, D., Davenport, J.N., Holt, M.S., Lynes, A., Mahieu, A., McCourt, B., Parker, J.G., Stephenson, R.R., Watkinson, R.J., and Wilkinson, T.G., 1987, Environmental effect of produced water from North Sea oil operations, Mar. Poll. Bull. 18(10):549-558.

Strathmann, M.F., 1987, Phylum Echinodermata, Class Echinoidea, *in*: "Reproduction and Development of the Marine Invertebrates of the Northern Pacific Coast: Data and Methods for the Study of Eggs, Embryos, and Larvae," M.F. Strathman ed., University of Washington Press, Seattle, Washington.

Tamm, S.L., 1989, Control of reactivation and microtubule sliding by calcium, strontium and barium in detergent-extracted macrocilia of *Beroe*, Cell Motil. Cytoskel. 12:104-112.

Tamm, S.L., and Tamm, S., 1990, Ca/Ba/Sr-induced conformational changes of ciliary axonemes, Cell Motil. Cytoskel. 17:187-196.

Timourian, H., and Watchmaker, G., 1970, Determination of spermatozoan motility, Dev. Biol. 21:62-72.

Tyler, A., 1953, Prolongation of life span of sea urchin spermatozoa and improvement of the fertilization reaction by treatment of the spermatozoa and eggs with metal chelating agents, Biol. Bull. 104:224-239.

Wells, P.G., 1972, Influence of Venezuelan crude oil on lobster larvae, Mar. Pollut. Bull. 3:105-106.

Wells, P.G., and Sprague, J.B., 1976, Effects of crude oil on American lobster (*Homarus americanus*) larvae in the laboratory, J. Fish. Res. Board Can. 33:1604-1614.

Werner, E.E., Gilliam, J.F., Hall, D.J., and Mittelbach, G.G., 1983, An experimental test of the effects of predation risk on habitat use in fish, Ecology 64(6):1540-1548.

SUBLETHAL EFFECTS OF PRODUCED WATER FROM CRUDE OIL TERMINALS ON THE CLAM *Donax faba*

Z. b. Din and A.b. Abu
Centre for Marine and Coastal Studies
Universiti Sains Malaysia
Penang, Malaysia

INTRODUCTION

Many earlier studies on the sublethal effects of pollutants on marine organisms dealt with single responses such as feeding or respiration rate. At times, these kinds of studies can be misleading. For example, on exposure to a certain kind of pollutant, an organism may reduce its feeding rate. However, if this were to be accompanied by a reduction in its respiration and/or excretion rate, as well as rate of fecal production, the original effect can be nullified.

Measurements of change in growth rate has been recommended as one of the procedures to study the overall effects of a particular toxicant on the physiology of an organism. This change is a fundamental component of the physiological fitness of an organism, and therefore represents an important index of environmental effect. Determination of the energy available for growth based on physiological analysis of the energy budget rather than direct measurement of growth itself, proves particularly useful in assessing the biological impact of an environmental stress. A measure of change in energy available for growth based on this concept is "Scope for Growth". Among those who have popularized this concept are Widdows (1985) and Bayne et al., (1985). Bayne and Worral (1980) demonstrated a close relationship between Scope for Growth (determined physiologically) and measured tissue growth in two populations of the blue mussel, *Mytilus edulis*.

Basically, Scope for Growth is a measure of the energy that is available to an organism for growth and gamete production. Of the food consumed by an animal, part of it is absorbed while the reminder is lost as feces. From the total energy absorbed, a major portion is used for respiration, and to a lesser extent, lost in excretion. What is left is termed Scope for Growth. The primary components of the energy budget can be described by the balanced equation of Winberg (1960):

$$C - F = A = R + U + P \quad (1)$$

or

$$P = A - (R + U)$$

where C = energy value of food consumed, F = energy lost as feces, A = energy absorbed from food, R = energy respired, U = energy excreted, P = energy incorporated into the biomass of the organisms (growth and gamete production).

Produced Water, Edited by J.P. Ray
and F.R. Engelhart, Plenum Press, New York, 1992

In this study, variations in Scope for Growth, along with changes in feeding and respiration rate were used as measures of sublethal effects of effluents from three crude oil terminals on the clam, *Donax faba.*

MATERIALS AND METHODS

Produced water regularly discharged from crude oil terminals of Sarawak Shell Berhad in Bintulu and Miri, Sarawak, Malaysia, and crude oil terminal of Sabah Shell Petroleum Company in Labuan, Malaysia, were used in this study. Except for the parameters listed in Table 1, no other chemical analyses were carried out on the effluents. Table 1 indicates that the effluents were similar in terms of pH and salinity but the dissolved oxygen in the Bintulu effluent was about 1.5 times higher than the other two. This can be attributed to the amount of total suspended organics (TSO) in the effluents; the amount of TSO in the Labuan and Lutong effluent was about 13 and 3 times that of the Bintulu effluent, respectively. The total suspended solids (TSS) in the Lutong effluent was more than twice that of Bintulu, while the effluent from Labuan had TSS value of about five times that of Lutong.

Crude oil and associated gases from offshore oilfields are piped onshore for storage and processing. At each of the above terminals, the formation water from the fields are separated and treated in effluent treatment systems, which consist of a combination of gravity separation (API, Corrugated Plate Interceptor) and induced air-floatation, with a final holding basin for final retention before discharge. The treated effluents are discharged through the outfall into the sea. Samples used in this study were collected prior to discharge into the sea.

Prior to effluent exposure, the test clams were acclimatized to laboratory conditions for a period of about 1 week. Bioassay experiments were conducted on these animals to determine the LC50 of the effluents. For these experiments, 20 clams were used in each glass tank containing 15 l of test solution. Three replicates were used for each effluent concentration. Procedures for the bioassay tests were in accordance with the United States Environmental Protection Agency (USEPA) protocol as suggested by Rand and Petrocelli (1985).

For the sublethal study, exposure concentrations were selected on the basis of LC_{50} values. Concentrations chosen corresponded to the LC_{10}, LC_{20}, and LC_{40} of the effluents. Use of similar lethal concentrations of the effluents makes the results more comparable since the effluents differed in toxicity. Actual effluent concentrations used together with their LC_{50} values are shown in Table 2. For these experiments, 120 clams were used in each of the fiberglass exposure tanks containing 150 l of test effluent.

After exposures to the effluents for a period of 4, 8, and 12 days, sets of four clams were randomly sampled from the tanks for the determination of the individual parameters for the calculations of Scope for Growth. These included (i) clearance rate, (ii) food absorption efficiency, (iii) rate of respiration and (iv) rate of ammonia excretion. Ammonia excretion was determined because it has been well documented that it forms the main component of excretion in aquatic invertebrates. These values were then converted to energy units as suggested by Bayne et al., (1985). Detailed calculations for Scope for Growth can be found in Bayne et al., (1985). For each of the parameters, duplicate readings were made.

Table 1. Some physico-chemical parameters of the effluent used

Effluent	pH	Dissolved Oxygen (ppm)	Salinity (ppt)	TSS (mg/l)	TSO (mg/l)
Bintulu COT	8.2	2.15	12	58	41
Lutong COT	8.5	1.45	16	137	114
Labuan COT	8.8	1.40	17	706	531

Table 2. Concentrations (ppm) of effluents used in the sublethal study. Corresponding 96-hr median lethal concentration (LC_{50}) values are also indicated

Effluent	LC_{10}	LC_{20}	LC_{40}	LC_{50}
Bintulu COT	8.6	11.1	14.7	15.3
Lutong COT	0.75	1.10	1.65	2.80
Labuan	0.08	0.11	0.14	0.20

Determination of Clearance Rate

Clearance rate is defined as the volume of water cleared of particles per hour by the test organism. These values were determined using the setup illustrated in Figure 1. Unfiltered seawater flows through the test chambers at a rate of about 200 ml/min. Actual flow rates were determined by measuring the amounts of seawater outflowing from the test chambers per unit time. The animals were left undisturbed in the chamber for about 60 minutes to allow their valves to open. Water samples were then collected every 15 minutes for 60 minutes from the outflow of each chamber. Particle numbers in the water samples were determined using a Model ZM Coulter Counter. Particle numbers from the control outflow represent the inflow particle numbers (Ci) while concentrations from the other outflows represent the outflow particle concentrations (Co). Clearance rate was then calculated as follows:

$$Clearance\ rate\ (\ l/h\ ) = \frac{Ci - Co}{Ci} \text{ x flowrate (l/h)} \qquad (2)$$

Determination of Food Absorption Efficiency

The feeding rate of the clam can be calculated using the following equation: Feeding Rate (mg h^{-1}) = Clearance Rate (l h^{-1}) x POM (mg l^{-1}) where POM = particulate organic matter in seawater. However, a large and variable portion of the POM is refractory and is not used as energy substrate. Thus, in order to estimate the amount of the POM that is actually utilized for energy, the clam's feeding rates need to be corrected by their food absorption efficiency (e).

There are several procedures that can be used to determine the values of food absorption efficiency. For this study, estimation of e was based on the Conover ratio (Conover, 1966):

$$e = \frac{F - E}{(1 - E)F} \qquad (3)$$

where:

e = absorption efficiency
F = ratio of ash-free dry weight : dry weight of food,
E = ratio of ash-free dry weight : dry weight of feces.

For the determination of dry weight, food and feces were collected and dried until constant weight (usually overnight) at a temperature of 90° C. For ash-free weights, the food and feces were ashed in a muffle furnace at a temperature of 500° C.

Determination of Respiration Rate

For the determination of respiration rate, test animals were placed in Model RC 400 Strathkelvin respiration chambers and changes in dissolved oxygen were determined using a Model 781 Strathkelvin oxygenmeter, attached with a Model 2030 Linear chart-recorder. Rate of oxygen consumption (respiration rate), was calculated using the following equation:

$$R\ (ml\ O_2/g\ dry\ ^{wt}/_h) = \frac{(DO_i - DO_f)\ x\ V}{M\ x\ t} \qquad (4)$$

where

R	=	respiration rate
DO_i	=	initial concentration of dissolved oxygen in respiration chamber (ppm)
DO_f	=	final concentration of dissolved oxygen in respiration chamber
V	=	volume of water in respiration chamber (l)
M	=	dry tissue weight of test clams (g)
t	=	period between reading DOi and DO_f (h)

Determination of Ammonia Excretion Rate

Four clams were placed in four beakers containing 200 ml of seawater that had been filtered through 0.45 μm filters. Another beaker with filtered seawater but with no animal, acted as control. After an incubation period of two hours, water samples were taken from each beaker for the analysis of ammonia using the method of Solorzano (1969). Ammonia excretion rate was determined using the following equation:

$$U\ (mg/\ dry\ tisue\ ^{wt}/_h = \frac{(A_f - A_i)\ x\ 0.2\ l}{M\ x\ 2\ h} \qquad (5)$$

where U = Ammonia excretion rate
A_f = Ammonia concentration at end of experiment
A_i = Ammonia concentration at beginning of experiment
M = Dry tissue weight of test clams

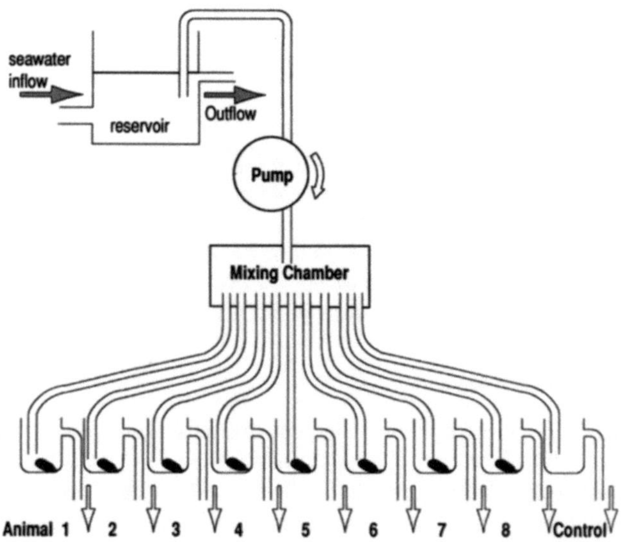

Figure 1. Schematic of set-up for the determination of clearance rate.

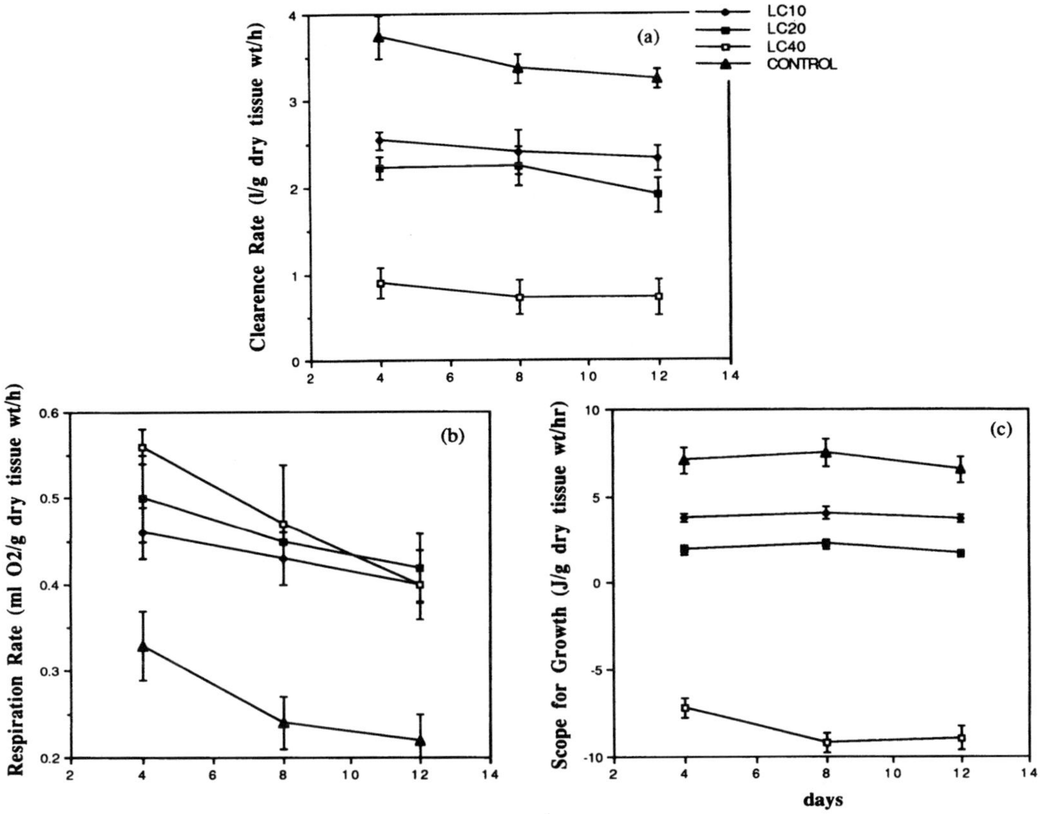

Figure 2. Effects of Bintulu Crude Oil Terminal effluent on various physiological parameters of the clam, *Donax faba*.

RESULTS

Sublethal Effects of the Bintulu COT Effluent on the Clam, *Donax faba*

Figure 2 shows the results of three kinds of sublethal effects that were determined for the clams on exposure to the Bintulu COT effluent. The feeding rate of the clam (as measured through the clearance rate) is shown to be inversely proportional to the effluent concentrations. On being exposed to 8.6% of the effluent (equivalent to its 96-hr LC_{10}) for four days, the clams were feeding at a rate of about 70% of the control animals. Those treated with 11.1% of effluent (which is equivalent to its LC_{20}) reduced their clearance rate to about 60% of the control, while animals exposed to the highest effluent level which was equivalent to the 96-hr LC_{40} were filtering food at a rate of between 20 to 25% of the control animals. These differences in feeding rates were similarly found after the eighth and twelfth day exposures.

The respiration rate of the clam was shown to change in direct proportion to the concentrations of the effluent used. There was no difference between the respiration rates of the "treated" clams ($p > 0.05$). However, the difference between the respiration rates of these animals and the control was very significant ($p < 0.001$); the treated animals were respiring at a rate of between 1.5 to almost twice that of the control.

The effects of the effluent on the overall growth physiology of the clams as measured by the Scope for Growth values follow the trend of the clearance rate, where an increase in effluent

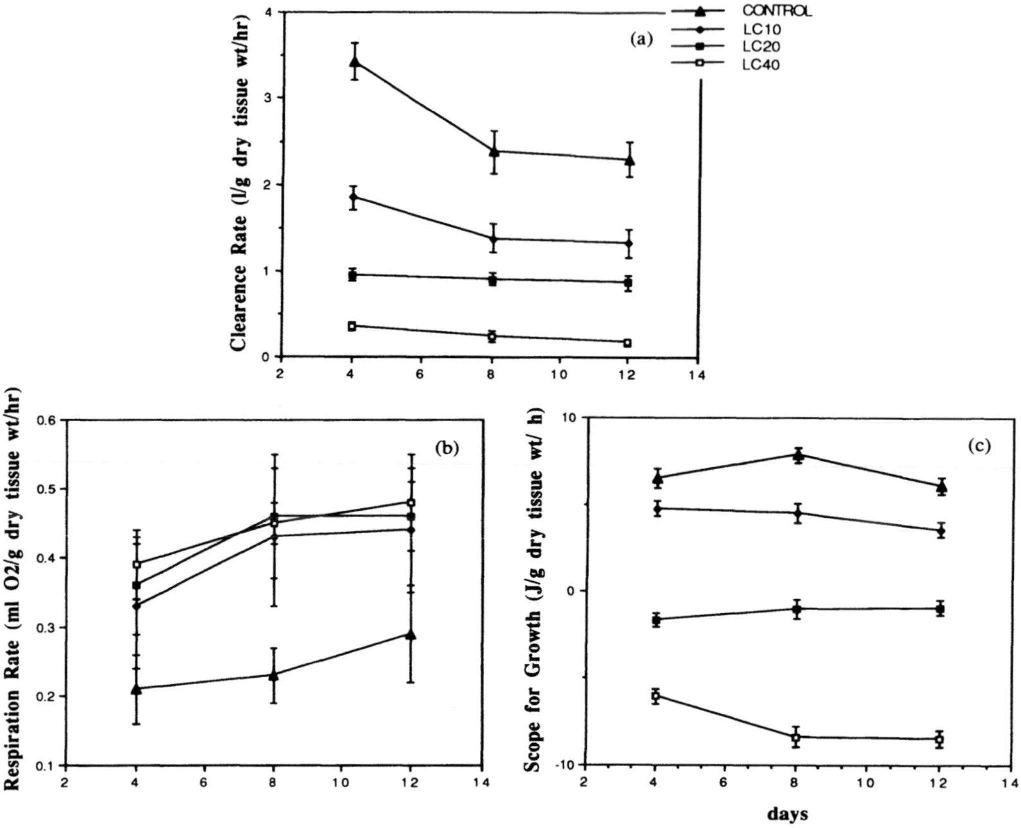

Figure 3. Effects of Lutong Crude Oil Terminal effluent on various physiological parameters of the clam.

concentration resulted in a decrease in the values of the Scope for Growth. Differences between the Scope for Growth of the clams treated with the effluent and that of the control were found to be quite similar for the three periods of exposure. The animals exposed to the lowest effluent concentration recorded Scope for Growth values ranging between 53 to 56% of the control while those exposed to the next effluent concentration had Scope for Growth values 25 to 30% of the amount available to the clams used as control. Negative values of Scope for Growth were recorded for the clams treated with the highest effluent concentration for all three periods of exposure. It is noteworthy that a negative value of Scope for Growth means that the animals were not assimilating enough food even to cater for their daily physiological requirements such as for respiration and excretion. The animals were probably burning some kind of stored energy to compensate for this deficiency.

Sublethal Effects of the Lutong COT Effluent on the Clam, *Donax faba*

The effects of the Lutong COT effluent on the clearance rate of the clams follow a trend similar to those exposed to the Bintulu COT effluent where an increase in effluent concentration resulted in a decrease in the rate of food filtration (Figure 3). However, the impact here was somewhat greater. When the animals were exposed to 0.75% of the effluent (equivalent to its 96-h LC_{10}), their clearance rate was reduced to between about 55 to 60% of the rate of the control. Equivalent exposure for the Bintulu COT effluent resulted in a decrease in clearance rate to only about 70% of the control. At the second level of effluent exposure (equivalent to its LC_{20}), the treated animals were filtering food at a rate of between 30 to 40% of the untreated ones, as compared to about 60 to 65% for the Bintulu effluent. When the clams were put in an effluent concentration equivalent to its LC_{40}, the clearance rate of the treated animals was reduced to about 10% of the untreated ones. The Bintulu effluent at this level caused a reduction in clearance rate to about 20 to 25% of the control.

The effects of the Lutong COT effluent on the respiration rate of the clam was found to follow a pattern similar to those exposed to the Bintulu COT effluent, where an increase in effluent concentration resulted in an increase in respiration rate. Here again, the differences between the respiration rates of the treated animals were not significant ($p > 0.5$) but the controls were respiring significantly less than them ($p = 0.04$). After 4 days of exposure to the effluent, the treated animals were respiring at a rate of about 1.5 times that of the control, and this increased to about 1.9 times for the animals exposed for 8 and 12 days. These differences were quite similar to those found for the animals treated with the Bintulu effluent.

As for the Scope for Growth, although the pattern of response for the clams exposed to this effluent was quite similar to the response of the animals exposed to the COT effluent from Bintulu, the impact here again was more severe. For the Bintulu effluent, negative values of Scope for Growth were only recorded for animals exposed to the highest effluent concentration equivalent to its LC_{40}. Here, even animals treated with an effluent level equivalent to its LC_{20} were recording negative values.

Sublethal Effects of the Labuan COT Effluent on the Clam, *Donax faba*

The responses of the clams when exposed to sublethal levels of the Labuan COT effluent follow a pattern similar to those exposed to the effluent from Lutong (Figure 4). This is rather surprising because, based on the results of the acute toxicity tests, the Labuan effluent was found to be about 9 times more toxic than the one from Lutong.

Detailed analysis of the individual responses reveal that the Labuan effluent was slightly more effective in influencing the clearance rate and Scope for Growth value of the test clams. On exposure to concentrations of the Labuan effluent equivalent to its LC_{10}, LC_{20}, and LC_{40}, the clearance rate of the treated animals was reduced to about 50, 26, and 7% of the control, respectively. The Scope for Growth of the animals treated with the lowest effluent concentration declined to about 20 to 50% of the untreated ones. By comparison, similar exposures to the Lutong COT effluent resulted in clearance rates of about 55, 32, and 9% of the control, respectively. The Scope for Growth value of the clams exposed to an equivalent LC_{10} concentration of the effluent was 20 to 50% of the untreated clams.

DISCUSSION

The results of this study indicate that the "growth" physiology of the clam *Donax faba* was affected when exposed to sublethal levels of the COT effluents. On the whole, the Labuan COT effluent had the greatest effect on the physiology of the clams. The Lutong effluent, although determined earlier (through acute toxicity tests) to be much less toxic than the Labuan effluent, had only a slightly lesser effect on the physiology of the test animals. The Bintulu effluent had least effects of the three.

The general trend in the effects of the three effluents on the physiology of the clams were similar. An increase in effluent concentration resulted in a decrease in clearance rate, an increase in rate of respiration, and a decrease in energy available for somatic growth and gamete production, measured in terms of Scope for Growth.

The effects of external stresses such as exposure to pollutants on marine organisms have been documented in many earlier studies. Gilfillan (1975) reported that when the blue mussel *Mytilus edulis* was exposed to low concentrations of oil (1% extract), its gross carbon consumption was elevated. However, the increase was nullified by an increase in carbon respiration. In the present study, exposure to all three effluents resulted in a decrease in food consumption; no stimulation of feeding activities was observed. The study by Gilfillan (1975) found that exposure to high concentrations of hydrocarbon reduced the carbon consumption of mussels. This is in agreement with the results of this study. Gilfillan (1975) indicated that the net carbon flux of the mussel was decreased by about 50% when exposed to 10% oil and to almost zero when the animals were put in 50% oil. A decrease in rate of clearance of food particles from suspension by the blue mussel when exposed to various concentrations of oil was reported by Widdows et al (1982). Similar results have been documented for the clam *Mercenaria* (Keck et al, 1978). Johnson (1977) suggested that a decrease in food clearance rate following exposure to hydrocarbons may be due to inhibition of the cilia involved in feeding.

The effects of the three COT effluents on the respiration rates of the clams, where elevation of respiration rates resulted from exposure to increasing concentrations of the pollutant, were quite different from results of several earlier reports on various species of bivalves. Fong (1976), Gilfillan et al (1976) and Stainken (1978) working with *Mya arenaria* and Figure 4. Effects of Labuan Crude Oil Terminal effluent on various physiological parameters of the clam, *Donax faba*.

Widdows et al (1982) working with the blue mussel *Mytilus edulis*, indicated that increased oxygen consumption was only observed at low concentrations of hydrocarbon; at higher levels of exposure, respiration rates declined. Decreased respiration rates following exposure to relatively high concentrations of oil were also reported by Gilfillan (1975) and Sabourin and Tullis (1981). These studies further suggested that the decrease in respiration rates was probably due to valve closure or narcotization of ciliary surfaces, or both. In the present study, we were surprised to find the treated animals respiring more than the controls. We expected the contrary because of reduced valve movements, as indicated by declines in clearance rates.

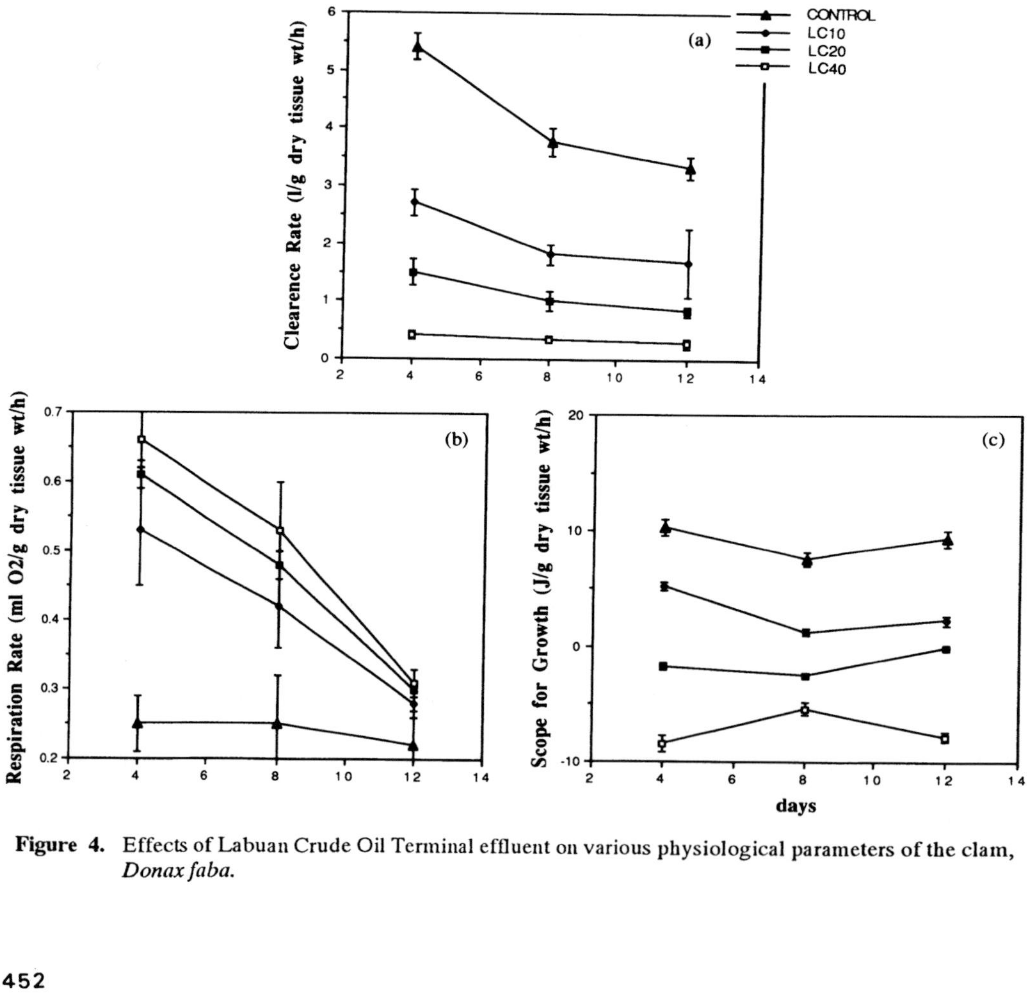

Figure 4. Effects of Labuan Crude Oil Terminal effluent on various physiological parameters of the clam, *Donax faba.*

The effects of pollutants on the growth rate of marine animals such as the bivalves have also been given a lot of attention. Published data have indicated that the rate of tissue growth or the potential for growth was depressed when animals were exposed to various kinds of pollutants. This is in agreement with the result of this study. In a series of observations on populations of *Mya arenaria* subjected to oil spills (the "Tamano" spill in Maine in 1972 and the "Arrow" spill in Nova Scotia in 1970), Gilfillan et al. (1976, 1977) and Gilfillan and Vandermeulen (1978) demonstrated a reduction in carbon flux (i.e., carbon available for growth) and tissue growth due to decreased feeding rates and increased respiration losses. Keck et al., (1978) documented reduced growth of juvenile hardshell clam *Mercenaria* when exposed to 60 µg total hydrocarbon per liter. They suggested that this reduction was due to decreased feeding rate as well as a declined assimilation.

The variations in values of Scope for Growth of the clams in this study seem to be in agreement with results from some of the earlier studies on the blue mussel, where an increase in external stress resulted in a decrease in the values of Scope for Growth (Widdows, 1978a, 1978b; Bayne, 1975; and Widdows et al., 1982). However, the range of values calculated in this study were found to be greater than those reported earlier, for example by Widdows et al., (1982) for the blue mussel. In this study, control clams were estimated to have Scope for Growth values of about 7 - 10 J/g dry tissue wt./h. These values declined to about -9 J/g dry tissue wt./h in animals exposed to effluent concentrations equivalent to their LC_{40}. Widdows et al (1982) documented that their control animals had Scope for Growth values of about 3.5 J/g dry mass/h.

In summary, this study indicates that the sublethal effects of the three kind of effluents were reflections of their toxicity, where most severe responses resulted from the Labuan effluent, followed in order by the Lutong and Bintulu effluents. The results of the study emphasize the importance of measuring several related responses on an organism when exposed to external stresses. These kinds of observations give a more complete picture of the effects of the stresses. In this study particularly, if respiration rate were to be the only response monitored, one might be misled to believe that the effluents acted as some sort of stimulant for the clams while in fact, on the overall, they inhibited the growth physiology of the animal.

ACKNOWLEDGEMENTS

The authors wish to thank Mohd Zambri Othman and Zubaidi Mokhtar for their assistance during the course of the study.

This study was primarily supported by a consultancy from PETRONAS and their contractors, Sarawak Shell Berhad and Sabah Shell Petroleum Company.

REFERENCES

Bayne, B.L., 1975, Aspects of physiological condition in *Mytilus edulis* (L.) with special reference to the effects of oxygen tension and salinity. *in:* "Proceedings of the 9th European Marine Biology Symposium," H. Barnes, Ed., Aberdeen University Press, Aberdeen, pp. 213-238.

Bayne, B.L., and Worrall, C.M., 1980, Growth and production of mussels *Mytilus edulis* from two populations. Mar. Ecol. Progr. Ser. 3:317.

Bayne, B.L., Brown, D.A., Burns, K., Dixon, D.R., Ivanovici, A., Livingstone, D.R., Lowe, D.M., Moore, M.N., Stebbing, A.R.D., and Widdows, J., 1985, "The Effects of Stress and Pollution on Marine Animals," Praeger, New York.

Conover, R.J., 1966, Assimilation of organic matter by zooplankton, Limnol. Oceanogr. 11:338.

Fong, W.C., 1976, Uptake and retention of Kuwait Crude Oil and its effects on oxygen uptake by the soft-shell clam *Mya arenaria*, J. Fish. Res. Bd Can 33, 2774.

Gilfillan, E.S., 1975, Decrease of net flux in two species of mussels caused by extracts of crude oil. Mar. Biol. 29:53

Gilfillan, E.S., Donovan, D., Hanson, S., and Mayo, D.W., 1976, Reduction in carbon flux in *Mya arenaria* caused by a spill of No. 2. fuel oil. Mar. Biol. 37:115.

Gilfillan, E.S., Mayo, D.W., Page, D.S., Donovan D., and Hanson, S., 1977, Effects of varying concentrations of petroleum hydrocarbons in sediments on carbon flux in *Mya arenaria*. in: "Physiological Responses

of Marine Biota to Pollutants," F.J. Vernberg, A. Calabrese, F.P. Thurberg, and W.B. Vernberg, eds., Academic, New York, pp. 299-314.
Gilfillan, E.S., and Vandermeulen, J.H., 1978, Alterations in growth and physiology in chronically oiled soft-shell clams, *Mya arenaria*, chronically oiled with Bunker C from Chedabucto Bay, Nova Scotia, 1970-76. J. Fish. Res. Bd Can. 35: 630.
Johnson, F.G., 1977, Sublethal biological effects of petroleum hydrocarbon exposures: Bacteria, algae and invertebrates, in: "Effects of Petroleum on Arctic and Subarctic Marine Environments and Organisms, Vol. II. Biological Effects," D.C. Malins, ed., Academic, New York.
Keck, R.T., Heess, R.C., Wehmiller, J., and Maurer, D.,1978, Sublethal effects of the water soluble fraction of Nigerian crude oil on the juvenile hard clam, *Mercenaria* L, Environ. Pollut. 15: 109.
Rand, G.M., and Petrocelli, S.R., 1985, "Fundamentals of aquatic Toxicology, Methods and Applications. Hemisphere.
Sabourin, T.D., and Tullis, R.E., 1981, Effects of three aromatic hydrocarbons on respiration and heart rate of the mussel *Mytilus californianus*. Bull. Environ. Contam. Toxicol. 26: 729 .
Solorzano, L., 1969, Determination of ammonia in natural waters by the phenolhypochlorite method, Limnol. Oceanogr. 14:799.
Stainken, D.M., 1978, Effects of uptake and discharge of petroleum hydrocarbons on the respiration of the soft-shell clam *Mya arenaria*. J. Fish. Res. Bd Can. 35: 637.
Widdows, J., 1978a, Combined effects of body size, food concentration, and season on the physiology of *Mytilus edulis*, J. Mar. Biol. Ass. U.K. 58:109.
Widdows, J., 1978b, Physiological indices of stress in *Mytilus edulis*, J. Mar. Biol. Ass. U.K., 58:125.
Widdows, J., 1985, Physiological responses to pollution. Mar. Poll. Bull. 16 (4): 129.
Widdows, J., Bakke, T., Bayne, B.L., Donkin, P., Livingstone, D. R., Lowe, D.M., Moore, M.N., Evans S.V., and Moore, S.L., 1982, Responses of *Mytilus edulis* L. on exposure to the water accommodated fraction of North Sea oil. Mar. Biol. 67:15.
Winberg, G.G., 1960, Rate of metabolism and food requirement of fishes, Fish. Res. Bd. Can. Transl. Ser., 194: 1.

A PROTOCOL FOR DETERMINING OIL AND GREASE IN PRODUCED WATERS

K. Simms[1], A. Zaidi[1], and O. Bhargava[2]

[1] Wastewater Technology Centre
Burlington, Ontario
Canada

[2] Omtek Inc.
Dundas, Ontario
Canada

INTRODUCTION

In the process and regulatory monitoring of produced waters from onshore and offshore oil recovery operations, oil and grease is widely used as a water quality parameter. The control of discharges of oil and grease in the brines from oil production is of concern to the oil industry because of: (i) potential deleterious effects of process upsets on downstream equipment for produced water recycle, (ii) plugging problems if the produced water is re-injected to subsurface formations, and (iii) regulatory requirements if the produced water is discharged to the marine environment. Currently, there is considerable controversy surrounding the issue of oil and grease analysis in produced waters. The controversy relates to several issues including: (i) the definition of oil and grease, (ii) the importance of distinguishing between petroleum hydrocarbon material and non-petroleum source extractable material, (iii) the identification of specific target compounds of concern in various oil and grease monitoring applications, and (iv) the analytical methods for determination of oil and grease in produced waters.

Presently, one of the key issues facing the oil industry is the appropriateness of the current methods being used to determine the oil and grease concentration in produced waters. Although a large number of analytical methods have been published for the determination of oil and grease, there is as yet no consensus among researchers, industry and regulatory agencies as to which method(s) is most suitable for application to produced waters. Difficulties relate to defining the specific components of oil and grease in the produced water that are of interest in a particular monitoring application. If the focus is on removal of oil and grease using conventional oil removal processes, then the suspended fraction is of primary concern; while the soluble oil and grease components may also be relevant if the produced water is to be discharged to the marine environment. Despite the different reasons for monitoring oil and grease and a wide variation in the methods being used for analysis of oil and grease, monitoring decisions are currently being made without due consideration of the limitations of the standard analytical techniques and the state of the art of produced water treatment technology.

This issue was recently brought to the forefront as a result of an international agreement (Montreal Protocol, 1990) to phase out the most commonly used extracting solvent in oil and grease analysis, namely CFC-113, by the year 2000. It therefore became necessary to modify the current standard and regulatory oil and grease methods for produced water analysis by that

Produced Water, Edited by J.P. Ray
and F.R. Engelhart, Plenum Press, New York, 1992

time. To understand the significance of such a change, it is important to recognize the method dependence of the oil and grease parameter. Oil and grease does not represent a specific chemical property or component of a wastewater sample; rather it is a gross parameter intended to provide a general indication of hydrophobic material in a sample. The specific components of a produced watersample quantified as oil and grease may include normal aliphatic hydrocarbons, carboxylic and naphthenic acids, complex aromatic compounds and sulphur and nitrogen containing hydrocarbon derivatives and hydrocarbon based treatment chemicals depending on the specific solvent and analytical technique employed. The introduction of an alternate solvent and/or technique will therefore impact both on the available database on oil removal equipment performance and on the offshore discharge regulations which have traditionally reflected equipment performance as well as environmental standards. Given the number of variables associated with oil and grease analysis, it is essential that a rigorous experimental program be conducted to generate hard data related to the suitability of any proposed alternative solvents and analytical techniques for application to produced waters.

Since 1985, the Wastewater Technology Centre has been involved in developing approaches for resolving issues related to oil and grease removal technologies. In 1990, a study was initiated to develop and recommend oil and grease analytical methods for monitoring onshore and offshore produced waters. This study is being conducted in two phases. In the initial phase of the work, a number of potentially suitable solvents and analytical techniques for the determination of total and suspended oil and grease will be examined under controlled laboratory conditions using simulated produced water samples. The test work will then be extended to actual wastewater samples both in-house and in the field, during the second phase of the study. It is expected that this work will culminate in the development of oil and grease analytical method(s) for monitoring of produced waters. The WTC study is focusing initially on offshore produced water applications because of the requirement for suitable regulatory control of marine discharges.This paper provides a brief review of concerns regarding current analytical techniques for oil and grease determination, presents the results of initial WTC work on the evaluation of solvents for total oil and grease analysis, and identifies various approaches for future work on the development of analytical methods for oil and grease analysis.

TECHNICAL BACKGROUND

Oil and Grease Analysis

Reviews of oil and grease analysis in petroleum contaminated wastewaters are available in the published literature (API, 1975; CONCAWE, 1984; Oldham, 1985;). The published methods are based on liquid/liquid extraction and generally comprise of sample pretreatment, serial extraction of the sample with an organic solvent (such as CFC-113) and quantitation of the oil in the solvent extract by either gravimetric or infrared spectrophotometric techniques. From Table 1, which shows some published methods used in North America and Europe for oil and grease analysis, it can be seen that there is considerable variation in all of the steps of the analytical protocols being used.

Sample pretreatment normally implies acidification with hydrochloric acid, however, nitrogen purging for removal of volatiles is also employed. In the gravimetric method, oil is recovered from the solvent extract by evaporation of the solvent under controlled conditions. This method will result in the loss of light hydrocarbons that are volatilized at the specified drying temperature. In the infrared (IR) spectrophotometric technique, oil content is determined by comparing the absorbance of the solvent extract at one, two or three wavelengths in the range of 2925 to 3030 cm^{-1}, against either a crude oil sample or a synthetic calibration standard. With IR analysis, higher oil and grease values would normally be obtained than with gravimetric determination. In another WTC study (Fraser, 1991), an exception to this trend was noted when solid phase extraction was tested for recovery of oil and grease in place of the standard liquid/liquid extraction procedure. At low concentration (40 mg/l), both IR and gravimetric analysis of the solid phase eluent gave similar results, possibly because the solid phase did not adsorb the lighter hydrocarbon fraction.

Table 1. Detail of some published oil and grease analytical methods (1 of 3)

ORGANIZATION METHOD DATE	U.S. EPA # 00556 1976	U.S. EPA # 413.2 1978	U.S. EPA #418.1 1978	APHA/WPCF/AWWA #5520 B 1990	APHA/WPCF/AWWA # 5520 C 1990
TITLE	Oil & Grease Total Recoverable, sep. funnel extraction	Oil & Grease Total Recoverable, IR spectrophotometric	Petroleum-hydrocarbon total recoverable IR, spectrophotometric	PARTITION-GRAVIMETRIC	PARTITION-INFRARED
MATERIAL DETECTED	CFC extractable not volatilized at 70° C	CFC extractable	CFC extractable not adsorbed by silica gel	CFC extractable not volatilized at 70° C	CFC extractable
SAMPLE volume bottle	1000 ml glass bottle	1000 ml glass bottle	1000 ml glass bottle	1000 ml glass bottle	1000 ml glass bottle
PRETREATMENT pH acid used	2 or lower 5 ml H_2SO_4 or HCl	2 or lower 5 ml HCl	2 or lower 5 ml HCl	2 or lower HCl	2 or lower HCl
EXTRACTION solvent volume (ml) container mixing by time	CFC-113 30/30/30 + 10/20 2 l sep. funnel hand shaking 2 min. each	CFC-113 30/30/30 + 10/20 2 l sep. funnel hand shaking 2 min. each	CFC-113 30/30/30 + 10/20 2 l sep. funnel hand shaking 2 min. each	CFC-113 30/30/30 + 10/20 2 l sep. funnel hand shaking 2 min. each	CFC-113 30/30/30 + 10/20 2 l sep. funnel hand shaking 2 min. each
DRYING filter paper drying agent quantity (g) collected in	solvent moistened sodium sulphate 1.0 g 100 ml vol. flask	solvent moistened sodium sulphate 1.0 g 100 ml vol. flask	solvent moistened sodium sulphate 1.0 g 100 ml vol. flask	solvent moistened sodium sulphate 1.0 g tared boiling flask (250 ml)	solvent moistened sodium sulphate 1.0 g tared boiling flask (250 ml)
ABSORBENT type quantity (g) method			silica-gel, 60-100 mesh, 3.0 g magnetic stirrer at least 5 min. Discard first 5-10 ml extract		
QUANTITAION method details	Gravimetric 70°C (covered water bath) for 15 min. Draw air for 1 min. Cool in dessicator for 30 min.	IR Scan standards and samples at 3200-2700 cm^{-1}. Measure absorbance at 2930 cm^{-1}	IR Scan standards and samples at 3200-2700 cm^{-1}. Measure absorbance at 2930 cm^{-1}. After IR add another 3 g silica gel	Gavimetric 70°C bath for 15 min. Draw air for 1 min. Cool 30 min. in desiccator.	IR Scan standards and samples at 3200-2700 cm_{-1}. Measure absorbance at 2930 cm_{-1}. After IR add another 3 g silica gel
Blank Range Detection Limit Precision	5-1000 mg/l 93% recovery ± 0.9mg	0.2-1000 mg/l 99% ± 1.4 mg/l	<1.0 mg/l 99% ± 1.4 mg/l	93% recovery s.d. = 1.4 mg.	99% recovery s.d. = 1.4 mg/l

Table 1. Detail of some published oil and grease analytical methods (2 of 3)

ORGANIZATION METHOD DATE	APHA/WPCF/AWWA # 5520 F 1990	ASTM #D3921 1987	ASTM # D3921 1987	ASTM # D4281 1987
TITLE	HYDROCARBONS	OIL GREASE & PETROLEUM HYDROCARBONS IN WATER	SILICA GEL TREATMENT	OIL & GREASE BY GRAVIMETRIC DETERMINATION
MATERIAL DETECTED	CFC extractable, not absorbed by silica gel	CFC extractable	CFC extractable, petroleum hydrocarbons	CFC extractable
SAMPLE volume bottle	 1000 ml glass bottle	 750 ml glass with teflon lined lid	 750 ml glass with teflon lined lid	 1000 ml glass/non-organic insoluble lid
PRETREATMENT pH acid used	 2 or lower HCl	 2 or lower H_2SO_4, $NaHSO_4$	 2 or lower H_2SO_4, $NaHSO_4$	 2 or lower HCl, H_2SO_4, $NaHSO_4$
EXTRACTION solvent volume (ml) container mixing by time	 CFC-113 30/30/30 + 10/20 2 l sep. funnel hand shaking 2 min. each	 CFC-113 30/30/30 2 l sep. funnel hand shaking 2 min. each	 CFC-113 30/30/30 2 l sep. funnel hand shaking 2 min. each	 CFC-113 60/60/60, 20 rinse 2 l sep. funnel hand shaking 2 min. each
DRYING filter paper drying agent quantity (g) collected in	 general purpose sodium sulphate 1.0 g tared boiling flask (250 ml)	 general ashless sodium sulphate 1.0 g 100 ml vol. flask	 general ashless sodium sulphate 1.0 g 100 ml vol. flask	 general phase sep. sodium sulphate 1.0 g tared boiling flask (250 ml)
ADSORBENT type quantity (g) method			 silica gel 100-200 mesh 3.0 g magnetic stirrer for 10 min.	 silica gel 60-200 mesh 3.0 g magnetic stirrer at least 5 min. Use redissolved residue
QUANTITATION method details	IR*/Gravimetric** *Scan at 3200-2700. Measure absorbance at 2930 cm_{-1}. , ** 70°C bath for 15 min. Draw air for 1 min. Cool 30 min. in desiccator.	IR Scan 3.13-3.70 μm Net absorbance at 3.41 μm Calibrate with known oil or reference (1:1 iso-octane:cetane)	IR Measure before and after Si gel. Scan 3.13-3.70 μm., Net absorbance at 3.41 um Calibrate with known oil or reference, (1:1 iso-octane:cetane)	Gravimetric steam bath Draw air for 5 min. Desiccate for 1 hr.
Blank Range Detection Limit Precision	 0.2 mg/l 99 % recovery, s.d. = 1.4 mg	none 0.5-100 mg/l 0.5 mg/l st=0.167x + 0.333, x = conc. (mg/l)	none 0.5-100 mg/l 0.5 mg/l st=0.160x + 0.329, x = conc. (mg/l)	 4-100 mg/l bias: 5% at 66 mg/l

Table 1. Detail of some published oil and grease analytical methods (3 of 3)

ORGANIZATION METHOD DATE	PARCOM 1987	FRENCH AFNOR 78503 1979	GERMAN DIN 38409 1981	SYENSK STANDARD SS 02 81 45 1979
TITLE	DETERMINATION OF OIL CONTENT OF MIXTURES	DETERMINATION OF TOTAL HYDROCARBONS	DETERMINATION OF HYDROCARBONS	OIL&GREASE IN WATER IR SPEC METHOD
MATERIAL DETECTED	Most light oil fractions		Petroleum hydrocarbons	CCl_4 extractable, not absorbed on Al_2O_3
SAMPLE volume bottle	 1000 ml glass bottle inert cap liner	 1000 ml	 500 ml	 1000 ml glass bottle
PRETREATMENT pH acid used other	 2 or lower HCl Bubble 30C N_2/air @ 1 l/min for 45s	 5 HCl (1 N) + 5 g NaCl	A: 1-2, H_2SO_4 B: 1-2, H_2SO_4 if <20 mg/l, add 100g Mg_2SO_4 if >20	 2 HCl (6N), or 50 g/l Mg Cl_2 for emulsion
EXTRACTION solvent volume (ml) container mixing by time	 CFC-113 or CCl_4 50/50 1 l sep. funnel hand shaking 2 min. each	 CCl_4 50 machine 15 min.	 CFC-113 25 1 l sep. funnel machine (centrifuge for emulsion) 10 min.	 CCl_4 50/10-20 rinsing 1-2 l sep. funnel machine/centrifuge extract @ 3000 rpm for emulsions
DRYING filter paper drying agent quantity (g) collected in	 medium grade	 yes sodium sulphate 1.0 g	 sodium sulphate 10 g	 Whatman 31 sodium sulphate 100 ml vol. flask
ADSORBENT type quantity (g) method	 Florisil	 Florisil MgO-SiO_2 column media	 Al_2O_3 10 g column chromatography	 Activated Al_2O_3 70-230 mesh 10 cm column (fig.) pass extract for separation of hydrocarbons
QUANTITATION method details	 IR Scan at 3200-2700. Measure absorbance at 2930 cm^{-1}. Calibrate with known crude or reference oil (n-hexadecane) 2-80 mg/l to at least 1000 mg/l.	 IR Measure at 3.42 μm Calibrate with known crude	 IR Measure at 2958, 2924, 3030 cm^{-1}. Use synthetic standards.	 IR Scan standards & samples at 3300-2700 cm_{-1}. Measure absorbance at 2925. Calibrate with reference (37.5% hexadecane 37.5% isooctane, 25% benzene).
Blank Range Detection Limit Precision	 0.2 mg/l 99% recovery s.d. = 1.4 mg		 0.1 mg/l 99% ± 1.4 mg/l	 0.1-100 mg/l 93% recovery s.d. = 1.4 mg.

Oldham (1985) has reviewed the published methods and recommends that regulatory methods for produced waters be compatible with field conditions, conducted by sufficiently trained personnel and selected to quantify only the petroleum hydrocarbon content. Sun et al (1987) have identified several anomalies and variations in the oil and grease analysis of petroleum wastewaters associated with the analytical procedures employed. For example, it was shown that a number of non-petroleum compounds such as phenolics and carboxylic acids, may contribute significantly to measured oil and grease concentrations in produced waters. Treatment of the solvent extract for removal of polar compounds or prefiltration of the produced water sample may provide some compensation for these soluble oil and grease compounds. However, solvent selection should be an initial consideration as each solvent will have a different selectivity and extraction efficiency for targeted as well as undesirable components of produced waters.

Solvent Selection

As indicated in Table 1, CFC-113 (1,1,2-trichloro-1,2,2-trifluoroethane) has been widely used for oil and grease analysis. First introduced as a replacement for n-hexane, CFC-113 is not flammable and relatively non-toxic, forms an organic layer below the aqueous phase, and can be used for both gravimetric and infrared spectrophotometric analysis. It has been shown, however, that heavy oil fractions such as asphaltenes and bitumen are not dissolved in CFC-113 (Gruenfeld, 1973) while non-petroleum sulphur compounds may be readily extracted (American Public Health Assoc., 1990). In addition, the high ozone depletion potential of CFC products will limit the availability of this solvent in the near future. Under the provisions of the Montreal Protocol (1990), the production and consumption of fluorinated solvents will be reduced by half by 1995 and will be completely phased out by January 1, 2000. The manufacturers of CFC have proposed the use of recently developed hydrochlorofluorocarbon products as temporary replacements for CFC-113. Recommended products include 1,1-dichloro, 2,2,2-trifluoroethane (HCFC-123) and an 80:20 wt% blend of 1,1-dichlorofluoroethane (HCFC-141b) and HCFC-123. These products have significantly less severe environmental impact than CFC-113 but will likely also be subject to future restrictions.

Several other options can be identified for potentially suitable solvents for oil and grease analysis in offshore produced waters on the basis of oil industry experience, ease of handling and safety considerations. Carbon tetrachloride has been used for infrared spectrophotometric determination of oil and grease in produced waters from offshore oil production. Mathavan and Viraraghavan (1990) recently compared CCl_4 with two CFC products for manual and instrumental IR spectrophotometric oil and grease analysis and found similar extraction efficiencies in all cases. However, carbon tetrachloride also falls under the agreements of the Montreal Protocol and should therefore be considered obsolete as an alternative to CFC-113 in any oil and grease method development work. Other solvents that could be used include dichloromethane, trichloroethylene, and tetrachloroethylene. All three of these solvents have been used for oil and grease analysis by various oil companies and are considered to have higher extraction efficiencies for certain crude oils than the extraction efficiency of CFC-113. Tetrachloroethylene is also suitable for IR spectrophotometric analysis.

The United States Environmental Protection Agency has recommended (USEPA, 1991) the use of a 20:80 vol% mixture of methyl tertiary butyl ether (MTBE) and n-hexane as an alternative to CFC-113 in the gravimetric determination of oil and grease. Both the ASTM and AWWA have also drafted methods based on the use of this solvent blend. The USEPA approach to solvent selection was to identify a solvent that was most comparable to CFC-113 in terms of solvating capabilities and boiling point (Kawahara, 1991) for a range of wastewater matrices. The main concern with this solvent blend, as with n-hexane, is its flammability. The mixture is also difficult to use because it does not form an organic layer at the bottom of the separatory funnel during the extraction procedure. In addition, MTBE is highly soluble in water (5% v/v) and solvent loss may result in low oil and grease values.

Solvent selection for oil and grease analysis is a complex problem and a systematic approach to solvent selection has been lacking to date. While the initial consideration for solvent selection should include selectivity and extraction efficiency, there is little data on what target organic components would be of concern for various produced water monitoring applications. Further criteria such as safety, ease of use and toxicity are also important. Even if specific compounds were defined as oil and grease and some prediction of solvent behaviour was made based on physical/chemical data such as dielectric constant and polarity index, the complexity and variability of the oilfield brine matrix would still mandate experimental evaluation of solvents to observe their dissolution characteristics.

SCOPE OF THE WORK

Two separate solvent test programs were conducted by the WTC as shown in Table 2. Test work was conducted using simulated produced water samples generated by addition of crude oils to brine solutions. Oil concentration levels and brine compositions were chosen to reflect expected characteristics of produced waters from Canadian offshore oil production platforms.

The first program was conducted in 1990 and focused on a comparison of two HCFC solvents with CFC-113 for gravimetric oil and grease analysis using a simulated produced water sample generated with Terra Nova crude. Some properties of the tested solvents are presented in Table 3.

The second study outlined in Table 2, was conducted in 1991 to evaluate four alternative solvents for CFC-113 for gravimetric oil and grease analysis and one alternative solvent for infrared analysis (tetrachloroethylene). In this study, simulated produced water samples were prepared using Terra Nova, Prudhoe Bay and Avalon crude oils.

Table 2. Scope of WTC test work

	HCFC Study	Six Solvent Study
Solvents	CFC-113, HCFC-123, HCFC 141b/HCFC-123 80:20 blend	CFC-113, MTBE:Hexane, n-Hexane, DCM, TCE, TTCE
Crude Oil	Terra Nova	Terra Nova, Prudhoe Bay, Avalon
Method	Gravimetric	Gravimetric, Infrared
pH	2	2 & 7
Concentration	100 mg/l (500 ml sample)	40 & 80 mg/l (1 l sample)
Replicates	4,5	3

The solvents tested were 1,1,2-trichloro-1,2,2-trifluoroethane (CFC-113), 20:80 % mixture of methyltertiarybutyl ether and n-hexane (MTBE:hexane), n-hexane, dichloromethane (DCM), trichloroethylene (TCE), and tetrachloroethylene (TTCE). Most of these solvents were selected because they have been used by the oil industry for produced water monitoring. The solvent mixture of MTBE:hexane was selected because it had been recommended for use by the USEPA. Table 4 presents some basic properties of the organic solvents included in this study. Due to the high boiling point of TTCE, this solvent could not be included in the gravimetric test program and was therefore compared with CFC-113 using the partition infrared technique. The other solvents absorb strongly in the IR range between 3.1 and 3.7 μm and were only used in the gravimetric analytical procedure.

A factorial experimental design was employed in the initial stages of the six solvent study. For Terra Nova crude oil, the six solvents were evaluated at two pH levels (2 and 7) and two concentration levels (low - 40 mg/l and high - 80 mg/l). For each test, two procedure blanks and three test replicates were employed. This allowed a determination of the relative significance of solvent type, pH level and concentration, on oil recovery. The subsequent test work using Prudhoe Bay and Avalon crude oils was limited to one level of conditions, namely pH 2 and low concentration of added oil. A concentration of 40 mg/l was chosen for this work because it is the current regulatory limit for oil in produced water discharge from Canadian offshore oil production platforms (COGLA, 1990). During this work, characteristics of some of the crude oils and solvent extracts were examined by gas chromatography to determine if this would be a suitable technique to identify components selectively extracted by each solvent.

Table 3. Properties of solvents used in HCFC study

Solvent	CFC-113	HCFC-141b	HCFC-123	80:20 Blend
Supplier	Dupont		Dupont	Allied Signal
Formula	$CF_2ClCFCl_2$	CH_3CCl_2F	$CHCl_2CF_3$	—
Mol. Wt.	187.4	116.9	152.9	—
Normal b.p.(°C)	47.6	32.0	27.8	31.0
Density (g/ml)	1.57	1.23	1.46	1.38
Vapour Pressure (mm Hg)	300	—	688	—
Flash Point	None	None	None	None
Ozone Depleting Potential	0.8	0.1	0.02	0.08
Greenhouse Potential	2.3	0.14	0.03	0.11
Toxicity-TLV (ppm)	500 - 1000	50 - 100	100	50 - 100

The Terra Nova crude was supplied by Petro-Canada Resources, Calgary. The Prudhoe Bay and Avalon crude oils were supplied by Environment Canada's Environmental Emergency Technology Division, Ottawa. Table 5 presents some characteristics of the three crude oil samples.

The criteria for comparison of the solvents was the accuracy, measured as weight percent recovery of added oil, and the precision, as determined by the ratio of standard deviation to mean recovery value (coefficient of variation or c.v.).

METHODS

The synthetic produced water samples were generated by adding a known weight of crude oil into a brine solution in a separatory funnel and shaking manually. The crude oil sample (known weight) was first dissolved and diluted in a volumetric flask in the specified solvent and then added to the brine using a pipette. While the target oil concentration levels were 40 and 80 mg/l, the actual concentrations varied between sets of samples. For example, at the low concentration level, the actual weight of oil added to the one litre brine solutions varied from 39 to 48 mg, while at the high level, the weight ranged from 79.3 to 116.8 mg. For the HCFC study, the brine solution was generated by addition of 25 grams of sodium chloride to 500 ml of distilled water. For the six solvent study, the brine solution consisted of one litre of distilled water with commercial sea salt at a concentration of 100 g/l. If pH adjustment was required, hydrochloric acid was added to the brine prior to addition of the crude oil. A one litre sample was serially extracted three times with 30 ml aliquots of the specified solvent in a separatory funnel. After removal of the entrained water through sodium sulphate and phase separating paper, the solvent extract was either: (i) diluted in a volumetric flask with the solvent for partition infra-red spectrophotometric measurements of oil and grease, determined by comparison of the absorbance at 2930 cm^{-1} of the sample extract with calibration standards, or (ii), evaporated at a specified temperature on a rotary evaporator in a tared boiling flask for partition gravimetric determination. Details of the WTC gravimetric and infrared procedures are listed in Table 6. The solvent-partition procedure was slightly more complicated with MBTE:hexane and n-hexane than with the other solvents because the solvent phase was at the top of the aqueous phase in the separatory funnel. The water had first to be drawn off before the solvent phase could be collected and the water then returned to the separatory funnel for further extractions. In the

HCFC study, a water bath temperature of 45°C was used for the HCFC solvents because of their low boiling points as compared to CFC-113. The drying temperature for the six solvent study was set at 70°C to generate comparable test results, although the temperature could potentially be optimized for each solvent. With trichloroethylene, however, a temperature of 90°C was required due to its high boiling point. The spectrophotometric calibration standards were prepared by addition of the test crude to the specified solvent.

Table 4. Properties of Solvents used in Six Solvent Study

Solvent	CFC-113	MTBE	n-Hexane	DCM	TCE	TTCE
Supplier	Caledon Lab. Ltd.	BDH Chemicals	Baxter Chemicals	Caledon Lab. Ltd.	Caledon Lab. Ltd.	Caledon Lab. Ltd.
Formula	$C_2Cl_3F_3$	$CH_3OC(CH_3)_3$	$CH_3(CH_2)_4CH_3$	CH_2Cl_2	$CHClCCl_2$	CCl_2CCl_2
Mol. Wt.	187.4	88.14	86.18	84.94	131.4	165.8
Normal b.p.(°C)	47.6	55.2	69	39.8	87.1	121
Vapour Pressure (mmhg) @ 20°C	300	313 @ 30°C	124	400	58	14
Density @ 20°C	1.57	0.741	0.66	1.33	1.47	1.62
%Solubility in Water @ 20°C	0.02	4.8	0.001	1.6	0.10	<0.1%
Flash Point	none	-28°C	-21.7°C	none	none	none
Toxicity- TLV (ppm)	1000	—	50	50	50	50
Polarity Index (P')	—	~2.5	0.1	3.1	1.0	—
Dielectric Constant E	2.41 @ 25°C	4.0 @ 20°C	1.89 @ 20°C	9.08 @ 20°C	3.4 @ 16°C	—

Table 5. Properties of test oils

Crude Type	Terra Nova	Prudhoe Bay	Avalon
Density (kg/m^3)	0.876	0.903	0.843
Gravity (°API)	30.0	25.2	36.4
Pour Point(°C)	-6	-12	9
Flash Point (°C)	0	-8	10
Wax Content (wt%)	7.9	3.9	6.1
Viscosity (mm^2/s) @ 20°C	35.3	47.3	145

RESULTS AND DISCUSSIONS

The extraction efficiencies determined for the various solvents and analytical methods are presented as percentage recovery of added oil. All reported data have been corrected for procedure blanks.

HCFC Study

Oil recoveries determined using CFC-113, an HCFC-141b /HCFC-123 blend and HCFC-123 are compared in Table 7. Mean oil recoveries are seen to range from 63.3 to 71.6%. Gravimetric analysis using HCFC-123 resulted in the highest mean recovery of the added Terra Nova crude oil at 71.6%; however, this solvent also exhibited the lowest precision, with the coefficient of variation exceeding 6%. There were also some difficulties in handling of HCFC-123 solvent due to its high volatility, which made volumetric measurement and metering of the solvent difficult. Statistical comparison of the mean recoveries, using the "F-test" and the "t-test" at the 95% confidence level, indicated that there was no significant difference between the results obtained with the HCFC-141b/HCFC-123 blend and HCFC-123, while these two solvents did differ significantly from CFC-113. It should be noted, however, that the difference of mean oil recoveries between CFC-113 and HCFC-123 is only 8.3%. At the test concentration of about 100 mg/l, this represents a difference of 8 mg/l between the highest and lowest mean recovery values. Although HCFC solvents would appear to be viable solvents for oil and grease analysis, test work on these solvents was not extended to actual produced water samples because of the likelihood of future restrictions on their use. Subsequent test work, under the six solvent study, focused on other solvents that had been identified as potentially suitable for determination of oil and grease in produced waters.

Table 6. WTC oil and grease analytical methods

Organization Method Date	WTC Gravimetric Method 1991	WTC Infrared Method 1991
Title	Partition Gravimetric	Partition Infrared Spectrophotometric
Material detected	Solvent Extractables	Solvent Extractables
Sample		
volume	500/1000 ml	1000ml
bottle	glass bottle	glass bottle
Pretreatment		
pH	2 or lower	2 or lower
acid used	HCl	HCl
Extraction		
solvent	Specified	Specified
volume(ml)	30/30/30 + 10/30	30/30/30 + 10/30
container	2 l separatory funnel	2 l separatory funnel
mixing by	hand shaking	hand shaking
time	2 min. each	2 min. each
Drying		
filter paper	1PS phase separatory paper	1PS phase separatory paper
drying agent	sodium sulfate	sodium sulfate
quantity(g)	1.0g	1.0g
collected in	tared boiling flask (125ml)	tared boiling flask (100ml)
Quantitation		
method	Gravimetric	IR known crude
details	70° C bath and rotavap	calibration
	(90°C for TCE, 45°C for HCFC's)	2930 cm^{-1}
	1 hour desiccation	

Table 7. Results of HCFC Study (Terra Nova crude)

Solvent	Mean Oil Recovery (%)	Coefficient of Variation (%)	n
CFC-113	63.3	3.5	4
HCFC-141b:HCFC-123	68.8	4.8	5
HCFC-123	71.6	6.1	4

Six Solvent Study

Effect of pH and Concentration. Tables 8 and 9 present the results of gravimetric analyses on the Terra Nova crude samples at two different pH (2 and 7) and concentration levels (40 and 80 mg/l). The mean oil recoveries ranged from 66.6% to 77.1% over all conditions tested. The highest recovery (77.1%) of Terra Nova crude oil was found using CFC-113 at pH 2 and low oil concentration. The lowest recovery was found using n-hexane, at pH 2 and at high concentration. A perusal of the mean recoveries indicates that pH and concentration do not affect the oil recovery to as great an extent as solvent type. This suggestion was confirmed by statistical analysis of variance. Solvent type was found to be the most significant factor affecting recovery of added oil, although the interaction effects between pH and concentration and between pH and solvent were also significant. No trend was detected as to the pH and concentration levels that resulted in the highest recoveries for all five solvents tested. The calculated coefficient of variation data, in Table 9, indicates that n-hexane yielded the highest precision under all pH and concentration conditions tested with an overall c.v. of 1.3 %. The data did not indicate any clear trend towards higher precision under certain pH or concentration conditions.

Higher precision was achieved with the infrared method than with the gravimetric technique as shown in Table 11, with TTCE yielding the most precise results. Coefficients of variation from the infrared oil and grease tests did not exceed 1.6% with CFC-113 or 0.7% with TTCE.

Table 8. Effect of pH, concentration, and solvent on recovery by gravimetric method

Solvent	Recovery of Added Oil (%)				
	pH 2		pH 7		Overall
	40 mg	80 mg	40 mg	80 mg	
CFC-113	77.1	75.7	74.0	73.7	75.1
MTBE:Hexane	69.7	71.5	69.6	72.5	70.8
n-Hexane	67.7	66.7	68.2	69.1	67.9
DCM	68.7	68.8	68.5	71.0	69.3
TCE	72.1	70.7	66.6	69.7	70.0
OVERALL	71.1	70.7	69.4	71.2	70.0

Table 9. Effect of pH, concentration and solvent on precision by gravimetric method

Solvent	Coefficient of Variation (%)				
	pH 2		pH 7		Overall
	40 mg	80 mg	40 mg	80 mg	
CFC-113	1.9	2.1	3.8	2.5	2.6
MTBE:Hexane	2.7	2.2	2.8	2.6	2.6
n-Hexane	1.4	1.8	0.7	1.3	1.3
DCM	2.8	2.7	2.8	2.2	2.6
TCE	2.5	1.6	1.3	3.6	2.3
OVERALL	2.3	2.1	2.3	2.4	2.3

Table 10. Effect of pH, concentration, and solvent on recovery by infrared method

Solvent	Recovery of Added Oil (%)				
	pH 2		pH 7		Overall
	40 mg	80 mg	40 mg	80 mg	
CFC-113	87.9	79.3	88.4	84.4	85.0
TTCE	95.0	94.2	97.1	91.1	94.4
OVERALL	91.5	86.8	92.8	87.8	89.7

Table 11. Effect of pH, concentration, and solvent on precision by infrared method

Solvent	Coefficient of Variation(%)				
	pH 2		pH 7		Overall
	40 mg	80 mg	40 mg	80 mg	
CFC-113	1.2	1.4	1.4	1.6	1.4
TTCE	0.0	0.3	0.7	1.0	0.5
OVERALL	0.6	0.9	1.1	1.3	1.0

These results demonstrate the strong dependence of the oil and grease parameter on the extracting solvent used for both partition gravimetric and partition infrared spectrophotometric analysis. The effects of pH, concentration and their interactions with solvent type were determined to be more significant when the infrared spectrophotometric method is employed than when the oil is quantified gravimetrically. It is likely that the effects of pH and concentration were confounded by the loss of light hydrocarbons during the solvent evaporation step of the partition gravimetric method. Clearly, spectrophotometry is the more powerful technique for the examination of solvent extraction efficiency and selectivity for targeted oil and grease components. However, it is important to establish that the calibration standards employed in the IR method (synthetic or crude oil) closely match the sample matrix. Some tests were also conducted to determine the effect of salt concentration on oil recovery using CFC-113 with a simulated produced water sample generated with Terra Nova crude oil. The results indicated that there was no significant difference at the 95% confidence level between oil recoveries obtained at 0 and at 100 g/l brine concentration.

Based on the results of these tests it was decided that further work on additional simulated produced water samples could be conducted at only one level of salt concentration, pH, and oil concentration conditions and still allow valid comparison of solvents for oil and grease determination. The salt concentration represents expected levels in offshore produced waters from the East coast of Canada. A pH of 2 was selected as it is normally recommended in the standard published methods and the low oil concentration level was selected as it reflects current regulatory limits on offshore produced water discharge. It is cautioned, however, that any test work on actual produced water samples should be designed so as to consider the pH and concentration effects and their interactions with solvent type.

Solvent Comparison. Table 12 presents the results of the oil and grease analyses conducted on all three simulated produced water samples using the WTC partition gravimetric method. Oil recovery ranged from 62.7% to 77.9% and procedure blanks ranged from 0.1 to 4.8 mg. These data are summarized for ease of interpretation in Table 13. Mean oil recoveries determined using the gravimetric method ranged from 65.4 to 77.1%. At a test concentration of 40 mg/l, this represents a difference of 5 mg/l between the highest and lowest mean recovery values. No one solvent was found to give the highest recoveries for all the types of crude oil tested. The solvent which produced the highest recovery of Terra Nova crude oil was CFC-113 with 77.1%

Table 12. Partition gravimetric test data

	Recovery of Added Oil (%)					
	CFC-113	MTBE: hexane	n-Hexane	DCM	TCE	RANGE
Blank	0.4 to 4.8 mg	0.8 to 2.4 mg	1.5 to 4.8 mg	0.1 to 2.0 mg	1.7 to 3.3 mg	0.1-4.8 mg
Terra Nova	75.4 77.9 77.9	68.6 68.7 71.9	66.9 67.4 68.7	66.5 69.3 70.2	70.6 71.5 74.1	66.5 - 77.9
Prudhoe Bay	64.3 68.3 69.3	67.4 68.0 68.9	65.3 65.8 67.9	68.4 68.7 72.2	70.3 70.8 74.2	64.3 - 74.2
Avalon	66.4 66.6 67.5	67.6 68.1 69.4	64.3 67.2 69.3	69.4 71.5 73.1	62.7 64.3 69.2	62.7 - 73.1
RANGE	64.3 - 77.9	67.4 - 71.9	64.3 - 69.3	66.5 - 73.1	62.7 - 77.9	62.7 - 77.9

Table 13. Comparison of partition gravimetric results

	Recovery of Added Oil (%)										
	CFC-113		MTBE: Hexane		n-Hexane		DCM		TCE		OVERALL
	mean	c.v.	mean	c.v.	mean	c.v.	mean	c.v.	mean	c.v.	mean
Blank	2 mg	—	1.3 mg	—	3.2 mg	—	1.1 mg	—	2.2 mg	—	2.0 mg
Terra Nova	77.1	1.8	69.7	2.7	67.7	1.3	68.5	2.8	72.1	2.5	71.0
Prudhoe Bay	67.3	3.9	68.1	1.2	66.3	2.1	69.8	3.0	71.8	2.9	68.7
Avalon	66.8	0..7	68.4	1.2	66.9	3.1	71.3	2.1	65.4	4.3	67.8
OVERALL	70.4	2.1	68.7	1.7	67.0	2.2	69.9	2.6	69.8	3.2	69.2

recovery, while TCE produced the highest recovery of Prudhoe Bay crude oil at 71.8% and DCM produced the highest recovery of Avalon crude oil at 71.3%.

Gas chromatography was used to compare the oil recovered by extraction of a Prudhoe Bay crude oil sample using the test solvents. Compounds lighter than n-nonane (C_9) were not distinguished due to interference from the carrier solvent. Examples of the simulated distillation curves generated from the GC data are shown in Figures 1 and 2 for CFC-113 and DCM respectively. The temperature range from 150° C to 500° C represents hydrocarbon compounds ranging from C_9 to C_{37}. Note that the oil from partition gravimetric analyses with CFC-113 and DCM shows a considerable loss of light ends. The results using CFC-113 parallel the original crude oil sample for the most part, however, some loss of light ends (b.p. < 200° C) can be seen with the CFC-113 extract. Comparison of this curve with a similar curve for DCM extract, shown in Figure 2, indicates that DCM may be less suitable than CFC-113 for extracting light oil components (b.p. < 270° C) of the Prudhoe Bay crude. It is suggested that GC/FID can be used to determine components lost during solvent evaporation and to compare solvent selectivity for groups of compounds in produced water samples. It would not be recommended, however, for targeting specific compounds that may be of concern in a certain monitoring application.

The overall data shown in Table 13 indicates that CFC-113 would be the preferred solvent in terms of oil recovery with an overall mean recovery of 70.4%. The lowest mean recovery was found using n-hexane at 67.0%. Both crude oil type, solvent type and their interaction were shown to be statistically significant factors at the 95% confidence level in the recovery of oil under the conditions tested.

A review of data in Table 13 does not reveal any marked difference between the procedure blanks or the test precision obtained with the five solvents used for gravimetric analysis. Mean procedure blanks ranged from 1.1 to 3.2 mg and the coefficients of variation ranged from 0.7 to 4.3%. It is interesting to note that the more complex manipulations required for solvent extraction with MTBE:hexane and n-hexane, did not result in lower precision under the conditions of this test program. As this work was conducted under controlled laboratory conditions, however, this analytical precision might not be maintained in field work with actual oilfield brine samples.

The test data from partition infrared analyses of the three simulated produced water samples using CFC-113 and TTCE as the extracting solvents are shown in Table 14. Substantially more accurate results were obtained with the infrared method as compared to the gravimetric method; with oil recovery values ranging from 86.7% to 99.7% in the IR tests. The calculated mean recovery values and coefficients of variation are shown in Table 15. With the infrared technique, TTCE gave more accurate results than CFC-113 for all three crude oils tested. The mean oil recoveries for Terra Nova, Prudhoe Bay, and Avalon crude oil using TTCE were 95.0%, 99.6% and 92.7% respectively. The lowest recovery (87.9%) was found with CFC-113 using Terra Nova crude oil. As with the gravimetric test results, the mean oil recovery values spanned a range of approx. 12%, which would represent a difference in measured oil and grease content of 5 mg/l at a concentration of 40 mg/l. Statistical analysis of variance was conducted on the

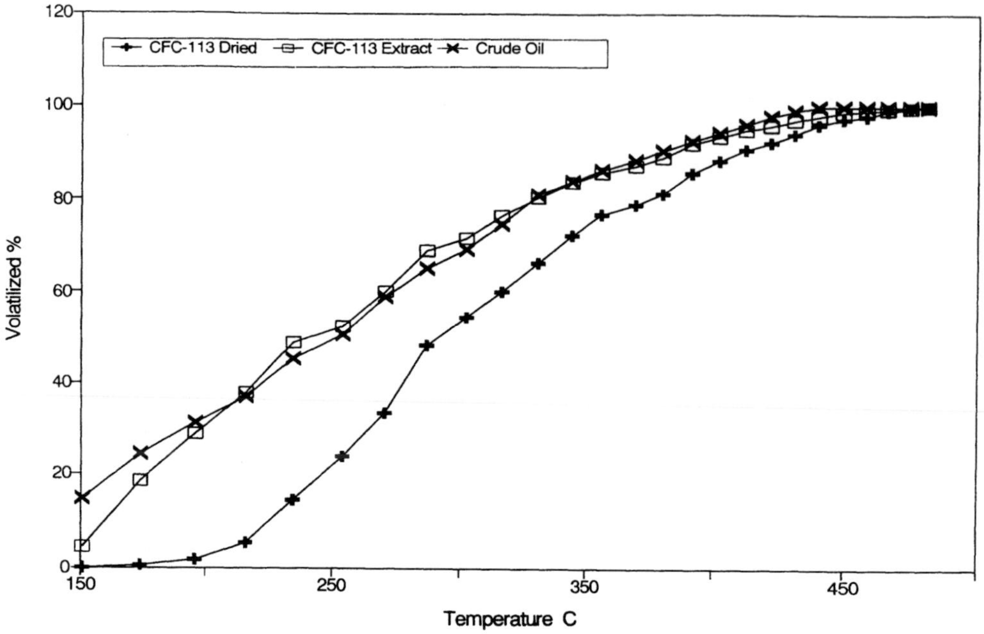

Figure 1. GC Simulated distillation curves (CFC-113).

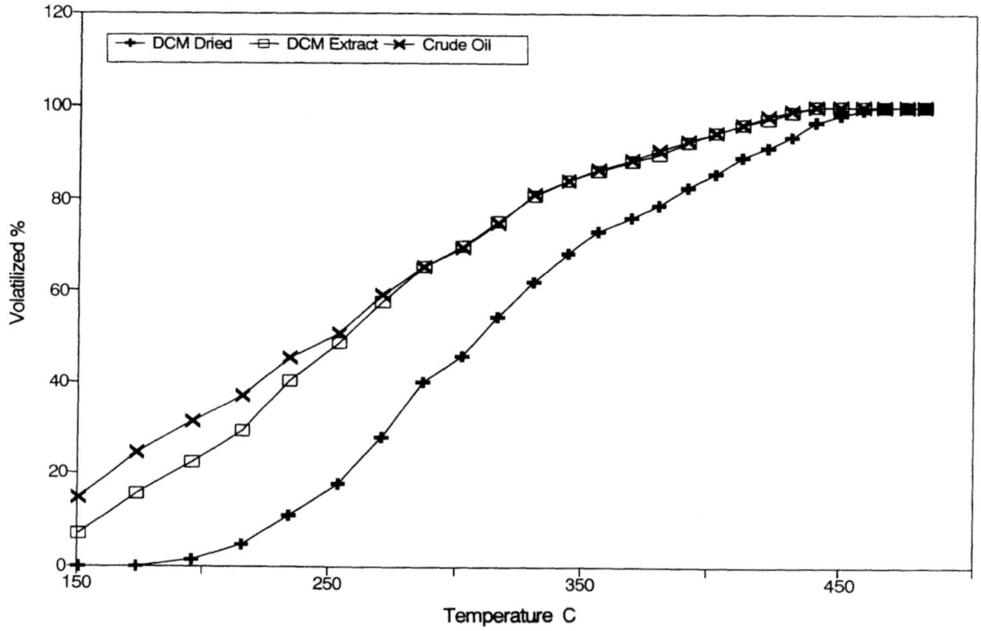

Figure 2. GC Simulated distillation curves (DCM).

data in Table 15 and each of solvent type, crude oil type and solvent-crude oil interaction effects were shown to be significant factors above the 95% confidence level in the recovery of added oil. It can be seen, from Table 15, that more precise results are obtained with the infrared method as compared to the gravimetric method, with TTCE giving more precise data than CFC-113. Calculated coefficients of variation ranged from 0.2% to 2.5% with CFC-113 and from 0% to 0.8% with TTCE.

Table 14. Partition infrared test data

	Recovery of Added Oil (%)		
	CFC-113	TTCE	RANGE
Blank	0.0 to 1.2 mg	-1.0 to 0.2 mg	-1 - 1.2 mg
Terra Nova	86.7 88.5 88.5	95.0 95.0 95.0	86.7 - 95.0
Prudhoe Bay	88.1 91.7 92.4	99.5 99.7 99.7	88.1 - 99.7
Avalon	89.0 89.0 89.4	92.0 92.7 93.3	89.0 - 93.3
RANGE	86.7 - 92.4	92.0 - 99.7	86.7 - 99.7

Table 15. Comparison of partition infrared results

	Recovery of Added Oil (%)				
	CFC-113		TTCE		OVERALL
	Mean	c.v.	Mean	c.v.	Mean
Blank	0.6 mg	—	-0.5 mg	—	0.05 mg
Terra Nova	87.9	1.1	95.0	0.0	91.5
Prudhoe Bay	90.7	2.5	99.6	0.1	95.1
Avalon	89.1	0.2	92.7	0.8	90.9
OVERALL	89.2	1.3	95.8	0.3	92.5

The results of the WTC study to date demonstrate the variations in oil recovery with solvent and crude oil type and the markedly lower oil recoveries obtained with the gravimetric technique due to the loss of low boiling compounds. For the gravimetric technique, no one solvent is obviously preferable in terms of oil recovery and precision for all three simulated produced water samples tested. A statistical comparison of means was conducted using a two sample t-test to determine the significance of the difference between the four alternative solvents and CFC-113 for gravimetric oil and grease analysis. CFC-113 was found to give significantly different recoveries for Terra Nova crude oil than each of MTBE:hexane, n-hexane, DCM and TCE. However, no significant difference at the 95% level was found between CFC-113 and the other solvents for the Prudhoe Bay and Avalon crude oil tests and for the overall results from gravimetric oil and grease analysis. Each of MTBE:hexane, n-hexane, DCM, and TCE can therefore be considered generally comparable with CFC-113 under the conditions tested. The infrared technique can be seen to yield significantly higher and more precise results than the gravimetric technique. For the results from infrared spectrophotometric determination of oil and grease, a significant difference at the 95% level was found between CFC-113 and TTCE, with TTCE being the preferred solvent for all three crude oils tested.

CONCLUSIONS

A number of conclusions can be drawn from the WTC work completed to date on the determination of oil and grease in oilfield brines.

1. The strong dependence of the oil and grease parameter on the characteristics of the extracting solvent and the nature of the crude oil in the sample was confirmed in this test work.
2. HCFC-123 solvent and the 80:20 HCFC-141b:HCFC-123 solvent blend were shown to yield significantly higher oil and grease values than CFC-113 when used for the gravimetric determination of Terra Nova crude in a salt water sample. The blended HCFC141b:HCFC123 solvent would likely be more suitable for routine laboratory use due to its lower volatility. This solvent may be suitable as an interim replacement for CFC-113 but will likely be subject to future restrictions due to environmental concerns.
3. For the three simulated produced water samples, the results suggest that each of MTBE:hexane, n-hexane, dichloromethane and trichloroethylene could potentially provide reasonably accurate and precise gravimetric oil and grease determinations as compared with CFC-113. If this is confirmed through more extensive tests using simulated and actual wastewater samples, it would suggest that other criteria for solvent selection such as toxicity, environmental impact, ease of use, safety, and cost must be given consideration in addition to the relevant solubility characteristics of the solvent.
4. Oil recoveries obtained with the partition infrared spectrophotometric analytical method were found to be substantially higher than those obtained with the gravimetric method. Results of the IR analyses indicate that tetrachloroethylene has a higher extraction efficiency and precision with all of the crude oil samples tested as compared to CFC-113.
5. The WTC solvent comparison test work conducted to date has focused on simulated produced water samples. While providing some comparative information regarding the suitability of the test solvents for determination of oil and grease in produced waters, it will not be prudent to expect a strong correlation between oil and grease data from simulated samples and actual wastewaters. The test work must therefore be extended to actual oilfield brine samples.
6. GC/FID analysis could be a useful tool to provide some indication of the groups of components in the produced water that are being extracted by the organic solvents and that are being lost during solvent evaporation. This technique could be employed with actual produced water samples to provide some comparison of solvent selectivity.

ACKNOWLEDGEMENTS

This paper was supported, in part, by the Federal Panel on Energy, Research and Development (PERD). The assistance of S. Kok is acknowledged with appreciation.

REFERENCES

American Public Health Association, "Standard Methods for the Examination of Water and Wastewater", 17th ed., Washington, D.C., (1990).

API, 1975, (American Petroleum Institute) - Task Group on Analytical Methods, "Review of Analytical Methods for Determination of Oil and Grease in Produced Waters from Oil and Gas Extraction Industry Operations," prepared for the API Committee on Environmental Conservation, Division of Protection, July 14 (1975).

COGLA, 1990, (Canada Oil and Gas Lands Administration), Ottawa, Ontario, "Offshore Waste Treatment Guidelines," Jan.(1990).

CONCAWE, 1984, "Determination of Hydrocarbons in Aqueous Effluents from the Oil Industry by Infrared Analysis", prepared by CONCAWE's Special Task Force on Guidelines on Tests and Standard Methods (1984).

Fraser, J., 1991, Wastewater Technology Centre, Burlington, Ontario, personal communication, November (1991).
Oldham, G.F., "Methods of Analysis and Sampling of Oil in Water," report prepared by B.P. International Ltd., presented at the Seminar on Aspects of the Prevention of Pollution in the Mediterranean Sea, organised by CONCAWE and E & P Forum in association with the Coordinating Unit for the Mediterranean Action Plan, UNEP, Athens, 11-13 February (1985).
Gruenfeld, M. and Frederick, R., 1973, Extraction of dispersed oils from water for quantitative analysis by infrared spectrophotometry, Environ. Sci. Technol. 7(7):636 (1973).
Kawahara, F., 1991, Office of Research and Development, USEPA, Cincinnati, Ohio, personal communication, February (1991).
Mathavan, G.N. and Viraraghavan, T., 1990, Comparison of I.R. determinations of oil and grease in water, Env. Tech.11:455 (1990).
Montreal Protocol (1990): UNEP, Regulations on freeze of production and consumption of CFC (chloro-fluoro-carbons), (1987), (1990).
Sun, P.T., C.L. Price, J.C. Raia, and Balderas, R.A., 1987, Anomalies in oil and grease analyses of petroleum wastewaters and their implications, 42nd Purdue University Industrial Waste Conference Proceedings, 151, (1987).
USEPA, 1991, (United States Environmental Protection Agency), "Oil and Grease, Total, Recoverable. Method 413.3. (Gravimetric, Separatory Funnel Extraction)," STORET NO. 00556, Proposed for NPDES, Issued August (1991).

INVESTIGATIONS INTO THE USE OF MODIFIED ZEOLITES FOR REMOVING BENZENE, TOLUENE, AND XYLENE FROM SALINE PRODUCED WATER

J.S. Janks[1] and F. Cadena[2]

[1] Texaco Inc., Exploration and Production Technology Department
Bellaire, Texas
USA

[2] Department of Civil, Agricultural and Geological
New Mexico State University
Las Cruces, New Mexico
USA

ABSTRACT

"Tailored" zeolites, made by adsorbing neutralized amines onto natural zeolites, were studied as an adsorption medium for the removal of BTEX from saline produced water. The major components in natural zeolites are smectite, clinoptilolite (or chabazite) and cristobalite. BTEX levels were reduced a minimum of 70 to 85 % in the produced water after treatment with ethyleneamine and methylamine zeolites. X-ray diffraction was used to analyze the natural and tailored zeolites.

INTRODUCTION

Produced water is the largest volume stream in the exploration and production operation. In the water are variable amounts of metals, salts, oil, radionuclides, and treatment chemicals. One specific group of organic compounds that may be present and potentially harmful to humans is benzene, toluene, ethyl-benzene, and *o*-, *m*- and *p*-xylene, collectively known as "BTEX." Produced water from oilfield operations can contain BTEX. Benzene typically ranges from 5.8 to 12.2 mg/l in gas operations to 1.3 to 8.7 mg/l in oil productions (Stephenson, 1991). However, that range can be exceeded. Benzene levels in produced water as high as 65 mg/l have been reported in the news media (Jackler, 1991). Produced water is currently exempt from federal rules on hazardous waste as part of the oilfield operations.

The type of disposal is critical, as it must adequately protect the environment and yet have the least cost to the operator. In many areas where there is no alternative, the water is simply reinjected into underground formations containing water unsuitable for human consumption. In other places, it is discharged directly into the environment. The potential for introducing BTEX into the environment is a concern, and this paper addresses our initial investigations into new technology that may reduce hazards to the environment by removing BTEX from water.

Produced Water, Edited by J.P. Ray
and F.R. Engelhart, Plenum Press, New York, 1992

The removal of BTEX from water

BTEX removal from water by altered clay minerals and granular activated carbon (GAC) has been known for years. GAC is very effective, but it tends to be nonspecific when adsorbing organic compounds. Most of the pollutant-removal studies have been performed on hazardous compounds, underground storage tank (UST) leachate, and water sources (Wolfe, et al., 1986; Cadena and Jeffers, 1987; Boyd, et al., 1988; Cadena, 1989). Smectite (montmorillonite) has been shown to effectively remove organic compounds, if the clay mineral's naturally occurring metallic cations are exchanged with organic ones (McBride, et al., 1977; Boyd, et al., 1988; Lee, et al., 1990). Removal of BTEX from produced water has not been as widely studied. Produced water has highly variable salt levels ranging from near salt-free conditions to several hundred thousand ppm, free and soluble hydrocarbons, and treatment chemicals, all of which may affect the sorbent-water interaction.

Clay minerals can effectively remove BTEX from water under certain conditions. All clays have relatively high surface areas that allow for the adsorption of organic compounds, but the smectite family, with its large surface area and charged interlamellar surface, is the most effective (Chou and Shoup, 1985). In water however, smectite is hydrophilic due to the hydration spheres around counterions such as Ca^{2+} and Na^{+}. Positively charged counterions are present in all smectites, and balance the negative structural charges inherent in this clay. The presence of metallic counterions generally hinders its ability to adsorb organic molecules from water.

The use of trimethylamine (TMA), tetraalkylamine (TAA), and hexadecyltrimethylamine (HDTMA) in the ammonium form have been particularly effective in removing BTEX from aqueous solutions up to about 60% of the clay's cation exchange capacity (CEC)(Cadena, 1988). Smectite-BTEX adsorption work has produced significant results that are applicable to USTs and contaminant removal, however, smectite presents two problems for BTEX removal from produced water: limited CEC and very low permeabilities.

Background and purpose of this paper

This paper presents the results of a study assessing the effectiveness for BTEX removal of modified ("tailored") commercially-available, natural zeolites, tested on laboratory solutions and produced water. "Natural" zeolites are those zeolites obtained from commercial mining operations without altering the mineral's chemistry. "Modified" zeolites are created by altering the natural chemistry via adsorption of various ammonium ions. The laboratory part of the study determines the relationship among BTEX components, the tailoring agent, and the natural zeolite. The produced water part of the study applies the data to field situations. Studies employing laboratory test solutions have encompassed most of the recent BTEX-zeolite studies (Gao et al., 1991). These experiments have increased our understanding of the complex nature of sorbents and BTEX removal, however these solutions did not approximate produced water. Moderate-to-high salt content, heavy metals and various soluble and insoluble hydrocarbon phases are factors which must be considered in moving the experiments from laboratory to field conditions. In this study, the produced water had 1% insoluble, straight-chain hydrocarbons, 860 mg/l soluble hydrocarbons and 70 mg/l BTEX.

Natural zeolites have not been widely studied for organic pollutant removal. Granular, natural zeolites have two advantages over clays: (1) higher CECs (they may be approximately 50% greater), and (2) very high permeabilities. The tailored zeolites have the additional benefit that unlike GAC, some are more specifically oriented toward certain BTEX species, particularly benzene (Janks and Cadena, 1991). Due to the regulation and limits on benzene, and to a lesser degree toluene, the specific removal of benzene is significant. By tailoring the zeolites, adsorption matrices can be prepared so that specific groups of pollutants (e.g., benzene) may be selectively removed, leaving relatively innocuous substances in solution.

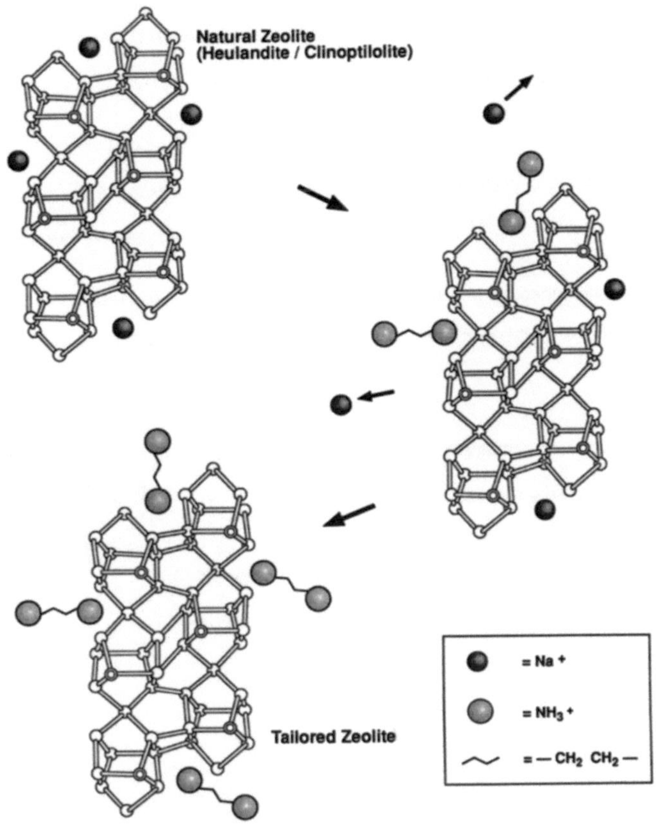

Figure 1. Cage-like structure of heulandite/clinoptilolite (zeolite) with the naturally occuring sodium counterions being replaced by the ammonium salt of ethylene diamine. The final product is a "tailored" zeolite. Zeolite structure taken from Gottardi and Galli (1985). Used with permission of Springer-Verlag (New York) and the authors.

Terminology

The term "smectite" is used throughout this paper as the generic name that encompasses all 100% expandable clays. In everyday usage it is interchangeable with "bentonite" and "montmorillonite." We retained the term "zeolite" for what is commercially sold as such, but is in fact a mixture of minerals. The true zeolite portion of the mixture is commonly less than 50%. The term "tailored" zeolite applies to natural zeolites that we have altered in the laboratory by adsorbing the ammonium salts of test amines into the cation exchange sites held by the natural counterions such as Na^+, Ca^{2+}, or K^+ (Figure 1). Any smectite present in the natural zeolite will also exchange cations. Natural zeolites are framework alumino-silicates whose structure contains channels filled with water and exchangeable cations. Many texts are available on the structure and chemistry of zeolites and those aspects will not be discussed here (Mumpton, 1977a; Gottardi and Galli, 1985).

EXPERIMENTAL

The amines studied were diethylenetriamine (DETA), ethylenediamine (EDA), hexadecyltrimethyl ammonium chloride (HDTMA), monoethanolamine (MEA), piperazine (PIP), tetramethylammonium chloride (TMA), triethylamine (TEA), triethylenepentamine (TEPA), and

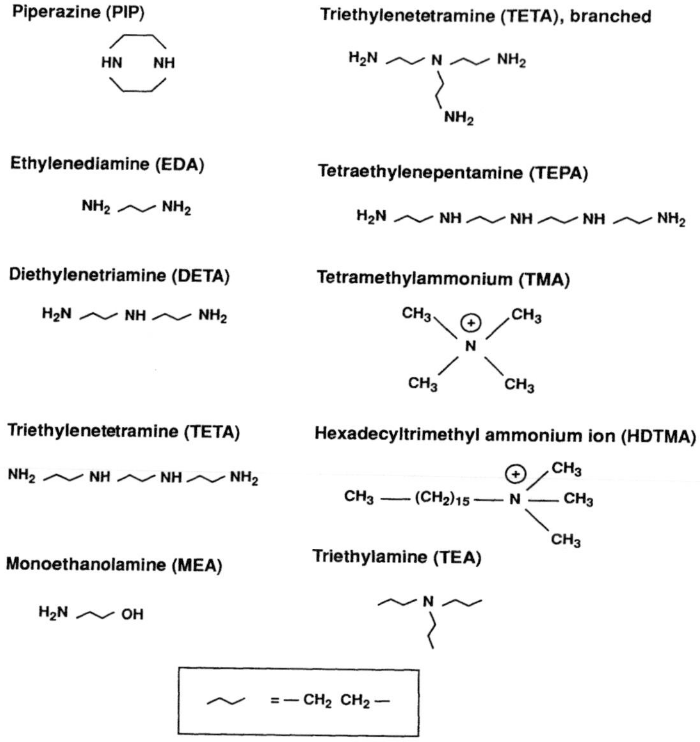

Figure 2. Structure and nomenclature of amines used.

triethylenetetramine (TETA) (Figure 2). Ammonium salts were created by mixing the amines 1:1 with de-ionized water, and neutralizing the solution to pH 3.0 with reagent grade HCl.

The natural zeolite was reacted with the tailoring agent in doses from 0 to 100% CEC. The CEC was determined by the ammonium absorption method and found to be 72 meq/100 g for chabazite and 80 meq/100 g for clinoptilolite (analysis performed by Soil, Water and Plant Testing Laboratory, New Mexico State University, Las Cruces, NM). Reactor vessels, 14 ml test tubes, were used to hold the zeolite suspension with the test water and reacted for 10 minutes. Preliminary studies indicated that this was sufficient time for adsorption to reach equilibrium. The suspensions were centrifuged at 3500 rpm for 10 minutes and the supernatant was then injected into the gas chromatograph port.

For the produced water tests, the ammonium salt solution was placed in contact with the natural zeolite for 24 hours in a 1000 ml beaker. Approximate stoichiometric quantities of ammonium salts and zeolites were calculated by determining the molecular weight of the amine and the theoretical CEC of the zeolite. After the 24-hour period the tailored zeolite was washed free of excess ammonium salts with de-ionized water and allowed to dry. The dried zeolite was placed in contact with the produced water at a ratio of 2.5 parts water to 1 part tailored zeolite. The experiment was run for four hours in a roller oven at 25° C. A produced water blank was run under the test conditions. After four hours, the produced water blank and the water in contact with the tailored zeolite were analyzed for BTEX by the purge and trap gas chromatography method. In all cases, both the photo ionization detector (PID) and flame ionization detector (FID) were used. FIDs were found to have lower detection limits. In all analyses the lowest detection limit value was used if the value was below detection. Inorganic metal analysis was performed by ion coupled plasma (ICP) analysis.

Table 1. X-ray diffraction data, expressed in percent

Sample	TOTC	QTZ	CRIS	CLIN	CARB
CL-1	45	—	22	32	tr
CL-2	52	2	20	24	4
CL-3	36	—	18	40	2
CL-T	38	10	21	20	8
CL-G	23	14	24	36	—
CL-S	10	20	7	61	—

TOTC = total clay content, QTZ = quartz, CRIS = cristobalite, CLIN = clinoptilolite, CARB = calcite + dolomite, "—" = not detected, tr = trace amounts

X-ray diffraction analysis of the natural and tailored zeolites were performed on a Siemens D-5000 diffractometer. The samples were ground to 325 mesh and scanned from 2° - 70° 2θ at a rate of 2° per minute using CuKα radiation. Semiquantitative estimates were obtained using peak areas and intensity factors (Schultz, 1964). An accuracy of +/- 10 % is generally assumed for all analyses. A sample of the whole zeolite was disaggregated with an ultrasonic vibrator and the < 2 μm fraction separated and placed onto glass slides (Drever, 1973).

These slides were examined under relative humidities of 50% and after exposure to ethylene glycol vapors at 60°C for 1 hour. The < 2 μm fraction was scanned from 2°- 48° 2θ.

RESULTS

X-ray diffraction analysis

The natural "zeolite" is a mixture of minerals (Table 1). The "zeolite" is composed of the major components cristobalite (a high temperature polymorph of silica), smectite, and clinoptilolite (a zeolite), and the minor components quartz, calcite, and dolomite. Minor phases were never more than a few percent of the total sample. The total clay fraction, which is almost universally smectite, ranged from about 10-50%. It is a major mineral phase in what is sold as natural "zeolite." Cristobalite is present in all samples analyzed, ranging from 7-24%; it is chemically inert. Clinoptilolite ranged from 20-61% of the T sample. Zeolite sources are highly

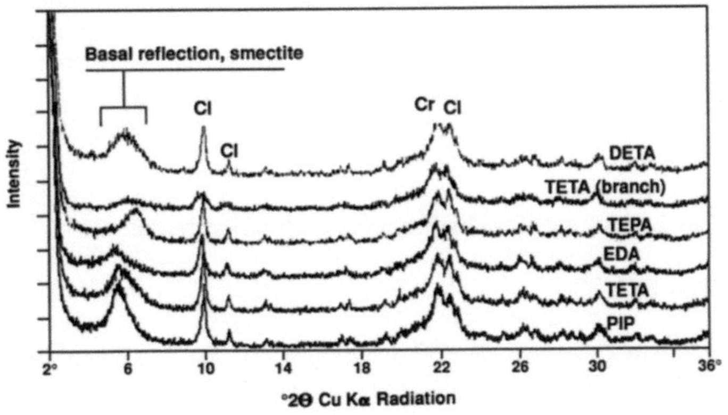

Figure 3. X-ray diffraction trace of the <2μm fraction for natural zeolites with ethyleneamines. The basal reflections (left) are from the smectite present in the natural "zeolite." The clinoptile and cristobalite peaks did not move.

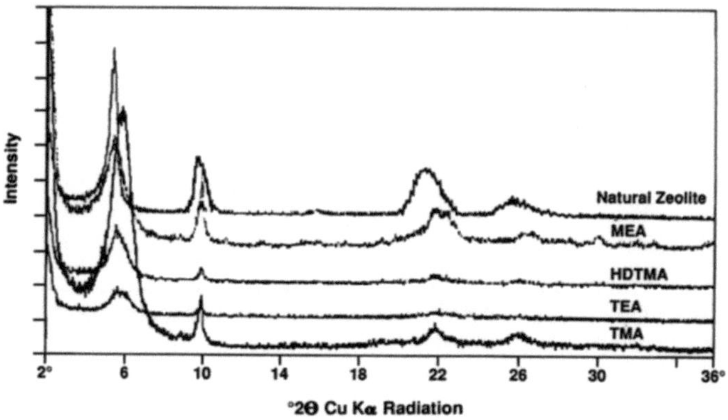

Figure 4. X-ray diffraction patterns, <2 μm fraction.

variable in the clay and clinoptilolite content, reflecting the source locality and mode of formation.

The basal reflection of smectite (typically found about 5° 2θ) is caused by clay platelets orienting themselves in parallel fashion. While the reflection is evident on the whole sample pattern, it is enhanced by the selection of the < 2 μm fraction, which concentrates the clay component. These patterns are shown in Figures 3 and 4. The basal reflection is characteristic for each tailored zeolite. The peak position is a measure of the distance between interlamellar planes which in turn is dependent upon the size of the ions present and their orientation. Well-defined peaks indicate the basal reflections are ordered sufficiently to produce constructive interference of the x-ray beams. In this study, there is a relationship between peak broadening and height and the molecular size of the intercalated ammonium ion. TMA, MEA and PIP produced the sharpest and best-defined peaks. The branched TETA produced the flattest pattern. EDA produced a semi-broad peak. TEA also produced relatively sharp peak with a large d-spacing. While the clays showed considerable variation and change, the associated clinoptilolite did not. Zeolite reflections remained unmoved by the adsorption of ammonium salts.

BTEX adsorption, laboratory experiments

A simplified linear version of the Freundlich equation was used to evaluate the equilibrium data obtained from adsorption isotherms. The Freundlich model was chosen as the most useful because our goal was to quantify the extent of adsorption. The Freundlich model "K" parameter numerically defines adsorption capacity, expressed as amount of sorbate adsorbed per unit weight of sorbent at equilibrium. The basic form of the equation is:

$$x/m = KC^{1/n}$$

when 1/n is equal to 1, the simplified linear partition version becomes:

$$x/m = KC$$

A plot of x/m versus C was constructed where x/m was the dependent variable and C was the independent variable. A best-fit line of the data points was created using least-squares linear regression forcing the calculated line through the origin. The slope of the line is the value of K for the adsorbent.

Two clinoptilolites ("T" and "S") and one chabazite, tailored with TMA, were tested to determine the effectiveness of benzene and toluene from laboratory test solutions as a function of their CEC. For benzene removal (Figure 5), maximum K values occurred at 0.92 l/g for chabazite,

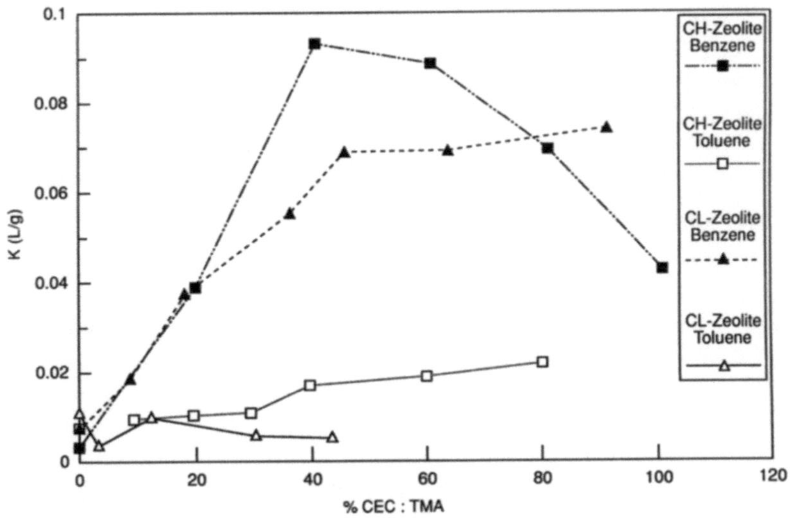

Figure 5. Benzene and toluene absorption for clinoptilolite (CL) and chabazite (CH) as a function of CEC. The tailoring agent was TMA.

and 0.77 l/g for clinoptilolite. The maximum value for chabazite occurs at 40-60 % CEC. For clinoptilolite, the slope of the K values changes markedly at about 40% CEC. The difference in K values between 90% CEC and 40% CEC is only about 0.003 l/g. Toluene removal efficiency is significantly lower than benzene. Maximum K values are 0.02 l/g for chabazite and 0.01 l/g for clinoptilolite.

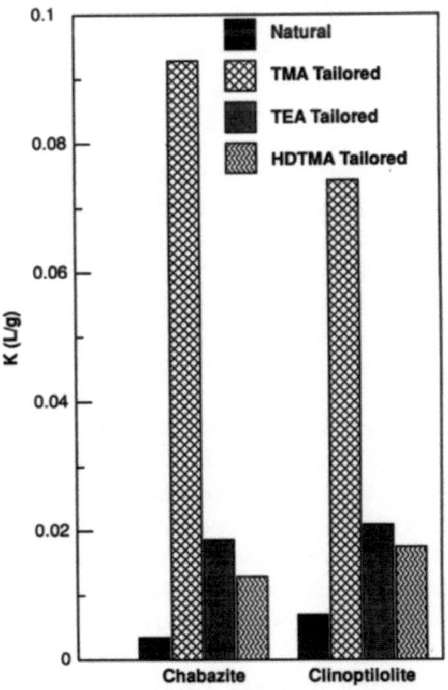

Figure 6. The effect of different tailoring agents on the removal of benzene from laboratory solutions. With both zeolites TMA was the most effective.

Several ammonium compounds were adsorbed onto the zeolites and the efficiency was compared between benzene and toluene. The results for benzene are shown in Figure 6. TMA-zeolites are clearly the most effective removers of benzene, followed by TEA, and HDTMA. The natural zeolite has very little effect (K values are under 0.01 l/g). The reverse trend is seen in toluene removal (Figure 7). The most effective is HDTMA, followed by TEA and TMA. For toluene removal, TMA is about as effective as the natural zeolite.

BTEX removal from produced water

The effects of increasingly complex ammonium salts, and the problems associated with field sample produced water were also examined. The amines ranged from the relatively simple EDA to the complex branched TETA. The produced water sampled contained variable amounts of benzene (15-40 mg/l), toluene (20-60 mg/l), ethyl-benzene (3-15 mg/l) and xylene (undifferentiated, 25-75 mg/l), as well as other hydrocarbons and metal ions. The water analysis is presented in Table 2 below:

Table 2. Produced water analysis. All values except pH are in mg/l. TDS is >10,000 mg/l (saline water)

Na	K	Ca	Mg	Ba	Sr
3,000	263	490	74	28.1	5.6
Fe	As	Cd	Cr	Cu	Ni
0.123	<0.004	<0.001	<0.01	0.14	0.098
Pb	Mn	Se	Ag	Zn	Hg
<0.002	1.54	<0.002	<0.001	0.034	<1.0
Cl^-	$SO_4^=$	HCO_3^-	Br^-	Si	pH
4,980	12	1,430	19.0	7.11	7.36

Total dissolved solids: 10,310
Total organic carbon (soluble): 860

The gas chromatogram of the produced water shows the presence of benzene, toluene, ethyl-benzene and xylene (Figure 8). In addition, there are significant compounds eluting after the BTEX components. They appear to be predominantly n-paraffins and possibly some other aromatics, however these were not analyzed. Stephenson (1991) points out that soluble aliphatic hydrocarbons ranging from decane through tetratriacontane have been found in produced water. In our study, soluble organics were 860 mg/l TOC (including BTEX). This did not include about 1% free hydrocarbons that were also present in the water. These free hydrocarbons were not separated prior to testing.

BTEX removal efficiency from the produced water was tested using TMA, EDA, PIP, TETA, TEPA, and MEA (Table 3). The results of the BTEX removal experiments for some ethyleneamines are shown in Figures 9 and 10. The tailored zeolites removed as little as 9.4% of the total BTEX (MEA) to as much as 85.2% (EDA). Some are highly specific to benzene (TMA), while others are more efficient at removing benzene and toluene than xylene (PIP). In the ethyleneamine group, the higher molecular weight amines were less efficient than the lower molecular weight ones in removing BTEX (Figure 9). Figure 10 shows specific BTEX component removal with different amines. TEPA is the least effective for toluene removal. Xylene is best removed by EDA. PIP however generally excluded xylene. MEA, while having a small molecular weight and the ability to expand smectite to about 16-17 Å, had almost no ability to remove BTEX. There is a correlation between those amines that produced the largest d-spacing (excluding MEA) in smectite and the effectiveness of the tailored zeolite to remove benzene (EDA, PIP and TMA).

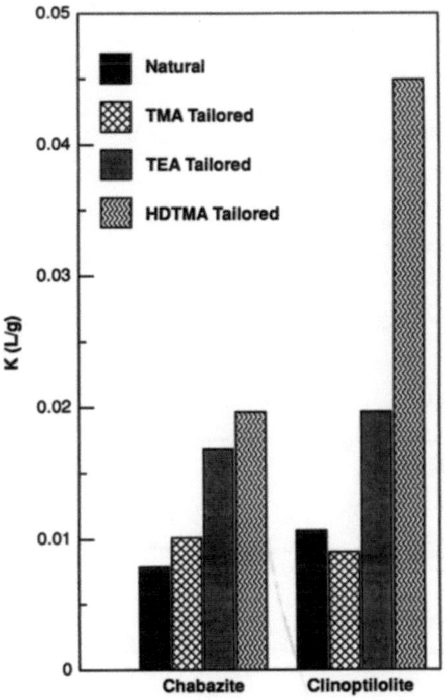

Figure 7. Toluene removal from laboratory solutions. The most effective tailoring agent is HDTMA.

Table 3. BTEX removal data. Upper value is the PID detector; lower value is the FID detector. All values are in mg/l

Sample	Benzene	Toluene	E-benzene	Xylene	BTEX
Untr.	13.8	24.1	<4.8	16.8	59.5
	15.5	24.2	4.9	26.2	70.9
TEPA	<4.8	<4.8	<1.0	<4.8	15.4
	4.4	6.0	3.0	7.3	21.4
TETA	<4.8	<4.8	<1.0	<4.8	15.4
	4.3	6.0	<3.0	5.1	18.4
PIP	<4.8	<4.8	<1.0	5.2	15.8
	4.3	2.3	<3.0	5.2	13.5
EDA	<4.8	<4.8	<1.0	<4.8	15.4
	<3.0	2.1	<3.0	2.4	10.5
TMA	<3.0	<4.8	<1.0	9.4	18.2
	1.5	3.0	<3.0	9.0	16.5
MEA	12.0	24.0	<4.0	16.0	56.0
	15.0	24.0	<3.0	25.0	67.0

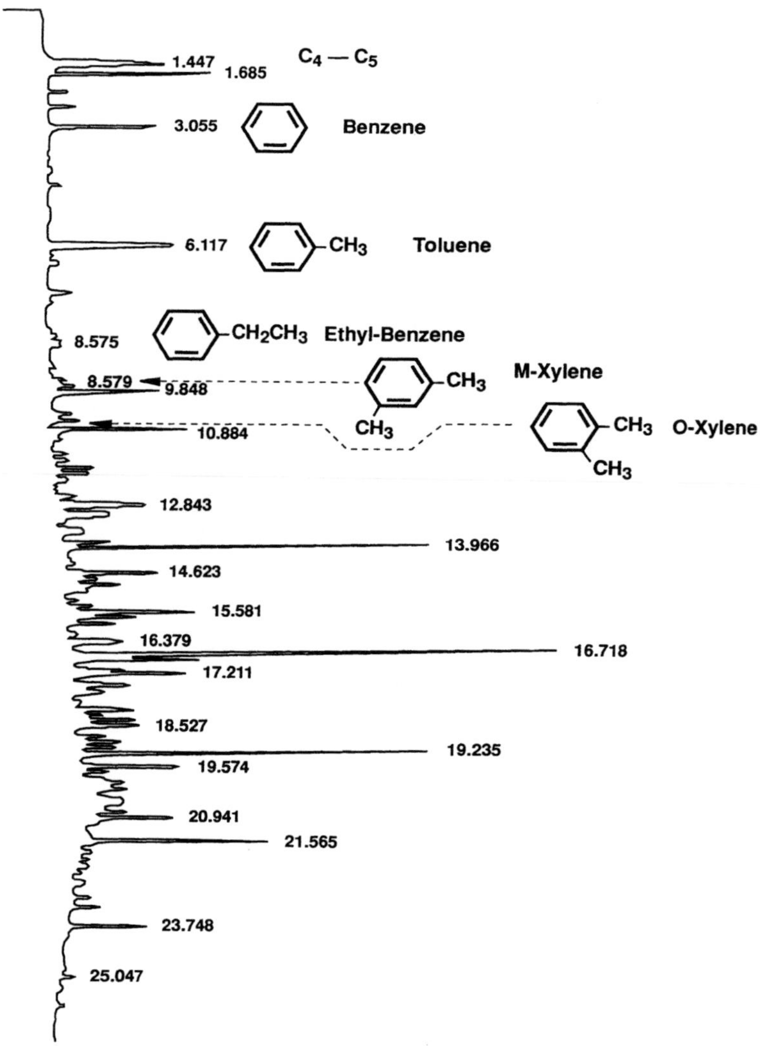

Figure 8. Representative gas chromatogram of produced water. BTEX components are labeled. The large peaks (e.g., 16.718 minutes) are probably n-paraffins. The x-axis is retention time, in minutes.

DISCUSSION

Smectite, chabazite and clinoptilolite actively exchange ions. Zeolites, unlike smectite, are cage-like structures that do not expand (Gottardi and Galli, 1985), meaning that their x-ray diffraction peaks will not shift. In zeolites, the external as well as the internal surfaces of the crystal is capable of cation exchange. While the ammonium cations were adsorbed, the effectiveness of the ammonium zeolite is speculative as there is no direct measurement technique to test its BTEX-removal ability. Indirect methods, such as varying smectite-zeolite ratios and using pure end members will have to be conducted to resolve the issue. The ability of chabazite to be more effective in benzene and toluene removal from laboratory solutions suggests that the type of zeolite as well as the amount of smectite intermixed with the zeolite is a factor. Chabazite tests clearly indicated that it significantly outperformed clinoptilolite in removal of both benzene and toluene. The ability of some zeolites, such as chabazite or clinoptilolite ("T") to perform well while others, particularly clinoptilolite "S" to perform poorly is an area of future investigation. It may be that the smectite concentration affects the ability of the zeolite to perform, and

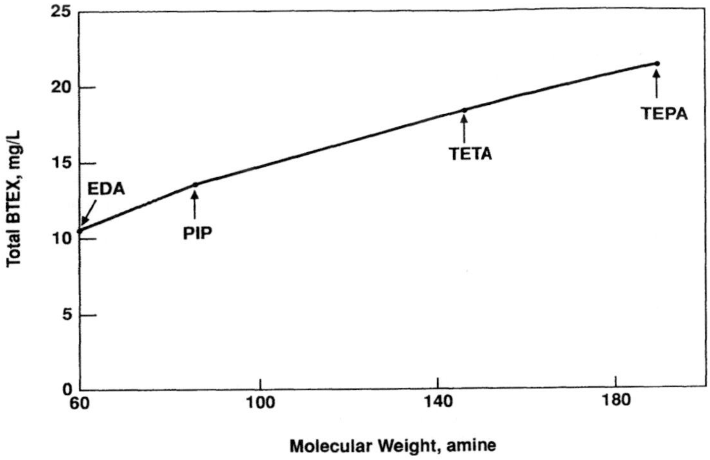

Figure 9. BTEX removal efficiency of four ethyleneamines.

that the combination of for example chabazite with the smectite is preferential over clinoptilolite with the smectite. Most perplexing is the maximum adsorption at approximately 40% of the zeolites CEC (Figure 5). This could be a function of the absolute amount of smectite in the sample, the efficiency of chabazite over clinoptilolite to remove benzene and toluene, or the surface area (a function of grain size). Clinoptilolite "S" was clearly the poorest remover of benzene and toluene. It is substantially richer in clinoptilolite and poorer in smectite than clinoptilolite "T." This may substantiate that the active agent in BTEX removal is smectite. Chabazite clearly has the highest efficiency, yet the amount of benzene and toluene adsorbed decreases after about 40% of the total CEC. It could be that the smectite component adsorbs benzene and toluene up to its maximum capacity, which is some function of the total CEC (zeolites have high CECs). After that maximum the benzene- and toluene-saturated smectite begins to lose its effectiveness. If this scenario is correct, the true zeolite portion of the natural zeolite simply functions as a dilutant for the smectite.

Smectite adsorbs ammonium cations and the effects of this adsorption are measurable. Changes in cation orientation and size affect the thickness and composition of the interlamellar space, as measured by changes in d-spacing of the basal reflections (Figure 11). BTEX compounds are trapped in these interlamellar spaces and removed from the water phase. We found an inverse relationship between the size of the molecule and the d-spacing; the largest

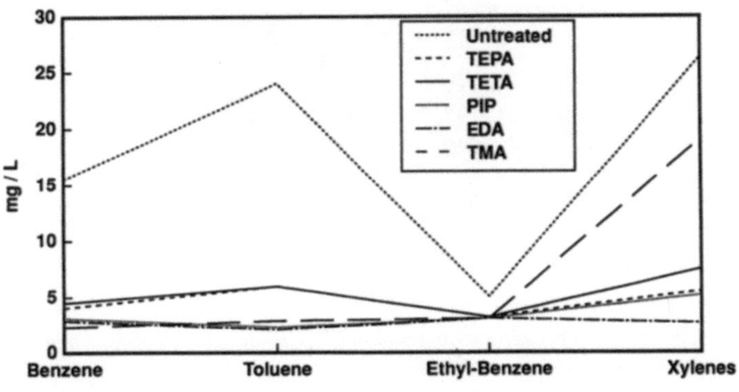

Figure 10. Specific component behavior with various amine-zeolites. TMA is highly efficient at removing benzene, but not so with xylene.

d-spacings belong to TMA, MEA and PIP. DETA and TEPA created much smaller basal d-spacings, despite the larger molecular size. The largest and most complicated molecule, the branched TETA, produced a broad, poorly defined peak. Its position suggests an interlamellar distance similar to DETA and TEPA.

With some tailored zeolites, benzene is preferentially adsorbed over toluene and xylene (Table 3). TMA- and PIP-tailored zeolites are effective in removing benzene, presumably by adsorbing it into the smectite interlamellar space, but forcing xylene to remain in solution. Three factors may affect the ability of aromatic hydrocarbons to be removed by the tailored zeolites: (1) the water solubility of the species (benzene > toluene > xylene), (2) size of the molecule and structural problems created by methyl groups attached to the aromatic ring, and (3), the interlamellar cation complexity may essentially "crowd out" any BTEX molecule attempting to enter the structure. Toluene and xylene, which are larger and less water soluble are partially excluded (Lee, et al., 1990). One possibility is that toluene and xylene may orient themselves parallel to the interlamellar space while benzene may orient itself vertically (Figure 11).

The relationship between BTEX adsorption and molecular weight (Figure 9) is similar to the relationship between benzene adsorption and the length of alkyl chains on an amine. The ability of a straight-chain tailoring agent to remove benzene from water decreases as the alkyl chain length increases (Cadena, et al., 1990). For branched tailoring agents this decrease was much more rapid and dramatic. In the ethyleneamines, longer molecules with multiple ammonium groups could adsorb into the interlamellar space becoming too crowded for BTEX access. Another possibility is that orientation of the ethyleneamine controls access to the interlamellar space. The smaller ethyleneamines such as PIP or EDA may orient perpendicular to the basal plane of the clay minerals, while the larger ethyleneamines may orient themselves parallel to the basal plane. Even if smaller amines were to align themselves parallel to the basal plane, they can produce multiple stacks between planes (Grim, 1968).

Ethyleneamines, methylamine and hydroxyamines (MEA) appear to behave different chemically. Monoethanolamine, with a -OH group attached to the ethylene group, is almost totally ineffective in removing BTEX, despite it small size and ability to expand smectite.

The experiments using produced water suggest that the presence of hydrocarbons (1% free and 860 mg/l soluble), under saline conditions may have affected the ability of the tailored zeolites only

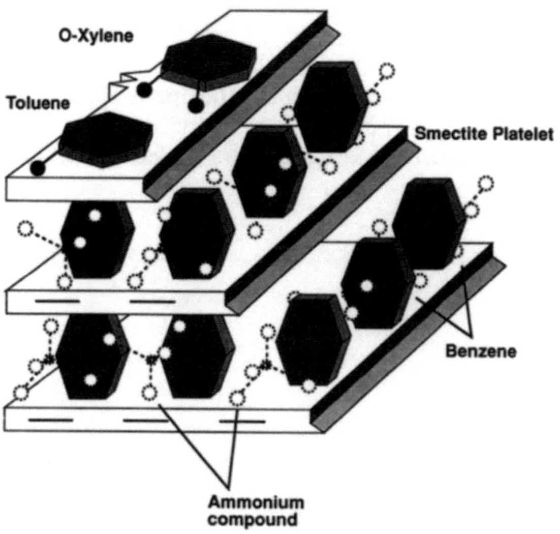

Figure 11. Proposed model for benzene absorption in a TMA-smectite. Toluene and xylene may orient themselves parallel to clay platelets.

Table 4. Ba levels in produced water after exposure to tailored zeolites. All values in mg/l

Untreated water	EDA	PIP	TETA	TEPA	DETA	TMA
28.1	28.0	17.3	25.8	23.9	27.5	27.0

slightly. The insoluble hydrocarbons were probably of such chain lengths that their incorporation into the smectite structure was minimal and their affects slight. The water-soluble organics' effects on the tailored zeolites was unknown, however they could be anionic as well as cationic and therefore have minimal affinity for the cation exchange sites in either the zeolites or the smectite (Dzidic, et al., 1988). Produced water (Table 2) can contain heavy metals. Zeolites have the ability to remove heavy metals and other undesirable cations from solution (Mumpton, 1977b). We tested the ability of the tailored zeolite to remove Ba, as a representative heavy metal, from the test produced water (Table 4). The data indicate that for the contact times and concentrations involved, the tailored zeolites are ineffective in removing this heavy metal. The sole exception to this was PIP, whose value of 17.3 mg/l is markedly lower than the other amines.

CONCLUSIONS

Laboratory tests examined the ability of tailored natural zeolites to remove BTEX from saline produced water containing approximately 70 mg/l BTEX as well as n-paraffins, other aromatics, and heavy metals. The natural zeolites were modified by the adsorption of the ammonium salts of commercially available amines. Natural "zeolites" are actually mixtures of clinoptilolite and chabazite, smectite, and cristobalite, with minor amounts of calcite, dolomite and pyrite. Natural "zeolites" often contain less than 50% true zeolite. The amines tested were monoethanolamine (MEA), trimethylamonium (TMA) and the ethyleneamines EDA (ethylenediamine), DETA (diethylenetetramine), TETA (triethylenetetramine), TEPA (triethylenepentamine) and PIP (piperazine). The tailored zeolites were analyzed by x-ray diffraction of the whole sample and the < 2 µm fraction. TMA, PIP and EDA zeolites produced the largest d-spacing in the basal reflections of the smectite that was associated with the natural zeolite. The zeolite, clinoptilolite, did not show any changes in x-ray diffraction behavior due to the adsorption of ammonium compounds. The produced water was analyzed by ICP and gas chromatography to analyze the inorganic and organic phases present.

The results of the study are:

(1) Tailored zeolites removed a minimum of about 70 to 85% of the BTEX present in the produced water,

(2) The smaller molecules TMA, PIP and EDA were the most effective in removing BTEX; larger ethyleneamines were less effective. TMA was the most effective tailoring agent found. Monoethanolamine, although a small molecule, was found to be ineffective in removing BTEX.

(3) The contaminant-removing behavior of the amine could be correlated to the d-spacing of the basal reflection of the smectite. In general, the larger the d-spacing the more efficient the removal of BTEX,

(4) Different amines have different properties associated with their BTEX removal abilities: e.g., PIP is effective in removing benzene and toluene but while excluding xylene; EDA is effective in removing all pollutants; TMA is selective to benzene,

(5) Other organic and inorganic contaminants present in the water had little apparent effect on the contaminant-removal ability of the tailored zeolites. These zeolites will be effective in removal of BTEX in field situations.

RECOMMENDATIONS FOR FUTURE WORK

This study represented a preliminary investigation into the complex relationship between BTEX removal and tailored zeolites. This relationship needs to be examined in depth in the following areas:

(1) delineation of the adsorption capacities of zeolites for various aromatics,
(2) determine the relationship between produced water salinity and BTEX removal,
(3) the effects of such parameters as particle size and contact time on BTEX removal,
(4) regeneration and disposal option studies.

ACKNOWLEDGEMENTS

The authors wish to thank Texaco Inc. for permission to publish this paper. We are indebted to the following Texaco Exploration and Production Technology Department personnel who performed analyses for us: C.F. Perdue, V.H. Schievelbein, and J.W. Wieser (gas chromatography), M.Q. Henley and N.R. Miranda (inorganic chemical analysis), L.R. Morrow, J.L. Sides and J.E. Tyler (total organic carbon), D. Tolopka (x-ray diffraction), W.C. Haltmar (infrared spectroscopy) and W. Tolopka (extraction and wet chemistry). I.T. Cutter drafted the illustrations and is responsible for their high quality. The authors wish thank the anonymous reviewers for their helpful comments and suggestions

REFERENCES

Boyd, S.A., Mortland, M.M., and Chiou, C.T., 1988, Sorption characteristics of organic compounds in hexadecyltrimethyl-ammonium smectite: Soil Sci. Soc. Am. J., v. 52, pp. 652-659.

Cadena, F., 1988, Treatment of water supplies contaminated with toxic pollutants using tailored soils: New Mexico Water Resources Res. Inst. Tech. Comp. Rept. Proj. No. 13445668, WRI Rep. No. 235, Las Cruces, NM, 54 p.

Cadena, F., 1989, Use of tailored bentonite for selective removal of organic pollutants: ASCE J. Envir. Eng., v 115, p. 757-767.

Cadena, F. and Jeffers, S.W., 1987, Use of tailored clays for selective adsorption for hazardous pollutants: Proc. 42nd Ind. Waste Conf., Purdue University, v. 42, p. 113-119.

Cadena, F., Garcia, R., and Peters, R.W., 1990, Adsorption of benzene from aqueous solutions by bentonite treated with quaternary amines, Envir. Prog. v. 9, p. 245-253.

Chiou, C.T. and Shoup, T.D., 1985, Soil sorption of organic vapors and effects of humidity on sorptive mechanisms and capacity: Envir. Sci. Tech., v. 19, p. 1196-1200.

Drever, J.I., 1973, The preparation of oriented clay mineral specimens for x-ray diffraction analysis by a filter membrane peel technique: Am. Mineral., v. 58, p. 553-554.

Dzidic, I., Somerville, A.C., Raia, J.C., and Hart, H.V., 1988, Determination of naphthenic acids in California crudes and refinery wastewaters by fluoride ion chemical ionization mass spectrometry, Anal. Chem., v. 60, p. 1318-1323.

Gao, F., Cadena, F., and Peters, R.W., 1991, Use of tailored zeolites for removal of benzene and toluene from water: 45th Purdue Indus. Waste Conf. Proc., Lewis Publ., Chelsea, Michigan, p. 509-516.

Gottardi, G. and Galli, E., 1985, "Natural Zeolites", Springer-Verlag, New York, 409 p.

Grim, R.E., 1968, Clay Mineralogy, McGraw-Hill, New York City, 596 p.

Jackler, R., 1991, Krueger Fears Change in Drilling Waste Rules, Houston Post, September 11, 1991, Houston, Texas.

Janks, J.S., and Cadena, F., 1991, Identification and properties of modified zeolites for the removal of benzene, toluene and xylene from aqueous solutions: Soc. Petrol. Eng. Pap. No. 22833, 66th Annual Tech. Conf., Dallas, TX, October 6-9, 1991, p. 153-167.

Lee, J.F., Mortland, M.M., Chiou, C.T., Kile, D.E., and Boyd, S.A., 1990, Adsorption of benzene, toluene and xylene by two tetramethylammonium-smectites having different charge densities Clays and Clay Minerals, v. 38, p. 113-120.

McBride, M.B., Pinnavaia, T.J., and Mortland, M.M., 1977, Adsorption of aromatic molecules by clays in aqueous suspension: *in* "Fate of pollutants in the air and water environments", I.H. Suffet (ed.), Part 1, 8, p. 145-154.

Mumpton, F.A., 1977a, Mineralogy and Geology of Natural Zeolites, Mineral. Assoc. America Short Course Notes, v. 4, Southern Printing Co., Blacksburg, Virginia, 233 p.

Mumpton, F.A., 1977b, "Utilization of natural zeolites: in Mineralogy and Geology of Natural Zeolites", F.A. Mumpton (ed.), v. 4, Southern Printing Co., Blacksburg, Virginia, p. 177-204.

Peters, R.W., Cadena, F., and Page, M.W., 1989, Containment of organic leachates from leaking underground storage tanks using tailored soils as liners: Symposium on Haz. Waste Envir. Man. in Gas Tech.; Gas Technology Institute, Chicago, IL, May 1-2, 1989.

Schultz, L.G., 1964, Quantitative interpretation of mineralogical composition from x-ray and chemical data for the Pierre Shale: U.S. Geol. Surv. Prof. Pap. No. 391-C, 31 p.

Stephenson, M.T., 1991, Components of produced water: a compilation of results from several industry studies, Soc. Petrol. Eng. Pap. No. 23313, First Intl. Conf. Health, Safety and Envir., Hague, Netherlands, November 10-14, 1991, p. 25-38.

Wolfe, T.A., Turgut, D., and Baumann, E.R., 1986, Adsorption of organic pollutants on montmorillonite treated with amines J. Water Pollut. Control Fed., 58, p. 68-75.

RECENT ADVANCES IN THE APPLICATION OF MEMBRANE TECHNOLOGY FOR THE REMOVAL OF OIL AND SUSPENDED SOLIDS FROM PRODUCED WATERS

A. Zaidi[1], K. Simms[1], S. Kok[1], and R. Nelson[2]

[1] Wastewater Technology Centre
Burlington, Ontario
Canada

[2] Alberta Oil Sands Technology and Research
Authority, Calgary, Alberta
Canada

INTRODUCTION

Membrane processes for produced water treatment, specifically microfiltration (MF) and ultrafiltration (UF), have been attracting considerable attention from the oil industry in Canada and the USA due to recent advances in MF/UF technology and increasing regulatory pressures (USEPA, 1991) for more effective produced water treatment. MF and UF processes offer the potential for generation of high quality effluent water on a more consistent basis than conventional technologies and may offer additional advantages in terms of space and weight savings and treatment costs. Although the ultimate fate of the produced water and the required effluent quality vary depending on the type of oil recovery operation and the regulatory environment, microfiltration and ultrafiltration will likely have an increasing role in the treatment of both onshore and offshore produced water.

Test work on MF and UF for produced water treatment has been conducted at both bench and pilot scale for onshore and offshore application and a few full scale installations are now in place. Although work in this field has been conducted by membrane and systems manufacturers, private test centres, government agencies, oil companies and oil industry organizations, much of this work has been conducted under arrangements that mitigate against the release of experimental data in the open literature. In addition, there have been no concerted efforts to systematically select the most cost-effective combination of membrane material, module and system for the produced water application from the wide range of membrane technology currently on the market or under development. This paper reviews the available information from recent studies of membrane technology for produced water treatment in order to identify the technical obstacles to the successful implementation of this technology, and define key technical areas where future work should focus.

Produced Water, Edited by J.P. Ray
and F.R. Engelhart, Plenum Press, New York, 1992

TECHNICAL BACKGROUND

Concept of Crossflow Membrane Filtration

The principle of crossflow membrane filtration (CFF) processes such as microfiltration and ultrafiltration is shown schematically in Figure 1. In crossflow membrane filtration processes, the liquid to be treated flows parallel to the surface of a porous membrane. Clean water permeates through the membrane while suspended matter (oil and solids) is selectively retained in the concentrate stream. The driving force for flow of permeate is the pressure difference across the membrane. Because of the positive barrier available for separating the oil from the water, the effluent generated by MF/UF processes is expected to be consistently free of suspended oil.

The flow rate of permeate per unit area of the membrane surface is termed the flux and is a key performance parameter for crossflow membrane filtration. Several variables affect the performance of crossflow membrane systems including; (i) crossflow velocity at the surface of the membrane, (ii) transmembrane pressure differential, (iii) feed temperature, (iv) feed characteristics (ie. contaminant concentration), and (v) membrane characteristics (ie. porosity, thickness of active layer, pore size and surface chemistry). There is as yet no consensus as to what comprises a target flux level for economic application of membrane technology to oilfield brine treatment. Any estimates must consider system cost, weight and space requirement, conventional equipment to be replaced, and, value of any incremental improvement in produced water quality, among other variables. To date, estimates ranging up to 1000 l/m^2h have been quoted (Zaidi et al., 1991). Another key performance parameter in crossflow membrane filtration is permeate quality. Membrane processes for the produced water application must consistently generate an effluent of acceptable quality to be considered technically viable.

Crossflow filtration processes use membranes in a wide range of pore sizes. For example, MF membranes have pores in the range of 0.1 to several microns while UF membranes have much finer pores (<0.01 µm). MF and UF membranes are made from a variety of organic polymers and inorganic materials. Some commercially available membranes that may be suitable for produced water applications are listed in Tables 1 and 2. While no one material or pore size has been proven most suitable for oilfield brine application, some desirable membrane characteristics would include: high porosity, surface hydrophilicity, adequate temperature and pressure tolerance, compatibility with required cleaning agents and common oilfield treatment chemicals, long operating lifetime and low cost. The membrane may be packed or configured in a variety of shapes and module types (tubular, plate and frame, capillary, hollow fibres and spiral wound) that represent some compromise between size of the feed flow channel, hydrodynamic flow conditions and packing density (membrane surface area per unit volume of the module).

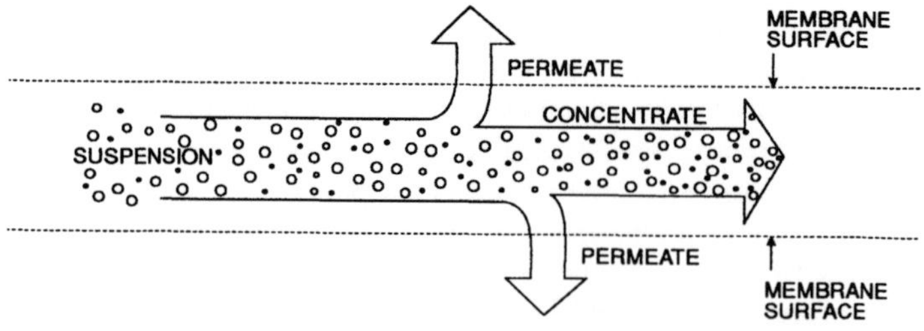

Figure 1. Operating principle of crossflow membrane filtration.

Table 1. Partial list of inorganic membranes suitable for oil/water separation

Manufacturer	Trade Name	Type	Configuration	Pore Size	Material
Asahi Glass	MPG	MF	tubular	0.1-1.5 μm	75-80% SiO_2, 10-20% B_2O_3,
		UF	tubular	8-30 nm	4-6%Al_2O_3, 96% SiO_2
Alcoa/SCT	Membralox	MF	tubular	0.2-5 μm	αAl_2O_3, $\gamma Al_2O_3/\alpha Al_2O_3$
		UF	tubular	4-100 nm	$ZrO_2/\alpha Al_2O_3$
DuPont/Carre		MF	tubular	0.1 μm	ZrO_2/stainless steel
Carbone Lorraine		MF	tubular	0.2 m, 0.8 μm,	carbon composite
		UF		4-100 nm	
CeraMem		MF	tubular	0.2 m, 0.5 μm,	$\gamma Al_2O_3/\alpha Al_2O_3$
		UF	tubular	5000 nm	
Cortane		MF	tubular	0.8 μm, 0.45 μm,	TiO_2/Al_2O_3
		UF		0.2 μm	
Millipore/Norton	Ceraflo	MF	tubular	0.2-1 μm	αAl_2O_3
Tech Sep/Rhône-Poulenc	Carbosep	MF	tubular	0.08-0.45 μm	ZrO_2/C
		UF		15,000-10,000 D	

In continuous large volume applications, such as produced water treatment, crossflow membrane filtration systems are usually operated in the feed and bleed mode, shown in Figure 2, where feed is continually supplied to the system while permeate is continually withdrawn and concentrate is bled off intermittently. A consistent feature of these systems is the requirement for recycle of a large volume of concentrate. Only a small volume of water permeates through the membrane with each pass and high flow volumes are required to maintain desired hydrodynamic conditions at the membrane surface. The cost of recycling lines, valves and pumps contribute significantly to the total capital cost of this type of system.

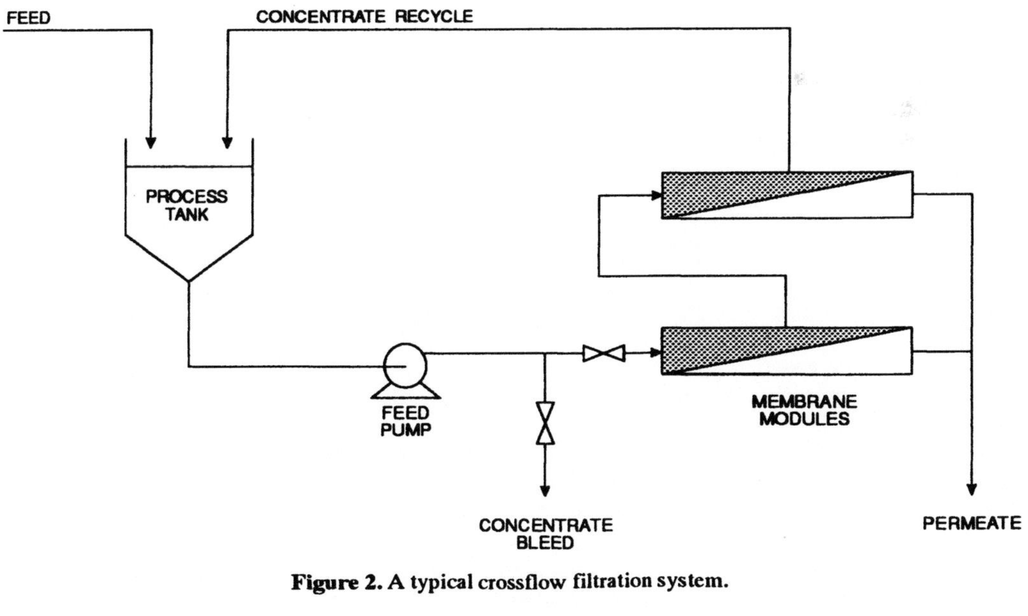

Figure 2. A typical crossflow filtration system.

Membrane Flux Decline

Common to all membrane processes is the phenomenon of flux decline or degradation, which represents a fundamental limitation to the economic viability of MF/UF in produced water treatment applications. While this flux decline is generally associated with fouling due to adsorption or accumulation of certain components of the oilfield brine on the membrane surface or in the membrane pores, the specific foulants and mechanisms have not been identified. When the permeate flux declines to an unacceptable level, detergent, solvents, caustic or acids may be used to clean the membrane. Strong and frequent cleaning can result in damage to certain membranes so that the cleaning regime is considered to be a major determinant of membrane life. In some cases, some materials may sorb directly in the membrane resulting in a structural change and subsequent flux decline. In this case, cleaning of the membrane surface does not recover the original flux rate. This type of fouling is termed irreversible.

Table 2. Partial list of organic polymeric membranes suitable for oil/water separation

Manufacturer	Trade Name	Type	Configuration	Pore Size or MWCO	Material
Amicon		UF	flat sheet, spiral wound	3,000 - 100,000 MWCO	PS
		MF	hollow fibre	0.1 μm	PS copolymer
Aqua Air Environmental		UF	spiral wound, hollow fibre		polymeric organic
Dow/DDS		MF	flat sheet	0.1 - 5 μm	PS, fluoropolymer
Desalination Systems		UF MF	spiral wound	1,000 - 1,5000 MWCO 0.02 - 3 μm	thin film polymer PTFE fluorocarbon
Enka		MF	capillary, tubular	0.2 μm	PP
Epoc/Exxflow		MF	tubular	0.5 - 5 μm	textile
Hoescht/Celanese	Nadir/ Molsep	UF	spiral wound tubular	4,000 - 200,000 MWCO 10,000 - 40,000 MWCO	PES modified PAN
Koch/Abcor		MF	spiral wound, tubular	0.1 - 4 μm	polymeric organic
Memtec	Memcor	MF	hollow fibre	0.2 μm	PP
Millipore		MF	flat sheet	0.1 - 0.5 μm	fluoropolymer
Nitto/Denko		UF	spiral wound, tubular	20,000 MWCO	PO
Osmonics	Osmo	UF	spiral wound	100 nm - 0.1 μm	modified PS, PI, fluoropolymer
Patterson Candy International		UF	tubular, flat sheet	4,000 - 200,000 MWCO	PS, PES, PVDF, PAN
Romicon		UF	hollow fibre	0.005 μm	PP, PVC
Sartorius		UF	flat sheet	20,000 MWCO	cellulose triacetate
Separation Dynamics Inc.	Extran	UF	hollow fibre	40,000 MWCO	polyglucose, PP, PVDF, PTFE
Stork Friesland	Wafalin	MF/UF	tubular		PS, PES, PAN, PVDF
Tech Sep/Rhône-Poulenc	Pleiade	MF	flat sheet	0.1 - 1.0 μm	fluoropolymer
X-Flow		MF	hollow fibre flat sheet	0.1 - 0.2 μm 0.05 - 1 μm	PS, PI
Zenon		UF	tubular	5,000 - 20,000 MWCO	PS, polymeric organic

Note: PAN - polyacrylonitrile, PES - polyether sulphone, PI - polyetherimide, PO - polyolefine, PP - polypropylene, PS - polysulphone, PVC - polyvinylchloride, PVDF - polyvinylidene fluoride.

A number of techniques may be employed to enhance flux rate and minimize fouling in microfiltration and ultrafiltration by maximizing mass transfer at the surface of the membrane. Table 3 presents a partial list of flux enhancement techniques. High crossflow velocity is most commonly maintained by operating MF/UF systems at a high feed velocity but alternative techniques can also be used to enhance the turbulence on the membrane. Other techniques include the use of backflushing with air or permeate or pulsing of the feed flow to dislodge any potential foulants. Operating strategies for minimizing fouling include the use of chemical pretreatment to generate precipitates which may either form a protective membrane layer, or continuously scrape the membrane surface to keep it clean. Table 4 lists some commercially available crossflow filtration systems and the flux enhancement techniques employed by these systems.

Table 3. Approaches to minimize flux degradation

Membrane Selection		
	Surface Modifications	
	Novel Materials	
Operating Strategies		
	Pretreatment	Prefiltration
		Addition of reagents
		Addition of abrasive particles
		Aeration
	Operating Conditions	Module design
		Transmembrane pressure
		Crossflow velocity
	Operating Modes	Batch
		Semi-batch
		Feed and bleed
Special Techniques		
	Hydrodynamic	Turbulence promoters
		Backwashing
		Periodic backpulse
		Pulsed flow
		Rotating/vibration membranes
		Rotating blade on membrane
	Others	Crossflow electrofiltration
		Ultrasonic enhancement

RECENT STUDIES

Table 5 summarizes some recent experimental and paper studies that have been conducted on MF/UF for produced water application and highlights the membrane systems that have been evaluated. Primarily two types of systems have been extensively demonstrated in produced water treatment applications. These are tubular ceramic microfiltration and tubular polymeric ultrafiltration membrane systems.

Systems based on Tubular Ceramic Membranes

Ceramic membranes offer particular advantages for oilfield brine applications due to their tolerance for high temperatures, strong cleaning agents and high concentrations of suspended solids in the feed. Currently, commercial ceramic membranes are only available in tubular

Table 4. Some commercial crossflow filtration systems

Membrane Type	Module Type	Flux Enhancement Techniques	Manufacturer
Flat sheet	Plate and frame	None	Dow/DDS, Tech-Sep
		Cross rotation	ABB/Flakt
		Vibration	New Logic
	Spiral wound	None	Nitto/Denko
Hollow fibres		None	Aqua Air Environmental, Amicon, Romicon, Seperation Dynamics, Inc.
		Periodic gas backpulse	Enka, Memtec
Tubular	Multichannel	Periodic liquid backpulse abrasive or large particles in feed	Alcoa/SCT, Cera Mem, Epoc/Exxflow
		None	Schenk
	Single tubes	None	Zenon, Kalle, Enka, Nitto, Tech-Sep, Stork-Friesland
		Rotation	Pall-Sulzer

modules. The large flow channels of the tubular configuration minimize the risk of tube plugging by solids, but this configuration also results in very low packing density (membrane surface area per unit volume).

The most widely tested system for treatment of produced water in North America is the Petrolox system manufactured by Alcoa Separation Technology Division (now U.S. Filter Corp.) of Warrendale, Pennsylvania. The Alcoa system is based on the use of ceramic tubular microfiltration membranes with feed pretreatment and backpulsing of the permeate for flux enhancement. The MF membranes comprise of thin layers of alumina with pore sizes ranging from 0.2 to 5 μm.

Table 5. Some recent studies on crossflow membrane filtration for produced water treatment

Sponsor	Scope	Systems
Orkney Water Test Centre	Tested systems using synthetic offshore oilfield brine	Alcoa, Shelda Delta,,SDI, Stork, X-Flow, Zenon
Offshore platforms in Holland/Denmark	Short term platform tests	Ceramic MF, Polymeric UF systems
Joint AOSTRA/Industry/WTC	Parallel heavy oilfield tests over 6 months	Alcoa, Zenon
USEPA development document	Paper study	Alcoa
Paragon study for OOC	Paper study	Alcoa, Stork, SDI X-Flow, Zenon
CH2M Hill study for Amoco/Shell/EM&R/E.C.	Paper study on produced water reuse technology	Overview
Arco/Marathon	Review of performance of one commercial and one pilot plant installation	Alcoa

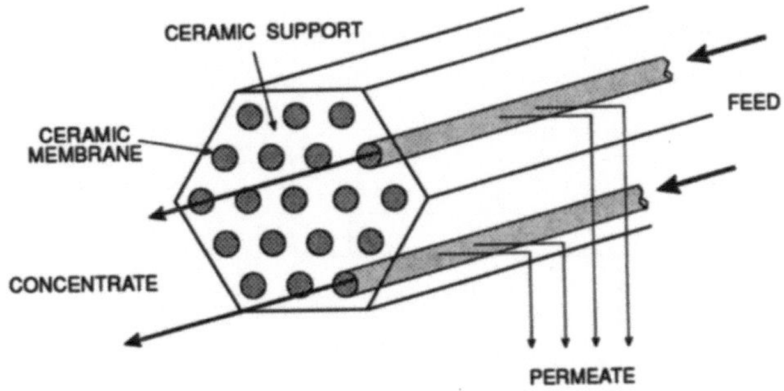

Figure 3. Schematic of Alcoa ceramic membrane filter.

The commercial membrane module (Alcoa Separation Technology Division, 1991), shown in Figure 3, is an 850 mm long porous ceramic monolith containing 19 flow channels (either 4 or 6 mm diameter) in which the membrane layer is cast. Stainless steel modules are available which contain one to 36 of the multichannel elements.

Figure 4 shows a schematic of the Alcoa pilot system (Chen et al., 1991) consisting of a feed tank, a pretreatment system, cleaning system, backpulse system and the membrane separation modules. Two complementary techniques are employed in the operation of the Alcoa membrane system to minimize fouling of the membrane. Pretreatment of the feed is effected by addition of ferric chloride and other chemicals in order to precipitate a ferric hydroxide floc. The mechanism of flux enhancement by pretreatment is not yet fully understood, however, it is theorized that the floc then coats the surface of the membrane, forming a dynamic membrane layer which prevents small particles from clogging the ceramic membrane pores. In addition, a periodic backpulse is applied by forcing permeate back through the membrane pores to briefly dislodge any loose deposits in the pores or at the membrane surface. At the same time, the feed flow is pulsed to sweep away any loose deposits. By backpulsing and fast flushing, the dynamic membrane is temporarily dislodged and its thickness is controlled. The Alcoa system is operated by maintaining a constant flux rate and crossflow velocity (between 1 and 5 m/s). As the membrane gradually becomes fouled, the pressure of the feed is increased until the transmembrane pressure exceeds some set value (usually 240 kPa) and the membrane must be cleaned. The cleaning sequence usually includes alternating caustic wash, fresh water rinse and acid wash. As shown in Figure 4, oily float and solids blowdown streams are occasionally withdrawn from the process feed tank. These streams along with spent cleaning fluids comprise the waste streams from the Alcoa ceramic microfiltration process.

To date, the ceramic tubular membrane system has been field tested at pilot scale for onshore and offshore produced water (the most extensive work being at heavy oilfield sites in Western Canada), and two commercial installations are now in place. Available data from a short test at an Alberta heavy oilfield (Separ Systems and Research Ltd, 1988) indicate a flux ranging from 800 to 1200 l/m2h. However, in longer term pilot tests, up to 800 hours (Zaidi and Kok, 1990), at another Alberta site, substantially lower stable flux rates were observed. In addition, during the long term tests, no correlation could be established between the use of chemical pretreatment and the permeate flux rate. While permeate quality was generally high during this test work, there was one incident of severe permeate degradation associated with changes in the produced water characteristics. This field test work on ceramic membranes was conducted using pore sizes of 0.8 μm, however, the rationale for pore size selection was unclear. In recent short term (up to 7 days) pilot tests at two Louisiana onshore and two Gulf of Mexico offshore sites (Chen et al., 1991), Alcoa reported that permeate flux rates of 1450 to 3400 l/m^2h

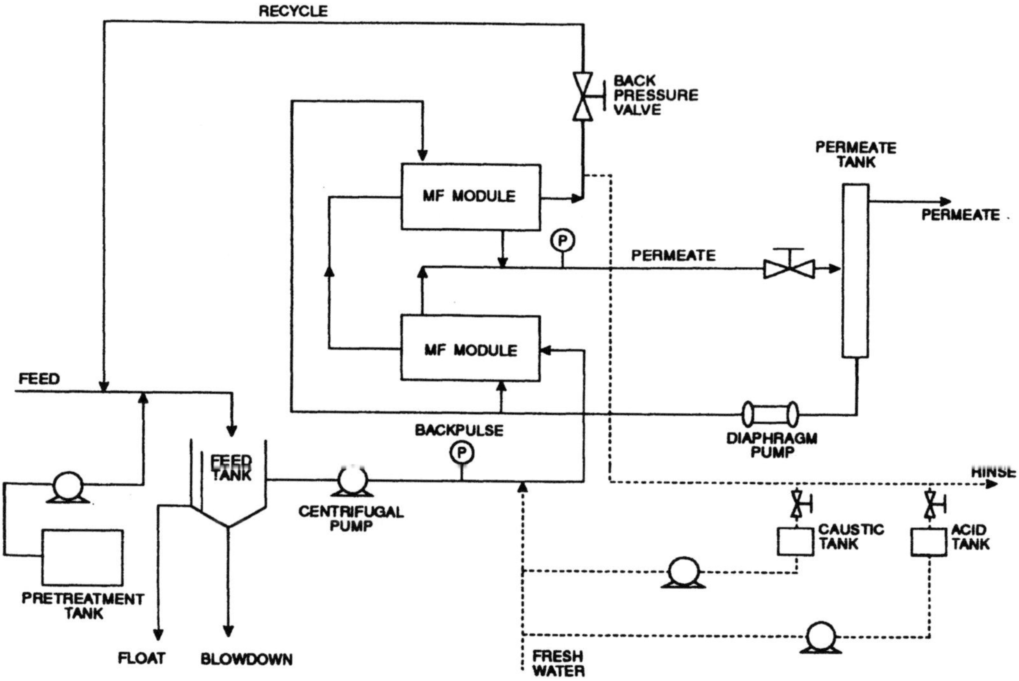

Figure 4. Alcoa Petrolox pilot unit.

could be sustained using chemical pretreatment and backpulsing with cleaning intervals greater than 66 hours. The total oil and grease level in the produced water was reduced from 27 - 583 mg/l in the feed to below 10 mg/l in the permeate. The volume of oily float stream from the process tank ranged from 1 to 10% of the total produced water volume treated.

Table 6 describes the two full scale ceramic membrane installations currently in place. During commissioning of the Marathon unit and at pilot tests at an Arco site (Yang and Meyer, 1991), the permeate flux rate was observed to be detrimentally affected by: low pH conditions, the presence of seawater, corrosion inhibitor, oil slugs and flow variations. Concerns were also raised with respect to the handling of the concentrate stream which may interfere with upstream treatment processes if recycled. Also, it has not been clearly established that ceramic microfiltration can generate a high quality permeate more suitable for re-injection than conventional processes. For example, at the Marathon platform, injection well plugging problems were not eliminated by use of ceramic microfiltration. The Marathon and Petro-Canada ceramic membrane installations represent the initial applications of this technology to produced water treatment at full scale. However, targeted performance criteria were not met with the original installations and substantial operational and design modifications have been required.

As shown in Table 1, a number of other ceramic membranes may also be suitable for the produced water application. For example, Stork offers MF/UF alumina ceramic membrane systems with backpulsing that have been tested for oily water application (Paragon Engineering Services, 1990). Zirconia or silica based membranes may also be of interest because of their high hydrophilicity relative to alumina. Tighter pore size ceramic membranes (in the UF range) may have the potential to minimize plugging due to entrained solids without the use of feed pretreatment although lower fluxes may also be expected with ultrafiltration membranes. Tech-Sep manufactures zirconia ultrafiltration membranes in 6 mm diameter tubular configuration that have been widely used for oil/water separation in the metal finishing industry (Tech-Sep, 1991). Ceramic tubular membranes are also amenable to an alternative flux enhancement technique termed crossflow electrofiltration (CFEF), where a continuously applied electrical field results in charged particles and colloids moving away from the membrane due to electrophoretic transport. Alternatively, the electric field may be periodically applied and material on the membrane surface that retains a charge can be removed. Some researchers (Visvanathan, 1990) have combined CFEF with a hydrodynamic flux enhancement method, namely rotation of the tubular membrane, and noted significant improvements in flux as compared to the conventional tubular systems. One experimental technique, termed direct electrical membrane cleaning (DMC), uses short pulses of current and relies on electrolytically generated bubbles to dislodge foulants from the membrane surface (Wakeman, 1986). To date, however, the application of CFEF has been limited to pilot scale or pre-industrial studies and the cost and complexity of these systems may be prohibitive for produced water treatment.

Common concerns with all ceramic membrane systems relate to the relatively high cost of the membrane, while the use of feed pretreatment advocated by Alcoa, raises concerns regarding handling of the waste streams generated. The use of polymeric membrane materials has been proposed to alleviate these problems.

Table 6. Alcoa Petrolox Installations

Company	Marathon	Petro-Canada
Date Installed	October, 1989	Fall, 1990
Location	Eugene Island Block 349 "B" platform Gulf of Mexico	Valhalla Field, Alberta
Design Capacity	1000 m^3/d	1000 m^3/d
Membrane Surface Area	45 m^2	—
Pore Size	0.8 μm	0.8 μm
Feed	Slipstream from precipitator effluent	5% from heater treater, 95% groundwater
Permeate Fate	Reinjection	Reinjection
Pretreatment	$FeCl_3$	$FeCl_3$

Systems based on Tubular Polymeric Membranes

One company that has attempted to develop polymeric membranes targeted to produced water is Zenon Environmental Inc. of Burlington, Ontario. Zenon has been investigating the application of tubular polymeric membrane to heavy oilfield produced water since 1983. One of the Zenon systems, shown in Figure 5, comprises of a feed tank, a prefilter, membrane modules and a cleaning system. The membranes tested to date were composed of proprietary polymeric material with 10,000 to 20,000 molecular weight cut off pore size, cast in 22 mm diameter tubes. Zenon's proprietary Z-8 membrane module features 8 tubes in series in a single module shell. Because these membranes have a tendency to become fouled upon exposure to free oil, prefiltration for removal of oily particles greater than 20 μm was used in the work conducted to date. The system is operated at a design pressure of 483 to 552 kPa and crossflow velocity of 3 to 4.6 m/s. Detergent or mechanical cleaning is initiated when the flux declines to an unacceptably low level.

To date, Zenon's work has incorporated off-site bench scale work and pilot tests at two heavy oil recovery operations in Western Canada. Extensive bench scale work, conducted on Zenon proprietary tubular membranes, focused on optimizing operating conditions and cleaning techniques using produced water samples from eight different enhanced oil recovery locations. While flux rates were found to vary dramatically with produced water type, the performance was not significantly affected by variations in the inlet oil concentration. Subsequent pilot testing in the field (Zaidi and Kok, 1991) was conducted in 1989/90 to evaluate the long term performance (over 11 months) of the most promising Zenon membranes. This work was conducted in parallel with tests on the Alcoa Petrolox system. In the Zenon test work, approximately 90% of the produced water was recovered as permeate. The permeate quality was generally high (< 20 ppm total O&G), but flux rates were significantly lower than predicted from bench scale work, ranging from 90 to 50 l/m^2h over 3 to 10 day test runs with 45 psi transmembrane pressure (Paragon Engineering Services, 1990). In addition, cleaning was not always effective in restoring original flux so that some component of the total flux decline was considered to be irreversible. As with the ceramic membranes, free oil was detected in the membrane permeate stream on one occasion.

Other polymeric UF/MF systems for oil/water separation are commercially available, as previously shown in Table 2. Stork offers tubular membranes in a range of polymeric materials. A number of systems have been installed for oily wastewaters and tests have been conducted on offshore produced water samples. According to one report (Paragon Engineering Services, 1990), a 0.1 m^3/hr unit was tested over two months on a Mobil gas platform and generated high quality permeate (< 3 ppm O&G) at feed oil concentrations of 100 ppm.

Since no full scale systems have yet been installed, integration of polymeric tubular systems into produced water treatment facilities is unproven. The concentrate bleed and any skim oil from the feed tank should be amenable to recycle to the oil separation train because treatment chemicals are not generally used with polymeric UF systems. Difficulty may arise, however, because of the substantial prefiltration requirement associated with polymeric ultrafiltration and lower flux rates would be expected with ultrafiltration than with microfiltration.

Other Systems

Alternative membrane configurations, such as capillary tubes and hollow fibres, have also been proposed for treatment of produced waters. In this case, the inner diameter of the tubes generally ranges from a few millimetres with capillary tubes to less than 0.5 mm with hollow fibres. Feed pretreatment is therefore essential to prevent clogging of these narrow membrane bores. This type of module configuration may be advantageous because the narrow tubes result in high crossflow conditions which may reduce fouling. In addition, capillary and hollow fibre membranes have much higher packing densities than tubular membrane. One manufacturer

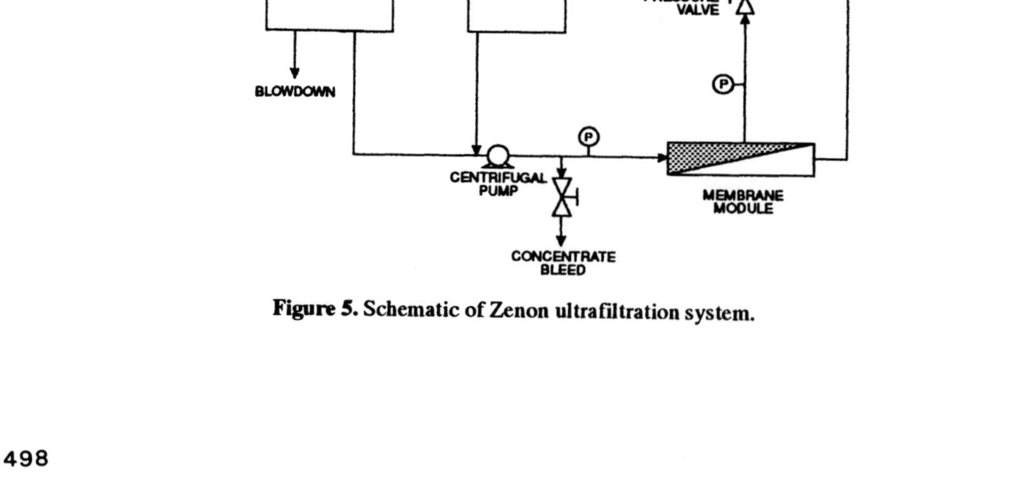

Figure 5. Schematic of Zenon ultrafiltration system.

(Taylor et al., 1991) has estimated that a floor area of only 24 m^2 would be required for a 7,950 m^3/d (50,000 BPD) hollow fibre system. By way of comparison, a similarly sized Alcoa ceramic microfiltration system (7,400 m^3/d) has been estimated to require approx. 160 m^2 (Paragon Engineering Services, 1990).

X-Flow BV of Holland has developed 1.45 mm diameter polyether sulphone capillary membranes with 0.2 µm pores for produced water application. These membranes have been incorporated into a 3S Teknik system which employs a high crossflow velocity (5 m/s) and backpulsing to minimize fouling. An 80 to 150 µm filter is recommended for feed pretreatment. Manufacturer's field data (X-Flow, 1991) indicate that X-Flow membranes can operate at 80 to 120 l/m^2h on oily wastewater at 200 kPa feed pressure and can generate permeate with < 10 ppm oil and grease. X-Flow membranes have been tested at bench and pilot scale to treat produced water from oil and gas recovery operations. Oil levels have reportedly been reduced from 1,000-2,000 mg/l to less than 40 mg/l with oil levels of up to 20% in the concentrate stream. A full scale unit, acquired by Maersk Oil and Gas Company, has a capacity of 160 to 320 m^3/d (1,000 to 2,000 BWPD) and has been reported to have operated for 2 months without cleaning (Paragon Engineering Services, 1990).

Another system that has been recently tested for produced water application is manufactured by Separation Dynamics Inc., Southfield, Michigan and employs inside-flow water-diffusion hollow fibres. Thousands of the fibres are fabricated into separation modules by first laying them in a parallel mat and then rolling the mat into a bundle which is then potted into a stainless steel or polyvinylchloride sleeve. The individual fibres are made of regenerated cellulose and have an inside diameter of 400 µm and can operate at up to 276 kPa transmembrane pressure. Separation of hydrocarbons is claimed to result from diffusion of water and water soluble compounds through the oleophobic membrane. Because of the small bore diameter of the fibres, prefiltration for removal of particles greater than 25 µm is recommended although a filter cut off of up to 50 µm may be adequate. SDI recently sponsored (Taylor et al., 1991) a five day test on simulated offshore produced water to examine the effect of crossflow velocity and solids loading on membrane performance. When a significant pressure drop across the module was observed, the feed flow was reversed to clear any plugged fibres. Measured permeate flow decreased by 20% under high solids loadings. Membrane flux rates were not disclosed, however, an average permeate flow rate of 400 l/h was obtained using a module of length 610 mm and diameter 152 mm, and the level of oil and grease in the permeate was 10 mg/l.

Other membrane systems that have been tested for produced water treatment include; the Aqua Air Environmental hollow fibre ultrafiltration system, the Epoc dynamic microfiltration system which uses textile membrane in a collapsible tube configuration and the New Logic vibration shear enhanced process (V-Sep) which enhances flux rates by vibrating a disk stack of membranes. However, hard data on the long term performance of these systems for treatment of produced water is not available.

CONCLUSIONS

A number of general conclusions and comments can be drawn from recent work on the removal of oil and suspended solids from produced water by MF/UF processes.

1. While this technology seems capable of reducing oil content in produced water to low levels (< 20 mg/l oil and grease) it remains to be seen if crossflow micro/ultrafiltration processes can maintain this performance on various types of produced water encountered at onshore and offshore oil production facilities. Also the long term cost-effectiveness of commercial MF/UF systems needs to be confirmed.

2. Although available data from produced water treatment tests indicate that in short term tests (days to weeks) membrane fluxes of up to 3,400 l/m^2h were achieved, in long term tests, the flux rates were generally one, or even two, orders of magnitude lower. In membrane crossflow filtration, permeate flux rates are sensitive to produced water characteristics and operating variables, and may be detrimentally affected by the variable flow, treatment chemicals and oil loading conditions normally seen at produced water treatment facilities.
3. Work to date has generated little fundamental or comprehensive understanding of the causes and mechanisms associated with the flux decline commonly seen in the membrane treatment of produced water.
4. Design targets for flux rate, permeate quality, and cleaning frequency have not been clearly defined in the application of membrane technology to produced water treatment.
5. Major technical obstacles to integration of membrane processes at existing produced water treatment facilities include the requirement for extensive prefiltration for polymeric membranes and the requirement for chemical pretreatment and resultant generation of intractable oily waste streams for ceramic microfiltration membranes.
6. Recent initiatives have focused on the use of dynamic membranes and various hydrodynamic techniques such as backpulsing or high shear operation to reduce membrane fouling, enhance flux rate, and minimize cleaning requirements.
7. To date, there has been a general lack of coordination amongst industry, research scientists and manufacturers to address key technical issues that prevent the successful implementation of microfiltration and ultrafiltration for produced water treatment. These key technical issues include low average flux and rapid flux decline, resulting in high capital and operating costs.

Given that the MF/UF technology comprises a wide variety of membrane materials, membrane modules, and systems that could be configured in various combinations, a rational objective approach would be needed to identify the most cost-effective membrane system for produced water application. Research is required is a number of areas including:

1. Review of conventional produced water treatment process performance and cost and anticipated treatment requirements to determine target applications for UF/MF and cost and performance criteria in each target application.
2. Identification of foulants and fouling mechanisms for membranes currently being used in oilfield brine and coordination of these studies with development of appropriate strategies to minimize fouling.
3. Comparative studies to evaluate alternative system designs and flux enhancement techniques on actual produced water samples.
4. Generation of performance data through long term parallel field evaluations of commercially available and experimental membranes potentially suitable for produced water.

ACKNOWLEDGEMENTS

This work was supported, in part, by the Federal Panel on Energy R & D (PERD).

REFERENCES

Alcoa Separations Technology Division, 1991, Warrendale, Pennsylvania, product literature.

Chen, A.S.C., Flynn, J.T., Cook, R.G., Casaday, A.L., 1991, Removal of Oil, Grease, and Suspended Solids from Produced Water with Ceramic Crossflow Microfiltration, SPE Production Engineering.

Paragon Engineering Services, Houston, Texas, 1990, Crossflow Membrane Separations Systems Study, Report prepared for the Offshore Operators Committee, New Orleans, Louisiana.

Separ Systems and Research Ltd., Calgary, Alberta, 1988, Crossflow Microfiltration of Produced Water, a report prepared for Environment Canada, Wastewater Technology Centre, Burlington, Ontario.

Taylor, J., Larson, R., and Scherer, B., 1991, Treatment of Offshore Produced Water: An Effective Membrane Process, Environment Northern Seas International Conference and Exhibition, Stavanger, Norway.

Tech-Sep, France, 1991, Product literature.

United States Environmental Protection Agency, 1991, Federal Register, Vol. 56, No. 49, Proposed Rules, March 13, 1991.

Visvanathan, C., Ben Aim, R., and Vigneswaran, S., 1990, Electrofiltration - State of the Art, Proceedings Vth International World Filtration Congress.

Wakeman, R., 1986, Electrofiltration: Microfiltration plus Electrophoresis, The Chemical Engineer, June.

X-Flow BV, Almelo, The Netherlands, 1991, Product literature.

Yang, J.C., and Meyer, J.P., 1991, Industry's Field Experience with Membrane Filtration Technology, Report prepared by Marathon Oil Company, Houston and ARCO Oil and Gas Company, Plano, Texas.

Zaidi, A., Kok, S., Simms, K., and Schmidt, J., 1991, Oil and Suspended Solids Removal from Produced Water Using Ultrafiltration and Microfiltration, presented at the UNITAR/UNDP Fifth International Conference on Heavy Oil and Tar Sands, Caracas, Venezuela, September,1991.

Zaidi, A., and Kok, S., 1990, Recent Advances in the Treatment of Produced Water from In-Situ Heavy Oil Recovery, presented at the Second Canada-China Heavy Oil Technology Symposium, Beijing, China, October, 1990.

SULPHATE SEPARATION FROM SEAWATER BY NANOFILTRATION

T. Bilstad

Rogaland University Centre
Stavanger, Norway

INTRODUCTION

An oilfield is comprised of oil, natural gas and what is known as formation water. Initially oil will rise up the well due to natural pressure within the reservoir. In order to maintain a high pressure and to enhance recovery of oil, seawater may be injected around the edges of a reservoir. Later in the life of a producing oilfield, a mixture of oil and water is produced from the well. This water can be a mixture of formation water and injected seawater. For some oilfields there is a problem when the seawater meets the formation water as the latter may have high concentrations of both barium and strontium which react with the sulphate in the seawater. The reaction products are precipitates which block the pores in the rock and scale the pipework on the platform. In some instances the problem can be so severe that the production wells become blocked.

Although a number of anti-scalants may look promising when tested in the laboratory, no products have proven successful in preventing barium sulphate scaling in the field; even with doses of anti-scalants up to 500 mg/l. There are numerous reasons for this seemingly anomalous situation: high barium concentrations in the formation water (2000 mg/l), high CO_2 in the oil (20 mole %), high reservoir temperature (125^0 C) and high reservoir pressure (50 MPa=500 bar). The most appropriate solution therefore seems to be the selective removal of sulphate from the injected seawater. It is important, however, that the bulk of the salinity remains in the seawater, as fresh water leads to swelling of the clays within the reservoir which again may block the wells.

The production of hydrogen sulfide by sulphate reducing bacteria (SRB) is also prevented by removing sulphate; i.e., their substrate. The souring of reservoirs and inherent corrosion effects are therefore limited.

Nanofiltration is excellent for treatment of fluids where monovalent ions (Na^+ and Cl^-) do not have to be separated. Compared to alternative separation processes for divalent ions; chemical precipitation, ion exchange and reverse osmosis, the investment and operational expenses are favorable for nanofiltration. The rejection is high for divalent and multivalent anions and organics with molecular weight above 200 - 300. Nanofiltration is therefore defined within the grey area between reverse osmosis and ultrafiltration.

For this work separation efficiencies for sulphate and other ions in seawater are determined for several different recoveries; i.e., permeate to feedflow ratio. Potential precipitates of hard to dissolve salts such as barium sulphate, strontium sulphate and calcium carbonate are negated by removing sulphate, calcium and carbonate from seawater prior to seawater injection into barium and strontium rich oil reservoirs.

Produced Water, Edited by J.P. Ray
and F.R. Engelhart, Plenum Press, New York, 1992

THEORY

The NF 40 membranes are crosslinked polyamides spiral wound with proprietary composition. The material is hydrophobic and negatively charged. This means:

- reduced clogging of membrane surfaces. Colloids, oil and proteins are usually hydrophobic and negatively charged and therefore repelled by the negatively charged membrane material;
- increased water flux due to favorable positioning of water-dipoles;
- increased and selective separation of electrolytes.

The electrically charged membrane groups repel anions. Due to electroneutrality, counter ions (cations) are also repelled. The ion separation efficiency is dependent on both charges and concentrations of ions.

The separation efficiency is increased for divalent and multivalent anions (SO_4=). At extremely high electrolyte concentration in the water phase, the charged membrane groups may be neutralized or masked by counter ions. This reduces membrane selectivity.

A consequence of the negatively charged membrane groups on the nanomembrane surface and the ions in the water phase is poor separation of monovalent ions; i.e., Na^+ and Cl^-. Separation efficiency for divalent anions, however, such as SO_4= and CO_3=, is excellent. Softening; i.e., Ca^{++}, Mg^{++} and Sr^{++} separation, is also significant.

Rejection of magnesium sulphate is above 90% by nano-separation. Comparative removal for sodium chloride is less than 50%. Since the NF membranes are negatively charged, it is the anion repulsion which mainly determines the solute rejection. For example the rejection of calcium chloride is about the same or lower than that of sodium chloride, while rejection of sodium sulphate is about the same as that for magnesium sulphate.

Figure 1 demonstrates the effect of ionic strength on separation in a carbonate system. Water with 10 g/l $NaCO_3$ is adjusted from pH 6.5 to pH 11.6. The carbonate equilibrium is pH dependent and the changing feed concentrations are reflected in Figure 1. The resulting permeate concentrations clearly demonstrate the changing removal efficiencies for monovalent bicarbonate and divalent carbonate; 0 - 50% for HCO_3- and above 90% for CO_3=.

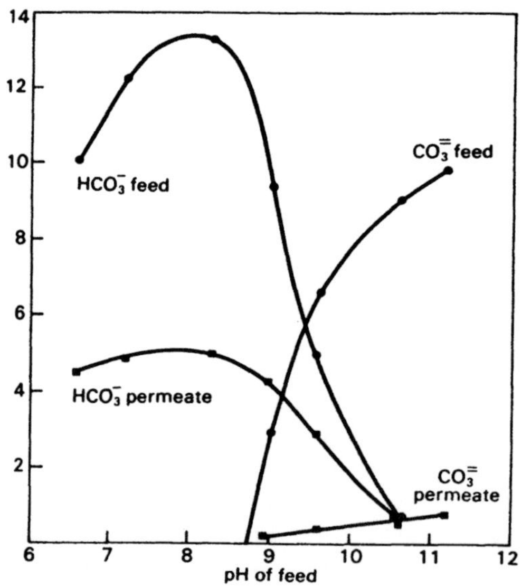

Figure 1. Carbonate/bicarbonate speciation in the feed and permeate as a function of feed pH for a 10 g/l Na^+ solution of sodium carbonate/bicarbonate (Simpson et al., 1987).

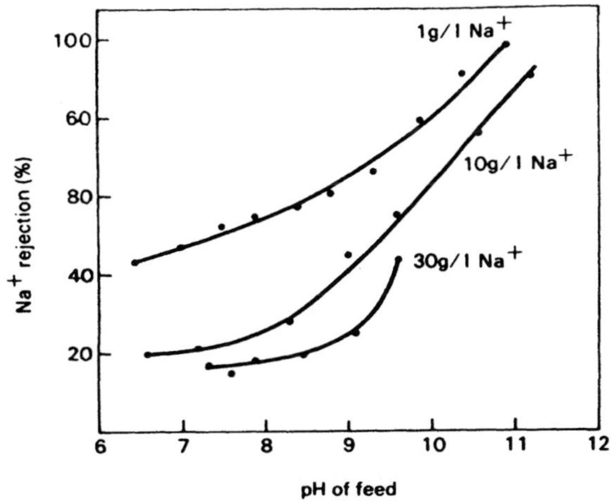

Figure 2. Membrane Na^+ rejection as a function of pH and the concentration of the sodium carbonate/bicarbonate feed (Simpson et al., 1987).

The feedwater represented in Figure 1 is without chloride. Sodium separation would be significant in a carbonate system because of electroneutrality in the concentrate and permeate. Figure 2 shows for different Na^+ concentrations how pH affects the Na^+ separation. A detrimental effect of increasing monovalent ion separation is an increase in osmotic pressure across the membrane. Thus, the water flux at constant operating pressure decreases with increasing pH as shown in Figure 3 for three different concentrations of sodium in the feed water.

Sodium separation is pH dependent in carbonate solutions. For sodium in solutions with sulphate or chloride, however, this pH dependency is minimal. Figure 4 summarizes efficiencies of sodium in different solutions:

1. Na^+ separation increases with increasing pH in a Na_2CO_3/$NaHCO_3$ solution.
2. Na^+ separation is efficient for all pH values in a Na_2SO_4 solution. This is because of efficient SO_4= separation, and electroneutrality requires Na^+ separation.

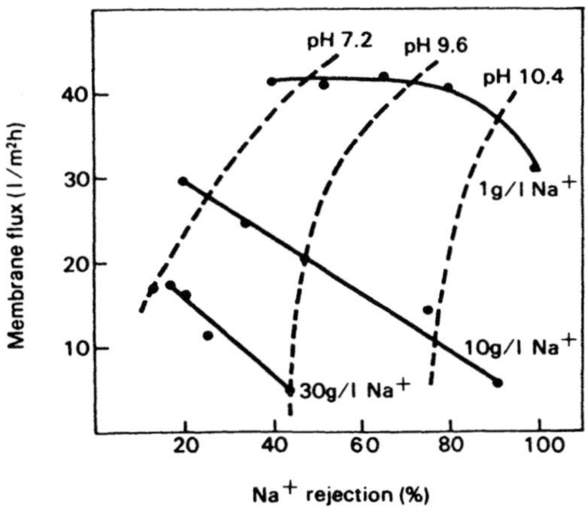

Figure 3. Membrane flux as a function of Na^+ rejection for sodium carbonate/bicarbonate solutions (Simpson et al., 1987).

3. Na^+ separation is poor in a NaCl solution and not pH dependent. Chloride is poorly separated due to its single valence and also electroneutrality negates efficient Na^+ separation.

PILOT STUDIES

A schematic of the pilot plant is depicted in Figure 5. The membrane employed was a Dow Separation Systems FilmTec NF 40-2540 thin film composite with 2.1 m^2 active spiral wound membrane area. The following operating parameters had to be adhered to: P < 4.1 MPa; T < 45° C; 2 < pH < 11; and Cl_2 < 0.5 ppm. Normal seawater from the Statfjord Field in the North Sea was the feedwater. The tank in Figure 5 acted as feed-reservoir for the nanoseparation plant. The recovery was regulated by flow control valves Vk in Figure 5.

Base line

In reference to Figure 5 the membrane system was initially flushed with fresh water at 15 l/min at P1 = 0.4 MPa for one hour. The baseline was thereafter established by feeding the system for 12 hours with seawater free of suspended particles. The recirculation loop was closed by adjusting valves Vr1 and Vr2. The concentrate flow was adjusted by turning valve Vk until the permeate recovery (F2) was at 5%.

Fifty Percent Recovery

At a flux rate of 27 l/m^2 • h (F2 = 57 l/h) at 23° C, the pressure P1 = 2.5 MPa. The raw water feed F1 = 115 l/h whereas the recirculation loop F4 = 1,105 l/h. The pressure loss in the feed to concentrate, P1 - P2, equals 0.3 MPa. Trials were performed for weeks without pressure and flux changes.

Seventy Five Percent Recovery

At a flux rate of 27 l/m^2 • h at 26°C the feed pressure P1 equals 3.0 MPa. The raw water feed F1 = 76 l/h and the internal concentrate recirculation F4 = 1,143 l/h. The concentrate to waste F3 = 19 l/h. The pressure loss P1 - P2 equals 0.3 MPa. Trials were performed for three weeks without pressure or flux changes.

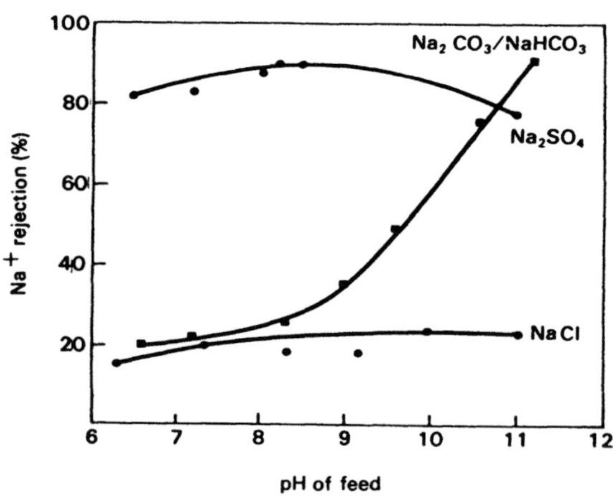

Figure 4. Membrane sodium rejection as a function of anion species at varying pH for solutions containing 10 g/l Na^+ (Simpson et al., 1987).

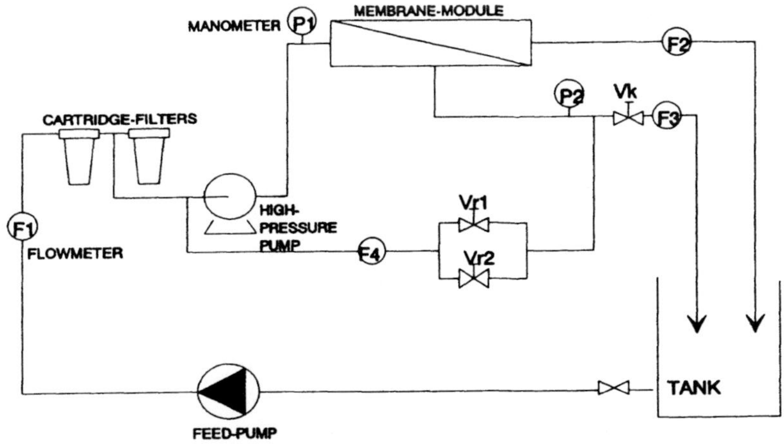

Figure 5. Schematics of pilot plant for nanoseparation of seawater.

Sampling and Analysis

Water samples were collected from the raw feed, the concentrate stream and the permeate. The samples were analyzed for several parameters. For sulphate, both turbidimetric and ion-chromatographic techniques were used. For chloride, the Mohr method employing silver nitrate as the titrant was used as well as potentiometric titration. For bacteria, both blood- and marine-agar were used as growth media at 15° C and 30° C.

RESULTS AND DISCUSSION

Figure 6 schematically represents the flow system for 75% recovery. Tables 1 and 2 summarize the analytical data for three different recoveries; 5%, 50% and 75%.

Sulphate separation is very efficient for all recoveries. For 5% recovery, without internal recirculation of concentrate, the sulphate separation is 99.5%. The separation efficiency drops to 98% for 75% recovery. The slight drop from 99.5% to 98% is likely due to increasing electrolyte concentration in the feedwater (F1 + F4) and a partial masking of the charged membrane surface.

The ion concentration in the concentrate for 75% recovery is shown in Table 1. The concentrate is continuously recirculated at 1,143 l/h and mixed with 76 l/h raw feed according to Figure 6 above. The feed concentration to the membrane is therefore constant over time.

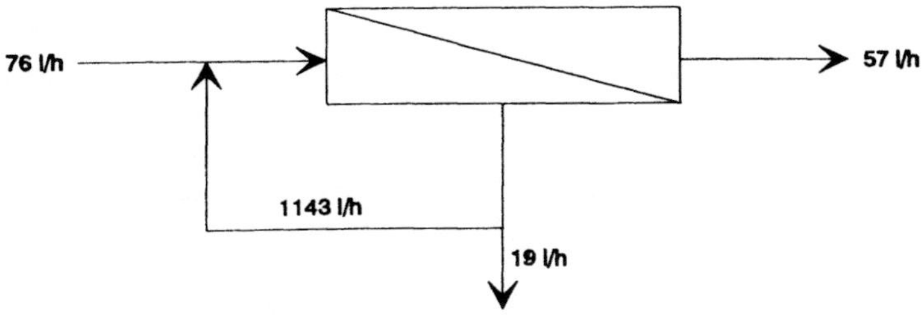

Figure 6. Mass balance of nanoseparation at 75% recovery.

Table 1. Chemical composition (mg/l) of seawater before and after nanoseparation with three different recoveries (%)

Parameter	Feed	Permeate			Concentrate
		75%	50%	5%	75%
Na^+	9400	8700	7900	6500	12000
K^+	358	330	297	248	428
Ca^{++}	328	202	163	108	701
Sr^{++}	6	3	2.5	1.5	14
Mg^{++}	1080	467	338	195	2850
Cl^-	16900	15700	14100	11300	21600
HCO_3-	122	77	65	37	277
SO_4=	2460	60	30	14	8330
pH	7.7	7.5	7.4	7.3	7.7
Conductivity (mS/m)	4065	3600	3300	2700	5200
Density	1.023	1.018	1.016	1.013	1.036
Bacteria /100 ml	10^3	0	0	0	10^3

Table 2. Concentration changes (%) from feed seawater to permeate and concentrate for different recoveries

Parameter	Permeate			Concentrate
	75%	50%	5%	75%
Na^+	7	16	31	28
K^+	8	17	31	20
Ca^{++}	38	50	67	114
Sr^{++}	50	58	75	133
Mg^{++}	57	69	82	164
Cl^-	7	17	33	28
HCO_3-	37	47	70	127
SO_4=	98	99	99	239
Bacteria	100	100	100	0

Sulphate is for all practical purposes the only di- and multi-valent anion in seawater. The 98% separation efficiency for sulphate is in accordance with theory for negatively charged nanomembranes. Electroneutrality for pH 7.7 feedwater also dictates efficient separation for multivalent cations in seawater; Ca^{++}, Sr^{++} and Mg^{++}. Table 2 indicates 67%, 75% and 82% separation at 5% recovery for Ca^{++}, Sr^{++} and Mg^{++}, respectively. With increasing recovery, the separation efficiencies decrease significantly.

The dissolved chemical constituents in normal seawater consists of 87% monovalent ions according to Table 1: mostly Na^+, K^+ and Cl^-. The separation efficiency for the three ions is approximately 30% without recirculation of concentrate; i.e., 5% recovery. At 75% recovery, the separation efficiency for the monovalent ions is around 7%. These results are in good agreement with Figure 4 which indicates about 20% Na^+ separation in a 10 g/l NaCl solution.

A very practical consequence of poor monovalent ion separation is the resulting low osmotic pressure of the separation system. This of course is the major advantage of nanoseparation compared to reverse osmosis. The natural osmotic pressure for seawater is about 3 MPa and typical reverse osmosis operating pressures are 6 - 8 MPa. For nanoseparation the operating pressure in our experiments were always less than 3 MPa. In Figure 3 it is clearly shown that the permeate flux decreases sharply with increasing Na^+ separation. This of course is natural as the osmotic pressure across the membrane (feed/permeate) increases enormously with increasing electrolyte separation. For nanoseparation the sulphate separation from seawater (Na^+

solution) is not coupled to Na^+ and Cl^- separation. Therefore, there is very little change in osmotic pressure across the nanomembrane.

The pH changes from seawater feed to permeate across the nanomembrane reflect the bicarbonate concentrations. Table 1 shows a HCO_3^- decrease from 122 mg/l in the raw seawater to 37 mg/l in the permeate at 5% recovery. The respective pH values are 7.7 and 7.3. The separation efficiencies of bicarbonate agree well with the literature data represented in Figure 1. Also, the values for conductivity and density in Table 1 are a consequence of the change in electrolyte concentrations across the nanomembrane.

DESIGN AND CONCLUSIONS

Charged nanomembranes are tailored for selective sulphate and bacteria separation from seawater. No other process separates 98% sulphate for as little investment cost and energy of operation as nanomembranes. From these studies a design flux of 25 $l/m^2 \cdot h$ at 8°C and 3 MPa seems appropriate.

Based on the above findings a full scale sulphate removal plant for 75% recovery of 11,500 m^3/24 h of raw seawater would have the following dimensions: 10 m long, 7.5 m wide and 5.9 m high. This includes cartridge filters, high pressure pumps, control panels, cleaning systems for membranes, membranes, piping and valves.

The power requirements for the above design would be 670 kw, with the dry weight of the plant at 96 tons.

When Agip's Tiffany field comes on stream in 1993 (Offshore Engineer, 1992) it will boast the first sulphate removal package to be installed on a new North Sea platform. Using the latest membrane separation technology, the process will selectively remove sulphate and other divalent ions from seawater prior to injection into the Tiffany reservoir. At the heart of the process are more than 900 special nanofiltration membranes, NF 40, as used in our work. The £7 million Tiffany system is to operate with 2 x 50% trains with an overall conversion of feed to product of 75%. The design output is 15,924 m^3/24 h (100,000 BWPD). The product quality is guaranteed to reduce sulphate level to less than 150 mg/l. The package dimensions are 10 m long, 7.5 m wide and 8.2 m high. The dry weight is 122 tons and the wet weight is 177 tons (Westgarth, 1992).

According to Offshore Engineer (1992), payback period for the treatment plant is around three years, based on the cost of chemical dosing which it replaces. Taking into account other operational costs associated with using chemicals, for example removing chemicals from produced water before reuse or disposal, the membrane route appears still more attractive.

REFERENCES

Offshore Engineer, 1992, Membranes tackle Tiffany scale, in: "Offshore Engineer Magazine," pages 40-41.

Simpson, A. E.; Kerr, C. A. and Buckley, C. A., 1987, The effect of pH on the nanofiltration of the carbonate system in solution, in: "Proceedings of the Third World Congress on Desalination and Water Reuse," Cannes, France.

Weir Westgarth Ltd pamphlet, 1992, Inverlair Avenue, Cathcart, Glasgow G44 4EX, UK.

INITIAL DESIGN CONSIDERATIONS FOR A COST EFFECTIVE TREATMENT OF STRIPPER OIL WELL PRODUCED WATER

M.A. Adewumi[1], J.E. Erb[2], and R.W. Watson[1]

[1] The Pennsylvania State University
University Park, Pennsylvania
USA

[2] Pennsylvania Department of Environmental Resources
Harrisburg, Pennsylvania
USA

BACKGROUND AND INTRODUCTION

The Appalachian Basin oil industry is composed primarily of small, independent producers operating oil wells in the state of New York, Pennsylvania, West Virginia, Ohio, Virginia, Kentucky and Tennessee. The wells operated by these independents produce very small quantities of crude oil. For example, the Independent Petroleum Association of America (1991) indicated that in 1989, Appalachian producers operated 101,274 oil wells (16.8% of the total number of operating domestic wells) while only producing 21,626,000 barrels of oil, less than one percent of the country's total 1989 production. Because of this small daily production, an average of less than one barrel of oil per day, almost all producing wells in the Appalachian Basin are considered stripper (economically marginal)wells. Because of the large number of wells needed to produce such low volumes of hydrocarbons, operators of these wells conduct business on the basis of very thin profit margins. The typical Appalachian oil well earned a mere $13.76 per day in 1989. The slim profit margins associated with operating these economically marginal oil wells make them extremely sensitive to increases in operating costs or decreases in prices paid for the commodity they produced.

Appalachian Basin oil wells typically produce along with the oil, small volumes of waste water-brine. (Produced water has constituents similar to those of sea water.) For example, 87% of the primary stripper oil wells produce less than one barrel of brine per day. This brine has been identified as a potential source of environmental degradation, and each oil well site where brines are produced is a potential candidate for regulatory control under the U.S. Environmental Protection Agency's National Pollutant Discharge Elimination System (NPDES) permitting program. Presently, brines produced from stripper oil well brine discharges are exempt from the Clean Water Act's zero-discharge effluent guideline and consequently, operators may treat and discharge brine into surface streams. In some states however, independent producers are also subject to separate state permit programs designed to address identical issues. In Pennsylvania, for example, oil well operators may use the stream discharge option for brine disposal if they are able to satisfy the NPDES permit requirements of the Clean Water Act as implemented in the Pennsylvania Department of Environmental Resources (DER) Part I permit process and the

Produced Water, Edited by J.P. Ray
and F.R. Engelhart, Plenum Press, New York, 1992

Part II Water Quality permit requirements of the state's Clean Streams Law. The two programs are administered and enforced by the DER, and stripper wells must be permitted and regulated under both simultaneously. State regulations implementing the two programs were adopted in 1978, but compliance with them has been virtually impossible for Pennsylvania's independent stripper oil well operators because of the excessive costs incurred in completing the appropriate permit applications. To apply for the necessary discharge permits, an operator must pay two application fees of $500 each, provide chemical analysis of the produced water quantifying no less than 29 parameters, and design an engineering plan for a treatment facility adequate to bring the produced water into compliance with water quality standards. The costs of the required analytic work for the NPDES permit ranges from $500 to $900, and the cost of the engineering study required by the Part II permit has, in the past, ranged between $5,000 and $50,000. The total costs incurred by a well operator, exclusive of his own administrative and clerical support, for each site required to comply with the NPDES and Part II permit requirements, therefore, equals a minimum of $6,500. This figure does not include the costs incurred by the operator in constructing, operating and monitoring the performance of the facility once approved by the DER. The financial burden imposed by the permit processes essentially prohibits the operator from complying with regulatory requirements.

OBJECTIVES OF THE RESEARCH

Given the economic realities existent in the operation of stripper oil wells, a research program to develop a cost effective treatment methodology of stripper oil well brine was initiated. The global objectives of this research effort, begun in 1988, are to permit the environmentally safe discharge of brines produced from stripper oil wells in Pennsylvania to the surface waters of the Commonwealth and to rapidly bring as many wells as possible into compliance with the law. The specific objective of this phase of the work is to evaluate some initial design considerations.

The work to achieve the above elucidated objectives had to take into consideration certain practical aspects of the field and economic nature of production from stripper oil wells. These design considerations included:

- the treatment unit is to be constructed of readily available materials such as plywood and plastic pipe;
- the treatment unit needed to be a gravity flow system since the units could be located at well sites remote from electric power; and
- the treatment design needed to be universal in application since stripper well production is from various sand formations which contain brines with a variety of constituent concentrations.

DEVELOPMENT OF THE PROCESS

To date, the research effort has led to development of a field scale pilot brine treatment facility which was constructed in 1989. The fabrication design used in the field scale pilot brine treatment facility evolved through a series of laboratory tests. These laboratory tests began in early 1988 with the evaluation of a brine treatment scheme developed by the Pennsylvania Oil and Gas Association's Environmental Committee with the assistance of the engineering firm, Baker/TSA, Inc. The laboratory evaluation of this brine treatment design indicated that the process design was inadequate. The model was subsequently modified, and the tests proved that the changes made to the process were promising with respect to the removal of dissolved iron.

In 1989, the second phase of the project was begun. It consisted of work on a computer program aimed at the development of the design parameters needed to scale-up the laboratory apparatus for a full-scale field demonstration; further laboratory work to refine the treatment process and to provide the reaction kinetics data needed to complete the computer program; and a field demonstration of the viability of the treatment process. A computer model that would be the basis for scaling up the laboratory scale model to the field scale was developed. This model was based on the data collected from laboratory experiments which considered the process flow path and the kinetics of the reactions taking place in the treatment process. The scale model of the treatment process was used to study the process flow path. A demonstration facility was designed and constructed using the knowledge obtained from the laboratory scale model and the computer simulator. This facility was located at Cabot Oil and Gas Corporation's Franklin Brine Treatment plant in Franklin, Pennsylvania. The facility was tested using brine produced from the Red Valley sandstone located in Venango County, Pennsylvania. The Red Valley brines are considered to be a "worst case" brine because of the comparatively high concentrations of dissolved iron and other metals. The results of this field testing indicated that the proposed treatment methodology should allow operators to satisfy the current federal and state brine disposal permit requirements.

THE LABORATORY-SCALE UNIT

System Description

In-depth analysis and study of the treatment process necessitated the building of a laboratory prototype, which allowed us to conduct parametric studies under controlled conditions. The field model is a larger replica of this model, even though the construction differs only for convenience and durability. Figure 1 shows the schematic diagram of the laboratory prototype.

The laboratory prototype is made up of four principal components, namely, the pH Adjustment Unit, the Aeration Unit, the Solid Separation Unit and the Sand Filtration Unit. The

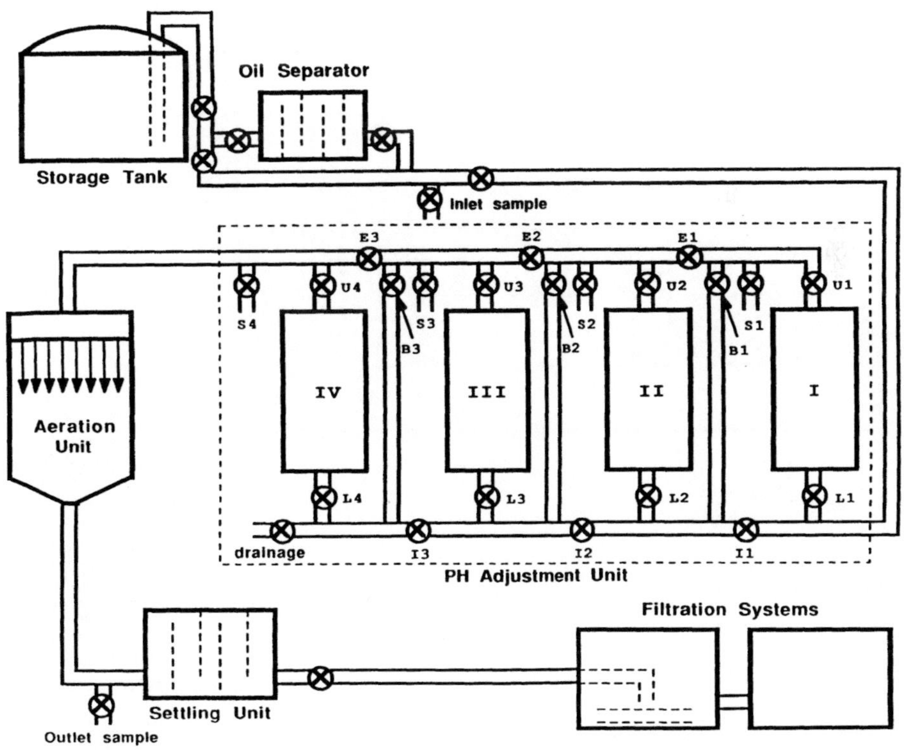

Figure 1. Schematic diagram of the laboratory unit of the Penn State/Program Brine Treatment Process.

effectiveness of the treatment process is predicated upon three essential elements. First, the pH of the brine being treated must be raised to a certain level necessary to initiate the oxidation reaction. Second, air contact with the brine must be adequate to ensure adequate availability of the oxygen needed for reaction. Third, sufficient retention time must be imposed to allow for settling of the precipitated solids. Finally, the fine solids that could not settle within a reasonable retention time must be removed. Each of these essential steps in the treatment process is effected in different units. Since this is a continuous process, all the units must be coupled together in such a way that no unit becomes a bottleneck that inhibits treatment rate. In this regard, comprehensive data collection, analysis and parametric studies must be conducted to determine how each unit effects and affects the rate and effectiveness of the treatment scheme.

The pH Adjustment Unit is made up of two packed-bed columns of limestone arranged in series. These columns are vertically oriented and fluid flow is from the bottom to the top so as to afford maximum contact between the fluid and the limestone chips. This has proved to be quite effective. Each column is made out of 5-inch O.D. PVC pipe and they are 26 inches tall. The effluent from the pH adjustment unit usually has a pH value of slightly above 7.0. This effluent then goes to the aeration unit where the oxidation process takes place. Here, for instance, iron in the ferrous state which is in solution, reacts with oxygen and is precipitated as a ferric salt.

The Aeration Unit is basically made up of a flat surface that is oriented at an angle to the horizontal. This angle is adjustable so as to ensure control of the liquid film thickness while it passes through this unit. The basic idea is to maximize air-brine contact and the best way to achieve this is to have brine flow on the bed of the aeration unit as a thin film. Since the total flow rate may vary, an adjustable angle permits the achievement of the thin-film principle. The

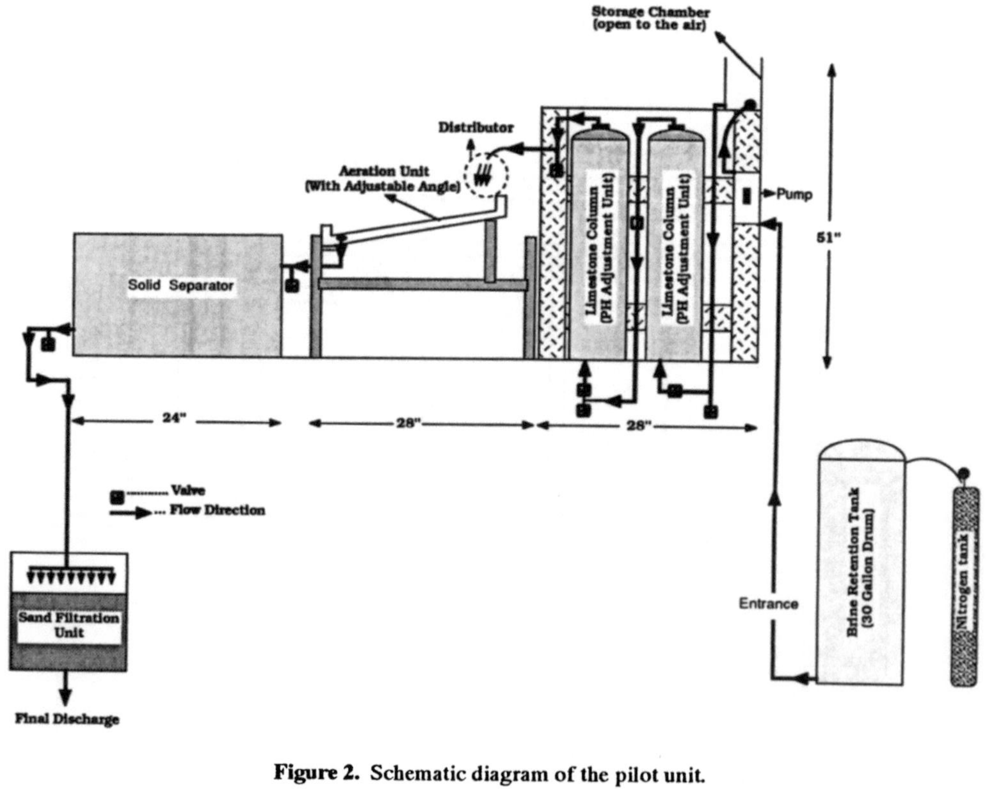

Figure 2. Schematic diagram of the pilot unit.

flat bed is 28 inches long and 12 inches wide. The liquid, being treated comes into the aeration unit through a horizontal distributor to ensure uniform flow over the bed. The bed, which is made out of plexiglass is roughened to prevent flow channeling, which was observed during the initial testing of this system.

The effluent from the aeration unit goes to the solids separator whose primary function is to provide the necessary avenue to effect separation of the solid precipitate from the treatment fluid. The unit has internal partitions that intersect the direction of flow. They are arranged in such a way that space is left below each partition to allow for flow and maximize the travel path and hence the retention time. There are two internal partitions in the present systems, thus giving us three compartments. This can be changed if need be but the system as it is quite effective. Generally, most settling takes place in the first compartment upstream and the least in the compartment closest to the exit point. In fact, the coloration of these compartments gives an indication of how much iron is being removed. The solid separator is made of plexiglass and it is 24 inches long, 12 inches wide and 12.5 inches deep.

The effluent from the settling unit still contains the finer particles, that need to be removed. This stream of fluid goes through a Sand Filtration Unit for removal of these fine particles. The filtration unit is basically a sand box, filled with sandstone. The box is 19 inches long, 19 inches wide and 14.5 inches deep and it is made out of plexiglass. The effluent from this unit is the final discharge from the treatment facility.

There is one unit that is not a part of the treatment process, but whose main use is for control purposes. This is the nitrogen-pressured brine storage tank. The whole idea is to control the oxygen environment in the system so that we do not initiate the reaction prematurely. The nitrogen is used to displace the air above the tank so as to prevent contact with oxygen before reaching the aeration unit.

The Field-Scale Unit

A schematic diagram of the field scale test unit is shown on Figure 2. The flow of brine through the system can be separated into a series of individual processes. These processes are: storage; oil, grease, basic sediments and water separation; pH adjustment; aeration; solids separation and filtration.

Brine to be processed by the treatment unit is delivered to the site by tank truck. This brine is then transferred to two fiber-glass lined 70 barrel storage tanks. The storage tanks were sized to accommodate the typical capacity of an over-the-road tanker which is 100 barrels. The storage tanks are lined with fiber-glass to mitigate any reactions between the stored brine and the steel walled tank. Brine then flows from the storage tank to the oil, grease, basic sediments and water separation unit. A siphon coupled to a flow regulator provides constant brine flow from the storage tanks to the separation unit. After the brine exits this unit, it enters the pH adjustment unit. The purpose of the pH adjustment is to raise the pH of the brine. The pH adjustment is accomplished by passing the brine through a chipped limestone pack. The limestone used in the pack was construction grade limestone obtained from a local quarry and is typical of the grade used by the Pennsylvania Department of Transportation in road construction and maintenance. After passing through the pH adjustment unit, the effluent flows to an aeration device. Aeration of the brine was achieved by distributing the flowing brine uniformly over a plywood surface coated with a water resistant epoxy. The effect achieved is a thin sheet of water flow, or a thin film. Thickness of the thin film is directly proportional to the velocity of the distributed liquid. Velocity and consequently, film thickness, was controlled by changing the slope of the plywood surface. To accommodate the change of slope, a hinge was attached to the base of the aeration unit. Stabilization of the chosen slope was accomplished using railroad tie cribbing. After aeration, the brine passes to a solids removal unit. This unit is similar in design to the oil, grease, basic sediments and water separation unit and is used to remove precipitated solids. This removal is accomplished using a series of weirs which permits flow of the brine through the device via a tortuous path. Effluent from the solids removal unit is then directed to a sand filtration unit.

The purpose of this sand filtration unit is to remove suspended solids which by-passed the solids removal unit. Experience with the system has indicated that a layer of limestone chips spread over the bottom of the sand filtration unit permitted an increase in pH to a value necessary to permit stream discharge.

TREATMENT PERFORMANCE EVALUATION

Experimental Procedure

One of the main objectives of the laboratory work was to isolate the effects of the various parameters on the effectiveness of the treatment process. In order to do this, the constituents of the fluid being processed had to be adequately controlled so as to ensure that experiments were being conducted under controlled conditions. The logical answer to this constraint was to use synthetic brine (i.e., brine whose constituents and their concentrations are well-defined). Large quantities of the brine were needed because the process is a continuous one, rather than batch. The synthetic brine was prepared with pre-specified ion concentrations and stored in a thirty-gallon plastic drum. The headspace was filled with nitrogen to prevent premature oxidation from taking place. All the experimentations carried out so far concentrate on a single element, iron. The iron-containing brine was prepared by adding iron salt [ammonium iron (II) sulfate hexahydrate] to the brine whose pH had been measured.

The prepared brine was transferred, on a continuous basis, to a small storage chamber mounted on a platform elevated higher than the rest of the system. This afforded a gravity feed system, similar to the expected field system. From there, the fluid flowed through the system in the sequence, already described (see Figure 1). Each unit was critical to the treatment process. For instance, the brine's pH must be pre-adjusted to a sufficiently high level; otherwise, little or no oxidation would take place in the aeration unit. Similarly, the fluid must flow through the latter as a thin film; otherwise the necessary contact with air would not occur and hence, little or no oxidation would take place. The precipitated solids must be given sufficient residence time for settling to take place in the solid separator. Finally, the remaining fine precipitates should be removed in the sand filtration system.

Effluent Testing Program

A comprehensive sampling program was essential to a successful monitoring of the system performance and unit contribution to the overall treatment process. The model was designed to permit sampling at various points along the flow path and sampling ports were located so that sampling could be effected without substantial disruption to the experiment. Furthermore, sampling assumed that the system was transient and hence each port was sampled at regular time intervals of thirty minutes. Changes in iron concentration with time were thus monitored and can provide useful information about how long it takes the system to achieve equilibrium state if the composition of the feed stream is fixed. Approximately 50 samples are collected for each experimental run. The samples were stored in 50 ml plastic bottles, to which hydrochloric acid was immediately added to halt further reactions. The samples were subsequently analyzed.

Sample Analysis

The three principal analytical instruments used were Plasma Emission Spectrometer, (PES), Atomic Absorption Spectrometer and the Optical System Spectrometer. Each of these have different capabilities and are used for different purposes. The Plasma Emission Spectrometer is used to detect the main ions in a brine sample. It is capable of detecting Zn, Fe, Ni, Cu, Mn, V, Cr, and Al. With the Atomic Absorption unit, three more ions, namely, Pb, Co,and Cd can be detected. Distinction of ferrous from ferric ion is achieved by using Optical System Spectrometer. It detects and measures only the ferrous concentration. By using mass balance and the total Fe concentration from the PES, the ferric ion concentration can be calculated.

Results and Analysis

Several runs have been made and results of only a few of the runs are included in this paper. Furthermore, this series of runs emphasizes the system's efficacy with respect to only one element, iron. The results included in this paper focus on the removal of this single element. Further efforts are underway to investigate in detail the other elements. It is pertinent to mention that one of the objectives of our work is to identify the key elements that would characterize the produced brines and perhaps investigate the possibility of using these to characterize the treatment effectiveness. Our hypothesis is based on equilibrium thermodynamic principles and this will be tested as this research evolves and more data are generated. All the results shown are obtained using primarily the atomic adsorption spectrometer since results from the various spectrometers have been found to be consistent within the limit of experimental error. All experiments are performed under similar conditions except where noted. The only parameters that are altered from run to run are the flow rate and inlet brine composition. These two are particularly significant in order to reflect the fact that brine production rate will differ from operation to operation and the concentration should be field-dependent. The general observation was that the system is very effective in removing iron from the brine and its effectiveness depends on the flow rate and inlet concentration. It is important to know these parameters so that they can be factored into the design process. For example, in two runs where the flow rates of 2.94 barrels per day and 1.91 barrels per day were used, 99% of the iron present was removed in the latter whereas only 90% was removed in the former. Both runs were made at the same inlet iron concentration.

Experiments show that the system reaches equilibrium within 2 to 3 hours. This, of course could be flow rate and concentration dependent but within the limit of the runs made so far, this was found to be the case. One other parameter of interest is the proportion of iron removed in each of the units in the system. This was carefully analyzed through appropriate sampling at various locations.

Table 1 shows the proportion of the iron removed in each of the process unit. It is clear that iron is removed from solution as the fluid progresses through the system, thus demonstrating the general effectiveness of the treatment system. Figure 3 graphically displays this trend.

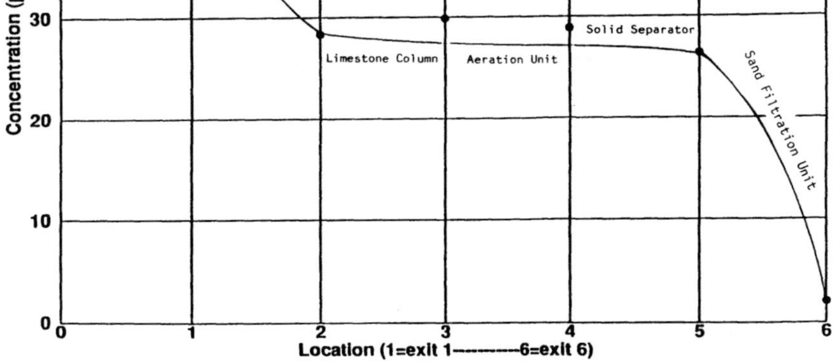

Figure 3. Concentration of total iron vs. location of sampling.

Similar data for different flow rates show the same trend, thus suggesting that this shape was characteristic for the treatment process. Consider for example Figure 3 for brine flow rate of 3.81 bbl/d. The initial iron concentration of 52 ppm was reduced to 32 ppm in the first limestone column. This was due to the fact that the pH was effectively adjusted there and instigates the oxidation process with the dissolved air. As the brine passed through the second limestone column, no significant reaction takes place, but the pH is increased. This makes major oxidation possible in the aeration unit but the iron is not separated until it reaches the solid separation unit where a substantial fraction of the precipitated iron is removed. The final treatment stage is the sand filtration unit which effectively removes the finer solid iron precipitates left. All elements of the system appear to contribute to the system's effectiveness, either for pH adjustment, oxidation or separation purposes.

Table 1. Percentage removal of total iron by different parts of the prototype model

Avg. inlet conc. = 50 ppm

Flowrate (bbl/d)	Exit 1 after the first limestone column	Exit 2 after the aeration unit	Exit 3 after the seperator	Exit 4 after the sand filteration
2.0	0.10%	1.11%	4.39%	100%
3.0	4.36%	0.64%	13.43%	46.62%
4.0	0.71%	1.63%	3.90%	32.87%

Avg. inlet conc. = 100 ppm

Flowrate (bbl/d)	Exit 1 after the first limestone column	Exit 2 after the aeration unit	Exit 3 after the seperator	Exit 4 after the sand filteration
2.0	3.80%	2.94%	2.38%	99.51%
3.0	7.90%	-0.11%	5.35%	43.14%
4.0	0.93%	1.40%	9.58%	31.16%

Avg. inlet conc. = 150 ppm

Flowrate (bbl/d)	Exit 1 after the first limestone column	Exit 2 after the aeration unit	Exit 3 after the seperator	Exit 4 after the sand filteration
2.0	0.28%	0.86%	2.30%	96.99%
3.0	2.11%	1.85%	6.11%	28.38%
4.0	1.84%	2.07%	3.90%	14.16%

Avg. inlet conc. = 200 ppm

Flowrate (bbl/d)	Exit 1 after the first limestone column	Exit 2 after the aeration unit	Exit 3 after the seperator	Exit 4 after the sand filteration
2.0	2.59%	1.13%	2.94%	91.16%
3.0	2.08%	0.45%	0.92%	24.78%
4.0	0.45%	1.57%	2.50%	12.19%

PROCESS MODELING

At this stage, development of the model is designed for predicting iron concentrations along the aeration unit. The kinetic equation formulation used in this model is coupled with a material balance equation by assuming constant temperature. The resulting equations are a system of differential equations which is solved numerically using the Implicit Runge-Kutta technique. For fine tuning and validating this model, the laboratory and field data are used. This model is coded in Fortran 77. Aeration is central to the treatment process as it is the means of providing the necessary oxygen for the reaction.

The reaction between oxygen and iron can be written as:

$$4\ Fe = 3\ O_2 \rightarrow 2\ Fe_2O_3$$

where ferric oxide, (Fe_2O_3) is the product of the oxidation reaction which will precipitate from the brine, as a solid phase. The rate and extent of this reaction depends on several factors such as retention time, brine flow rate, pH of the brine, brine composition, temperature conditions etc. Since these properties and conditions vary from field to field, the equipment geometry used for the aeration technique will be different also for different fields. In order to achieve optimum performance of the aeration process, a precise reactor design is therefore necessary. A kinetic model which has been developed in our study is used for designing this reactor to suit the particular properties and field conditions. The capability of this model will be expanded to handle multi-element systems where several chemical components need to be removed from the brine and/or where temperature changes occur. The experimental work provides the necessary kinetic data for determining the reaction constants.

Kinetics Model

For a homogeneous reaction, the reaction rate, r_i, is a function of temperature and concentration of the reactants and products. In an irreversible reaction:

$$aA + bB \rightarrow cC + dD \tag{1}$$

the rate of reaction of A may be written.

$$r_A = -\frac{dC_A}{dt} = K\,C_A^{\alpha}\,C_B^{\beta} \tag{2}$$

where α is the order of reaction with respect to A, and β is the order with respect to B, K is the reaction rate constant which is independent of concentrations. It is not necessary that the order of reaction be equal to the stoichiometric number. In fact, these constants can be determined from extensive kinetic experiments.

Description of the Process Model

For any reactor, the conservation principle requires that the mass of species "i" in element of reactor volume V obeys the following statement:

$$\left[\frac{\text{Rate of i into}}{\text{Volume Element}}\right] - \left[\frac{\text{Rate of i out of}}{\text{Volume Element}}\right] + \left[\frac{\text{Rate of Production of i}}{\text{within the Volume Element}}\right] =$$

$$\left[\frac{\text{Rate of Accumulation of i}}{\text{within the Volume Element}}\right] \tag{3}$$

By appropriately defining the various terms of the preceding equation and making reasonable assumptions, the governing system of differential equations is obtained. For one-dimensional flow system, this system of equations becomes:

$$\frac{\partial}{\partial t}(C_i) = K C_i^{\alpha} - \left(v_x \frac{\partial C_i}{\partial x} + C_i \frac{\partial v_x}{\partial x}\right) \tag{4}$$

In this study, the numerical method of lines is used for solving equation (4) which is categorized as a hyperbolic partial differential equation. For discretizing the spatial differential terms, the backward-finite-difference scheme is used. The resulting system of ordinary differential equations can be written as:

$$\frac{dC_i}{dt} = K C_i^{\alpha} - v_{io} \frac{C_i - C_{io}}{\Delta x} - C_{io} \frac{v - v_{io}}{\Delta x} \tag{5}$$

where

C_i = Concentration of species "i" at outlet
C_{io}= Concentration of species "i" at inlet
v = velocity at outlet
v_o = velocity at inlet
Δx = grid size in x-direction

Equation 5 can be written in a compact form as:

$$\frac{\partial}{\partial t}(C_i) = E(C_i) \tag{6}$$

Numerical Solution

The general formulation of the implicit Runge-Kutta method used to solve Equation (6) follows:

$$B(C_i)\frac{\partial C_i}{\partial t} - E(C_i) = \vec{0} \tag{7}$$

$$\left\{B(C_i) - \beta_1 \Delta t \left[\frac{\partial E}{\partial C_i}\right]_n\right\}\vec{K}_1 = \vec{E}(C_i) \tag{8}$$

$$\left\{B(C_i) - \beta_1 \Delta t \left[\frac{\partial E}{\partial C_i}\right]_n\right\}\vec{K}_2 = \vec{E}(C_i + \beta_2 \Delta t \vec{K}_i) \tag{9}$$

$$\vec{C}^{(n+1)} - \vec{C}^{(n)} = \Delta t\,(C_i \vec{K}_1 + C_2 \vec{K}_2) \tag{10}$$

where

$\beta_1 = 0.292893219$, $\beta_2 = 0.485281374$, $C_1 = 0.426776696$
$C_2 = 0.573223304$, K = slope of equation

The component variable $E(C_i)$ in equation 7 is defined as

$$E(C_i) = k\,C_i^{\alpha} - v_x \frac{\partial C_i}{\partial x} \tag{11}$$

By assuming that velocity, v_x is constant across the system, Equation (11) becomes.

$$E(C_i) = k\,C_i^{\alpha} - v_o \frac{\partial C_i}{\partial x} \tag{12}$$

In the difference form, Equation (12) can be written as

$$E(C_i) = k\,C_i^{\alpha} - v_o \frac{(C_i)_j - (C_i)_{j-1}}{\Delta x} \tag{13}$$

The derivative of E(C) with respect to C_i is

$$\frac{dE}{dC_i} = (k\,\alpha)\,C_i^{\alpha-1} - v_o \frac{1}{\Delta x} \tag{14}$$

For example, using a 5-grid-point system with uniform grid size, Δx would yield the following coefficient matrices:

$$[B] = \begin{bmatrix} 1 & 0 & 0 & 0 & 0 \\ 0 & 1 & 0 & 0 & 0 \\ 0 & 0 & 1 & 0 & 0 \\ 0 & 0 & 0 & 1 & 0 \\ 0 & 0 & 0 & 0 & 1 \end{bmatrix} \tag{15}$$

$$\left[\frac{\partial E}{\partial C_i}\right] = \begin{bmatrix} \frac{\partial E_1}{\partial C_i} & 0 & 0 & 0 & 0 \\ 0 & \frac{\partial E_2}{\partial C_i} & 0 & 0 & 0 \\ 0 & 0 & \frac{\partial E_3}{\partial C_i} & 0 & 0 \\ 0 & 0 & 0 & \frac{\partial E_4}{\partial C_i} & 0 \\ 0 & 0 & 0 & 0 & \frac{\partial E_5}{\partial C_i} \end{bmatrix} \tag{16}$$

$$\left[E_i(C_i)\right] = \begin{bmatrix} E_1(C_i) \\ E_2(C_i) \\ E_3(C_i) \\ E_4(C_i) \\ E_5(C_i) \end{bmatrix} \tag{17}$$

Using the Iannelli-Baker method, this system of ordinary differential equations can be solved, and this model can then be used for designing and predicting reactor performance in the aeration process. At this stage, the single reaction between iron species and oxygen is considered. The reaction model which is described in equation (2) is used in this model by assuming that the value of equals zero. In addition, the temperature in the system is assumed to be constant and its effect is negligible. The constants K and α on other hand, are determined experimentally from laboratory data.

CONCLUSIONS

A rather simple treatment process has been developed for treating oil-well brine. The system is proven, both by laboratory experiments and field-scale pilot tests, to be capable of removing iron from the brine stream. In addition, the process shows good promise in terms of removing other heavy metals but more work needs to be done to quantify the effectiveness of the process in this regard. Extensive kinetics data have been developed to form the basis for the kinetics model needed to develop the process flow model. Fundamental modeling of the process has been undertaken to afford the operator the necessary tool to custom-design the process to meet his own particular needs. This work is progressing and efforts are continuing at extending the study to multi-element systems and also to address brine from gas wells.

ACKNOWLEDGEMENTS

Leksono Mucharam, Shifeng Tian, Scott Heckman and Siamak Safargar made various contributions to the research work and their contributions are gratefully acknowledged. Grants from POGAM, Ben Franklin Partnership Program of Pennsylvania and Pennsylvania Energy Development Authority supported the research efforts. Steve Rhoads, president of POGAM initiated the effort leading to this work and his contributions are very much appreciated.

REFERENCES

Independent Petroleum Association of America, 1991, The oil and natural gas producing industry in your state, 1990-1991, Washington D.C.

Kohler, T. E., Heckman, S. W., Watson, R. W., Adewumi, M. A., and Rabosky, J. G., 1989, The design, development, and testing of a standardized treatment system for oil field brines, Proc. SPE Eastern Reg. mtg. (SPE # 19319).

Smith, J. M., 1981, Chemical Engineering Kinetics, Mc Graw-Hill Book Co., New York.

BIORECLAMATION OF OILFIELD PRODUCED WASTEWATERS: CHARACTERIZATION AND FEASIBILITY STUDY

Gilbert T. Tellez, and N. Nirmalakhandan

New Mexico State University
Las Cruces, New Mexico
USA

ABSTRACT

This research evaluates the feasibility of bioreclamation of hydrocarbon contaminated brine waters produced during oil and gas recovery. Samples from five typical oil field waste storage pits and separatory tanks in southeastern New Mexico were evaluated. Studies focused on physical-chemical characterization and biodegradability of the samples. A commercially available culture, PETROBAC-S (Polybac Corporation, 1990) was selected for the biodegradability evaluation. Batch biotreatability studies using a computer interfaced respirometer were conducted to evaluate the degradation of petroleum hydrocarbons under acclimated and non-acclimated biomass conditions.

Results indicated that biological degradation using a non-acclimated culture was inhibited by produced waters at dilutions of 5:1 and 10:1, with inhibition being more predominant in the 5:1 dilution. However, with acclimated cultures reductions of 65 to 98% for total petroleum hydrocarbon were obtained under the same dilution ratios.

INTRODUCTION

Treatment methods based on biological oxidation of oilfield wastewaters are unique and rarely used. This unfamiliarity may be attributed to the limited amount of previous and/or existing research and development studies needed to properly design conventional biological treatment processes. With the high amounts of dissolved organics (e.g., organic acids and phenols), presently used physical treatment such as gravity-separation and flotation are not designed to remove these components. Additional treatment processes, such as biological oxidation, may be needed to remove soluble organic constituents in order to produce a less hazardous, organic free effluent.

For microbial degradation of the organic materials present in oilfield produced wastewaters, inhibition of possible toxic inorganic compounds must be oppressed. Toxic constituents may include chromium, chromates, zinc, arsenic, copper, cyanides, sodium and chlorides. Many of these may be present in produced waters at elevated conditions, and could potentially hamper microbial activity.

Sodium chloride for example (as Na^+ or Cl^-), at elevated concentration can cause considerable environmental stress, produce microbial lysing effects, promote loss of biomass, and increase the concentration of soluble organic materials. Work has been done to investigate the effect of brines or salts on biological treatment processes during both steady state and shock conditions. Individual species can be acclimated to high salt concentrations. However, a common organism such as *Escherchia coli* loses its acclimation within a short time after resuspension in fresh water. In contrast, when an acclimation is changed from a low "fresh-water" concentration to a higher concentration, stimulation of biomass growth takes place (Gaudy and Gaudy, 1980).

The specific objectives of the following research were to characterize southeastern New Mexico's oilfield produced wastewaters and to evaluate the feasibility of bioremediation as a cleanup technology. Biological treatability studies were conducted to determine the most economical dilution ratio required to ensure maximum organic substrate removal. Abiotic testing was also made to evaluate organic substrate removal mechanisms other than by biological means.

Produced water samples from five different sites were evaluated. A commercial culture PETROBAC-S, capable of growing in saline environments, was chosen for this study. Through laboratory scale batch reactors, biodegradation using a wastewater acclimated commercial seed, was evaluated for the removal of petroleum hydrocarbons, even at elevated brine concentrations.

Produced Wastewater Collection and Characterization

Produced water samples from various types of separator tanks and pits were collected from five oil producing sites in southeastern New Mexico. All samples were preserved and tested in accordance with Standard Methods for the Examination of Water and Wastewaters (American Public Health Association, 1985).

Physical-chemical testing of the five oilfield produced wastewater samples is listed in Table 1. Test results were evaluated with respect to parameters set by the State of New Mexico's Water Quality Control Commission Regulations for discharges onto or below the surface of the ground (New Mexico Water Control Commission, 1988).

The organic content of the New Mexico produced water samples were measured using three different quantitative techniques; soluble total organic carbon (TOC), oil and grease (O&G) determination, and total petroleum hydrocarbon measurements (TPH). Soluble TOC concentration for the wastewaters ranged from 150 to 750 mg/l. The separatory tanks maintained the lower end of the range and the waste pits the upper. This discrepancy in TOC concentrations could be attributed to various constituents that are contained in waste pits and not in separatory tanks, such as drilling muds, waste oil rig lubricants, drilling additives, etc. Pits are also exposed to organic materials which exist in the natural surroundings (i.e. soils, blow materials, animals waste, etc.), which could contain high concentration of organic material. When compared to state ground water stipulations, all samples exceed the TOC standard of 20 mg/l, and all are characterized as a high strength domestic wastewater. Roswell and Lovington pits represent the higher concentrations, with values of 722 and 694 mg/l, respectively.

Samples containing oil and grease exhibit traits of high strength domestic wastewater. The Roswell separator waste stream contained the largest O&G concentration of 1,540 mg/l. State of New Mexico's ground and surface water regulations do not stipulate a O&G requirement. However, other states such as Alaska and Texas impose a ground and surface water limit of 15 mg/l. All tested produced waters exceed this limitation.

Total petroleum hydrocarbon analysis, unlike oil and grease determination, measures only mineral oils (e.g., (C_6 - C_{14}) petroleum hydrocarbon chains). TPH values of the produced waters ranged from approximately 300 to 1,000 mg/l. The organic materials extracted have infrared (IR) absorptions near 2,900 cm^{-1} (alkyl). This spectrum includes the IR absorption of benzene and its derivative groups such as toluene, xylene, ethylbenzene, and isobutylbenzene. State of New Mexico's ground and surface water regulations do not stipulate a TPH requirement.

Table1. Major chemical constituents found in typical southeastern New Mexico oilfield produced water

Parameter	Roswell Tank	Artesia Pit	Hobbs Tank	Roswell Pit	Lovington Pit
pH	6.8-7.2	6.7-7.2	6.8-7.4	7.0-7.5	7.0-7.3
Dissolved oxygen, mg/l	0.1	0.71	0.06	1.1	1.4
Solids, mg/l					
Total Residue	250,000	64,900	60,000	75,700	6,100
Total Dissolved	225,000	64,220	59,800	74,300	5,700
Total Suspended	25,000	700	200	1,400	400
Total Volatile	40,000	8,550	20,000	12,900	2,200
Total Fixed	210,000	56,350	40,000	62,800	3,900
Total organic carb., mg/l	250	218	155	722	694
Oil & Grease, mg/l	1,540	768	1,040	1,259	714
Total Petroleum Hydrocarbon, mg/l	375	363	886	640	442
Total Nitrogen, mg/l	24-30	10-15	20-25	0-5	5-10
Hydrogen sulfide, mg/l	600	407	265	459	195
Alkalinity as $CaCO_3$, mg/l	565	166	740	160	265
Chloride, mg/l	122,450	34,900	34,000	15,250	1,198
Cation, mg/l					
Ca	725	660	1,086	1,750	570
Mg	293	840	2,160	460	47
Na	59,900	21,300	16,000	8,000	460
K	1,650	550	38	337	8.0
Total Phosphorus, mg/l	6.0	0.65	2.0	2.70	<0.1
Sulfate, mg/l	1,750	1,560	1,011	1,640	650
Heavy Metals, mg/l					
As	<0.001	<0.001	<0.001	<0.001	<0.001
Ba	0.1	0.2	0.1	0.3	0.1
Cd	<0.005	<0.005	<0.005	<0.005	<0.005
Cr	0.8	<0.02	<0.02	0.2	<0.2
Pb	<0.5	<0.5	<0.5	<0.5	<0.5
Hg	0.0071	0.0027	0.0018	0.0002	0.0002
Se	<0.001	<0.001	0.001	<0.001	<0.001
Ag	<0.02	<0.02	0.02	<0.02	<0.02

The combined sodium and chloride (Na/Cl) concentrations, in the form of Na^+ and Cl^- respectively, are the major inorganic ions found in the produced wastewaters tested. High sodium and chloride concentrations greater than 2,000 mg/l have been reported to hinder biological growth due to lysing or osmotic destruction of the microorganism (Gaudy and Gaudy, 1980). All samples, for the exception of Lovington pit, maintained a combined sodium and chloride concentration well above 2,000 mg/l.

Heavy metals were present in all produced waters tested, but in trace quantities. For example, concentrations ranged from 0.02 to 0.8 mg/l for chromium, and 0.001 mg/l for arsenic. All values are well below State of New Mexico ground and surface water standards. Trace metals have been known to stimulate microbial growth and increase substrate metabolism.

Biomass Cultures

Based on above chemical analysis, a bacterial culture with the ability to withstand high brine concentrations would be required for effective degradation of many produced waters. Therefore, a commercial biomass PETROBAC-S (supplied and manufactured by POLYBAC Corporation), was selected as the aerobic host bacterial culture in the treatability and biodegradation studies. PETROBAC-S is a microbial hydrocarbon degrader specifically formulated to degrade crude or refined hydrocarbons in saline environments. It was used in the non-acclimated and acclimated forms throughout this study.

Acclimation of the host culture was established by aerating the suspended growth culture with a 50 ml daily dose of Hobbs separator waste (medium strength). A two liter batch reactor was maintained at a mix liquor suspended solid (MLSS) concentration of 1,300 mg/l, with an eight day sludge detention time. Nutrient water containing elements such as calcium, magnesium, and phosphorus etc., was supplied to the suspended culture at a rate of 200 ml per day. A biomass volume of 250 ml was wasted per day. The Hobbs produced water was the only source of organic substrate available for biological consumption.

Biological Treatability and Degradation Analysis

An automated eight station COMPUT-OX respirometer (N-CON Systems Company, Inc., Larchmont, NY) was utilized for all studies performed. Respirometric test runs lasted five days at a temperature of 20° C. Substrate removal was determined by evaluation of the five day BOD curves with respect to the control as well as by TOC & TPH reductions.

Treatability Studies. Produced water treatability tests were initiated on the Hobbs medium strength produced water. Wastewater strength is in terms of its combined sodium and chloride (Na/Cl) concentrations. Test objectives were to inoculate this produced water with the Hobbs acclimated and PETROBAC non-acclimated cultures, to determine if biological degradation is feasible with either of the two forms. If effective, an optimum and most economical dilution ratio to sufficiently degrade the organic substrate can be determined. Dilution ratios (water:waste) of 50:1, *10:1, 5:1,* 1:1 and 0:1 (or no dilution) were chosen from the non-acclimated culture testing, The acclimated treatability testing utilized the dilution ratios of 10:1, 5:1 and 0:1.

Results were evaluated by comparing the five day biochemical oxygen demand (BOD) curves for each of the dilutions in relation to that of its inoculated control (acclimated or non-acclimated). The control reactors consisted of a 20 ml non-acclimated or acclimated biomass and 200 ml of nutrient water, for a 10% volume per volume (V/V) concentration.

Biodegradation Studies. Degradation testing for each individual waste also used the acclimated biomass at a concentration of 10% v/v (20 ml inoculum per 200 ml waste/nutrient water). Testing was performed using the optimum dilution ratio determined from the above treatability analysis. Inhibitory characterization according the waste's five day TOC and TPH removal efficiencies was performed. Inhibition was determined by comparison to the inoculated control reactor.

Soluble TOC measurements were performed on the four produced water samples before and after five-day respirometric testing. A change in TOC would determine the effectiveness (or efficiency) of an acclimated culture in removing the soluble carbon substrate from produced waters. Testing was in accordance with section 505 of Standard Methods for the Examination of Water and Wastewater (American Public Health Association, 1985).

TPH testing was employed to determine the overall removal efficiency of total petroleum hydrocarbons during the five day biooxidation period. TPH concentrations were determined using a 100 ml extraction of each sample before and after the five day respirometric period. Testing was per EPA Method 418.1 (Spectrophotometric, Infrared) Petroleum Hydrocarbon, Total Recoverable.

Abiotic Studies. Petroleum hydrocarbon substrate removal measured and analyzed on the basis of respirometric techniques, may be misleading due to abiotic removal mechanisms, such as adsorption, volatilization, and chemical oxidation. Therefore, testing to determine the extent of abiotic removal after the five day biological oxidation period was performed. To distinguish between abiotic and biological removal, a biocide was added to each wastewater respirometer bottle before respirometric testing. Respirometric analysis was performed on two of the four waste streams; Roswell produced water pit and Hobbs tank separator.

Initially, one gram of $HgCl_2$ biocide was placed into each respirometer bottle containing the optimum dilution for each respective waste stream. The $HgCl_2$ was used as a biocide which prevents biomass growth and eliminates the biodegradation mechanism. Therefore, only abiotic mechanisms of removal are represented after the five day testing period. The extent of abiotic removal was analyzed and quantified by TPH analysis.

Results and Discussion

Biological treatability and degradation studies were evaluated by using one or more of the following methods; respirometric measurements, TOC analysis, and TPH scanning, Test results are discussed below.

Treatability Studies. Characterization of southeastern New Mexico's produced waters has shown to contain high concentrations of total dissolved solids with sodium and chloride being the major constituents. The magnitude of the combined sodium and chloride concentrations (Na/Cl) ranged from 180,000 to 1,500 mg/l. At these elevated values, inhibition could possibly effect the non-acclimated heterogeneous biomass from degrading the produced waters. This is illustrated in Figure 1, which shows insignificant biological degradation of the Hobbs produced water even at dilutions of 50:1. This would suggest that the non-acclimated culture is

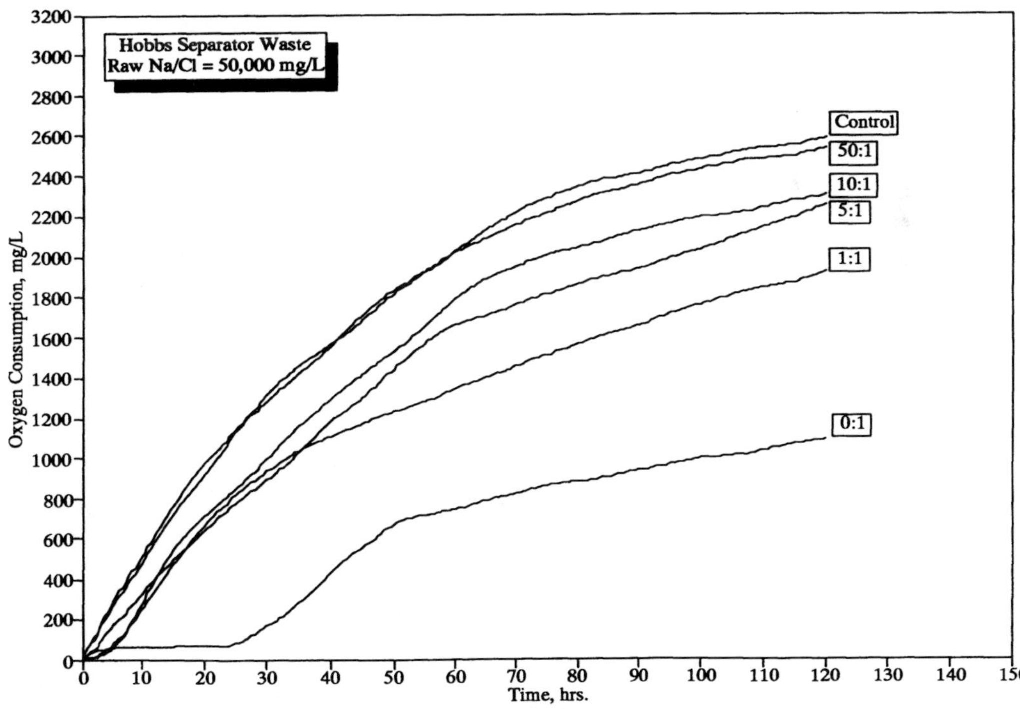

Figure 1. Respirometric five day BOD treatability test using a non-acclimated PETROBAC-S inoculum.

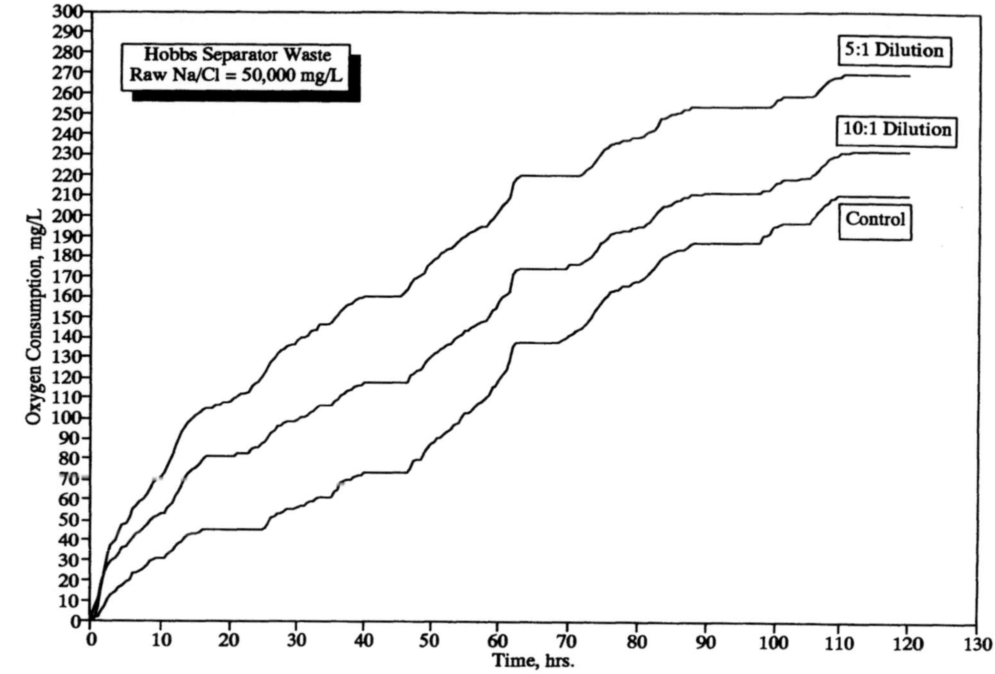

Figure 2. Respirometric five day BOD treatability test using an acclimated PETROBAC-S inoculum.

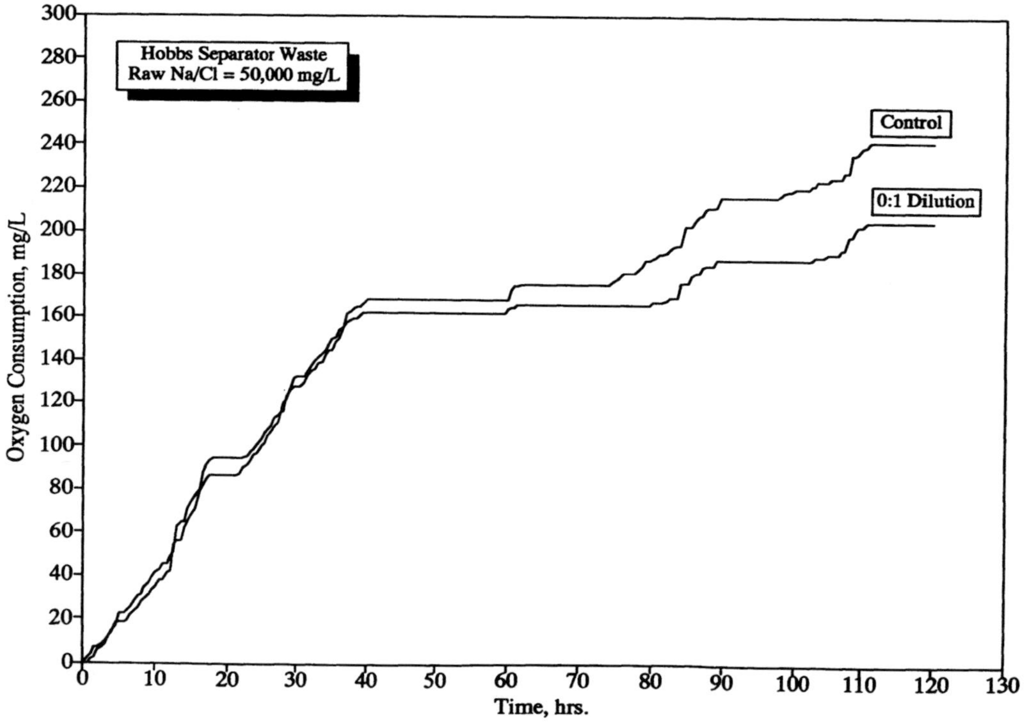

Figure 3. Respirometric five day BOD treatability test using an acclimated PETROBAC-S inoculum.

being inhibited by constituents present in the produced wastewater to degrade the available organic substrate.

Hence, a treatability experiment was performed to determine if introducing an inoculum which has been acclimated to the Hobbs produced water, would overcome the inhibition. As seen in respirometer plots Figure 2, this hypothesis regarding acclimation was proven to be correct. Both dilution factors tested (5:1 and 10:1) now exceeded the BODs control reactor and exhibit biodegradation characteristics.

The inhibitory effect of the waste between dilutions was also overcome by acclimation. In the previous treatability test using the non-acclimated culture, the 10:1 dilution BOD_5 curve maintained a degradation potential greater than that of the 5:1. However in contrast, the BOD_5 curve for the 5:1 ratio surpassed that of the 10:1 while using the acclimated culture. This would suggest that higher removal efficiencies are obtained by the acclimated culture when introduced to higher strength produced water.

Acclimated culture testing was also performed on a undiluted raw produced water as illustrated in Figure 3. However, biological inhibition increased substantially compared to the dilutions of 5:1 and 10:1 in Figure 2. The 0:1 dilution exhibited no biodegradation traits.

Upon review of the treatability test results, it was determined that for economic restraints on dilution water, and substantial increased biological inhibition, dilutions of 5:1 is recommended. Respirometric evaluations of all oilfield waste streams were there after conducted at dilutions of 5:1, while employing the Hobbs acclimated inoculum.

Biodegradation Studies. Total soluble organic carbon analysis was performed on the four waste samples before and after respirometric testings, Figure 4 depicts TOC removal efficiencies for the acclimated inoculum. Upon review of the figure, all produced water samples showed considerable removal efficiencies, ranging from 96% (Roswell separator) to 83% (Roswell waste pit).

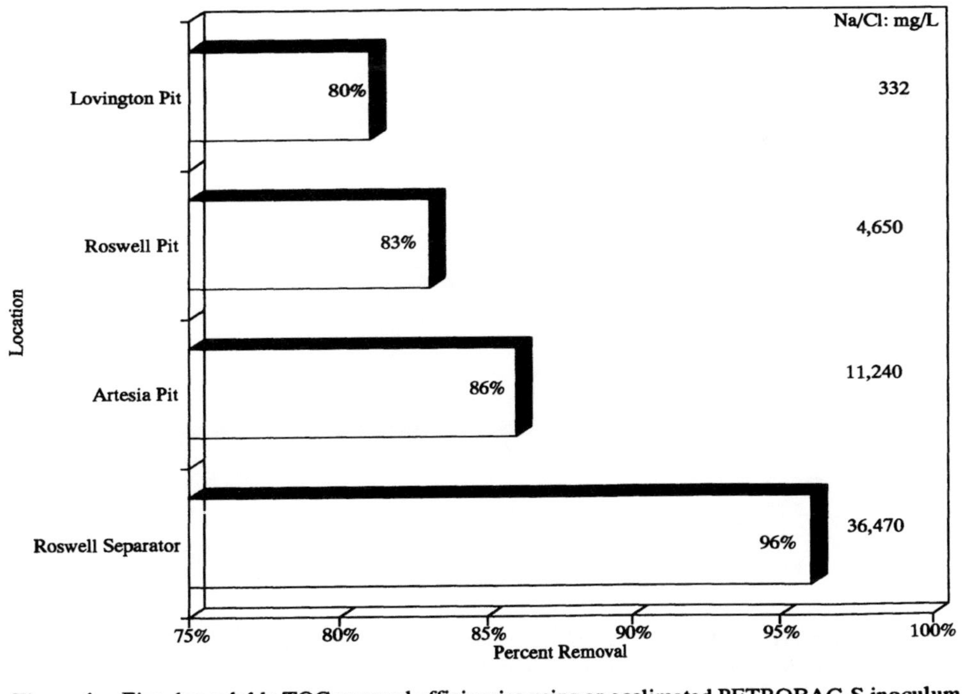

Figure 4. Five day soluble TOC removal efficiencies using an acclimated PETROBAC-S inoculum, introduced to each wastestream diluted 5:1.

Total petroleum hydrocarbon analysis as illustrated in Figure 5 represents TPH removal efficiencies before and after respirometric five day testing. TPH testing reflects the findings of that performed in the TOC analysis. As the combined sodium and chloride (Na/Cl) concentrations increased from the Lovington pit to the Roswell separator, biomass activity is stimulated, which in turn produces higher removal efficiencies. TPH removals ranged from 92% for the Roswell separator to 65% for the Lovington waste pit.

As exhibited by both Figures 4 and 5, the combined sodium and chloride concentrations for each produced water in relation to their respective removal percentages, parallels the work performed by Gaudy (Gaudy and Gaudy, 1980). Microbial species can be acclimated to a high salt concentration, but when introduced to a fresher water environment, acclimation is lost and biological degradation is curtailed. It is also evident that when the acclimated culture is introduced to a stronger TDS or combined sodium and chloride water environment biological degradation of the produced water is stimulated.

Abiotic Studies. Test results performed during abiotic testing is illustrated in Figure 6. As each respirometer waste sample was treated with 1 gram of the $HgCl_2$ biocide, TPH analysis was performed before and after the five day respirometric period. The BOD_5 curves clearly show that oxygen consumption for both the Hobbs separator and Roswell pit concluded after 15 hours. The total BOD_5 for both wastes were approximately 50 mg/l. Results would indicate that biodegradation of the organic material present was eliminated after the 15 hour period, leaving only the abiotic mechanisms accountable for substrate removal.

With biological degradation eliminated by the biocide, abiotic mechanisms accounted for 5 to 15% of the total petroleum hydrocarbons from the Roswell pit and Hobbs separator respectively. Refer to Figure 7 for illustration. These low removal percentages would suggest

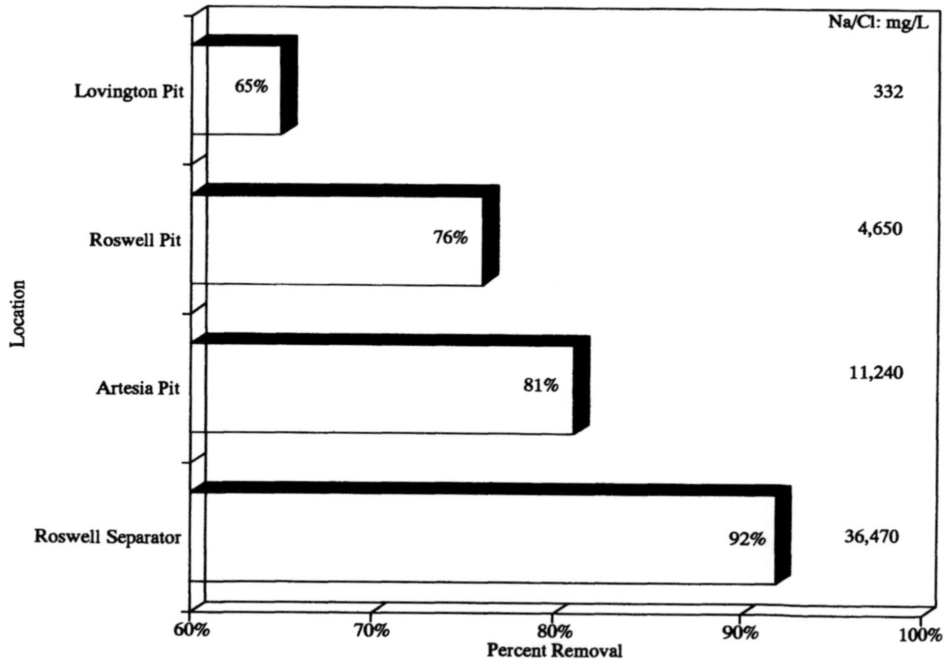

Figure 5. Five day soluble TOC removal efficiencies using an acclimated PETROBAC-S inoculum, introduced to each wastestream diluted 5:1.

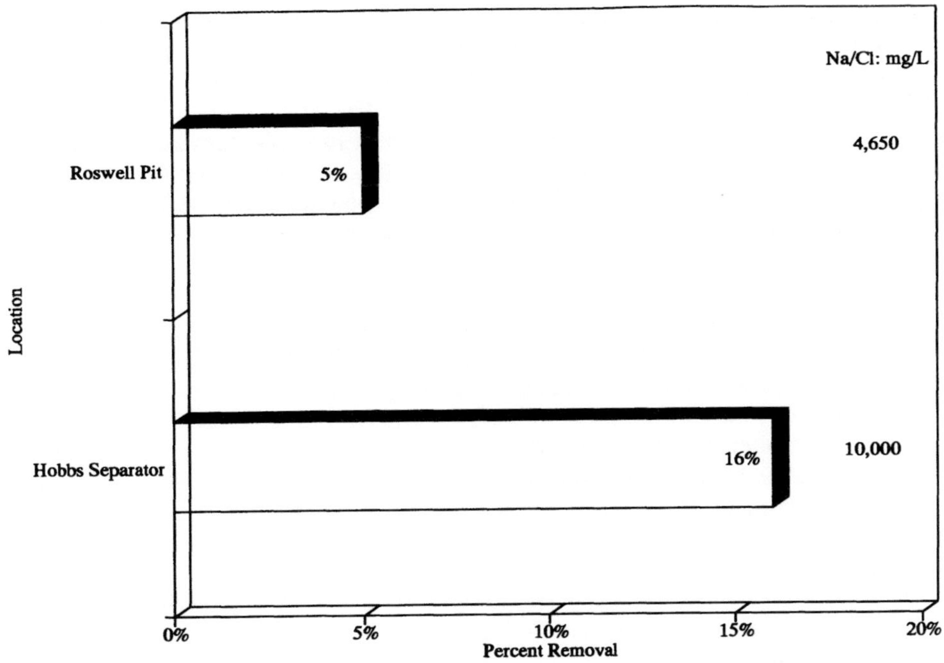

Figure 6. Respirometric five day test using an acclimated PETROBAC-S inoculum with biocide.

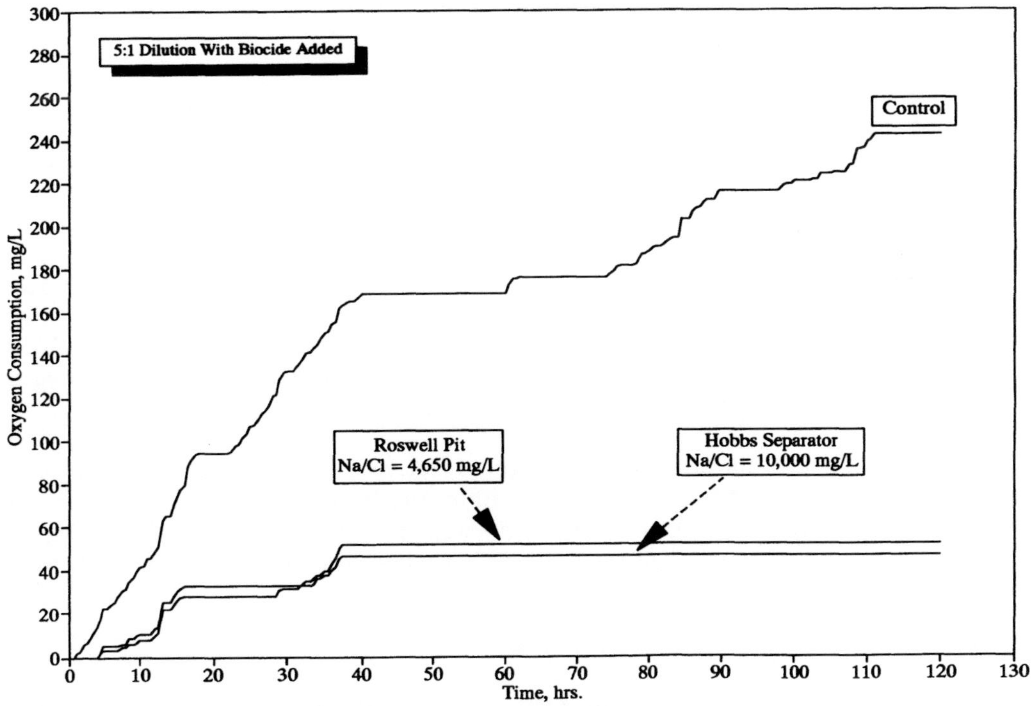

Figure 7. Five day abiotic TPH removal efficiencies using an acclimated PETROBAC-S inoculum, introduced to each wastestream diluted 5:1.

that the organic removal potential for abiotic mechanisms are minimal and that TPH removals exhibited in previous respirometric testing was mainly due to biological degradation.

Summary and Conclusions

As environmental problems increase due to present storage and disposal methods used for oilfield produced wastewaters, awareness to their quality, contamination potential, and treatability is essential Water quality as well as the biological treatability and degradation characteristics for the tested southeastern New Mexico's oilfield produced waters are summarized below.

Water Quality Analysis

1. Combined sodium and chloride (Na/Cl) concentrations, makeup approximately 80% of the total dissolved solids, with values between 180,000 to 1,500 mg/l.
2. Total petroleum hydrocarbon (TPH) analysis produced values of 300 to 1,000 mg/l. This includes the IR absorption of benzene and its derivative groups such as toluene, xylene, ethylbenzene, and isobutylbenzene.
3. Total organic carbon (TOC) concentration ranged from 150 to 750 mg/l This exceeds the TOC standard of 20 mg/l set by the State of New Mexico.

Treatability Studies

1. Non-acclimated cultures exhibit no biological degradation capabilities of the produced waters even at a dilution ratio of 50:1.
2. An inoculum which has been acclimated to the produced water maintains biological degradation properties at dilution of 5:1 and 10:1.
3. Higher removal efficiencies are acquired when the acclimated culture was introduced to a higher TDS produced water than that of its acclimation water. When a fresher water environment was treated, acclimation and biodegradation was lost.

Biodegradation Studies

1. Use of the acclimated inoculum produced TOC removals efficiencies, ranging from 96% to (Roswell separator) to 83% (Roswell waste pit).
2. Total petroleum hydrocarbon removals ranged from 92% (Roswell separator) to 65% for the Lovington waste pit.

Abiotic Studies

1. Abiotic mechanisms accounted for 5 to 15% of the total petroleum hydrocarbons from the Roswell pit and Hobbs separator respectively.
2. Total petroleum hydrocarbon removal potential for abiotic mechanisms were determined minimal, and TPH removals obtained in previous testing was mainly due to biological degradation.

In accordance with the research results provided, biological treatment of petroleum hydrocarbons present in produced waters is feasible. Biodegradation is substantial even at elevated TDS conditions of 5:1, or a combined sodium and chloride concentration of 36,000 mg/l However, a produced water acclimated culture is recommended for all biological treatment activities.

REFERENCES

American Public Health Association, 1985, Standard Methods for the Examination of Water and Wastewater, 16th.ed.

Gaudy, A.F., Gaudy, E.T., 1980, "Microbiology For Environmental Scientists and Engineers," McGraw-Hill Book Company, New York, New York, 1980.

New Mexico Water Control Commission, 1988, "Water Quality Control Commission Regulations," November 1988.

Polybac Corporation, 1990, "Product Information Booklet," Biomass Engineering, 1990.

Tellez, Gilbert T., 1991, "Bioreclamation of New Mexico Oilfield Produced Wastewaters: Characterization and Feasibility Study," MS Thesis, New Mexico State University May.

USE OF SOLAR PONDS TO RECLAIM SALT PRODUCTS FROM BRINE WATERS FROM OIL AND GAS WELL OPERATIONS IN NEW YORK

J.F. Atkinson[1], M.R. Matsumoto[1], M.D. Bunn[2], and D.S. Hodge[3]

[1] Department of Civil Engineering
[2] Marketing Department, School of Management
[3] Department of Geology
State University of New York at Buffalo
Buffalo, New York
USA

INTRODUCTION

New York State is a relatively minor producer of oil and gas. Presently, there are approximately 4,300 active gas wells and 4,600 oil wells owned by about 600 operators in New York State. The majority of active oil wells are in Allegany and Cattaraugus Counties, and the majority of active gas wells are in Chautauqua County. In 1990, 0.42 million barrels of crude oil and 25,400 million cubic feet of natural gas were produced (DEC, 1990). Current annual brine generation rates, estimated by the New York State Division of Mineral Resources (DMN), are about 300,000 barrels (12.6 million gallons) from gas production and about 2.1 million barrels (88.2 million gallons) from oil production.

The composition of the generated brine varies with location and well operation. Major constituent characteristics of brines from New York oil and gas producing geologic zones are presented in Table 1. In general, these brines contain a high concentration of dissolved solids (salt), and varying amounts of heavy metals, oil and grease, and trace organics.

In the past (prior to 1984) brine was generally disposed in a brine pit adjacent to the well. Separated brine was placed in the brine pit and allowed to infiltrate into soil and/or evaporate. However, brine pits are no longer considered to be environmentally acceptable because of the high potential for groundwater contamination, and are no longer allowed for disposal.

In a draft generic environmental impact statement (GEIS) on the oil, gas, and solution mining industry prepared by the New York State Department of Environmental Conservation (DEC) in 1988, three (3) brine disposal methods currently used were identified: 1) road spreading of brines for deicing and road stabilization, 2) discharge to surface waters through permitted facilities, and 3) underground injection.

Since the passage of the New York State Environmental Quality Review Act in 1977, the DEC has sought to establish rational bases and criteria for environmental review which can be used to recommend changes in existing regulations to strengthen their mandated role to protect the state's environment. The primary vehicle for reviewing a broad regulatory program, such as the oil and gas regulatory program, is preparation of a generic environmental impact statement (GEIS). A GEIS for the oil and gas industry was prepared by the DEC in January, 1988. In this document, all aspects of the oil and gas industry and their impact on the environment were examined.

Produced Water, Edited by J.P. Ray
and F.R. Engelhart, Plenum Press, New York, 1992

Table 1. Major constituent characteristics of brines from New York State's gas and oil producing formations[a] (NYSDEC, 1988)

	Formation					
Constituent	Potsdam/ Theresa	Queenston	Medina	Oriskany	Bass Island	Deicing Road Salt
Sodium	76,710	73,500	68,900	45,460	60,750	59,500
Calcium	31,260	36,600	37,120	33,680	56,400	1,090
Magnesium	3,500	2,890	2,770	5,170	3,160	210
Strontium	—	—	—	—	—	0.2
Barium	750	—	—	—	—	0.01
Potassium	3,370	1,120	—	1,310	—	530
Iron	20	200	680	220	20	2
Chloride	183,700	187,420	181,300	145,440	203,000	120,000
Bromide	1,420	1,120	1,720	1,690	—	—
Iodide	10	10	20	10	—	—
Sulfate	20	—	740	60	180	—
Bicarbonate	90	—	20	200	50	—
TDS	299,190	302,870	292,730	232,740	323,560	200,800

[a]All units in mg/l.

Because preparation of a GEIS is the key step in developing criteria for recommending changes in existing regulations, it is anticipated that new regulations related to brine disposal may be proposed in the future. Any new proposed regulations may affect the current methods of brine disposal. Therefore, in anticipation of future changes in brine disposal regulations and the possibility of developing a commercial brine disposal enterprise to service the oil and gas producers in New York State, the New York State Energy Research and Development Authority (NYSERDA) and the Gas Research Institute (GRI), in conjunction with the State University of New York at Buffalo, are conducting a research study investigating various alternatives for brine disposal in New York State.

The overall objectives of this study are to 1) develop options for disposal of brine from current and future oil and gas production operations, 2) examine the technical and economic feasibility of these options, and 3) assess the associated environmental impacts. A major part of this study focuses on the development of a resource recovery facility to produce marketable salt products from the oil/gas brine. A marketing study is also part of the study to determine the market potential for recovered brine salt products. The brine reclamation facility outlined in this paper utilizes an evaporation system to recover salt. Several heat energy sources are being investigated. It is hypothesized that a reclamation system using solar ponds for process heat supply could provide a cost effective method for brine disposal.

In this paper, an overview of the brine disposal situation in New York state, and costs associated with brine disposal is presented. Greater detail is provided regarding the feasibility of developing resource reclamation facility, utilizing solar ponds as a primary energy source.

OVERVIEW OF BRINE DISPOSAL ALTERNATIVES

There are two basic strategies for brine disposal: direct disposal and resource recovery. Brine disposal alternatives for both of these categories are presented in Figure 1. From an environmental management/protection viewpoint, resource recovery strategies are more desirable than direct disposal methods. However, selection of the actual brine disposal method depends primarily on economic and regulatory considerations.

Under the current regulatory and economic situation in New York, the general order of brine disposal preference in practice is: 1) surface water discharge with no treatment, 2) road

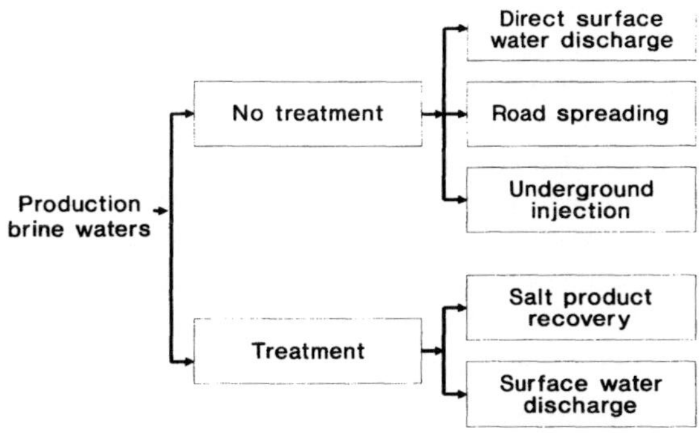

Figure 1. Overview of brine disposal alternatives.

spreading with no treatment, 3) surface water discharge with treatment, and 4) underground injection. Brief descriptions of these alternatives are presented in the following sections.

Surface Water Discharge With No Treatment

In New York all point discharges to surface and ground waters are permitted through the State Pollutant Discharge Elimination System (SPDES), which is administered by the DEC. Permit requirements for such discharges include limitations of total dissolved solids (TDS), oil and grease, and any other pollutants of water quality concern.

Because of the long history of waterflooding operations at shallow oil wells in New York, the salt concentration of the produced waters from these shallow oil wells is relatively dilute. Direct discharge of these dilute produced waters without treatment is permitted in New York. Nearly 1.8 million barrels of these dilute produced waters are discharged annually into surface waters (DEC, 1990).

Road Spreading

The use of salt for road maintenance operations such as deicing and road stabilization of unpaved roads is commonly practiced in New York state. Sodium chloride (NaCl) and calcium chloride ($CaCl_2$) are the principal salts used in these operations.

Produced waters typically contain a significant percentage of both sodium and calcium chloride, and the characteristics of the more concentrated brines are similar to those of commercial road salts (see Table 1), making them attractive to local highway departments for road maintenance purposes. Until recently, municipal and county highway departments in the oil and gas producing areas of New York state used commercial salts for these maintenance operations. However, as a result of the symbiotic need of producers and highway departments, the more concentrated brines, from gas wells and oil wells from the Bass Island formations, have been utilized increasingly over the past five years as a substitute. The DMN estimates that about 170,000 barrels of brine were disposed by road spreading in 1989 (DEC, 1990).

Highway departments generally accept brines directly from producers as long as certain minimum requirements are met (e.g. no visible oil/grease and calcium chloride concentrations above about 5%). Producers are not charged for "disposal" and the highway departments do not pay for the "salt solution". To help minimize transportation costs, small storage facilities are located at various locations throughout the county. Producers take the brine to the nearest storage facility for "disposal". The highway departments fill up road spreading equipment with "salt solution" from the storage facilities.

Despite the apparent success of road spreading, there are environmental concerns (DEC, 1988). As with the use of brine pits, road spreading is a potential source for contamination of water supply aquifers. Although the chemical properties of oil and gas brines are similar to those of commercial salts, the fact that brine is in liquid form presents potential problems of runoff and spills if the brine is not properly applied. The number of applications needed for dust control using brine is 2.5 times that needed for commercial solutions of salt for dust control. Thus, the potential for runoff and spills is also increased.

Based on the typical concentrations of salt in the brine, production brine has a freezing point of about -60° F. As a consequence it is suitable for use as an ice control agent, although its liquid form presents a potential problem of runoff from the road surface during application. Solid granules are thought to be more effective because they stay on the road surface better and penetrate through the ice better with traffic to break the ice bond with the pavement.

Surface Discharge With Treatment

Another method of disposal is to treat brines in a wastewater treatment facility to remove suspended matter, heavy metals, and oil and grease prior to a surface water discharge. Both municipal wastewater treatment plants and commercial brine treatment facilities have been used. One major concern at municipal wastewater treatment plants is the potential disturbance of the biological treatment processes caused by osmotic shock (high salt gradients). Thus, only a very small quantity of brine (21,000 bbl in 1990) has been disposed through a municipal wastewater treatment plant by adding the brines in very small amounts to the domestic wastewater flow.

Two commercial brine treatment facilities (capacity - about 200,000 gal/day each) are currently operating in northwestern Pennsylvania, within 50 miles of the principal oil and gas fields of New York, one in Warren and the other in Franklin. Brine is delivered in 2,000 or 5,000 gallon tanker trucks, treated through a treatment system as shown in Figure 2, and then discharged into a receiving water. Current treatment fees are approximately $1.25 to $1.50 per barrel. During 1990, approximately 300,000 bbl were hauled out of state to these two facilities.

Underground Injection

Although underground injection has been widely used in other states, the use of this technique for disposal of brine wastes has been discouraged in New York. In fact, National Fuel Gas, during in-house studies of brine disposal options, did not consider the use of underground injection as a viable long-term option because of regulatory concerns.

Similar to surface discharges, a SPDES permit is required for an injection disposal well. In addition, state regulations require a permit from the state Division of Mineral Resources

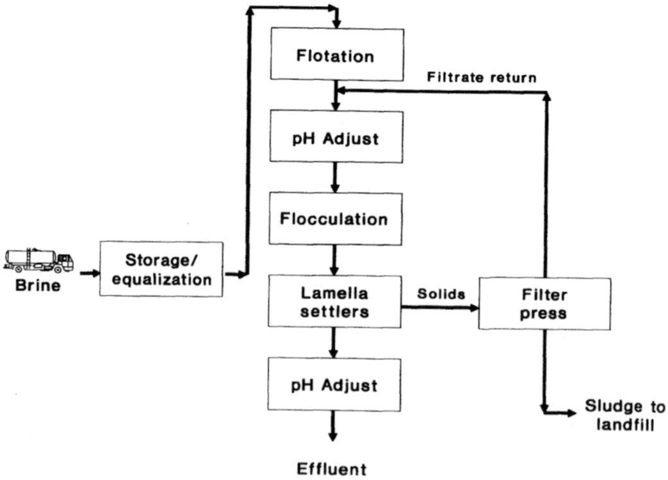

Figure 2. Flowsheet for commercial brine treatment facilities.

(DMN) and a federal Underground Injection Control (UIC) permit. The primary environmental consideration in obtaining injection well permits is the protection of groundwater resources. Stringent requirements on the casing and cementing of the injection well, ability of the formation to accept injection fluids, pressure limitations on injection, and groundwater monitoring and reporting deter most operators from seeking to develop an underground injection well.

At present, there are only two injection wells in Chautauqua and Cattaraugus counties, near the oil and gas fields. Of these two injection wells, only one is operational. Approximately 58,000 bbl of brine was injected into this well. There are three additional injection wells outside of the two major oil and gas producing counties. Total brine injected into all five of these wells was about 118,000 bbl in 1990. Only one of these injection wells is available for commercial operations; neither of them are in Chautauqua or Cattaraugus counties. The remaining four wells are reserved solely for the needs of the owner/producers. Disposal charges for the one commercial injection well is comparable to the brine treatment charges at Warren and Franklin.

RESOURCE RECOVERY OPTION

From strictly an environmental management viewpoint, resource recovery is a much more desirable alternative than direct disposal into the environment. However, under the current regulatory situation in New York, there is little incentive for producers to consider this alternative and little research has been conducted to assess this possibility. For this reason, a major emphasis of this study was to investigate the technical and economic feasibility of developing a salt reclamation facility utilizing produced waters as a feedstock.

Produced waters contain a number of inorganic salts which can be used for industrial, municipal or agricultural purposes. The principal constituents with marketable value are sodium chloride (NaCl) and calcium chloride ($CaCl_2$). Based on an average characterization of brines (see Table 1), produced waters (other than those from waterflooding operations) contain approximately 14 % NaCl and 7% $CaCl_2$. Therefore, the 600,000 bbl of highly saline produced waters generated annually contain about 16,000 tons of NaCl and 8,000 tons of $CaCl_2$.

Marketing Analysis

To determine the feasibility of selling salt products reclaimed from production waters, a marketing study was conducted. The focus of the study was on characterizing the competitive supply markets for the products (NaCl and $CaCl_2$), identifying the potential applications for the reclaimed brine minerals, studying the buyer behavior for the products in each application, and describing the appropriate marketing elements for the profitable sale of the products in each market. A summary of the preliminary results from that study are presented below.

Uses of sodium and calcium chloride, and the fraction of the total amount used for each purpose are presented in Table 2. The annual usage of NaCl in the U.S. is approximately 37 million tons. Forty-two (42)% of the NaCl is used as a brine solution (28% NaCl), 40% as rock salt, and 10 % as granulated or powdered salt. Annual usage of calcium chloride is about 600,000 tons. The vast majority of $CaCl_2$ is sold in solution form; a very small fraction is sold as a solid flake or pellet.

The primary user markets for the sodium chloride reclaimed from brine are highway usage, animal feedstocks, and water softening uses. There are a significant number of potential users in proximity to our study area and medium coarse rock salt is the general form of the product used.

The primary user markets for calcium chloride will be for highway usages and a variety of industrial processing. Similar to sodium chloride there are a significant number of potential users of calcium chloride in proximity to our study area. The principal form of calcium chloride product is solution (40+%).

Depending on the final product form, prices for sodium and calcium chloride vary considerably, from \$15 to \$90 per ton for sodium chloride and \$120 to \$300 per ton for calcium chloride. For the forms of product that are projected from a proposed brine reclamation facility, medium coarse rock salt and calcium chloride solution, average market prices are $65 and $120 per ton, respectively. Assuming that the quality of the salt products reclaimed from produced waters meets buyer criteria, it is estimated that the revenue potential is nearly $2 million per year. Thus, after an initial analysis it would seem that resource recovery may be economically feasible.

Table 2. Uses of sodium and calcium chloride in the United States

Sodium Chloride			Calcium chloride		
Use	Amount (tons)	Percent of total	Use	Amount (tons)	Percent of total
Chemical products	19,400,000	53.0	Highway de-icing	212,000	35
Highway use	8,400,000	8.4	Dust control	133,000	22
Animal feeds	1,300,000	3.5	Industrial uses	121,000	21
Food processing	1,200,000	3. 3	Oil recovery	61,000	10
Industrial uses	1,800,000	4.9	Concrete treating	30,000	5
Water softening	1,000,000	2.7	Tire ballasting	24,000	4
Other	3,600,000	9.7	Other	24,000	4
TOTAL	36,700,000	100.0	TOTAL	605,000	100.0

Resource Recovery Facility

For the reclaimed products to be marketable, they must be of high quality, free of oil/grease, potentially toxic impurities (heavy metals), and aesthetic impurities (suspended matter). To produce a product that is of the highest quality possible, a preprocessing system to remove these impurities is required. A treatment system similar to that shown in Figure 2 will provide the desirable preprocessing. During preliminary testing conducted in conjunction with this study, after treatment, heavy metal and oil/grease concentrations in the treated brine were less than 1 mg/l (each); suspended solids concentrations were less than 20 mg/l; and turbidity was less than 5 NTU.

The recovery of commercial salts from brines is a well-developed practice. Multiple effect evaporation is the most common method of producing solidified salt and is proposed for this investigation. The overall salt recovery system will include a rotary drum filter, rotating cylinder dryer, and screens. Various pumps, recycle tanks and piping are also necessary, as well as a packaging/storage facility for the final products. As previously noted, it is anticipated that two primary products will be reclaimed from the production waters, vacuum pan evaporated salt (NaCl) with a minimum purity of 99.5% and a concentrated calcium chloride solution (40+ %) with a 95% purity.

Solar Ponds as an Energy Source for Evaporators

Because of the relatively high amount of energy that is required to evaporate the brine, conventional heat sources such as fuel oil and gas may preclude development of a brine recovery facility. It is hypothesized that the use of solar ponds to provide process heat for an evaporation system will be more cost effective than the conventional heat sources and that with solar ponds, brine evaporation/recovery will be a viable, cost-effective disposal option. On this basis, a study was conducted to determine the technical and economic requirements for developing solar ponds as a means of providing the principal energy for the evaporative system for generating the salt products. Other energy sources are being evaluated for the full study.

In Figure 3, a proposed flowsheet for a brine evaporation system which incorporates a salt gradient solar pond (SGSP) is presented. Heat energy generated from the solar pond will be

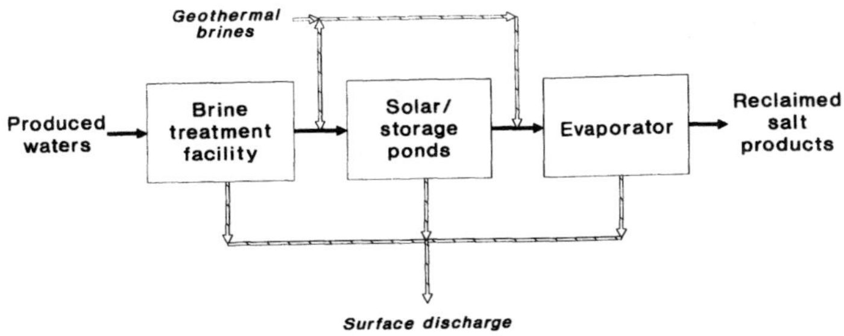

Figure 3. Flowsheet for resource recovery system.

used preheat brine as it enters the evaporation system, thereby reducing significantly the amount of auxiliary energy and cost for evaporation.

The phenomenon of relatively warm water being generated at the bottom of a shallow body of water was described near the turn of the century (Kaleczinsky, 1902), though it was not until much later that this concept was proposed for development as a useful solar energy collector (Tabor, 1963). There are several collector designs which are known generally as solar ponds, but the one which has received the most attention is the salt gradient solar pond (SGSP). This is the design of interest for this resource recovery project.

The pond surface area should be large enough to avoid excessive edge losses and not so large that piping and pumping considerations become a problem. Existing ponds in the U.S. are typically on the order of 0.5 to 1 acre in size. One pond in Israel has a surface area greater than 50 acres. Even larger ponds have been proposed, notably for the Salton Sea in southern California (French and Meitlis, 1980), but with the total area divided into a number of smaller sections, or modules.

Total depth is usually about 10 feet and is divided into the three regions shown in Figure 4. The bottom convecting zone serves as a combination collector and storage unit. Solar radiation reaching this layer warms the water there which is prevented from convecting upwards by the overlying density gradient in the gradient zone provided by salinity stratification. The stabilizing effect of the added salt on density is greater than the destabilizing effect of temperature and this is the main operating principle for a SGSP. An upper convecting zone exists near the top of the pond. Though it is desirable to minimize the depth of this layer to increase collection efficiency, it is impossible to eliminate a small mixed region near the surface. The depth of the upper zone depends on wind-mixing, convective mixing, and possible through-flow and diffusion (Atkinson and Harleman, 1987). Although the theoretical mechanics are not completely understood, practical experience from a number of ponds in the U.S., Israel and Australia has shown that surface mixing may be controlled with the installation of a floating net cover. These and other aspects of SGSPs are reviewed in some detail by Hull and Nielsen, 1986.

Heat is extracted from the pond either by pumping a working fluid through a heat exchanger at the bottom of the pond or by withdrawing fluid directly from the bottom layer and into an external heat exchanger. Both of these options have been used successfully, though the latter option is usually preferable for larger ponds where the size of an internal heat exchanger would be unmanageable. An external heat exchanger is used in the large Israeli pond and also in the pond operated by the University of Texas at El Paso (UTEP), which supplies process heat for a canning plant and powers a desalination unit producing 5,000 gallons of potable water per day.

SGSPs are relatively simple to install and are considerably less expensive than other solar collector systems, depending on the availability of land, water, and salt, and on whether a bottom liner is needed. Installation costs must be determined on a site-specific basis, but installed costs of $5 per square foot of surface area are reasonable. One of the main reasons for this low cost

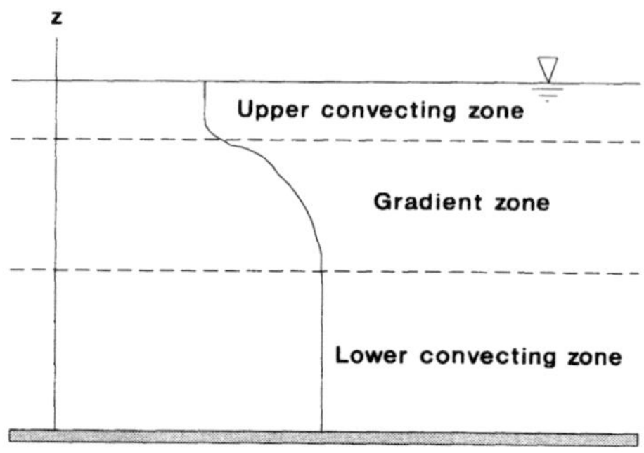

Figure 4. Typical stratification in solar ponds.

is the advantage of combining the collection and storage processes in a common pond. However, there are several possible drawbacks that should be considered when designing a system with a SGSP.

Perhaps the most significant problem of a SGSP is its relatively low collection efficiency. Depending on water clarity, depth, bottom/edge losses and meteorology, the collection efficiency is typically around 20 percent. Thus, large areas are often needed. The pond is also limited in the maximum temperatures that may be obtained. In a hot climate temperatures in the bottom layer may reach boiling if sufficient heat is not extracted. This condition is probably not a significant problem in New York, but has occurred in a pond in New Mexico, causing excessive mixing and destroying the gradient (Weeks and Bryant, 1981). This temperature limitation affects heat transfer efficiency (in the heat exchangers) and also limits the theoretical efficiency of any sort of engine that may be driven by the pond. It is important to design a good match between the pond's thermal output and the end use to maximize the usefulness of the pond.

To optimize the output of a SGSP it is necessary to maintain the density gradient in the gradient zone. This process usually involves a periodic adjustment by selectively withdrawing and/or injecting fluid of desired density at specific locations using a thin diffuser plate. This procedure was originally proposed by Zangrando (1987) for initially filling large ponds and has since been adopted in almost every current SGSP for gradient maintenance. Practical experience with these diffusers has shown that it is possible to maintain the gradient under closely controlled conditions. However, except under high heating or high wind conditions, the relatively low diffusivity of salt allows a pond to be operated with only occasional adjustments in the gradient.

Most SGSPs in the U.S. have been built for basic research on the fundamental processes of importance in a pond. These have provided some important operating experience over the past ten years. Several ponds have also been developed for practical use, notably the UTEP pond previously mentioned. A pond in Miamisburg, Ohio was built to heat a municipal swimming pool (Wittenberg and Harris, 1981), ponds have been used for agricultural purposes (grain drying, greenhouse heating, etc.) (Short et al, 1979) and ponds have been built in California for driving desalination units (Engdahl, 1984). In addition, a number of ponds in Israel and Australia (and other countries) have been built for similar purposes and operated satisfactorily.

As envisioned for this brine reclamation facility, a series of 3-acre solar ponds, about 10 feet in depth, would be constructed. Each pond would have the capacity to provide enough heat to process approximately 100,000 bbl of brine per year. Initially, pretreated brines would be used to create the dense, saline layer at the bottom of the solar pond. Pretreatment is necessary to improve clarity of the brines to enable proper operation of the solar pond and also to insure that quality standards are met for the desired salt product. After reaching a predetermined depth, about 5 feet, lower density water, with a low salt concentration, will be introduced over the saline layer to create the salt gradient. Based on the climatic conditions in western New York, solar pond heating will be effective for about six months each year, from the beginning of May to the end of October. For the remaining months of the year, brine will have to be stored until the next heating season, or an alternative energy source used for the other half of the year.

Three different operating schemes have been examined for heating pretreated brines using energy from the solar ponds. In the first scheme, the hot saline water from the lower layer of the solar pond is withdrawn and sent through an external heat exchanger to preheat preprocessed brine prior to its introduction into the evaporator. Saline water from the solar pond passing through the heat exchanger is recirculated back into the pond. As previously noted, this is the most common method of extracting heat from a solar pond.

In the second heating scheme, preprocessed brine is introduced directly into the bottom layer of the solar pond for heating. Brine is withdrawn directly from the bottom layer and introduced into the evaporator. There are two significant advantages with this operating scheme: 1) no external heat exchanger is needed and 2) by allowing the depth of the saline layer to fluctuate, winter storage can be provided within the solar ponds themselves, reducing or eliminating the need for separate storage facilities.

There is a significant geothermal brine source in western New York. Geothermal brines with a temperature of about 75° C are available in the region where the brine reclamation facility is proposed. In the third heating scheme, geothernal brine is sent through an external heat exchanger to heat the preprocessed brine as it is introduced into the solar pond bottom layer. The geothermal brines are similar in quality to the preprocessed brines expected from the treatment system of the reclamation facility. As such, they can be processed similarly to generate additional salt product. Significantly higher brine temperatures can be obtained using this operating scheme. However, the size of the evaporator will need to be increased to accommodate the higher overall flow rates.

To determine expected temperatures, the operation of a solar pond for the proposed system was simulated using the numerical model described by Atkinson and Harleman (1983, 1987) and recently modified by Atkinson (1990). This model considers primarily one-dimensional processes (vertical direction), though it also incorporates horizontal inflow and outflow. It solves the conservation equations for heat, salt and momentum and includes a detailed evaluation of surface heat exchange and heat loss to the ground. Simulations were carried out for a two year period, using 40-year daily-averaged meteorological conditions for the Buffalo, New York area. Each simulation starts on January 1 with a constant pond temperature of 10° C (50° F). Because of transient start-up responses, the second year results are of greater interest, as they are approaching steady conditions. Each of the three brine heating schemes was simulated to determine their effect on solar pond performance. In addition, a control condition, in which heat extraction or brine inflow/outflow is not employed, was simulated. These simulations for a three acre solar pond are shown in Figure 5. Note that the following designations were employed.

Case 1: Control condition, passive pond; no heat extraction, no inflow or outflow.

Case 2: Heat extracted through an external heat exchanger to preheat process stream; 100,000 bbl treated between May 1 and October 31.

Case 3: Brine inflow directly into the pond; 100,000 bbl processed between May 1 and October 31. Assumed inflow brine temperature =10° C.

Case 4: Brine inflow is heated with geothermal brine before being introduced into the pond with additional geothermal brine added; 100,000 bbl processed between May 1 and October 31. Assumed geothermal brine temperature = 75° C with a flow rate of 40 gal/min.

The first twelve months of the simulation represent the startup of the solar pond are not reflective of the long-term temperatures expected in the bottom layer of the solar pond. The second twelve months of the simulation are more representative of the "steady-state" operating conditions. The difference in temperature between the control condition (Case 1), and the Case 2 and 3 heating schemes is proportional to the amount of energy that will be extracted from the solar pond to heat the process brine. The difference in temperature between the control condition and the Case 4 heating scheme is reflective of the net heat gained as a result of preheating with geothermal brine.

As can be seen in Figure 5, the Case 3 heating scheme produces the lowest temperatures in the solar pond, between 55° and 60° C. These temperatures, however, are sufficient for solutions introduced into single stage evaporators. As long as solution temperatures are above about 50° C, the rate of evaporation is primarily controlled by the application of a vacuum, which lowers the boiling point of the solvent (water). Higher solution temperatures permit less vacuum pressures to be applied.

Because a typical evaporator is needed to increase the solution temperature, rather than for applying the vacuum, very little energy savings will be achieved using the Case 2 or Case 4 heating schemes. For this reason and the advantages discussed previously, Case 3 is the most appropriate of the three heating schemes.

Economic Analysis

To determine the economic feasibility of reclaiming salt products from production water, a preliminary design and cost analysis was performed. Design conditions for this preliminary analysis were as follows:

Design flow = 600,000 bbl/yr
Preprocessing treatment system operation = year round
Solar pond/evaporator system operation = 6 months/year, Case 3 heating scheme

The preliminary design for each major component of the reclamation facility was conducted separately. In addition, because of the differences between operation of the preprocessing treatment system and the solar pond/evaporator system, a storage pond was included as part of the preliminary design. Cost estimates for both capital, and operation and maintenance costs are listed in Table 3. These estimates are very conservative and may be considered to be an upper limit.

Based on the preliminary designs and cost analyses, the payback period for a reclamation facility using solar ponds as the primary energy source for the multiple effect evaporators will be on the order of 4 to 5.5 years, depending on whether there is a charge for brine disposal imposed by the reclamation facility owner/operators. Thus, the project would seem to be economically feasible, although the payback period may be longer than many companies like to consider.

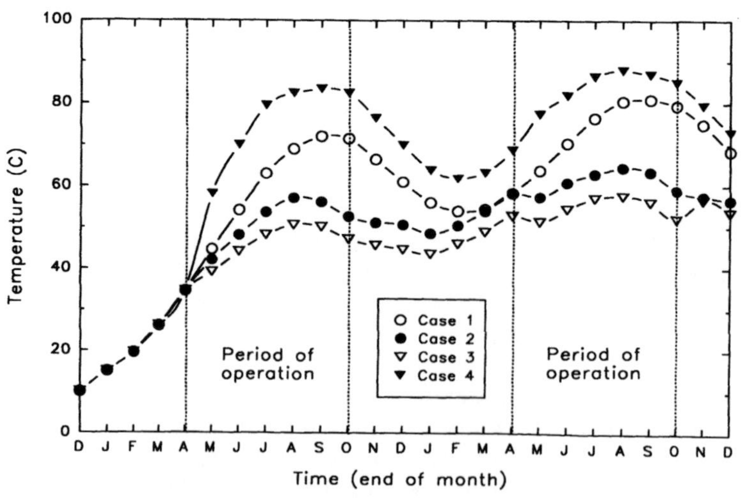

Figure 5. Computer simulation of temperature in solar pond bottom layer.

Table 3. Cost estimates for brine reclamation facility based on preliminary design

Component	Capital Cost, $	Annual Operating Cost, $
Preprocessing Treatment	1,300,200	581,100
Solar Ponds (18 acres)	585,000	40,000
Salt Recovery System	892,000	728,540
Component Subtotal	2,777,200	1,349,640
Non-Component Costs		
Land acquisition	140,000	
Site work	160,310	
Electrical and instrumentation	365,930	
Piping	315,020	
Miscellaneous	120,000	
Contractor	199,710	
Engineering	326,100	
Contingencies	406,590	
Non-component Subtotal	2,033,660	
Total Construction Costs	4,810,860	454,150 (a)
Annual Operating Costs		1,349,640
Total Annual Cost		1,803,790
Annual Revenue from Salt Products		2,000,000
Annual Revenue from Disposal Charges(b)		819,000
Net Annual Profit Potential(with no disposal charges)		196,210
Net Annual Profit Potential (including disposal charges)		1,015,210

(a) 20 year project life, 7% interest
(b) Disposal charge = $ 0.0325/gal

SUMMARY AND CONCLUSIONS

New York's oil and gas industry make a substantial contribution to the local economies of southwestern and central New York. It is estimated that for every $1.00 of well output, $1.40 is added to the state economy (DEC, 1988). However, because of the relatively low market prices for oil and gas, the added burden of increased disposal costs will result in the shutdown of many existing oil and gas wells (those that generate brine) and severely curtail drilling of new wells.

The current costs for brine disposal, other than by direct surface water discharges or road spreading, are between $1.50 to $3.00 per barrel (includes the cost for transporting the brine to the disposal sites).

In an effort to ascertain whether more cost effective and environmentally acceptable methods are available to oil and gas producers in New York, a study is currently underway, sponsored by NYSERDA and GRI. One alternative disposal option, discussed in this paper, involves treatment and reclamation of salt products from the brine utilizing solar ponds as a principal energy source. Conclusions from this study pertaining to this disposal option are as follows:

- Sodium and calcium chloride are the principal salt products that can be generated from oil and gas brines.
- Based on a preliminary economic analysis, a reclamation process consisting of three major processing steps: pretreatment of brine to remove suspended and dissolved impurities, preliminary heating and storage of brine in solar ponds, and recovery of salt using a multiple-effect evaporator appears to be economically viable.
- A brine reclamation facility could generate a net revenue from sales of reclaimed salt products and brine disposal chargers between $0.2 million and $1.0 million per year. An initial investment of about $5 million would be required and would have a payback period of about 3.3 to 7 years.
- A reclamation facility that is centrally located to the oil and gas fields in New York could result in significant cost savings to producers. Depending on the economic goals of an owner/operator of the reclamation facility and whether disposal fees are charged, overall costs for disposal could be reduced between 15 and 90 percent.
- Solar ponds provide a cost effective means of heating and storing brines prior to salt reclamation. They are equal to, or considerably less expensive than, the use of natural gas, electrical power, or geothermal wells in heating brines.

Recommendations based on this study are:

- Assuming that roadspreading of brines is restricted in the future and that brine volumes generated increase or maintain their present level in New York State, development of a salt reclamation facility should be pursued.
- Because of the competitiveness of the salt markets, development of a salt reclamation facility should be carried out in partnership with one of the current principal salt producers.

ACKNOWLEDGMENTS

The authors would like to express their appreciation to Mahmoud Hammoud, Robert J. Seyfried, and Ketan M. Shah for their invaluable parts in this study. In addition, the authros gratefully acknowledge the work of Dr. Burton Krakow of NYSERDA and Mr. James Evans of GRI for their interest in our project.

This study was co-sponsored by the New York State Energy Research and Development Authority, the Gas Research Institute, and the State University of New York at Buffalo. Opinions, findings, and recommendations expressed in this paper do not necessarily reflect the views of NYSERDA, GRI, or SUNY Buffalo.

REFERENCES

Adamiak, R., 1987, Franklin brine treatment plant. Paper presented at the Ohio Oil and Gas Association Technical Meeting, Columbus, Ohio, October 13, 1987.

Atkinson, J.F., 1990, Effects of surface ice on solar pond performance, Solar Energy, v. 45 (4).

Atkinson, J.F., and Harleman, D.R.F., 1987, Wind-mixing experiments for solar ponds, Solar Energy, v. 38, 389-403.

Engdahl, D., 1984, California Department of Water Resources, personal communication.

Fillo, J.P. et al., 1987, Production waters associated with the production, processing, transmission, and storage of natural gas: a literature survey. A report to the Gas Research Institute prepared by ERT, A Resource Engineering Company, Pittsburgh, PA.

French, R.L., and Meitlis, I., 1980, Salton sea solar pond project, Proceedings, AIAA 15th Intersociety Energy Conversion Engineering Conf., v. 2, 1430-1431.

Hull, J.R., and Nielsen, C.E., 1986, Solar Ponds, Chap. 11, "DOE Solar Thermal Energy Conversion Technology Status and Assessment Project," ed. Francis de Winter.

Kaleczinsky, A., 1902, Uber die ungarischen warmen und heissen Kochsalzseen als naturaliche Warmeaccumulatoren sowie uber die Herstellung von warmen Salzseen und Warmeaccumulatoren, Annalen der Physik, v. 7, 408-416.

New York State Department of Environmental Conservation (DEC), 1988, "Draft Generic Environmental Impact Statement on the Oil, Gas and Solution Mining Regulatory Program," Volumes 1-3, Division of Mineral Resources.

New York State Department of Environmental Conservation (DEC), 1988, New York State Oil and Gas Drilling and Production, 1988, Division of Mineral Resources.

New York State Department of Environmental Conservation (DEC), 1989, New York State Oil and Gas Drilling and Production, 1989, Division of Mineral Resources.

New York State Department of Environmental Conservation (DEC), 1990, New York State Oil and Gas Drilling and Production, 1990, Division of Mineral Resources.

Short, T.H., Badger, P.C., and Roller, W.L., 1979, The development and demonstration of a solar pond for heating greenhouses, "Sun II, Proc. Int. Solar Energy Soc. Cong.," ed. K.W. Boer and B.H. Glenn, v. 2, 1021-1025.

Tabor, H., 1963, Solar ponds: large area collectors for power production, Solar Energy, v. 7, 189-194.

Weeks, D.D., and Bryant, H.C., 1981, "What happens when a solar pond boils?, Solar Rising, Proc. Annual Meeting Am. Solar Energy Soc.," ed. B.H. Glenn and G.E. Franta, v. 4.1, 768-771.

Wesolowski, D., et al., 1987, "Characterization of produced waters from natural gas production operations, Volumes 1 and 2," a report to the Gas Research Institute prepared by ERT, A Resource Engineering Company, Pittsburgh, PA.

Wittenberg, L.J., and Harris, M.J., 1981, Construction and startup performance of the Miamisburg salt gradient solar pond, J. Solar Energy Eng., v. 103, 11-16.

Zangrando, F., 1980, A simple method to establish salt gradient solar ponds, Solar Energy, v. 25, 467-470.

SUBSURFACE DISPOSAL OF PRODUCED WATERS: AN ALBERTA PERSPECTIVE

R. J. Cox

Energy Resources Conservation Board
Calgary, Alberta
Canada

INTRODUCTION

Alberta's oil and gas industry was born in the early 1900's, and currently (1990) produces in the order of 60 million cubic meters (378 million barrels) per year of light, medium and heavy crude oil and 90 billion cubic meters (3194 billion cubic feet) of natural gas. This level of hydrocarbon production, coupled with the mature stages of Alberta's oilfields, results in significant volumes of produced water that must be reinjected for disposal or pressure maintenance purposes. Additional volumes of equivalent saline or waste fluids also result from a variety of industrial and oilfield processes.

The management of these fluids in a safe, effective, environmentally acceptable manner is a problem both industry and regulators are constantly dealing with. These fluids are either stored on the surface, treated and returned to the surface watershed, or most often, disposed to a subsurface geological formation.

The purpose of this paper is to illustrate the areas of regulatory concern in the subsurface disposal of produced water and other liquid wastes in Alberta, and to highlight the technical factors that should be considered by both operators and regulators of subsurface disposal operations. Many of the matters discussed will also apply to wells injecting produced water for reservoir pressure maintenance purposes, hereafter referred to as injection wells.

DISPOSAL/INJECTION HISTORY

Subsurface disposal of fluids is not new to Alberta. The first disposal occurred in the 1920s with the injection of produced water. A further development occurred in 1951 with the disposal of liquid process wastes (industrial) and finally in 1957, the first water injection for the purpose of pressure maintaining an oil pool for improved hydrocarbon recovery.

In 1990, there were some 4,462 injection and disposal wells operating in the province of Alberta, and these wells injected a total of 233 million cubic meters (1467 million barrels) of water and waste fluids. An additional 43 million cubic meters (271 million barrels) of fresh and recycled produced water was injected as steam for the recovery of crude bitumen in Alberta's Oil Sands Areas. The number of wells used for water injection or pressure maintenance have increased steadily the past 30 years as have the injected volumes. In 1990, a total of 3,655 wells contributed to the injection of approximately 108 million cubic meters (682 million barrels) of water, approximately half of which was fresh water. By comparison, the number of wells used for disposal of produced water and waste fluids have increased sharply since the early 1970's, as have the annual disposal volumes as shown in Figure 1. In 1990, a total of 807 wells

Produced Water, Edited by J.P. Ray
and F.R. Engelhart, Plenum Press, New York, 1992

contributed to the disposal of approximately 125 million cubic meters (785 million barrels) of produced water and waste fluids.

Disposal rates typically range from 25 to 500 cubic meters (150 to 3,150 barrels) per day, but in rare cases can exceed 5,000 cubic meters (31,500 barrels) per day. Many of the disposal wells accept fluid on vacuum (i.e., without applied wellhead pressure) but most require a wellhead pressure to achieve disposal at the required rates.

GEOLOGICAL SETTING

Alberta overlies the Western Canadian Sedimentary Basin where oil and gas deposits are trapped by numerous shale barriers and fault systems. The nature of the basin provides for sedimentary deposits ranging in depth from 6,000 meters (~20,000 feet) along the edge of the foothills in the west to zero in the northeastern tip of the province. The stratigraphy provides trapping mechanisms for oil and gas, and subsequently provides much opportunity for stratigraphic isolation of disposed fluids. A cross-section of central Alberta is shown in Figure 2.

Significant sources of groundwater, suitable for domestic consumption, agriculture, and numerous industrial uses are available in varying quantities throughout the province. Alberta's Department of Environment has defined usable groundwater as any groundwater with a total dissolved solids content of less than 4,000 milligrams per liter (approx. parts per million). Exceptions to this criterion occur where shallow groundwaters exceed this value or where depths of groundwaters containing less than 4,000 milligrams per liter are excessively deep. Depths of groundwater protection in Alberta range from 65 meters (215 feet) to 600 meters (1,970 feet) in some areas (Alberta Environment, 1991). Alberta's Department of Environment is currently mapping depths for groundwater protection across the province. The approximate base of groundwater protection in the area represented by the cross-section is also shown in Figure 2.The dominant disposal formations as depicted in Figure 2 are located in the Lower Cretaceous system in the northeast, with a trend towards the Mississippian and Upper Devonian systems at the southwest portion of the cross-section. Although there are certainly some exceptions throughout the province, Figure 2 illustrates the favorable extent of vertical separation and numerous shale barriers between the common disposal formations and usable groundwaters.

Approved disposal operations in Alberta range in depth from 178 meters to 4,314 meters (584-14,155 feet) with the majority lying between 750 - 1,250 meters (2,460-4,100 feet). Figure 3 shows a distribution of well depths for disposal operations.

REGULATORY ENVIRONMENT

The Energy Resources Conservation Board (ERCB) is responsible for the approval of all applications for subsurface disposal of produced water or waste liquids. All such applications are reviewed by technical staff prior to approval. ERCB review is intended to ensure adequate formation characteristics, competent wellbore completion, and safe injection pressures. Matters of waste suitability and groundwater protection are forwarded to the Alberta Department of Environment for review. A summary of the ERCB's technical review follows.

Formation Integrity

The prime objective in subsurface disposal operations is to ensure that fluids are injected into a reservoir or formation that is capable of containing the disposal fluids both initially and for the long term. This can be evaluated from the perspective of whether or not the proposed formation exhibits stratigraphic isolation, and by review of the formation to estimate whether sufficient pore volume capacity exists to meet the disposal requirements.

Stratigraphic isolation is considered adequate when it can be determined that layers of confining formation rock are located both above and below the disposal horizon. Where disposal

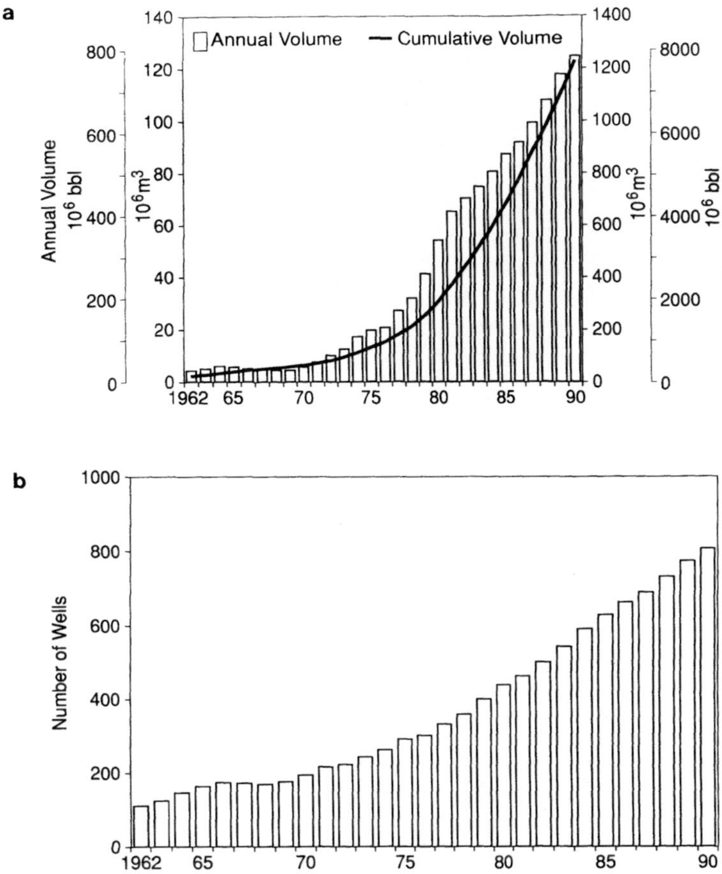

Figure 1. Volumes and number of wells for produced water and waste disposal.

is into a known oil or gas bearing formation it can be concluded that sufficient trapping mechanisms exist to contain the disposal fluid. Otherwise, the ERCB may review the application and related geology to evaluate the presence and extent of overlying shales as caprock, and underlying shales as base rock.

With respect to disposal capacity, the total compressibility of formation rock and fluid is typically low so substantial pore volume must be available to store significant volumes of fluid with an acceptable pressure increase. Regional geology is reviewed to evaluate potential formation disposal capacity. Rigorous analysis is not usually conducted, rather formation performance and capacity is indirectly controlled by setting a maximum wellhead injection pressure on the disposal well. Formation plugging or reservoir fill-up would result in increasing injection pressures to dispose of the necessary volumes, and by relying on an approved maximum injection pressure the need for rigorous compatibility tests and pore volume estimations is alleviated. The onus is therefore left to the operator to evaluate pore volume and water compatibility.

Where oilfield or industrial wastes are disposed, formation performance and capacity may be evaluated in a more direct manner through monitoring of static formation pressure, typically on an annual basis. The ERCB does not currently impose limitations on static formation pressures.

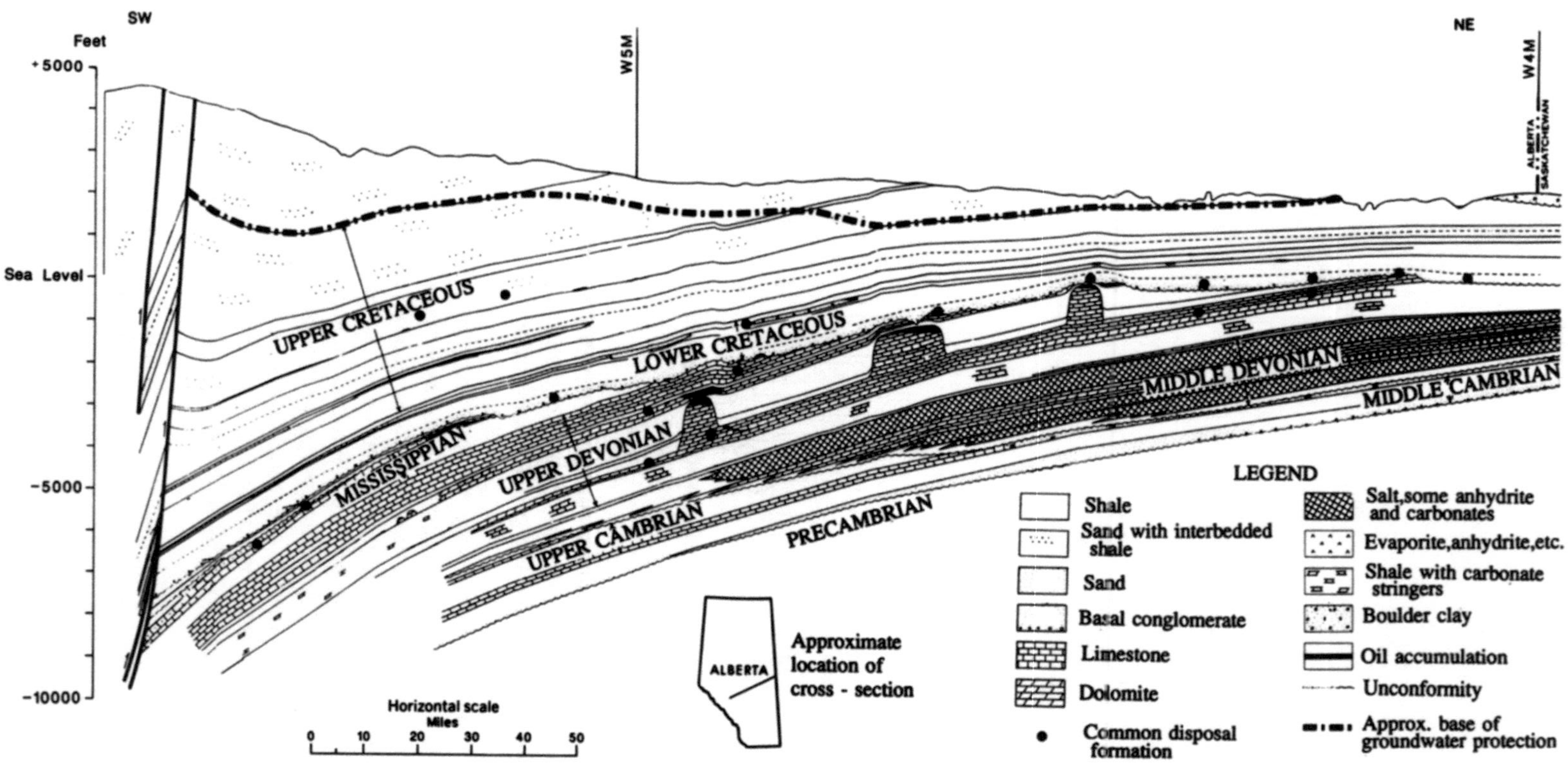

Figure 2. Schematic cross-section of the Western Canada Sedimentary Basin showing common disposal formations and the approximate depth of groundwater protection. Based on map adapted by J.R. Pow from Maps 1558A and 1559A (1981), Geological Survey of Canada.

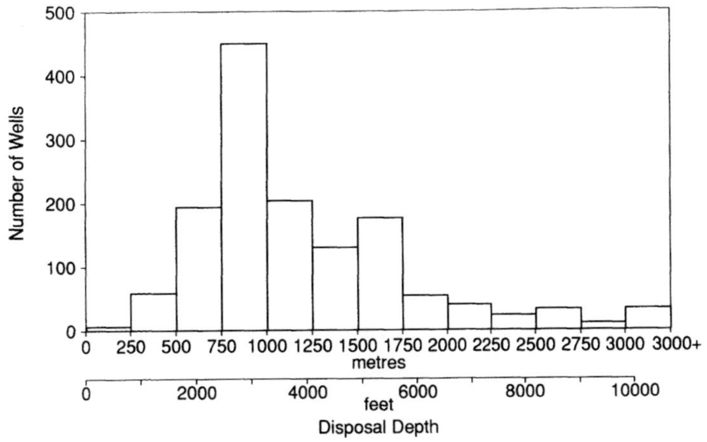

Figure 3. Historical depths for produced water and waste disposal.

Formation Fracture

Maximum wellhead injection pressures are generally restricted to result in a bottomhole pressure below the fracture pressure of the formation. To some extent, increased injectivity may be achieved by initiating and propagating fractures within the designated disposal zone. Depending on cap and base rock or cement qualities, potential exists for propagation of the fracture outside the disposal zone, thereby jeopardizing stratigraphic isolation and containment.

The consequences of fracture include potential contamination of usable groundwaters as well as potential detriment to hydrocarbon recovery in adjacent zones. In the absence of site-specific or reliable offset fracture data, the ERCB uses a conservative fracture gradient based on statistical analysis of fracture data of approximately 5,700 wells across the province. Determining wellhead pressures on the basis of an average or typical gradient may not be appropriate, as by nature of its mathematical definition an average gradient implies that approximately half the wells approved by this means could be injecting at pressures exceeding fracture pressure. To illustrate, Figure 4 shows the composite fracture data for Alberta. As shown, an "average" or typical gradient of 18 kPa/m (0.80 psi/ft) would result in many wells injecting or disposing above the fracture pressure. In absence of well-specific or reliable offset/regional fracture data, fracture pressures are determined for a well by this composite data such that there is a 90% probability that the formation will not fracture at that pressure. The conservative fracture gradient used by the ERCB varies with depth, but is approximately 14 kPa/m (0.62 psi/ft). A further safety factor is applied to the estimated fracture pressure from which an equivalent wellhead injection pressure is determined and specified on the approval. If the ERCB-estimated fracture pressure and resulting maximum wellhead injection pressure prove too restrictive to the operator, opportunity exists for the operator to justify a higher pressure. Step-rate injectivity tests are generally required to determine the actual fracture pressure of the disposal zone at the well, or alternatively to illustrate a pressure at which disposal can occur without initiating a fracture.

In some instances, fracture pressure may be exceeded for produced water disposal where formation thicknesses are large and rock characteristics are favorable (i.e., high quality cap and base rock and some indication of effective barriers to fracture propagation), or where communication with adjacent zones would have no adverse impacts on groundwaters or hydrocarbon recovery. Disposal above fracture pressure must certainly be considered an exception to the rule at the present time.

Wellbore Integrity

The wellbore must be completed in a manner that ensures the disposed fluid is isolated from all other formations, both on route to the intended formation and also after its arrival. It is imperative that migration of the disposed fluids along the wellbore does not occur. Failure to contain the fluids in and along the wellbore means a failure of the disposal operation. Such a failure could more importantly give rise to potential pollution problems through contamination of groundwater or hydrocarbon producing horizons.

A "text-book" disposal well is completed with a string of surface casing typically set at 200-300 meters (650-985 feet) and cemented to surface. This is followed by a longer production casing string set at the total depth of the well and cemented to surface. The production casing is then perforated opposite the desired formation, or where geological conditions permit, an open-hole completion across the zone of interest may be used.

The aforementioned completion would be typical of a well drilled specifically for the purpose of water or waste disposal. However, disposal wells are more often proposed as conversions of existing producing wells in which production casing may be cemented to a pre-determined depth, generally above the highest hydrocarbon-bearing formation. Production casing cement requirements are specified at the time of well licensing and may vary depending on the location. In cases where less than 180 meters (590 feet) of surface casing has been run, or surface/intermediate casing is not set and cemented from more than 25 meters (82 feet) below any usable water aquifer to surface, the production casing must be cemented full length. Establishment of these requirements has been an evolutionary process such that existing wells may or may not meet these requirements at the time of proposed conversion to disposal. As a minimum in any disposal operation, there must be sufficient cement across the disposal formation to provide hydraulic isolation of the disposal fluid.

In cases where surface or intermediate casing does not cover usable groundwaters, the Department of Environment may require groundwater monitoring wells adjacent to the disposal well or an assessment of the risk to groundwater of the proposed wellbore completion.

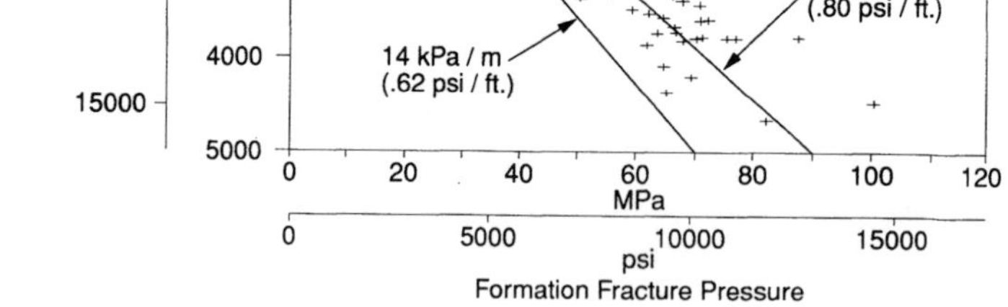

Figure 4. Formation fracture pressures (bottomhole), based on initial shutin pressures obtained from fracture treatments.

Given this general completion scenario, wellbore integrity can be achieved through the following completion practices.

Tubing and packer. All wells injecting fluids other than fresh water or steam are required to be completed with a tubing and packer. Injection through a tubing string with a production packer protects the production casing from exposure to the disposal fluid and any significant pressures applied to achieve the desired injection rates. Protection of the production casing from corrosion is essential as the casing and associated exterior cement serve to isolate all uphole formations and ensure stability of the wellbore. Tubing corrosion, although undesirable, can be tolerated as the tubing can be pulled and repaired while casing, on the other hand, is cemented in place and can only be repaired with some difficulty. The packer should be placed as close as practical to the top of the perforations to minimize the length of casing exposed to disposal fluid and ensure zonal isolation below the packer in the event of casing corrosion.

Inhibited annulus. As an additional measure of protection, the tubing/casing annulus of disposal or injection wells (except those injecting steam or fresh water) must be filled with a corrosion inhibited fluid. This protects both the exterior of the tubing and the interior of the casing from deterioration by corrosion, and the fluid in the annulus also serves as a medium through which casing or tubing leaks can be detected.

Casing integrity logs. Logging of the well casing to establish the presence and extent of corrosion gives an indication of the general casing condition, in particular the wall thickness and potential trouble spots. This operation is particularly necessary when converting an old well to injection or disposal, and can also be used for detecting or locating potential leaks. Casing integrity logs are currently required when converting a well of age greater than five years to disposal or injection. These logs must be capable of determining internal and external corrosion and detecting holes and pits, and accordingly caliper logs and logs capable of indicating only an average wall thickness are not acceptable.

Cement integrity logs. Cement logging of the wellbore gives an indication of the quality of the cement-to-casing bond and overall effectiveness of the cement with respect to hydraulic isolation. The function of the cement with respect to a disposal operation is to provide hydraulic isolation between porous zones, particularly between the disposal zones and adjacent formations. The mere presence of cement behind the casing may not be sufficient to provide hydraulic isolation and the logs must be interpreted carefully to ensure that vertical migration of the disposal fluid will not occur. For wells disposing at depths greater than 600 meters (1,970 feet), cement bond or cement evaluation logs are currently required across the disposal zone and the surface casing shoe where production casing is cemented full length, and across the disposal zone and cement top where production casing is not cemented full length. For wells injecting at depths of less than 600 meters (1,790 feet), both cement bond and evaluation logs are required from total depth to 25 meters (82 feet) inside the surface casing.

A typical wellbore completion, in compliance with current regulations, is shown in Figure 5. The aforementioned requirements, specifically regarding casing and cement integrity logs, are currently under review by the ERCB. Proposed changes are discussed later in this paper.

It is important to note that adjacent wellbores could potentially jeopardize the safe and effective operation of a subsurface disposal scheme. Improperly abandoned or poorly cemented wellbores in communication with a disposal formation can allow for vertical migration of the disposal fluid. Consideration must therefore be given to the wellbore completion and subsequent integrity of adjacent wellbores. No specific policies or requirements for review exist at present, although this matter has been addressed in specific applications where it was believed warranted.

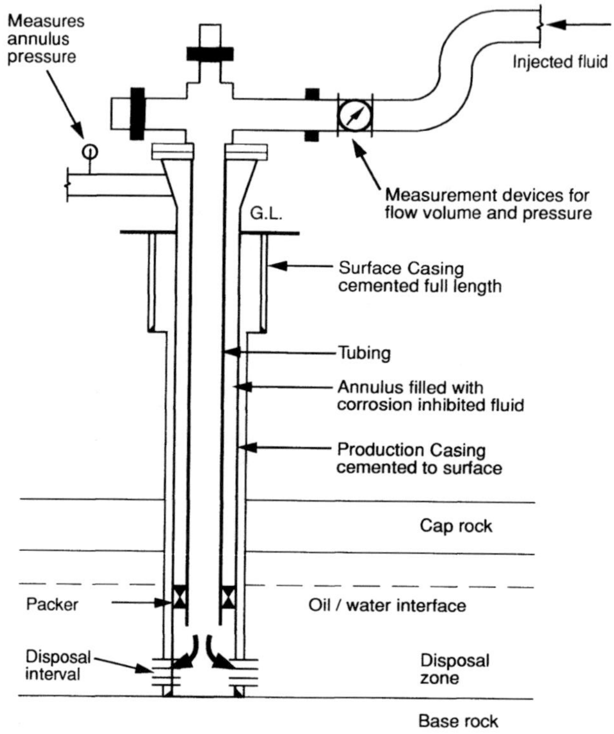

Figure 5. Preferred disposal well completion schematic diagram.

OPERATION AND MONITORING REQUIREMENTS

Given a properly completed wellbore as described in the previous section, and a competent and suitable disposal formation, the disposal well should initially perform safely and efficiently. However, the components of a wellbore completion are susceptible to failure, whether through equipment design failures, manufacturing faults, or time-related aspects such as corrosion. A system of monitoring must therefore be implemented to ensure continued wellbore integrity. With respect to pollution control, failures in the subsurface environment hold potential for going undetected for an appreciable length of time.

Failure Analysis

Subsurface failure of a disposal operation may result from any one of a number of possible failure modes. The consequences of a failure in relation to environmental pollution are variable.

From the required wellbore completion as shown in Figure 5 it is apparent that, given a competent formation and cement job, the potential for a sudden or catastrophic failure of the subsurface system is low as it would require simultaneous failure of two or more components of the completion. That is, failure of a single component such as tubing, casing, or packer would not result in a threat to groundwaters or adjacent hydrocarbons as in all cases, a second line of defense exists.

For the purpose of this discussion, this type of failure would be termed of "secondary consequence" due to the backup protection. A failure of "primary consequence", by contrast, would be one in which no backup protection exists. With respect to the required wellbore completion, Table 1 summarizes the possible failure modes and associated consequences.

It is obvious from this analysis that the cementing of the well and the selection of the formation itself are key parameters in a successful disposal operation.

Monitoring Parameters

Monitoring and surveillance procedures should be applied to subsurface disposal schemes that are consistent with both the nature and the sensitivity of the scheme in general. For instance, high pressure injection of a waste fluid should command more attention than a well disposing of near-potable water at very low pressure or even vacuum conditions. With the aforementioned in mind, monitoring of the following parameters may be required in an ERCB approval for produced water or waste disposal.

Table 1. Summary of disposal well failure modes and consequences

Failure Mode	Consequence	Backup Protection
Tubing	Secondary	Casing/Cement
Packer	Secondary	Casing/Cement
Casing	Secondary	Requires tubing or packer failure
Cement	Primary	None
Formation(integrity)	Primary	None
Formation (fracture)	Primary	None

Disposal Volume. Continuous measurement of all fluids injected through a well to an underground formation is required by the ERCB. This allows for evaluation of reservoir disposal capacity and, where appropriate, waste front migration. When considered as an injection rate it can also be used as an indicator, in conjunction with injection pressure, of well integrity. Any changes in the injectivity may indicate wellbore or formation integrity problems.

Tubing Injection Pressure. The tubing pressure is a measure of the resistance of the formation and system to accept the disposal volumes. As such, it is an indicator of injectivity problems, component failure, or formation failure when compared to corresponding disposal rates. Generally speaking, a sudden drop in the injection pressure with all other factors being equal (particularly disposal rates) is an indication that a combined casing and packer/tubing failure, or reservoir fracture may have occurred. Conversely an increase in tubing pressure at constant injection rates alludes to formation plugging or reservoir fillup, and can therefore be used as an indicator of potential problems. All new disposal wells, waste or produced water, are subject to a maximum wellhead injection pressure to avoid formation fracture. Waste disposal wells may also require some continuing monitoring of tubing injection pressure as a condition of the disposal approval.

Casing-tubing Annular Pressure. Monitoring the pressure of the liquid in the casing-tubing annular space provides an indication of tubing, packer, and casing integrity. Changes in the annular pressure, other than those expected due to tubing expansion and contraction, are indications of possible failure of either of these components. The specific response is dependent on the nature of the fluid gradients and injection pressures and the corresponding pressure contrasts at the point of failure. Monitoring the fluid level in the annular space is a common alternative to pressure measurements. Annular pressure or fluid level monitoring is often required for waste disposal wells as a condition of the disposal approval, and may be specified for produced water disposal where continuing casing integrity is suspect.

Annular pressure monitoring is not without its limitations, as the annular pressure or fluid level is expected to fluctuate due to temperature and pressure changes causing tubing expansion or contraction. Pressure or fluid level fluctuations must therefore be evaluated in light of changing injection conditions before a failure can be concluded.

In addition to monitoring this parameter under normal operating conditions, a test of packer and tubing/casing annular integrity is required at least annually by applying an external pressure to the annulus. A negative test does not always identify the component that has failed, but rather that one of the components (casing, tubing, packer) has failed.

Static Formation Pressure. Regular bottomhole static pressure surveys along with the pressure falloff behaviour allows for analysis of reservoir performance, particularly with respect to formation quality or damage and future reservoir disposal capacity. In large-volume waste disposal schemes and those of significant sensitivity, annual bottomhole pressure surveys may be required as a condition of the ERCB approval. They are not required for all waste disposal schemes and are typically not required for produced water disposal schemes.

Formation failure can be considerably more difficult to detect, and generally requires analysis of injectivity and bottomhole pressure data. Regarding formation fracturing, monitoring of tubing pressures and injectivity can be used as an indicator, as fracturing typically requires a steadily building pressure to initiate the fracture followed by a sudden reduction in pressure (increased injectivity) while the fracture propagates. In the absence of continuous injectivity monitoring, however, formation failures may be evidenced only through the presence of the disposal fluids in adjacent formations. Such failures can be avoided by injecting at a pressure below the fracture pressure. If it is necessary to approach the fracture pressure then appropriate monitoring of injectivity should be performed.

The previous discussion has assumed that cement integrity is such that hydraulic isolation will not allow vertical migration of disposal fluids beyond the injection interval. In absence of hydraulic isolation and depending on the injection pressure vs. the resident wellbore gradients, vertical migration could go undetected for a significant length of time. Unfortunately, as with a suspected formation failure, monitoring of this problem is difficult and generally the failure may not be discovered until some measure of damage has already been done. Accordingly, this reinforces the need for prudent cementing practices and conscientious evaluation of hydraulic isolation.

PERFORMANCE

The ERCB has in the past denied some applications for subsurface disposal based on some of the items addressed in this paper. It also has seen numerous instances in which wells have failed to maintain injectivity due to formation plugging or poor formation choice. In the Board's experience with administering subsurface disposal wells, there have been relatively few instances of formation or wellbore failure that have resulted in communication with formations other than the intended disposal zone or escape of fluids to surface. Documented cases of releases to surface or contamination of adjacent formations due to subsurface disposal are extremely low. Two documented cases which did result in a return of fluids to surface revealed the causes of failure to be a lack of hydraulic isolation of the disposal zone due to poor cementing and injection pressures in excess of the formation and/or cement fracture pressure. This helps to emphasize the need for prudent management of wellhead injection pressures. Other injection-related failures have occurred in pressure maintenance or steam injection schemes as evidenced by lack of voidage replacement and charging of adjacent zones. Similar causes were suspected here, but again the frequency of documented failures of these types are extremely low.

Historical data suggests a failure rate of annular integrity tests in the order of 10 percent, representing a combination of tubing, packer, and casing failures. This percentage may be misleadingly high as emphasis in inspection and surveillance of these tests is placed on wells where problems may be suspected. Monitoring programs have been generally successful in detecting tubing, packer, or casing failures in time to take remedial action before significant damage is done.

CURRENT ISSUES AND FUTURE DIRECTIONS

It is suggested that further evaluation of some primary-consequence failures may be appropriate, specifically in the areas of hydraulic isolation and formation integrity. The ERCB is satisfied that present operating and monitoring practices are capable of preventing or detecting other failure modes. The following discusses recent developments in the ERCB review process, and identifies issues for future consideration.

Hydraulic Isolation

With respect to hydraulic isolation, the ERCB is in the process of moving away from inferential techniques such as cement bond or evaluation logs towards a direct measurement of flow behind casing such as radioactive tracers and temperature surveys, or suitable alternatives such as oxygen activation. Limitations of bond logs are well recognized, and the merits of requiring a more direct measurement are reinforced by mechanical integrity testing conducted by Thornhill et al (1990) which documents the inability of cement logs to detect certain known channels in a cemented wellbore. Cement logs would therefore become an option rather than a requirement, although circumstances may well warrant cement logging as the preferred diagnostic tool when isolation of uphole zones behind casing are a concern. It has further been proposed that maximum wellhead injection pressures be determined as the pressure at which the successful tracer/temperature survey was run. In wells injecting produced water, this could assist in assessment of wellhead injection pressure with respect to fracture pressure, and further evaluate vertical fracture propogation (if any) and consequence.

Although this represents a more definitive evaluation of hydraulic isolation, it is an initial review of cement integrity and does not address the issue of long-term wellbore integrity. Matters of current interest related to hydraulic isolation include the potential for cement degradation upon continued exposure to produced water or waste streams, and the ability of wellbore cement to withstand high injection pressures.

Formation Integrity

The other large unknown in produced water disposal operations is that of long-term formation integrity. It can be surmised that if a formation was capable of trapping water or hydrocarbons initially, it should be capable of continued isolation under disposal operations. Currently, no limit is specified for the maximum static formation pressure in a disposal scheme, rather the issue of declining formation capacity is handled by way of maximum wellhead injection pressures and the operator's injectivity needs.

Well Classifications

Industry, government, and regulators have been conducting open conversations recently regarding disposal well practices, policies, and procedures. It has been concluded that a system of well classifications is necessary to provide for operating and monitoring parameters that are commensurate with the nature of the fluid injected.

A classification system has recently been proposed which is similar, but not identical to the United States Environmental Protection Agency classifications. At the time of writing this paper the proposal was being prepared for industry and government review. Similar nomenclature is being proposed, designating produced water disposal wells as Class II.

SUMMARY AND CONCLUSIONS

Subsurface disposal of produced water and wastes through deep wells is a safe and responsible practice where wellbore and formation integrity can be achieved, both initially and on an ongoing basis. Subsurface formations must be selected that are capable of containing the disposal fluids and of providing stratigraphic isolation of the fluids from usable groundwaters or recoverable hydrocarbons. Wellhead injection pressures must also be managed so as to preserve formation integrity, typically by limiting injection pressures to below the formation fracture pressure. Wellbore integrity requires prudent cementing practices to confine the disposal fluid to the host formation, while completion with tubing and packer provides an additional level of protection as well as a means of monitoring wellbore mechanical integrity. Monitoring programs provide for continued wellbore integrity and should be consistent with the sensitivity of the disposal scheme. Monitoring programs have been largely successful in detecting wellbore failures prior to significant release of injected fluid to surface or adjacent zones.

ACKNOWLEDGMENTS

The author gratefully acknowledges the assistance in preparation of data and figures for this paper by Mr. Kent Geddes of the Energy Resources Conservation Board's Drilling and Production Department.

REFERENCES

Alberta Department of Environment, Groundwater Protection Branch 1991, personal conversations

Thornhill, J. T., and Benefield, B.G., 1990, "Injection Well Mechanical Integrity", Robert S. Kerr Environmental Research Lab., Ada, OK

WATER REUSE TESTING AT SHELL'S PEACE RIVER COMPLEX

G. D. Giesbrecht[1], F. Chang[2], and R. Nelson[3]

[1]CH2M Hill Engineering Ltd.
[2] Shell Canada Ltd.
[3] Alberta Oil Sands Research and Technology Authority

INTRODUCTION

Shell Canadá Ltd. has operated a heavy oil recovery plant near Peace River in northern Alberta since 1979. The 8° API gravity heavy oil is recovered using the pressure cycle steam drive process. Typical heavy oil and produced water rates are 10,000 bbl/d and 40,000 bbl/d, respectively. The main plant facilities consist of produced fluid treating, source water treating, steam generation, and an asphalt plant. Water for steam production has been obtained from Cadotte Lake, which is near the plant, and from Peace River. Deoiled produced water has been injected down onsite disposal wells.

The Alberta Energy Resources Conservation Board has a policy of requiring serious efforts toward reuse of produced water at all commercial production batteries with significant water consumption. In addition, Shell recognized that there is an economic incentive to reuse produced water to raise steam because of the reduced raw water pumping and treatment and the reduced produced water disposal costs.

WATER REUSE CONCERNS

The typical produced water composition from the Peace River Complex (PRC) is shown in Table 1. In common with other produced water from steamfloods, it is very saline, hard, and sour. It has a total silica concentration of 200 to 300 mg/l, which is typical of high-pressure steam floods. One unusual characteristic of the produced water is its very high alkalinity. While there is evidence that some of the alkalinity is due to organic acids rather than carbonates, which are the usual source, there is no doubt that the total carbonate concentration is quite high. The other unique feature relating to water reuse is the practice at PRC of separating the liquid and vapor phases of the steam so as to inject dry steam.

Feedwater quality specifications for high-pressure, once-through steam generators have changed very little since they were first published. Typical feedwater specifications are presented in Table 2. The specified concentration maxima for oxygen, hardness, and oil have not changed significantly since oil field steam generators came into common use. Silica specifications, on the other hand, began at 50 mg/l SiO_2 and have increased to 150 mg/l subject to certain conditions. There have been some unpublished indications that a feedwater silica specification of 300 mg/l was feasible. Increased silica concentration limits have been accompanied by much lower limits on hardness and on minimum ratios of alkalinity to silica. The basis for increased limits have been brief pilot tests and theoretical considerations of the solubilities of various

Produced Water, Edited by J.P. Ray
and F.R. Engelhart, Plenum Press, New York, 1992

silicate minerals. There have been examples of reasonably successful operations at low steam quality (60 percent) and low steam pressure (1,200 psi) but no sustained operations at high steam pressure and quality.

Table 1. Produced water quality

	Shell PRC	Typical Range[1]
Calcium	1,800	1200 - 11,200
Sodium	70	40 - 920
Magnesium	20	10 - 460
Chloride	1,900	1,400 - 20,000
Alkalinity (as HCO_3^-)	1,700	450 - 915
Sulphate	50	110 - 225
Silica (as SiO_2)	200	50 - 175
pH (pH units)	8.5	6.7 - 7.7
Total Dissolved Solids	5,700	3,500 - 33,000

(1) All units mg/l as ion except as noted.

The only commercial treatment process to reduce silica from several hundreds of parts per million to below 50 or so is lime softening. Lime softening is costly even under favorable conditions. At PRC it is even more expensive and produces a much larger amount of sludge than at other locations because of the high alkalinity. Large amounts of lime must be added to overcome the buffering effect of the alkalinity in addition to that required for the softening reactions.

Consideration of the high cost of treating its produced water and of the progressive increase in feedwater silica specifications prompted Shell to undertake a program to determine the feasibility of reusing produced water at PRC without removing silica.

Table 2. Feedwater quality specifications

Total Hardness	< 0.5 mg/l as $CaCO_3$
Total Silica	< 50 mg/l as SiO_2
Oxygen	< 0.04 mg/l
Free Oil	< 1 mg/l
Sulphide	absent
Iron	< 1.0 mg/l
Total Dissolved Solids	< 8,000 mg/l

Pilot Plant Investigation

In 1986 May Shell, with support from the Alberta Oil Sands Technology and Research Authority (AOSTRA), commissioned CH2M HILL ENGINEERING LTD. to design, build, and operate a pilot facility that would answer a number of questions about reuse of PRC-produced water. Specific objectives of the pilot plant study were:

- Determine the feasibility of operating a once-through steam generator on up to 100% produced water at the dissolved solids and silica concentrations experienced at PRC.
- Obtain operating experience under field conditions with produced water reuse using the treatment processes selected.
- Investigate the various operating characteristics of ion exchange systems treating PRC produced water without prior lime softening.

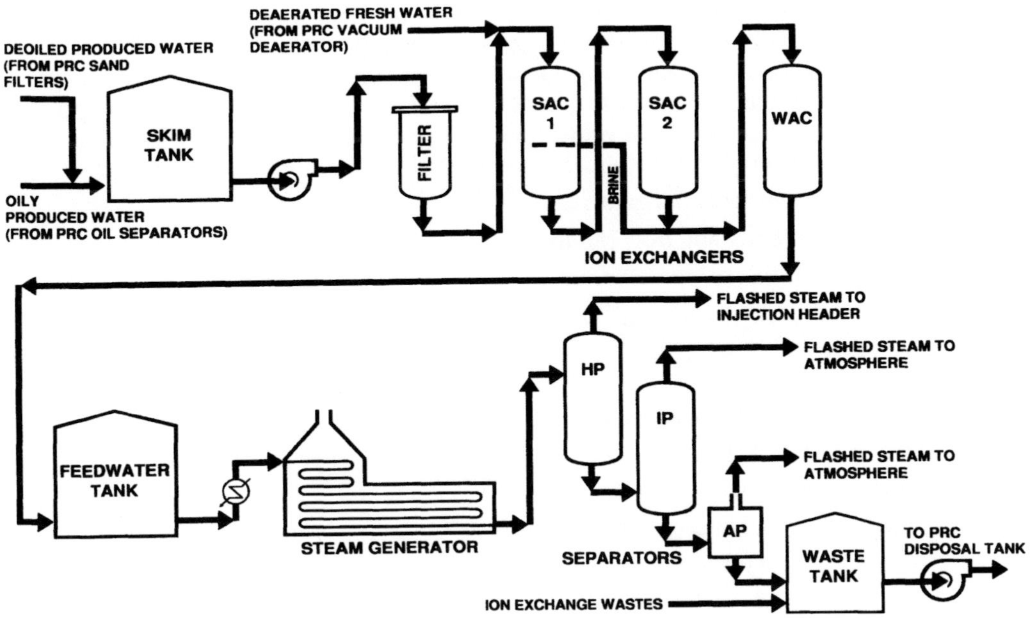

Figure 1. Shell/AOSTRA pilot plant.

- Investigate certain aspects of material performance in water treatment and steam separation equipment.

Only the first objective will be discussed in the remainder of this paper.

The pilot facilities consisted of a 1610 kW (5.5 MBtu/hr) high-pressure once-through steam generator that was specifically designed and built for the study, an ion exchange system to soften the water, chemical systems for ion exchange regeneration and to modify the feedwater chemistry, tankage and utilities, and an extensive data acquisition and control system. The steam pressure and quality were set at PRC operating conditions, 16 MPa (2300 psi) and 80%, respectively. A schematic of the pilot plant is shown in Figure 1.

Produced water was piped from two locations in the main plant. The first location was downstream from the final stage of oil removal before disposal and was the main produced water source. The second was from the water leg of the production treaters and was used to add a small amount of free oil for specific short tests.

The total silica concentration was controlled by blending produced water with fresh water and then adding sodium chloride and sodium bicarbonate to approximate the TDS and alkalinity found naturally in the produced water. Filtered and deaerated water was piped in from the plant fresh water system for blending and for utility water use.

Softening was carried out in a three-stage system. The first two stages were salt regenerated strong acid cation resin operating in the sodium cycle. The third stage was weak acid cation exchange resin operating in the sodium cycle. All three stages would not necessarily be required in order to obtain soft water, but one aspect of the investigation was to acquire data that could be used to optimize the process sequence for subsequent full-scale designs. A small cartridge filter was installed upstream of the ion exchangers to guard against oil upsets from the main plant. A PC-based data acquisition system was used to collect process data and carry out basic calculations.

The steam generator was specially designed for the program by Struthers TIW of Calgary. Extensive computer modeling was carried out to ensure the pilot generator was as representative of a full-scale unit as possible. Complete geometric and thermal similarity between two different sized pieces of equipment is impossible without using a different fluid. Since use of produced water was fundamental to the study, critical parameters were matched to full-scale units and others were allowed to vary. Parameters that were held to full-scale values included:

- Waterside pressure and temperature;
- Waterside velocity and flow regime;
- Average heat flux in each section;
- Steam quality profile through the waterside path;
- Fireside temperature.

Parameters that could not be held equal to full-scale units included:

- Radiant section L/D ratio;
- Relative roughness of the tubes;
- Tube spacing and return bend radius;
- Volumetric heat release;
- Water-side residence time.

The thermal design was such that boiling began at the beginning of the radiant section. Details of the pilot steam generator are shown in Figure 2. The steam generator was heavily instrumented with flowmeters, pressure transmitters, and thermocouples on all available process streams and thermocouples on the fireside of a number of convective and radiant tubes.

The test plan called for a number of runs of one month in duration at a particular feedwater silica concentration to be followed by an inspection and decision with respect to the next run. A diagram of the decision-making process is shown in Figure 3. The first run was based on fresh water in order to ensure satisfactory mechanical operation and to provide baseline data for the

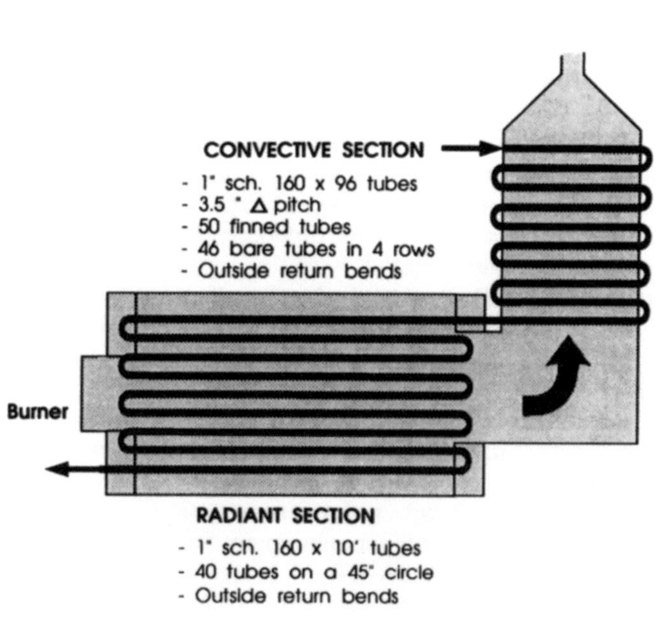

Figure 2. Pilot steam generator configuration.

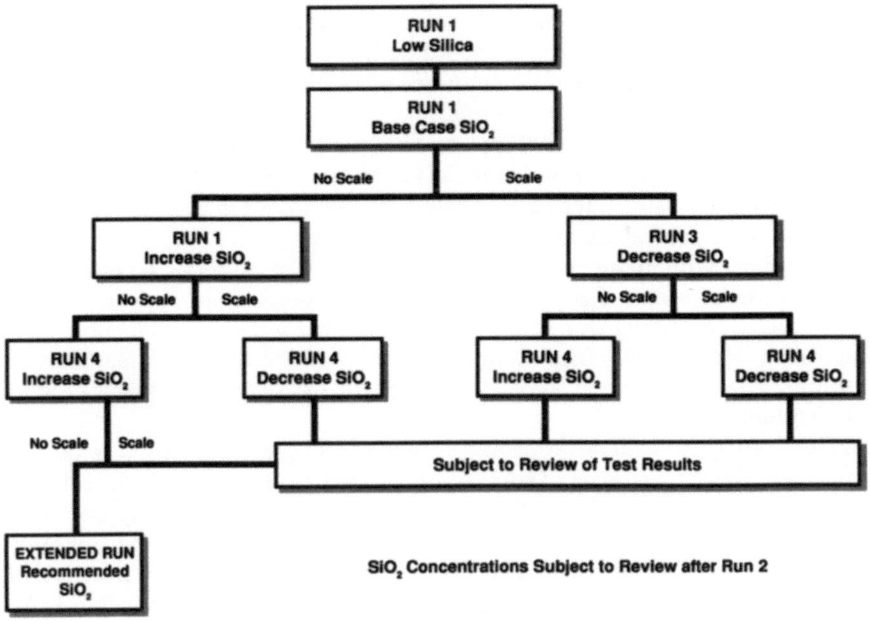

Figure 3. Silica concentration decision tree.

subsequent analysis. The second run was based on a conservative feedwater silica concentration comparable to that which would have been selected for a full-scale plant in the absence of any test work. As the program proceeded, the feedwater silica concentration was increased until it was clear that the associated scale deposition rate was excessive. Following the short runs, an extended five month run at the final recommended feedwater silica concentration was carried out to confirm the long-term acceptability of the operation.

After each run the steam generator was inspected in order to determine the operating conditions for the next run. The details of the inspection varied as knowledge and experience were gained. A typical inspection would consist of removing a number of return bends and inspecting the adjacent tubes visually with a boroscope. Selected tubes would then be removed and sectioned for detailed analysis of deposited scale. Samples of scale were analyzed by X-ray diffraction and X-ray fluorescence to determine the identity of the minerals present and gravimetrically to determine mass loading.

Impact of Silica on Steam Generator Operation

Observations made during this study are specific to the produced water composition and steam generator conditions used. Similar results may not be found at other steam conditions or with different produced water chemistry.

Scale is formed in boiling mixtures because the solubility of the scale-forming compounds is exceeded. Since the concentration of dissolved species in the liquid phase increases as the steam quality increases along the water path, there was an expectation that scale would form at some point along the water path that was related to the local liquid phase silica concentration. Furthermore, the point at which scale formation began would be expected to move back along the water path as the feedwater silica concentration increased. The rate of scale deposition would be expected to increase as the feedwater silica concentration increased.

In actual operation it was found that the rate of scale formation did increase as feedwater silica concentration increased, but that when scale appeared, it did so almost uniformly throughout the radiant section. A graph of observed scale deposition rate versus feedwater silica concentration is presented in Figure 4. It should be noted that the data on which this graph is based are somewhat limited in that different tubes were inspected for different runs and some of them were in service for more than one run.

Local variations in heat flux did not appear to have any influence on the deposition of scale. Scale deposits were approximately the same thickness on the fire side (facing the flame) as on the dark side (facing the wall) of the tubes. Scale was also found in comparable amounts in the return bends that were outside the radiant section tube sheet.

Silica combines with many cations to form hundreds of minerals in a variety of crystal structures. Analysis of scales deposited in the steam generator showed only nine different minerals with acmite, a sodium/iron substituted silicate being the most common. All minerals found were ribbon or chain structures. Cations present were iron, sodium, aluminum, calcium, and magnesium. The scales varied somewhat in physical structure from dense hard layers to more porous structures.

During the course of one of the inspections it was found that a significant amount of erosion had taken place on the outer tangent of some of the return bends in the radiant section. Not only was there no scale in these areas, but there was considerable metal loss. The cause was believed to be abrasion by fine silicate particles carried in the liquid phase of the wet steam. After completion of this work Esso Resources Canada reported that it had seen return bend erosion at its Cold Lake Alberta production complex. Esso had discovered particles of amorphous silica being carried in the liquid phase without being deposited on the tube wall.

Scale is considered highly undesirable in most fired equipment because it is a thermal insulator and causes a rise in the metal temperature and loss of strength in the pressure tubes. Thermocouples were installed on the fire side of many radiant and convective tubes in order to monitor the skin temperature and possibly detect the onset of scale formation. In practice skin temperatures were not a good indication of scale formation. The thermocouple itself affected heat transfer and, therefore, the measured temperature in the immediate area. The severe conditions in the radiant section resulted in a high failure rate of the thermocouples.

Temperature gradient calculations were done using a value for the thermal conductivity of the scale layer in the range of published values for several silicate minerals. They showed that the tube temperature should increase by about 10° C for a 400 micron thick layer of scale. This thickness is comparable to scales observed in the course of the study.

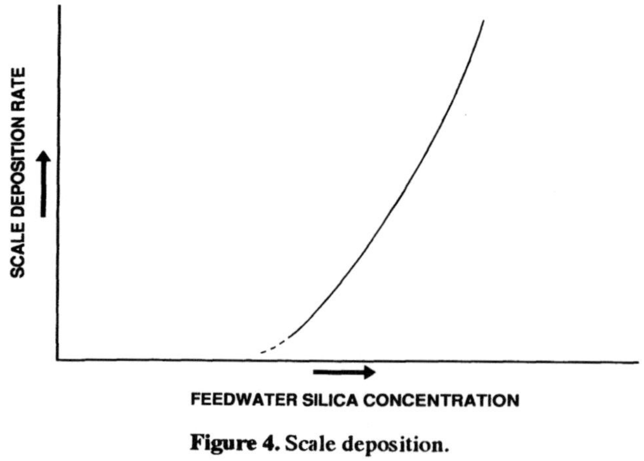

Figure 4. Scale deposition.

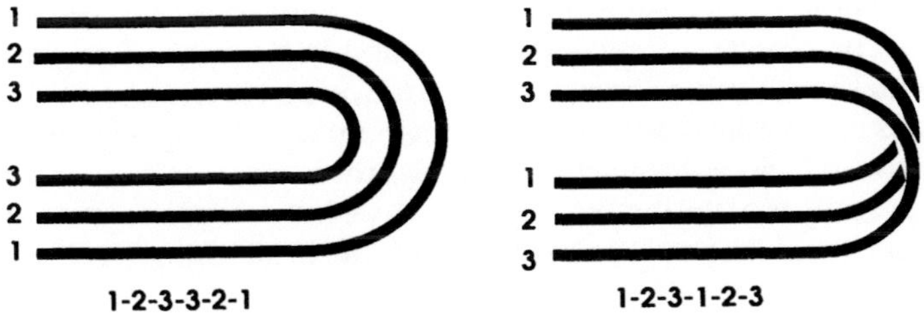

Figure 5. Return bend layout.

At one point in the program when a significant quantity of scale had accumulated, a chemical cleaning of the water side was undertaken. The cleaning agent was a mixture of hydrochloric acid and ammonium fluoride at an elevated temperature. This mixture produces hydrofluoric acid, which is one of the few effective means of dissolving silicate minerals. The cleaning procedure was only moderately effective. Most of the scale was removed; however, over 20 to 30% of the area, a portion of the original scale was left behind. The pilot steam generator had a much larger surface-to-volume ratio and a much larger ratio of metal to internal volume than a full-scale unit. These two factors made it difficult to maintain cleaning solution at the desired temperature which is believed to have reduced its effectiveness.

Impact on Future Plant Design

The pilot program showed that when using current water treatment technology, the maximum tolerable silica concentration was lower than the untreated produced water concentration.

Return bend erosion is a potential problem that operators will need to be aware of and plan for. Regular inspections of the tubes would be required as a minimum. In a large three path steam generator, long radius bends, as in a 1-2-3-1-2-3 radiant tube pattern, are preferred to a 1-2-3-3-2-1 pattern (Figure 5).

Provision for regular internal cleaning of the tubes would need to be made. If chemical cleaning is the method of choice, connection points, dead ends, and material selection all need to be considered in the equipment design.

Future Research Needs

Operation of high-pressure steam generators at 80% quality using high silica produced water and current feedwater treatment technology without silica removal will probably result in relatively high rates of scale deposition and attendant risk of equipment failure. Silica reduction processes that do not cost as much or produce as much solid waste as lime softening require continuing research.

This work was done with specific reference to the produced water composition found at Shell Peace River. Improved knowledge of the effect of alkalinity on the solubility of silicate minerals would help in deciding whether or not alkalinity should be modified at PRC or other locations. Similarly, research on all aspects of the interaction of the various species present in produced water, the mechanism of scale formation, and the cause of metal loss would help mitigate the problems discussed.

Day-to-day control of feedwater hardness is currently based on the colorimetric EDTA titration method. The detection limit for this method is 0.5 mg/l. The color level in many produced waters makes analysis in the region of 0.5 to 1.0 mg/l even more challenging. Improved low-level hardness measurement in a plant environment would help a great deal when treating high silica waters with high-performance softening processes.

Actual measurements of the thermal conductivity of silicate scales would help in determining the risk inherent in a given rate of scale deposition and in adjusting mechanical design details to reduce that risk.

If tube cleaning is to become a more common occurrence, there is ample scope for developing internal cleaning methods that are more effective, faster, cheaper, and involve less dangerous chemicals than current methods use.

ACKNOWLEDGEMENTS

The support of the project sponsors, Shell Canada Limited, Pecten Canada Limited, and the Alberta Oil Sands Technology and Research Authority are gratefully acknowledged. The complete project report is available from AOSTRA through its technology transfer program.

Numerous individuals within Shell Canada, both in Calgary and in Peace River, provided essential technical and operational input to the study.

REFERENCES

CH2M HILL Canada Ltd., 1983, Industry/AOSTRA Produced Water Treatment Study.

Rubinstein, I., 1990, Canadian Heavy Oil Association.

OVERVIEW OF THE PRODUCED WATER SYSTEM AT THE PRUDHOE BAY UNIT - ALASKA, NORTH SLOPE

T. A. Denman and S. Starr

BP Exploration
Anchorage, Alaska
USA

ABSTRACT

The produced water system at the Prudhoe Bay Unit (PBU) is one of the largest in the world, currently designed to handle in excess of one million barrels of produced water per day. Due to the large scale of the PBU production facilities, a major effort has gone into all the aspects of planning and operating the produced water system.

During the early stages of production from Prudhoe Bay the relatively low volumes of produced water were disposed to a shallow, permeable formation at low pressure. As the fields production matured sea water injection was initiated for pressure maintenance. Water- flooding also utilizes produced water which has continued to increase relative to oil production.

In some areas, as the produced water rates increased, injection water quality decreased due to separation equipment limitations and solids buildup in vessels. Expansion to the produced water system has included additional pumps for greater injection capacity as well as additional separation equipment to reduce oil and solids in injection water. The separation equipment expansion includes skim tanks, hydrocyclones and gas flotation cells. Additionally, sand jetting equipment has been installed in some upstream facilities.

Several other significant issues are currently being addressed relative to the PBU produced water handling system, including the impacts of increasing bacteria levels and the disposal capacity for water not reinjected to the producing formation. There is a significant effort underway to determine the impact of bacteria on injection water quality. Potential operating constraints and possible additional regulatory restrictions on produced water disposal are prompting an evaluation of options for handling the produced water not reinjected to the producing formation.

FIELD DESCRIPTION

Location

The Prudhoe Bay oil field is located on the Beaufort Sea coast of Alaska, approximately 250 miles north of the Arctic Circle in the region known as the North Slope (Figure 1). Prudhoe Bay is the largest oil field discovered in North America, with original oil in place of approximately 22 billion barrels. There are a number of other significant oil fields in this region. Three of these fields, Kuparuk, Lisburne and Endicott have in excess of one billion barrels of oil in place.

Produced Water, Edited by J.P. Ray
and F.R. Engelhart, Plenum Press, New York, 1992

Key to the Prudhoe Bay Unit and all other production from the North Slope is the Trans Alaska Pipeline System (TAPS) which carries all the regions production to the Valdez oil loading terminal 800 miles to the south. The 48 inch pipeline has a capacity of approximately two million barrels of oil per day.

Environment

The Prudhoe Bay Unit is located in what is considered to be an environmentally sensitive environment. The entire field area consists of tundra and shallow lakes underlain by permafrost extending to a depth of 1,800 feet. The region has a variety of wildlife, but the waterfowl and caribou which inhabit the area during the summer months are the subject of the highest degree of environmental study.

Operating Conditions

Ambient temperatures can range from less than -60° F (-51° C) to in excess of 80° F (27° C) with an annual mean of approximately 10°F (-12°C). Most process equipment is located inside heated buildings or modules to protect equipment and plant operators from the adverse weather conditions.

Reservoir Description

The Prudhoe Bay Unit produces from the Sadlerochit formation, a prolific sandstone and gravel structure. The Permian-Triassic age Sadlerochit reservoir is approximately 9,000 feet deep in the area of the Prudhoe Bay Unit. There are numerous complications to the Sadlerochit reservoir but generally it is a structural fault trap with a well developed gas cap area.

The initial gas expansion and gravity drainage production mechanisms have been supplemented with gas cap gas reinjection, waterflood, and miscible fluid injection. The miscible injection program involves alternating the injection of water with slugs of miscible injectant generated from the processing of the fields natural gas.

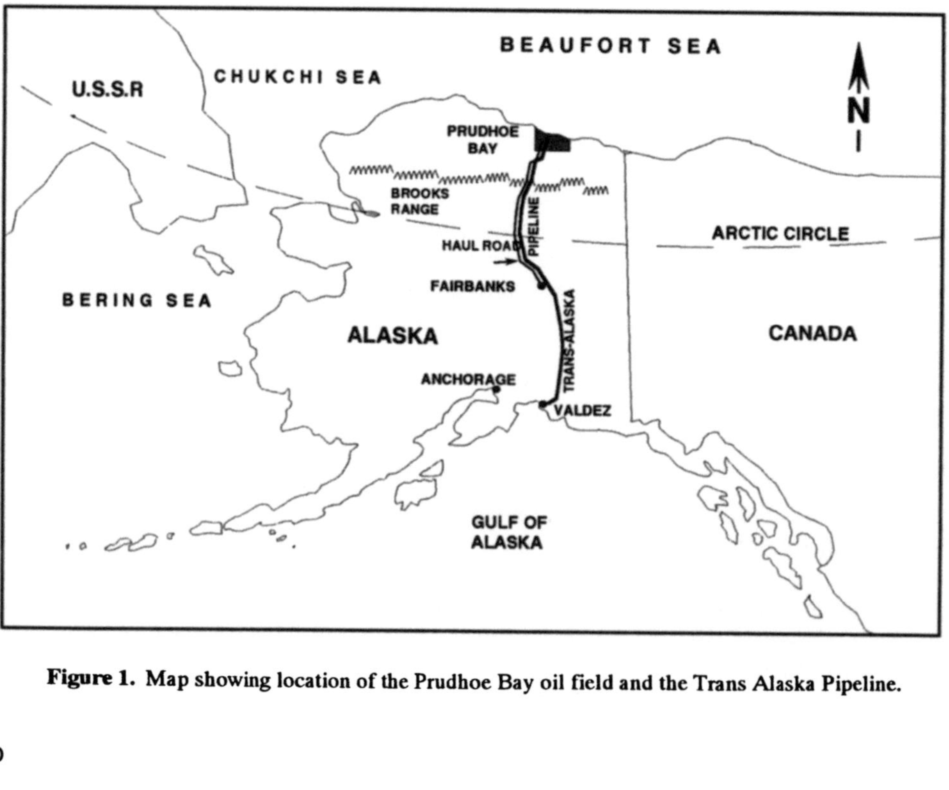

Figure 1. Map showing location of the Prudhoe Bay oil field and the Trans Alaska Pipeline.

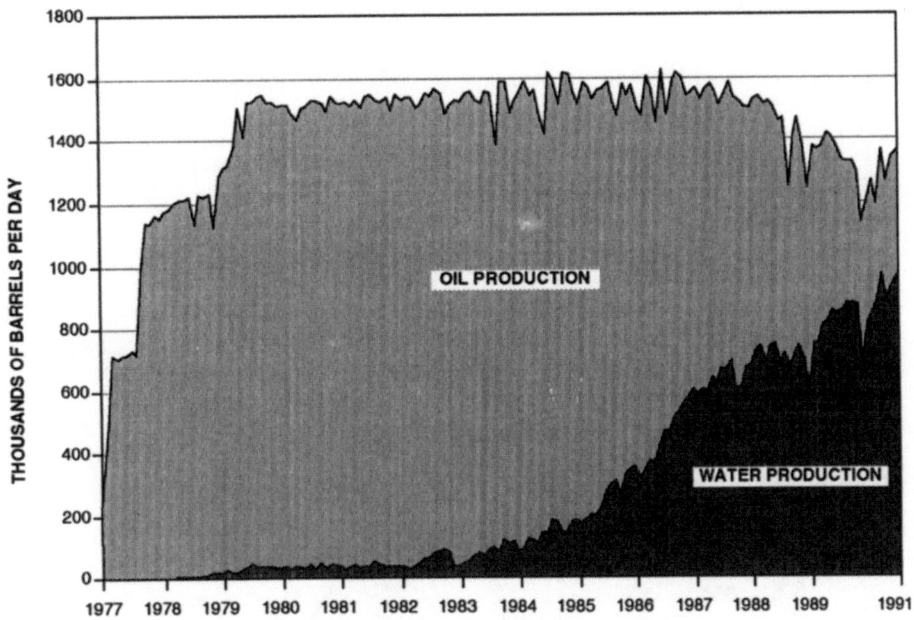

Figure 2. PBU daily oil and water production curves.

Current Production

From the maximum field oil production rate of 1.6 million barrels per day in 1987, production has declined to a level of 1.3 million barrels of oil per day as of October 1991. The associated water and gas production (for 1991) was 1.0 million barrels per day and 4,982 million cubic feet per day respectively. Figure 2 is a field production curve illustrating the daily rates of oil and water production from PBU.

FIELD LAYOUT

The Prudhoe Bay Unit encompasses approximately 350 square miles and is divided into the two operating areas. The western operating area (WOA) is operated by BP Exploration Alaska Incorporated (BPX) and the eastern operating area (EOA) is operated by ARCO Alaska Incorporated (AAI). The nomenclature differs in the two operating areas for facilities having the same functions. There are approximately 1200 wells in the Unit including both production and injection wells. The primary production facilities include well pads(WOA)/drill sites(EOA), gathering centers(WOA)/flow stations(EOA), produced water handling facilities, sea water handling plants, and gas handling facilities (Figure 3).

Wells are drilled from well pads or drill sites which can be considered analogous to offshore platforms in that multiple wells are drilled in proximity and directed significant horizontal distances to intersect target areas. The well pads/drill sites include manifolding which directs production and injection fluids between the individual wells and the separation facilities. There are thirty four well pad/drill sites in the PBU.

Primary Separation Facilities

Across the field there are six primary separation facilities referred to as gathering centers/flow stations. These facilities contain the equipment which separates produced fluids from the well pad/drill sites into the oil, water and gas production streams which exit the gathering

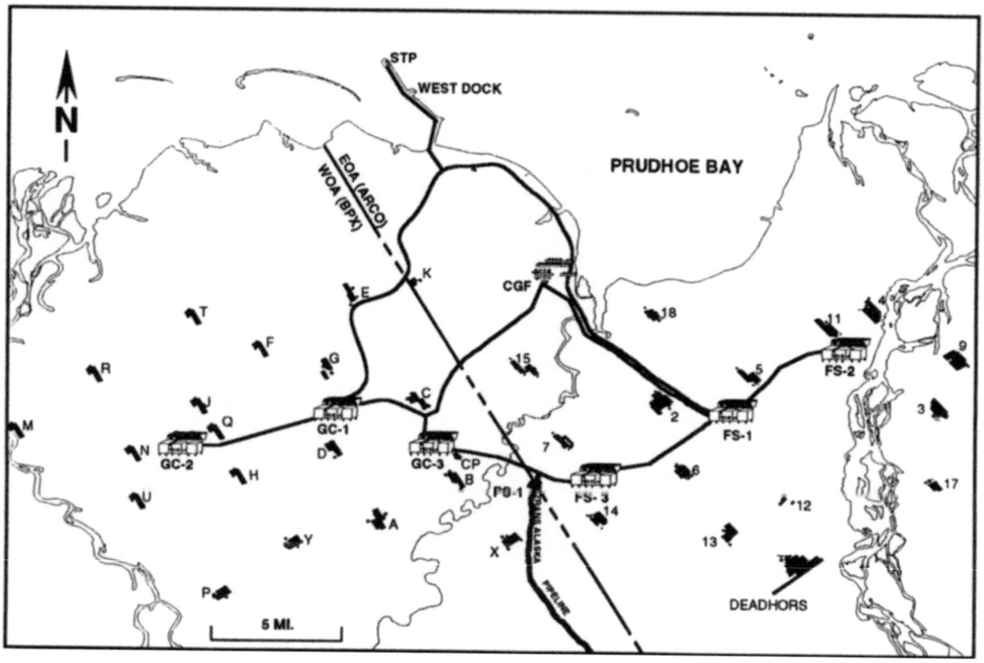

Figure 3. Prudhoe Bay field map showing operating facilities and EOA-WOA boundary.

centers/flow stations. These facilities also contain the equipment necessary for produced water reinjection or disposal.

The dehydrated oil flows to the custody transfer facilities located at Pump Station One, the start of the Trans Alaska Pipeline System. Produced water exiting the gathering centers/flow stations is transferred to facilities for pressurizing and distribution to injection wells for enhanced oil recovery. Gas is transferred to a central gas facility or utilized for artificial lift.

Water Handling Facilities

Located on each side of the Prudhoe Bay Field are facilities which increase the pressure on the produced water following the separation process. The pressurized produced water is distributed to well pads/drill sites where it is manifolded to individual wells for reinjection.

Seawater is taken from the Beaufort Sea at the Seawater Treatment Plant located offshore north of the Prudhoe Bay field. The treated seawater is transferred to facilities on each side of the field for pressurizing and distribution to injection wells. Seawater treatment consists primarily of removing the naturally occurring solids and deoxygenation.

Gas Handling Facilities

The gas produced from the Prudhoe Bay field which is not used for artificial lift or burned for fuel, travels to the Central Gas Facility where it is processed to extract natural gas liquids and miscible fluid for enhanced oil recovery. The natural gas liquids are combined with the produced oil prior to custody transfer at Pump Station One. The remaining gas exiting the Central Gas Facility is pressurized at the Central Compression Plant and reinjected to the fields gas cap.

PRODUCED WATER SYSTEM DEVELOPMENT CHRONOLOGY

The Prudhoe Bay field was discovered in 1968 but the first production was not delivered until completion of the Trans Alaskan Pipeline in the summer of 1977. Initial oil production included minimal associated produced water which was disposed of to a shallow "water bearing zone" referred to as the Cretaceous interval.

In the mid-eighties water injection was initiated in select areas of the field for secondary recovery and pressure maintenance. Significant volumes of seawater were necessary initially to provide adequate injection volumes. Produced water was used to supplement seawater injection. Cretaceous zone disposal of produced water was employed to handle operational swings in the produced water handling system and to avoid reinjection of poor quality water which could result in damage to injection wells.

As the Prudhoe Bay waterflood matured and produced water rates increased, injection wells were converted from seawater injection to produced water injection. The key issue associated with the conversion from seawater to produced water is the potential reduction in injectivity or damage which can occur in injection wells due to plugging material injected with produced water.

The increasing rates of water production required the installation of additional separation equipment to provide adequate separation of oil, solids and water. This separation process is key to oil sales as operators must maintain water-in-oil levels below the pipeline specification. The separation of oil and solids from the downstream produced water is important to minimize the level of oil and solids reinjected with produced water.

In addition to separation equipment it has also been necessary to add facilities to transfer and pressurize produced water following the separation process. Natural gas fueled turbines power pumps which deliver high volumes of produced water at up to 2600 psi for reinjection.

WATER SYSTEM FACILITIES OVERVIEW

The original facilities installed at Prudhoe Bay were designed to handle the low volumes of initial water production. Prior to the higher volumes of water associated with waterflooding produced water handling consisted of primary oil-water separation and disposal by injection to the shallow Cretaceous interval.

During 1990 and 1991 major water treatment facilities were commissioned at four of the six separation centers, servicing the waterflood areas of the field. Facilities were justified based on reserve benefits at risk due to injection well plugging. Water specification targets were determined to be 50 ppm oil in water and 10 mg/l suspended solids. The largest of the new treatment facilities is installed at Flow Station 2 where produced water rates are projected to peak at 650 MBWPD. Primary treatment utilizes new skim tanks followed by mechanical induced-gas flotation cells which utilize small gas bubbles to facilitate oil and solid separation from produced water.

The next largest facilities are at Gathering Centers 2 and 3 where water rates are expected to peak at 330 and 230 MBWPD respectively. Expanded skim tank capacity coupled with new gas flotation cells provides treatment capabilities. Figure 4 is a summary level process flow diagram illustrating the induced-gas flotation cell system design.

The most recent treatment facility installation is at Flow Station 3. It is designed to handle anticipated peak rates of 240 MBWPD. A hydrocyclone system was installed which uses centripetal force to encourage oil-water separation. Figure 5 presents a simplified process flow diagram for the hydrocyclone system. A two stage hydrocyclone treatment philosophy was adopted to meet the targeted water specifications. Water from the slug catchers flows through the first stage cyclones, into the skim tanks and on to the second stage cyclones prior to reinjection.

All separation centers have the capability to divert produced water to the Cretaceous disposal system. Each facility location has at least one disposal well which is capable of between 30 to 80 MBWPD each. Ideally, as produced water rates increase, seawater injection wells are gradually converted to produced water injection, allowing Cretaceous disposal to maintain some minimum swing or surge rate. In the interest of effective resource management some seawater injection is maintained in lieu of produced water reinjection in areas where produced water reinjection may reduce waterflood recovery.

Developing the relationship of produced water quality to waterflood efficiency is one of the key areas of current study. Degradation of water quality increases as water rates increase due to a number of variables, including facility retention time, chemical treatment and bacterial activity.

PRODUCED WATER MANAGEMENT ISSUES

Prudhoe Bay secondary recovery projects are experiencing typical problems associated with major waterflood programs, ones that are not new to the oil industry. What is atypical about Prudhoe Bay are the solutions. High production and injection rates coupled with the arctic environment and logistic problems make for very costly solutions to "typical" oil field problems. Sensitive environmental issues further complicate the problems and strongly impact the solutions.

Injection Well Performance

The potential negative impacts of poor quality injection water is being actively studied across the Prudhoe Bay field due to the potential for reduced overall injection and vertical conformance. Reduced injection capacity and loss of vertical conformance can both negatively impact waterflood efficiency.

Vertical conformance is proving to be a major issue in management of the waterflood programs, particularly in the multi-zone flood areas. Based on the range of permeabilities that exist within a given zone and between different zones, vertical conformance problems exist naturally. Poor water quality tends to compound the problem. To further complicate the issue one zone, a conglomerate, acts as a thief zone and is virtually undamageable. Prior to installation of water treatment facilities injection well treatment programs had limited short term success and minimal long term benefit since water quality could not be controlled or improved. Installation of water treatment facilities provides renewed interest. Work is ongoing to identify damaged injection wells and implement a treatment program to restore injectivity and vertical conformance.

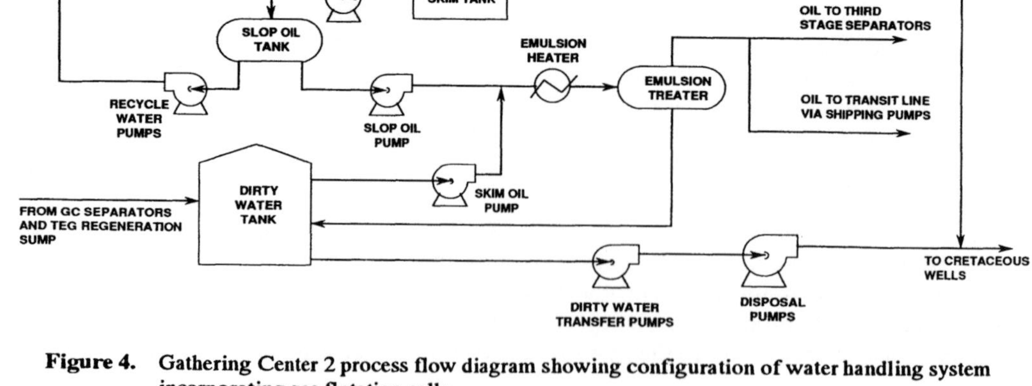

Figure 4. Gathering Center 2 process flow diagram showing configuration of water handling system incorporating gas flotation cells.

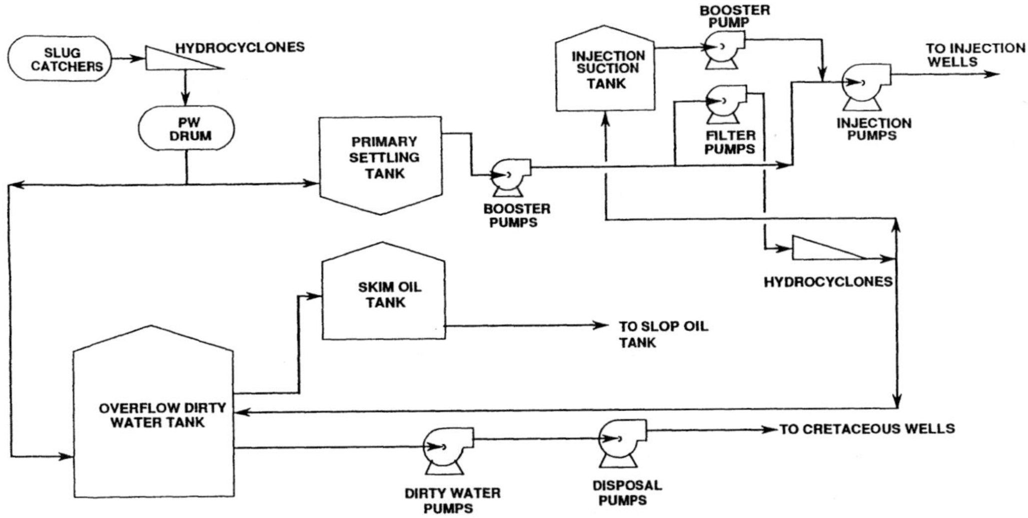

Figure 5. Flow Station 3 process flow diagram showing configuration of water handling system incorporating hydrocyclones.

Development of a successful treatment program will take into account the key variables that contribute to injection well plugging. An aggressive sampling and analysis program is underway to identify the composition and origin of the plugging material. Surface and downhole samples of a black, sludgy substance have been obtained. Buildup is so extreme in some wells that wireline activities are severely hampered. Bacterial activity and chemical compatibility problems are two major items suspected of contributing to this sludge presence.

Another complicating factor in evaluating injection well performance is the presence of thermal and hydraulic fractures and their impact on vertical conformance and reserve benefits. Seawater, injected at 80° F into a reservoir of 180° F to 220° F, has a significant impact on the fracture gradient. Produced water is injected at temperatures in the range of 160°F and realizes little benefit from thermal fracturing. Immediate injectivity increases are realized when wells are converted from produced water to seawater due to thermal fractures induced from the difference in water temperatures.

Water Treatment and Reinjection

As with all waterflood projects, a decision must be made on disposition of produced water. The most viable options for Prudhoe Bay are reinjection to the Sadlerochit formation or disposal to the shallow water bearing Cretaceous formation. Reinjection in the waterflood area has been considered the most feasible, cost effective and environmentally sound option. The alternative of long term Cretaceous disposal is potentially unattractive from an economic and environmental perspective.

The issue of produced water treatment has been considered for over a decade. Injectivity tests conducted in the early 1980's were inconclusive as to the effects of increased oil and suspended solids on injection well performance. In the absence of clearly identified incentives the major capital investment for produced water treatment was deferred until impacts on injection well performance could be demonstrated.

Water quality gradually deteriorated at the separation centers servicing the waterflood areas. Water quality worsened as water handling rates increased primarily due to reduced retention time in separation equipment. The decrease in water quality became most notable at FS-2 which handled production from the largest waterflood area. Oil in water concentrations of 1,000 ppm to 2,000 ppm were typical and frequent excursions as high as 5,000 ppm were experienced. Declines in injection well performance were observed when wells were converted from seawater injection to produced water injection.

During the later half of the 1980s, the issue of water quality was again addressed in an attempt to determine long term effects of injecting poor quality water. Injection well performance was monitored in several Prudhoe Bay wells after their conversion from seawater to produced water service. Results of these tests indicated reserve losses could be expected at oil in water concentrations greater than 25-50 ppm and suspended solids levels in excess of 10 mg/l. Reserve loss estimates were able to justify the high cost of water treatment facilities.

Additional Produced Water Treatment

As previously discussed, four of the six major separation centers have produced water treatment facilities and are using produced water for reinjection in waterflood project areas. The remaining two facilities, Flow Station 1 and Gathering Center 1, employ Cretaceous disposal as their primary means of handling produced water. The issue at hand regarding these facilities is justification and timing for produced water treatment equipment.

These two facilities service primarily the main gravity drainage area of the field. At this time only a handful of waterflood patterns exist at each facility. Field depletion plans may eventually include a major waterflood project for this area which may provide the justification for additional separation equipment.

Water quality specifications will be addressed in an overall evaluation of injection well performance. Injection well performance benefits at the current water specifications of 50 ppm oil in water and 10 mg/l suspended solids have not yet been demonstrated. There are plans to assess the benefits of the produced water treatment facilities recently installed at Prudhoe Bay. Injection well monitoring programs are under development to investigate injection well performance with treated produced water. The issue of thermal fracturing will also be investigated under this program. Results of this evaluation will provide the basis for refining water quality requirements and understanding the reservoir benefits of treatment.

Alternatives to costly treatment equipment will be investigated. Redistributing the water to one of the other Gathering Centers or Flow Stations may be more cost effective, given the uncertainty of a major waterflood program.

Bacteria Related Issues

There is increasing evidence that bacterial activity present in the Prudhoe Bay surface facilities contributes to operational problems and may potentially impact waterflood recovery. Bacteria related impacts on water quality, facility performance, corrosion and hydrogen sulfide generation are the key areas of concern.

Some impacts of bacterial activity are well documented while other aspects are very difficult to grasp. Microbiologically induced corrosion accompanied by hydrogen sulfide and iron sulfide generation are common products of bacterial activity and have been observed in the major waterflood areas of Prudhoe Bay. However, the possible relationship between high levels of bacteria in the production facilities and reduced produced injection water quality is not clearly understood. It is reasonable to assume that high levels of viable bacteria and other associated biomass contribute to reduced water quality by stabilizing emulsions and increasing oil, solids and sludge accumulation in the system.

The primary bacteria control option under investigation is biocide treatment. An effective biocide program must be able to address both surface and downhole issues. Field trial programs assessing biocide effectiveness and benefits are being planned for implementation at several of the main separation centers.

Alternative bacterial control options are also under investigation. Studies are planned to assess the impacts of nutrient levels and temperature on bacteria growth and the economic feasibility of implementing this type of control.

Produced Water Composition

The composition of produced water has changed as the PBU waterflood has matured. Prior to waterflooding, the small volumes of produced water were connate or original aquifer water. With the breakthrough of injected seawater and increased produced water volumes the composition of produced water has expectedly shifted towards seawater composition. The scaling tendency for barium sulfate scale has increased with the modified composition of produced water as mixing of connate and seawater continues in the reservoir and in surface facilities.

The shift in produced water composition has also had an impact on the development of bacteria populations throughout the producing system. Neither connate or sea water contain all the components conducive to rapid bacteria growth. However, when the waters are mixed the result is an attractive medium for bacteria growth. A significant effort is underway to monitor the levels of bacteria in the producing system as well as determining the production impacts and options for control.

Chemical Incompatibility

An operational issue often overlooked is one of chemical incompatibility and its impact on water quality. Water breakthrough has exacerbated operational problems such as emulsions, increases in corrosion rates, bacterial activity and the deposition of scale and solids. Chemical treatment is used to combat or mitigate the problems. Emulsion breakers, scale inhibitors, corrosion inhibitors and biocides are either injected continuously or periodically throughout the system. The potential operational impacts of this chemical soup are currently being studied.

Understanding all the variables that impact produced water quality is critical to the ongoing evaluations of Prudhoe Bay produced water handling. Chemical incompatibility is suspected of playing a key role in some of the major water quality issues.

Chemical incompatibility contributes to emulsion stability problems and decreased efficiency of separation equipment. Additionally, the chemicals used throughout the system are suspected of adding nutrients which promote bacteria growth. It is likely the chemical additives also contribute to the accumulation of the sludgy substance in the produced water system.

Test programs are underway to specifically define chemical compatibility problems at two of the separation centers. Monitoring the impacts of chemical additives in the produced water system was incorporated in the performance test program for the gas flotation cells commissioned in 1991.

Water Disposal

As discussed previously, water disposal to the shallow Cretaceous formation has been ongoing at Prudhoe Bay since the field started production.

Given the high cost of produced water handling and the potential reserves at risk due to reinjection of produced water the alternatives to produced water treatment and reinjection are under continual review. The alternative of flooding with seawater requires a long term disposal option for produced water. Disposal of high volumes of produced water is not considered feasible for a number of reasons.

The primary obstacle to any produced water handling strategy calling for disposal of large volumes is the uncertainty of future regulations which could prohibit such practices or make them cost prohibitive. Regulatory issues regarding long term Cretaceous disposal may play a significant part in the water handling decisions. In the absence of major waterflood expansion projects the potential justification for additional water handling expansions may be based primarily on the need to minimize Cretaceous disposal to accommodate regulatory changes.

SUMMARY

The first fourteen years of production from the Prudhoe Bay Unit have seen significant change in the volumes, composition and relative importance of water produced in association with oil and gas. There is a significant effort by the operators to anticipate future water handling requirements in the interest of making cost effective decisions which are operationally and environmentally sound.

One issue fundamental to the consideration of produced water handling requirements is the relationship between produced water quality and reserve recovery from the PBU water flood. The current effort to address this issue will further develop an understanding of the impacts of oil and solids levels in produced water on injection well performance.

The potential impacts of produced water quality on waterflood recovery will along with potential regulatory changes influence future decisions regarding the level of produced water disposal and seawater injection.

As the PBU waterflood matures and water rates continue to increase it is likely that additional facility modifications will be required. The modifications may be additional equipment for oil and water separation or may provide the ability for greater flexibility in routing produced water to facilities with available capacity.

Higher rates and varying compositions may also require some modifications to operating practices. The potential impacts of chemical incompatibility on produced water quality are being studied along with the bacteria control needs.

Note:

Techniques and or conclusions are those of the authors and are not necessarily shared by the other Prudhoe Bay Unit working interest owners.

PRODUCTION WATER/WASTE TREATMENT AND SITE REMEDIATION TECHNOLOGY DATABASE FOR THE NATURAL GAS INDUSTRY

J. T. Tallon[1], J. P. Fillo[1] and J. M. Evans[2]

[1] ENSR Consulting and Engineering
Pittsburgh, Pennsylvania
USA

[2] Gas Research Institute
Chicago, Illinois
USA

INTRODUCTION

Natural gas industry operations generate various production waters and wastes. These materials are generally regulated and managed in accordance with the specific disposal method used. Depending upon the manner in which production waters/wastes currently are and have been managed, treatment may be required, prior to ultimate discharge or disposal, to comply with applicable regulations. Instances have also occurred in the past where these waters and wastes were not managed in accordance with current required practices. In these instances, what is now considered to be improper management of these releases could have resulted in chemical contamination of surface and subsurface soils and/or ground water. Thus, remediation could be required for the resulting chemical contamination.

In order to help the natural gas industry address their needs for ongoing production water/waste management and soil/ground water remediation, the Gas Research Institute (GRI) sponsored development of a treatment/remediation technology computer database. The intent of this effort was to consolidate in one place data on the treatment of constituents and streams/media relevant to the natural gas industry, as well as the associated costs for achieving treatment. To ensure the quality of the database, only data from published reports and literature were utilized and referenced in the database. The primary product of this research was a user-friendly database compiled with Paradox 2[1] database software and a GRI topical report.

The purpose of this database is to provide natural gas industry environmental managers and others with a means to rapidly identify and screen potential solutions to treatment/remediation problems. Although there is much published, reviewing the many reference sources to target applicable technologies can be an arduous task. This database provides environmental managers with a consolidated source of treatment/remediation performance and economic data, and thus facilitates the initial screening of relevant technologies on a constituent-, matrix-, and technology-specific basis. Results from such a screening evaluation can then be used to target more in-depth engineering and treatability investigations on a technology-specific basis.

---

[1] Paradox is a trademark of Borland International, Inc.

Produced Water, Edited by J.P. Ray
and F.R. Engelhart, Plenum Press, New York, 1992

This paper describes the scope of the database and provides an overview of its contents, capabilities, and use. It summarizes the technologies, waste streams, and constituents for which data are present in the database, and provides basic information on use of the database. The paper also provides the information necessary to obtain the database and associated topical report. Note that the report contains detailed discussions of the scope and approach to this work, the database structure and its operation, case studies, and contains printed copy of all entries in the database (Tallon et al., 1990).

SCOPE OF THE DATABASE

This database focuses on constituents in typical production water/waste streams from natural gas industry operations, and management options for remediating soil and ground water contamination resulting from former waste disposal practices. Production waters/wastes associated with industry operations and former disposal practices were initially identified, based on data developed in GRI and other programs, to guide the investigation. Constituents of interest related to the various production water/waste streams, along with potential treatment options, were subsequently identified to focus the evaluation. The constituents, streams, and technologies targeted for this database were reviewed with GRI and gas industry representatives prior to the initiation of this effort in order to ensure accuracy and completeness.

Production Water/Waste Streams

Natural gas industry production water/waste streams considered in this evaluation are listed in Table 1. Twenty-seven water/waste streams were identified from published literature and GRI project efforts. These streams provided a firm basis for defining potential constituents of interest and treatment/remediation technologies.

Constituents of Interest

Chemical constituents that may be present in natural gas industry waters and wastes and may require remediation in environmental media were identified from published literature and ongoing GRI programs. Constituents were selected for consideration if they were: (1) known to be present in production water/waste streams and typically required treatment to ensure compliance with environmental regulations; or (2) identified at elevated concentrations during prior analytical characterization and were, therefore, suspected to require treatment prior to final disposition. The target constituents are listed in Table 2. Materials that may have been used previously at a facility, such as diesel oil, were also included because they may require remediation if found in soil and/or ground water.

Treatment Technologies

Candidate treatment technologies fall into four categories and include:

- physical treatment, which includes methods that rely on physical means such as differences in specific gravity, vapor pressure, adsorption, particle size, etc.;
- chemical treatment, which includes methods that chemically alter contaminants to produce inert constituents, or reduce their toxicity or mobility;
- biological treatment, which includes treatment procedures that use microorganisms to biodegrade contaminants; and
- thermal treatment, which includes processes that employ elevated temperatures to destroy or remove contaminants.

Table 1. Production water/waste streams associated with natural gas industry facility types

	Natural Gas Industry Facility Type			
Production Water/Waste Stream	Field Production/ Gathering	Processing Conditioning	Transmission	Underground Storage
Production Water	•			•
Dehydration Water	•			•
Flare Knockout Water	•			
Surface Runoff	•			
Cooling Tower Blowdown		•	•	•
Hyrostatic Test Water	•		•	
Pipeline Condensate and Drips	•		•	
Claus Waste Heat Boiler Blowdown		•		
Boiler Blowdown		•	•	•
Scrubber Water/Waste		•	•	•
Washdown Water/Floor Drains	•	•	•	•
Water Softner Backwash/Blowdown		•	•	•
Condensed Stripping Steam		•		
Pipeline Cleaning Receiver Wastes[(a)]	•	•	•	•
Acid Gas Removal Blowdown (Amines)	•	•		•
Waste Glycol	•	•		•
NGL Separation & Processing Wastes		•		
Spent Lube Oils	•	•	•	•
Sulfur Recovery Unit Wastes[(a)]	•	•		
Workover Fluids	•			
Process Filters[(a)]	•	•		•
Gas Filters[(a)]	•	•	•	•
Waste Iron Sponge[(a)]	•			
Spent Catalyst[(a)]		•		•
Spent Dehydration Media[(a)]		•		•
Sludge/Tank Bottoms[(a)]	•	•	•	•
Drilling Muds[(a)]	•			

[(a)]Solid Wastes

Table 2. Production water/waste constituents of interest

Conventional Parameters	
pH	Total Orgainic Carbon (TOC)
Conductivity	Chemical Oxygen Demand (COD)
Total Dissolved Solids (TDS)	Biochemical Oxygen Demand (BOD)
Total Suspended Solids (TSS)	Oil and Grease

Inorganic Constituents				
Metals			Other Inorganics	
Arsenic	Lead	Sodium	Ammonia	Phosphate
Barium	Lithium	Strontium	Chloride	Sulfate
Cadmium	Magnesium	Vanadium	Nitrate	Sulfide
Calcium	Mercury	Zinc		
Chromium	Nickel			
Copper	Potassium			
Iron	Selenium			

Organic Constituents				
Volatile Aromatics	Aliphatic Hydrocarbons		Halogenated Hydrocarbons	
Benzene	n-Decane (C10)	n-Docosane (C22)	Chlorobenzene	Tetrachloroethylene
Toluene	n-Dodecane (C12)	n-Tetracosane (C24)	Dichloroethane	Trichlororthylene
Xylenes	n-Tetradecane (C14)	n-Hexacosane (C26)	Dichloroethylene	1,1,1- Trichloroethane
Ethylbenzene	n-Hexadecane (C16)	n-Octacosane (C28)	Methylene Chloride	
	n-Octadecane (C18)n-Triacotane (C30)			
	n-Eicosane (C20)			

Other Hydrocarbons	Phenolics	Polynuclear Aromatic Hydrocarbons	
Acetone	Phenol	Acenaphthene	Phenanthrene
Ketones (MEK, 2-Butanone)	2-Methylphenol	Anthracene	Pyrene
	4-Methylphenol	Benzo(a)pyrene	2-Methylnaphthalene
	2,4-Dimethylphenol	Fluorene	Naphthalene

Other Organics		
Anthraquinone	Diesel Oil	Amines (DEA,MEA,MDEA)
Kerosene	Polychlorinated Biphenyls (PCB)	Mercaptans
Glycols (EG,TEG)	Disulfonic Acid	Used Oil
Methanol		Gasoline

Table 3. Water/waste treatment technologies

Treatment Technologies	Production Water/Waste		Site Remediation			
			Aqueous Liquids		Solids/Semisolids	
	Aqueous Liquids	Solids/ Semisolids	In Situ	Ex Situ	In Situ	Ex Situ
Physical Treatment						
Coagulation/Flocculation	•			•		
Oil-Water Separation	•			•		
Air Stripping	•			•		
Steam Stripping	•			•		
Carbon Adsorption	•			•		
Filtration	•			•		
Reverse Osmosis	•			•		
Sedimentation	•			•		
Evaporation	•			•		
Distillation	•					
Solvent Extraction		•				•
Freeze Crystallization	•					
Centrifugation		•				•
Sonic Treatment	•			•		
Soil Venting					•	•
Chemical Treatment						
Neutralization	•			•		
Precipitation	•			•		
Ion Exchange	•			•		
Chemical Oxidation	•			•		
Photolysis (Ultraviolet Light)	•			•		
Wet Air Oxidation	•			•		
Stabilization/Solidification		•			•	•
Acidification with Stripping	•					
Dechlorination					•	•
Soil Washing					•	•
Biological Treatment						
Activated Sludge	•			•		
Activated Sludge with Powdered Activated Carbon	•			•		
Aeration Tank	•			•		
Fixed Film	•			•		
Fluidized Bed	•			•		
Rock Reed Filter	•			•		
Sequencing Batch Reactor	•			•		
Trickling Filter	•			•		•
Composting			•		•	
In Situ Biodegradation					•	•
Land Treatment	•	•				
Liquid/Solid Systems	•			•		
Anaerobic				•		
Thermal Treatment						
Incineration		•				•
Thermal Desorption		•				•
Vitrification						•

Table 3 lists relevant technologies for each treatment category. This table notes the potential technology application for production water/waste and site remediation as it relates to the management of aqueous liquids and solids/semi-solids that may be encountered. Ultimately, data from 44 treatment technologies were compiled in the database.

Database PROGRAM

Database Capabilities

Treatment technology performance and economic data were compiled using the computer database program Paradox2. This program can be used to search for and display any number of keywords, parameters, and descriptive fields. Paradox was selected because it is a user-friendly relational database that can manage and display the large fields of information required for this application. Essential features of the program include:

- ability to run on IBM and compatible computer systems;
- rapid search and output capabilities;
- easy access to data from one or multiple tables;
- ability to search one or multiple fields without the need for exact identification; and
- display and output capabilities for any number of field headings.

These features allow quick screening and searching through the database to prepare reports tailored to specific technology interests, such as identifying options for treating a particular constituent or compound.

System Requirements

The Paradox2 database system can operate on an IBM XT, AT, PC, and Personal System-2 computer or a 100 percent compatible system. The program requires a minimum of 512 K bytes of available random access memory (RAM) and DOS version 3.1 or a more recent version. Paradox2 can be run from a two floppy disk drive system. However, due to the large data files in the GRI database, both a hard disk and a floppy disk drive are required for this application. The hard disk should have a minimum of 3 MB of available storage for the program and data files.

Enhanced options, while not necessary, allow the database to run more efficiently. These include an 80286 or 80386 microprocessor, 640 K bytes of RAM and an expanded memory adapter. These options greatly enhance performance because Paradox 2 takes advantage of RAM by allocating a temporary storage area and information cache in the expanded memory. Because RAM access is much faster than disk access, the more RAM available, the better.

The Paradox2 program can also be run from a network system with one of the following configurations:

- Novell with Novell Advanced Netware 2.0A;
- 3Com 3Plus with 3Com 3Plus operating system 1.0;
- IBM Token Ring or PC Network with IBM PC Local Area, Network Program 1.12;
- Torus Tapestry Network Version 1.4;
- AT&T Starlan Network with AT&T P6300 1.1; or
- a 100 percent compatible network with one of the above.

To operate Paradox2™ from a network, the system must use DOS 3.1 or a later version.

LEVEL I DATA

Information in the database is presented in two levels of detail. Level I presents a general overview of each technology, and Level II presents performance and economic information for specific technology applications.

Level I overviews each technology and describes their general design, performance characteristics, and application. Information in the Level I assessment includes the following technology-specific information:

- a summary description;
- pertinent design parameters;
- traditional performance characteristics;
- constituent removal ranges;
- relative economics; and
- key regulatory and environmental health and safety (EH&S) issues.

Table 4 summarizes the specific information presented under each data field in the Level I technology overview, with examples of field-specific information.

LEVEL II DATA

The Level II evaluation is divided into two sections: performance data and economics data. These two sections are configured to provide data related to specific applications of each technology. Information in Level II provides published performance data from the treatment of gas industry production water/waste streams or other relevant streams/matrices and constituents. Similarly, economic data were compiled from published reference sources for natural gas industry or other relevant applications.

Approximately 730 treatment/remediation performance entries are included in the database. Performance information included in each data field is listed in Table 5. To assist in evaluation, treatment effectiveness, the operating rate, influent and effluent concentrations, and percent removal for each constituent are included, where available. As noted in Table 5, additional information is provided for water/waste source, stream type, pretreatment, literature reference, and other performance-related comments.

Approximately 240 entries for capital, operating and/or annualized costs are included in the database. Specific costs cited in the database were drawn from published case studies where the technology was used alone, or in conjunction with other treatment methods.

The contents included in each database field are described in Table 6. Expenditures are itemized by initial capital investment; operating, utility, chemical, and annualized costs; and year in which cost data were developed. Generally, these data are compiled as reported in the specific literature reference.

DATA REFERENCING

All data entries included in the Level I and II evaluations are referenced to the specific literature source from which they were obtained. The documents included are referenced by code numbers or STIC numbers to provide a concise referencing method. The STIC numbers used in the Natural Gas Industry Environmental Health and Safety (EH&S) Information System (Muenchow et al., 1991) are also used to reference documents referred to in the Level I and Level II evaluations. The Natural Gas Industry EH&S Information System includes more than 900 (approximately 500 added during this evaluation) documents related specifically to the natural gas industry and water/waste treatment technologies.

Table 4. Level I treatment technology information

Data Field	Contents/Text
Process Name	Name of the Technology
Treatment Category	Chemical, Physical, Biological, Thermal, Isolation
Description	
Process	Describe the unit operations employed and other relevant information describing the process
Operation & Design Parameters	Identifies key operational and design variables such as flow, temperature, contact time, etc.
Equipment	Identifies major process equipment.
Chemicals	Identifies chemicals used.
Developmental Status	Bench-scale, pilot, demonstration, commercial.
Industry Application	Lists industry application (i.e., general industry, natural gas industry, steel manufacturing, etc.)
Performance	
Streams	Lists waste or water streams that could potentially be treated using this technology.
Constituents	Lists constituents of interest that are treated by this technology.
Effectiveness	Identifies process limitations, degree to which treatment is irreversible, etc.
Reliability	Identifies general performance record, ability to operate technology (e.g., need for highly specialized personnel),etc.
Economics	
Capital Cost	Cost identified by relative ranking of high, medium, or low.
Operating Cost	Cost identified by relative ranking of high, medium, or low.
Annualized Cost	Cost identified by relative ranking of high, medium, or low.
Regulatory/EH&S Issues	
Air	Identifies (1) the need for additional permitting, monitoring, and treatment, and (2) secondary environmental issues, such as types and quantities of residual materials (e.g., by-product waste requiring treatment and disposal), and the degree of hazard associated with any residues.
Water	As above.
Waste	As above.
References	Lists key references used according to the Natural Gas Industry EH&S Information System STIC Numbers and other sources of data that may be helpful to understand the use of the technology.
Comments	Lists pertinent comments regarding the selection or use of this technogy.

Table 5. Level II treatment technology performance information

Data Field	Contents/Text
Process Name	Name of the Technology
Developmental Scale	Bench, Pilot, Demonstration, Commercial
Operating Rate	Process operating rate at the time the performance data was obtained.
Operating Units	Operating units (i.e., GPM, tons/day, etc.).
Water/Waste Source	Industrial source of the water/waste stream (i.e., natural gas industry, manufactured gas, etc.).
Stream Type	Identifies the name of the stream treated. If applicable, identifies associated water/waste streams that were combined to form the stream (i.e., 75% production waste, 10% dehydration water, and 15% cooling water blowdown).
Pretreatment	Lists the order of any pretreatment steps (i.e., coagulation/flocculation followed by sedimentation) prior to treatment by this technology.
Constituent	Lists the constituent treated or managed by the technology. Each constituent comprises an entry or row (i.e., record) in the data table.

Table 5. Level II treatment technology performance information. (Continued)

Data Field	Contents/Text
Matrix	Unique factors related to the water/waste that may affect treatment.
Influent Concentration	Identifies the initial concentration of the constituent prior to treatment by the technology (i.e., after pretreatment steps).
Influent Units	Concentration units (i.e., mg/l, g/kg, etc.).
Effluent Concentration	The concentration after treatment.
Effluent Units	The discharge or effluent concentration units.
% Removal	Lists the removal efficiency achieved through treatment.
Reference	Source of information which is referenced according to the Natural Gas Industry EH&S Information system STIC Number.
Comments	Pertinent comments regarding the selection or use of technology in the specific application cited.

DATABASE FILES AND REPORTS

Level I includes a table that describes the 44 identified technologies, introduces the user to the technologies, and provides summary descriptions, design parameters, and other pertinent information. Level II includes two tables, with one containing performance data and the other containing economic data. The file names for Level I and Level II tables are:

Table		File Name
Level I	Description	GRI1
Level II	Performance	GRI2P
Level II	Economics	GRI2E

There are many ways to manipulate Paradox2 data to obtain information. However, the more specific your request for information is, the less information you will retrieve. Therefore, when searching concentrations or removal efficiencies, ranges are recommended, rather than specific values, especially in the initial search. A series of report formats are pre-defined in this package for the three tables. Level I has one standard report format. Tables GRI2P and GRI2E both have one standard report format and three additional formatted reports.

Nine predesigned mini reports have been developed for use with the database for printing performance and economics data. The mini reports, which correspond to Tables GRI1, GRI2P, and GRI2E, are subsequently listed.

The mini reports were developed to contain specific information that would typically be asked for in a database query. A list of the mini reports and their appropriate fields for Table

Table	Corresponding Mini Reports
GRI1	Standard
GRI2P	Const Operate Source Standard
GRI2E	Annual Economic Mechanic Standard

Table 6. Level II treatment technology economic information

Data Field	Contents/Text
Process Name	Name of the Technology
Developmental Scale	Pilot, Demonstration, Commercial, Conceptual
Operating Rate	Operating rate used to develop the economic data.
Operating Units	Operating units (i.e., GPM, tons/day, etc.).
Water/Waste Source	See descriptions in Table 5.
Stream Type	See descriptions in Table 5.
Capital Cost, Units, Yr.	Lists the capital cost, units (i.e., $XXXX/10 MGD plant and the year in which the cost was calculated.
Equipment	Lists the equipment included in the capital cost estimate.
Operating Cost, Units, Yr.	Process operating cost, units (i.e., $XXX/1,000 gallons per 10 MGD plant), and the year in which the cost was calculated.
Utility Cost, Units, Yr.	Process utility cost, units (i.e., $XXX/1,000 gallons per 10 MGD plant), and the year in which the cost was calculated.
Utilities	Lists the utilities included in the cost.
Chemical Cost, Units, Yr.	Process chemical cost, units (i.e., $XXX/1,000 gallons per 10 MGD plant), and the year in which the cost was calculated.
Chemicals	Lists the chemicals included in the cost.
Annualized Cost, Units, Yr.	Annualized cost, units (i.e., $XXX/1,000 gallons per 10 MGD plant), and the year in which the cost was calculated.
Annualized Cost Basis	Lists critical assumptions related to the cost developed, including project life, interest rate, depreciation schedule, etc.
References	Identifies the source of information which is referenced according to the Natural Gas Industry EH&S Information System STIC Number.
Comments	Other pertinent economic comments.

GRI2P and Table GRI2E appears in Tables 7 and 8, respectively. The standard reports for Table GRI1 consist of all the fields.

DATABASE UPDATE

Currently, the database is being revised to significantly expand and update its information content. The list of technologies will be expanded to include data from emerging and innovative technologies. Moreover, constituents of interest will be added to the database to address new parameters coming under regulatory consideration within the natural gas industry and present in additional streams relevant to natural gas industry operations. As part of this effort, current technology performance and economic data are being solicited from individuals and research organizations, along with suggestions for improving the database. The revision and expansion of the database is planned for completion in 1993.

Table 7. Corresponding Fields for GRI2P Mini Reports

	GRI2P Mini Reports			
Fields	Standard	Const	Operate	Source
Process Name	•	•	•	•
Developmental Status	•	•	•	•
Operating Rate	•		•	

Table 7. Corresponding Fields for GRI2P Mini Reports (Continued)

	GRI2P Mini Reports			
Fields	Standard	Const	Operate	Source
Operating Units	•		•	
Water/Waste Source	•			•
Stream Type	•			•
Pretreatment	•		•	
Constituents	•	•	•	•
Matrix	•		•	
Influent Concentration	•	•		•
Influent Units	•	•	•	
Effluent Concentration	•	•	•	
Effluent Units	•	•	•	
% Removal	•	•	•	•
References	•	•	•	•
Comments	•		•	

Table 8. Corresponding Fields for GRI2E Mini Reports

	GRI2E Mini Reports			
Fields	Standard	Annual	Economic	Mechanical
Process Name	•	•	•	•
Developmental Status	•	•	•	•
Operating Rate	•	•	•	•
Operating Units	•	•	•	•
Water/Waste Source	•			
Stream Type	•			
Capital Cost	•		•	
Capital Units	•		•	
Capital Year	•		•	
Equipment	•			•
Operating Cost	•		•	
Operating Units	•		•	
Operating Year	•		•	
Utility Cost	•		•	
Utility Units	•		•	
Utility Year	•		•	
Utilities	•			•
Chemical Cost	•		•	
Chemical Units	•		•	

Table 8. Corresponding Fields for GRI2E Mini Reports (Continued)

Chemical Year	•		•	
Chemicals	•			•
Annualized Cost	•	•	•	
Annualized Units	•	•	•	
Annualized Year	•	•	•	
Annualized Cost Basis	•			
Reference	•	•	•	•
Comments	•			•

SUMMARY

In order to assist the natural gas industry to address potential treatment/remediation needs, a report and database have been developed by GRI to facilitate screening appropriate environmental management options. To accomplish this objective, state-of-the-art data on performance and economics of 44 treatment technologies suitable for managing production water/waste streams and remediating site-related contamination, have been consolidated and organized in a database format. The technology evaluation report and database were developed for use by gas industry managers who are responsible for addressing environmental issues, and by others who are responsible for addressing waste treatment and remediation.

The information is compiled on Paradox 2 database software to facilitate examination of technology-specific data. Data are presented in two levels of detail; Level I provides a general overview of each technology, and Level II compiles performance and economic information for specific technology applications. In all, approximately 730 treatment performance entries are included, along with approximately 240 economics entries. The companion technical report presents the scope and approach to the study, describes the database's structure and operation, and contains a printed copy of all entries in the database. Case studies are included in the report to demonstrate the various uses and capabilities of the system.

ACKNOWLEDGEMENT

Funding for this investigation was provided by the Gas Research Institute under Contract Numbers 5084-253-1117 and 5086-245-1334. The topical report and database were developed by Remediation Technologies, Inc. and ENSR Consulting and Engineering.

ORDERING INFORMATION

To receive a free copy of the report "Evaluation of Treatment Technologies in the Natural Gas Industry; Production Waste/Water Management and Site Remediation", order codes GRI-89/0263.1, GRI-89/0263.2, GRI-89/0263.3, contact James Evans, GRI Senior Project Manager, Environmental and Health Research, Gas Research Institute, 8600 West Bryn Mawr Avenue, Chicago, IL 60631 [Telephone (312) 399-8329; Fax (312) 399-8170]. Indicate if the computerized database is also desired, along with the size and density of diskette desired.

To receive a free copy of the Natural Gas Industry EH&S Information System, contact James Evans at the above address. The database can be obtained by requesting a copy of the report "Users Manual for the Gas Research Institute, Natural Gas Industry Environmental, Health and Safety Information System", order code GRI-91/0007.

REFERENCES

Muenchow, C., S. Koraido, J. Balik and Fillo, J.P. 1991, "User's Manual for the Gas Research Institute Natural Gas Industry Environmental, Health and Safety Information System." GRI-91/0007. February

Tallon, J.T., L.A. Peach, J.P. Fillo, and Halapin, T., 1989, "Evaluation of Treatment Technologies in the Natural Gas Industry: Production Water/Waste Management and Site Remediation, Volumes I, II, and III," GRI Report Numbers GRI-89/0263.1, GRI-89/0263.2 and GRI-89/0263.3, October.

TREATMENT AND DISPOSAL OF WASTEWATERS PRODUCED WITH COALBED METHANE BY REVERSE OSMOSIS

B. F. Simmons
RMI Environmental Services, Inc.
Farmington, New Mexico
USA

Due to the high water production rates associated with coalbed methane (CBM) production, several water disposal alternatives are currently being evaluated. Most of these methods have been used and explored on a full-scale basis in active producing areas, but either economic or environmental limitations have presented insurmountable problems. The research of this paper evaluates the technical feasibility of utilizing reverse osmosis (RO) to dispose of water produced from CBM. Extensive study and testing indicates that a reverse osmosis process is feasible from a technical and economical standpoint and provides a highly attractive alternative. Research has included a combination of water quality characterization, RO computer modeling and bench testing, and the operation of a full-scale pilot for a six-month period. The treatment technologies of the RO process can be divided into two basic categories which include, 1) pretreatment, and 2) reverse osmosis. The first stage of the pretreatment consists of an enhanced filtration process to remove suspended solid contaminants that may harm and/or affect the performance of the subsequent treatment processes. Next, the feed waters are chemically treated to condition the water to maximize the performance and efficiency of the RO equipment. The final treatment stage utilizes reverse osmosis to reduce the total dissolved solids to a decreased level allowing surface use or discharge. During this stage of the process, approximately 75% of the produced water is treated to meet EPA permit requirements for surface discharge and the remaining 25% becomes a concentrated brine containing 90+% of the original dissolved solids. The concentrated brine can then be disposed of by evaporation, either in an open pond utilizing solar energy, or in a direct-fired thermal evaporator utilizing natural gas, or by the more conventional method of underground injection in a class II injection well. The successful development and demonstration of this RO treatment process introduces a viable new disposal alternative for produced waters. It is proven to be both technically and environmentally sound. Additional applications of the process under various production conditions will ultimately determine the overall flexibility and usefulness of the process.

INTRODUCTION

The available tax credits on natural gas produced from coal seams has made this expanding energy source increasingly important to gas producing companies (Soot, 1988). Projected development in the San Juan Basin of northwestern New Mexico and southwestern Colorado is estimated to exceed more than 2700 new wells by the end of 1990. These wells are included in a projected estimate of more than 9000 CBM wells to be drilled in the U.S. (Rountree, 1990),

Produced Water, Edited by J.P. Ray
and F.R. Engelhart, Plenum Press, New York, 1992

and with the extension of the tax credit for an additional two years, CBM production could exceed three billion cubic feet of gas per day by early 1993.

One unique characteristic of coalbed methane production relative to conventional production in most of the actively drilled basins is the large amount of water produced initially and during the production life of the gas wells. As it is produced, the water is unsuitable for discharge because its high dissolved mineral and residual hydrocarbon content exceed the governing regulatory guidelines. Investigation and research over the past two years, both on bench and full pilot scale, has demonstrated that it is both economically and technically feasible to treat and dispose of CBM wastewater using reverse osmosis (RO). This process produces a clean permeate stream representing 70-80% of the original feed stream volume, and a highly concentrated brine consisting of 20-30% of the original feed stream volume. This will allow for the discharge of the majority of the water (the permeate stream) to the surface, which is highly attractive since it avoids the potential environmental hazards of groundwater contamination by reinjection and provides a large amount of potable water for surface use. The quantity of produced water treated by the RO process on the Southern Ute Indian Reservation of southwestern Colorado has been projected to reach 67,000 acre-feet per well (approx. 520 million barrels) over a ten year period (Bauch, 1990). Some of the potential uses in this arid southwest region include the development or enhancement of farming, ranching, and/or wildlife wetland habitat. Other advantages of the treatment process include reclamation of the minerals, elimination of the risk associated with drilling and operating a successful injector (well), and flexibility of disposal capacity and location. Public relations between the operator and the private sector are also enhanced due to the natural resource recycling and positive environmental achievement.

The concentrated effluent can be disposed of by several methods available to the gas producers. The two primary methods of concentrate disposal include deep well injection and evaporation, which have typically been used for disposal of the untreated produced waters. Evaporation can be accomplished using thermal or vapor recompression evaporators, or solar energy with the use of an evaporation pond. One-hundred percent evaporation of the brine will result in the production of a dry salt product which may have a marketable value. Depending upon the concentrations of the various dissolved ions in the produced water, different salt by-products can be recovered for use in local industrial activities. Sodium carbonate and sodium chloride represent the majority of the salts potentially recoverable from the concentrated effluent.

The primary objective of this paper will be to evaluate the technical feasibility of utilizing reverse osmosis and other complementary water treatment technologies, as an integrated process, for the treatment and disposal of waters produced from coalbed methane wells. Results from the research that has been performed by RMI Environmental Services will be presented and discussed as well as the advantages and disadvantages of treatment versus alternative methods of CBM water disposal.

WATER DISPOSAL ALTERNATIVES

Water disposal alternatives are highly dependent upon the producing area in question. The analyses and comparisons discussed herein are based on the environmental conditions of the San Juan Basin of northwestern New Mexico and southwestern Colorado. Although these conditions are somewhat unique to this area, the disposal methods and associated problems are very similar in other producing basins of the United States.

Historically, oil and gas producers have attempted to rely solely on well-established conventional water disposal methods to resolve the problem. However, the abnormally high water production associated with CBM and the recent increase of environmental awareness, has required operators to look beyond the older disposal methods to new and more technologically advanced disposal alternatives and/or combinations of the old and the new. Each alternative method possesses advantages and disadvantages and should be evaluated on a site specific basis.

The primary conventional water disposal method utilized by the majority of the oil and gas producers in the U.S. is deep well injection. Deep well injection typically represents the simplest, most cost-effective disposal method and remains the most widely used in the energy industry today. However, several factors can greatly affect or even prohibit the use of deep well injection.

The increased demand for shallow fresh water aquifers in the southwest has focused regulators' attention on protecting water resources and water quality. This has forced producers to drill deeper, more costly injectors (wells), in order to continue deep well disposal of the produced waters. Not only does this inflate the capital expenditures, but the risk associated with targeting highly porous and permeable disposal zones is also increased. The primary injection zones range in depth from 7,000 to 10,000 feet and consist of discontinuous sandstones (i.e., the Dakota, Bluff, and Burro Canyon formations; Kelso and Wicks, 1988). The permeability and porosity can change rapidly with stratigraphic pinchouts or clay plugging within the formation. All these factors combine to make deep well injection a highly scrutinized, risky, and expensive disposal method. However, the operators are very familiar with injection drilling and operations, and associate this method with limited future environmental liabilities.

Another common disposal alternative is the use of evaporation ponds. Generally, in the southwestern United States, evaporation ponds can be utilized to dispose of produced waters, but the high water volumes and large land areas required with CBM production and/or environmental limitations can prohibit their use. Some of the disadvantages include the requirement for large flat-surface land areas, expensive construction costs, permitting requirements and delays, inconsistent evaporation rates, and the ultimate removal and disposal of the dry solids. Land area requirements for an evaporation pond in the Four Corners area are estimated to equal approximately one acre/100 barrels of water per day (BWPD) based on an evaporation rate of 3.0 gal/min/acre (Public Service Company, 1989). However, operators will experience peak rates in the summer and reduced rates in the winter months which must be taken into account. An evaporation pond must also be permitted with all the regulatory agencies having jurisdiction in the area. Factors such as the impacts to the local wildlife, groundwater resources, and the regional aesthetics must be considered. The protection of these various areas results in increased construction costs (i.e., expensive liners, leak detection systems, fencing, and netting). Ultimately, the operator is still faced with the disposal of the solid salts, and must devise a method to remove the salts without damaging the liner, or face costly replacement of the liner. The consideration and resolution of these variables, in addition to the high construction costs, typically prohibits the use of evaporation ponds for CBM produced water disposal.

The final alternative which will be discussed is utilized by most of the operators in conjunction with one of the other methods. Water truck hauling is used by most operators to transport a portion or all of the produced water to a company owned, or commercially operated injector or evaporation disposal facility. Truck hauling is usually limited to producing sites with lower water volumes or remote locations. The cost to haul produced water can triple disposal costs. Therefore, in areas where it is feasible, operators construct a pipeline gathering system to collect the produced water. Also, hauling produced water to another site does not eliminate the operator's liabilities as they relate to that final disposal site.

WATER TREATMENT AND DISPOSAL

Due to the environmental concerns and economic parameters associated with commonly used water disposal alternatives, the treatment by RO and surface discharge of the majority of the water is highly attractive. Even though permitting a treatment site can still be complicated, transporting the water can be reduced or eliminated and associated adverse environmental impacts are greatly reduced. Public relations are also enhanced with the reclamation of the produced water in a relatively dry area of the southwest.

The following treatment methods were identified as required components of an integrated reverse osmosis treatment process for the effective and efficient recovery and disposal of produced CBM water. This research focused on the evaluation and implementation of this integrated system at a full-scale facility located on an operating CBM production site.

- flotation/aeration
- solids filtration
- chemical addition/treatment
- reverse osmosis

These technologies can be divided into two basic categories and will be grouped together as follows for further discussion and analysis. A) Pretreatment: utilizing flotation/aeration, solids filtration, and chemical addition and treatment; and B) total dissolved solids (TDS) reduction; utilizing reverse osmosis.

DEFINITION OF REVERSE OSMOSIS

The key treatment technology of the process is the use of reverse osmosis for the reduction and removal of the majority of the dissolved solids. Reverse osmosis has been developed in direct competition with distillation processes for the desalination of seawater and/or wastewater. A major advantage of reverse osmosis lies in the fact that it requires no heat and limited operating energy. Referring to Figure 1, the normal direction of water flow across a membrane is down the concentration gradient. Therefore, when an ideal semi-permeable membrane (one that is permeable to solvent but not solute) is placed between two compartments, one containing pure solvent and the other containing a solution (solvent plus solute), the solvent passes through the membrane to the solution side. This phenomenon is called "osmosis".

Transport through the membrane occurs due to the chemical potential driving force which is caused by the presence of the solute. If a pressure (Δp) is applied to the solution side just equal to the pressure difference between the solute and the solvent ($\Delta\pi$), water flow ceases, and an osmotic equilibrium occurs. The exact pressure that must be applied to the solution side to stop the solvent flux is called the "osmotic pressure" (Khan, 1986; Parekh, 1988).

"Reverse osmosis" is a process in which a pressure greater than the osmotic pressure ($\Delta p > \Delta\pi$) is applied to the solution to reverse the flow and drive solvent from the solution side to the pure solvent side (see Figure 1.3).

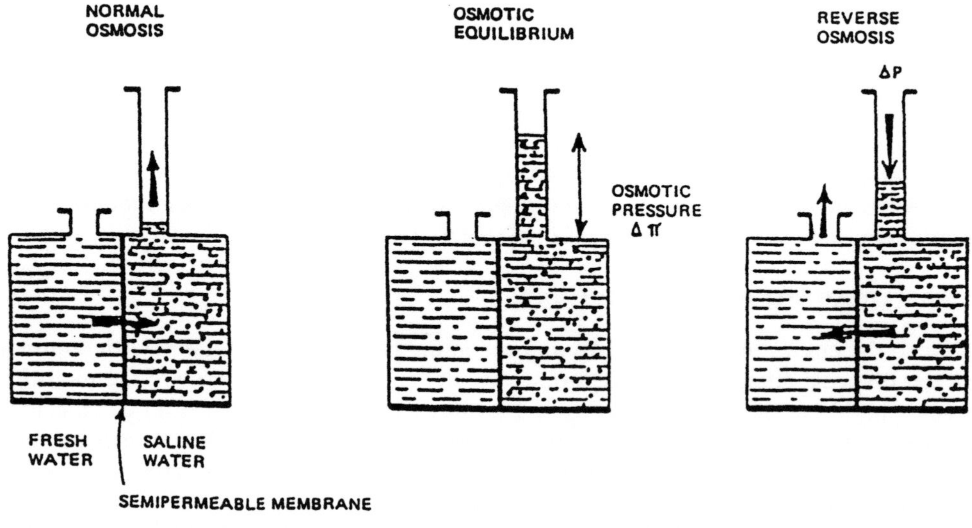

Figure1 The Principle of the Reverse Osmosis Process **Source:** Backish, R., 1973. Chap. 2 *in* "Practice of Desalination", Ed. R. Backish. New Jersey: Noyes Data Corporation.

Several driving forces are possible in membrane transport including pressure, concentration, electrical potential, and temperature. However, the primary driving forces are pressure and concentration, which lead to the flux of solvent and solute.

Most waters treated by reverse osmosis have relatively high osmotic pressures (e.g., for seawater as high as 500 psi). However, the CBM waters typically have osmotic pressures ranging from 50 to 350 psi due to much lower TDS concentrations. As water is removed from the feed solutions during the RO process, the osmotic pressure of the concentrate increases. This can be a very important design consideration in preventing membrane scaling and fouling in the final RO membranes in series as flow rates decrease and concentrations increase.

Another important design consideration that will be discussed is the membrane permeate "flux" or the volume of permeate produced per unit surface area of membrane per unit time (i.e., gal/ft^2/day) (Parekh, 1988).

The limits of the flux rate for a particular membrane are typically provided by the membrane manufacturer. Manufacturers rate the packaged membranes by measuring the module's flux under standard test conditions. This productivity is the recommended maximum obtainable. The water flux is defined by equation 1 (Dhawan, 1990):

$$Fw = A\,(\Delta p - \Delta\pi) \tag{1}$$

Where: A = water permeability constant (gal/cm^2 sec Atm)
Δp = pressure differential applied across the membrane (Atm)
$\Delta\pi$ = osmotic pressure differential across the membrane (Atm)

It is important to follow the membrane manufacturers' recommendations and not exceed the maximum element (membrane) flux. These recommendations are based on element size and the standard quality of feed water.

Other key measurements that need to be monitored are the membrane "rejection" and the membrane "recovery". Membrane recovery is defined as the ratio of permeate flow to feed flow for that particular membrane. Recovery can be adjusted by varying the flow rate and operating pressures; however, caution should be taken not to exceed the performance limits of the membranes and the chemical concentrations that would contribute to scaling.

Maintaining proper operating conditions is the key preventative step to minimize membrane fouling. Membrane rejection is defined as the amount of solute that does not pass through the membrane (C2) relative to the feed concentration (C1). Mathematically this is defined by the equation (2) and is expressed as a percentage (Dhawan, 1990).

$$\text{Membrane Rejection (\%)} = (C1\text{-}C2/C1) \times 100 \tag{2}$$

Membrane rejection should also be consistently monitored to prevent rapid deterioration and early replacement of the membranes. The membrane rejection should not exceed the manufacturers specifications to maintain the RO system performance.

The TDS concentration of CBM waters in the San Juan basin ranges from as low as 1000 ppm to levels as high as 30,000 ppm. The primary dissolved ions include Na, Cl, and HCO_3. A typical water analysis is shown in Table 1.

The feed water quality directly impacts the primary factors that determine the effectiveness of the treatment process and those monitored operating levels as discussed above. The initial dissolved solids concentration and composition determines the necessary operating pressures of the RO unit and controls the design of an effective pretreatment system.

Water flux will decrease as feed water salinity increases due to the efficiency of the given membrane. Salt rejection will also eventually decline as water flux becomes insufficient for proper operation (Dhawan, 1990).

Other water characteristics which will affect the RO process include the water temperature and pH. Flux increases almost linearly with temperature; however, salt rejection declines with increasing temperature. Ideal operating temperatures vary with different membrane modules, but 75 - 85° F is typically the most desirable temperature range. Salt rejection and flux are relatively constant over a broad pH range, however, the pH of the feedwater can greatly affect scaling tendencies and extreme pH levels will damage most membranes. Typically, produced water temperatures and pH values of CBM waters in the San Juan Basin, as well as other gas producing basins, fall in or can be easily modified to the desirable range of values.

There are two principal types of RO membranes which have evolved over the last 50 years to capture virtually the entire water purification market. They are the thin film composite (TFC) spiral wound membrane and the polyamide hollow-fiber membrane. The TFC spiral wound membrane was selected for CBM water treatment because of its exhibited structural stability at higher pressures, superior solute rejection, resistance to bacterial attack, and overall flexibility to a wide range of operating conditions. The thin film composite material is formed in large flat sheets and then wound into concentric rings to maximize the surface area of the membrane. The standard brackish water TFC spiral wound membrane will effectively reject the dissolved solids contained in CBM waters, however, the key to consistent water treatment is more dependent upon sufficient pretreatment steps to protect the membrane performance.

Table 1. Typical coalbed methane produced water analysis

Constituent	mg/l (except as noted)	meq/l
Na^+	3530.0	153.546
K^+	16.5	0.422
Ca^{++}	30.9	1.542
Mg^{++}	11.2	0.921
Total Fe^{++} & Fe^{+++}	4.7	0.252
Positive Sub-Total	3593.3	156.684
Cl^-	680.0	19.180
CO_3=	180.0	6.000
HCO_3-	9120.0	149.466
OH^-	0.0	0.000
SO_4=	<10.0	0.000
Negative Sub-Total	9980.0	174.674
Total Dissolved Solids	9180.0	
pH	8.3	
Conductivity (μho/cm)	11000.0	
Hardness as $CaCO_3$	149.0	
Alkalinity as $CaCO_3$	7780.0	
Silica as SiO_2	33.4	
Field Temperature °C	36	

Source: RMI Pilot Research Data. Water Analyses: March-August 1990, RMI Environmental Services, Inc., Farmington, NM

PRETREATMENT DEVELOPMENT

Feedwater pretreatment is probably the most important factor in contributing to a successful RO plant. If pretreatment is inadequate, membranes will become fouled, because in addition to removing dissolved solids, the elements will also reject suspended solids, colloids, and metals. The result will be a reduction in the water flux. In order to maintain the permeate recovery, the RO unit will have to operate at higher pressures, which are potentially damaging and more costly. Pretreatment can also control and decrease the tendency of scaling resulting from the precipitation of salts on the membrane surface. Precipitation can be minimized by increasing the solubility of the constituents before the feed water reaches the membranes. If acid is added, as will be shown later, the pH is lowered and the concentration of the scaling minerals will also be lowered.

Another pretreatment variable which must be considered is the control and prevention of bacterial fouling. Any feedwater which contains sufficient nutrients, in particular CBM produced water, will promote the rapid growth of organisms. This growth can lead to the deposition of bacterial slime on the membrane surface which can cause irreversible fouling and loss of performance.

The pretreatment for the CBM water treatment process was primarily tested and developed from the research that took place both at RMI's laboratory and operations in the field at a full scale pilot plant. The pilot plant was placed in southwestern Colorado at a CBM producing facility and was operated by RMI from March through July of 1990.

The treatment facility was designed to process approximately 3000 barrels of produced water a day with a TDS concentration of approximately 3500 ppm. The test methods utilized during the treatment site research were developed in an attempt to maintain steady-state uninterrupted operations. Fixed operating guidelines were implemented in order to amplify the effects and gross relationships of the various pretreatment and RO controls and their impacts over time on the system operations. The controlling factors of the system operations research included the following variables, both individually and combined:

- feed pump pressure;
- RO operating pressure;
- pressure drop across each set of membranes;
- pressure drop across the multi-media filter;
- pressure drop across the cartridge filters;
- feed water TDS;
- feed water flow rate;
- feed water temperature;
- recovery and rejection rates;
- pH (HCl injection rate) ;
- biocide injection.

The system operations research was designed to maintain the values of selected parameters within a limited range, to focus the results over an extended time frame on a single variable. The test protocol required the continuous monitoring and recording of all of the above variables at predetermined time intervals. The results were periodically reviewed and evaluated in order to monitor relative changes and make adjustments to the controlling factors and/or the measurement time interval. This method of isolating a single control variable and monitoring the system response enabled the evaluation of all of the various pretreatment and reverse osmosis components.

Initially, research and bench tests indicated that tank settling and 5 micron cartridge filtration would be adequate as pretreatment steps with some chemical pretreatment to control bacterial growth in the RO system. However, within the first two weeks of the pilot operation, the RO membranes began to foul. The use and replacement of 5 micron cartridge filters increased dramatically, almost doubling within the second week. Operating pressures increased 25% and a marked decrease in the flux and recovery of the system was noted (RMI, 1990).

The cartridge filters exhibited red-orange staining indicating iron fouling. The membranes were also removed from the RO unit and inspected. Reddish-orange scaling and an oily slime buildup with a recognizable hydrocarbon odor was found on the surfaces of the membranes.

The fouling evidenced in the system indicated hydrocarbon contamination from oil and grease and iron scaling from iron precipitation. A very low iron concentration was reported in the initial water samples, however, due to the large amount of iron piping and components in the system, the iron concentrations had increased. Iron in the ferrous (Fe^{++}) soluble form was being concentrated and oxidized in the RO unit to the ferric form and fouling was occurring from the precipitation of $Fe(OH)_3$. Also, the presence of oil and grease had been noted in water samples, but the continuous flow of the large volumes of water and the multiple sources of water caused an accumulation of hydrocarbons on the membrane surfaces. Research was initiated to identify effective pretreatment technologies to eliminate the iron and residual hydrocarbons.

Through laboratory bench tests it was determined that the addition of an induced air flotation (IAF) unit would oxidize the iron, and, it could be subsequently removed prior to reaching the RO unit. In diffused aeration systems, air bubbles are formed at an orifice from which they break off and rise through the water, finally bursting at the liquid surface. The size of the bubbles varies from coarse to extremely fine, depending on the specific aeration device. The IAF unit is capable of removing suspended solids, metals, and emulsified oil by the physical method of inducing fine air bubbles through the water which migrate to the surface (Eckenfelder, 1989). Solids and oil droplets adsorb to the bubbles and are carried to the water's surface. The dissolved iron concentrations are reduced by oxidation which also forms on the bubbling water surface in the form of $Fe(OH)_3$.

A multi-media filter was added after the IAF unit to filter out any suspended solids, remaining hydrocarbons, and ferric iron that might still be in the feed stream. The multi-media filter consisted of a dual bed of sand and anthracite.

The addition of the IAF and multi-media filter to the RO treatment facility solved the initial problems of iron scaling and hydrocarbon fouling. However, after operating the system for several more weeks an increase in the operating pressures and an overall drop in the performance of the RO unit again indicated a scaling or fouling problem. The RO unit was disassembled and the membranes were inspected to evaluate the cause of the deterioration of the RO performance. It was determined that scaling caused by the precipitation of calcium carbonate and/or bicarbonate, and magnesium hydroxide was causing the deterioration of the RO unit's performance. This diagnosis was consistent with the slow buildup and subsequent decline of the flux and recovery of the RO unit. It was also noted that the majority of the scaling had taken place in the last few membranes in series which is consistent with salt precipitation from the increase in concentration and the reduced flow across the membrane surfaces in the final stages of the RO unit.

In order to control and prevent salt scaling within the RO, the decision was made to adjust the pH of the feedwater by chemical addition. By lowering the pH of the feedwater, the solubility of the scaling salts would be increased which would also decrease the tendency of the salts to precipitate. Acid injection was implemented and proved to be both technically and economically feasible for the control and prevention of salt scaling.

The final pretreatment variable that had to be resolved, in order to guarantee the efficient operation of the reverse osmosis, process, is that of biological fouling. It is common with surface water that micro-organisms have a tendency to grow on most surfaces that are immersed in waters which do not contain biocides (Osta and Bakheet, 1987). The amount of growth is dependent upon factors related to the water supply such as type and degree of microbial contamination, nutrients present, temperature, and intensity of water flow.

Over the several months in which the pilot facility was operated, bacterial growth was evident both as a black slime on the membrane surfaces, and in the overall performance of the RO unit. It was noted during the pilot operation that bacterial fouling caused an increase in salt passage raising the permeate TDS as well as decreasing the permeate recovery. A gradual increase in the operating pressure was also experienced. A strong H_2S odor was noted on several occasions upon opening the pressure vessels to remove the membranes, in particular, after periods of temporary shutdown. In order to control biological fouling, a biocide was added continuously to the feed stream, and during downtime the membranes were stored in a dilute solution of biocide for preservation.

TDS REDUCTION - REVERSE OSMOSIS

In order to evaluate the feasibility of utilizing reverse osmosis as the basis for a CBM water treatment process, extensive research and a series of tests had to be conducted. Initially, RMI compiled data on the ranges of the coalbed water quality to determine if the produced waters would fall within the limits of membrane treatment technology. This initial investigation was focused primarily on identifying cations or anions which could prevent the use of RO, rather than the TDS levels of the selected water samples. Research revealed the CBM waters to be relatively benign containing low concentrations of calcium, manganese, barium, iron, strontium, sulfates, nitrates, ammonia, silica, and hydrogen sulfide.

The next step taken was preliminary computer modeling based on average analyses of the produced waters from various regions of the basin. Virtually all membrane manufacturers have developed software to model the efficiency of their products under various conditions. The flexibility of the model allows the user to maximize the RO system efficiency and accurately predict treatment results without performing expensive bench tests or timely mathematical calculations. Most membrane manufacturers will not warrant their RO elements unless the guidelines generated by computer analysis are closely followed.

The computer model used to perform the RO System Analysis is a product of Filmtec Membranes, a subsidiary of the Dow Chemical Company. It is identified as the "Reverse Osmosis System Computer Analysis for the FT-30 Series Membranes" (RMI, 1990). The key variables that must be input into the model include the selection of the membrane module (element), pressure vessel array, projected optimum recovery, projected flow rate, water temperature, and the actual TDS or analysis of the water to be treated. After the information is entered, the model generates a step-by-step analysis of the permeate and reject (concentrate) water quantity and quality from each element, the average flux and recovery of each element, the osmotic pressure and RO operating pressures within the system, and the final product water quality and quantity.

In the preliminary models, the assumption was made that the CBM waters would meet the suspended solid and specific ion concentration limits that are set forth by the manufacturer. The modeling verified that it would be technically feasible to treat the produced CBM waters with RO, and that the recoveries and rejection rates would be adequate to save the CBM producers money by discharging the majority of the water to the surface. The model input value for the feedwater TDS was entered at 3,000 ppm; the permeate recovery calculated by the model equaled 75.8%; and the TDS of the permeate was projected to average 165 ppm. Based on the guidelines of the EPA issued National Pollutant Discharge Elimination System (NPDES) permit [RMI; NPDES Permit, 1990], which regulates the discharge of the treated effluent, the 165 ppm is well below the thirty day average discharge limit of 500 ppm TDS, and also, well below the highest one time discharge limit of 1000 ppm.

After computer modeling verified the viability of RO, bench tests were performed to simulate the treatment and recovery of a typical sample of produced water. The results of the bench tests of various samples of CBM water yielded recoveries similar to those projected by the computer modeling.

Based on the results of the combined initial research efforts, RMI decided to proceed with a full scale pilot test to evaluate the technical and economical feasibility of a RO treatment process for the disposal of CBM waters. The research for the CBM water treatment process was continued and basically proven to be viable at RMI's pilot treatment facility in southwestern Colorado. In this stage of the research, approximately 75% of the produced water is treated to meet EPA permit discharge requirements and the remaining 25% becomes a concentrated brine containing 90+% of the original dissolved solids. A synopsis of the results and operating parameters of the full-scale pilot research is summarized in Table 3. Of the four million gallons (approx. 96,000 barrels) treated, 40% was processed over the last twenty days of actual pilot operations. The thirty day average measurement of the permeate (discharge) TDS was 443 ppm with no measurable oil and grease, and/or no detectable hydrocarbons. As was demonstrated by whole effluent toxicity (WET) testing (see Table 2), the full strength effluent proved to have a higher survival rate than the control sample (RMI, 1990). The WET test measures the overall toxicity of the water, as compared with a benign control, by allowing sensitive organisms to live in the samples for a monitored time period.

The RO unit operated and performed much as predicted after an effective and efficient pretreatment system was implemented to precondition the water for the highly sensitive membrane separation process.

The final key that solidified the technical feasibility of operating an RO unit to treat CBM waters was the development and consistent performance of thorough and complete membrane cleanings. Membranes can be cleaned using cleaning solutions approved by the membrane manufacturer. Because of the TFC membrane's combination of pH stability and temperature resistance, cleaning can be done very effectively. Both acidic and alkaline cleaners can be used at temperatures to 50° C (Khan, 1986). Acid cleaning to remove mineral scale is best done with an acidic solution at a pH of 2. For alkaline cleaning, sodium lauryl sulfate generally gives the best results. The development of an effective cleaning schedule is very important to the maintenance of the RO unit and to limiting downtime. Factors that should be considered include the rate of fouling/scaling, design of an in-line cleaning system, use of the most effective cleaners/acids, flow rates during various cycles, and the economics of the schedule considering both time and materials.

Table 2. Abstract of Whole Effluent Toxicity (WET) Test Results CBM Water Treatment Pilot

Test Concentration: Control, 100%, 50%, 25%, 12.5%, 6.25%
Number of organisms in each concentration: 20
Replicates at each concentration: 4

	Ceriodaphnia	Fathead Minnows
Test vessel size:	30 ml	260 ml
Exposure volume:	15 ml	200 ml
LC_{50}'s:	100%	100%
Temp. range during test (°C):	20.2-23.0	20.0-25.0
Dissolved oxygen range(ml/l):	6.4-7.6	5.5-7.3
pH range during test:	8.3-8.5	8.0-8.4
	Control	100%
Alkalinity (mg/l as $CaCO_3$):	83	277
Hardness (mg/l as $CaCO_3$):	103	10
Total residual chlorine (mg/l):	ND	ND
Total ammonia in effluent (mg/l):	ND	ND

Results - No significant toxicity was measured for either test species. Survival in the tests ranged from 85-95% for *Ceriodaphnia* and 85-100% for the fathead minnows. In the former test, survival was 90% in the controls and 85% in the whole strength effluent. Ninety percent of the control fish survived compared to 95% in the 100% effluent.
Source: RMI Pilot Research Data. Water Analyses: March-August, 1990

ECONOMIC ANALYSIS

When evaluating the economics of water treatment versus other disposal methods, controlling cost factors can fluctuate considerably in their importance from one area to another.

Therefore, the economics of implementing and operating a CBM water treatment facility are almost exclusively dependent upon the local area and the environmental factors affecting that specific area. Some of the key variables to consider are the produced water quality characteristics, initial volumes and production decline of the water, well groupings and the associated gathering systems, access and topography, and depth and continuity of the locally approved injection zones. As can be expected, making general analyses concerning the economic feasibility of water treatment can be vague and inconclusive. A detailed economic analysis must be performed on a site-by-site basis and should consider all of the above factors.

Although this paper does not explore a detailed economic analysis, the reverse osmosis process has enabled the reduction of disposal costs in various specific applications.

Table 3. Reverse Osmosis Unit Operation Results for the CBM Water Treatment Pilot

Operating Parameter	Results	EPA Discharge Limitations
Rejection rates	87-93 %	
Recovery rates	68-85 %	
Average Flow Rate	69 gpm	
Permeate TDS	443 ppm	500
Feed water TDS	3,400 ppm	
pH (untreated)	7.9	
pH (treated)	7.1	6.5-9.0
temperature	92° F	
Operating Pressure	320 psi	
Total Treated Water Volume	4,020,600 gallons	

Source: RMI Pilot Research Data. Water Analyses: March-August 1990, RMI Environmental Services, Inc., Farmington, NM

CONCLUSIONS

Based on the research and operations conducted by RMI, it has been proven that the treatment and disposal of water produced with CBM by reverse osmosis is both technically and economically feasible. Because of the flexibility of the process, production unknowns such as those previously discussed, can be addressed by making basic modifications to the process and by developing a specifically engineered treatment system. Produced water disposal by this process also has many advantages, such as:

- Eliminating the risks associated with drilling and operating an injection well;
- Providing flexibility to relocate the treatment equipment and utilize at a different location, especially considering the high water production decline rates;
- Process creates clean water for use by local consumers and to replenish runoff;
- Dissolved solids may be recycled and could offset treatment costs.

However, due to the complexity of the process, several factors must be evaluated when considering reverse osmosis for a disposal alternative. All information must be evaluated both technically and economically, on a site-by-site basis, before a commitment is made to use this fast growing water treatment technology; reverse osmosis. Additional applications of the process under various production conditions will ultimately determine the overall flexibility and usefulness of the process.

ACKNOWLEDGMENTS

Special appreciation is extended to Stan R. Lueck, Pat Steenburg, and Hollis E. Brown of RMI for their invaluable assistance and support on this project and in compiling this report.

REFERENCES

Dhawan, G. K.: 1990, "Design and Operation Maintenance of Reverse Osmosis Systems", Seminar - Applied Membranes Inc., February 1990, San Marcos, CA, pp. 1-1 through 9-33.

Eckenfelder, W. W., 1989, Industrial Water Pollution Control: New York, NY, McGraw-Hill, Inc., 400 p.

Kelso, Bruce S., and Wicks, D. E., 1988, "A geologic analysis of the Fruitland Formation coal and coalbed methane resources of the San Juan Basin, southwestern Colorado and northwestern New Mexico." Geology and Coalbed Methane Resources of the Northern San Juan Basin of Colorado and New Mexico. Ed. James E. Fassett. Denver, CO: Rocky Mountain Association of Geologists, p. 69-70.

Khan, A. H., ed., 1986, Desalination Processes and Multistage Flash Distillation Practice: New York, NY, Elsevier Science Publishing Co. Inc., 596 p.

Osta, T. K., and Moh'd Bakheet, L, 1987, "Pretreatment System in Reverse Osmosis Plants." Desalination, v. 63, pp. 71 - 80.

Parekh, B. S. ed. 1988. Reverse Osmosis Technology. New York: Marcel Dekker, 516 p.

Public Service Company of New Mexico. 1989 Telephone conversation, 11 November. San Juan Power Plant, Shiprock, New Mexico.

RMI National Pollutant Discharge Elimination System Permit File. 1990. Unpublished. CBM water treatment site, southwestern Colorado, (1990). RMI Environmental Services, Inc., Farmington, NM.

RMI National Pollutant Discharge Elimination System Permit File. 1990. Unpublished. CBM water treatment site, southwestern Colorado, (1990). RMI Environmental Services, Inc., Farmington, NM.

Rountree, R., ed., 1990, Report Forcasts Coal Seam Rush, Western Oil World, 53(1), (July 1990)

The Durango Herald. 1990. Permit drafted for waste water discharge. 23 October. Bauch, Chris., p. 2.

CONVERSION FACTORS

To convert from:	To:	Multiply by:
Acre	Hectare	0.40468564
	Square kilometer	4.046856×10^{-3}
Bar	Atmosphere	0.9869233
Barrel (petroleum)	Cubic meter	0.1589873
	Gallon (U.S.)	42
	Liter	158.9873
Becquerel	Curie	2.702703×10^{-11}
	Picocurie	27
Centimeter	Foot	0.03280840
	Inch	0.3937008
	Yard	0.01093613
Centimeter/second	Foot/second	0.03280840
Cubic Centimeter	Gallon (U.S.)	2.641721×10^{-4}
	Milliliter	1
Cubic foot	Cubic meter	0.028316847
	Cubic yard	0.03703704
	Gallon (U.S.)	7.480519
	Liter	28.316847
Cubic inch	Cubic meter	1.6387064×10^{-5}
	Cubic yard	2.143347×10^{-5}
	Gallon (U.S.)	4.329004×10^{-3}
	Liter	0.016387064
Cubic meter	Barrel (petroleum)	6.289811
	Cubic centimeter	1×10^{6}
	Cubic yard	1.307951
	Gallon (U.S.)	264.1721
	Liter	1000
Cubic yard	Gallon (U.S.)	201.9740
	Liter	764.5549
Curie	Becquerel	3.7×10^{10}
Disintegrations per minute (DPM)	Pico curies	0.45
Foot	Meter	0.3048
	Mile (Nautical)	1.645788×10^{-4}
	Mile (Statute)	1.893939×10^{-4}
Foot/second	Kilometer/hour	1.09728
	Knot	0.5924838
	Meter/second	0.3048
Gallon (U.S., liquid)	Barrel (petroleum)	0.02380952 (1/42)
	Cubic centimeter	3785.412
	Cubic foot	0.13368056

To convert from:	To:	Multiply by:
Gallon (U.S., liquid) (continued)		
	Cubic yard	4.951132×10^{-3}
	Liter	3.785412
Gallon (U.S.)/minute	Liter/second	0.06309020
Hectare	Acre	2.471054
	Square kilometer	0.01
	Square meter	10000
Inch	Centimeter	2.54
Kelvin (temp. interval)	Celsius	1
	Fahrenheit	1.8
Kilogram	Pound	2.2046226
Kilogram/cubic meter	Pound/cubic foot	0.06242796
Kilometer	Mile (nautical)	0.5399568
	Mile (statute)	0.6213712
Kilometer/hour	Foot/second	0.9113444
	Knot	0.5399568
	Meter/second	0.2777778
Knot	Kilometer/hour	1.852
	Meter/second	0.5144444
Liter	Cubic meter	0.001
	Cubic yard	1.307951×10^{-3}
	Gallon (U.S.)	0.26417205
	Quart (U.S., liquid)	1.056688
Megapascal	Bar	10
Meter	Foot	3.2808399
	Yard	1.093613298
Meter/s	Kilometers/hour	3.6
	Knots	1.94
	Miles per hour	2.237
Mho (ohm^{-1})	Siemens	1
Mile (nautical)	Kilometer	1.852
Mile (statute)	Kilometer	1.609344
Mile/hour	Kilometer/hour	1.609344
	Knot	0.8689762
Pint (U.S., liquid)	Liter	0.4731765
	Quart (U.S., liquid)	0.5
Pound (avoirdupois)	Gram	453.59237
Pound/square inch (psi)	Pascal	1.047198×10^{3}
	Bar	0.0689746
	Grams/square centimeter	70.306958
Quart (U.S., liquid)	Cubic centimeter	946.35295
	Liter	0.94635295
Rad	Gray	0.01
Square foot	Acre	2.295684×10^{-5}
	Square meter	0.09290304
Square Kilometer	Hectare	100
	Square mile	0.38610216
Ton (metric)	Pound	2204.6226
	Ton (long)	0.98420653
	Ton (short)	0.1.1023113
	Kilogram	1000
Yard	Meter	.9144

AUTHOR INDEX

SUBJECT INDEX

MIX
Papier aus verantwortungsvollen Quellen
Paper from responsible sources
FSC® C105338

FSC
www.fsc.org

If you have any concerns about our products,
you can contact us on
ProductSafety@springernature.com

In case Publisher is established outside the EU,
the EU authorized representative is:
Springer Nature Customer Service Center GmbH
Europaplatz 3, 69115 Heidelberg, Germany

Printed by Libri Plureos GmbH
in Hamburg, Germany